T0210909

Lecture Notes in Computer Science   8584

Commenced Publication in 1973
Founding and Former Series Editors:
Gerhard Goos, Juris Hartmanis, and Jan van Leeuwen

Beniamino Murgante   Sanjay Misra
Ana Maria A.C. Rocha   Carmelo Torre
Jorge Gustavo Rocha   Maria Irene Falcão
David Taniar   Bernady O. Apduhan
Osvaldo Gervasi (Eds.)

Computational Science and Its Applications – ICCSA 2014

14th International Conference
Guimarães, Portugal, June 30 – July 3, 2014
Proceedings, Part VI

 Springer

Volume Editors

Beniamino Murgante, University of Basilicata, Potenza, Italy
E-mail: beniamino.murgante@unibas.it

Sanjay Misra, Covenant University, Ota, Nigeria
E-mail: sanjay.misra@covenantuniversity.edu.ng

Ana Maria A.C. Rocha, University of Minho, Braga, Portugal
E-mail: arocha@dps.uminho.pt

Carmelo Torre, Politecnico di Bari, Bari, Italy
E-mail: torre@poliba.it

Jorge Gustavo Rocha, University of Minho, Braga, Portugal
E-mail: jgr@di.uminho.pt

Maria Irene Falcão, University of Minho, Braga, Portugal
E-mail: mif@math.uminho.pt

David Taniar, Monash University, Clayton, VIC, Australia
E-mail: david.taniar@infotech.monash.edu.au

Bernady O. Apduhan, Kyushu Sangyo University, Fukuoka, Japan
E-mail: bob@is.kyusan-u.ac.jp

Osvaldo Gervasi, University of Perugia, Perugia, Italy
E-mail: osvaldo.gervasi@unipg.it

ISSN 0302-9743   e-ISSN 1611-3349
ISBN 978-3-319-09152-5   e-ISBN 978-3-319-09153-2
DOI 10.1007/978-3-319-09153-2
Springer Cham Heidelberg New York Dordrecht London

Library of Congress Control Number: 2014942987

LNCS Sublibrary: SL 1 – Theoretical Computer Science and General Issues

Typesetting: Camera-ready by author, data conversion by Scientific Publishing Services, Chennai, India

Printed on acid-free paper

Springer is part of Springer Science+Business Media (www.springer.com)

Welcome Message

On behalf of the Local Organizing Committee of ICCSA 2014, it is a pleasure to welcome you to the 14th International Conference on Computational Science and Its Applications, held during June 30 – July 3, 2014. We are very proud and grateful to the ICCSA general chairs for having entrusted us with the task of organizing another event of this series of very successful conferences.

ICCSA will take place in the School of Engineering of University of Minho, which is located in close vicinity to the medieval city centre of Guimarães, a UNESCO World Heritage Site, in Northern Portugal. The historical city of Guimarães is recognized for its beauty and historical monuments. The dynamic and colorful Minho Region is famous for its landscape, gastronomy and vineyards where the unique *Vinho Verde* wine is produced.

The University of Minho is currently among the most prestigious institutions of higher education in Portugal and offers an excellent setting for the conference. Founded in 1973, the University has two major poles: the campus of Gualtar in Braga, and the campus of Azurém in Guimarães.

Plenary lectures by leading scientists and several workshops will provide a real opportunity to discuss new issues and find advanced solutions able to shape new trends in computational science.

Apart from the scientific program, a stimulant and diverse social program will be available. There will be a welcome drink at Instituto de Design, located in an old Tannery, that is an open knowledge centre and a privileged communication platform between industry and academia. Guided visits to the city of Guimarães and Porto are planned, both with beautiful and historical monuments. A guided tour and tasting in Porto wine cellars, is also planned. There will be a gala dinner at the Pousada de Santa Marinha, which is an old Augustinian convent of the 12th century refurbished, where ICCSA participants can enjoy delicious dishes and enjoy a wonderful view over the city of Guimarães.

The conference could not have happened without the dedicated work of many volunteers, recognized by the coloured shirts. We would like to thank all the collaborators, who worked hard to produce a successful ICCSA 2014, namely Irene Falcão and Maribel Santos above all, our fellow members of the local organization.

On behalf of the Local Organizing Committee of ICCSA 2014, it is our honor to cordially welcome all of you to the beautiful city of Guimarães for this unique event. Your participation and contribution to this conference will make it much more productive and successful.

We are looking forward to see you in Guimarães.

Sincerely yours,

Ana Maria A.C. Rocha
Jorge Gustavo Rocha

Preface

These 6 volumes (LNCS volumes 8579-8584) consist of the peer-reviewed papers from the 2014 International Conference on Computational Science and Its Applications (ICCSA 2014) held in Guimarães, Portugal during 30 June – 3 July 2014.

ICCSA 2014 was a successful event in the International Conferences on Computational Science and Its Applications (ICCSA) conference series, previously held in Ho Chi Minh City, Vietnam (2013), Salvador da Bahia, Brazil (2012), Santander, Spain (2011), Fukuoka, Japan (2010), Suwon, South Korea (2009), Perugia, Italy (2008), Kuala Lumpur, Malaysia (2007), Glasgow, UK (2006), Singapore (2005), Assisi, Italy (2004), Montreal, Canada (2003), and (as ICCS) Amsterdam, The Netherlands (2002) and San Francisco, USA (2001).

Computational science is a main pillar of most of the present research, industrial and commercial activities and plays a unique role in exploiting ICT innovative technologies, and the ICCSA conference series has been providing a venue for researchers and industry practitioners to discuss new ideas, to share complex problems and their solutions, and to shape new trends in computational science.

Apart from the general track, ICCSA 2014 also included 30 workshops, in various areas of computational sciences, ranging from computational science technologies, to specific areas of computational sciences, such as computational geometry and security. We accepted 58 papers for the general track, and 289 in workshops. We would like to show our appreciation to the workshops chairs and co-chairs.

The success of the ICCSA conference series, in general, and ICCSA 2014, in particular, was due to the support of many people: authors, presenters, participants, keynote speakers, workshop chairs, Organizing Committee members, student volunteers, Program Committee members, Advisory Committee members, international liaison chairs, and people in other various roles. We would like to thank them all.

We also thank our publisher, Springer–Verlag, for their acceptance to publish the proceedings and for their kind assistance and cooperation during the editing process.

We cordially invite you to visit the ICCSA website http://www.iccsa.org where you can find all relevant information about this interesting and exciting event.

June 2014

Osvaldo Gervasi
Jorge Gustavo Rocha
Bernady O. Apduhan

Organization

ICCSA 2014 was organized by University of Minho, (Portugal) University of Perugia (Italy), University of Basilicata (Italy), Monash University (Australia), Kyushu Sangyo University (Japan).

Honorary General Chairs

Antonio M. Cunha	Rector of the University of Minho, Portugal
Antonio Laganà	University of Perugia, Italy
Norio Shiratori	Tohoku University, Japan
Kenneth C. J. Tan	Qontix, UK

General Chairs

Beniamino Murgante	University of Basilicata, Italy
Ana Maria A.C. Rocha	University of Minho, Portugal
David Taniar	Monash University, Australia

Program Committee Chairs

Osvaldo Gervasi	University of Perugia, Italy
Bernady O. Apduhan	Kyushu Sangyo University, Japan
Jorge Gustavo Rocha	University of Minho, Portugal

International Advisory Committee

Jemal Abawajy	Daekin University, Australia
Dharma P. Agrawal	University of Cincinnati, USA
Claudia Bauzer Medeiros	University of Campinas, Brazil
Manfred M. Fisher	Vienna University of Economics and Business, Austria
Yee Leung	Chinese University of Hong Kong, China

International Liaison Chairs

Ana Carla P. Bitencourt	Universidade Federal do Reconcavo da Bahia, Brazil
Claudia Bauzer Medeiros	University of Campinas, Brazil
Alfredo Cuzzocrea	ICAR-CNR and University of Calabria, Italy

Marina L. Gavrilova   University of Calgary, Canada
Robert C. H. Hsu   Chung Hua University, Taiwan
Andrés Iglesias   University of Cantabria, Spain
Tai-Hoon Kim   Hannam University, Korea
Sanjay Misra   University of Minna, Nigeria
Takashi Naka   Kyushu Sangyo University, Japan
Rafael D.C. Santos   National Institute for Space Research, Brazil

Workshop and Session Organizing Chairs

Beniamino Murgante   University of Basilicata, Italy

Local Organizing Committee

Ana Maria A.C. Rocha   University of Minho, Portugal (Chair)
Jorge Gustavo Rocha   University of Minho, Portugal
Maria Irene Falcão   University of Minho, Portugal
Maribel Yasmina Santos   University of Minho, Portugal

Workshop Organizers

Advances in Complex Systems: Modeling and Parallel Implementation (ACSModPar 2014)

Georgius Sirakoulis   Democritus University of Thrace, Greece
Wiliam Spataro   University of Calabria, Italy
Giuseppe A. Trunfio   University of Sassari, Italy

Agricultural and Environment Information and Decision Support Systems (AEIDSS 2014)

Sandro Bimonte   IRSTEA France
Florence Le Ber   ENGES, France
André Miralles   IRSTEA France
François Pinet   IRSTEA France

Advances in Web Based Learning (AWBL 2014)

Mustafa Murat Inceoglu   Ege University, Turkey

Bio-inspired Computing and Applications (BIOCA 2014)

Nadia Nedjah   State University of Rio de Janeiro, Brazil
Luiza de Macedo Mourell   State University of Rio de Janeiro, Brazil

Computational and Applied Mathematics (CAM 2014)

Maria Irene Falcao   University of Minho, Portugal
Fernando Miranda   University of Minho, Portugal

Computer Aided Modeling, Simulation, and Analysis (CAMSA 2014)

Jie Shen   University of Michigan, USA

Computational and Applied Statistics (CAS 2014)

Ana Cristina Braga   University of Minho, Portugal
Ana Paula Costa Conceicao
  Amorim   University of Minho, Portugal

Computational Geometry and Security Applications (CGSA 2014)

Marina L. Gavrilova   University of Calgary, Canada
Han Ming Huang   Guangxi Normal University, China

Computational Algorithms and Sustainable Assessment (CLASS 2014)

Antonino Marvuglia   Public Research Centre Henri Tudor,
                            Luxembourg
Beniamino Murgante   University of Basilicata, Italy

Chemistry and Materials Sciences and Technologies (CMST 2014)

Antonio Laganà   University of Perugia, Italy

Computational Optimization and Applications (COA 2014)

Ana Maria A.C. Rocha   University of Minho, Portugal
Humberto Rocha   University of Coimbra, Portugal

Cities, Technologies and Planning (CTP 2014)

Giuseppe Borruso   University of Trieste, Italy
Beniamino Murgante   University of Basilicata, Italy

Computational Tools and Techniques for Citizen Science and Scientific Outreach (CTTCS 2014)

Rafael Santos   National Institute for Space Research, Brazil
Jordan Raddickand   Johns Hopkins University, USA
Ani Thakar   Johns Hopkins University, USA

Econometrics and Multidimensional Evaluation in the Urban Environment (EMEUE 2014)

Carmelo M. Torre   Polytechnic of Bari, Italy
Maria Cerreta   University of Naples Federico II, Italy
Paola Perchinunno   University of Bari, Italy
Simona Panaro   University of Naples Federico II, Italy
Raffaele Attardi   University of Naples Federico II, Italy

Future Computing Systems, Technologies, and Applications (FISTA 2014)

Bernady O. Apduhan   Kyushu Sangyo University, Japan
Rafael Santos   National Institute for Space Research, Brazil
Jianhua Ma   Hosei University, Japan
Qun Jin   Waseda University, Japan

Formal Methods, Computational Intelligence and Constraint Programming for Software Assurance (FMCICA 2014)

Valdivino Santiago Junior   National Institute for Space Research (INPE), Brazil

Geographical Analysis, Urban Modeling, Spatial Statistics (GEOG-AN-MOD 2014)

Giuseppe Borruso   University of Trieste, Italy
Beniamino Murgante   University of Basilicata, Italy
Hartmut Asche   University of Potsdam, Germany

High Performance Computing in Engineering and Science (HPCES 2014)

Alberto Proenca   University of Minho, Portugal
Pedro Alberto   University of Coimbra, Portugal

Mobile Communications (MC 2014)

Hyunseung Choo   Sungkyunkwan University, Korea

Mobile Computing, Sensing, and Actuation for Cyber Physical Systems (MSA4CPS 2014)

Saad Qaisar   NUST School of Electrical Engineering and
                           Computer Science, Pakistan
Moonseong Kim   Korean Intellectual Property Office, Korea

New Trends on Trust Computational Models (NTTCM 2014)

Rui Costa Cardoso   Universidade da Beira Interior, Portugal
Abel Gomez   Universidade da Beira Interior, Portugal

Quantum Mechanics: Computational Strategies and Applications (QMCSA 2014)

Mirco Ragni   Universidad Federal de Bahia, Brazil
Vincenzo Aquilanti   University of Perugia, Italy
Ana Carla Peixoto Bitencourt   Universidade Estadual de Feira de Santana
                                  Brazil
Roger Anderson   University of California, USA
Frederico Vasconcellos Prudente   Universidad Federal de Bahia, Brazil

Remote Sensing Data Analysis, Modeling, Interpretation and Applications: From a Global View to a Local Analysis (RS2014)

Rosa Lasaponara   Institute of Methodologies for Environmental
                              Analysis National Research Council, Italy
Nicola Masini   Archaeological and Monumental Heritage
                              Institute, National Research Council, Italy

Software Engineering Processes and Applications (SEPA 2014)

Sanjay Misra   Covenant University, Nigeria

Software Quality (SQ 2014)

Sanjay Misra   Covenant University, Nigeria

Advances in Spatio-Temporal Analytics (ST-Analytics 2014)

Joao Moura Pires   New University of Lisbon, Portugal
Maribel Yasmina Santos   New University of Lisbon, Portugal

Tools and Techniques in Software Development Processes (TTSDP 2014)

Sanjay Misra   Covenant University, Nigeria

Virtual Reality and its Applications (VRA 2014)

Osvaldo Gervasi   University of Perugia, Italy
Lucio Depaolis   University of Salento, Italy

Workshop of Agile Software Development Techniques (WAGILE 2014)

Eduardo Guerra   National Institute for Space Research, Brazil

Big Data:, Analytics and Management (WBDAM 2014)

Wenny Rahayu   La Trobe University, Australia

Program Committee

Jemal Abawajy	Daekin University, Australia
Kenny Adamson	University of Ulster, UK
Filipe Alvelos	University of Minho, Portugal
Paula Amaral	Universidade Nova de Lisboa, Portugal
Hartmut Asche	University of Potsdam, Germany
Md. Abul Kalam Azad	University of Minho, Portugal
Michela Bertolotto	University College Dublin, Ireland
Sandro Bimonte	CEMAGREF, TSCF, France
Rod Blais	University of Calgary, Canada
Ivan Blecic	University of Sassari, Italy
Giuseppe Borruso	University of Trieste, Italy
Yves Caniou	Lyon University, France
José A. Cardoso e Cunha	Universidade Nova de Lisboa, Portugal
Leocadio G. Casado	University of Almeria, Spain
Carlo Cattani	University of Salerno, Italy
Mete Celik	Erciyes University, Turkey
Alexander Chemeris	National Technical University of Ukraine "KPI", Ukraine
Min Young Chung	Sungkyunkwan University, Korea
Gilberto Corso Pereira	Federal University of Bahia, Brazil
M. Fernanda Costa	University of Minho, Portugal
Gaspar Cunha	University of Minho, Portugal
Alfredo Cuzzocrea	ICAR-CNR and University of Calabria, Italy
Carla Dal Sasso Freitas	Universidade Federal do Rio Grande do Sul, Brazil
Pradesh Debba	The Council for Scientific and Industrial Research (CSIR), South Africa
Hendrik Decker	Instituto Tecnológico de Informática, Spain
Frank Devai	London South Bank University, UK
Rodolphe Devillers	Memorial University of Newfoundland, Canada
Prabu Dorairaj	NetApp, India/USA
M. Irene Falcao	University of Minho, Portugal
Cherry Liu Fang	U.S. DOE Ames Laboratory, USA
Edite M.G.P. Fernandes	University of Minho, Portugal
Jose-Jesus Fernandez	National Centre for Biotechnology, CSIS, Spain
Maria Antonia Forjaz	University of Minho, Portugal
Maria Celia Furtado Rocha	PRODEB and Universidade Federal da Bahia, Brazil
Akemi Galvez	University of Cantabria, Spain
Paulino Jose Garcia Nieto	University of Oviedo, Spain
Marina Gavrilova	University of Calgary, Canada
Jerome Gensel	LSR-IMAG, France

Maria Giaoutzi	National Technical University, Athens, Greece
Andrzej M. Goscinski	Deakin University, Australia
Alex Hagen-Zanker	University of Cambridge, UK
Malgorzata Hanzl	Technical University of Lodz, Poland
Shanmugasundaram Hariharan	B.S. Abdur Rahman University, India
Eligius M.T. Hendrix	University of Malaga/Wageningen University, Spain/Netherlands
Hisamoto Hiyoshi	Gunma University, Japan
Fermin Huarte	University of Barcelona, Spain
Andres Iglesias	University of Cantabria, Spain
Mustafa Inceoglu	EGE University, Turkey
Peter Jimack	University of Leeds, UK
Qun Jin	Waseda University, Japan
Farid Karimipour	Vienna University of Technology, Austria
Baris Kazar	Oracle Corp., USA
DongSeong Kim	University of Canterbury, New Zealand
Taihoon Kim	Hannam University, Korea
Ivana Kolingerova	University of West Bohemia, Czech Republic
Dieter Kranzlmueller	LMU and LRZ Munich, Germany
Antonio Laganà	University of Perugia, Italy
Rosa Lasaponara	National Research Council, Italy
Maurizio Lazzari	National Research Council, Italy
Cheng Siong Lee	Monash University, Australia
Sangyoun Lee	Yonsei University, Korea
Jongchan Lee	Kunsan National University, Korea
Clement Leung	Hong Kong Baptist University, Hong Kong
Chendong Li	University of Connecticut, USA
Gang Li	Deakin University, Australia
Ming Li	East China Normal University, China
Fang Liu	AMES Laboratories, USA
Xin Liu	University of Calgary, Canada
Savino Longo	University of Bari, Italy
Tinghuai Ma	NanJing University of Information Science and Technology, China
Sergio Maffioletti	University of Zurich, Switzerland
Ernesto Marcheggiani	Katholieke Universiteit Leuven, Belgium
Antonino Marvuglia	Research Centre Henri Tudor, Luxembourg
Nicola Masini	National Research Council, Italy
Nirvana Meratnia	University of Twente, The Netherlands
Alfredo Milani	University of Perugia, Italy
Sanjay Misra	Federal University of Technology Minna, Nigeria
Giuseppe Modica	University of Reggio Calabria, Italy

José Luis Montaña   University of Cantabria, Spain
Beniamino Murgante   University of Basilicata, Italy
Jiri Nedoma   Academy of Sciences of the Czech Republic,
                                       Czech Republic
Laszlo Neumann   University of Girona, Spain
Kok-Leong Ong   Daekin University, Australia
Belen Palop   Universidad de Valladolid, Spain
Marcin Paprzycki   Polish Academy of Sciences, Poland
Eric Pardede   La Trobe University, Australia
Kwangjin Park   Wonkwang University, Korea
Ana Isabel Pereira   Polytechnic Institute of Braganca, Portugal
Maurizio Pollino   Italian National Agency for New
                                       Technologies, Energy and Sustainable
                                       Economic Development, Italy
Alenka Poplin   University of Hamburg, Germany
Vidyasagar Potdar   Curtin University of Technology, Australia
David C. Prosperi   Florida Atlantic University, USA
Wenny Rahayu   La Trobe University, Australia
Jerzy Respondek   Silesian University of Technology, Poland
Ana Maria A.C. Rocha   University of Minho, Portugal
Humberto Rocha   INESC-Coimbra, Portugal
Alexey Rodionov   Institute of Computational Mathematics and
                                       Mathematical Geophysics, Russia
Cristina S. Rodrigues   University of Minho, Portugal
Octavio Roncero   CSIC, Spain
Maytham Safar   Kuwait University, Kuwait
Chiara Saracino   A.O. Ospedale Niguarda Ca' Granda - Milano,
                                       Italy
Haiduke Sarafian   The Pennsylvania State University, USA
Jie Shen   University of Michigan, USA
Qi Shi   Liverpool John Moores University, UK
Dale Shires   U.S. Army Research Laboratory, USA
Takuo Suganuma   Tohoku University, Japan
Sergio Tasso   University of Perugia, Italy
Ana Paula Teixeira   University of Tras-os-Montes and Alto Douro,
                                       Portugal
Senhorinha Teixeira   University of Minho, Portugal
Parimala Thulasiraman   University of Manitoba, Canada
Carmelo Torre   Polytechnic of Bari, Italy
Javier Martinez Torres   Centro Universitario de la Defensa Zaragoza,
                                       Spain
Giuseppe A. Trunfio   University of Sassari, Italy
Unal Ufuktepe   Izmir University of Economics, Turkey
Toshihiro Uchibayashi   Kyushu Sangyo University, Japan

Mario Valle	Swiss National Supercomputing Centre, Switzerland
Pablo Vanegas	University of Cuenca, Equador
Piero Giorgio Verdini	INFN Pisa and CERN, Italy
Marco Vizzari	University of Perugia, Italy
Koichi Wada	University of Tsukuba, Japan
Krzysztof Walkowiak	Wroclaw University of Technology, Poland
Robert Weibel	University of Zurich, Switzerland
Roland Wismüller	Universität Siegen, Germany
Mudasser Wyne	SOET National University, USA
Chung-Huang Yang	National Kaohsiung Normal University, Taiwan
Xin-She Yang	National Physical Laboratory, UK
Salim Zabir	France Telecom Japan Co., Japan
Haifeng Zhao	University of California at Davis, USA
Kewen Zhao	University of Qiongzhou, China
Albert Y. Zomaya	University of Sydney, Australia

Reviewers

Abdi Samane	University College Cork, Ireland
Aceto Lidia	University of Pisa, Italy
Afonso Ana Paula	University of Lisbon, Portugal
Afreixo Vera	University of Aveiro, Portugal
Aguilar Antonio	University of Barcelona, Spain
Aguilar José Alfonso	Universidad Autónoma de Sinaloa, Mexico
Ahmad Waseem	Federal University of Technology Minna, Nigeria
Aktas Mehmet	Yildiz Technical University, Turkey
Alarcon Vladimir	Universidad Diego Portales, Chile
Alberti Margarita	University of Barcelona, Spain
Ali Salman	NUST, Pakistan
Alvanides Seraphim	Northumbria University, UK
Álvarez Jacobo de Uña	University of Vigo, Spain
Alvelos Filipe	University of Minho, Portugal
Alves Cláudio	University of Minho, Portugal
Alves José Luis	University of Minho, Portugal
Amorim Ana Paula	University of Minho, Portugal
Amorim Paulo	Federal University of Rio de Janeiro, Brazil
Anderson Roger	University of California, USA
Andrade Wilkerson	Federal University of Campina Grande, Brazil
Andrienko Gennady	Fraunhofer Institute for Intelligent Analysis and Informations Systems, Germany
Apduhan Bernady	Kyushu Sangyo University, Japan
Aquilanti Vincenzo	University of Perugia, Italy
Argiolas Michele	University of Cagliari, Italy

Athayde Maria Emília Feijão Queiroz	University of Minho, Portugal
Attardi Raffaele	University of Napoli Federico II, Italy
Azad Md Abdul	Indian Institute of Technology Kanpur, India
Badard Thierry	Laval University, Canada
Bae Ihn-Han	Catholic University of Daegu, South Korea
Baioletti Marco	University of Perugia, Italy
Balena Pasquale	Polytechnic of Bari, Italy
Balucani Nadia	University of Perugia, Italy
Barbosa Jorge	University of Porto, Portugal
Barrientos Pablo Andres	Universidad Nacional de La Plata, Australia
Bartoli Daniele	University of Perugia, Italy
Bação Fernando	New University of Lisbon, Portugal
Belanzoni Paola	University of Perugia, Italy
Bencardino Massimiliano	University of Salerno, Italy
Benigni Gladys	University of Oriente, Venezuela
Bertolotto Michela	University College Dublin, Ireland
Bimonte Sandro	IRSTEA, France
Blanquer Ignacio	Universitat Politècnica de València, Spain
Bollini Letizia	University of Milano, Italy
Bonifazi Alessandro	Polytechnic of Bari, Italy
Borruso Giuseppe	University of Trieste, Italy
Bostenaru Maria	"Ion Mincu" University of Architecture and Urbanism, Romania
Boucelma Omar	University Marseille, France
Braga Ana Cristina	University of Minho, Portugal
Brás Carmo	Universidade Nova de Lisboa, Portugal
Cacao Isabel	University of Aveiro, Portugal
Cadarso-Suárez Carmen	University of Santiago de Compostela, Spain
Caiaffa Emanuela	ENEA, Italy
Calamita Giuseppe	National Research Council, Italy
Campagna Michele	University of Cagliari, Italy
Campobasso Francesco	University of Bari, Italy
Campos José	University of Minho, Portugal
Cannatella Daniele	University of Napoli Federico II, Italy
Canora Filomena	University of Basilicata, Italy
Cardoso Rui	Institute of Telecommunications, Portugal
Caschili Simone	University College London, UK
Ceppi Claudia	Polytechnic of Bari, Italy
Cerreta Maria	University Federico II of Naples, Italy
Chanet Jean-Pierre	IRSTEA, France
Chao Wang	University of Science and Technology of China, China
Choi Joonsoo	Kookmin University, South Korea

Choo Hyunseung   Sungkyunkwan University, South Korea
Chung Min Young   Sungkyunkwan University, South Korea
Chung Myoungbeom   Sungkyunkwan University, South Korea
Clementini Eliseo   University of L'Aquila, Italy
Coelho Leandro dos Santos   PUC-PR, Brazil
Colado Anibal Zaldivar   Universidad Autónoma de Sinaloa, Mexico
Coletti Cecilia   University of Chieti, Italy
Condori Nelly   VU University Amsterdam, The Netherlands
Correia Elisete   University of Trás-Os-Montes e Alto Douro,
                                         Portugal
Correia Filipe   FEUP, Portugal
Correia Florbela Maria da
   Cruz Domingues   Instituto Politécnico de Viana do Castelo,
                                         Portugal
Correia Ramos Carlos   University of Evora, Portugal
Corso Pereira Gilberto   UFPA, Brazil
Cortés Ana   Universitat Autònoma de Barcelona, Spain
Costa Fernanda   University of Minho, Portugal
Costantini Alessandro   INFN, Italy
Crasso Marco   National Scientific and Technical Research
                                         Council, Argentina
Crawford Broderick   Universidad Catolica de Valparaiso, Chile
Cristia Maximiliano   CIFASIS and UNR, Argentina
Cunha Gaspar   University of Minho, Portugal
Cunha Jácome   University of Minho, Portugal
Cutini Valerio   University of Pisa, Italy
Danese Maria   IBAM, CNR, Italy
Da Silva B. Carlos   University of Lisboa, Portugal
De Almeida Regina   University of Trás-os-Montes e Alto Douro,
                                         Portugal
Debroy Vidroha   Hudson Alley Software Inc., USA
De Fino Mariella   Polytechnic of Bari, Italy
De Lotto Roberto   University of Pavia, Italy
De Paolis Lucio Tommaso   University of Salento, Italy
De Rosa Fortuna   University of Napoli Federico II, Italy
De Toro Pasquale   University of Napoli Federico II, Italy
Decker Hendrik   Instituto Tecnológico de Informática, Spain
Delamé Thomas   CNRS, France
Demyanov Vasily   Heriot-Watt University, UK
Desjardin Eric   University of Reims, France
Dwivedi Sanjay Kumar   Babasaheb Bhimrao Ambedkar University,
                                         India
Di Gangi Massimo   University of Messina, Italy
Di Leo Margherita   JRC, European Commission, Belgium

Di Trani Francesco   University of Basilicata, Italy
Dias Joana   University of Coimbra, Portugal
Dias d'Almeida Filomena   University of Porto, Portugal
Dilo Arta   University of Twente, The Netherlands
Dixit Veersain   Delhi University, India
Doan Anh Vu   Université Libre de Bruxelles, Belgium
Dorazio Laurent   ISIMA, France
Dutra Inês   University of Porto, Portugal
Eichelberger Hanno   University of Tuebingen, Germany
El-Zawawy Mohamed A.   Cairo University, Egypt
Escalona Maria-Jose   University of Seville, Spain
Falcão M. Irene   University of Minho, Portugal
Farantos Stavros   University of Crete and FORTH, Greece
Faria Susana   University of Minho, Portugal
Faruq Fatma   Carnegie Melon University,, USA
Fernandes Edite   University of Minho, Portugal
Fernandes Rosário   University of Minho, Portugal
Fernandez Joao P   Universidade da Beira Interior, Portugal
Ferreira Fátima   University of Trás-Os-Montes e Alto Douro,
                                   Portugal
Ferrão Maria   University of Beira Interior and CEMAPRE,
                                   Portugal
Figueiredo Manuel Carlos   University of Minho, Portugal
Filipe Ana   University of Minho, Portugal
Flouvat Frederic   University New Caledonia, New Caledonia
Forjaz Maria Antónia   University of Minho, Portugal
Formosa Saviour   University of Malta, Malta
Fort Marta   University of Girona, Spain
Franciosa Alfredo   University of Napoli Federico II, Italy
Freitas Adelaide de Fátima
   Baptista Valente   University of Aveiro, Portugal
Frydman Claudia   Laboratoire des Sciences de l'Information et des
                                   Systèmes, France
Fusco Giovanni   CNRS - UMR ESPACE, France
Fussel Donald   University of Texas at Austin, USA
Gao Shang   Zhongnan University of Economics and Law,
                                   China
Garcia Ernesto   University of the Basque Country, Spain
Garcia Tobio Javier   Centro de Supercomputación de Galicia
                                   (CESGA), Spain
Gavrilova Marina   University of Calgary, Canada
Gensel Jerome   IMAG, France
Geraldi Edoardo   National Research Council, Italy
Gervasi Osvaldo   University of Perugia, Italy

Giaoutzi Maria	National Technical University Athens, Greece
Gizzi Fabrizio	National Research Council, Italy
Gomes Maria Cecilia	Universidade Nova de Lisboa, Portugal
Gomes dos Anjos Eudisley	Federal University of ParaÃba, Brazil
Gomez Andres	Centro de Supercomputación de Galicia, CESGA (Spain)
Gonçalves Arminda Manuela	University of Minho, Portugal
Gravagnuolo Antonia	University of Napoli Federico II, Italy
Gregori M. M. H. Rodrigo	Universidade Tecnológica Federal do Paraná, Brazil
Guerlebeck Klaus	Bauhaus University Weimar, Germany
Guerra Eduardo	National Institute for Space Research, Brazil
Hagen-Zanker Alex	University of Surrey, UK
Hajou Ali	Utrecht University, The Netherlands
Hanzl Malgorzata	University of Lodz, Poland
Heijungs Reinout	VU University Amsterdam, The Netherlands
Henriques Carla	Escola Superior de Tecnologia e Gestão, Portugal
Herawan Tutut	University of Malaya, Malaysia
Iglesias Andres	University of Cantabria, Spain
Jamal Amna	National University of Singapore, Singapore
Jank Gerhard	Aachen University, Germany
Jiang Bin	University of Gävle, Sweden
Kalogirou Stamatis	Harokopio University of Athens, Greece
Kanevski Mikhail	University of Lausanne, Switzerland
Kartsaklis Christos	Oak Ridge National Laboratory, USA
Kavouras Marinos	National Technical University of Athens, Greece
Khan Murtaza	NUST, Pakistan
Khurshid Khawar	NUST, Pakistan
Kim Deok-Soo	Hanyang University, South Korea
Kim Moonseong	KIPO, South Korea
Kolingerova Ivana	University of West Bohemia, Czech Republic
Kotzinos Dimitrios	Université de Cergy-Pontoise, France
Lazzari Maurizio	CNR IBAM, Italy
Laganà Antonio	Department of Chemistry, Biology and Biotechnology, Italy
Lai Sabrina	University of Cagliari, Italy
Lanorte Antonio	CNR-IMAA, Italy
Lanza Viviana	Lombardy Regional Institute for Research, Italy
Le Duc Tai	Sungkyunkwan University, South Korea
Le Duc Thang	Sungkyunkwan University, South Korea
Lee Junghoon	Jeju National University, South Korea

Lee KangWoo	Sungkyunkwan University, South Korea
Legatiuk Dmitrii	Bauhaus University Weimar, Germany
Leonard Kathryn	California State University, USA
Lin Calvin	University of Texas at Austin, USA
Loconte Pierangela	Technical University of Bari, Italy
Lombardi Andrea	University of Perugia, Italy
Lopez Cabido Ignacio	Centro de Supercomputación de Galicia, CESGA
Lourenço Vanda Marisa	University Nova de Lisboa, Portugal
Luaces Miguel	University of A Coruña, Spain
Lucertini Giulia	IUAV, Italy
Luna Esteban Robles	Universidad Nacional de la Plata, Argentina
Machado Gaspar	University of Minho, Portugal
Magni Riccardo	Pragma Engineering SrL, Italy, Italy
Malonek Helmuth	University of Aveiro, Portugal
Manfreda Salvatore	University of Basilicata, Italy
Manso Callejo Miguel Angel	Universidad Politécnica de Madrid, Spain
Marcheggiani Ernesto	KU Lueven, Belgium
Marechal Bernard	Universidade Federal de Rio de Janeiro, Brazil
Margalef Tomas	Universitat Autònoma de Barcelona, Spain
Martellozzo Federico	University of Rome, Italy
Marvuglia Antonino	Public Research Centre Henri Tudor, Luxembourg
Matos Jose	Instituto Politecnico do Porto, Portugal
Mauro Giovanni	University of Trieste, Italy
Mauw Sjouke	University of Luxembourg, Luxembourg
Medeiros Pedro	Universidade Nova de Lisboa, Portugal
Melle Franco Manuel	University of Minho, Portugal
Melo Ana	Universidade de São Paulo, Brazil
Millo Giovanni	Generali Assicurazioni, Italy
Min-Woo Park	Sungkyunkwan University, South Korea
Miranda Fernando	University of Minho, Portugal
Misra Sanjay	Covenant University, Nigeria
Modica Giuseppe	Università Mediterranea di Reggio Calabria, Italy
Morais João	University of Aveiro, Portugal
Moreira Adriano	University of Minho, Portugal
Mota Alexandre	Universidade Federal de Pernambuco, Brazil
Moura Pires João	Universidade Nova de Lisboa - FCT, Portugal
Mourelle Luiza de Macedo	UERJ, Brazil
Mourão Maria	Polytechnic Institute of Viana do Castelo, Portugal
Murgante Beniamino	University of Basilicata, Italy
NM Tuan	Ho Chi Minh City University of Technology, Vietnam

Nagy Csaba	University of Szeged, Hungary
Nash Andrew	Vienna Transport Strategies, Austria
Natário Isabel Cristina Maciel	University Nova de Lisboa, Portugal
Nedjah Nadia	State University of Rio de Janeiro, Brazil
Nogueira Fernando	University of Coimbra, Portugal
Oliveira Irene	University of Trás-Os-Montes e Alto Douro, Portugal
Oliveira José A.	University of Minho, Portugal
Oliveira e Silva Luis	University of Lisboa, Portugal
Osaragi Toshihiro	Tokyo Institute of Technology, Japan
Ottomanelli Michele	Polytechnic of Bari, Italy
Ozturk Savas	TUBITAK, Turkey
Pacifici Leonardo	University of Perugia, Italy
Pages Carmen	Universidad de Alcala, Spain
Painho Marco	New University of Lisbon, Portugal
Pantazis Dimos	Technological Educational Institute of Athens, Greece
Paolotti Luisa	University of Perugia, Italy
Papa Enrica	University of Amsterdam, The Netherlands
Papathanasiou Jason	University of Macedonia, Greece
Pardede Eric	La Trobe University, Australia
Parissis Ioannis	Grenoble INP - LCIS, France
Park Gyung-Leen	Jeju National University, South Korea
Park Sooyeon	Korea Polytechnic University, South Korea
Pascale Stefania	University of Basilicata, Italy
Passaro Pierluigi	University of Bari Aldo Moro, Italy
Peixoto Bitencourt Ana Carla	Universidade Estadual de Feira de Santana, Brazil
Perchinunno Paola	University of Bari, Italy
Pereira Ana	Polytechnic Institute of Bragança, Portugal
Pereira Francisco	Instituto Superior de Engenharia, Portugal
Pereira Paulo	University of Minho, Portugal
Pereira Ricardo	Portugal Telecom Inovacao, Portugal
Pietrantuono Roberto	University of Napoli "Federico II", Italy
Pimentel Carina	University of Aveiro, Portugal
Pina Antonio	University of Minho, Portugal
Pinet Francois	IRSTEA, France
Piscitelli Claudia	Polytechnic University of Bari, Italy
Piñar Miguel	Universidad de Granada, Spain
Pollino Maurizio	ENEA, Italy
Potena Pasqualina	University of Bergamo, Italy
Prata Paula	University of Beira Interior, Portugal
Prosperi David	Florida Atlantic University, USA
Qaisar Saad	NURST, Pakistan

Quan Tho	Ho Chi Minh City University of Technology, Vietnam
Raffaeta Alessandra	University of Venice, Italy
Ragni Mirco	Universidade Estadual de Feira de Santana, Brazil
Rautenberg Carlos	University of Graz, Austria
Ravat Franck	IRIT, France
Raza Syed Muhammad	Sungkyunkwan University, South Korea
Ribeiro Isabel	University of Porto, Portugal
Ribeiro Ligia	University of Porto, Portugal
Rinzivillo Salvatore	University of Pisa, Italy
Rocha Ana Maria	University of Minho, Portugal
Rocha Humberto	University of Coimbra, Portugal
Rocha Jorge	University of Minho, Portugal
Rocha Maria Clara	ESTES Coimbra, Portugal
Rocha Maria	PRODEB, San Salvador, Brazil
Rodrigues Armanda	Universidade Nova de Lisboa, Portugal
Rodrigues Cristina	DPS, University of Minho, Portugal
Rodriguez Daniel	University of Alcala, Spain
Roh Yongwan	Korean IP, South Korea
Roncaratti Luiz	Instituto de Fisica, University of Brasilia, Brazil
Rosi Marzio	University of Perugia, Italy
Rossi Gianfranco	University of Parma, Italy
Rotondo Francesco	Polytechnic of Bari, Italy
Sannicandro Valentina	Polytechnic of Bari, Italy
Santos Maribel Yasmina	University of Minho, Portugal
Santos Rafael	INPE, Brazil
Santos Viviane	Universidade de São Paulo, Brazil
Santucci Valentino	University of Perugia, Italy
Saracino Gloria	University of Milano-Bicocca, Italy
Sarafian Haiduke	Pennsylvania State University, USA
Saraiva João	University of Minho, Portugal
Sarrazin Renaud	Université Libre de Bruxelles, Belgium
Schirone Dario Antonio	University of Bari, Italy
Schneider Michel	ISIMA, France
Schoier Gabriella	University of Trieste, Italy
Schutz Georges	CRP Henri Tudor, Luxembourg
Scorza Francesco	University of Basilicata, Italy
Selmaoui Nazha	University of New Caledonia, New Caledonia
Severino Ricardo Jose	University of Minho, Portugal
Shakhov Vladimir	Russian Academy of Sciences, Russia
Shen Jie	University of Michigan, USA
Shon Minhan	Sungkyunkwan University, South Korea

Shukla Ruchi	University of Johannesburg, South Africa
Silva J.C.	IPCA, Portugal
Silva de Souza Laudson	Federal University of Rio Grande do Norte, Brazil
Silva-Fortes Carina	ESTeSL-IPL, Portugal
Simão Adenilso	Universidade de São Paulo, Brazil
Singh R K	Delhi University, India
Soares Inês	INESC Porto, Portugal
Soares Maria Joana	University of Minho, Portugal
Soares Michel	Federal University of Sergipe, Brazil
Sobral Joao	University of Minho, Portugal
Son Changhwan	Sungkyunkwan University, South Korea
Sproessig Wolfgang	Technical University Bergakademie Freiberg, Germany
Su Le Hoanh	Ho Chi Minh City Technical University, Vietnam
Sá Esteves Jorge	University of Aveiro, Portugal
Tahar Sofiène	Concordia University, Canada
Tanaka Kazuaki	Kyushu Institute of Technology, Japan
Taniar David	Monash University, Australia
Tarantino Eufemia	Polytechnic of Bari, Italy
Tariq Haroon	Connekt Lab, Pakistan
Tasso Sergio	University of Perugia, Italy
Teixeira Ana Paula	University of Trás-Os-Montes e Alto Douro, Portugal
Teixeira Senhorinha	University of Minho, Portugal
Tesseire Maguelonne	IRSTEA, France
Thorat Pankaj	Sungkyunkwan University, South Korea
Tomaz Graça	Polytechnic Institute of Guarda, Portugal
Torre Carmelo Maria	Polytechnic of Bari, Italy
Trunfio Giuseppe A.	University of Sassari, Italy
Urbano Joana	LIACC University of Porto, Portugal
Vasconcelos Paulo	University of Porto, Portugal
Vella Flavio	University of Rome La Sapienza, Italy
Velloso Pedro	Universidade Federal Fluminense, Brazil
Viana Ana	INESC Porto, Portugal
Vidacs Laszlo	MTA-SZTE, Hungary
Vieira Ramadas Gisela	Polytechnic of Porto, Portugal
Vijay NLankalapalli	National Institute for Space Research, Brazil
Villalba Maite	Universidad Europea de Madrid, Spain
Viqueira José R.R.	University of Santiago de Compostela, Spain
Vona Marco	University of Basilicata, Italy

Sponsoring Organizations

ICCSA 2014 would not have been possible without the tremendous support of many organizations and institutions, for which all organizers and participants of ICCSA 2014 express their sincere gratitude:

Universidade do Minho
Escola de Engenharia
Universidade do Minho
(http://www.uminho.pt)

University of Perugia, Italy
(http://www.unipg.it)

University of Basilicata, Italy (http://www.unibas.it)

 MONASH University   Monash University, Australia
(http://monash.edu)

Kyushu Sangyo University, Japan
(www.kyusan-u.ac.jp)

Associação Portuguesa de Investigação Operacional
(apdio.pt)

Table of Contents

The Modelery: A Collaborative Web Based Repository

Rui Couto, António Nestor Ribeiro, and José Creissac Campos

University of Minho,
Braga, Portugal
{rui.couto,anr,jose.campos}@di.uminho.pt
http://di.uminho.pt

Abstract. Software development processes are known to produce a large set of artifacts such as models, code and documentation. Keeping track of these artifacts without supporting tools is not easy, and making them available to others can be even harder. Standard version control systems are not able to solve this issue. More than keeping track of versions, a system to help organize and make artifacts available in meaningful ways is needed. In this paper we review a number of alternative systems, and present the requirements and the implementation of a collaborative web repository which we developed to solve this issue.

Keywords: Model Driven Development, Models repository, Web collaborative repository.

1 Introduction

Research into software development processes typically produces a large amount of artifacts, from documentation and different kinds of models to the actual code. Organizing and sharing those artifacts has shown to be somehow a difficult task, due to the lack of effective support. We are particularly interested in the development of tools and techniques to support software engineering and re-engineering (c.f. [1–3]), and the problems faced by teams applying them. The amount of produced artifacts when using these tools, and (in many cases) the distributed nature of the teams, begs the question of how to adequately store, catalog, archive and share such artifacts. It becomes all too easy to lose track of existing versions, the relations between artifacts, and even the artifacts themselves.

The use of standard version control systems (such as Subversion (SVN)) has shown to be inadequate [4]. In fact, it is not our objective to have a system with version control capabilities, as delta updates. Instead, we aim towards a repository for a diversity of artifacts. By artifacts, we are referring to the inputs and outputs of a software (re)engineering process, but mostly models. Example of artifacts include different types of models, test cases, pattern catalogs, processes descriptions, software prototypes, meta-models, or database schemes.

B. Murgante et al. (Eds.): ICCSA 2014, Part VI, LNCS 8584, pp. 1–16, 2014.

Three main functionalities are considered relevant in this context: repository functionalities (archive, catalog, categorize, search, explore and share capabilities); social functionalities (research groups support, associating groups with artifacts); scientific publications support (management and association with scientific publications). We classify such platform as a collaborative Web repository. On the one hand, it allows multiple researchers to collaborate in a project through a Web environment. On the other hand, it provides archiving capabilities (i.e., a repository). We consider a Web portal to be the best solution to access this type of system. It ensures that the users will be able to access it from almost any device with a Web browser, without the need to install any software. Some Web 2.0 functionalities, such as dynamic content and user supported contents (i.e., forums), improve both the interaction of the users with the platform, and among them.

In this paper we present and discuss the implementation of the proposed platform. Section 2 reviews related work, with the analysis of a number of similar tools. Section 3 builds on that to present the requirements for the platform. In section 4 the tool is described. Finally, Section 5 presents some discussion about what has been achieved, and Section 6 concludes the paper with some pointers for further work.

2   Related Work

A study was carried out to analyze the state of the art for collaborative repository tools.

It covered, not only software oriented repositories, but also other platforms, such as business process repositories and books cataloging systems. The analyzed repositories can be categorized into two main approaches. First, there are the data repositories, common among the database research communities. They are the extension of a database management system, with emphasis on metadata management. The repository consists in a "shared database of information about engineered artifacts produced and used by an enterprise" [5]. Model management systems are also related with data repositories, addressing problems of models representation and processing [6]. Second, there are the process model repositories, based in workflow and conceptual modeling. They provide a repository and execution environment for those models [7].

The analysis of the related work produced two major outcomes. First, it allowed evaluating how suitable for our purposes existing software systems were. Second, it provided valuable input regarding the requirements for this type of platform. A contribution of this work is the table presented in Section 5. It presents the comparison of the discussed platforms regarding their functionalities. This section presents the most relevant tools.

2.1   Repository for Model Driven Development (ReMoDD)

ReMoDD[1] is a Web platform developed by the Colorado State University Department of Computer Science and Engineering [8]. This platform aims to support the Model Driven Development (MDD) community by providing an easy and complete way to share models, informations, case studies and knowledge among multiple audiences, for instance teachers, researchers and students.

The tool provides a Web portal to interact with the repository. It supports browsing the repository by listing the models sorted by multiple attributes: name, description, categories, author(s) or update data. The only organization criteria is this sorting functionality. Other than that, it is only possible to open a model and view its full description (there is no search feature). By opening a model it is possible to visualize its details, post comments and download it. To perform further interactions user registration is required. This platform provides also a group (or forum) functionality, where registered users are able to interact.

As artifacts' discovery is relevant to us, the lack of search and list functionalities presents a big limitation. We consider that viewing the models' informations is also one of the most relevant functionalities, which is very limited in the ReMoDD, providing only general information and lacking the authoring tool, scope and version (among other informations). The group functionality is a forum-like functionality, but lacks a deep integration with the rest of the framework. Finally, at the moment, the platform is not accepting registrations.

2.2   ECOBAS

ECOBAS is an information system which supports online modeling and simulation. It is designed for ecology and environmental sciences [9]. This tool offers some interesting repository functionalities. It provides both a Web interface and local client. The Web interface allows users to search a model by name, by subject or by free-text. Viewing the models' informations is similar to other repositories. It is possible to select a model from a list, and its details are presented.

The focus of this platform on ecological and environmental context makes it unsuitable for our purposes. However, analyzing the tool made us aware of the importance of having an open platform. A flexible platform should provide support for a large variety of models, regardless of their application area. Another limitation of ECOBAS is the information shown about each model, which despite being detailed misses some relevant informations such as a visual representation.

2.3   Apromore

Apromore is a Business Process Model repository [7]. While it was possible to test a first version of the repository, that version has since then been deprecated and taken offline. A new version of the tool is under development but is currently unavailable to test. Hence the current analysis referes to the deprecated version.

---

[1] http://www.cs.colostate.edu/remodd/v1/ (visited January 30, 2014).

Apromore provides model storage and management funcionalities (both view and create/edit). The models' discovery functionalities are adequate, as they support listing, searching and filtering of models (by criteria). All the models' details are available, and it supports rating the models. This tool provides an intuitive user interface for model management. However, groups are not supported, and all models exist at the same level, being available to all users (there is no visibility concept). This platform is closer to a repository than to a collaborative environment. Additionally, it support only the storage of models created directly in the platform. Hence, the tool is too restrictive to be considered a generic collaborative platform.

2.4   Shelfari

We consider model organization, storing and categorization as the core of the a model repository. Such functionalities are found in books managing systems, as is the case of Shelfari. This platform provides a digital library to store and organize books. Book entries can be searched, listed, added, removed and rated. Cataloging is done through several aspects, such as subject, author and tags. The concept of group is also present, where a set of users sharing the same interests about a particular subject can discuss it. While not directly usable for our needs, the tool provides useful hints for developing a new platform, as the task of cataloging artifacts shares some concepts with cataloging books.

2.5   Other Tools

A number of other tools were identified and analyzed. Due to space constraints they are not discussed in-depth here. Briefly, from these tools, we may highlight the following ones.

- ATL Zoo[2], which presents a list of artifacts in a Web page, accessible also from the eclipse Integrated Development Environment (IDE).
- ARIS[3], an enterprise architecture management tool with repository functionalities.
- Adonis, a commercial platform focused in business process management [10].
- Colex, a model repository that targets model versioning and versioning conflicts [11].
- ModeleR, a model knowledge base for experts, targeting environmental model execution [12].

2.6   Discussion

None of the analyzed tools was found suitable for our purposes. Briefly, it is possible to say that the tools are either for a specific domain, for a specific

---

[2] http://www.eclipse.org/atl/atlTransformations/ (visited February 26, 2014).
[3] http://www.aris.com/ (visited January 31, 2014).

language, are closed (for registration), or are too limited in functionalities. A platform that seems promising is ReMoDD. This platform could fulfill our needs and solve our problems. However, a set of limitations (not the least of which is the fact that it is currently not accepting further registrations) made this platform inadequate for our objectives. Additionally, the platform lacks Web 2.0 functionalities to encourage collaboration between researchers [11].

While we found no suitable tool to cover our need for a collaborative artifacts repository, the analyzed tools were able to provide us with insights on what functionalities to implement to achieve an usable and adequate artifacts repository. This is discussed in the next section.

3   Requirements for a Collaborative Web Repository

As none of the analyzed tools is adequate for our purposes, we propose to create a new collaborative Web repository. Combining our need with the informations extracted from the tools' analysis allowed us to define a set of requirements to guide us in the development of a new tool. We present in this section these requirements, considered essential for our repository. Our objective goes towards the development of a Web platform supported (i.e., the artifacts are provided) by the community.

To start with, the platform will require what in [7] is designated as the standard repository functionalities, which include data storage, access control, and simple search queries. Those requirements are not enough when developing a new system, if we want it to be better than existing solutions. We decided to include some other functionalities, such as advanced search functionalities.

3.1   Artifacts Repository

One of the main functionality that we look forward in a repository, is the artifact archiving and cataloging. Archiving artifacts will help keep track of them, store them in a centralized platform and share them with third persons. Cataloging the artifacts allows to store them in a meaningful way, and later to ease the process of finding them. The cataloging enables also the possibility of other people finding models. We consider that multiple approaches should be possible when browsing the models, namely textual search, criteria listing and criteria browsing. Also, multiple criteria for cataloging should exist in order to ease the browsing process.

Searching artifacts by text should support finding models either by name or description. This is the most direct way to perform searches, since textual forms are the most common way found nowadays. Criteria listing represents a search which allows the user to select from a set of predefined criteria. With this approach it is possible to filter the models to a subset containing only the relevant criteria selected by the user. Criteria browsing is a refined search approach, which allows searching the models by reducing the number of results as we select criteria, related with previous selections. This approach will raise the probability of finding artifacts within the repository.

Models are prone to changes and updates, and such factor is essential when developing a repository. In order to support such behavior we propose supporting several versions of the same model, sorted in a meaningful way.

The decision of making an artifact public (accessible to everyone) or private is left to the user. Hence, the user might decide to kept an artifact private, for instance while in development, or only available to a subset of users. If an artifact is public, it should be accessible by anyone, allow to add comments, ratings, and even keep track of it. This is where the collaborative functionalities start, in the sense that other users may collaborate in the development or improvement of an artifact. If an artifact is private, only the author should be able to see and modify it. Lastly, in order to support collaboration, an artifact must be able to be restricted to a group.

3.2   Publications Management

Developing tools and works in academic context results in a large set of scientific publications. The publications arise in several computer science areas, and sometimes they are related with artifacts. Hence, in this context is makes sense to manage references to scientific publications, associating them with the artifacts. As an artifact might also be referred in several articles, we propose a bidirectional relationship between artifacts and publications. With this functionality it should then be possible to search models related with specific publications, or otherwise, search publications related with specific models.

3.3   Social Functionalities

It is common for the research process to involve interaction among several persons and ideas as well as previous works. The collaboration and sharing of information improves the research results. From multiple people, different approaches emerge and sometimes best results are found by combining several persons' ideas. This is the basis of the collaborative platforms [13].

An improvement on the repository would be to deeply integrate the social functionalities with the artifacts. The group concept, allied with the forum functionalities seems an appropriate requirement. By creating groups where the users could discuss ideas, and associate artifacts to them, it would allow a collaborative comportment.

In the same way that the models have a visibility option, it makes sense to have the same option for the groups. Hence, it should be possible to make a group (as well as its artifacts) restrict to a set of users. With this approach only the subset of persons related with the project would have access to the information. This is specially useful for private projects, projects in development, or simply by convenience. When an artifact is part of a group, it would be adequate to allow both the author and the members of the group to update it.

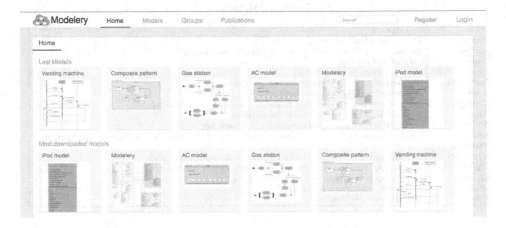

Fig. 1. Modelery main page

3.4 Levels of Sharing

Not all the artifacts and groups are developed for the same purpose. Some of them are intended to be public, other restricted to a subset of persons (and able to be updated by all these persons, or only the author) and other completely private. Also the groups may either be public or private (selecting the persons which should belong to them).

The distinction between all these visibility levels is crucial to cover a broader audience of developers. Also, an author might decide to keep a model private while developing it, and make it public once finished. It provides some more control over the development process.

It is easy to think in version control functionalities (e.g. for models) as adequate for such platform. However, at this point, such functionality will not be considered. Firstly, implementation of version control functionalities is known as a hard task [4]. Then, we produce models in many languages (some of them are not even standard), which results in known versioning problems [14]. By merging these two factors we face a complex problem that we decided not address at the moment. Furthermore we are more interested in cataloging artifacts (where the artifacts should be more stable and ready to be used by other users), than in a centralized development tool as is the case of control version systems.

4 The Modelery

In order to solve the inability of the analyzed tools to fulfill our needs we have developed the Models Refinery (Modelery)[4]. Our platform combines the proposed functionalities in a single Web environment, accessible through the browser as depicted in Figure 1. Here we present the decisions which lead to our tool, as

---

[4] http://modelery.di.uminho.pt

Table 1. Meta-data

Item	Description
Name	The name of the artifact
Author	The author of the artifact, automatically associated
Date	Date of submission
Description	A description of the artifact
Institution	Institution where the artifact was produced
Tool	Tool which originated this artifact
Category	Category where the artifact belongs, i.e., the area of knowledge
Tags	A set of tags, associated with the artifact
Scope	The artifact scope, within the area of knowledge
Language	The language in which the artifact was created (for instance, programming language)
Publications	List of publications associated with the artifact
Visibility	Visibility of the artifact: Only to author, to group, or public
Updatable	Whom may update the artifact: only the author, or the group
Group	The group which the artifact may belong
Image	An image representing the artifact
File	The artifact file itself

well as the description of the functionalities. It was developed according with a model driven methodology, and used the Modelery itself to keep track of the source models.

4.1   Artifacts Repository

The artifacts repository functionality was our major concern. We are interested in storing not only the artifacts, but also their meta-data. This meta-data constitutes the artifact's entry, provided by the user when submitting it to the repository, and it is essential information to provide when displaying an artifact. Table 1 summarizes an artifact's attributes. These are further discussed below. Figure 2 presents the corresponding Web page.

Collaborative functionalities are achieved by supporting interaction between the users, trough the artifacts in the platform. This interaction fosters the artifact's evolution, due to the users feedback. Indeed, registered users may interact with an artifact by adding comments (which may help the author or other users). Also, the users might rate it, expressing its satisfaction with the artifact, with a value from 1 to 5. The artifacts' author is able to both update the model (by submitting a new version - the previous version is kept on record), and to edit the artifacts' meta-data.

Our platform relies on artifacts created by the users. Hence, an artifact must always have an author. While any user might search and view (public) artifacts, registration is required in order to create a new one. When creating the artifact, the user should specify all the details, as well as group, publication and visibility options. The artifact file should be also specified, and it is then uploaded and stored online in the platform.

As the artifacts belong to a specific context, we provide two ways to specify it. First, we allow an artifact to be part of a group. This possibility enables us, not only to aggregate a set of artifacts in a specific group, allowing for their categorization, but also to restrict its access to a set of persons which may view or update it, the members of the group. Second, we provide also a means to

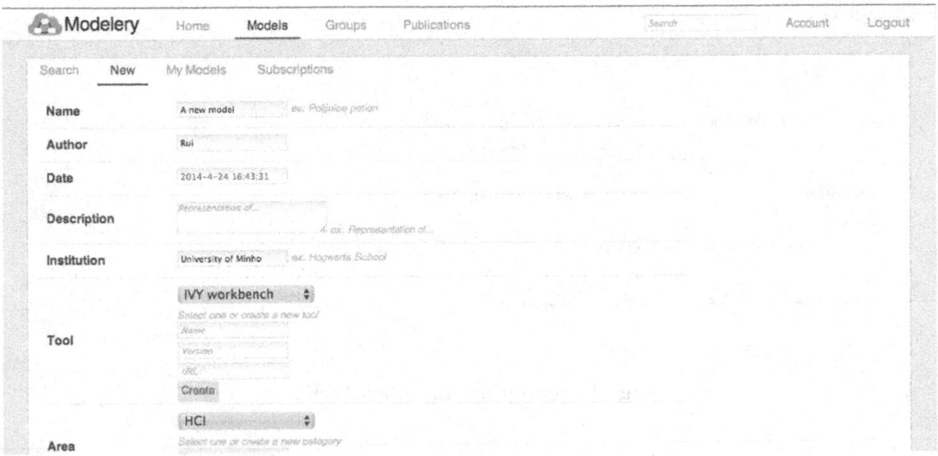

Fig. 2. Adding an artifact

identify the publications (for instance articles, papers or posters) in which an artifact is be involved. This constitutes a further dimension though which to classify and access artifacts.

The artifacts have different purposes, and some of them (such as models) may belong to different development phases. As such, the artifacts' visibility level defines if it should be visible to everyone, visible to the group members, or visible only to the author. This enables users to keep artifacts private during their development phase (or permanently, of course), or opt to share them with a restricted group of people who might comment of actively collaborate on its development. The visibility level allows also to define which users might update the model. Here, the owner of an artifact may let a group update it, or restrict updates to himself/herself. The visibility level and who may update an artifact are independent properties, since it may be visible to the group, but only the author might have permission to update it.

The platform provides several ways to search and explore artifacts. By selecting the search option, a listing of the existent artifacts is presented, as depicted in Figure 3. We provide the possibility, also, for an author to view a list of his/her own artifacts. The user may then input some text, and the listing will start to be filtered, by presenting only the artifacts whose name or description match the text being input. This is the more natural search approach, common in most repositories.

Alternatively to the textual search, the user has the possibility to browse the artifacts. Browsing differs from searching by presenting the user with a set of predefined criteria: the tool in which the artifact was developed, the language in which it was written, its category and its author. With this approach it is possible for a user to select all the artifacts containing the specified properties. This supports a rigorous filtering of the artifacts, and allows also the filtering of artifacts by several criteria at the same time.

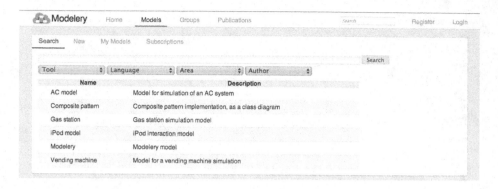

Fig. 3. Searching for an artifact

Fig. 4. Adding a publication

A specific artifact (or set artifacts) might be of interest to a user. In order to ease the user access to those relevant artifacts, we provide the possibility to "track" an artifact. This means that a user may choose to follow the progress of a specific artifact, keeping a reference for it. It is also possible for a user to list the artifacts that he is following.

4.2   Publications Management

As already mentioned, Modelery supports the possibility to create publication entries. The publications are registered with their name, abstract and URL for the article location, as shown in Figure 4. Contrary to what is provided for artifacts, publications management does not allow uploading the publication itself into the platform. We consider this to be a more efficient approach, as the platform's focus is not publications' management. Since publications may have more than one author, they are not automatically associated with the user which created them. Information of the authors is in the publication document itself.

Fig. 5. Searching groups

The relation between the artifacts and the publications can be explored starting from either entity. On the one hand, publications may refer a specific tool or artifacts, and it is possible to list the artifacts associated with a publication. On the other hand, an artifact may be referred in multiple publications, and it is possible to view all its associated publications. This functionality provides a convenient way to explore publications along with artifacts, and at the same time provides more information for a given artifact. It allows also exploration of the practical results (i.e. artifacts) of the publications.

As a large set of publications might be added, we provide also search functionalities for them. The textual search functionality is provided for publications. Along with textual search we provide the possibility to browse the publications, by their kind, date of publication or publisher.

4.3 Collaborative Functionalities

Since we are developing a collaborative repository, Web 2.0 functionalities are essential to promote interaction among users [12]. Once registered, users are automatically associated with any group, artifact, comment or update that they create. This allows other users to know who is the author of a given artifact, or the owner of a specific group. Users are responsible for managing the groups that they creates, by selecting which other users should be part of the group. In Figure 5 is presented an overview of the groups search page. A functionality which is essential for promote collaborative behaviors is the possibility of users to exchange messages inside the platform. The Modelery supports both personal one-to-one messages, and more public messages in a discussion group (or forum).

Other functionalities include the dynamic main page, which presents information such as the last submitted artifacts and most downloaded artifacts, and a tag cloud. This provides an overview of the contents of the repository, emphasizing most relevant artifacts.

4.4 Usability Improvements

Due to the relevance of usability considerations for the platform's success, an effort was made to create a responsive user interface (for instance avoiding to reload full pages for small requests) in order to improve the experience of the users. This was mainly achieved recurring to Ajax, by performing modular page loadings. This also enables us to provide more lightweight Web pages and reduced bandwidth usage. Resorting to a combination of HTML5[5], Cascading Style Sheets version 3 (CSS3) and jQuery, we are able to improve the user interface by, for instance, providing early error detections when filling field in the Web page, and better feedback (including animations when performing changes to the page contents).

Nevertheless, the Web interface was developed taking in mind compatibility with old browsers. Even if the visual aspect is not kept (mainly due to CSS3 compatibility), all the functionalities remain usable.

4.5 Implementation Status

The Modelery was developed according to a multi-layer architecture, using a model driven approach. The presentation layer was implemented using Java Server Pages (JSP) and servlets over the business layer. Following a multi-layer approach allows us to easily improve or change some of the platform components. For instance, it would be simple to add Web-services over the business layer.

At the moment, the tool is fully functional. All the described functionalities, including access control are available. It is also possible to access public artifacts and information without registration.

5 Discussion

Regarding our initial goal of a platform to support the archiving and exchange of models and other type of artifacts, relevant to research into software engineering methods and tools, we have now achieved a fully functional prototype, which we consider implements the more relevant functionalities identified.

An alternative approach to achieve a similar platform would had been to conjugate several other platforms into a single environment, for instance a Concurrent Version System (CVS) (such as SVN or GIT) for artifacts management, along with an online forum (such as phpbb) for discussion issues. However, the approach taken presents advantages over the integration of multiple platforms. First, CVS system are mainly used and optimized for textual documents (such as source code). They lack model targeted functionalities, and it is harder to add functionalities (such as an online model editor) later on. Furthermore, CVS systems are not targeted for sharing and cataloging. Using an online forum for our objectives suffers from similar issues as the usage of a CVS for the models, with the inability to provide specific functionalities. Integrating visibility levels

---

[5] http://www.w3.org/TR/html5/ (visited January 30, 2014).

in a CVS, or groups, managed by the users, in the forum, would have been a very hard and time consuming tasks. Combining these functionalities to collaborate together, by providing a platform as coherent and as practical as ours would have been more costly than developing this one. Finally, a poor integration of these technologies might easily lead to an unpractical platform, and result in a project failure.

Some of the repositories discussed in Section 2 offer online models' editing. That is an interesting functionality. However, not suitable for our repository at the moment. Since we allow any kind of artifact (therefore any models) in our the repository, it would require either a restriction on the type of supported artifacts (by imposing a metamodel, for instance), or selecting a subset of artifacts with online editing functionalities. We have chosen to ignore this functionality for now, since it would not lead to a solid and robust editor.

Comparing our platform against other repositories, it is possible to draw some conclusions. There are some similarities between our tool and ReMoDD, since our objectives are somehow similar. However, we provide some improvements over Modelery. First, our platform provides a larger group of functionalities without requiring registration. An unregistered user is free to explore all the public information, from groups to models and publications. ReMoDD is considerably more restricted in model browsing. The only way to search content in the site (any kind of content) is by textual search. Another possibility is to list all of the models. The platform provides also a forum, however completely disconnected from the models. Finally, it provides a workshop catalog system, once again, disconnected from the models. Viewing a model's information is very limited, since only few informations are displayed. ReMoDD claims to be a repository for model driven development, however our platform might provide a better support for model driven methodologies by overcoming some of ReMoDD shortcomings.

ECOBAS is targeted to different purposes, being aimed at a specific area and focusing on modelling and simulation. In what concerns management of models, ECOBAS is somewhat limited in terms of search functionality, since it only supports the listing of models by name, or performing a textual search. Opening a model's entry provides a large amount of information, but lacks some of the details we consider relevant, such as a visual representation of the model or the author. ECOBAS lacks also other functionalities such as publications management and discussion groups. From this point of view, Modelery provides a more complete environment as a model repository.

The Apromore platform shares some of our objectives, but is currently in a preliminary phase of development. The platform allows public models' submission only, limiting the models' scope. The model entries do not provide very complete information, since apart from its name, it is only possible to view their language, domain, ranking, version and author. The platform offers an interesting online model editor. However that editor is language specific, allowing only to edit one kind of model. Also, Apromore provides no other functionalities than a model repository. At the moment, this platform has limited browser support. Modelery provides a more usable option, since it is ready for use. Users are free

Table 2. Comparison of the analyzed repositories

Tool	Fully Web	List	View	Comments	Download	Public access	Groups	Advanced search	Open platform	Software oriented
ReMoDD	✓	✓	☐	✓	✓	×	☐	☐	×	✓
ECOBAS	×	✓	☐	×	☐	✓	×	✓	✓	×
Apromore (prev.)	✓	✓	✓	×	×	✓	×	✓	✓	×
Shelfari	✓	×	✓	✓	×	✓	✓	✓	✓	×
Modelery	✓	✓	✓	✓	✓	✓	✓	✓	☐	✓

to register (contrary to Apromore), and submit any artifact (not only models), as well as their relevant informations.

Table 2 summarizes the comparison of the platforms.

6   Conclusions

In this paper we have described a collaborative repository for software artifacts, with special focus in models, patterns and catalogs. We presented the Modelery, a platform which combines an online artifact repository, publication management and social functionalities. The presented functionalities came mainly from our needs to store, manage, catalog and make the artifacts we produce during our research projects, available online. Also, with this platform we have created a new means to discuss the artifacts within discussion groups.

This platform corresponds to a prototype developed in order to fulfill our need for a collaborative model repository. The implemented functionalities represent a first approach, and as such there are many planned improvements as future work. Our mainly outreach with this platform is the academic community. This is due to our platform nature, developed within an academic and research context. In long term, we intent to outreach other areas, such as general research purposes (including research in business contexts), and even to support model driven development and artifacts sharing for enterprise contexts.

We have started using the repository for our own needs[6]. This has allowed us to test the repository and made possible minor adjustments. Our immediate next objective is to make it publicly available and encourage other research groups to adhere to it.

In the longer run. we take also in account the possibility to include other functionalities in the platform. Namely, the possibility of integrating editors or the generation of graphical representations for particular modelling languages, and also integration with verification and validation tools (e.g. for certification purposes). The integration with other tools can be achieved by means of Web

---

[6] http://modelery.di.uminho.pt

services. We will study the possibility to include such functionality, through Simple Object Access Protocol (SOAP) or REpresentational State Transfer (REST) technologies.

Acknowledgments. This work was carried out in the context of project LATiCES: Languages And Tools for Critical rEal-time Systems (Ref. NORTE-07-0124-FEDER-000062) is financed by the North Portugal Regional Operational Programme (ON.2 - O Novo Norte), under the National Strategic Reference Framework (NSRF), through the European Regional Development Fund (ERDF), and by national funds, through the Portuguese funding agency, Fundação para a Ciência e a Tecnologia (FCT).

References

1. Couto, R., Ribeiro, A.N., Campos, J.C.: A Patterns Based Reverse Engineering Approach for Java Source Code. In: 2012 35th Annual IEEE Software Engineering Workshop (SEW), pp. 140–147 (2012)
2. Campos, J., Saraiva, J., Silva, C., Silva, J.: GUIsurfer: A Reverse Engineering Framework for User Interface Software. In: Telea, A. (ed.) Reverse Engineering - Recent Advances and Applications, pp. 31–54. InTech (2012)
3. Campos, J.C., Harrison, M.D.: Interaction engineering using the IVY tool. In: ACM Symposium on Engineering Interactive Computing Systems (EICS 2009), pp. 35–44. ACM, New York (2009)
4. France, R., Bieman, J., Cheng, B.: CRI: Collaborative Project: Repository for Model Driven Development (ReMoDD). Colorado State University (2006)
5. Bernstein, P.A., Dayal, U.: An Overview of Repository Technology. In: Proceedings of the 20th International Conference on Very Large Data Bases, VLDB 1994, pp. 705–713. Morgan Kaufmann Publishers Inc., San Francisco (1994)
6. Dolk, D.R., Konsynski, B.R.: Knowledge Representation for Model Management Systems. IEEE Transactions on Software Engineering SE-10(6), 619–628 (1984)
7. La Rosa, M., Reijers, H.A., van der Aalst, W.M.P., Dijkman, R.M., Mendling, J., Dumas, M., García-Bañuelos, L.: APROMORE: An advanced process model repository. Expert Syst. Appl. 38(6), 7029–7040 (2011)
8. France, R., Bieman, J., Cheng, B.H.C.: Repository for model driven development (ReMoDD). In: Kühne, T. (ed.) MoDELS 2006. LNCS, vol. 4364, pp. 311–317. Springer, Heidelberg (2007)
9. Cavalcanti, M.C., Mattoso, M., Campos, M.L., Llirbat, F., Simon, E.: Sharing scientific models in environmental applications. In: Proceedings of the 2002 ACM Symposium on Applied Computing, SAC 2002, pp. 453–457. ACM, New York (2002)
10. Karagiannis, D., Kühn, H.: Metamodelling Platforms. In: Bauknecht, K., Tjoa, A.M., Quirchmayr, G. (eds.) EC-Web 2002. LNCS, vol. 2455, p. 182. Springer, Heidelberg (2002)
11. Brosch, P., Langer, P., Seidl, M., Wieland, K., Wimmer, M.: Colex: a web-based collaborative conflict lexicon. In: Proceedings of the 1st International Workshop on Model Comparison in Practice, IWMCP 2010, pp. 42–49. ACM, New York (2010)

12. Pérez-Pérez, R., Benito, B.M., Bonet, F.J.: ModeleR: An enviromental model repository as knowledge base for experts. Expert Syst. Appl. 39(9), 8396–8411 (2012)
13. Wang, H., Johnson, A., Zhang, H., Liang, S.: Towards a collaborative modeling and simulation platform on the Internet. Adv. Eng. Inform. 24(2), 208–218 (2010)
14. France, R., Rumpe, B.: Model-driven Development of Complex Software: A Research Roadmap. In: 2007 Future of Software Engineering, FOSE 2007, pp. 37–54. IEEE Computer Society, Washington, DC (2007)

MetamorphosIS: A Process for Development of Mobile Applications from Existing Web-Based Enterprise Systems

Itamir de Morais Barroca Filho[1] and Gibeon Soares de Aquino Junior[2]

[1] Informatics Management Office (SINFO)
[2] Department of Informatics and Applied Mathematics
Federal University of Rio Grande do Norte Natal, Brazil
itamir@info.ufrn.br, gibeon@dimap.ufrn.br
http://www.info.ufrn.br, http://www.dimap.ufrn.br

Abstract. Considering the mobile computing era we realized that information systems are experiencing a process of metamorphosis to enable users to use new ways to access information from mobile devices. This is mainly due to the increased popularity of these devices such as smartphones and tablets. Driven by this new computing scenario, which is changing old habits and creating new ways to access information that previously was only accessible via traditional computers, there are growing demands for mobile enterprise applications (MEA). This increase is caused by the companies' need to ensure their customers new forms of interactions with their services. Thus, this paper aims to introduce a process called MetamorphosIS, which provides a set of activities subdivided into three phases: requirements, design and deployment, to assist on the development of mobile applications from existing information systems. The article also describes the SIGAA Mobile application, which is a case study of the use of the activities described in this process in an existing web information system called SIGAA.

Keywords: Mobile computing, mobile enterprise applications, process, MetamorphosIS, SIGAA mobile.

1 Introduction

Mobile computing is becoming more present in daily life. Nowadays, smartphones and tablets have processing power that until a while ago only existed in "modern" computers, with large memory and processing capacity. According to [1], 1.75 billion people have mobile phones with advanced capabilities and points to further growth in the use of such technology in the coming years. Thus, the information became accessible from powerful mobile devices in terms of resources and lower sizes. There is a global trend towards increasing the number of users connected to the network via mobile devices, which in turn will produce an increasing demand for information systems, applications and content for such equipment.

B. Murgante et al. (Eds.): ICCSA 2014, Part VI, LNCS 8584, pp. 17–30, 2014.

Moreover, as a result of the diversity of features and capabilities offered by such devices, we can observe a large increase in their sales in recent years. [2] estimates that 1.9 billion mobile phones will be shipped in 2014, a 5% increase in comparison to 2013. As a result of smartphones and tablets sales there is also an increase demand for new applications. This can be seen by the growing number of application downloads on mobile application markets, such as Google Play and Apple AppStore. About this fact, [3] estimates that the number of downloads of mobile applications will grow from 10.9 billion in 2010 to a total of 76.9 billion in 2014.

Thus, driven by this new computing scenario, that is changing old habits and creating new ways to access information that until now were only accessible via traditional computers, there is a growing demand for enterprise mobile applications (MEA). [4] predicts that developers will create apps for virtually every aspect of a mobile user's personal and business lives that will 'appify' just about every interaction between physical and digital worlds. There is a natural tendency for companies that have web information systems to begin to adapt them to fit this new computing scenario. This is an essential strategy for such systems to continue attracting and serving the needs of its users. According to [5] 90% of 250 IT managers had plans to develop new mobile apps within their company by the end of 2011 and there is a considerable interest in MEA and willingness to invest in these technologies.

Therefore, we realize that traditional information systems are experiencing a process of metamorphosis to fit this new computing context and new way of accessing information that is being made possible by today's mobile devices. However, it is important to note that according to [6], the development of these new types of systems involves several activities, such as: the construction of MEA considering development plataform (like Android, iOS, Web Mobile); integration with exclusive services on these devices, such as GPS, SMS, NFC and other telephony services; development or evolution of existing web and embedded systems; and integration between these systems and MEAs. We also can't fail to consider constraints that this metamorphosis should analyze, such as screen size and connectivity of mobile devices. It's important to review our knowledge of software development, particularly in processes, methods, techniques, patterns and architectural solutions for applications to fit this new computing environment.

Finally, this article aims to introduce a process named MetamorphosIS to assist in MEAs development from existing web information systems. This process is based on a set of activities subdivided into three phases: requirements, design and deployment, that will be presented in section 3. In Section 2 we discuss the characteristics of MEAs and in section 4 will be presented SIGAA Mobile, a case study of MetamorphosIS process utilization. Then, in section 5 we present conclusions of this article and future works.

2   Mobile Enterprise Applications (MEAs)

The MEAs are mobile applications developed with the purpose of providing enterprise information obtained from web information systems. In [5], an interview with six experts where everyone agreed with the definition of MEA was made and they added that, considering the enterprise context, the company aggregates value by increasing productivity and/or reducing cost.

Also in [5], the experts believe that the potential of the MEA is on customer support: sales and service applications. This may be seen at mobile applications markets (Google Play and Apple App Store) where there are a considerable range of customer support applications, such as: applications for airline check-in services, enterprise applications to consult deliveries and applications for sales. This kind of MEA is defined in [5] as business to private end consumer (B2C) whose goal is the client integration with the company's business information. [7] presents some scenarios of using mobile applications in business and enterprises such as: intelligent advertising, costumers profiling and loyalty and collective intelligence.

In this new computing scenario, mobile devices become faster access to information, which can be decisive given the competitiveness of the business sector. Moreover, this new way of interaction brings convenience to users of web information systems. Since the MEAs are created from existing information systems, it is important to know during the development which activities must be performed to ensure that they fit the constraints of the mobile computing scenario. Among these constraints we have: small screen size, connectivity, display Resolution, data entry methods [10] that complicate typing long texts. It is also essential that the MEAs offer the user a good experience. To assist with these issues a process called metamorphosis, whose aim is to provide a set of activities that should be performed to create a MEA will be presented in section 3.

3   MetamorphosIS Process

The metamorphosis process is a set of activities categorized into three phases: requirements, design and deployment, that should be considered during the development of a MEA from an existing web information system. This process is driven by two macro actions:

1. Selection and adjustment of existing functionality created for web information system to become accessible from mobile devices; Creation of native mobile application on platforms such as: Android, iOS or Web mobile; developing Restful Web Services and integration with the existing web information system that implements the business rules.
2. Development of specific functionalities for mobile devices using embedded technologies such as GPS and NFC.

As seen in section 1, MEAs development involve a new computing scenario and requires the use of new approaches for the different facets of software development process. Thus, driven by these two macro actions and new approaches, the metamorphosIS phases are:

Requirements: business approaches, involving activities that are related to the scope and stakeholders.

Design: technical approaches, related to the application's source code activities, architectural solutions [11], technologies, frameworks, design patterns and best practices.

Deployment: activities related to MEA deployment.

These phases are detailed in sections 3.1, 3.2 and 3.3.

3.1   Requirements

At the beginning of the project, it is very important to plan which functionalities are relevant in the context of mobile environments. The process of creating a mobile application from existing web enterprise system is not a direct mapping of functionality-to-functionality. This kind of simplification is a common mistake and must be treated carefully. Mobile devices have some intrinsic restrictions such as screen size, difficulties to type long texts and no guarantee of network access availability. Moreover, it has a different mode of user interaction with touch support, gesture events and rapid actions. According to [9], mobile applications tend to provide relevant advantages to their users in terms of design and usability. For this reason, it is important to follow some specific strategies on projects involving mobile application development.

Thus, the first activity of requirements phase, presented on figure 1, is to analyze which features of the existing web information system would be important to mobile context. For this, four practices shoud be considered:

1. Choose popular functionalities;
2. Avoid long-steps functionalities or long-fill forms;
3. Adapt existing functionalities;
4. Create specific functionalities for mobile application.

The artifact of this activity is a list of pre-selected functionalities that will be validated with the stakeholders, which is the second activity of this phase. If the pre-selected list is not validated with stakeholders, we need to go back to the previous activity, which is the analysis of the existing system's functionalities. Otherwise, we pass to the next activity, which is the evaluation of the functionalities pre-selected considering the mobile context restrictions. In this activity, we verify the sizes of the forms and the amount of steps on these functionalities. If we consider the functionality suited for mobile context, it passes to the second artifact of this phase, which is a list of selected functionalities for development. Otherwise, if a pre-selected funcionality is not suited for mobile context, we evaluated the possibility of adaption of the functionality, which can result in a new design for the operation, reducing the amount of fields and steps.

If such a change is possible in functionality, it is selected for development. Otherwise, it goes to another artifact of this phase, which is a list of functionalities not selected for development.

For each functionality selected for development, we evaluate the need to work offline. This is important because mobile devices are often connected to a network using wireless connections, whose availability may be low. Moreover, the connectivity may be unstable while the user is interacting with the system. The necessity to run offline directly impacts on the design phase.

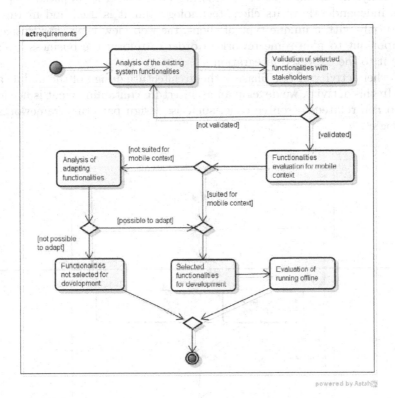

Fig. 1. MetamorphosIS requirements phase

3.2   Design

A challenge that needs to be addressed is how to integrate the MEA with the existing system. Moreover, we need to be concerned about how to reuse its already implemented business components. Nowadays, the use of the layer pattern [8] is particularly common in web information systems. For this reason, a generic approach that can be used to integrate with the existing web system is the definition of a new separated layer providing a set of services that must be used by the mobile application. This new layer integrates with the existing business rules layer using the already implemented and stable code and it is developed in design phase, presented on figure 2, where there is an activity to design this service layer.

Although the integration with the business layer seems so simple, it may require some refactoring in the existing code. It is caused by a habitual phenomenon called "software architecture erosion", where the violation of architecture principles during the system maintenance occurs without a malicious intent [12]. Usually the business layer on existing web system is prepared to provide services independently of its client technology, but it is used and matured integrated only with a unique type of client, the web view layer. For this reason, it is important to plan some rework in order to reform the business layer and prepare it to the expected integration.

The other activity of this phase is the architecture design of the mobile application. In this activity, we develop an architecture containing what is needed for MEA to run related to source code, such as: design patterns, frameworks and technologies.

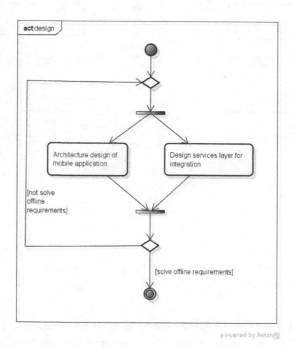

Fig. 2. MetamorphosIS design phase

Once developed the service layer on existing web system and MEA architecture, which are artifacts of design phase, we need to verify if our solutions solve the necessity of running offline detected on requirements phase. If it does, we are ready for the development of MEA selected funcionalities, otherwise we need to alter these artifacts.

3.3   Deployment

After the development of MEA, we enter in the deployment phase, presented on figure 3. The first activity of this phase is to evaluate the need for publishing the MEA on mobile applications markets, such as Google Play and Apple App Store. According to [5], there are companies that operate their own mobile app stores in order to distribute MEA to their employees, customers and partners. This is called in-house or corporate mobile app stores. So it is important to discuss with stakeholders the need to publish it on in-house or tradicional mobile applications markets.

If it has to be published on the mobile application market, one of the most important activity of this phase is the publicity around the MEA. The potential user must know this new style of enterprise system exists and they may be motivated to try this new way of accessing the system. Only publishing the application on a platform store, e.g. Android Play, Apple App Store, is not enough to make it known among users. For this reason, its existence should be well communicated to the target audience. So as artifacts of this phase, we have the MEAs publishing and publicity.

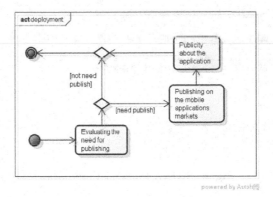

Fig. 3. MetamorphosIS deployment phase

4   SIGAA Mobile: A Estudy Case of MetamorphosIS Process Utilization

This section is organized into three subsections: 4.1, 4.2 and 4.3. In subsection 4.1, we present SIGAA (Integrated System for Management of Academic Activities), which is an enterprise web system to provide students and professors academic information from Federal University of Rio Grande do Norte (UFRN).

In subsection 4.2, we present the MetamorphosIS process application on SIGAA for SIGAA Mobile development. And finally, in subsection 4.3, we present SIGAA Mobile, which is a MEA to provide SIGAA users a new way to access academic information contained in it.

4.1 SIGAA

SIGAA (Integrated System for Management of Academic Activities) is a corporative web system developed by UFRN that computerizes the business process of the academic area through the modules that composes it: graduation; masters degree (strictu and latu sensu); technical study; elementary and high school; research projects submission and control; research grants; academic extension actions control; teaching projects submission and control (monitoring and innovation); registration and reports of professors' academic production; and a virtual teaching place called "virtual class". These modules are presented in figure 4 and are enabled according to user profiles.

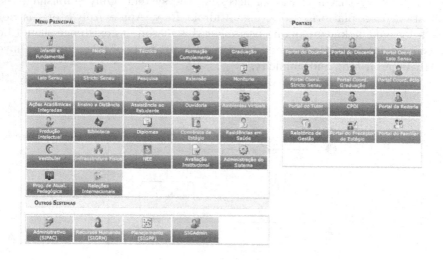

Fig. 4. SIGAA functional modules

Nowadays, SIGAA has 41,397 users, divided into the following profiles: students, professors and technicians. This web system was developed using open technologies such as Java, Hibernate, JavaServer Faces, Richfaces, Struts, EJB and Spring. It uses PostgreSQL as DBMS and is deployed into JBoss Application Server.

In terms of physical metrics, the SIGAA has 646,382 lines of code, 4,750 classes and 1,135 tables divided into 40 schemes. In terms of functional metrics, it contains 1,858 functionalities, accounting a total of 22,369 function points based on the NESMA account method. It depends on a software architecture also developed by UFRN, serving as an infrastructure to this corporative system.

This software architecture is composed by four projects presented in figure 5. The relationships between these projects are defined into three layers: data access, presentation and business:

Data access layer is responsible for persistence and data queries into database and contains Data Access Object, Hibernate and Java Data Base Connectivity (JDBC).

Presentation layer is responsible for controlling interaction between users and system and contains JSPs, web services and JSF.

Business layer is responsible for centralizing the domain business system logic. This layer contains EJB Commands.

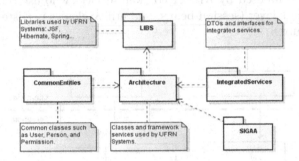

Fig. 5. UFRN Software Architecture and SIGAA

With the SIGAA's success in managing UFRN's academic activities, this system has been released for others Brazilian federal universities since 2009. Nowadays, nineteen universities are using SIGAA.

4.2   MetamorphosIS Process Utilization on SIGAA

For the development of SIGAA Mobile we used MetamorphosIS process, detaild in section 3, on SIGAA. The utilization of the activities of this project is presented in this section.

Starting from requirements phase, an analysis of the functionality SIGAA web was made. As we saw in section 4.1, SIGAA is a big information system regarding to the quantity of functionalities and it makes no sense to develop each of these features for the mobile context. Thus, as described section 3.1, we minimized the scope for the mobile version, selecting only the most popular and most suitable features for this context. After such analysis, we tried to involve several professors and students (stakeholders) to raise their validations, expectations and feedbacks about pre-selected functionalities for SIGAA Mobile. After the validation of the functionalities that were pre-selected for development with the stakeholders, we had a set of functionalities that concerned mostly in SIGAA's virtual class. Then the pre-selected functionalities were evaluated considering the constraints of the mobile context described in section 2. Some of them did not fit this context due to the amount of fields and steps. The others, that fit this context, were selected for development. Those that did not fit were analyzed to be adapted. All were able to be adapted and were selected for development. After having selected the functionalities for development, we evaluated the need for offline operation. Some operations such as register frequency of students were evaluated as necessary to work offline. At the end of the requirements phase, we had a list of operations selected for development and which ones needed to work offline.

Then, we pass to the design phase, where SIGAA Mobile architecture and SIGAA web services layer were developed. After these developments, we analyzed if the SIGAA Mobile architecture attended offline requirements. The services layer developed composed by RESTFUL web services who use the components of SIGAA business layer (session beans, and utility classes persistent entities) is presented in figure 6 .

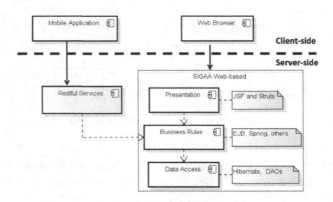

Fig. 6. Solution to integrate existing web-based SIGAA with new mobile application

As warned in section 3.2, it was necessary to perform refactorings to enable this integration through service layer, because specific objects from web frameworks were found beyond the business layer boundary. SIGAA business layer was designed for interaction with the presentation layer only through web browsers and the introduction of this new type of client (mobile devices) demanded that we removed these incorrect invasions performing refactorings. At the end of this phase we had an architecture for mobile application and a services layer in SIGAA web.

After the design phase, SIGAA Mobile was developed and we passed to Memorphosis deployment phase described in section 3.3. We evaluated the need to publish this application on Google Play and decided to use this app market instead of using an in-house solution. The reason for this decision was mainly that the cost of maintaining in-house solution was higher than the cost of publication in the applications market. In addition, we had a lot of potential users.

Then, after SIGAA Mobile release on Google Play, the AGECOM (Communication Agency of UFRN) did an intensive publicity about it and the numbers of installations grew significantly.

4.3 SIGAA Mobile

SIGAA Mobile, main screen presented in figure 7, was developed using the Android platform, using the activities of the Metamorphosis process, described in section 3, that resulted in the architecture presented in figure 8.

Fig. 7. SIGAA Mobile main screen

The View layer contains all Android activities, i.e. components which interacts directly with the user presenting the graphical interface and manipulating the user events. It uses a layer of abstraction for the business rules that are hosted in the server called Business Delegate. This layer uses a communication channel which implements the REST pattern to communicate with the server-side of the system. Moreover, it manipulates the Cache Data Access Layer, storing and recovering data, depending of the connectivity status, to allow the offline use capability.

From Android Developer Console, which is the tool for distributing Android applications, we will present some statistics about SIGAA Mobile uses extracted considering the following informations: active installations, country and application version. Nowadays, SIGAA Android is installed in about 2,400 devices (active installations), whose growing is presented on figure 9. It was uploaded do Google Play at 04/25/2012 and the download number is increasing ever since.

Observing country installations statics, it is perceived that Brazil dominates the number of downloads and installations, but analyzing figure 10 we realize that there are other countries on the statistics, like United States of America,

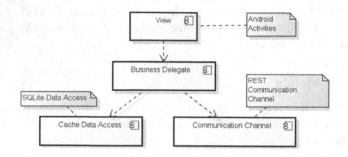

Fig. 8. Software Architecture of Android SIGAA Application

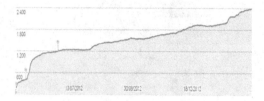

Fig. 9. SIGAA Android active installations

China, Spain, France and Portugal. This fact happens because undergraduate and postgraduate UFRNs students are participating on international programs related to the academic mobility and inter-institutional relations.

Continuing on Android Developers Console analysis, the third version of the application contains the higher active installation number, with 2,327 installations. SIGAA Mobile has 3 versions (1, 2 and 3). The version 1 has 15 installations, while version 2 has 58 installations.

Fig. 10. Countries active installations

After the active installations analysis with Developer Console, we used SIGAA log infrastructure to discover the average daily access and most used functionalities. SIGAA has a data base log controlled by an asynchronous process of persisting information about every user's operations. Analyzing daily access logins into SIGAA Android, we discovered that since its release, this application has 124,643 logins and an average of about 410 daily logins. We noted the increasing in login numbers at the beginning and the end of the academic activities period, as presented on the graph at figure 11. For example, in the month of vacation the number of access decreases, but immediately before and immediately after it, the number of access increases.

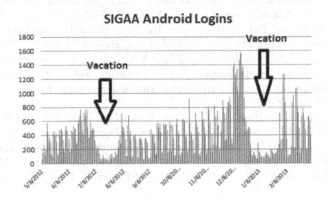

Fig. 11. SIGAA Android logins

5   Conclusions and Future Works

The activities of the metamorphosIS process proved useful in terms of MEAs development from existing web information systems. However, to be more confident of their efficiency, it is important to apply the same in other similar projects. Based on these experiences we intend to improve this process presenting more information, such as codes and how a particular functionality can be adapted for a MEA. Including providing details of how inputs can be adapted, possible differences in the workflow in both versions, affected components and other architectural details related to this process.

We believe that, as more information is collected by applying the MetamorphosIS process, we will be able to develop architectural solutions to the context of MEAs based on existing web systems, specifically regarding to design: architectural patterns, frameworks and reusable components.

References

1. Gartner Says Worldwide Mobile Phone Sales Declined 1.7 Percent in 2012, http://www.gartner.com/newsroom/id/2335616
2. Gartner Says Worldwide Traditional PC, Tablet, Ultramobile and Mobile Phone Shipments On Pace to Grow 7.6 Percent in 2014, http://www.gartner.com/newsroom/id/2645115
3. IDC Forecasts Worldwide Mobile Applications Revenues to Experience More Than 60% Compound Annual Growth Through 2014, http://www.idc.com/about/viewpressrelease.jsp?containerId=prUS22617910
4. Worldwide and U.S. Mobile Applications, Storefronts, and Developer 2010-2014 Forecast and Year-End 2010 Vendor Shares: The "Appification" of Everything, http://www.idc.com/research/viewdocsynopsis.jsp?containerId=225668
5. Giessmann, A., Stanoevska-Slabeva, K., Visser, B.: Mobile Enterprise Applications - Current State and Future Directions. In: 45th Hawaii International Conference on System Science, pp. 1363–1372. IEEE Press, Hawaii (2012)
6. Aquino Jr., G., Barroca Filho, I.: SIGAA Mobile – A sucessful experience of constructing a mobile application from an existing web system. In: 25th International Conference on Software Engineering & Knowledge Engineering, Boston, pp. 510–516 (2013)
7. Kolici, V., Xhafa, F., Pinedo, E., Nunez, J.: Analysis of Mobile and Web Applications in Small and Medium Size Enterprises. In: Eighth International Conference on P2P, Parallel, Grid, Cloud and Internet Computing, Compiegne (2013)
8. Buschmann, F., Meunier, R., Rohnert, H., Sommerlad, P., Stal, M.: Pattern-Oriented Software Architecture: A System of Patterns. John Wiley & Sons, New York (1996)
9. Nayebi, F., Desharnais, J.-M., Abran, A.: The state of the art of mobile application usability evaluation. In: 25th IEEE Canadian Conference on Electrical & Computer Engineering, Montréal (2012)
10. Nosseir, A., Flood, D., Harrison, R., Ibrahim, O.: Mobile Development Process Spiral. In: 7th International Conference on Computer Engineering & Systems, Cairo (2012)
11. Bass, L., Clements, P., Kazman, R.: Software Architecture in Practice, 3rd edn. Addison-Wesley (2013)
12. De Silva, L., Balasubramaniam, D.: Controlling software architecture erosion: A survey. Journal of Systems and Software 85, 132–151 (2012)

Link Prediction in Online Social Networks Using Group Information

Jorge Carlos Valverde-Rebaza and Alneu de Andrade Lopes

Departamento de Ciências de Computação
Instituto de Ciências Matemáticas e de Computação
University of São Paulo
Caixa Postal 668
13560-970 São Carlos, SP, Brazil
{jvalverr,alneu}@icmc.usp.br

Abstract. Users of online social networks voluntarily participate in different user groups or communities. Researches suggest the presence of strong local community structure in these social networks, i.e., users tend to meet other people via mutual friendship. Recently, different approaches have considered communities structure information for increasing the link prediction accuracy. Nevertheless, these approaches consider that users belong to just one community. In this paper, we propose three measures for the link prediction task which take into account all different communities that users belong to. We perform experiments for both unsupervised and supervised link prediction strategies. The evaluation method considers the links imbalance problem. Results show that our proposals outperform state-of-the-art unsupervised link prediction measures and help to improve the link prediction task approached as a supervised strategy.

Keywords: Link prediction, social networks, communities, social network analysis, graph mining.

1 Introduction

Online social networks are Web platforms that offer to their users the possibility of meeting and networking individuals with similar interests and behaviors [14]. Online social networks such as Flickr, LiveJournal, Orkut and Youtube have become part of the daily life of millions of people around the world who maintain and create new social relationships and interest groups [11]. This fact implies the growth and quick changes in underlying structures (nodes and links) of the social networks over time [8].

The boundless growth of online social networks has resulted in several research directions that examine structural and other properties of large-scale social networks. One of the most relevant research in social networks is the link prediction [8], [10], [14], [13], [9].

B. Murgante et al. (Eds.): ICCSA 2014, Part VI, LNCS 8584, pp. 31–45, 2014.
© Springer International Publishing Switzerland 2014

Link prediction addresses the problem of predicting the existence of missing relations or new ones [8], [10]. Detection of hidden social relationships is a friendship suggestion mechanism used by some online social networks and constitute one of the main application of link prediction. In such case, hidden relationships may consist in existing social ties that have not been established yet in a social network or in social ties missed during the social network evolution [14], [8].

Several methods have been proposed to cope with the link prediction problem. These methods can be divided into two different strategies: unsupervised [8], [13], [14] and supervised [5], [9], [1]. Furthermore, the strategy employed for a specific method influences how its performance will be evaluated [15].

Unsupervised methods assign a score for each pair of nodes with base on neighborhood nodes (local) or path (global) information. The state-of-the-art unsupervised link prediction methods are compared in [8], [10]. According to these experimental results on real networks, global methods usually achieve higher accuracy than local methods. Nevertheless, global methods are very time-consuming and usually infeasible for large-scale networks.

On the other hand, methods based on supervised strategy consider the link prediction problem as a classification problem [5], [1]. Thus, network information such as the structural ones and nodes attributes are used to build a feature vector for each pair of nodes. Then, these vectors are used to train different classifiers to determine the link existence or not between a pair of nodes.

Most proposals have focused on exploiting either the local or the global structural information of the networks. However, other information, such as the behavior of users in social communities, are not properly used. Thus, with the aim of improving the accuracy of link prediction, different hybrid methods using local information and community information have been proposed [18], [13], [12], [7]. These hybrid methods consider that the existence of high concentration of links within communities, as well as the low concentration of links between these communities, is a important property to be exploit in the link prediction problem.

Hybrid methods using community information have a better performance than most of local methods. Notice that these hybrid methods consider that a node belongs to just one community. However, in online social networks users usually belongs to more than one community.

In this paper we propose three new measures for link prediction considering user participation in multiple groups. We compare experimentally the most popular link prediction local methods with our proposals in both unsupervised and supervised strategies.

The remainder of this paper is organized as follows. In Section 2, we present the link prediction problem and state-of-the-art link prediction measures. In Section 3, we present and explain our three proposals. In Section 4, we present experimental results obtained from four online social networks. Finally, in Section 5, we summarize the main findings and conclusions of this work.

2   The Link Prediction Problem

The link prediction problem can be approached in two different strategies: unsupervised and supervised. The evaluation process, i.e., the performance of the link prediction task, therefore, must consider such strategies. Next, we describe both strategies.

2.1   Unsupervised Strategy

Given a network $G = (V, E)$, where V and E are sets of nodes and links respectively. Multiple links and self-connections are not allowed. If G is a directed network, consider the universal set, denoted by U, containing all $|V|(|V| - 1)$ potential directed links between pair of nodes in V, where $|V|$ denotes the number of elements in V. If G is an undirected network, the universal set U contains $\frac{|V|(|V|-1)}{2}$ links. The fundamental link prediction task in the unsupervised context is to find out the missing links (future links) in the set $U - E$ (set of nonexistent links) assigning a score for each link in this set. The higher the score, the higher the connection probability, and vice versa [10], [17], [13], [14].

Most existing unsupervised link prediction methods use node neighborhood (local) or path (global) information. In this work, we use the undirected and directed definitions as in [15] for five local measures: Common Neighbors (CN), Adamic Adar (AA), Jaccard Coefficient (Jac), Resource Allocation (RA) and Preferential Attachment (PA). Afterwards, they are referred to as base measures.

Two standard evaluation measures are used to quantify the prediction accuracy considering the link imbalance problem [10]: AUC (area under the receiver operating characteristic curve) and precision. The AUC is interpreted as the probability that for a randomly chosen link correctly predicted is given a higher score than for a randomly chosen link wrongly predicted. Thus, for n independent comparisons, if n' times for the links correctly predicted are given higher scores than for links wrongly predicted whilst n'' times for both correctly and wrongly predicted links are given equal scores. Thus, the AUC is approximately 0.5 when all the scores are generate from an independent and identical distribution. Therefore, the degree to which the value exceeds 0.5 indicates how better than by pure chance the algorithm performs. The AUC is defined by Eq. 1.

Different from AUC, precision only focuses on the L links with highest scores. Thus, if for the L top-ranked links there are L_r correctly predicted links. The precision is defined by Eq. 2. Clearly, higher precision means higher prediction accuracy.

$$AUC = \frac{n' + 0.5n''}{n} \qquad (1) \qquad precision_U = \frac{L_r}{L} \qquad (2)$$

2.2   Supervised Strategy

Supervised strategy considers the link prediction problem as a classification problem. Thus, network information such as the structural ones and nodes attributes are used to build a set of feature vectors for linked and not linked pairs of nodes

[5], [1], [15]. Classifiers are able to capture important interdependence relationships between nodes since feature vectors are formed based on unsupervised link prediction measures that capture different structural information sources of networks [9].

Using the supervised strategy is possible to use different validation processes, such as k-fold cross-validation [5]. Thus, we can use the traditional evaluation measures to compare classifiers performance. In this work, we use four standard evaluation measures [4]: accuracy (acc), precision, recall and f-measure (F). These measures are defined as follows:

$$acc = \frac{|tp| + |tn|}{|tp| + |tn| + |fp| + |fn|} \quad (3) \qquad precision_S = \frac{|tp|}{|tp| + |fp|} \quad (4)$$

$$recall = \frac{|tp|}{|tp| + |fn|} \quad (5) \qquad F = \frac{2 \times precision_S \times recall}{precision_S + recall} \quad (6)$$

where $|tp|$, $|tn|$, $|fp|$ and $|fn|$ represent true positives, true negatives, false positives and false negatives rates, respectively.

It is important to notice that the precision for unsupervised strategy is calculated differently than for supervised strategy but in both cases indicates the number of existent links correctly predicted with respect to a set of analyzed links. Furthermore, unsupervised evaluation measures are applied directly on results of link prediction measures but supervised evaluation measures are applied on results of classifiers [15].

3   Proposals

For a network G, we denote by $L_{x,y}$ and $\overline{L}_{x,y}$ the class variables of link existence and nonexistence, respectively, for a pair of nodes $(x, y) \in V$. The prior probabilities of $L_{x,y}$ and $\overline{L}_{x,y}$ are calculated according to Eq. 7 and 8, respectively.

$$P(L_{x,y}) = \frac{|E|}{|U|} \quad (7) \qquad P(\overline{L}_{x,y}) = \frac{|U| - |E|}{|U|} \quad (8)$$

Furthermore, in the network G there exist $M > 1$ groups identified by different group labels $g_1, g_2, \ldots, g_M$. Each node $x \in V$ belongs to a set of node groups $\mathcal{G} = \{g_a, g_b, \ldots, g_p\}$ with size P. Thus, $P > 0$ and $P \leq M$. Each $g_i \in \mathcal{G}$ is a group of nodes, whose elements share interests and behaviors. With M groups in G is possible to form N different sets of groups $\mathcal{G}_\alpha, \mathcal{G}_\beta, \ldots, \mathcal{G}_N$. When the node x belongs to a set of node groups $\mathcal{G}_\alpha$, this node is represented as $x^{\mathcal{G}_\alpha}$. A node belongs to just one set of node groups.

Considering the structural similarity, for undirected networks, the basic structural definition for a node $x \in V$ is its neighborhood $\Gamma(x) = \{y \mid (x, y) \in E \vee (y, x) \in E\}$ which denotes the set of neighbors of x. For directed networks, the set of nodes formed by directed links from x is different from the set of nodes formed by directed links from them to x. Thus, $\Gamma_{out}(x) = \{y \mid (x, y) \in E\}$ is defined as outgoing neighborhood and $\Gamma_{in}(x) = \{y \mid (y, x) \in E\}$ is defined as incoming neighborhood [17]. Also, the set of all common neighbors of the pair of disconnected nodes (x, y) is defined as $\Lambda_{x,y} = \Gamma(x) \cap \Gamma(y)$. When there is directionality, we have $\Lambda_{x,y}^{in} = \Gamma_{in}(x) \cap \Gamma_{in}(y)$ and $\Lambda_{x,y}^{out} = \Gamma_{out}(x) \cap \Gamma_{out}(y)$ [15]. Considering these definitions and based on the approach showed in [13], we propose three new link prediction measures.

3.1   Common Neighbors Within and Outside of Common Groups

For an undirected network, according to Bayesian theory [6], the posterior probabilities of the link existence and nonexistence between a pair of nodes $(x^{\mathcal{G}_\alpha}, y^{\mathcal{G}_\beta})$, given its set of all common neighbors $\Lambda_{x,y}$, are defined by Eq. 9 and 10, respectively.

$$P(L_{x,y}|\Lambda_{x,y}) = \frac{P(\Lambda_{x,y}|L_{x,y})P(L_{x,y})}{P(\Lambda_{x,y})} \qquad P(\overline{L}_{x,y}|\Lambda_{x,y}) = \frac{P(\Lambda_{x,y}|\overline{L}_{x,y})P(\overline{L}_{x,y})}{P(\Lambda_{x,y})}$$

$$(9) \hspace{5cm} (10)$$

Considering that $\mathcal{G}_{\alpha,\beta} = \mathcal{G}_\alpha \cap \mathcal{G}_\beta$, we define the set of all common neighbors such as $\Lambda_{x,y} = \Lambda_{x,y}^{WCG} \cup \Lambda_{x,y}^{OCG}$, where $\Lambda_{x,y}^{WCG} = \{z^{\mathcal{G}_\gamma} \in \Lambda_{x,y} \mid \mathcal{G}_{\alpha,\beta} \cap \mathcal{G}_\gamma \neq \varnothing\}$ is the set of common neighbors within common groups (WCG), i.e., the common neighbors of x and y belonging to at least one group to which both x and y belong to. The complement, $\Lambda_{x,y}^{OCG} = \Lambda_{x,y} - \Lambda_{x,y}^{WCG}$ is the set of common neighbors outside of the common groups (OCG), i.e., the common neighbors of x and y belonging to any group except to one group to which both x and y belong to. Clearly, $\Lambda_{x,y}^{WCG} \cap \Lambda_{x,y}^{OCG} = \varnothing$.

Hence, to estimate the probability of the common neighbors $\Lambda_{x,y}$ given the connection between $x^{\mathcal{G}_\alpha}$ and $y^{\mathcal{G}_\beta}$, we have to consider the number of common neighbors within common groups by the number of all common neighbors, as stated in Eq. 11. Similarly, to estimate the probability of the common neighbors $\Lambda_{x,y}$ given a disconnection between $x^{\mathcal{G}_\alpha}$ and $y^{\mathcal{G}_\beta}$, we have to consider the number of common neighbors outside of the common groups by the number of all common neighbors, as stated in Eq. 12.

$$P(\Lambda_{x,y} \mid L_{x,y}) = \frac{|\Lambda_{x,y}^{WCG}|}{|\Lambda_{x,y}|} \qquad (11) \qquad P(\Lambda_{x,y} \mid \overline{L}_{x,y}) = \frac{|\Lambda_{x,y}^{OCG}|}{|\Lambda_{x,y}|} \qquad (12)$$

In order to compare the existence likelihood between $x^{\mathcal{G}_\alpha}$ and $y^{\mathcal{G}_\beta}$, in Eq. 13, we define the likelihood score, $s_{x,y}$, of a node pair (x, y) as the ratio between Eq. 9 and 10.

$$s_{x,y} = \frac{P(\Lambda_{x,y} \mid L_{x,y})P(L_{x,y})}{P(\Lambda_{x,y} \mid \overline{L}_{x,y})P(\overline{L}_{x,y})} \qquad (13)$$

Substituting Eq. 11 and 12, we have the final score referred to as the **common neighbors within and outside of common groups (WOCG)** measure, defined as:

$$s_{x,y}^{WOCG} = \frac{|\Lambda_{x,y}^{WCG}|}{|\Lambda_{x,y}^{OCG}|} \times \Omega \qquad (14)$$

where $\Omega = \frac{P(L_{x,y})}{P(\overline{L}_{x,y})} = \frac{|E|}{|U|-|E|}$ is a constant for a network and its computation can be disregarded.

For a directed network, we consider $\Lambda_{x,y}^{WCG_{in}} = \{z^{\mathcal{G}_\gamma} \in \Lambda_{x,y}^{in} \mid \mathcal{G}_{\alpha,\beta} \cap \mathcal{G}_\gamma \neq \varnothing\}$, $\Lambda_{x,y}^{WCG_{out}} = \{z^{\mathcal{G}_\gamma} \in \Lambda_{x,y}^{out} \mid \mathcal{G}_{\alpha,\beta} \cap \mathcal{G}_\gamma \neq \varnothing\}$, $\Lambda_{x,y}^{OCG_{in}} = \Lambda_{x,y}^{in} - \Lambda_{x,y}^{WCG_{in}}$ and $\Lambda_{x,y}^{OCG_{out}} = \Lambda_{x,y}^{out} - \Lambda_{x,y}^{WCG_{out}}$. Thus, WOCG is defined based on the link direction: $s_{x,y}^{WOCG_{in}} = \frac{|\Lambda_{x,y}^{WCG_{in}}|}{|\Lambda_{x,y}^{OCG_{in}}|}$ and $s_{x,y}^{WOCG_{out}} = \frac{|\Lambda_{x,y}^{WCG_{out}}|}{|\Lambda_{x,y}^{OCG_{out}}|}$.

3.2 Common Neighbors of Groups

For an undirected network, considering a pair of nodes $(x^{\mathcal{G}_\alpha}, y^{\mathcal{G}_\beta})$, we define the set of common neighbors of groups $\Lambda_{x,y}^{\mathcal{G}} = \{z^{\mathcal{G}_\gamma} \in \Lambda_{x,y} \mid \mathcal{G}_\alpha \cap \mathcal{G}_\gamma \neq \varnothing \vee \mathcal{G}_\beta \cap \mathcal{G}_\gamma \neq \varnothing\}$. Thus, we define a score referred to as **common neighbors of groups** (CNG), as stated in Eq. 15.

$$s_{x,y}^{CNG} = |\Lambda_{x,y}^{\mathcal{G}}| \tag{15}$$

The CNG measure refers to the size of the set of common neighbors of x and y belonging to at least one group to which x or y belongs to.

For a directed network, we define the set of incoming common neighbors of groups $\Lambda_{x,y}^{\mathcal{G}_{in}} = \{z^{\mathcal{G}_\gamma} \in \Lambda_{x,y}^{in} \mid \mathcal{G}_\alpha \cap \mathcal{G}_\gamma \neq \varnothing \vee \mathcal{G}_\beta \cap \mathcal{G}_\gamma \neq \varnothing\}$ and the set of outgoing common neighbors of groups $\Lambda_{x,y}^{\mathcal{G}_{out}} = \{z^{\mathcal{G}_\gamma} \in \Lambda_{x,y}^{out} \mid \mathcal{G}_\alpha \cap \mathcal{G}_\gamma \neq \varnothing \vee \mathcal{G}_\beta \cap \mathcal{G}_\gamma \neq \varnothing\}$. Thus, CNG is defined based on the link direction: $s_{x,y}^{CNG_{in}} = |\Lambda_{x,y}^{\mathcal{G}_{in}}|$ and $s_{x,y}^{CNG_{out}} = |\Lambda_{x,y}^{\mathcal{G}_{out}}|$.

3.3 Common Neighbors with Total and Partial Overlapping of Groups

For an undirected network, we formulate a new proposal. Thus, according to Bayesian theory, the posterior probabilities of link existence and nonexistence between a pair of nodes $(x^{\mathcal{G}_\alpha}, y^{\mathcal{G}_\beta})$, given its set of common neighbors of groups $\Lambda_{x,y}^{\mathcal{G}}$, are defined by Eq. 16 and 17, respectively.

$$P(L_{x,y}|\Lambda_{x,y}^{\mathcal{G}}) = \frac{P(\Lambda_{x,y}^{\mathcal{G}}|L_{x,y})P(L_{x,y})}{P(\Lambda_{x,y}^{\mathcal{G}})} \tag{16}$$

$$P(\overline{L}_{x,y}|\Lambda_{x,y}^{\mathcal{G}}) = \frac{P(\Lambda_{x,y}^{\mathcal{G}}|\overline{L}_{x,y})P(\overline{L}_{x,y})}{P(\Lambda_{x,y}^{\mathcal{G}})} \tag{17}$$

Consider that $\Lambda_{x,y}^{\mathcal{G}} = \Lambda_{x,y}^{TOG} \cup \Lambda_{x,y}^{POG}$, where $\Lambda_{x,y}^{TOG} = \{z^{\mathcal{G}_\gamma} \in \Lambda_{x,y}^{\mathcal{G}} \mid \mathcal{G}_\alpha \cap \mathcal{G}_\gamma \neq \varnothing \wedge \mathcal{G}_\beta \cap \mathcal{G}_\gamma \neq \varnothing\}$ is the set of common neighbors with total overlapping of groups (TOG), i.e., the common neighbors of group of x and y belonging to at least one group of nodes to which x and y belong to. The complement, $\Lambda_{x,y}^{POG} = \Lambda_{x,y}^{\mathcal{G}} - \Lambda_{x,y}^{TOG} = \{z^{\mathcal{G}_\gamma} \in \Lambda_{x,y}^{\mathcal{G}} \mid \mathcal{G}_\alpha \cap \mathcal{G}_\gamma \neq \varnothing \veebar \mathcal{G}_\beta \cap \mathcal{G}_\gamma \neq \varnothing\}$ is the set of common neighbors with partial overlapping of groups (POG), i.e., the common neighbors of groups of x and y belonging exclusively to at least one group of nodes to which x or y belong to. Clearly, $\Lambda_{x,y}^{TOG} \cap \Lambda_{x,y}^{POG} = \varnothing$.

Using the same process presented in Section 3.1, we can estimate the probability of the common neighbors of groups $\Lambda_{x,y}^{\mathcal{G}}$ given the probability of link existence and nonexistence between $x^{\mathcal{G}_\alpha}$ and $y^{\mathcal{G}_\beta}$ as stated in Eqs. 18 and 19.

$$P(\Lambda_{x,y}^{\mathcal{G}}|L_{x,y}) = \frac{|\Lambda_{x,y}^{TOG}|}{|\Lambda_{x,y}^{\mathcal{G}}|} \quad (18) \qquad P(\Lambda_{x,y}^{\mathcal{G}}|\overline{L}_{x,y}) = \frac{|\Lambda_{x,y}^{POG}|}{|\Lambda_{x,y}^{\mathcal{G}}|} \quad (19)$$

In order to compare the existence likelihood between $x^{\mathcal{G}_\alpha}$ and $y^{\mathcal{G}_\beta}$, we define the likelihood score of a node pair (x,y) as the ratio between Eq. 16 and Eq. 17. Substituting Eq. 18 and Eq. 19, we have the final score called as the **common neighbors with total and partial overlapping of groups** (TPOG) measure, defined as:

$$s_{x,y}^{TPOG} = \frac{|\Lambda_{x,y}^{TOG}|}{|\Lambda_{x,y}^{POG}|} \times \Omega \qquad (20)$$

where $\Omega = \frac{P(L_{x,y})}{P(\overline{L}_{x,y})} = \frac{|E|}{|U|-|E|}$, in the same way that for WOCG, is a constant for a network and its computation can be disregarded.

For a directed network, we can consider $\Lambda_{x,y}^{TOG_{in}} = \{z^{\mathcal{G}_\gamma} \in \Lambda_{x,y}^{\mathcal{G}_{in}} \mid \mathcal{G}_\alpha \cap \mathcal{G}_\gamma \neq \varnothing \wedge \mathcal{G}_\beta \cap \mathcal{G}_\gamma \neq \varnothing\}$, $\Lambda_{x,y}^{TOG_{out}} = \{z^{\mathcal{G}_\gamma} \in \Lambda_{x,y}^{\mathcal{G}_{out}} \mid \mathcal{G}_\alpha \cap \mathcal{G}_\gamma \neq \varnothing \wedge \mathcal{G}_\beta \cap \mathcal{G}_\gamma \neq \varnothing\}$, $\Lambda_{x,y}^{POG_{in}} = \{z^{\mathcal{G}_\gamma} \in \Lambda_{x,y}^{\mathcal{G}_{in}} \mid \mathcal{G}_\alpha \cap \mathcal{G}_\gamma \neq \varnothing \vee \mathcal{G}_\beta \cap \mathcal{G}_\gamma \neq \varnothing\}$ and $\Lambda_{x,y}^{POG_{out}} = \{z^{\mathcal{G}_\gamma} \in \Lambda_{x,y}^{\mathcal{G}_{out}} \mid \mathcal{G}_\alpha \cap \mathcal{G}_\gamma \neq \varnothing \vee \mathcal{G}_\beta \cap \mathcal{G}_\gamma \neq \varnothing\}$. Thus, TPOG is defined based on the link direction: $s_{x,y}^{TPOG_{in}} = \frac{|\Lambda_{x,y}^{TOG_{in}}|}{|\Lambda_{x,y}^{POG_{in}}|}$ and $s_{x,y}^{TPOG_{out}} = \frac{|\Lambda_{x,y}^{TOG_{out}}|}{|\Lambda_{x,y}^{POG_{out}}|}$.

4   Experiments

We consider a scenario where new links of four online social networks must be predicted. Due to the fact that in online social networks users participate freely in different user groups, in each one of these social networks we use this natural group information to assign group labels to each node. We also compare the performance of our proposals to the base measures.

4.1   Datasets

Social network graphs considered in our experiments are Flickr, LiveJournal, Orkut and Youtube. These graphs, available in [11], are among the most popular social networking sites. On the other hand, these graphs have information both of links between users and of friendship groups to which each user belongs.

Each online social network have different features. Flickr[1] is a photo-sharing network to organize images using tags and allows users to form groups of common photography interests. LiveJournal[2] is a popular blogging site whose users form a social network and create custom user groups for posting discussion. Orkut[3] is

---

[1] http://www.flickr.com
[2] http://www.livejournal.com
[3] http://www.orkut.com

a social networking site run by Google considered a pure social network since it has the sole purpose of friendship networking and allows users to create groups of users with similar interests. Youtube[4] is a popular video-sharing site that includes a social network that allows users to create groups of users with similar video preferences.

Table 1. High-level topological features of our four social network graphs

	Flickr	LiveJournal	Orkut	Youtube
Number of nodes	$1,846,198$	$5,284,457$	$3,072,441$	$1,157,827$
Number of links	$22,613,981$	$77,402,652$	$223,534,301$	$4,945,382$
Average degree per node	12.24	16.97	106.1	4.29
Fraction of links symmetric	62.0%	73.5%	100.0%	79.1%
Average path length	5.67	5.88	4.25	5.10
Diameter	27	20	9	21
Average clustering coefficient	0.313	0.330	0.171	0.136
Average assortativity coefficient	0.202	0.179	0.072	−0.033
Number of node groups	$103,648$	$7,489,073$	$8,730,859$	$30,087$
Average number of groups membership per node	4.62	21.25	106.44	0.25
Average group size	82	15	37	10
Average group clustering coefficient	0.47	0.81	0.52	0.34

High-level topological features of the four social network graphs are presented in Table 1. From this table, we observe that by the high number of nodes and links these networks are considered as large-scale networks. The average degree per node indicates the average of number of neighbors per user. The fraction of links symmetric denotes the degree in which directed links from a source to a destination have an endorsement of the destination by the source. For instance, with the exception of Orkut, which is an undirected network (with 100% of links symmetric), the other networks (directed networks) have a significant degree of symmetry, i.e., many of the target of the links reciprocate. Furthermore, independent of the causes, the symmetric nature of social links affects the structure of large scale social networks, mainly by increasing the overall connectivity of the network and reducing its diameter [11].

Also, Table 1 shows global topological features of networks. The average path length is the average number of steps along the shortest paths for all possible node pairs and the diameter is defined as the maximum shortest path between any two nodes. In absolute terms, the average path lengths and diameters for all four networks are remarkably shorter compared with average path length and diameter of the Web graph (16.12 and 905, respectively) [2]. The average clustering coefficient is the degree to which nodes in a network tend to cluster together. A high average clustering coefficient suggests the presence of strong local community structure, i.e., in friendship social networks, users tend to be introduced to other users via mutual friends. The average assortativity coefficient indicates the likelihood for nodes to connect to other nodes with similar degrees. When this coefficient tends to 1, means that nodes likely are connected to nodes with similar degrees, and the opposite when the value tends to -1.

---

[4] http://www.youtube.com

Between the group features, we observe that the four networks have a high amount of node groups and that each user belongs on average to more than four groups (except Youtube). Thus, all groups of all four networks have a minimum of 10 users. It is important to note here that for each network each user can belongs to more than one group. Also, note that users in a group do not necessarily need link to each other in the network and user groups represent tightly clustered communities of users in social networks. This can be seen from the average group clustering coefficients, which is defined as the average of clustering coefficients of the subgraphs of the network consisting of only the users who are members of each group [11].

4.2 Experimental Setup

For the network preprocessing, for a network G, the set E is divided into the training set E^T and the testing set E^P. From the set E, for selecting the links for E^P, we take randomly two-third of the links formed by nodes whose number of neighbors is two times greater than the average degree per node. The remaining links, except those formed by nodes whose number of neighbors is less than two-third of the average degree per node, constitute the training set E^T. This evaluation method is widely used in the link prediction literature [14], [15], [16], [17].

After that, the link prediction process is initiated. This process includes both unsupervised and supervised strategies. In unsupervised strategy, for each pair of nodes from E^T, the connection likelihood is calculated based on the link direction, choosing the highest score between its *in* and *out* scores as final and unique score, e.g., by vertex pair (x, y) if $s_{x,y}^{out} > s_{x,y}^{in}$ then $s_{x,y} = s_{x,y}^{out}$, otherwise, $s_{x,y} = s_{x,y}^{in}$.

In supervised strategy, we use decision tree (J48), naive Bayes (NB), multilayer perceptron with backpropagation (MLP) and support vector machine (SMO) classifiers from Weka[5]. Previously, for each network, we compute a set of feature vector formed by randomly selected pair of nodes from E^T. If the pair of nodes taken from E^T is also in E^P then the feature vector formed by this pair of nodes takes the positive class (existent link), otherwise takes the negative class (nonexistent link). To avoid the links imbalance problem, the set of feature vectors for each network have 50% with positive class and 50% with negative class. Table 2 shows the number of instances by class and the total of instances for each social network.

For each network, we create five different data sets in ARFF format. Each data set is formed by features which combine different link prediction measures. Thus, VLocal is the data set whose feature vectors are formed by CN, AA, Jac, RA and PA. VGroup is the data set whose feature vectors are formed by WOCG, CNG and TPOG. VTop is the data set whose feature vectors are formed by the three best base measures from the literature, i.e., CN, AA and RA, and the two best measures based on group information, i.e., CNG and TPOG (see Section

---

[5] http://www.cs.waikato.ac.nz/ml/weka/

Table 2. Number of instances by class for all networks

	Existent	Non-existent	Total
Flickr	500001	500001	1000002
LiveJournal	300001	300001	600002
Orkut	1500001	1500001	3000002
Youtube	20001	20001	40002

4.3). Similarly, VTop2 is the data set whose feature vectors are formed by the five overall best link prediction measures (see also Section 4.3), i.e., TPOG, CNG, AA, WOCG and CN. VTotal is the data set whose feature vectors are formed by all base measures and all our proposals, i.e., CN, AA, Jac, RA, PA, WOCG, CNG and TPOG.

The experiments were carried out in a computer with 99 GB of RAM using Linux operating system.

4.3   Results

In order to validate our results, we use the evaluation measures presented in Section 2 both for unsupervised strategy and supervised strategy. For results of our unsupervised link prediction process, we employ AUC and precision to validation. Table 3 summarizes the prediction results measured by AUC, with $n = 5000$, for the four networks. Each AUC value is obtained by averaging over 10 run over 10 independent partitions of training and testing sets.

Table 3. The prediction results measured by AUC

	WOCG	CNG	TPOG	CN	AA	Jac	RA	PA
Flickr	0.637 (5.0)	0.728 (1.0)	0.728 (2.0)	0.674 (3.0)	0.656 (4.0)	0.431 (8.0)	0.616 (6.0)	0.566 (7.0)
Livejournal	0.596 (4.0)	0.611 (3.0)	0.665 (1.0)	0.582 (5.0)	0.580 (6.0)	0.624 (2.0)	0.565 (7.0)	0.542 (8.0)
Orkut	0.649 (2.0)	0.621 (3.0)	0.651 (1.0)	0.572 (7.0)	0.620 (4.0)	0.575 (6.0)	0.566 (8.0)	0.602 (5.0)
Youtube	0.434 (7.0)	0.723 (5.0)	0.555 (6.0)	0.834 (4.0)	0.928 (1.0)	0.217 (8.0)	0.892 (3.0)	0.917 (2.0)
Average rank	4.50 (4.0)	3.00 (2.0)	2.50 (1.0)	4.75 (5.0)	3.75 (3.0)	6.00 (7.5)	6.00 (7.5)	5.50 (6.0)

From Table 3, each value in parentheses represents the ranking of each link prediction measure for each network. In general, our proposals perform better than the base measures in Flickr, LiveJournal and Orkut. In Youtube, TPOG and CNG have their lowest performance and WOCG has the overall worst performance. This can be explained by the fact that Youtube has the lowest values of average clustering coefficient and average group clustering coefficient, i.e., friends of a user does not necessarily become friends and user groups are weakly dense. Also, Youtube has a negative value of average assortativity coefficient, i.e., there is a tendency of friendship relations between users that share few common interests and behaviors. Among the base measures, highlighted as best CN and AA and, surprisingly, PA has a better performance than Jac and RA.

To analyze the difference between all link prediction measures, based on results of Table 3, we perform the Friedman and Nemenyi post-hoc tests [3]. The critical value of the F-statistics with 7 and 21 degrees of freedom at 95 percentile is 2.49. Thus, according to the Friedman test using the F-statistics, the null-hypothesis that all link prediction measures evaluated behave similarly should not be rejected. According to the Nemenyi statistics, the critical difference (CD) for comparing the mean-ranking of two different link prediction measures at 95 percentile is 5.25.

Results from Nemenyi test are present in Figure 1, where we show the critical difference value on the top of the diagram. In the axis of the diagram are plotted the average rank of measures (whose values are explicit in the last row of Table 3). In the axis, the lowest (best) ranks are in the left side. Thus, the null-hypothesis that all link prediction measures have a similar behavior should not be rejected, i.e., all measures analyzed have no significant difference, so they are connected by a black line in the diagram. Although there is no significant difference among them, we observe that our proposals have the first, second and fourth best overall accuracy. The base measure best positioned is AA, which is third, and CN, which is fifth.

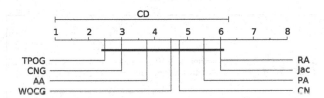

Fig. 1. Post-hoc test for results from Table 3 with CD = 5.25

Figure 2 shows the prediction quality measured by precision on all social networks analyzed. Different values of L are used. For Flickr and LiveJournal, all link prediction measures have a similar precision performance, highlighting AA and RA as the best overall measures in all L values, but reaching their maximum performance when $L = 100$. For Orkut, also all link prediction measures have a similar precision performance, highlighting WOCG and TPOG as the best overall measures in all L values. However, WOCG reaches the highest overall performance when $L = 1,000$. For Youtube, we observe a declining performance in all link prediction measures after $L = 100$, highlighting CNG and TPOG as the best overall measures in all L values. Furthermore, CNG reaches the highest overall performance when $L = 100$. Also, we observe that PA and Jac are the worst overall performance in all the four networks.

For results of our supervised link prediction process, accuracy and f-measure are employed to validate the quality of the classifiers in VLocal, VGroup, VTop, VTop2 and VTotal data sets for each social network. Tables 4 and 5 respectively show Accuracy and F-Value average values for four different classifiers after using 10-fold cross validation. For both Tables 4 and 5, values emphasized in

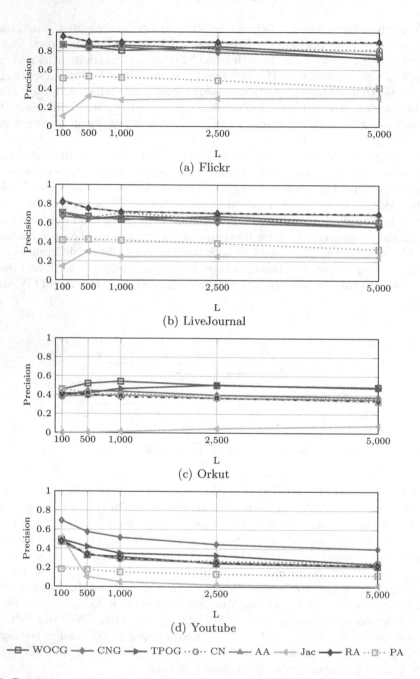

Fig. 2. Precision results on four social networks. Different values of L are used to select the top-L highest scores for predicting links.

black correspond to the highest result among the evaluated data sets for each classifier. Results highlighted in gray indicate that a classifier get best results in data sets formed by feature vectors using our proposals than VLocal data set, which is formed by feature vectors using only the base measures.

Table 4. Correctly classified instances (in percent)

	J48	NB	MLP	SMO		J48	NB	MLP	SMO
Flickr Vlocal	77.70	57.94	**71.29**	66.88	Orkut VLocal	82.53	71.91	80.01	76.75
Flickr VGroup	70.66	**62.50**	69.91	67.98	Orkut VGroup	78.11	69.35	77.32	74.14
Flickr VTop	72.35	59.08	71.13	68.60	Orkut VTop	79.86	72.99	77.01	76.12
Flickr VTop2	72.20	60.96	70.89	68.08	Orkut VTop2	79.32	73.09	77.31	76.09
Flickr VTotal	**77.72**	60.37	71.20	**68.97**	Orkut VTotal	**82.59**	**74.14**	**80.13**	**77.32**
LiveJournal VLocal	**79.70**	70.66	**78.85**	**77.67**	Youtube VLocal	82.35	59.73	**73.08**	62.01
LiveJournal VGroup	76.94	71.01	76.94	75.34	Youtube VGroup	67.24	61.16	67.09	61.45
LiveJournal VTop	79.14	71.63	78.76	77.50	Youtube VTop	78.94	60.55	72.43	64.71
LiveJournal VTop2	79.12	70.92	78.15	77.46	Youtube VTop2	78.03	**63.67**	71.69	64.39
LiveJournal VTotal	**79.70**	**71.77**	78.59	77.61	Youtube VTotal	**82.66**	62.38	72.33	**65.20**

Table 4 shows results for J48, NB, MLP and SMO classifiers. In most cases, the best accuracy is obtained by VTotal data set, i.e., the data set formed by feature vectors using all our proposals. For MLP classifier the best result is by using the VLocal data set, i.e., the data set formed only by the base measures. For Orkut, the best performance of MLP classifier is using the VTotal data set. Besides, we observe that using NB, for all the four networks, and using SMO, for Flickr and Youtube, the performance of classifiers in data sets formed by feature vectors using our proposals (VGroup, VTop, VTop2 and VTotal) is markedly better than in data sets formed by feature vectors using only base measures (VLocal).

Table 5 shows for J48, NB and SMO classifiers that the best f-measure results are obtained by the VTotal data set or in any other data set whose feature vectors use our proposals. For MLP classifier the best result is using the VLocal data set.

Table 5. Average of f-measure on four social networks

	J48	NB	MLP	SMO		J48	NB	MLP	SMO
Flickr VLocal	**0.777**	0.507	**0.713**	0.651	Orkut VLocal	0.825	0.702	0.800	0.764
Flickr VGroup	0.706	**0.583**	0.699	0.668	Orkut VGroup	0.781	0.676	0.773	0.737
Flickr VTop	0.724	0.525	0.711	0.676	Orkut VTop	0.799	0.720	0.77	0.759
Flickr VTop2	0.722	0.558	0.709	0.669	Orkut VTop2	0.793	0.722	0.773	0.758
Flickr VTotal	**0.777**	0.548	0.712	**0.680**	Orkut VTotal	**0.826**	**0.731**	**0.801**	**0.771**
LiveJournal VLocal	**0.797**	0.687	**0.788**	**0.774**	Youtube VLocal	0.823	0.531	**0.73**	0.565
LiveJournal VGroup	0.768	0.698	0.768	0.750	Youtube VGroup	0.658	0.563	0.655	0.567
LiveJournal VTop	0.791	0.700	0.787	0.772	Youtube VTop	0.789	0.543	0.724	0.617
LiveJournal VTop2	0.79	0.691	0.781	0.772	Youtube VTop2	0.780	**0.600**	0.717	0.613
LiveJournal VTotal	**0.797**	**0.702**	0.786	**0.774**	Youtube VTotal	**0.826**	0.577	0.723	**0.623**

In general, we observe that the performance of a classifier measured by accuracy is similar that measured by f-measure. Thus, from entries highlighted in gray in Tables 4 and 5 we observe that classifiers perform better in data sets formed by feature vectors that include our proposals. This happens mainly when all base measures and all our proposals are combined into a feature vector, i.e., in the VTotal data set.

5   Conclusions

We proposed three new link prediction measures, referred to as WOCG, CNG and TPOG measures. These measures use information about the groups to which the nodes belong. Differently from the link prediction measures based on group information described in the literature, our proposals consider that a node can belong to more than one group, as usually occurs in real social networks. Thus, WOCG divides the common neighbor set of two nodes and use the neighborhood intersection information. CNG defines the set of common neighbors of two nodes belonging to at least one group to which these nodes belong and use the size of this set as a link prediction measure. TPOG uses the same schema that WOCG but using the set of common neighbors of groups defined by the CNG measure.

Since for applying measures based on group information is need, in a previous phase, partitioning the network into groups, researchers use different community detection algorithms that have low computational cost and that improve the link prediction performance [14], [13], [7], [12]. This makes the performance of link prediction measures based on group information strongly depend of clustering quality. Thus, to eliminate this dependence and spend less time executing a community detect algorithm, in social networks domain we can use the natural group information, i.e., the information from friendship groups to which users belong to.

To evaluate our proposals, we use both unsupervised and supervised strategies on four real and large-scale online social networks: Flickr, LiveJournal, Orkut and Youtube. When an unsupervised strategy is performed, the performance of our proposals compared to other measures was better under the AUC criterion. When analyzing precision, highlight AA, RA, WOCG, TPOG and CNG but there is no clear winner. It is important to note here that the performance of our proposals is influenced by the topological structure of the analyzed network.

When a supervised strategy is performed, our results show that combining measures based on local information and based on group information improves the performance of classifiers. But the improvement may not be significant because the selection processes for generating feature vectors of data sets are diverse, so how to select the most appropriate links for a supervised strategy is a challenging problem.

In summary, our experiments suggest that communities to which users belong convey relevant clues about user's interest and behavior. Hence, our proposals improve the performance of the link prediction task by considering mainly the information of common groups to which users belong to.

Acknowledgment. This work is partially supported by Grants 2011/22749-8 and 2013/12191-5 from São Paulo Research Foundation (FAPESP) and 151836/2013-2 from National Council for Scientific and Technological Development (CNPq).

References

1. Benchettara, N., Kanawati, R., Rouveirol, C.: A supervised machine learning link prediction approach for academic collaboration recommendation. In: RecSys 2010, pp. 253–256 (2010)
2. Broder, A., Kumar, R., Maghoul, F., Raghavan, P., Rajagopalan, S., Stata, R., Tomkins, A., Wiener, J.: Graph structure in the web. Comput. Netw. 33(1-6), 309–320 (2000)
3. Demsar, J.: Statistical comparisons of classifiers over multiple data sets. JMLR 7, 1–30 (2006)
4. Fatourechi, M., Ward, R.K., Mason, S.G., Huggins, J., Schlogl, A., Birch, G.E.: Comparison of evaluation metrics in classification applications with imbalanced datasets. In: ICMLA 2008, pp. 777–782 (2008)
5. Al Hasan, M., Chaoji, V., Salem, S., Zaki, M.: Link prediction using supervised learning. In: SDM 2006 Workshop on Link Analysis, Counterterrorism and Security (2006)
6. Hastie, T., Tibshirani, R., Friedman, J.: The elements of statistical learning: data mining, inference and prediction, 2nd edn. Springer (2009)
7. Hoseini, E., Hashemi, S., Hamzeh, A.: Link prediction in social network using co-clustering based approach. In: WAINA 2012, pp. 795–800. IEEE (2012)
8. Liben-Nowell, D., Kleinberg, J.: The link-prediction problem for social networks. JASIST 58(7), 1019–1031 (2007)
9. Lichtenwalter, R.N., Lussier, J.T., Chawla, N.V.: New perspectives and methods in link prediction. In: ACM SIGKDD KDD 2010, pp. 243–252. ACM (2010)
10. Lü, L., Zhou, T.: Link prediction in complex networks: A survey. Physica A: Statistical Mechanics and its Applications 390(6), 1150–1170 (2011)
11. Mislove, A., Marcon, M., Gummadi, K.P., Druschel, P., Bhattacharjee, B.: Measurement and analysis of online social networks. In: ACM SIGCOMM IMC 2007, pp. 29–42. ACM (2007)
12. Soundarajan, S., Hopcroft, J.: Using community information to improve the precision of link prediction methods. In: WWW 2012, pp. 607–608. ACM (2012)
13. Valverde-Rebaza, J., de Andrade Lopes, A.: Link prediction in complex networks based on cluster information. In: Barros, L.N., Finger, M., Pozo, A.T., Gimenénez-Lugo, G.A., Castilho, M. (eds.) SBIA 2012. LNCS (LNAI), vol. 7589, pp. 92–101. Springer, Heidelberg (2012)
14. Valverde-Rebaza, J., de Andrade Lopes, A.: Structural Link Prediction Using Community Information on Twitter. In: CASoN 2012, pp. 132–137. IEEE (2012)
15. Valverde-Rebaza, J., de Andrade Lopes, A.: Exploiting behaviors of communities of Twitter users for link prediction. Social Network Analysis and Mining 3(4), 1063–1074 (2013)
16. Yin, D., Hong, L., Davison, B.D.: Structural link analysis and prediction in microblogs. In: CIKM 2011, pp. 1163–1168 (2011)
17. Zhang, Q.-M., Lü, L., Wang, W.-Q., Zhu, Y.-X., Zhou, T.: Potential theory for directed networks. PLoS ONE 8(2), e55437 (2013)
18. Zheleva, E., Getoor, L., Golbeck, J., Kuter, U.: Using friendship ties and family circles for link prediction. In: Giles, L., Smith, M., Yen, J., Zhang, H. (eds.) SNAKDD 2008. LNCS, vol. 5498, pp. 97–113. Springer, Heidelberg (2010)

Improving OLAM with Cloud Elasticity

Guilherme Galante[1,2], Luis Carlos Erpen De Bona[1], and Claudio Schepke[3]

[1] Federal University of Paraná - UFPR
Curitiba, PR, Brazil
[2] Western Paraná State University - Unioeste
Cascavel, PR, Brazil
[3] Federal University of Pampa - UNIPAMPA
Alegrete, RS, Brazil
`guilherme.galante@unioeste.br, bona@inf.ufpr.br,`
`claudioschepke@unipampa.edu.br`

Abstract. Elasticity is considered one of the fundamental properties of cloud computing, and can be seen as the ability of a system to increase or decrease the computing resources allocated in a dynamic and on demand way. This feature is suitable for dynamic applications, whose resources requirements cannot be determined exactly in advance, either due to changes in runtime requirements or in application structure. A good candidate for using cloud elasticity is the Ocean-Land-Atmosphere Model (OLAM), since it presents a significant load variation during its execution and due to online mesh refinement (OMR), that causes load unbalancing problems. In this paper, we present our efforts to adapt OLAM to use the elasticity offered in cloud environments to dynamic allocate resources according to the demands of each execution phase, and to minimize the load unbalancing caused by OMR. The results show that elasticity was successfully used to provide these features, improving the OLAM performance and providing a better use of resources.

Keywords: Ocean-Land-Atmosphere Model, Cloud Computing, Elasticity, Load Balancing, Cloudine Framework.

1 Introduction

Scientific Computing is the field of study concerned with modeling and analyzing natural and engineered processes using computational techniques. The evolution of scientific computing methodology is fast and the construction of infrastructures for computational science lies at the front edge of technical development. During the last decades, scientific applications have been executed over high performance infrastructures, including supercomputers, clusters and grids.

Recently, cloud computing emerged as an alternative execution environment, attracting attention from industry and academic worlds, becoming increasingly common in the literature to find cases of cloud adoption by companies and research institutions. Cloud computing refers to a flexible model for on-demand access to a shared pool of configurable computing resources (such as networking, servers, storage, platforms and software) that can be easily provisioned as needed [1]. In addition, a new feature emerged on cloud computing: *elasticity*.

B. Murgante et al. (Eds.): ICCSA 2014, Part VI, LNCS 8584, pp. 46–60, 2014.

Elasticity is the ability of a system to increase or decrease the computing resources allocated in a dynamic way [2]. It implies that the virtual environment used to execute an application may be changed over time, without any long-term indication about the future resources demands. This feature is suitable for dynamic applications, whose resources requirements cannot be determined exactly in advance, either due to changes in runtime requirements or in application structure [3].

An application with this characteristics is the Ocean-Land-Atmosphere Model (OLAM) [4]. OLAM is a global weather and climate prediction model that consists essentially of a finite volume representation of the full compressible non-hydrostatic Navier-Stokes equations over the planetary atmosphere. Recently, the model was modified to support multilevel parallelism using MPI and OpenMP and online mesh refinement (OMR) [5]. OLAM is a good candidate for using cloud elasticity, since it presents a significant load variation during its execution and due to OMR, that causes load unbalancing problems.

In this paper, we present our efforts to adapt OLAM to support elasticity and to use this feature to provide elastic provisioning of resources (virtual processors and memory) and load balancing. The elastic allocation of resources enables OLAM to use the cloud resources efficiently, improving the shared use of resources and the performance. In addition, we propose an approach for load balancing in which elasticity is used to automatically distribute the available virtual processors among the nodes to equalize the processing time and reduce the total running time. To the best of our knowledge, there are no works on literature addressing the use of cloud elasticity to provide similar capabilities to parallel applications as presented in this paper.

The remainder of the paper is organized as follows. Section 2 presents some related works. Section 3 introduces the cloud elasticity and describes the Cloudine, the framework used to provide elasticity features in this work. Section 4 describes OLAM in details. In Section 5 we present how elasticity is employed in OLAM. Section 6 presents some tests and results of elasticity-aware OLAM. Finally, Section 7, concludes the paper.

2   Related Work

Many elasticity solutions have been implemented by public cloud providers and by academy [2], enabling the development of elastic scientific applications.

In the work of Byun et al. [6] is presented the exploration of elasticity in astronomy, seismology and genetics applications. The goal of the study is to use elasticity to find the minimum set of resources for application workflow that meet the given deadline.

Raveendran et al. [7] describe an initial work towards making existing MPI applications elastic for clouds. The elasticity is achieved by terminating the application execution and restarting a new one using a different number of virtual instances. They test the proposed framework using Jacobi and conjugate gradient iterative solvers.

Rajan et al. [8] presented Work Queue, a framework for the development of elastic master-slave applications. There are several applications developed with this framework, such as, simulations of replica exchange (molecular dynamics), genome processing and protein folding [9].

Galante and Bona [10] propose a platform, called Cloudine, for exploring elasticity in scientific applications. The platform is based on the concept of elasticity primitives, enabling the development of tailor made elasticity controllers for scientific applications. Based on this work, the authors also propose a mechanism to provide elasticity to OpenMP applications [11]. Cloudine is used in the present paper to provide the elasticity features to OLAM.

Dynamic memory allocation in scientific applications is addressed in the work of Moltó et al [12]. The authors present a mechanism for adapting the memory size of the VM based on the memory consumption pattern of the application.

The use of cloud computing in the execution of climate and atmospheric models are presented in literature. The work of Evangelinos and Hill [13] is one of the first works towards the use of cloud computing in this subject. They present some experiments with the MIT General Circulation Model on Amazon EC2 in order to determine the feasibility of running of the application on clouds. A similar study is presented by Martinez et al. [14], where the authors evaluate the execution of the Weather Research and Forecasting (WRF) code on an on-premise virtualized environment.

Kumar et al. [15] present CloudCast, a mobile application for personalized short-term weather forecasting (up to 15 minutes in the future for areas as small as $100m^2$). CloudCast uses instances from commercial and research cloud services to execute a short-term weather forecasting application based on a Nowcasting algorithm.

The work of Mattmann et al. [16] present the efforts to infuse cloud computing and virtualization in the Regional Climate Model Evaluation System (RCMES). The authors describe the utilization of cloud computing for elastic data ingestion and querying in data warehousing, and the use of virtual machines as a delivery mechanism for the complex software layer of data analysis toolkit.

Using of computational clouds to execute climate and atmospheric codes is not a novelty, however, to the best of our knowledge, the exploration of elasticity features is not addressed in this type of application. In this sense, we present our efforts to adapt OLAM to support elasticity and take advantage of this feature to provide dynamic allocation of resources and load balancing.

3   Cloud Elasticity

Elasticity is considered one of the fundamental properties of the cloud [17]. It can be defined as the ability to adaptively scale resources up and down in order to meet varying application demands. It implies that the resources that compose the virtual environment may be added or removed on-the-fly and without service interruptions. Ideally, to the consumer, the resources available for provisioning often appear to be unlimited and can be purchased in any quantity at any time [1].

Elasticity can be classified as horizontal and vertical, depending on the granularity in which the resources are provisioned. Horizontal elasticity consists in adding/removing complete virtual machines (VM) from user virtual environment, and is the most widely used method to provide elasticity. In vertical elasticity, processing, memory and storage resources can be added/removed from a running virtual instance.

Currently, cloud elasticity has been used for scaling traditional web applications to handle unpredictable workloads, and enabling companies to avoid the downfalls involved with the fixed provisioning (over and under-provisioning) [18]. In the scientific scenario, the use of cloud computing is discussed in several studies [19], but the use of elasticity in scientific applications is a subject that is starting to receive attention from research groups [2].

Some academic researches have addressed the development of elastic scientific applications. It is possible to find works focusing on scientific workflows [6], MapReduce [20], MPI [7] and master-slave applications [8]. In this paper, we use Cloudine framework to include elasticity features to OLAM. The platform is described in details in next section.

3.1   Cloudine Framework

Cloudine is a framework for development and execution of elastic scientific applications in clouds [10]. The framework focuses on parallel and distributed applications that runs directly over the VM operating system of IaaS clouds.

In Cloudine, the elasticity control is based on the concept of elasticity primitives, which are basic functions inserted into source code that allow applications to perform requests for allocation or deallocation of resources. The primitives enable the dynamic allocation and deallocation of resources in several levels, ranging from nodes of a virtual cluster, to virtual processors and memory of a node. Primitives for collecting information and monitoring data from the cloud system are also available. Cloudine supports C/C++ languages and provides a set of primitives for dynamic allocation of VCPUs, memory and virtual machines. The primitives list can be found in the work of Galante and Bona [10].

The communication between elastic applications and the cloud is done by Cloudine platform. It receives the requests of resources generated by elasticity primitives and converts them to cloud specific commands using protocols such as OpenNebula XML-RPC[1], OCCI[2] or EC2[3].

Figure 1 illustrates an example of Cloudine operation. The programmer inserts the `clne_add_vcpu(2)` primitive in the original source code. This primitive is used to request the addition of two VCPUs to the virtual machine in which the application runs. It sends a request to the Cloudine platform, which converts

---

[1] http://opennebula.org/documentation:rel4.0:api
[2] http://occi-wg.org/
[3] http://docs.aws.amazon.com/AWSEC2/latest/UserGuide/using-query-api.html

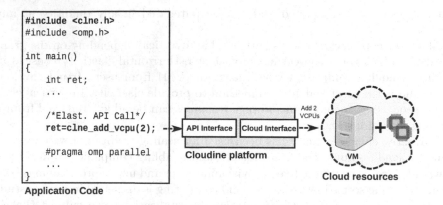

Application Code

Fig. 1. Cloudine framework: elasticity primitives operation

the message and interacts with the cloud, requesting the resources through the interface provided by the cloud provider.

The use of the elasticity primitives approach enables to control explicitly the application elasticity. Thus, programmers can construct tailor made elasticity controllers that take into account the application particularities. These features are specially interesting for dynamic applications, whose resources requirements cannot be determined exactly in advance, either due to changes in runtime requirements or in application structure, as is the case of OLAM.

4   Ocean-Land-Atmosphere Model - OLAM

Ocean-Land-Atmosphere Model (OLAM) [4] was chosen as background to the development of this work. This model is used to forecast weather and climate in research and forecast centers.

The model consists essentially of a finite volume representation of the full compressible non-hydrostatic Navier-Stokes equations over the planetary atmosphere with a formulation of conservation laws for mass, momentum, and potential temperature, and numerical operators that include time splitting for acoustic terms [21]. The finite volumes are defined horizontally by the global triangular-cell grid mesh and subdivided vertically through the height of the atmosphere forming vertically-stacked prisms of triangular bases.

Local grid or mesh refinement can be applied to cover specific geographic areas with higher resolution. The mesh points that represent these areas are subdivided cyclically while the expected mesh resolution is not achieved. Each cyclical division doubles the resolution. The global grid and its refinements define a single grid, as opposed to the usual nested grids schemes of regional models. The grid points, which represent a more refined area, do not overlap the grid points that represent the global domain, but substitute them.

An example of local mesh refinement is illustrated in Figure 2, where the resolution is exactly twice that of the original resolution. This is achieved by subdividing each previously triangle into 2 smaller triangles. For this purpose, auxiliary edges were inserted at the boundary between the original and refined regions to preserve the rule of the three neighboring triangles for each triangle.

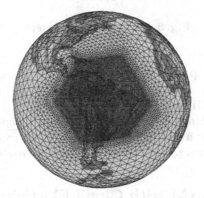

Fig. 2. Local mesh refinement applied to a selected part of the globe

OLAM was originally developed in FORTRAN 90 and parallelized with Message Passing Interface (MPI) under the Single Program Multiple Data (SPMD) model. In this paper, the OLAM was re-implemented using C language and parallelized with using MPI, for inter-node parallelism, and OpenMP for intra-node parallelism [5]. Online Mesh Refinement (OMR) support was also included in this version [22]. The OMR implementation allows local mesh refinement at execution time, increasing the resolution of a discrete representation of part of a domain. This solution provides higher mesh resolution for atmospheric models with low impact in the execution time, providing also better numerical results.

The execution of OLAM involves several steps and can be divided in three major parts: the parameter initialization, the atmosphere time state calculation and results output, as illustrated in Figure 3. The first part of the code involves the pre-processing, where settings are read and applied for memory allocations and the processing of the information of terrain, vegetation, soil and sea.

The remainder of the algorithm consists of an iterative step, involving the physical parametrization. The physical parametrization includes the radiation transfer, micro-physics, bio-physical schemes, turbulence and convective clouds like cumulus clouds.

During the iterative steps, it is possible to call OMR. In this case, the execution is stopped and the mesh is refined on specific points or Earth regions, according to a climatological condition. After the conclusion of an online mesh refinement call, the iterative execution proceeds normally.

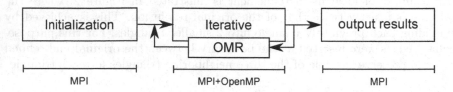

Fig. 3. OLAM execution steps

After the iterative step, and before the end of the program, some results are written in specific files, storing values of the physical conditions of the atmosphere to a determined time.

All steps are executed in parallel by MPI processes. OpenMP is used in iterative steps to execute the loops from OLAM's highest hotspot routine, named *progwrtu*, which is responsible for up to 70% of the CPU time of each timestep [23].

5   Improving OLAM with Cloud Elasticity

In this section, we present our efforts to adapt OLAM to use the elasticity offered in cloud environments. We employed elasticity for three purposes: (1) dynamic allocation of virtual processors (VCPUs), (2) load balancing and (3) dynamic allocation of memory.

Considering the OLAM phases, only the iterative part is parallelized using OpenMP, while the others parts are executed using a single VCPU. Using elasticity provided by Cloudine, we modified OLAM to starts with a single VCPU and dynamically allocate additional processors in the iterative step, where they are used by OpenMP threads, as presented in Figure 4. It enables the use of cloud resources efficiently, allocating the resources only when demanded and avoiding overprovisioning and idle resources in the initialization and output phases. The released resources can be used by other users or applications.

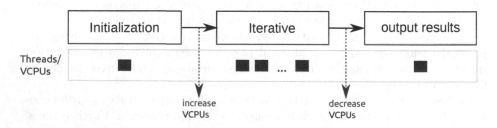

Fig. 4. Dynamic VCPUs allocation in OLAM

Other feature included in OLAM is load balancing, originally not addressed in the model. The online mesh refinement brings unbalanced load distribution after it is called, since refinement results in the addition of new elements to some process. Figure 5 illustrates a mesh with resolution of 50 Km before (200,002 triangles) and after the OMR (275,386 triangles). The mesh refinement was done from the Cartesian coordinate 0,0 with radius of 2,000 km.

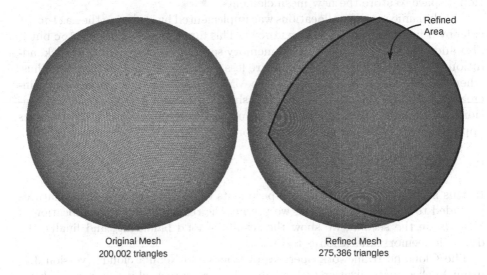

Original Mesh
200,002 triangles

Refined Area

Refined Mesh
275,386 triangles

Fig. 5. Original mesh and refined mesh

As consequence of the load unbalancing, the execution time of each iteration (of iterative part) is determined by the time of the slower process, affecting the application performance as a whole. We used the Cloudine elasticity primitives to enable the application to distribute automatically the available VCPUs among the nodes according to the workloads of each MPI process in order to equalize the processing time of all processes involved, and consequently, reduce the iterations time.

Compared to traditional load balancing strategies, the proposed approach requires few modifications in original source code (few dozens of lines) and do not require stopping the application or redistributing the load among the processes involving data exchanges [24][25].

At last, we also included support for elastic memory allocation, taking advantage of the ballooning technique available in most of modern hypervisors (e. g., KVM and Xen). The ability to dynamically modify the memory of a VM at runtime without any service disruption represents an important capability for applications with dynamic memory requirements. This means to automatically adapt the underlying virtual computing platform (i.e., the VM or the set of VMs) to the runtime profile of memory consumption of the application [12].

Considering that OLAM simulations can be executed with different resolutions and refinements, to choose the most appropriate amount of memory is not a simple task. Thus, we implemented a mechanism to allocate memory to the VM according to the application demands. The mechanism was implemented to allocate the memory needed by application and keeping free the initial amount of the VM memory (to be used by operating system and other services). This feature is also useful in the cases where the OMR require the allocation of additional space to store the new mesh elements.

The dynamic memory allocations was implemented by replacing the `malloc()`[4] calls by a function called `elast_alloc()`. This function resembles `malloc` but it also adds to the VM the requested memory space if necessary. In this work, additional memory is allocated if there are less than 512 MB free in the VM. Thus, when a data structure is allocated, the VM memory space is automatically increased, avoiding the occurrence of thrashing [12](which impact on the applications performance) without the need of memory over-provisioning, and prevents application aborting due to lack of memory.

6 Tests and Results

In this section we present three experiments to validate the elasticity features provided to OLAM. In the first, we present the results of dynamic allocation of VCPUs, in the second, we show the results of load balancing, and finally, the dynamic memory allocation is tested.

The Cloudine framework operates over an OpenNebula cloud[5] (version 3.4) using Xen[6] as virtualization technology. The hardware used is a 24-core workstation equipped with three AMD Opteron 6136 at 2.40 GHz and 98 GB RAM. We used 4 cores and 10 GB RAM for dom0 and the remaining CPUs and memory for the other VMs. The operating system in all tests is Ubuntu Server 12.04 with kernel 3.2.0-29.

6.1 Dynamic VCPUs Allocation

We tested the dynamic VCPUs allocation feature using a mesh with global resolution of 50 Km, using 1,440 iterations before and 1,440 after OMR, simulating 2 days of real time (each iteration represents 60 seconds of real time). Two virtual cluster configurations were employed: with two and with four nodes.

In the first scenario, the two nodes have initially a single VCPU and other seven are added in iterative stage, released in OMR and reallocated in the next iterative stage, according to demand of the parallel regions created by OpenMP. The allocation of VCPUs are illustrated in Figure 6. The total running time was 4,396 seconds, approximately 2% slower than the execution using eight VCPUs per process during whole execution (4,304 seconds).

---

[4] The function `malloc` is used to allocate a certain amount of memory during the execution of a program.

[5] http://www.opennebula.org

[6] http://www.xen.org

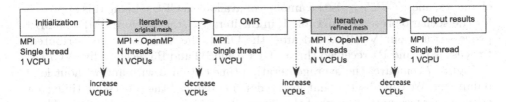

Fig. 6. Dynamic allocation of VCPUs

Similarly, in the second scenario, the execution starts with four nodes with one VCPU and other 3 are allocated dynamically for iterative steps. The total running time was 4,016 seconds, approximately 1,7% slower than the execution using four VCPUs per process during whole execution (3,950 seconds).

The results show that is possible to employ successfully the elasticity to allocate resources more efficiently, improving the shared used of cloud infrastructures, with small overhead.

6.2 Load Balancing

Other use for Cloudine primitives in OLAM is load balancing, originally not addressed in OLAM model. The online mesh refinement causes load unbalancing after it is called, since the new created elements are not distributed uniformly among all processes. Table 1 shows the load distribution before and after the OMR, in the simulation using the mesh of 50 Km resolution and four processes. It is possible to observe that processes P1 and P4 were affected significantly by OMR.

Table 1. Load unbalancing caused by OMR

Process	Before OMR	After OMR	Diff. from P0
P0	50,556	50,556	–
P1	50,812	50,812	0.5%
P2	51,110	88,875	75%
P3	50,886	85,143	68%

Considering this problem, we employed the elasticity primitives to construct a load balancer. Differently from traditional load balancing techniques that are based on the load redistribution, in our approach we distribute automatically the available VCPUs (16 in this experiment) proportionally to the workload of each MPI process, using the formula:

$$VCPU_{Pi} = \lceil NE_{Pi}/NE_{total} \times VCPU_{total} \rceil$$

where $VCPU_{Pi}$ is the number of VCPUs allocated to process Pi, NE_{Pi} is the number of mesh elements assigned to Pi, NE_{total} is the total number of mesh elements, and $VCPU_{total}$ is the amount of available VCPUs. Using this approach for the scenario presented in Table 1, in the iterative stage (before OMR) all MPI processes use four VCPUs, and after the OMR, the VCPUs are redistributed. Processes P0 and P1 received three VCPUs and P2 and P3 received five VCPUs.

Figure 7 compares the average iteration time obtained with and without load balancing. We can observe that load balancing reduced the processing time (not considering message passing) of the stage after OMR, on average, in 24%. These results represent a decrease of 17% in the execution time of the stage (1,999 to 1,720 seconds), and 7% of total OLAM execution time (3,950 to 3,677 seconds). Note that the performance gains was achieved using the same amount of resources.

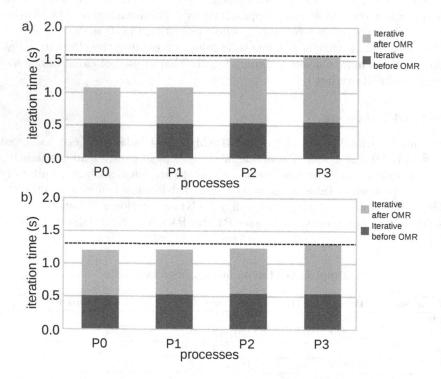

Fig. 7. Iteration time. a) Without load balancing. b) With load balancing.

The use of Cloudine primitives in the construction of the load balancing feature in OLAM proved to be an interesting option. To the best of our knowledge, there are not elasticity mechanisms that enable the development of load balancing using the approach presented in this section. Unfortunately, OLAM does not implement load balancing features and thus it was not possible to compare results.

6.3 Dynamic Memory Allocation

We tested the dynamic memory allocation a mesh with global resolution of 40 Km, using 120 iterations for each iterative phases (before and after OMR), simulating 4 hours of real time. We used two virtual machines with 2 GB memory. Figure 8 shows the memory allocation for processes P0 and P1. The proposed mechanism successfully allocated memory to keep the application data in resident memory and preserving memory for the operating system and other services running on the machine.

Note that the peak memory use (~55 minutes) is almost 8 times greater than in the beginning of simulation. This large variation makes difficult to predict the memory to be allocated at VM instantiation. Thus, the use of elasticity is of great value in this type of scenario.

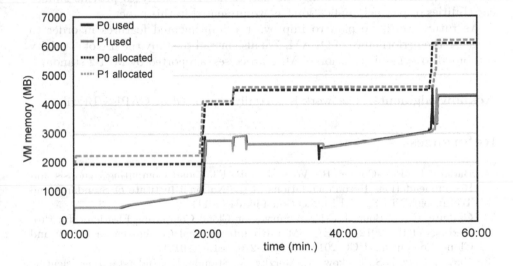

Fig. 8. Dynamic allocation of memory in OLAM

We compared the results with the results of a monitoring-based solution mechanism proposed in the work of Moltó et al. [12]. This mechanism failed in allocating the memory for the workload peak and application aborted after approximately 20 minutes of simulation, since the mechanism is not able to allocate memory fast enough.

7 Conclusion

The elasticity offered by cloud computing enable the development of applications with capabilities that cannot be implemented in traditional environments, such as clusters and grids. In this paper, we presented our efforts to adapt OLAM to use the elasticity offered by cloud environments. Elasticity was successfully employed to provide dynamic allocation of VCPUs and memory, and to add load balance support to OLAM.

The dynamic allocation of VCPUs was employed to improve the use of shared resources, enabling OLAM to allocate the processors on-demand and resulting in rational use of resources. Dynamic memory allocation worked very well, making possible to adapt the VMs to provide memory enough to applications, maintaining the data on resident memory and avoiding application errors due lack of resources. The use of elasticity to provide load balancing to OLAM also presented good results, considering that was possible improve the performance of the model using the same amount of resources and without needing additional inter-process communication.

Considering the state-of-art, the use of clouds to execute climate and atmospheric codes is not a novelty. However, to the best of our knowledge, there are no cases in literature describing the use of cloud elasticity to provide similar capabilities to parallel applications as we presented in this paper.

As future work, we plan to improve the implemented features in order to increase the performance of OLAM. We also intend to study the use of elasticity combined to dynamic creation of MPI processes, supported by MPI-2 standard.

Acknowledgments. This work is partially supported by CAPES/Brazil.

References

1. Badger, L., Patt-Corner, R., Voas, J.: DRAFT Cloud Computing Synopsis and Recommendations Recommendations of the National Institute of Standards and Technology. Nist Special Publication 146, 84 (2011)
2. Galante, G., de Bona, L.C.E.: A Survey on Cloud Computing Elasticity. In: Proceedings of the 2012 IEEE/ACM Fifth International Conference on Utility and Cloud Computing, UCC 2012, pp. 263–270. IEEE (2012)
3. Jha, S., Katz, D.S., Luckow, A., Merzky, A., Stamou, K.: Understanding Scientific Applications for Cloud Environments. In: Buyya, R., Broberg, J., Goscinski, A.M. (eds.) Cloud Computing: Principles and Paradigms, pp. 345–371. John Wiley & Sons (2011)
4. Walko, R.L., Avissar, R.: The Ocean-Land-Atmosphere Model (OLAM). Part I: Shallow-Water Tests. Monthly Weather Review 136(11), 4033–4044 (2008)
5. Schepke, C., Maillard, N., Schneider, J., Heiss, H.U.: Online Mesh Refinement for Parallel Atmospheric Models. International Journal of Parallel Programming 41(4), 552–569 (2013)
6. Byun, E.K., Kee, Y.S., Kim, J.S., Maeng, S.: Cost Optimized Provisioning of Elastic Resources for Application Workflows. Future Gener. Comput. Syst. 27(8), 1011–1026 (2011)
7. Raveendran, A., Bicer, T., Agrawal, G.: A Framework for Elastic Execution of Existing MPI Programs. In: Proceedings of the International Symposium on Parallel and Distributed Processing Workshops and PhD Forum, IPDPSW 2011, pp. 940–947. IEEE (2011)

8. Rajan, D., Canino, A., Izaguirre, J.A., Thain, D.: Converting a High Performance Application to an Elastic Cloud Application. In: Proceedings of the 3rd International Conference on Cloud Computing Technology and Science, CLOUDCOM 2011, pp. 383–390. IEEE (2011)
9. Rajan, D., Thrasher, A., Abdul-Wahid, B., Izaguirre, J.A., Emrich, S., Thain, D.: Case Studies in Designing Elastic Applications. In: 13th IEEE/ACM International Symposium on Cluster, Cloud and Grid Computing, CCGrid 2013, pp. 466–473. IEEE (2013)
10. Galante, G., Bona, L.C.E.: Constructing Elastic Scientific Applications Using Elasticity Primitives. In: Murgante, B., Misra, S., Carlini, M., Torre, C.M., Nguyen, H.-Q., Taniar, D., Apduhan, B.O., Gervasi, O. (eds.) ICCSA 2013, Part V. LNCS, vol. 7975, pp. 281–294. Springer, Heidelberg (2013)
11. Galante, G., Bona, L.C.E.: Supporting Elasticity in OpenMP Applications. In: Proceedings of 22nd Euromicro International Conference on Parallel, Distributed and Network-Based Processing, PDP 2014 (to appear, 2014)
12. Moltó, G., Caballer, M., Romero, E., de Alfonso, C.: Elastic Memory Management of Virtualized Infrastructures for Applications with Dynamic Memory Requirements. Procedia Computer Science 18, 159–168 (2013)
13. Evangelinos, C., Hill, C.N.: Cloud Computing for parallel Scientific HPC Applications: Feasibility of running Coupled Atmosphere-Ocean Climate Models on Amazon's EC2. In: Cloud Computing and Its Applications, CCA 2008 (2008)
14. Martinez, J.C., Wang, L., Zhao, M., Sadjadi, S.M.: Experimental Study of Large-scale Computing on Virtualized Resources. In: Proceedings of the 3rd International Workshop on Virtualization Technologies in Distributed Computing, VTDC 2009, pp. 35–42. ACM (2009)
15. Krishnappa, D.K., Irwin, D.E., Lyons, E., Zink, M.: CloudCast: Cloud computing for short-term mobile weather forecasts. In: Proceedings of the 31st IEEE International Performance Computing and Communications Conference, IPCCC 2012, pp. 61–70. IEEE (2012)
16. Mattmann, C., Waliser, D., Kim, J., Goodale, C., Hart, A., Ramirez, P., Crichton, D., Zimdars, P., Boustani, M., Lee, K., Loikith, P., Whitehall, K., Jack, C., Hewitson, B.: Cloud computing and virtualization within the regional climate model and evaluation system. Earth Science Informatics, 1–12 (2013)
17. Han, R., Ghanem, M.M., Guo, L., Guo, Y., Osmond, M.: Enabling Cost-aware and Adaptive Elasticity of Multi-tier Cloud Applications. Future Generation Computer Systems 32, 82–98 (2014)
18. Chieu, T.C., Mohindra, A., Karve, A.A., Segal, A.: Dynamic Scaling of Web Applications in a Virtualized Cloud Computing Environment. In: Proceedings of the 2009 IEEE International Conference on e-Business Engineering, ICEBE 2009, pp. 281–286. IEEE (2009)
19. Wang, L., Zhan, J., Shi, W., Liang, Y.: In Cloud, Can Scientific Communities Benefit from the Economies of Scale? IEEE Trans. Parallel Distrib. Syst. 23(2), 296–303 (2012)
20. Iordache, A., Morin, C., Parlavantzas, N., Riteau, P.: Resilin: Elastic MapReduce over Multiple Clouds. Rapport de recherche RR-8081. INRIA (October 2012)
21. Marshall, J., Adcroft, A., Hill, C., Perelman, L., Heisey, C.: A Finite-Volume Incompressible Navier-Stokes Model for Studies of Ocean on Parallel Computers. Journal of Geophysical Research 102(C3), 5753–5766 (1997)

22. Schepke, C., Maillard, N., Schneider, J., Heiss, H.U.: Why Online Dynamic Mesh Refinement is Better for Parallel Climatological Models. In: Proceedings of the 2011 23rd International Symposium on Computer Architecture and High Performance Computing, SBAC-PAD 2011, pp. 168–175. IEEE (2011)
23. Osthoff, C., Grunmann, P., Boito, F., Kassick, R., Pilla, L., Navaux, P., Schepke, C., Panetta, J., Maillard, N., Silva Dias, P., Walko, R.: Improving performance on atmospheric models through a hybrid openmp/mpi implementation. In: 9th International Symposium on Parallel and Distributed Processing with Applications, ISPA 2011, pp. 69–74. IEEE (2011)
24. Xu, C., Lau, F.C.: Load Balancing in Parallel Computers: Theory and Practice. Kluwer Academic Publishers (1997)
25. Dorneles, R.V., Rizzi, R.L., Diverio, T.A., Navaux, P.O.A.: Dynamic Load Balancing in PC Clusters: An Application to a Multi-Physics Model. In: Proceedings of the 15th Symposium on Computer Architecture and High Performance Computing, SBAC-PAD 2003, pp. 192–198. IEEE (2003)

Estimating and Controlling the Traffic Impact of a Collaborative P2P System

Pedro Sousa

Centro Algoritmi/Department of Informatics,
University of Minho, Braga, Portugal
pns@di.uminho.pt

Abstract. Nowadays, P2P applications are commonly used in the Internet being an important paradigm for the development of distinct services. However, the dissemination of P2P applications also entails some important challenges that should be carefully addressed. In particular, some of the important coexistence problems existing between P2P applications and Internet Service Providers (ISPs) are mainly motivated by the inherent P2P dynamics which cause traffic to scatter across the network links in an unforeseeable way.

In this context, this work proposes a collaborative framework of a BitTorrent like system. Using the proposed framework and based on the exchange of valuable information between the application and network levels, some novel techniques are proposed allowing to estimate and control the traffic impact that the P2P system will have on the links of the underlying network infrastructure. Both the framework and the presented techniques were tested resorting to simulation. The results clearly corroborate the viability and effectiveness of the formulated methods.

Keywords: P2P, Traffic Engineering, BitTorrent, Collaborative systems.

1 Introduction

P2P overlays [1] can be considered as self-organized systems operating on top of a given network infrastructure. Such systems usually adopt specific protocols and peering strategies which may significantly change the traffic profiles observed in the network, thus also posing new problems to the Internet Service Providers (ISPs). BitTorrent [2] is just an example of a widely used P2P protocol over which many applications rely to exchange considerable large resources among a significant number of users, being also responsible by a considerable amount of the Internet traffic [3,4]. However, several coexistence problems between ISPs and P2P applications emerged in the last years, being this motivated by several factors. In fact, P2P dynamics cause traffic to scatter across the network links in an unforeseeable way. As consequence, P2P approaches are not always consistent with ISP economic models, as specific links from the underlying network might be under excessive and unpredictable traffic loads and some unnecessary inter-domain traffic could also be generated [5,6]. Another important issue is that ISPs

B. Murgante et al. (Eds.): ICCSA 2014, Part VI, LNCS 8584, pp. 61–75, 2014.

many times use several Traffic Engineering (TE) techniques for tasks such as capacity planning, resilience improvements, routing optimization, among others. One of the critical inputs for TE tasks is the estimation of the traffic matrix of the network infrastructures. In this perspective, P2P overlay networks make complex the demand matrix estimation, and such estimation errors will also affect all the other TE related tasks [7,8].

Given all the above mentioned, this work presents a contribution to attain a BitTorrent-like P2P system architecture with the ability to collaborate and help network level entities to better deal with the P2P traffic generated by the application level. For that purpose, the devised framework includes methods allowing to estimate the traffic impact that the P2P system will have on the underlying network links. Such qualitative impact estimation values are able to provide a preliminary view about the traffic patterns that will traverse the network infrastructure, thus being an important asset from the ISP point of view. Additionally, within the proposed framework, the P2P tracker ruling the P2P swarm behavior is able to be dynamically configured in order to make an effort to protect specific network links from the generated P2P traffic. The presented framework corroborates the advantages of pursuing collaborative efforts in this area, as also highlighted by other works (e.g. [9,10]). In this case, the proposed solution raises the P2P application level with mechanisms allowing to estimate and control the P2P traffic impact, making such enhanced methods available to network level entities through specific configuration interfaces. As compensation, ISPs are expected to provide such collaborative P2P systems with a privileged traffic treatment, in contrast with more aggressive techniques used to punish other nonconforming P2P approaches.

The paper is organized as follows: Section 2 presents the proposed framework rationale, also explaining the devised mechanisms for estimating the network impact of P2P swarms and for protecting specific links from P2P traffic. Section 3 describes the implemented simulation platform and presents illustrative results corroborating the effectiveness of the devised mechanisms. Finally, Section 4 concludes the presented work.

2   P2P System Architecture and Devised Methods

This work proposal focus on a BitTorrent-like framework having some enhanced features. In the devised system the network level is able to obtain estimations about the traffic impact that the P2P system will have on the network infrastructure and, if required, control how such traffic traverses the network domain. For this end, the framework depicted in Figure 1 assumes a collaborative perspective between the application level (e.g. Service Providers) and network level entities (e.g. ISPs) with the exchange of valuable information. The framework might be used in distinct contexts. For instance, it could allow ISPs to offer Internet users a friendly P2P system behaving in a collaborative way with the network level. In a different perspective, it can be also used by Service Providers to develop specific services involving their clients, such as the upload of large resources (e.g. generic data files, media, software packages, etc.) to all of their customers in a P2P fashion, also benefiting from specific agreements made with the network provider. This exchange might occur in previously scheduled time periods, having end users to notify the Service Provider that they intend to integrate the corresponding P2P swarm on such allocated time slots.

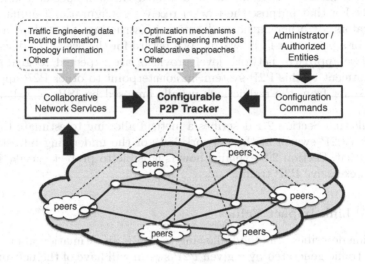

Fig. 1. General view of the envisioned P2P framework and participating entities

The operation of a BitTorrent like system usually implies that interested peers establish a contact with the P2P tracker controlling a specific P2P swarm. As consequence, the P2P tracker sends a random sample of peers already present in the swarm to the contacting peer. With that, the newly arrived peer attempt to establish network connections with other peers in order to exchange the pieces of the file. From that point on, the BitTorrent defines several rules that affect the data transfer and the choke/unchoke processes among the swarm peers [1,2,11]. In addition, periodically, peers are allowed to contact the tracker to obtain a renewed sample of peers. The framework depicted in Figure 1 assumes that the P2P tracker fully controls the peering information provided to clients, which means that clients are not allowed to exchange peering information between them. Most of the classical BitTorrent based P2P systems can also behave in this way by the use of specific options conveniently defined in the *.torrent* file.

As also visible in Figure 1, the framework adopts the use of configurable P2P trackers making possible that distinct configurations could be made on the tracker, also allowing the tracker interaction with other external entities.

The tracker internal architecture is not the focus of this work, but could follow similar directives as the presented in [10,12]. Moreover, if required, the tracker may also resort to several optimization mechanisms, including mechanisms from the field of computational intelligence (e.g. [16], [15]). The configurable P2P tracker integrated in Figure 1 is then able to receive valuable network level information from collaborative network services, such as topological, routing and other traffic engineering related inputs. The tracker can also be programmed with internal methods that might be activated through specific configuration commands. For that purpose the tracker receives configuration commands from administrators, or other authorized entities, instructing it to adopt a specific behavior for a particular P2P swarm. As a reward for the use of the devised P2P collaborative approach, network level providers are expected to give a better traffic treatment to this P2P system, in counterpoint to other P2P approaches that will suffer from the restrictions usually imposed by ISPs (e.g. bandwidth throttling).

The following Section 2.1 describes a method allowing to estimate the traffic impact of a P2P swarm in the network links of the underlying infrastructure. Following that, Section 2.2 explains how is possible to protect specific network links from excessive P2P traffic.

2.1 P2P Link Impact Values

This section describes a method allowing to attain an estimation about the impact that traffic generated by a given P2P swarm will have of the network links, when involving a considerable number of peers. Thus, for a given swarm composition and assuming the tracker behaving in the classical mode, the objective is that such qualitative link impact information could be provided to the ISP.

Lets assume a classical mathematical representation of a network, with the graph $G = (N, L)$ expressing a network domain (e.g. an ISP network), were N is a set of the network nodes/routers and L a set of the interconnecting network links, for which routing link weights are also considered for shortest path computation. Part of the network nodes/routers might also be viewed as Points of Presence (PoP) to end-users areas having peers interested to participate in a given P2P swarm. For convenience, the location of such end-users areas is denoted by the corresponding ISP network router, a, with $a \in A$ and $A \subseteq N$.

Within the scope of the proposed mechanism, several graph measures (e.g. [13,14]) could constitute valuable inputs, in particular the concept of betweenness centrality in a graph, here adapted and extended to provide estimations of the P2P traffic link impact. The devised impact estimation metric combines distinct factors that could present a preliminary snapshot of the traffic patterns exchanged within a large P2P swarm. For a specific ISP link, l, and a pair of end-users areas, $i, j \in A$, we consider the ratio between the number of shortest paths from i to j, $sp_{i,j}$, and the number of such paths that effectively pass through link l, $sp_{i,j}(l)$. By this way, each link l is assigned with a partial impact value of

$\frac{sp_{i,j}(l)}{sp_{i,j}}$ for the case of peering adjacencies between areas i, j. When accounting all possible area adjacencies this metric will present higher values for links which integrate a higher number of shortest paths among the areas, thus having such links higher probabilities of being traversed by the P2P swarm traffic. A second weighting factor, $w_{i,j}$, is also considered for case of P2P swarms where end-user areas have an unbalanced distribution of peers. This factor considers the ratio between the number of peers involved in the peering adjacencies of areas i, j over the total number of peers involved in all possible adjacencies, favoring the importance of shortest paths connecting areas involving higher number of peers.

The above mentioned rationale can be further enhanced taking into account some characteristics of the TCP protocol that is used in the data transfers among BitTorrent peers. In fact, in such protocolar approach, peers often have a higher probability to establish peering connections with nearest peers in the network, taking advantage of lower network round-trip times (RTT). Thus, for shortest paths between areas i and j a preference value[1] ($p_{i\leftarrow j} \in [0,1]$ with $\sum_{j\in A, j\neq i} p_{i\leftarrow j} = 1$) is assigned to such adjacencies, implicitly denoting how close are areas j and i. Considering all the above mentioned reasoning, and for the case of a tracker returning random samples to contacting peers, Equation 1 presents the devised normalized P2P link impact value ($P2P_{LIV}$) value for link l, within the interval $[0,1]$. The tracker may announce these estimations to network services or administrators which in turn are able to instruct the tracker to protect specific links from the infrastructure.

$$P2P_{LIV}(l) = \sum_{\substack{i,j \in A \\ i \neq j}} [(|A| - 1) \cdot p_{i\leftarrow j}] \cdot \frac{sp_{i,j}(l)}{sp_{i,j}} \cdot w_{i,j} \quad l \in L \quad (1)$$

The metric presented by Equation 1 has the major objective of gathering a preliminary snapshot of which links are expected to be traversed by higher amounts of P2P traffic. The objective is that the comparison between the $P2P_{LIV}$ values of two links can be used to foresee which one will be traversed by higher amounts of P2P traffic, i.e. that the order relations between $P2P_{LIV}$ values could also somehow express the order relation between the P2P traffic that will flow over such links.

In order to validate the correctness of such impact estimations, the function $f(l, z)$ (presented in Equation 2) is defined for two distinct links $l, z \in L$. As observed in Equation 2, the function $f(l, z)$ might return two alternative values $\{0, 1\}$ according with the estimated $P2P_{LIV}$ metrics and the traffic that effectively traverses such links (function $T(l)$) after running a real/simulated experiment of the framework. If the $P2P_{LIV}$ order relations also express the $T(l)$ order relations the value returned by $f(l, z)$ is 1, otherwise 0. For the particular

---

[1] This value is then multiplied by the total number of distinct external areas adjacencies that could be made by peers in a given area, i.e. $|A| - 1$, for normalization purposes.

case of links having exactly equal $P2P_{LIV}$ values a small deviation (controlled by the γ variable) is accepted when comparing the observed traffic on each link.

$$f(l,z) = \begin{cases} 1 \text{ if } \left(P2P_{LIV}(l) > P2P_{LIV}(z)\right)\&(T(l) > T(z)) \\ 1 \text{ if } \left(P2P_{LIV}(l) < P2P_{LIV}(z)\right)\&(T(l) < T(z)) \\ 1 \text{ if } \left(P2P_{LIV}(l) = P2P_{LIV}(z)\right)\&(T(l) \in [T(z) \cdot (1-\gamma), T(z) \cdot (1+\gamma)]) \\ 0 \text{ otherwise} \end{cases}$$

(2)

Based on the $f(l,z)$ function, Equation 3 defines now the function $\psi(l)$ expressing the order conformity of the $P2P_{LIV}$ impact value of link l. Thus, $\psi(l)$ represents the average $f(l,z)$ values obtained when directly comparing link l with all the other links of a given network topology. Therefore, $\psi(l)$ values will vary within the interval $[0,1]$, with values close to 1 expressing that most of the order relations among $P2P_{LIV}$ values also express the order relations between the P2P traffic that effectively traverses the links. Thus, function $\psi(l)$ will be used to assess the quality of the $P2P_{LIV}$ results obtained in the experimental part of this work.

$$\psi(l) = \frac{\sum_{z \in L \setminus \{l\}} f(l,z)}{|L| - 1} \quad l \in L$$

(3)

2.2 Protecting Links from P2P Traffic

As previously explained, links having higher $P2P_{LIV}$ values are expected to be traversed by larger amounts of traffic. In this perspective, we now explore a possible method allowing the tracker to control the P2P swarm traffic distribution in the network domain, namely by protecting specific links of the network from P2P traffic. The devised method allows the tracker to reduce the link impact values of specific network links by conveniently manipulating the peer samples returned to the contacting peers.

Algorithm 1 presents the pseudo-code of the proposed method. As inputs it receives the swarm identification, an ordered set of the protected links and collaborative information provided by the network level. The methods starts by considering a set with all the area pairs combinations of the network (X_s, line 2), where each pair (a_i, a_j) means that when contacted by a peer from area a_i the tracker is able to include in the random sample peers from the area a_j. After that, and for each protected link $link_l$, the algorithm uses the topology and routing information provided by collaborative network entities to construct a subset Y containing the (a_i, a_j) pairs for which the shortest paths connecting such areas traverse $link_l$ (line 4). In the next step the algorithm verifies if is possible to remove a specific (a_i, a_j) entry from X_s in order to reduce the impact of the P2P swarm traffic on such link. The function $swarm\_totally\_connected()$ (in line 6) verifies if the swarm is still totally connected when considering that the tracker will not include peers from area a_j in peer samples sent to peers

---

Algorithm 1. *protecting_links_from_P2P_Traffic (s, K, data)*

1: {**Comment:** *s*- *a swarm identification;* ***K***- *a decreasingly ordered set with all $link_l \in L$ protected links (ordered by priority);* ***data***- *auxiliary information provided by collaborative services (topology, routing, etc.)*}

2: $X_s \leftarrow$ decreasingly ordered set with all (a_i, a_j) area pairs having peers from swarm s, $a_i, a_j \in A$ {**Comment:** X_s *is a* $w_{i,j} * p_{i \leftarrow j}$ *ordered set*}

3: **for all** $link_l \in K$ **do**

4:    $Y \leftarrow$ decreasingly ordered subset of X_s with (a_i, a_j) pairs which shortest paths include $link_l$ {**Comment:** Y *is a* $w_{i,j} * p_{i \leftarrow j}$ *ordered set*}

5:    **for all** $(a_i, a_j) \in Y$ **do**

6:       **if** *swarm_totally_connected*$(s, X_s \setminus \{(a_i, a_j)\}) = TRUE$ **then**

7:          $X_s \leftarrow X_s \setminus \{(a_i, a_j)\}$

8:       **end if**

9:    **end for**

10: **end for**

11: *update_tracker*(s, X_s)

---

from area a_i. The swarm is assumed to be totally connected if all peers have the opportunity to contact one of the swarm seeds, or contact other peers that directly or indirectly have access to the pieces sent by a seed. Otherwise the swarm is considered to be partitioned and some peers will never receive all the pieces of the shared file. In the case that the swarm would not become partitioned the (a_i, a_j) pair is effectively removed from X_s (line 7).

Algorithm 2. *get_peer_sample(peer p, swarm s)*

1: *peer_area* $\leftarrow$ get_peer_location(p)

2: **if** swarm_in_initial_state(s) **then**

3:    *peer_sample* $\leftarrow$ random_sample(s)

4: **else**

5:    *peer_sample* $\leftarrow$ random_sample_from_X_s $(s, peer\_area, X_s)$ {**Comment:** X_s *was previously computed by the tracker using Algorithm 1*}

6: **end if**

7: update_swarm_info(p, s)

8: return$(peer\_sample)$

As result, at the end of Algorithm 1, the set X_s will contain all area pairs that the tracker should consider to build the random peers samples. The considered peering adjacencies are sufficient to build a totally connected swarm, having also the minimum possible traffic impact on the considered protected links. After the computation of the final X_s set, the tracker will adopt Algorithm 2 to return a peers sample whenever contacted by any peer. As illustrated in Algorithm 2 in the initial state of the swarm, i.e. during a short initial period over which only few peers have contacted the tracker, the tracker behaves in the classical mode, i.e. a random peer sample is build considering all the available peers[2]. After that period, the tracker takes into account the area of the contacting peers and builds random peer samples constrained by the allowed peering adjacencies expressed in the X_s set returned by Algorithm 1 (line 5 of Algorithm 2).

3   Simulation Testbed and Results

Figure 2 presents the modules that were implemented in a simulation platform (ns-2 [17]) and the selected network topology to present illustrative results. A patch implementing the dynamics of the BitTorrent protocol [18] was used as the baseline over which other components from the devised framework were added. Specific modules were built for the implementation of the configurable tracker as well as the collaborative network services module providing network level information to the tracker. A specific interface was devised, allowing to interact with the tracker and provide configuration commands to activate some implemented methods. The link impact estimation and the link protection methods (explained in Sections 2.1 and 2.2) were programmed in the tracker internal logic, being able to be activated whenever triggered by specific configuration commands.

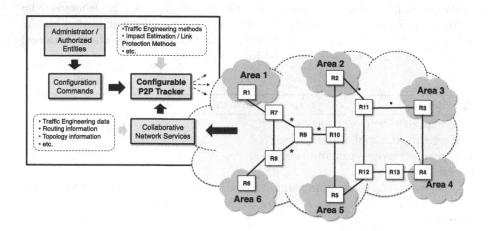

Fig. 2. Implemented framework and illustrative topology used in the experiments

To present illustrative simulation results the network topology depicted in Figure 2 was used. It illustrates a network topology consisting of several end-users areas interconnected by several links and core routers (some of which can also be viewed as possible Points of Presence (PoPs) of the ISP). The depicted

---

[2] This limited initial period will generate an almost negligible volume of traffic on the protected links, but contributes for assuring an improved initial distribution of some file pieces among all the areas.

topology intentionally includes several topological characteristics making more challenging the test of the devised mechanisms, such as: areas connected by paths with distinct distances, equal cost paths between some areas, critical links which failure will originate a partition on the network, single and multi-homed areas, etc. For routing purposes it is assumed that the ISP shortest paths are the ones having a small number of hops between a given source/destination pair (i.e. routing link weights of 1 for all links). The scenario assumes that peers participating in the P2P swarm are distributed along six areas, being each area composed by a second level of routers/links. In the developed simulation platform, several parameters can be configured, including the number of peers and seeds per area, the file size, the chunk size, among others.

The examples presented in the following sections assume a total number of 300 peers in the swarm, exchanging a file of 50 MB and operating with a chunk size of 256 KB. The parameters P_D and S_L might be used to control the distribution of the peers and seeds in the distinct network areas, respectively. By default, the peer sample returned by the tracker includes 25 peer contacts. At each area the peers have an upload capacity of 1 Mbps and a download capacity of 8 Mbps, thus simulating common residential scenarios where users have higher download capacities. To force some heterogeneity within each area, propagation delays of the users access links randomly vary within the interval [1, 50] ms. Due to the collaborative nature of the devised P2P system, the scenario also assumes that the ISP allows on each link a share of 50 Mbps exclusive for P2P traffic generated by the proposed P2P system, and the propagation delays of such links are at least two times higher than the end users access links. In the following sections, for each one of the described experiments, five simulations were made and the corresponding mean values were taken for analysis.

3.1   P2P Traffic Impact - Link Impact Values ($P2P_{LIV}$)

Based on the scenario depicted in Figure 2 several results are now presented regarding the tracker method to estimate the P2P impact on the network links. In the provided examples several scenarios were considered for distinct combinations of peers distribution in the network, P_D, and seed locations, S_L, and the obtained results are shown in Figure 3. The scenarios vary from an uniform distribution of peers in the network areas (first row of Figure 3 with all areas having 50 peers, i.e. $P_D = (50, 50, 50, 50, 50, 50)$) to other scenarios where a higher density of peers is considered to exist in specific parts of the network. The results of such additional peer distributions are presented in the other rows of Figure 3, assuming that the left, right, upper and bottom sides of the topology of Figure 2 have a higher density of peers, respectively. In addition, for each of the mentioned P_D distributions, three distinct seed positioning scenarios are considered: i) all areas having one seed; ii) a single seed positioned in area 1 and iii) a single seed positioned in area 4 (first, second and third columns of Figure 3).

Each graph of Figure 3 presents the results obtained on each particular scenarios (five independent simulations runs were made for each one and the plotted results are averaged values). For comparative analysis, on each graph, the

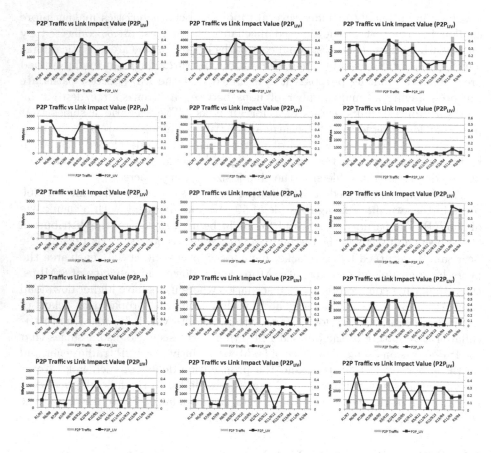

Fig. 3. P2P traffic on links vs $P2P_{LIV}$ values for distinct P_D and S_L values. Row1: $P_D=(50,50,50,50,50,50)$; Row2: $P_D=(70,70,10,10,70,70)$; Row3: $P_D=(10,70,70,70,70,10)$; Row4: $P_D=(90,90,90,10,10,10)$; Row5: $P_D=(10,10,10,90,90,90)$; Column1: S_L=all ; Column2: $S_L=A_1$; Column3: $S_L=A_4$.

cumulative P2P traffic which traversed each link during the swarm lifetime is represented by gray filled columns (in MBytes), being the previoulsy computed $P2P_{LIV}$ impact metrics[3] for each link (Equation 1) represented by a black line-plot representation (normalized values within $[0,1]$). A detailed analysis of Figure 3 allows to verify that in all of the considered scenarios both the $P2P_{LIV}$ link values and the overall P2P traffic on each link follow a similar trend. This constitutes a preliminary indication that $P2P_{LIV}$ metric could in fact denote the relations between the P2P traffic traversing each link during the swarm lifetime.

In order to verify the correctness of the $P2P_{LIV}$ metrics, the link impact order conformity metric (function $\psi(l)$ in Equation 3) was evaluated for each one of

---

[3] As in real scenarios the tune of $p_{i \leftarrow j}$ values is difficult, in the experiments only nearest areas are differentiated ($p_{i \leftarrow j}=0.4$), the remaining areas have values of 0.15.

the topology links within each one of the simulated scenarios. The obtained $\psi(l)$ values are summarized in Table 1[4]. As observed the link impact metrics obtained high order conformity values. In fact, in most of the presented scenarios and independently of the peers distribution and seed locations the $\psi(l)$ averaged values fall within the interval $[0.89, 97]$. This means that, for an expressive majority of the cases, the $P2P_{LIV}$ link impact values computed by the tracker also denote the foreseeable order relations between the P2P traffic traversing each link. In that way, $P2P_{LIV}$ values can effectively be used to have a preliminary view about which links will suffer higher impact from the P2P swarm traffic, thus being this information a valuable asset for ISPs and network administrators.

Table 1. Link Impact Value Order Conformity $\psi(l)$ on the Simulated Scenarios (for each simulated instance of Figure 3)

Scenar.		Link Impact Value Order Conformity $\psi(l)$															
P_D	S_L	R_1 R_7	R_6 R_8	R_7 R_8	R_7 R_9	R_8 R_9	R_9 R_{10}	R_2 R_{11}	R_{10} R_5	R_2 R_{11}	R_5 R_{12}	R_{11} R_{12}	R_{12} R_{13}	R_{13} R_4	R_{11} R_3	R_3 R_4	Avg $\overline{\psi(l)}$
50,50,	all	0.86	0.86	1.00	1.00	1.00	0.93	0.93	0.93	0.79	1.00	1.00	1.00	1.00	0.86	0.71	**0.92**
50,50,	A_1	0.93	0.93	1.00	1.00	1.00	1.00	0.93	0.93	0.86	1.00	1.00	1.00	1.00	0.93	0.93	**0.96**
50,50	A_4	0.86	0.86	1.00	1.00	1.00	0.93	0.93	0.93	0.86	1.00	1.00	1.00	1.00	0.86	0.79	**0.93**
70,70,	all	0.79	0.79	0.79	0.93	0.93	0.79	0.79	0.86	1.00	0.93	1.00	1.00	1.00	0.93	0.93	**0.90**
10,10,	A_1	0.79	0.79	0.86	0.93	0.93	0.86	0.86	0.86	1.00	0.93	1.00	1.00	1.00	1.00	0.93	**0.91**
70,70	A_4	0.79	0.79	0.86	0.86	0.86	0.86	0.86	0.86	1.00	0.93	0.86	0.93	0.93	1.00	0.93	**0.89**
10,70,	all	0.93	0.93	1.00	0.93	0.93	0.86	1.00	1.00	1.00	1.00	1.00	0.93	0.93	0.93	0.93	**0.95**
70,70,	A_1	0.93	0.93	1.00	0.93	0.93	0.79	0.93	0.93	1.00	0.86	0.93	0.93	0.93	1.00	1.00	**0.93**
70,10	A_4	0.93	0.93	1.00	0.93	0.93	0.79	1.00	0.93	1.00	0.93	0.93	0.93	0.93	1.00	1.00	**0.94**
90,90,	all	0.86	0.93	0.93	1.00	0.93	0.86	0.86	1.00	1.00	1.00	1.00	1.00	1.00	1.00	0.93	**0.95**
90,10,	A_1	0.86	0.93	0.86	1.00	0.93	0.86	0.86	1.00	1.00	0.93	1.00	1.00	1.00	1.00	0.93	**0.94**
10,10	A_4	0.86	0.93	0.93	1.00	0.93	0.86	0.86	1.00	1.00	1.00	1.00	1.00	1.00	1.00	0.93	**0.95**
10,10,	all	1.00	0.93	0.93	0.93	1.00	0.93	0.93	1.00	1.00	1.00	1.00	0.93	0.93	1.00	0.79	**0.95**
10,90,	A_1	1.00	0.93	0.93	0.93	1.00	0.93	1.00	1.00	1.00	1.00	1.00	1.00	1.00	0.93	0.93	**0.97**
90,90	A_4	1.00	0.93	0.93	0.93	1.00	0.93	1.00	0.79	1.00	0.93	1.00	0.93	0.93	1.00	1.00	**0.95**

P_D - Peers distribution in the network $(A_1,...,A_6)$, S_L - Seeds location in the network

3.2   Protecting Network Links from P2P Traffic

This section illustrates the tracker configuration mode explained in Section 2.2, namely in Algorithm 1, where some specific network link(s) are protected from the traffic generated by the P2P swarm.

---

[4] For $\psi(l)$ computation (Eq. 2) variable γ was assigned with a value of 0.025, i.e. only allowing a traffic deviation of 2.5% when comparing links with equal $P2P_{LIV}$ values.

In the first example the tracker was instructed to protect the links $R_7 \to R_9$, $R_8 \to R_9$ and $R_9 \to R_{10}$ from the network topology (links identified with a ⋆ mark in Figure 2) considering the scenario with a balanced distribution of peers in the network and one seed in all the network areas. The resulting traffic behavior is plotted in Figure 4, which compares the traffic observed in the network when the tracker behaves in the classical mode, Figure 4 a), and when configured with Algorithm 1 to protect the mentioned links, Figure 4 b). As observed in Figure 4 b) with the devised mechanism the cumulative P2P traffic traversing the selected links is almost imperceptible[5], comparatively with the scenario where the tracker assumes the classical behavior and a significant amount of traffic is observed in links $R_7 \to R_9$, $R_8 \to R_9$ and $R_9 \to R_{10}$ (plotted in Figure 4 a)), i.e. 2175, 1087 and 1087 MBytes, respectively. In this example, the protection of the links is obtained as Algorithm 1 computes a X_s set that only maintains area adjacencies pairs in two independent groups. In the first group, peers from areas A_1 and A_6 are not allowed to receive peers samples involving peers from other areas (i.e. A_2, A_3, A_4 and A_5), and in the second group peers from areas A_2, A_3, A_4 and A_5 are not able to receive peers samples integrating peers from areas A_1 and A_6. In this way all the P2P swarm traffic that would intersect the protected links is avoided by the tracker computed peering constraints. Thus, in the example of Figure 4 b) the tracker has computed the following allowed adjacencies:

$$X_s = \{(A_1, A_1), (A_1, A_6), (A_6, A_1), (A_6, A_6), (A_2, A_2), (A_2, A_3), (A_2, A_4), (A_2, A_5), (A_3, A_2), (A_3, A_3),$$
$$(A_3, A_4), (A_3, A_5), (A_4, A_2), (A_4, A_3), (A_4, A_4), (A_4, A_5), (A_5, A_2), (A_5, A_3), (A_5, A_4), (A_5, A_5)\}$$

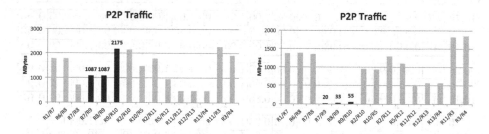

Fig. 4. Scenario with P_D=(50, 50, 50, 50, 50, 50), S_L=*all* a) classical tracker configuration; b) tacker configured to protect $R_7 \to R_9, R_8 \to R_9, R_9 \to R_{10}$ using Algorithm 1

The example presented in Figure 4 could be considered as having lower complexity due to the fact that distinct seeds were considered to exist on each network area. Thus, the behavior of Algorithm 1 could be considered has somehow foreseeable, not having to deal with possible swarm partitioning problems that could occur in more complex scenarios. In this perspective, a second example is now presented with a more challenging task. This case assumes the same peer distribution as in Figure 4, but considering now that only a single seed in area

---

[5] The residual values observed are due to the first phase of Algorithm 2 were no peering adjacencies constraints are considered.

A_1 exist, for the same set of links to be protected. As consequence, Algorithm 1 returns in this case a slightly distinct solution to the tracker, also integrating the (A_6, A_4) areas pair in the X_s set of the previous example. Otherwise, without such pair, a partition will occur in the swarm[6]. Figure 5 b) plots the results for this new scenario. As observed, this time the links $R_8 \to R_9, R_9 \to R_{10}$ have been traversed by some traffic from the P2P swarm, which is required to preserve the swarm totally connect (traffic exchanged between areas A_6 and A_4). Nevertheless, as Algorithm 1 tries to minimize traffic on protected links, there is still a significant traffic reduction even in such links, as observed when comparing Figures 5 a) and b). In fact, traffic on link $R_9 \to R_{10}$ is now five times lower than in the classical configuration, traffic on link $R_8 \to R_9$ is nearly two and a half times lower and traffic on link $R_7 \to R_9$ only presents residual values.

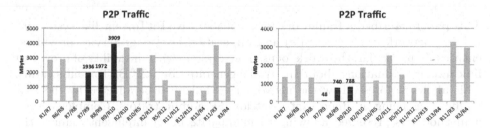

Fig. 5. Scenario with $P_D=(50, 50, 50, 50, 50, 50)$, $S_L=A_1$ a) classical tracker configuration; b) tacker configured to protect $R_7 \to R_9, R_8 \to R_9, R_9 \to R_{10}$ using Algorithm 1

The last example presented assumes that the tracker was instructed to protect the links $R_2 \to R_{11}$ and $R_{11} \to R_3$ (identified with a ● mark in Figure 2), for the same scenario as in Figure 5. The results presented in Figure 6 a) and b) corroborate again the effectiveness of the proposed link protection approach, as only residual traffic values are observed in the protected links. For this specific example the tracker has computed the following allowed adjacencies:

$$X_s = \{(A_1, A_1), (A_1, A_2), (A_1, A_5), (A_1, A_6), (A_2, A_1), (A_2, A_2), (A_2, A_5), (A_2, A_6), (A_3, A_3), (A_3, A_4),$$
$$(A_4, A_3), (A_4, A_4), (A_4, A_5), (A_5, A_1), (A_5, A_2), (A_5, A_4), (A_5, A_5), (A_5, A_6), (A_6, A_1), (A_6, A_2),$$
$$(A_6, A_5), (A_6, A_6)\}$$

A more depth analysis of the X_s computed by the tracker allows to verify that, in this example, peers from area A_3 are very constrained in peering opportunities, only being allowed to contact peer in the same area or in area A_4. However, peers in area A_4 are allowed to contact peers in area A_5 which, in turn, have directly or indirectly access to the pieces sent by the seed in area A_1. In this perspective, once again the computed X_s solution ensures the integrity of the P2P swarm and the protection of the considered links.

---

[6] Note that area A_6 is able to contact peers from Area A_1 where the seed is located. Thus, the (A_6, A_4) peering adjacency now added to X_s indirectly allows that areas A_2, A_3, A_4 and A_5 have also access to all pieces of the files exchanged in the swarm.

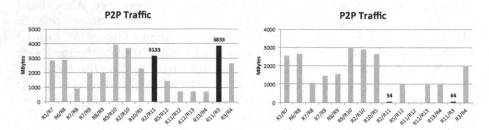

Fig. 6. Scenario with $P_D=(50,50,50,50,50,50)$, $S_L=A_1$ a) classical tracker configuration; b) tacker configured to protect $R_2{\rightarrow}R_{11},R_{11}{\rightarrow}R_3$ using Algorithm 1

4   Conclusions

This paper described a framework for a collaborative BitTorrent-like system involving network level (e.g. ISPs) and application level (e.g. Service providers) entities. In particular, this work focused on a system with the ability of providing link impact estimations about the traffic generated by P2P BitTorrent swarms. This allows to foresee how the network level links will be affected by the P2P traffic, thus being an important asset for ISP administrators. Complementary, a method was presented allowing to manipulate in an intelligent manner the peering information sent by the trackers. As consequence, the P2P tracker can be informed about which link(s) it should protect from the P2P swarm, generating for that purpose an optimized set of the allowed peering adjacencies, still ensuring the full connectivity of the swarm.

As a proof of concept, both the framework modules as well the devised methods were implemented in a simulation platform. The preliminary results obtained clearly corroborate that the mechanisms for P2P link impact estimations and for the protection of links from P2P traffic presented acceptable behavior. As future work, we intend to pursue the study on the effectiveness of the proposed mechanisms, analyzing additional complementary scenarios and configuration parameters. In a similar way, it is also intended to further enrich the proposed framework with other intelligent mechanisms that could benefit the integration of collaborative P2P applications in current networking environments.

Acknowledgments. This work has been supported by FCT - Fundação para a Ciência e Tecnologia within the Project Scope: PEst-OE/EEI/UI0319/2014.

References

1. Lua, K., Crowcroft, J., Pias, M., Sharma, R., Lim, S.: A survey and comparison of peer-to-peer overlay network schemes. IEEE Communications Surveys & Tutorials 7(2), 72–93 (2005)
2. Choen, B.: Incentives build robustness in BitTorrent. In: Proceedings 1st Workshop on Economics of Peer-to-Peer Systems, Berkeley (June 2003)

3. Karagiannis, T., et al.: Is p2p dying or just hiding? In: Proceedings of GLOBE-COM, Dallas, USA (November 2004)
4. Schulze, H., Mochalski, K.: Internet Study 2007: The Impact of P2P File Sharing, Voice over IP, Skype, Joost, Instant Messaging, One-Click Hosting and Media Streaming such as YouTube on the Internet. Technical Report (2007)
5. Xie, H., Krishnamurthy, A., Silberschatz, A., Yang, Y.R.: P4P: explicit communications for cooperative control between P2P and network providers (2008), http://www.dcia.info/documents/P4P_Overview.pdf
6. Seetharaman, S., Ammar, M.: Characterizing and mitigating inter-domain policy violations in overlay routes. In: Proceedings of IEEE International Conference on Network Protocols, ICNP (2006)
7. Keralapura, R., Taft, N., Chuah, C., Iannaccone, G.: Can ISPs take the heat from overlay networks? In: Proceedings of HotNets-III, San Diego, CA (November 2004)
8. Qiu, L., Yang, Y.R., Zhang, Y., Shenker, S.: On selfish routing in Internet-like environments. In: Proceedings of SIGCOMM, Karlsruhe, Germany (August 2003)
9. Xie, H., et al.: P4P: Provider Portal for Applications. In: Proceedings of ACM SIGCOMM 2008, Seattle, Washington, USA, August 17-22 (2008)
10. Sousa, P.: Context Aware Programmable Trackers for the Next Generation Internet. In: Oliver, M., Sallent, S. (eds.) EUNICE 2009. LNCS, vol. 5733, pp. 78–87. Springer, Heidelberg (2009)
11. Legout, A., et al.: Clustering and Sharing Incentives in BitTorrent Systems. In: Proceedings of ACM SIGMETRICS 2007, San Diego, USA, June 12-16 (2007)
12. Sousa, P.: Flexible Peer Selection Mechanisms for Future Internet Applications. In: Proceedings of BROADNETS 2009 - Sixth International ICST Conference on Broadband Communications, Networks and Systems, Madrid, Spain (2009)
13. Opsahl, T., Agneessens, F., Skvoretz, J.: Node centrality in weighted networks: Generalizing degree and shortest paths. Social Networks 32(3), 245–251 (2010)
14. Narayanan, S.: The betweenness centrality of biological networks. MSc Thesis, Faculty of the Virginia Polytechnic Institute and State University (2005)
15. Rocha, M., Sousa, P., Rio, M., Cortez, P.: QoS constrained internet routing with evolutionary algorithms. In: Proceedings of IEEE Congress on Evolutionary Computation, pp. 2720–2727 (2006)
16. Sousa, P., Rocha, M., Rio, M., Cortez, P.: Efficient OSPF Weight Allocation for Intra-domain QoS Optimization. In: Parr, G., Malone, D., Ó Foghlú, M. (eds.) IPOM 2006. LNCS, vol. 4268, pp. 37–48. Springer, Heidelberg (2006)
17. ns-2 (The Network Simulator). Sources and Documentation, http://www.isi.edu/nsnam/ns/
18. Eger, K., Hoßfeld, T., Binzenhofer, A., Kunzmann, G.: Efficient Simulation of Large-Scale P2P Networks: Packet-level vs. Flow-level Simulations. In: Proceedings of 2nd Workshop on the Use of P2P, GRID and Agents for the Development of Content Networks (2007)

Making 3D Replicas Using a Flatbed Scanner and a 3D Printer

Vaclav Skala[1], Rongjiang Pan[2], and Ondrej Nedved[1]

[1] Faculty of Applied Sciences, University of West Bohemia
CZ 30614 Plzen, Czech Republic
[2] School of Computer Science and Technology, Shandong University
250101 Jinan, China

Abstract. This paper describes a novel approach to making 3D replicas of near-ly flat objects using a flatbed scanner and a 3D printer. The surface reconstruc-tion is based on the fact that the light in a flatbed scanner shines under a given constant angle and the CCD sensor records different intensities depending on the angle between a local normal vector of a micro-facet and the vector towards the light source position. The scanned object is rotated by 90° and thus four dif-ferent images are obtained. It enables normal vector estimation followed by a surface reconstruction based on analogy with solution of partial differential eq-uations. 3D replicas are produced using a 3D printer based on the data from the surface reconstruction. Due to high resolution of the flatbed scanner, resulting replicas are of a high precision as well. This method can be used e.g. in making replicas of archaeological parts.

Keywords: computer graphics, 3D surface reconstruction, 3D printing, digital archaeology.

1 Introduction

Surface reconstruction methods are used in many applications. They are based on laser scanning or other similar techniques, like deflectometry. Most of the devices have difficulties in scanning small objects or objects with fine details.

This paper presents a new approach for a surface reconstruction and 3D print of nearly flat small objects using a flatbed scanner. The approach is based on normal vectors map computation from the scanned image followed by a surface reconstruc-tion and thereafter construction of a final 3D object representing the reconstructed surface. The size of the object is limited by the scanning area – usually A4 or "letter" format. Due to high hardware resolution of a scanner, usually up to 4800 dpi, scan-ning precision is high and it enables to reconstruct a surface with very fine details. It is expected that the given approach can be used especially in producing 3D replicas of nearly flat objects, like coins or rough fabrics etc.

B. Murgante et al. (Eds.): ICCSA 2014, Part VI, LNCS 8584, pp. 76–86, 2014.
© Springer International Publishing Switzerland 2014

2   Related Work

As for the related work, there are several papers to be mentioned. M.K.Johnson et al. [5] from MIT extended his former work cooperating with Adelson [4] and proposed a texture independent method for getting a surface with a fine resolution. On the other hand a sophisticated device for using the mentioned method is required. The former work of Johnson and Adelson [4] itself uses a special elastomer, which is „pushed into a relief", to get a texture independent surface, which is then reconstructed using a well-known photometric stereo algorithm. Chen, Goesele and Seidel [2] proposed a method for intermediate surface detail reconstruction from specularity. As for the prior partially related work, Liu et al. [6] introduced algorithms for synthesis of bidirectional texture functions (BTF) for gaining the texture information and thus material dependence. Our approach does not require any specially constructed hardware, can be used for reconstruction of either fine cloths or fabrics, as in the mentioned papers, and well as for glossy objects like polished coins too.

3   Scanning

In photometric stereo, at least three images of the surface are required [12]. Typically, more than the minimum three images are used in practice when considering the noise. Digital cameras can be used to take multiple images from the same view point illuminated with different light sources of known position. Multiple images of an object at different orientations can also be scanned with a flatbed scanner. To make the process easier and robust, we take scans rotated by 90 degrees. In order to make the orientation of the object as accurate as possible along the side of the scanner platen, we utilize a square clamp to fix the scanned object. A planar checkerboard pattern is glued on the surface of the clamp and utilized in the multiple scans registration. The checkerboard corners are automatically extracted in subpixel precision. We find an optimal rigid transformation using the corresponding corners in each scan. After aligning the other three scans to the first scan, we cut out the image regions that are out of interest. Fig.1 shows four registered images of an 'Yi Jiao' coin at different orientations along the flatbed scanner.

| (a) 0 degree | (b) 90 degree | (c) 180 degree | (d) 270 degree |

Fig. 1. Four scans of an 'Yi Jiao' coin

4   Normal Vectors Reconstruction

The light source of a flatbed scanner is linear and placed at a fixed angle α with respect to the scanner platen. According to [9], the angle α is approximately $\pi\,/\,6$. Each point on the scanned surface has a normal n and albedo ρ. For the scan taken at zero degree orientation, we define a left-handed coordinate system as shown in Fig.2, whose origin is at the surface point considered. The x-axis is parallel to the CCD sensor array at the surface point considered, xy-plane is the scanner platen and z-axis points straight down. The light source is approximated by a line segment extending from $-l$ to $+l$ in the x-axis direction and is offset by a in the y-direction and b in the z-direction. We define $n = \left[n_x, n_y, n_z\right]^T$ as a normalized surface normal at the point and $l = (x, a, b)^T / \sqrt{x^2 + a^2 + b^2}$ the normalized lighting direction vector along the linear light source.

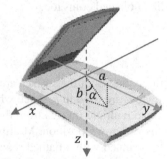

Fig. 2. Coordinate system at orientation zero degree

The observed intensity of such a surface point is then:

$$I_0 = \varrho \int_{-l}^{l} \langle n, l \rangle dx = \varrho \int_{-l}^{l} \frac{n_x x + n_y a + n_z b}{\sqrt{x^2 + a^2 + b^2}} dx = \tag{1}$$

$$\varrho\left(n_y a + n_z b\right) \int_{-l}^{l} \frac{1}{\sqrt{x^2 + a^2 + b^2}} dx = \varrho s\left(n_y a + n_z b\right)$$

where:

$$s = \int_{-l}^{l} \frac{1}{\sqrt{x^2 + a^2 + b^2}} dx = 2 ln \frac{l + \sqrt{l^2 + a^2 + b^2}}{\sqrt{a^2 + b^2}} \tag{2}$$

as:

$$\int_{-l}^{l} \frac{n_x x}{\sqrt{x^2 + a^2 + b^2}} dx = 0 \tag{3}$$

is an odd (anti-symmetrical) function on the interval $\langle - l, l \rangle$.

In the same way, scanning the same surface point with the object rotated by 90, 180 and 270 degrees, we get:

$$I_{90} = \varrho s(-n_x a + n_z b) \tag{4}$$

$$I_{180} = \varrho s(-n_y a + n_z b)$$

$$I_{270} = \varrho s(n_x a + n_z b)$$

We arrange the four equations (1) and (4) into a matrix equation which can be solved using linear least-squares. Its solution is:

$$\varrho s b n_x = (I_{270} - I_{90})/(2\,tan\alpha) \tag{5}$$

$$\varrho s b n_x = (I_0 - I_{180})/(2\,tan\alpha)$$

$$\varrho s b n_y = (I_0 + I_{90} + I_{180} + I_{270})/4$$

where: $tan\,\alpha = a\,/\,b$. Finally, the normal n at this point of the surface is obtained and normalized.

Fig.3 shows the computed normal maps of the Yi Jiao coin, a game coin and a piece of a fine fabric. To visualize the normal map, n_x, n_y and n_z values of the surfaces' normals are mapped to RGB components respectively, i.e., n_x maps from $(-1.0, 1.0)$ to red $(0, 255)$, n_y maps from $(-1.0, 1.0)$ to green $(0, 255)$ and n_z maps from $(0.0, 1.0)$ to blue $(0, 255)$.

Fig. 3. Normal map of two coins and a piece of a fine fabric

When the normal map is computed, there is a question of how to reconstruct a surface of the object.

5   Surface Reconstruction

Surface reconstruction is usually based on points given by scanning etc. However, we have a slightly different task as we have a map of normal vectors and a boundary of the given object. This is a classical formulation of the boundary problem known from the partial differential equations, namely from computational physics field, e.g. heat diffusion computation etc.

It can be seen that for each row or column we have values at the boundary of the object and inside of such an interval we have only normal vectors, e.g. derivatives.

Derivatives can be estimated using centered difference as:

$$\frac{\partial f}{\partial x} \approx \frac{f(x+h,y) - f(x-h,y)}{2h} \tag{6}$$

$$\frac{\partial f}{\partial y} \approx \frac{f(x,y+h) - f(x,y-h)}{2h}$$

where h is an increment in the x and y axis and in our case it is set as $h = 1$.

When using a direct solution, we can reformulate the problem as:

$$2f_x = f(x+1,y) - f(x-1,y) \tag{7}$$

$$2f_y = f(x,y+1) - f(x,y-1) \tag{8}$$

Let us consider a normal map of a size $M \times N$. Avoiding normals at the border leaves us $(M-2) \times (N-2)$ inner points to compute the surface for. There are two equations for each inner point, adding the boundary conditions gives us an overdetermined system of linear equations with approximately $2 \cdot (M \cdot N)$ equations for $M \cdot N$ points to be computed.

Even though the size of the linear system is very large, it can be solved efficiently considering that the matrix is sparse. It can be seen that for this linear system of equations to be solved, least square error estimation must be applied, while minimizing:

$$\min \sum |s_x - s_y|^2 \tag{9}$$

Where s_x is height of the surface according to the first condition from (7) and s_y height of the surface according to the second condition from (8).

6   3D Print

Surface generation itself is quite simple as the surface reconstruction produces a $2\frac{1}{2}D$ height field on a rectangular grid. Therefore tessellation to a triangular mesh is straightforward.

For 3D print a closed volume is actually needed. It means that additional surfaces have to be added, i.e. a surface representing a border of the given object of a specified height and also bottom surface of the object. It should be noted that such final surface has to be print-ready, i.e. consistent orientation of facets, with no holes etc.

Now the scanned object is represented as a triangulated surface having its volume and can be printed out on a 3D printer.

7 Experimental Results

Normal vectors reconstruction from scanned objects on a flatbed scanner was verified on several types of objects. For a surface reconstruction a simple approach based on an analogy with partial differential equations (PDE) with boundary condition formulation was used.

Our approach to surface reconstruction using a flatbed scanner was verified on different coins. Fig.4, 5 and 6 show our results, from left to right the original scan, a normal map, a reconstructed $2\frac{1}{2}D$ surface and a resulting up-scaled copy printed on a 3D printer.

Fig. 4. Results of a 'Yi Jiao' coin. Left to right: original, normals, reconstruction, 3D print.

Fig. 5. Results of a '2 Kč' coin. Left to right: original, normals, reconstruction, 3D print

Fig. 6. Results of a '20 Kč' coin. Left to right: original, normals, reconstruction, 3D print

It should be noted that we set $h = 1$ in the difference estimation for simplicity of computation. The actual height of the reconstructed surface is to be adjusted manually.

Table 1. Experimental data for a coin reconstruction with 1683×1557 points

	Normal map computation	Surface reconstruction
64 bits	68 [ms]	18 959 [ms]
32 bits	110 [ms]	24 289 [ms]

For experiments a standard PC was used – 3.16 GHz 12 GB RAM 64 bits, Windows 8 Professional x64 and a standard scanner with only 600 dpi was used. As for the computation times, as shown in Table 1, the normal map acquisition is of an algorithmic complexity $O(N)$, where N is the number of elements of an image, and therefore very fast in contrast with the surface computation, where a linear equations system of $N \sim 2\ mil.$ is solved. A 32 bit version of the software is available too.

For 3D print the ZPrinter 650 powder based 3D printer was used with 0.1 mm particle size.

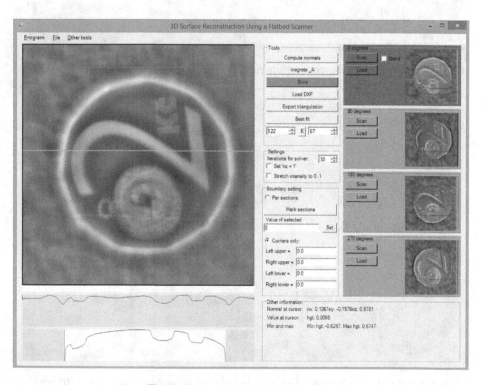

Fig. 7. The implemented user interface

Fig.7 presents user interface of the developed application. After loading 4 scanned images, each rotated by 90°, the normal map is computed followed by surface reconstruction and data export for 3D print. A comparison of a slice with an exact measurement is available too.

8   Comparison of the Reconstruction with Original

Additional experiments were made to compare the reconstructed surface to the real surface. Exact slices of coins used for this experiment were measured with a precision of 5 micrometers, which is sufficient for the used scanner having 600 dpi resolution.

Fig.8 to Fig.11 show a comparison of slices of the original coins and slices of their corresponding reconstructed models. An average error for all four coins can be found in Table 2.

Table 2. Average absolute error of the reconstruction compared to the original coin

	2 Kč	20 Kč	5 Kč	10 Kč
Average error	0.0342 mm	0.0399 mm	0.0409 mm	0.0270 mm

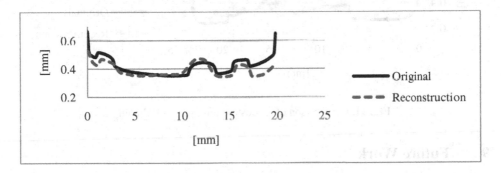

Fig. 8. Comparison of reconstruction of a '2 Kč' coin

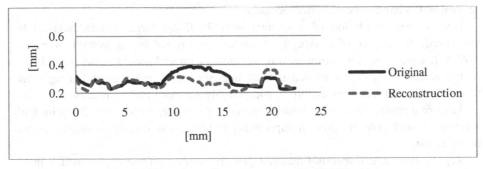

Fig. 9. Comparison of reconstruction of a '20 Kč' coin

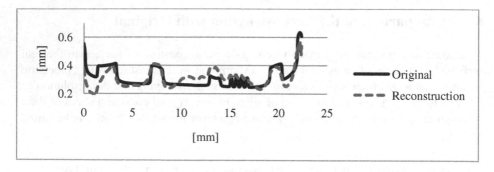

Fig. 10. Comparison of reconstruction of a '5 Kč' coin

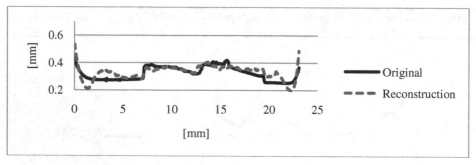

Fig. 11. Comparison of reconstruction of a '10 Kč' coin

9   Future Work

As experiments proved there is a simple method for making duplicates of nearly flat objects on a 3D printer. However, for a practical commercial or non-commercial use several issues have to be solved, especially:

Due to high resolution of a scanner over 4800 dpi large images have to be processed. An object of a size $4 \times 4\ inches$ means solving a system of nearly $369\ mil.$ equations. Resolution used is a critical factor and should be set accordingly to the scanned object. Computed normal map should be preprocessed using some filters to avoid unnecessary distortion and possibly get more precise results.

Texture mapping from the scanned images to the model with a final 3D print with a texture would give the user an opportunity to utilize the already acquired material information.

Representing and respecting material properties of a scanned object, which influences the normal map computation, e.g. surface roughness, could give better results.

10   Conclusion

A novel approach for 3D surface reconstruction and making 3D nearly flat objects replicas using 3D printer is presented. Experiments made proved expected basic properties of the proposed approach. Experiments also proved high precision of reconstruction even when a cheap commodity flatbed scanner is used.

Several types of objects were successfully reconstructed, including glossy coins or fine fabrics.

The presented approach is applicable to $2\,{}^1\!/_2$ dimensional, nearly flat objects only and therefore cannot be extended to higher dimensions.

Presented algorithms were implemented on .NET platform for 32 and 64 bits.

Future work will be concentrated on application of Radial Basis Function approximation and optimization of computation.

Acknowledgments. The authors express their thanks to colleagues at the University of West Bohemia and Shandong University for their help and to anonymous reviewers for their critical comments and advices.

The research was supported by the Ministry of Education of the Czech Republic, projects No.LH12181, ME10060, SGS-2013-029 and by China-Czech Scientific and Technological Collaboration Project (40-8), Shandong Natural Science Foundation of China (Grant No. ZR2010FM046).

References

1. Blinn, J.F.: Simulation of wrinkled surfaces. SIGGRAPH Comput. Graph. 12(3), 286–292 (1978)
2. Chen, T., Goesele, H.P.: Seidel: Mesostructure from Specularity. In: 2006 IEEE Computer Society Conference on Computer Vision and Pattern Recognition (CVPR 2006), vol. 2, pp. 1825–1832. IEEE (2006), doi:10.1109/CVPR.2006.182
3. Clarkson, W., Weyrich, T., Finkelstein, A., Heninger, N., Halderman, J.A., Felten, E.W.: Finger printing Blank Paper Using Commodity Scanners. In: Proceedings of the 2009 30th IEEE Symposium on Security and Privacy (SP 2009), pp. 301–314. IEEE Computer Society, Washington, DC (2009)
4. Johnson, M.K., Adelson, E.H.: Retrographic sensing for the measurement of surface texture and shape. In: 2009 IEEE Conference on Computer Vision and Pattern Recognition, pp. 1070–1077. IEEE (2009), doi:10.1109/CVPRW.2009.5206534
5. Johnson, M.K., Cole, F., Raj, A., Adelson, E.H.: Microgeometry capture using an elastomeric sensor. ACM Trans. Graph. 30(4), Article 46, 8 pages (2011)
6. Liu, X., Hu, M.Y., Zhang, H.P.J., Tong, X., Guo, B., Shum, H.-Y.: Synthesis and rendering of bidirectional texture functions on arbitrary surfaces. IEEE Transactions on Visualization and Computer Graphics 10(3), 278–289 (2004), doi:10.1109/TVCG.2004.1272727
7. Myronenko, A., Song, X.B.: On the closed-form solution of the rotation matrix arising in computer vision problems. CoRR abs/0904.1613 (2009)
8. Pan, R., Skala, V.: Surface Reconstruction with higher-order smoothness. The Visual Computer 28(2), 155–162 (2012) ISSN 0178-2789

9. Pintus, R., Malzbender, T., Wang, O., Bergman, R., Nachlieli, H., Ruckenstein, G.: Photo Repair and 3D Structure from Flatbed Scanners Using 4- and 2-Source Photometric Stereo. In: Ranchordas, A., Pereira, J.M., Araújo, H.J., Tavares, J.M.R.S. (eds.) VISIGRAPP 2009. CCIS, vol. 68, pp. 326–342. Springer, Heidelberg (2010)
10. Skala, V.: Projective Geometry and Duality for Graphics, Games and Visualization - Course SIGGRAPH Asia 2012, Singapore (2012) ISBN 978-1-4503-1757-3
11. Toler-Franklin, C., Finkelstein, A., Rusinkiewicz, S.: Illustration of complex real-world objects using images with normals. In: Proceedings of the 5th International Symposium on Non-Photorealistic Animation and Rendering (NPAR 2007), pp. 111–119. ACM, New York (2007)
12. Woodham, R.J.: Photometric stereo: A reflectance map technique for determining surface orientation from image intensity. In: Proc. 22nd SPIE Annual Technical Symposium, vol. 155, pp. 136–143 (1978)
13. Woodham, R.J.: Photometric Method for Determining Surface Orientation from Multiple Images. Optical Engineering 19(1), 139–144 (1980)

Computational Tool for the Energy Performance Assessment of Horticultural Industries – Case Study of Industries in the Centre Inner Region of Portugal

Diogo Neves[1], Pedro D. Gaspar[1,*], Pedro D. Silva[1], José Nunes[2],
and Luís P. Andrade[2]

[1] University of Beira Interior, Electromechanical Engineering Department,
Rua Marquês d'Ávila e Bolama 6200 Covilhã, Portugal
diogopneves@gmail.com, {dinis,dinho}@ubi.pt
[2] Politechnic Institute of Castelo Branco,
Av. Pedro Álvares Cabral n°12 6000 Castelo Branco, Portugal
{jnunes,luispa}@ipcb.pt

Abstract. Food processing and conservation represent decisive factors for the sustainability of the planet given the significant growth of the world population in the last decades. Therefore, the cooling of any food products has been subject of study and improvement in order to ensure the food supply with good quality and safety. A computational tool for the assessment of the energy performance of agrifood industries that use refrigeration systems was developed. It aims to promote the improvement of the energy efficiency of this industrial sector. The computational tool for analysis of the energy profile is based on a set of characteristic parameters used for the development of correlations, including the amount of raw material, annual energy consumption and volume of cold stores. In this paper, the developed computational tool was applied to companies in the horticultural sector, specifically to resale-type companies. The results obtained help on the decision making of practice measures for the improvement of the energy efficiency.

Keywords: Computational tool, Energy performance, Energy efficiency, sustainability, perishable products, horticultural, cold storage, Matlab, GUIDE.

1 Introduction

There are several topics that are targets of research or scientific studies, and both sustainability and food safety are no exception. In the early twentieth century the world population was about 1500 million people, however just over a century the number of habitants increased to approximately 7000 million [1]. Thus, it is clear that the demand for food is increasing and so it becomes imperative to find short-term solutions for the sustainability of the planet, hence putting an enormous strain on the production

---

* Corresponding author.

B. Murgante et al. (Eds.): ICCSA 2014, Part VI, LNCS 8584, pp. 87–101, 2014.

and conservation sectors of the food chain. In this context, cooling plays an important role, allowing food storage in times of increased production when the market has no capacity for product flow, or just make it available when needed. The refrigeration is a process with the ability to preserve perishable products ensuring that they retain their chemical, physical and nutritional properties but it is also indispensable in the processing of perishable foods such as meat, fish, milk, fruit and so on. The demand of chilled and frozen food has increased substantially, which leads to new requirements for the efficiency of refrigeration systems [2]. Energy represents a factor of greatest importance not only for country's economy, but especially for the well-being of its citizens. In 1971 there was a global electricity consumption of around 439 $Mton_e$, and in 2010 this figure rose to 1 536 $Mton_e$ [3]. This increase is related to the population growth, being the industry responsible for about 35.2% of energy consumption worldwide in 2010. The energy consumed globally for the cooling process accounts for 15% of energy consumption worldwide [1].

Horticultural products, i.e., fruits and vegetables are important elements of a balanced diet and are associated with a healthy lifestyle, due to vitamins, fiber, minerals and sugar that can be obtained from it. Besides that, the high intake of this kind of food helps to reduce the risk of chronic diseases such as diabetes [4], a few cancers (mouth, larynx, oesophageal, stomach, pharynx and lungs) [5] and cardiovascular health problems [6,7,8].

According to the European Food Information Council (EUFIC), there are different definitions of fruit and vegetables since some countries might consider dry fruits as the same kind of food (statistically). In a general way, potatoes are not considered as a vegetable to statistical results. Note that the World Health Organization (WHO) recommends eating more than 400 grams per day of fruits and vegetables, excluding potatoes and other starchy tubers such as cassava [9]. Unfortunately there is no information regarding the consumption of this food group, however based on the vegetable supply in Europe, it can be stated that its consumption has increased over the past four decades (FAO). Actually, the average fruit and vegetables intake (excluding juices) in Europe is 220 grams per day and 166 grams per day respectively, which leads to a total of 386 grams per day [10]. Austria and Portugal are the countries that show highest intakes of fruit and vegetables, while Spain and Iceland show the lowest values.

Thus, is important to develop studies and tools to improve the efficiency of these processes in order to ensure a better sustainability of this industrial sector. Although it can be highlighted that there are a few studies or projects developed in this area, more specifically in the creation of computational tools for the analysis of several points related to the cooling processes (whether the level of energy consumption, environmental impact, among others). In this regard, Gogou *et al.* [11] describe the FRISBEE (*Food Refrigeration Innovations for Safety, Consumers Benefit, Environmental impact and Energy*) project, which considers the development of a software tool for the evaluation of quality, energy and environmental impact of the European cold chain.

This program has the ability to predict the temperature of chilled or frozen products on certain circumstances and to calculate the validity/quality of particular food at different stages of the cold chain. An internet platform was developed to conduct the surveys which has more than 5500 records [11]. The data entered on this platform is organized according to the following fields: phase of the cold chain, temperature range from food storage, characterization of food, type of food, food product, packaging, and country of origin. This computational tool is innovative not only by their own concept of allowing simulating certain behaviours of the cold chain in Europe, but mainly because it has three very important fields in cooling: chilled product quality, energy consumption and environmental impact. FRISBEE covers five categories of foods, including fruit, meat, fish, dairy products and vegetables [12] and its main objective aims to collect data of different stages of the cooling process using the online platform. The FRISBEE CCP (Cold Chain Predictor) [13] is the computational tool developed, which performs simulations on specific conditions defined by the user, constructing graphs representing the variation of the temperature over time and estimating the remaining shelf life of the product. These simulations are performed based on the method of Monte Carlo [11] generating distributions of time/temperature for every stage of the cold chain and the selected product. The results represent realistic behaviour scenarios of food products and based on these it is possible to take corrective actions in order to optimize the efficiency of the cold chain ensuring product quality. In this same context, Foster et al. [14] describes the ICE-E (*Improving Cold Storage Equipment in Europe*) project [15] devoted to the creation of tools with the same goals: reduction of energy consumption and greenhouse gases emissions through the application of more efficient equipment, taking into account the energy and environmental standards of the EU [16]. The database of this project includes not only small businesses but also large multinationals, with data collected through an online platform. The ICE-E does not analyse the quality of chilled products with the respective indicators of safety and quality [17] being this the main difference between the presented projects. It should be noted that both projects developed freely computational tools to be used by any owner or employee of a company in the sector in order to perform an energy characterization of his company and to determine the achievable savings if their consumption is exaggerated. In addition, Eden et al. [15] describes the CHILL-ON [18], which is a project developed in the same area and with similar concerns. However, it is noteworthy that this project focuses on the quality of chilled products, mainly fish and birds, making a deep study in terms of microbiology. It includes a computational tool, QMRA (*Quantitive Microbial Risk Assessment*) [19], which is combined with the principle of the Hazard Analysis and Critical Control Points (HACCP) to enable the improvement of quality and safety of food in a preventive approach. More specifically, this tool allow to estimate the risk of pathogens growth based on the temperature, chemical and nutritional characteristics of food. As result, the SLP (*Shelf Life Predictor*) [20] was developed with the ability to predict the remaining shelf life of foods. Apart from the quality and safety of foods, this tool allows to track (traceability) products in order to know its real time location in so that the consumer can make use of reliable information about the foods origin and to over-look the manufacturing process from the beginning to the end. Furthermore, the development of new technologies in the field of refrigeration is also covered by the CHILL ON project such as the introduction of "smart labels": TTIs (*Time Temperature Indicators* or *Integrators*) [19]. Combining this technology with RFID (*Radio Frequency Identification*), it became possible to send an electronic and optical signal through a wireless connection directly to a software that calculates the remaining shelf life.

Butler *et al.* [21] describes another European project, NIGHT WIND, which can also be related to these topics being the reduction of power consumption its main objective. The concept of this project is very simple, and it consists in making use of the existing cold storages as "batteries" that will store cold air [21], in other words, energy. It is proposed the creation of a control software of cold stores temperature taking into account the price of electricity and the daily consumption profile of the company. It can be stated that the compressors will work during periods in which electricity is cheaper (usually at night), accumulating energy as cold air (temperature below -20°C) to use it rationally during peak hours [22]. However, this procedure is executable only in the facilities with enough cold storage capacity to store the energy demand of the next day.

2   The Energy Performance Assessment in Portuguese Industries

The abovementioned projects covered the development of computational tools aimed to improve the energy efficiency of cold stores. Although, Portuguese companies had not been included in the databases. In this context, a project was developed in Portugal directed to the identification of the energy consumption profiles of the agrifood industry and the promotion and development of actions that contribute to a real improvement of energy efficiency and the competitiveness of this sector. The work developed and presented in this paper is part of an activity of the project and its main objective is the development and implementation of an analysis algorithm to be validated with companies outside the visited/studied sample of companies. It is considered a tool to support strategic decisions in the companies allowing the estimation of their energy performance and pointing practice measure that lead to an effective improvement of energy efficiency. Note that this project is not aimed for the energy characterization of the cold stores in general, but to the characterization by sector, including meat, fish, milk, horticultural (fruit and vegetables), wine and vineyard and distribution sectors in order to obtain real data that can be inputted in the model/algorithm.

It should be noted that in the case of Portugal, in 2011, the number of small and medium enterprises was around 1.112.000, providing employment to about 78.5% of the employed Portuguese population [23], a factor that characterizes the Portuguese economy. Furthermore, according to a study conducted by the OECD (Organization for Economic Cooperation Development), Portugal was considered in 2005 as one of the countries whose population had fewer qualifications, being positioned at the same level as Brazil and Turkey. The study surveyed individuals between 25 and 65 years of age belonging to the labour force, and 59% of them had qualifications below the sixth grade [24], while in Denmark, also a European country, this figure was only 1%. This situation was taken into account for the development of computational tools, given the lack of literary skills of part of the Portuguese population. Thus, the computational tool created is extremely simple, intuitive and easy to understand in order to be accessible to all employees in the refrigeration industry regardless of their qualifications.

After an intensive collection of field data of a given sample of companies, analytical correlations were developed based on the work of Nunes [25] and Nunes *et al.* [26-28] to predict the average energy performance in the horticultural industry in Portugal. Considering the correlations for different sectors of agrifood industry, a computational tool was developed that allows the estimation of the energy consumption of a particular company against the national average. Depending on the results, the tool also provides practice measures to improve the energy efficiency. Thus, it can be used on energy-decision making process. This work introduces the tool in the horticultural (fruit and vegetables) sector, more specifically, for the industry of fruit resale.

3   Computational Tool for the Prediction of Energy Performance

3.1   Overview

The computational tool, Cool-OP (*Cooling Optimization Program*), was developed in MATLAB using the GUIDE (*Graphical User Interface Design Environment*) tool [29] that allows the creation of multiple windows that graphically illustrate the results. This toolbox is used by advanced programmers giving them the ability to create the graphical layout of the program more easily than using the old methods. It consists in usual programming code to define the position of the GUI. The GUIDE has facilitated graphical programming allowing to reduce the working time to develop the graphical interface. Using this application it is possible to create windows of the desired size, adding various interactive icons of user interface with the program, from buttons to static boxes or dynamic text, and many others as shown in Figure 1.

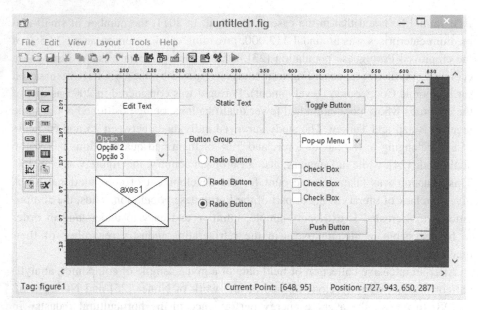

Fig. 1. GUIDE interface software in MATLAB

It is important to note that all files created by GUIDE are .fig format and they must have an associated .m file where a function is assigned to each graphical component. Some of the features of this toolbox allow the programmer to display static text and other components such as images, graphs or tables, without the need to have any dynamic function. By other hand, editable text boxes can have two distinct functions, such as reading a value or text that is entered by the user, or displaying information based on the .m file associated. Regarding the axes, these are always used as a reference to insert an image or to plot a graph. Besides that, the programmer can make use of many other graphical components such as radio buttons, checkboxes, pop-up menus or list. Finally there are the usual buttons, which are normally associated with a link to a different window or action such as closing a window, for example. Each component has always a *tag*, or a label that will be identified in the source file (.m) by creating a *callback*. Thus, every GUI component has a *callback* that consists in a function that will control its behaviour according to the respective programming code when the event occurs. This event may be the click of a button, the selection of an option from a menu or inserting a text or number. For example, consider a .fig file with a button that starts the plotting of a graph. When the user clicks the button, the software reads the *callback* function associated to the button, performing its action according to the code written.

It should be noted that all GUI components are configurable through the Property Inspector (see Figure 2). Consider a push button as an example. As expected, it is possible to change the button size, type, font colour and size, text alignment, background colour, among others. But beyond that there are some editable programming parameters such as the value assumed as default of *pushbuttons*, the setting of short-cut keys to trigger buttons, the name that identifies the .m file, the automatic or manual creation of *callbacks*, and a few others.

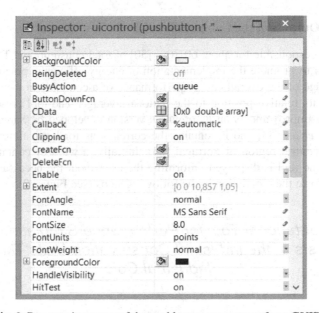

Fig. 2. Property inspector of the pushbutton component from GUIDE

The Cool-OP windows here developed using the abovementioned programming concepts. Besides these programming details, it is important to highlight that the code was converted to an executable file (.exe) that can run in any computer.

This computational tool predicts the current state of the energy consumption in a company and suggests a few practice measures to reduce this consumption. The correlations used in the tool were developed statistically and are focused on several key parameters that characterize companies of agrifood sector, including: the quantity of raw material processed, power consumption and the total volume of cold stores. The energy consumption parameter includes not only the energy consumed for cooling or processing products, but also other energy consumption related with lighting, heating, office, etc. Since the correlations, that are the core of the algorithm, were obtained through real values collected in companies, it becomes imperative to define the variation domain of each parameter to provide valid predictions. These restrictions on utilization of the computational tool are given in Table 1. As it can be seen, the energy consumption parameter is not limited to a domain, since this is the main evaluation parameter which characterizes the company.

Table 1. Application domain of each of the parameters of the computational tool

Industry	Raw Material [ton]	Chambers volume [m³]
Fruit Resale	100-1800	50-1000

3.2   Inputs/Outputs of the Computational Tool

Cold stores are responsible for about 60-70% [30] of total consumption. This computational tool aims to promote the implementation of energy efficiency practice measures by the knowledge of the overall energy performance of a company in the agrifood sector, as well as its relative position to the national average. The tool presented in this study considers improvements on its previous version developed by Santos *et al.* [31-32] and Neves *et al.* [33] and it contains the correlations for horticultural companies located in the central region of Portugal [25]. Initially, a window containing a brief description of the tool is displayed, informing the user about the necessary input data (inputs) to perform the simulation in the following steps (see Figure 3).

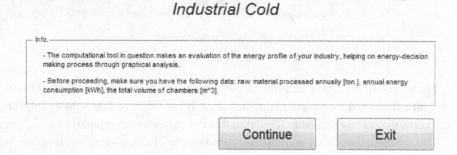

Computational Tool for Performance Simulation and Analysis of the Industrial Conservation Units Through Industrial Cold

┌─ Info. ───┐

- The computational tool in question makes an evaluation of the energy profile of your industry, helping on energy-decision making process through graphical analysis.

- Before proceeding, make sure you have the following data: raw material processed annually [ton.], annual energy consumption [kWh], the total volume of chambers [m^3].

| Continue | | Exit |

Fig. 3. Cool-OP: Opening window

In the next menu, the user can select which industry fits the company to be analysed (see Figure 4-a)), including meat, fish, horticultural (fruit and vegetables) and dairy products. Within each of the different sectors there are many subcategories. In the specific case of the horticultural sector or Fruit and Vegetables industry, there are (1) fruit resale warehouses and (2) fruit plants as shown in Figure 4-b). This paper focus on the former subcategory.

Thereafter a window appears in which are introduced the values of the parameters that characterize the company: amount of raw material processed annually [ton], annual electricity consumption [MWh] and the total volume of cold stores [m³]. Notice that it's always possible to return to the previous menu or close the window using the respective buttons for navigation. Any of the subcategories of agrifood industries opens a similar window requesting the same data, changing only the title (Figure 4-c)).

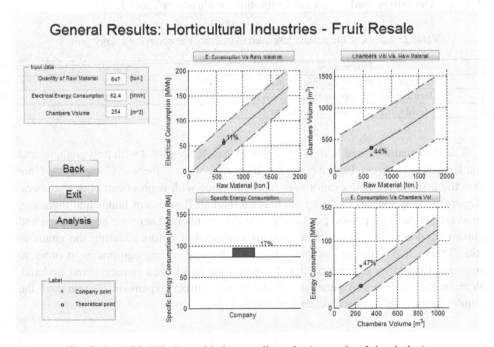

Industries	**Horticultural**	**Horticultural industries - Fruit Resale**
Meats		Quantity of Raw Material 0 [ton.]
	Fruit resale	Electrical Energy Consumption 0 [MWh]
Fishes		Chambers Volume 0 [m^3]
Fruit and vegetables	Fruit Plants	
Milk products		Continue
Exit	Back Exit	
		Back Clear Exit

a) Menu window.   b) Menu window of Horticultural sector.   c) Input window of the parameters of the company.

Fig. 4. Cool-OP: Windows to input the company parameters

After the correct introduction of the data according to the International System of Units (SI) and within the valid variation domain of each parameter, pressing the button "Continue" gives access to the general results that summarize the current state of energy performance of the company. In this window (see Figure 5) the tool immediately processes the information previously entered, generating graphs that relate the evaluation parameters of the company.

Fig. 5. Cool-OP: Window with the overall results (example of simulation)

All graphs have a green shade which represents a confidence level of 95% of the experimental values used to create the statistical correlations. In addition, it is also shown in each graph the percent deviation of the point under consideration (Company point – red dot) against the Portuguese national average (black dot). If the user wants to analyse a particular graph, he/she can press the button above it to open a new larger window.

4   Case Study

4.1   Company Presentation

A company was randomly selected to validate the tool, hereinafter referred to as Company A for reasons of business confidentiality. The company has not been part of the statistical data used in the development of the correlation.

This company is in the fruit resale sector and began operating in 2010. It has 10 workers and according to the Portuguese legislation it is classified as a small enterprise. Annually it processes about 900 tons of horticultural products, particularly vegetables (200 ton), pear and apple (350 ton each). Its facilities are equipped with cold rooms with a total volume of approximately 294 m^3 and its total covered area is 400 m^2. The values used to run the simulation are shown in Table 2.

Table 2. Values of the parameters used to evaluate the company energy profile

Parameters	Values
Quantity of raw material [ton]	900
Electrical Energy consumption [MWh]	46.8
Cold Chambers total volume [m^3]	294

The cold chambers are built with sandwich panels, insulated with polyurethane and the floor is in concrete. Figure 6 shows one of the cold chambers of Company A. Note that the insulation of the evaporators tubing in made with neoprene in both chambers. Regarding heat sources, lighting is obtained through fluorescent lamps and there are, in average, about 10 person per hour inside the chambers. There are both manual and automatic doors that allows access to an antechamber before entering the chamber itself. Since it is a recent company it hasn't invested in any equipment in order to improve the energy efficiency and the refrigerant fluid used is recent and updated. With regard to the refrigeration system it is a direct expansion system and the condensers are defrost by an electrical resistor.

Fig. 6. Case study: Inner vision of a cold chamber of Company A

4.2   Analysis and Discussion of Results

The predictive values of energy performance of Company A are shown in Figure 7. It is stated that the company has a specific energy consumption below average, characterizing it as competitive business.

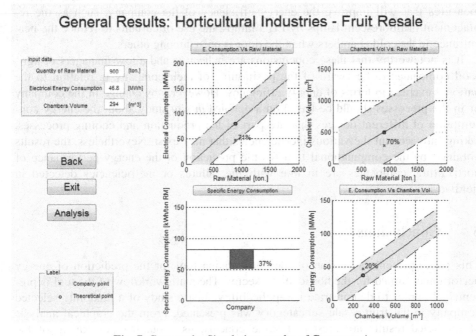

Fig. 7. Case study: Simulation results of Company A

The first graph (left upper side of the tool window of Figure 7) compares the electrical consumption with the raw material processed by the company. Its value is 71% under the national average, which is a good result meaning that most of similar companies with the same energy consumption can only process about 600 tons of raw material. The second graph (right upper side) shows a similar result, although it compares the cold chambers volume with the quantity of raw material. Despite a value 70% under the average, which might lead the user to think that the chambers are under dimensioned, it means that the company as a stable volume of business all over the year making good use of the chambers storage space. The third graph (left bottom side) is related to the Specific Energy Consumption (SEC) and it can be considered as the most conclusive graph of the computational tool. It relates the energy consumption with the quantity of processed raw material. In this particular company, the SEC is 37% under the average, approximately 52 kWh/ton, allowing the user to conclude that it is a competitive and well sustained company with advantage against the majority of the other companies in the same business sector. The fourth and last graph indicates an energy consumption value 20% above the average regarding the chambers volume. There are many factors related to the variation of this relationship. In this case study, it is due to the fact that there are in average ten person per hour inside the cold chambers (significant heat load), and despite having an air conditioned antechamber, there are always significant losses that lead to an increased energy consumption since the compressors will work more time in order to compensate the temperature increase.

It is important to highlight that all predictive values are within the confidence interval of 95% (shaded green). However, in the fieldwork were identified practice measures that will improve the energy efficiency of the company, such as the replacement of fluorescent lamps by LED lamps, the use of curtains to reduce the heat entrance in the cold chambers when the door is open, among others.

It is noteworthy that these conclusions are indicative and allow managers of agrifood companies to be aware of the positioning of their company in relation to the national average in terms of energy efficiency. However, any change in the company or in the processes should be based on a detailed *in-situ* study about the energy consumption of different devices that are part of the production and cooling processes, taking into account the various inefficiencies that may exist. Nevertheless, the results obtained by the computational tool for the prediction of the energy performance of horticultural companies are in line with the failures or inefficiencies detected in fieldwork.

5   Conclusions

This paper presents a version of the computation tool for the prediction of energy performance aimed for the horticultural sector. The main workflow of the tool is presented. In order to demonstrate its applicability, a case study of a randomly selected company in the fruit resale subcategory was presented. From the graphical analysis of predicted results are presented some conclusions about the positioning of the company's performance in relation to the national average. The Cool-OP computational tool allows to perform an evaluation of the energy performance of companies in the agrifood sector comparing parameters such as raw materials processed, energy consumption and volume of cold stores with national averages so that the user has the ability to conclude which are the possible weaknesses or strengths of its company. The graph of specific energy consumption is very conclusive since it relates the energy consumption per ton of raw material. However it should be mentioned that the developed tool only aids in the analysis of the energy performance, so it is necessary that the user has sensitivity to identify possible problems of technical origin on the company facilities. Thus, the analysis does not eliminate the need for a more detailed study to determine the particular conditions that can be improved. This conclusion arises from the comparison of results obtained from simulations and results from audits to the respective companies. It became clear that they had adequate working conditions and a proper maintenance of the equipment, such as the compressors, the refrigerant pipes and their proper insulation, and the quality and maintenance of refrigeration chambers.

The current state of computational tool allows the user to manually enter the data of annual energy consumption, raw material processed annually and volume chambers. With these performance predictions, the user can decide how to improve the energy performance of his company. The practical application of this tool demonstrates its usefulness in helping decision-making in the implementation practice measures for the improvement of energy efficiency. This tool includes other correlations for other agrifood sectors, such as meat, fish, milk, vine and wine and distribution. Moreover, taking into account the sector of the company under analysis, it will provide automatic suggestions for the improvement of energy performance. Changing the linear correlation values, this tool can be used in any country, since the production process and storage conditions may differ depending on the food product.

Acknowledgments. This work is part of the anchor-project "InovEnergy – Eficiência Energética no Sector Agroindustrial" of the Action Plan of InovCluster: Associação do Cluster Agro-Industrial do Centro. The study was funded by the National Strategic Reference Framework (QREN 2007-2013) - COMPETE/POFC (Operational Competitiveness Programme), SIAC (Collective action support system): 01/SIAC/2011, Ref: 18642.

References

1. Pachai, A.C.: From Crandle to table – cooling and freezing of food. In: 2nd IIR International Conference on Sustainability and the Cold Chain (ICCC 2013), Paris, France, pp. 2–4 (April 2013)
2. Laguerre, O., Hoang, H.M., Flick, D.: Experimental investigation and modelling in the food cold chain: Thermal and quality evolution. Trends in Food Science & Technology 29(2), 87–97 (2013)
3. IEA: 2012 Key World Energy Statistics. International Energy Agency (IEA) (2012)

4. Harding, A.H., Wareham, N.J., Bingham, S.A., Khaw, K., Luben, R., Welch, A., Forouhi, N.G.: Plasma vitamin C level, fruit and vegetable consumption, and the risk of new-onset type 2 diabetes mellitus: the European prospective investigation of cancer–Norfolk prospective study. Archives of Internal Medicine 168(14), 1493–1499 (2008)

5. World Cancer Research Fund (WCRF) Panel: Food, Nutrition, Physical Activity, and the Prevention of Cancer: A Global Perspective. World Cancer Research Fund: Washington, DC (2007)

6. Mirmiran, P., Noori, N., Zavareh, M.B., Azizi, F.: Fruit and vegetable consumption and risk factors for cardiovascular disease. Metabolism 58(4), 460–468 (2009)

7. Hung, H.C., Joshipura, K.J., Jiang, R., Hu, F.B., Hunter, D., Smith-Warner, S.A., Colditz, G.A., Rosner, B., Spiegelman, D., Willett, W.C.: Fruit and vegetable intake and risk of major chronic disease. Journal of the National Cancer Institute 96(21), 1577–1584 (2004)

8. Rissanen, T.H., Voutilainen, S., Virtanen, J.K., Venho, B., Vanharanta, M., Mursu, J., Salonen, J.T.: Low intake of fruits, berries and vegetables is associated with excess mortality in men: the Kuopio Ischaemic Heart Disease Risk Factor (KIHD) Study. Journal of Nutrition 133(1), 199–204 (2003)

9. World Health Organization: WHO European Action Plan for Food and Nutrition 2007-2012. WHO: Copenhagen, Denmark (2008)

10. European Food Safety Authority: Concise Database summary statistics - Total population (2008),
http://www.efsa.europa.eu/en/datexfoodcdb/datexfooddb.htm

11. Gogou, E., Katsaros, G., Derens, E., Li, L., Alvarez, G., Taoukis, P.: Development and applications of the European Cold Chain Database as a tool for cold chain management. In: 2nd IIR International Conference on Sustainability and the Cold Chain (ICCC 2013), Paris, France, pp. 2–4 (April 2013)

12. Food Refrigeration Innovations for Safety, consumers Benefit, Environmental impact and Energy optimisation along the cold chain in Europe

13. Stahl, V., Alvarez, G., Derens, E., Hoang, H.M., Lintz, A., Hezard, B., El Jabri, M., Lepage, J.F., Ndoye, F.T., Thuault, D., Gwanpua, S.G., Verboven, P., Geeraerd, A.: An essential part of the frisbee software tool: Identification and validation of models quantifying the impact of the cold chain on the RTE pork products. In: 2nd IIR International Conference on Sustainability and the Cold Chain (ICCC 2013), Paris, France, pp. 2–4 (April 2013)

14. Foster, A., Zilio, C., Corradi, M., Reinholdt, L., Evans, J.: Freely available cold store energy models. In: 2nd IIR International Conference on Sustainability and the Cold Chain (ICCC 2013), Paris, France, pp. 2–4 (April 2013)

15. ICE-E Improving cold storage equipment in Europe (2013)

16. Evans, J.A., Hammond, E.C., Gigiel, A.J., Foster, A.M., Reinholdt, L., Fikiin, K., Zilio, C.: Assessment of methods to reduce energy consumption of food cold stores. Applied Thermal Engineering 62(2), 697–705 (2014)

17. Evans, J.A., Foster, A.M., Huet, J.-M., Reinholdt, L., Fikkin, K., Zilio, C., Houska, M., Landfeld, A., Bond, C., Scheurs, M., van Sambeeck, T.W.M.: Specific energy consumption values for various refrigerated cold stores. Energy and Buildings 74, 141–151 (2014)

18. Eden, M., Colmer, C.: Improved Cold Chain Management: efficiency and food safety through international project results. Food Safety and Tecnhology 24(1), 30–32 (2010)

19. Welcome to the Chill-on Project, http://www.chill-on.com/

20. Colmer, C., Kuck, M., Lohmann, M., Bunke, M.: Novel Technologies to improve safety and transparency of the chilled food supply chain. Innovation in the Nordic Marine Sector (May 2009)

21. Butler, D.: Fridges could save power for a rainy day. Nature (2007)
22. Saint Trofee – Refrigeration Tools,
    http://www.nightwind.eu/night-wind.html
23. IAPMEI, PME em números, http://www1.ionline.pt/conteudo/37754-populacao-activa-portuguesa-e-das-que-tem-menos-habilitacoes-literarias-
24. Ionline, http://www1.ionline.pt/conteudo/37754-populacao-activa-portuguesa-e-das-que-tem-menos-habilitacoes-literarias-
25. Nunes, J.: Energetic efficiency evaluation in refrigeration systems of agrifood industries in the Beira Interior region. PhD Thesis, University of Beira Interior, Covilhã, Portugal (2014)
26. Nunes, J., Silva, P.D., Andrade, L.P., Gaspar, P.D.: Characterization of specific energy consumption of electricity of Portuguese sausages industry. Energy and Sustainability 2014 - WIT Transactions on Ecology and the Environment 186 (2014)
27. Nunes, J., Silva, P.D., Andrade, L.P.: Energetic efficiency evaluation in refrigeration systems of meat industries. In: 23rd International Congress of Refrigeration, ICR 2011, Prague, Czech Republic, August 21-26 (2011)
28. Nunes, J., Silva, P.D., Andrade, L.P., Gaspar, P.D., Domingues, L.C.: Energetic evaluation of refrigeration systems of horticultural industries in Portugal. In: 3rd IIR International Conference on Sustainability and Cold Chain (ICCC 2014), London, United Kingdom, June 23-25 (2014)
29. Matworks: MATLAB Creating Graphical User Interfaces. The Mathworks, Inc., USA (2013)
30. Evans, J.A., Hammond, E.C., Gigiel, A.J., Reinholdt, L., Fikiin, K., Zilio, C.: Improving the energy performance of cold stores. In: 2nd IIR International Conference on Sustainability and the Cold Chain (ICCC 2013), Paris, France, April 2-4 (2013)
31. Santos, R., Nunes, J., Silva, P.D., Gaspar, P.D., Andrade, L.P.: Ferramenta computacional de análise e simulação do desempenho de unidades de conservação de carne através de frio industrial. In: VI Congreso Ibérico y IV Congreso Iberoamericano de Ciencias y Técnicas del Frío (CYTEF 2012), Madrid, Spain (February 2012)
32. Santos, R., Nunes, J., Silva, P.D., Gaspar, P.D., Andrade, L.P.: Computational tool for the analysis and simulation of cold room performance in perishable products industry. In: 2nd IIR International Conference on Sustainability and the Cold Chain (ICCC 2013), Paris, France, April 2-4 (2013)
33. Neves, D., Gaspar, P.D., Silva, P.D., Andrade, L.P., Nunes, J.: Computational tool for the energy efficiency assessment of cheese industries - Case study of inner region of Portugal. In: V Congreso Iberoamericano de Ciencias y Técnicas del Frío (CYTEF 2014), Tarragona, Spain, June 18-20 (2014)

hLCS. A Hybrid GPGPU Approach for Solving Multiple Short and Unbalanced LCS Problems

Pedro Valero-Lara[1,2]

[1] Numerical Simulation Unit, Research Center for Energy,
Environment and Technology, CIEMAT, Madrid, Spain
pedro.valero@ciemat.es
[2] School of Computer Science, The Complutense University of Madrid, Spain

Abstract. The "Longest Common Subsequence" is one of the most widely used and well-known methods within the similarity search community, applicable to a wide range of fields. Currently, modern multicore CPU and GPU-based systems offer an impressive cost/performance ratio and are an attractive test platform to accelerate response time and increase the number of problems solved per second. The use of GPUs for carrying out sequences alignment is widely extended for bioinformatics applications. However, we focus on the use of this algorithm applied to other problems which supposes a new and different approach. In particular, the most important difference is found in the pattern of the sequences. While, on one hand, the size of the biological sequences are large and similar, on the other hand, the sequences in other applications are short and unbalanced. Furthermore, this work aims to use one multicore CPU and GPU system for computing multiple problems simultaneously instead of computing only one. The main contribution of this work is a new hybrid approach which combines the two classical parallel techniques for our problem in two different phases. This new implementation is up to 80× and 25× faster, in terms of speedup, over the sequential and multicore counterpart respectively for our particular problem, that is, solving multiple "Longest Common Subsequence" problems on short and unbalanced sequences with a high ratio of problems solved per second.

Keywords: Longest Common Subsequence, parallel algorithms, multicore and GPU.

1 Introduction

The "Longest Common Subsequence" (LCS) algorithm is a widely used method within the similarity search community for carrying out sequence processing. In this work, we have focused on the LCS as this is considered as the source of the sequence processing algorithms, while numerous other algorithms are extension from it. Therefore, most optimizations carried out on the LCS can also be applied to its extensions as well. This algorithm has a high computational cost in terms of resources (memory) and computing (response time). A large number of studies have been carried out [4] based on the early pioneering work in this field [5].

B. Murgante et al. (Eds.): ICCSA 2014, Part VI, LNCS 8584, pp. 102–115, 2014.

This type of algorithms is applicable to a wide range of applications [6]. Recently, the use of sequence processing algorithms has focused on bioinformatic (see for example [7]), in which the similarity of 2 large DNA sequences is computed. However, it is important not to disregard the importance of these algorithms in other fields, such as data compression [8], files comparison [9], intrusion and virus detection [10]. In these fields, in which the response time is a very important factor, the sequence processing algorithms are used to compare several short sequences with one bigger sequence.

The *LCS* algorithm is one of the main problems into the stringology, the science of sequences processing. Given two different sequences A (left sequence) and B (top sequence) with a size equal to M and N respectively, the LCS obtains the similarity between both sequences. An example of the sequential LCS algorithm is shown in Figure 1-left. A $(M+1) \times (N+1)$ matrix is necessary which stores the information of the matches between the elements of the sequences. From this information, it is possible to know the similarity (last element of the matrix), the common sequence/s and all occurrences between the two sequences. To solve the (i, j) matrix element, it is necessary to know the value of three different elements $(i-1, j-1)$, $(i-1, j)$ and $(i, j-1)$. Thus, this algorithm presents a high data dependency which is the most important feature from computing point of view. An example of a LCS problem is illustrated in Figure 1-right.

The era of single-threaded processor performance increasing has come to an end due to the limitation of the current Very Large Scale Integration (VLSI) technology. In response, the most hardware manufactures are designing and developing multicore processors and/or specialized hardware accelerators [1, 2]. Programs will only increase in performance if they use and exploit the new parallel characteristics of new architectures. The recent appearance of GPUs for general purpose computing platforms offers powerful parallel processing capabilities at a low cost. Although the Graphics Processing Units (GPUs) are traditionally

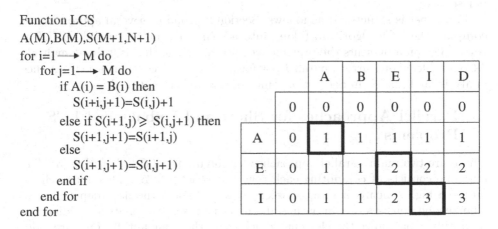

Fig. 1. LCS algorithm (left) and example (right)

associated to interactive applications involving high rasterization performance, they are also widely used to accelerate much more general applications (now called General Purpose Computing on GPU (GPGPU) [14]) which require an intense computational load and present parallel characteristics. The main feature of these devices is a large number of processing elements integrated into a single chip, which reduces significantly the cache memory. These processing elements can access to a local high speed external DRAM memory, connected to the computer through a high speed I/O interface (PCI-Express). Overall, these devices can offer a higher main memory bandwidth and can use data parallelism to achieve a higher floating point throughput than CPUs [16]. On the other hand, current multicore processors are becoming another interesting parallel platform, due to their low cost and the absence of the constraints of GPU architecture. Both types of processors offer a very interesting performance/cost ratio, and so they are increasingly used in parallel computing.

It is possible to find numerous works in which the GPU is used to accelerate the response time of the LCS algorithm [19–26]. However, all these works are focused on a single application which is the biological sequence alignment. This application consists of computing the LCS or other extensions on 2 large DNA sequences, where both sequences share a similar size. In contrast, this work focuses on other applications [8–10]. These applications share a common pattern which is different with respect biological sequences alignment. While, on one hand, the sequences size are large and very similar for biological approach, on the other hand, short and unbalanced sequences are used in the other applications. Furthermore, we compute several LCS problems on a set of input sequences, instead of executing one LCS on two large biological sequences. The main contribution of the cited works consists of minimizing the memory requirements to align 2 large sequences by using the maximum number of threads possible. However, our target is different which consists of studying the use of multicore CPU and GPU platforms to obtain the maximum ratio of LCS problems solved per second.

The paper is structured as follows: Section 2 proposes several approaches to compute the LCS algorithm taking into account the multicore and GPU features; Section 3 presents the problem we wish to tackle, that is, solving multiple LCS problems on current parallel platforms. Section 4 contains a performance analysis, and finally, in Section 5, some conclusions are outlined.

2   Parallel Approaches for Short and Unbalanced LCS Problems

There are two main parallel approaches for solving the LCS algorithm [3]. The first one consists of computing each row in parallel ($rLCS$). This approach requires a high synchronization between rows to respect the data dependencies. The second consists of computing all the elements of the matrix antidiagonal by antidiagonal, since the elements which share the same antidiagonal are independent between them ($aLCS$). Both approaches are graphically illustrated in Figure 2.

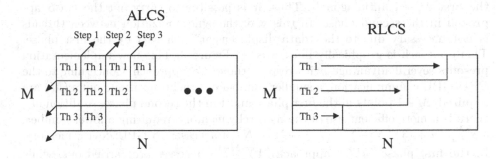

Fig. 2. Parallel LCS approaches, antidiagonal-LCS (left) and row-LCS (right)

Next, a set of GPU based approaches to implement the LCS algorithm are presented for the particular case of short and unbalanced sequences. First of all, it is discussed the implementation of the classical parallel approaches, $rLCS$ and $aLCS$. The $rLCS$ approach presents several impediments concerning GPU computing. Due to the SIMD (Single Instruction Multiple Data) architecture of the current GPUs, the synchronization between threads reaches a starvation state in which all threads are checking if the upper thread has finished. On the other hand, the $aLCS$ approach can be carried out on the current GPUs. As shown in the previous section, all the elements of the same antidiagonal are independent from each other, and so can be computed in parallel by using GPUs. However, a high number of CPU-GPU synchronization points (one per antidiagonal) and accesses to global memory are necessary. Therefore, for a problem size equal to $M \times N$, this approach needs at least $4 \times M \times N$ accesses to GPU device memory and $(M - 1) + N$ synchronization points between CPU and GPU.

A new hybrid-LCS ($hLCS$) approach is proposed in order to mitigate these disadvantages by combining both parallel approaches, $rLCS$ and $aLCS$, in two different phases. The first phase consists of computing the $aLCS$ approach on

HLCS

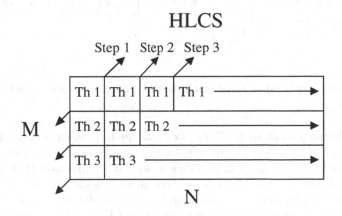

Fig. 3. Parallel hybrid LCS

the first $M - 1$ antidiagonals. Thus, it is possible to carry out the $rLCS$ approach in the second phase. In this way the synchronization between threads is not necessary due to the "data displacement" achieved by the first phase. This approach is graphically illustrated in Figure 3. This new implementation presents several advantages with respect the $aLCS$ approach. Concerning to the CPU-GPU communication, a smaller number of synchronization points (M) are required, $M - 1$ points in the first phase and 1 in the second phase. Additionally, there is a more efficient use of the hierarchy memory, requiring a lower number, $4\sum_{i=1}^{M-1} i + (3MN - \sum_{i=1}^{M-1} i) + M + N$, of accesses to GPU device memory. In the first phase ($aLCS$ approach) $4\sum_{i=1}^{M-1} i$ accesses are carried out, while the rest $(3MN - \sum_{i=1}^{M-1} i) + M + N$ are performed in the second phase ($rLCS$ approach). In last phase, each "i-row" thread reads the i element from the left sequence once (M). Also, it is possible to exploit the use of shared memory or L1 cache in the accesses to the elements of the top sequence which have to be used by all threads (N). Finally, the number of accesses to the matrix elements correspond to $3MN - \sum_{i=1}^{M-1} i$. As shown, the main bottleneck in terms of memory management is found in the accesses to matrix elements. Therefore, henceforth we focus on the memory management paying special attention to matrix elements.

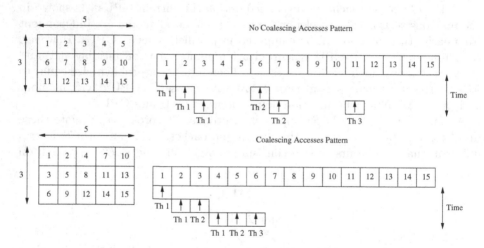

Fig. 4. No coalescing (top) and coalescing (bottom) memory locations for the elements of the matrix

Memory management in GPU devices has an impressive impact in performance due to the high latency suffered by the device memory. In order to manage the bandwidth of memory in an efficient way, it is important that all threads into a block of threads access to consecutive memory locations. In this favorable case, all memory accesses are carried out into a consolidated access (coalescing access to memory). In both approaches, $aLCS$ and $hLCS$, the accesses to the left and top sequences are performed by contiguous threads on contiguous locations of memory and so the memory bandwidth is efficiently exploited. In contrast, the most number of accesses to the device memory are carried out on the matrix elements and do not coalesce. Therefore, a new memory locations order on the elements of the matrix is proposed in order to achieve consecutive threads access to continuous locations of memory at the same period of time. Basically, the new approach consist of accessing to the matrix elements in an antidiagonal wise order instead of a row wise order. This new approach is a very profitable scheme for both implementations, $aLCS$ and $hLCS$, as illustrated in Figure 4. It implies a much more difficult indexing process over the no-coalesce approach. However, taking into account the high latency suffered by the device memory, the overhead of this new indexing can be much lower with respect no-coalesce access pattern.

3   Multiple LCS Problems

This section presents a set of different strategies to solve multiple LCS problems in parallel on both platforms, multicore and GPU, in order to achieve a high ratio of problems solved per second.

The multicore approach carries out a coarse-grain parallelism by mapping a set of continuous LCS problems on each core which are solved sequentially by using the $rLCS$ algorithm. This distribution is well balanced and the use of the memory is optimized taking into account that in C style programming languages, information is stored in row wise order. The set of problems can be easily parallelized with this approach annotating some of its loops with OpenMP [28] pragmas.

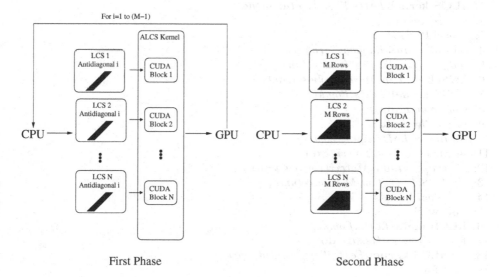

First Phase   Second Phase

Fig. 5. HLCS approach

On the other hand and concerning to GPU implementations, an effective fine-grained strategy is carried out in numerous studies [19–26], in which each thread computes either one matrix element or a small portion of the matrix (part of a row) and several blocks of threads (CUDA blocks) solve one big problem which saturates the power of the GPU by itself. However, our problems are much smaller with respect DNA sequences alignment. Therefore, the computational resources would be underutilized by using the aforementioned strategy and each problem would be computed separately, reducing the ratio of LCS problems solved per second. Additionally, R. Uribe et al. [11–13] proposed to compute a similar method to obtain the similarity between word as part of range search on metric spaces. They use a coarse-grained implementation where each thread computes the Euclidean distance for 2 different words. This is an efficient approach due to the small size of the words (15 maximum). Unfortunately, the size of the sequences used in this work are much bigger and this strategy is not profitable for our purposes.

Here, we propose other approach which maximize the use of GPU resources for our particular problem, that is, solving multiple "medium" LCS problems simultaneously in the same GPU. This consists of a medium-grained parallelism which distributes each LCS problem on one CUDA block. We used the same CUDA block mapping on both approaches, $aLCS$ and $hLCS$. In this way, the threads within the same CUDA block can work together by exploiting the fastest region of the memory hierarchy efficiently (registers, shared memory, L1 cache). The number of threads per CUDA block is given by the size of the left sequences

---

Algorithm 1. hLCS kernels

1: **aLCS_kernel**($Lefts,Top,M,antidiagonal$)
2: $row,column,ind_M$
3: $row = threadIdx$
4: $column = antidiagonal - row$
5: $ind_M = indexMatrix(row, column)$
6: $LCS(Left, Top, M, row, column, ind_M)$
7: **rLCS_kernel**($Lefts,Top,M$)
8: $row,column,ind_M$
9: $row = threadIdx$
10: $column = Lefts.size - row$
11: **while** $column < Top.size$ **do**
12:      $ind_M = indexMatrix(row, column)$
13:      $LCS(Left, Top, M, row, column, i)$
14:      $syncThreads$
15: **end while**
16: **hLCS_main**($Lefts,Top,M$)
17: **for** $i = 1 \rightarrow Lefts.size$ **do**
18:      **aLCS_kernel**($Lefts,Top,M,antidiagonal$)
19: **end for**
20: **rLCS_kernel**($Lefts,Top,M$)

(M). Although, the two approaches share the same block mapping pattern, different features are noted. As introduced in Section 2, it is necessary an iterative process to carry out the $aLCS$ approach on GPU. In each iteration, all the elements of the given antidiagonal are computed in all LCS problems. In the most of the iterations, the number of threads per CUDA block is M. However, in the first and last $M - 1$ iterations the size of CUDA block changes, increasing in the first iterations and decreasing in the last iterations. The $hLCS$ approach is composed by two phases. The first one consist of computing the $aLCS$ approach on the first $M - 1$ antidiagonals in all problems. The last one consists of computing the $rLCS$ approach on the rest of the matrix elements in all LCS problems.

In order to avoid *race* conditions, it was necessary to resort to barriers for guaranteeing coherent executions in the second phase ($rLCS$). This barriers were implemented through the call to *syncthreads* functions which force to all threads of the same CUDA block do not advance up to all of them have achieved the same synchronization point (barrier). In particular we placed the barrier at the end of every iteration after computing LCS. The CUDA block mapping and the pseudo-code are graphically illustrated in Figure 5 and algorithm 1 respectively.

4   Performance Evaluation

This section presents the performance evaluation in which all proposed approaches and both platforms, multicore and GPU, are evaluated in terms of execution time, speedup and number of LCS problems solved per second. The set of tests were carried out on a heterogeneous platforms [27], whose main features are summarized in Table 1. To control the GPU devices and manage memory, we have used in the present work the high level programming language CUDA [18] introduced by NVIDIA.

As introduced in Section 1, this problem is used in a large number of applications, therefore rather than evaluating a particular test case, an exhaustive number of synthetic tests are performed focusing on the requirements of the applications and the limitations of the computational resources. The sequences are randomly initialized. It is important to note that the maximum number of threads per CUDA block for the GPU system used is 1024. However, this number

Table 1. Platform

Platform	Xeon E5520	Fermi C2050
Cores	4	512
on-chip Mem	L1 32KB (per core)	SM 16/48KB
	L2 256KB (unified)	L1 48/16KB
	L3 8MB (unified)	L2 768KB (unified)
Memory	16 GB DDR3	3 GB GDDR5
Bandwidth	25.6 GB/s	148 GB/s
OS	Linux Red Hat 2.6.32-220 amd-64	
Compiler	gcc 4.4.6	nvcc 4.1

has been increased recently in the new GPU architecture (Kepler). This section is divided into 3 different parts, providing an analysis of the different approaches and platforms by evaluating the most important performance aspects.

In the first one, the influence on performance of the most important parameters is evaluated, i.e. the sizes of the left sequences and top sequence, and the number of *LCS* problems. Therefore, the tests have consisted of increasing the size of each parameter, while the other remain constant. The influence of the memory transfers overhead in the GPU approaches is not included here, but will be discussed later. Figure 6 shows the speedup achieved by 6 different parallel implementations: multicore, *aLCS* and *hLCS* with and without coalescing memory locations for the matrix elements. The L1 level and shared memory can be combined in two different ways, 16KB shared memory and 48KB L1 or 48KB shared memory and 16KB L1. Therefore, for the case of *hLCS* with coalescing memory locations, two different approaches have been carried out which consist of using 48KB shared memory and 16KB L1 (*hLCS(Coalescing-SM)*) or 48KB L1 cache memory and 16KB shared memory (*hLCS(Coalescing-L1)*). While, in the *hLCS(Coalescing-SM)* approach the elements of the top and left sequences are stored in shared memory, in the *hLCS(Coalescing-L1)* the sequences are stored in L1 cache memory. Three graphs are shown in Figure 6 which show the speedup, execution time of the sequential approach over the execution time of the set of parallel implementations, achieved by the different porposed approaches. The "A" graph shows the trend of the speedup by increasing the size of left sequences, keeping constant the size of the top sequence (2048) and the number of *LCS* problems (32). The "B" graph shows the performance by increasing the size of the top sequence and keeping constant the size of the left sequences (32) and the number of problems (32), and finally, the last ("C") graph shows the trend of the speedup by increasing the number of problems and keeping constant the size of the left sequences (32) and top sequences (2048). As expected, a better result is obtained by using of those approaches which exploit the coalescing memory locations on matrix elements, being the *hLCS(Coalescing-L1)* approach the fastest. Therefore, the exploitation of L1 cache memory is more efficient than the use of shared memory. Although, the shared memory shows much better features in terms of access time, this can suffer from bank memory conflict. On the other hand, the multicore approach shows a good performance, obtaining a speedup around 3.4 in all test cases by using 4 cores. As shown, the GPU approaches achieve a higher speedup by increasing the size of the left sequences and the number of *LCS* problems. On the other hand, the speedup remains constant with a higher size for the top sequence. Obviously, better results are achieved on the cases where an increase of computational cost, (number of *LCS* problems and size of left sequences) supposes a higher parallelism degree (number of CUDA blocks or number of threads per CUDA block).

As well known, the use of GPU platforms requires an additional cost due to the transfers between CPU and GPU memory. In our case the transfers are carried out at the beginning and the end of the problem. Firstly, it is transferred the

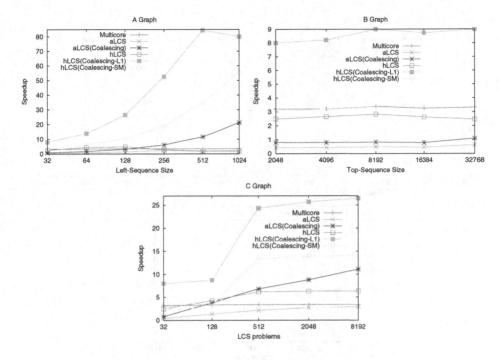

Fig. 6. Speedup achieved

set of left sequences and the top sequence from CPU memory to GPU memory. In contrast, the last elements (similarity) of the matrices are transfered from GPU memory to CPU memory at the end. In the second part of this section, the overhead of the data transfer on the fastest approach, *hLCS(Coalescing-L1)*, is analyzed. Figure 7 shows the trend of the overhead (% with respect the total execution time) caused by memory transfers. The overhead achieved does not suppose a high cost over the total execution time. It is important to note that, the memory transfers overhead increases in the cases where the speedup increases as well. Obviously, the memory transfers can not be parallelized and so the overhead caused by these is higher when a increase of memory requirements supposes a higher speedup (increasing either the size of left sequences or the number of LCS problems). Otherwise, on the case where the speedup remains constant by increasing the memory requirements (increasing the size of the top sequences) the memory transfers overhead is constant as well.

Mainly, the parallelization of this algorithm was carried out to obtain a high ratio of problems solved per second. Hereafter, the overhead of the memory transfer is taken into account and the only GPU implementation used corresponds to *hLCS(Coalescing-L1)*. A higher computational cost by increasing the size of the left sequences and top sequence and keeping constant the number

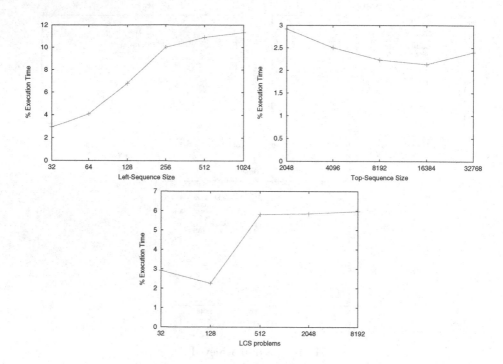

Fig. 7. Memory transfers overhead

of LCS problems shows an important decrement in the ratio. Both parameters show an impressive impact in performance, being the size of the top sequence the most influential. However, the behavior is different by increasing the number of problems and keeping the size of the sequences constant. Figure 8 shows the trend for this case where the sequential and multicore approaches achieve a constant ratio while the benefit of the *hLCS(Coalescing-L1)* approach increases as the number of *LCS* problems increases.

Finally, an overall analysis is presented which consists of showing the performance by increasing the value of all parameters, i.e. the size of the sequences and the number of problems. From the first test with a configuration for the size of left sequences, top sequence and number of problems equal to 64, 256 and 512 respectively, the parameters values for the next test are the double bigger than in previous test. Although the *hLCS* approach achieves high speedups, an important decrement in performance in terms of number of problems solved per second is performed by increasing the cost per problem.

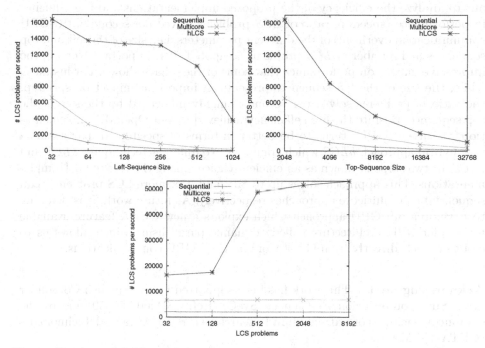

Fig. 8. Number of *LCS* problems per second achieved by the sequential, multicore, and GPU (*hLCS*) implementations

Table 2. Execution time in ms, speedup and number of problems solved per second (Ratio)

Sequential		Multicore			hLCS		
Time	Ratio	Time	Speedup	Ratio	Time	Speedup	Ratio
62.50	8192	19.23	3.25	26625	1.72	36.33	297674
531.25	1927	156.25	3.4	6553	8.69	61.13	117836
4093.75	500	1218.75	3.35	1680	49.4	82.86	41457

5   Conclusions

It is possible to find numerous problems which need a high ratio of *LCS* problems solved per second. In particular, some of them, such as, instruction detection, data compression, files comparison, . . . , use the same sequences pattern, that is, short and unbalanced. Taking into account the particularities of the sequences and the aim of computing multiple problems simultaneously, in order to perform a high ratio of problems solved per second, a set of parallel approaches have been proposed and studied on current heterogeneous platforms, multicore CPU and GPU, focusing on different parallel techniques and memory control. An exhaustive performance study composed by numerous synthetic tests has been carried out to analyze the efficiency of the proposed implementations and the influence in performance (speedup, ratio of LCS problems solved per second, CPU-GPU communication overhead) of the different parameters, i.e., size of the left and top sequences and number of LCS problems computed. It is important to note, the impressive impact on performance that some of them have shown. For instance, while, the size of the left sequences presents an important impact on speedup, the ratio of problems solved per second is highly influenced by the size of the top sequence. Despite the high efficiency achieved by our OpenMP implementation, the use of GPUs turns to be better in terms of speedup. In particular, a new hybrid approach $hLCS$ which combines two different techniques, $aLCS$ and $rLCS$, in two steps is shown as an efficient choice among the rest of GPU implementations. This approach solves up to 83× and 25× more LCS problems than sequential and multicore approaches respectively. As future work, it is planning to develop a full GPU approach which exploits a new CUDA feature available for Kepler GPU architecture called "dynamic parallelism" which allows us to call to kernels directly from GPU avoiding CPU-GPU communications.

Acknowledgments. This work has been supported by the Spanish Consolider grant Supercomputación y e-Ciencia (SyeC) (Ref: CSD2007-00050) and by the computing facilities of Extremadura Research Centre for Advanced Technologies (CETA-CIEMAT).

References

1. Asanovic, K., Bodik, R., Christopher, B., Gebis, C.J.J., Husbands, P., Keutzer, K., Patterson, D.A., Plishker, W.L., Shalf, J., Williams, S.W., Yelick, K.A.: The landscape of parallel computing research: A view from berkeley. Technical Report UCB/EECS-2006-183, EECS Department, University of California, Berkeley (2006)
2. Geer, D.: Chip markets turn to multicore processors. Computer 38(5), 11–13 (2005)
3. Xiao, L., Song, H., Zhu, M., Kuang, Y.: A PC cluster based Parallel Algorithm for Longest Common Subsequences Problems. In: The International Conference on Bioinformatics and Biomedical Engineering, ICBBE 2008 (2008)
4. Bonizzoni, P., Vedova, G.D., Mauri, G.: Experimenting an Approximation Algorithm for the LCS. Discrete Applied Mathematics (1998)
5. Dowling, G., May, P.: Aproximate String Matching. Journal ACM Computing Surveys (1980)
6. Navarro, G.: A Guided Tour to Approximate String Matching. Journal ACM Computing Surveys (1999)
7. Ning, K., Ng, H.K., Wai, L.H.: Finding Patterns in Biological Sequences by Longest CommonSubsequences and Shortest Common Subsequences. In: IEEE Symposium BioInformatics and BioEngineering, BIBE (2006)
8. Bentley, J., Mcilroy, D.: Data Compression using Long Common Strings. In: IEEE Data Compression Conference (1999)
9. Heckel, P.: A technique for isolating differences between files. Communications of ACM (1978)

10. Kumar, S., Spafford, E.: A pattern-matching model for intrusion detection. In: National Computer Security Conference (1994)
11. Uribe-Paredes, R., Valero-Lara, P., Arias, E., Sánchez, J.L., Cazorla, D.: A GPU-Based Implementation for Range Queries on Spaghettis Data Structure. In: Murgante, B., Gervasi, O., Iglesias, A., Taniar, D., Apduhan, B.O. (eds.) ICCSA 2011, Part I. LNCS, vol. 6782, pp. 615–629. Springer, Heidelberg (2011)
12. Uribe-Paredes, R., Valero-Lara, P., Arias, E., Sánchez, J.L., Cazorla, D.: Similarity search implementations for multi-core and many-core processors. In: The 2011 International Conference on High Performance Computing & Simulation (HPCS), pp. 656–663 (2011)
13. Uribe-Paredes, R., Arias, E., Sánchez, J.L., Cazorla, D., Valero-Lara, P.: Improving the Performance for the Range Search on Metric Spaces Using a Multi-GPU Platform. In: Liddle, S.W., Schewe, K.-D., Tjoa, A.M., Zhou, X. (eds.) DEXA 2012, Part II. LNCS, vol. 7447, pp. 442–449. Springer, Heidelberg (2012)
14. GPGPU. General-purpose computation using graphics hardware, http://www.gpgpu.org
15. Rost, R.J.: OpenGL Shading Language. Addison-Wesley (2005)
16. Feng, W.C., Manocha, D.: High-performance computing using accelerators. Parallel Computing 33, 645–647 (2007)
17. Mark, W.R., Glanville, S.R., Akeley, K., Kilgard, M.J.: Cg: a system for programming graphics hardware in a C-like language. In: SIGGRAPH 2003: ACM SIGGRAPH 2003 Papers, pp. 896–907. ACM Press (2003)
18. NVIDIA. NVIDIA Compute Unified Device Architecture (CUDA) Programming Guide, http://docs.nvidia.com/cuda/cuda-c-programming-guide/index.html
19. Manavski, S.A., Valle, G.: CUDA compatible GPU cards as efficient hardware accelerators for Smith-Waterman sequence alignment. BMC Bioinformatics (2008)
20. Kloetzli, J., Strege, B., Decker, J., Olano, M.: Parallel Longest Common Subsequence using Graphics Hardware. In: Eurographics Symposium on Parallel Graphics and Visualization (2008)
21. Flavius, E., Alba, S., de Melo, C.M.A.: CUDAlign: Using GPU to Accelerate the Comparison of Megabase Genomic Sequences. In: Principles and Practice of Parallel Programming, PPoPP (2010)
22. Singh, J., Aruni, I.: Accelerating Smith-Waterman on Heterogeneous CPU-GPU Systems. In: International Conference on Bioinformatics and Biomedical Engineering (iCBBE), p. 14 (2011)
23. Ligowski, L., Rudnicki, W.: An efficient implementation of Smith Waterman algorithm on GPU using CUDA, for massively parallel scanning of sequence databases. In: IEEE International Symposium Parallel & Distributed Processing (IPDPS), p. 18 (2009)
24. Liu, Y., Maskell, D.L., Schmidt, B.: CUDASW++: optimizing Smith-Waterman sequence database searches for CUDA-enabled graphics processing units. BMC Res. Notes 2, 73 (2009)
25. Du, Z., Yin, Z., Bader, D.: A tile-based parallel Viterbi algorithm for biological sequence alignment on GPU with CUDA. In: IEEE International Symposium on Parallel & Distributed Processing Workshops (IPDPSW), pp. 1–8 (2010)
26. Striemer, G.M., Akoglu, A.: Sequence alignment with GPU: Performance and design challenges. In: IEEE International Symposium on Parallel & Distributed Processing (IPDPS), pp. 1–10 (2009)
27. CETA-CIEM, AT, http://www.ceta-ciemat.es/index.php?lang=en
28. OpenMP, http://www.openmp.org

CONCLAVE: Ontology-Driven Measurement of Semantic Relatedness between Source Code Elements and Problem Domain Concepts

Nuno Ramos Carvalho[1], José João Almeida[1], Pedro Rangel Henriques[1], and Maria João Varanda Pereira[2]

[1] University of Minho, Braga, Portugal
{narcarvalho,jj,prh}@di.uminho.pt
[2] Polytechnic Institute of Bragança, Bragança, Portugal
mjoao@ipb.pt

Abstract. Software maintainers are often challenged with source code changes to improve software systems, or eliminate defects, in unfamiliar programs. To undertake these tasks a sufficient understanding of the system (or at least a small part of it) is required. One of the most time consuming tasks of this process is locating which parts of the code are responsible for some key functionality or feature. Feature (or concept) location techniques address this problem.

This paper introduces CONCLAVE, an environment for software analysis, and in particular the CONCLAVE-MAPPER tool that provides a feature location facility. This tool explores natural language terms used in programs (e.g. function and variable names), and using textual analysis and a collection of Natural Language Processing techniques, computes synonymous sets of terms. These sets are used to score relatedness between program elements, and search queries or problem domain concepts, producing sorted ranks of program elements that address the search criteria, or concepts. An empirical study is also discussed to evaluate the underlying feature location technique.

1 Introduction

Reality shifts, bug fixes, updates or introducing new features often require source code changes. These software changes are usually undertaken by software maintainers that may not be the original writers of the code, or may not be familiar with the code anymore. In order to carry out these changes, programmers need to first understand the source code [43]. This task is probably the main challenge during software maintenance activities [9].

Software reverse engineering is a process that tries to infer how a program works by analyzing and inspecting its building blocks and how they interact to achieve their intended purpose [8,30]. Many of these techniques rely on mappings between human oriented concepts and program elements [35]. These are often used to locate which parts of the program are responsible for addressing specific

B. Murgante et al. (Eds.): ICCSA 2014, Part VI, LNCS 8584, pp. 116–131, 2014.

domain concepts [3], and are usually referred in the literature as feature location techniques [12].

Programming languages unambiguous grammars limit the sentences that can be used to write software. Still some degree of freedom is given to the programmer to use natural language terms (e.g. program identifiers, constant strings or comments). These terms can give clues about which concepts the source code is addressing, and the meaningfulness of these terms can have a direct impact on future program comprehension tasks [24]. Most of the programming communities promote the use of best practices and coding standards that usually include rules and naming conventions that improve the quality of terms used (e.g. the *"Style Guide for Python Code"* [1]). Feature location techniques that exploit such elements are typically described as textual analysis, often combined with static analysis [12].

This paper introduces CONCLAVE[2], a system of tools for software analysis, with special focus on CONCLAVE-MAPPER[3], a tool that provides a technique to measure semantic relatedness between source code elements, and elements supplied by the maintainer as query searches. It allows the creation of mappings between source code and real world concepts, facilitating feature location activities. The main goal of this system is to provide programmers with insight and information about software packages to enhance program understanding activities and ease software maintenance tasks.

CONCLAVE-MAPPER uses source code static analysis to extract data from source code (e.g program identifiers, function definitions). The extracted data is loaded to an ontology that represents the program. Other ontologies can be added to the system if available (e.g. the problem domain ontology, dynamic traces information). Using a set of Natural Language Processing (NLP) techniques and textual analysis, *kind-of* Probabilistic Synonymous Sets (kPSS) are computed for every element present in the ontologies, and a scoring function is used to measure the semantic relatedness[4] between them. The main output of this tool are ranks – sorted by relevance – of program elements that are prone to address some specific real world domain concept. This tool also provides a Domain Specific Language (DSL), for writing search queries.

In the next section related work, and some state-of-the-art feature location techniques are discussed; Section 3 gives a brief overview of the CONCLAVE environment; and Section 4 describes in more detail the CONCLAVE-MAPPER tool. Section 5 describes the experimental validation held to evaluate the feature location technique, including results discussion. Finally, Section 6 includes some final remarks and trends for future work.

---

[1] http://www.python.org/dev/peps/pep-0008/ (Last accessed: 29-01-2014).

[2] http://conclave.di.uminho.pt (Last accessed: 27-01-2014).

[3] http://conclave.di.uminho.pt/mapper (Last accessed: 27-01-2014).

[4] In ontologies the term similarity is used to refer how similar two concepts are, and is usually based on a hierarchy of *is-a* relations, in the context of this work concepts can be related in many ways, hence the adoption of the term relatedness.

2   Related Work

Program Comprehension (PC) is a field of research concerned with devising ways to help programmers understand software systems. In the context of PC, feature (or concept) location is the process of locating program elements that are relevant to a specific feature implementation. This is typically the first step a programmer needs to perform in order to devise a code change [3, 28, 35].

Feature location techniques are usually organized by types of analysis: (a) dynamic analysis, which is based in software execution traces, and examines programs runtime (e.g. [1, 13, 40, 44]); (b) static analysis, based on static source code information, such as slicing, control or data flow graphs (e.g. [7, 27, 39]); and (c) textual analysis, explore natural language text found in programs like comments or documentation. This last type can be based on Information Retrieval (IR) methods (e.g. [4,5,26]), NLP (e.g. [18,41]), or pattern matching (sometimes also referred as grep-*like*) based approaches (e.g. [14]). For more details about different trends and other approaches please refer to surveys [12] and [44].

The CONCLAVE-MAPPER underlying feature location technique uses a combination of static and textual analysis, and ontologies. Examples of other approaches that explore the same combination of analysis include: in [47], Zhao *et al* use a static representation of the source code named BRCG (branch-reserving call graph) to improve connections between features and computational units gathered using an IR technology; in [17], Hill *et al* present a technique that exploits the program structure and also program lexical information; in [36], Ratiu and Florian establish a formal framework that allows the classification of redundancies and improper naming of program elements, which is used as a based to represent mappings between the code and the real world concepts in ontologies; in [16], Hayashi *et al* proposed linking user specified sentences to source code, using a combination of textual and static analysis domain ontologies. Other applications of ontologies in software engineering in [15].

State-of-the-art feature location approaches involve combining techniques taking advantage of having data produced from different types of analysis (e.g. [23, 26]).

3   CONCLAVE Overview

The CONCLAVE environment provides a set of tools to perform software analysis. The main system workflow is divided in three main stages: (a) collecting data; (b) processing collected data and loading ontologies; and, (c) reasoning about data in the ontologies and providing views of computed information. Figure 1 illustrates this workflow, and the next sub-sections describe in more detail the different stages. All the tools implemented in the context of this system are modular (or work as plugins), and some provide web-services, so that they can be used as standalone applications, or composed together to create more complex applications or workflows (like the one illustrated in Fig. 1).

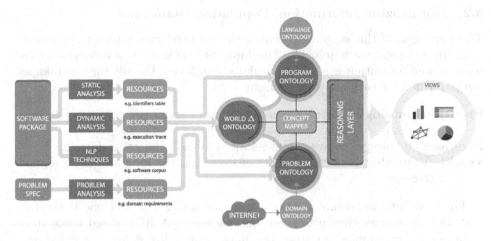

Fig. 1. Overview of the major stages of the CONCLAVE system workflow

3.1 Collecting Data

This is the first stage of the main workflow; its goal is to collect data from a software package, and any kind of problem specification if available. It takes as input the complete package (and other available documents) and produces as output an heterogeneous collection of resources. The processing tools involved in this stage can use different type of analysis: static source code analysis (e.g. parsing code to extract identifiers and static call graphs), dynamic analysis (e.g. execution traces), etc.

Any analysis can be used to collect information, and produce a resource. In the context of this work, some tools were implemented to provide some initial data to the system and contribute to PC in general, here are some examples:

CONC-CLANG: is a static analysis tool, based on the clang compiler library [22] for gathering identifiers and static functions calls information for C/C++ programs;

CONC-ANTLR: is a static analysis tool, based on the ANTLR parser generator framework [31], for gathering program identifiers information for Java programs;

CONCLAVE-IDSPROCESSOR: provides a tool for splitting program identifiers, mainly because programmers tend to use abbreviations and word combinations to name program elements (like variables or functions), these need to be split and expanded to improve feature location techniques [10, 11].

The heterogenous set of tools used during this stage produce a multitude of resources in distinct formats. In order to take advantage of all these results all this information needs to be conveyed to a common format, more suitable for querying and processing. Ontologies were adopted as a common target format. Building ontologies from collected data is done during the second stage, which is discussed in the next section.

3.2   Normalizing Information, Populating Ontologies

The main goal of this stage is to convey the data collected during the previous stage into the system ontologies. The input of this stage is a collection of resources, and the output is a set of populated ontologies. Usually three ontologies are populated for each software package:

Program Ontology: abstract representation of some key program elements
(e.g. methods, functions, variables, classes);
Problem Ontology: concepts and relations in the application domain;
World △ Ontology: runtime effects of executing the program (e.g. program
run traces).

There are two important details about this stage. The first one is the format and technology chosen to store the ontologies. A RDF based triple-store technology was adopted to store the data. This allowed for a scalable and efficient method for performing storing and querying operations, and also allows to export the data in several community accepted ontology formats (e.g. OWL, RDF/XML, Turtle) [19, 21]. Querying facilities are also readily available; for instance, SPARQL is a querying domain specific language for RDF triple-stores [32, 34].

Although these technologies provide scalable and efficient environments for handling information, development wise, they are far from the abstraction desired by the applications level implementation. To overcome this problem the Ontology ToolKit (OTK)[5] was implemented, which provides an abstraction layer on top of the RDF technology, to develop ontology-*aware* applications. In practice, when applications developers want to perform an ontology related operation, instead of using triple-store low level primitives, they can use the abstraction layer. To motivate for the development of this abstract framework, consider the modern Object-Relational Mappers (ORM) in the context of relational databases. Which provide an abstraction layer and interface for programming languages to handle data (stored in databases) as objects, allowing the development of applications regardless of the underlying database technology used [20].

The second important detail is the data semantic shift. Resources tend to produce raw data, but the data stored in the ontologies conveys a richer semantic. Most resources require a specific tool to read the resource data, and translate it to information that is ready to store in the ontology, i.e. follows the semantic defined by the ontology. OTK has also proven useful to implement this family of tools.

A simple example to illustrate the previously discussed details follows. Imagine the CONC-CLANG tool was used to process a C source code file, included in a software package. The raw output of this tool is something like[6]:

```
Function,source.c::add::6,add,,source.c,6,8
```

---

[5] Implemented as a set of libraries for the Perl programming language.
[6] For more examples please visit the tool website:
http://conclave.di.uminho.pt/clang (Last accessed: 27-01-2014).

This line by itself conveys small to none semantic of the data being included in the final resource. In loose english this line states that: *"in the 'source.c' file there is a 'Function' definition which has a identifier represent by 'add' that starts in line '6' and ends in line '8'"*, and this is the kind of semantic that needs to be conveyed to the ontological program representation of the program. The Program Ontology has a class to represent instances of elements that are functions in the source code, another for identifiers, and the line numbers are stored as data proprieties [7]. To illustrate the use of OTK, the following snippet illustrates a simplified version of the required code to load this information to the Program Ontology.

```
use OTK;
my $ontology = OTK->new($pkgid, 'program');
$ontology->add_instance('add', 'Function');
$ontology->add_instance('add', 'Identifier');
$ontology->add_data_prop('add', 'hasLineBegin', 6, 'int');
$ontology->add_data_prop('add', 'hasLineEnd', 8, 'int');
$ontology->add_obj_prop('add', 'inFile', 'source.c');
```

The Program Ontology used is in line with other authors' proposed descriptions (e.g. [37, 45, 46]). This also eases future integration processes with other tools that followed those same approaches. Figure 2 illustrates a subset of the class hierarchy exported to OWL. Once all the data is stored in the ontologies, the reasoning layer can be used to relate informations gathered from different

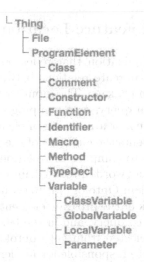

Fig. 2. Program ontology sub-set of the class hierarchy

---

[7] Although a triple-store RDF approach is used to store the actual information, we are using OWL vocabulary and specification to make clear the aimed semantics for the program representation [2].

elements and domains to build semantic bridges between elements. More details about this stage are discussed in the next section.

3.3   Reasoning and Views

During this stage more knowledge about the system is build and provided to the system end-user. The tools in this stage use as input the ontologies built during the previous stage, and generally fall in one of the two categories, either they: (a) process information to compute new information and knowledge about the system – usually in this case the tool output is new content added to the ontologies; or (b) information or knowledge suitable for visualization is built – in this particular case the final output of the tool is a view for the package system.

Querying the ontology, and adding information if necessary, can easily be done using the OTK framework. Also note that the tools in this stage are language agnostic, in the sense that data about the source code (language dependent) has already been gathered, and OTK tools do not depend anymore on the source language. For example, if a tool processes identifiers, to get a list of the program identifiers simply query the Program Ontology using OTK, as follows:

```
use OTK;
my $ontology = OTK->new($pkgid, 'program');
my @identifiers = $ontology->get_instances('Identifier');
```

CONCLAVE-MAPPER, described in detail in the next section, is an example of tools that are used during this stage.

4   CONCLAVE-MAPPER Feature Location Approach

CONCLAVE-MAPPER is an application that relies on data computed by other tools (see Sec. 3.1 and 3.2), to create relations between elements of any of the ontologies available for a given package. The input for this application is a set of ontologies, and either a search query, or a mapping query; and the output is a sorted rank of element relations, or a mapping of element relations respectively. The Program Ontology represents the elements of the program, a software maintainer can ask the application to compute the relations between elements in the program and either a set of keywords provided in a search query, or elements in other ontologies (e.g. Problem Ontology) using a mapping query. In the first case, the result is a sorted rank of the program elements that are related with the keywords provided in the search query, and in the latter a matrix of relatedness score between the elements selected from both ontologies. Both can be used to find which parts of the code are responsible for implementing a domain concept – feature location.

Before discussing function implementations, a formal description of the data types required for creating each output follows[8]:

---

[8] *Haskell* syntax is used to describe the data types and functions discussed; some details have been slightly simplified to improve readability.

type *Rank* = [*Entry*]
data *Entry* = *Entry* { *score* :: *Float*, *element* :: *Element* }

This defines a *Rank* as a collection of *Entry*, where each *Entry* contains the semantic relatedness (*score*), between the element (*element*) and the query search (more details about how this score is calculated in Sec. 4.2). The type *Element*, represents an instance in any ontology (if elements of the Program Ontology are being used all other data is also available: source file, begin and end line, identifier, etc.).

data *Map* = *Map* { *rows* :: [*Element*], *cols* :: [*Element*], *cells* :: [*Cell*] }
data *Cell* = *Cell* { *row* :: *Int*, *col* :: *Int*, *score* :: *Float* }

This defines a *Map* as a matrix, with an *Element* for each row and column; each *Cell* in the matrix (besides its position information) contains the semantic relatedness measure (*score*) for the corresponding elements (more details on how score is calculated in Sec. 4.2).

The application implements two main functions to compute each one of the available output types. The *locate* function creates a *Rank* and is defined as:

```
locate    :: Query → Rank
locate q = let elements = getElements q
               entries  = [ Entry score e | e ← elements, computeScore q e ]
           in Rank entries
```

This function, given a *Query*, computes a *Rank*, by iterating over all the elements being analyzed (defined by the search query), and for each element computing a semantic relatedness score, and adding it to the *Rank* as a new *Entry*. The element set being searched and the scoring function are defined by the search query (see Sec. 4.2 for details).

The *mapping* function creates a *Map* and is defined as:

```
mapping          :: Query → Query → SFunction → Map
mapping q1 q2 f = let rows = getElements q1
                      cols = getElements q2
                      i    = 0
                      j    = 0
                      cells = [ Cell i j (f r c) |
                                    r ← rows, c ← cols,
                                    i ← i + 1, j ← j + 1]
                  in Map rows cols cells
```

This function, given two queries (of type *Query*), and a scoring function (*SFunction*), calculates a matrix of elements where each cell includes the relatedness score between the corresponding row and column element. This provides a matrix of relations between all selected (program, application domain, etc.) elements, that can be sorted by relevance. Figure 3 illustrates a possible view of these mappings, highlighting the best relevance ranking between the application domain and functions.

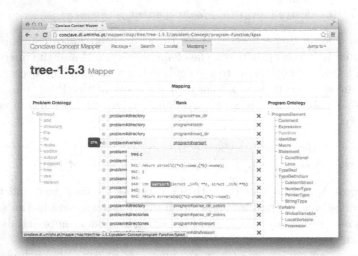

Fig. 3. A mapping produced by CONCLAVE-MAPPER: on the left the Problem Ontology can be used to constrain the concepts being searched, on the right the Program Ontology can be used to constrain the range of which program elements are being analyzed, and in the center the resulting rank sorted by relevance

The *Query* type used before describes a query supplied by the user (a predefined set of queries is also available via the system interface). A DSL was developed to describe these queries (either search or mapping); this language is discussed in the next subsection.

4.1   The Query Language

In order to compute rankings and mappings at least one query is required. This section describes the DSL that was devised to create such queries. Each query has at least three main components: (a) keywords; (b) domain and range constrains (e.g. search only functions, or variables); and, (c) the scoring function used to compute the relatedness score between the elements (some of these have default values).

The DSL query language allows two major types of queries: (a) simple strings; or (b) complex queries where a set of proprieties that define the query are supplied in the form of `PropriertyName=PropriertyValue`[9]. To distinguish simple queries from complex, the latter are enclosed in square brackets (`[ ]`). To introduce possible queries and proprieties a set of query examples are presented next.

---

[9] Keywords are only used for search queries by the locate function, and are ignored when creating maps.

The simplest possible query that can be written is a simple string, where the words that represent features or concepts being search are concatenated together using blank spaces, for example:

```
"color"
"color schema"
```

These queries search for elements that are closely related with the words `"color"` (or also `"schema"`). By default, a score is computed for all the Program Ontology elements (functions, variables, classes, etc.), and the scoring function used is `kpss` (scoring functions are discussed in detail in Sec. 4.2).

The program maintainer when initially addressing a possible bug fix, may be interested in searching functions (or methods) only. The `class` propriety can be used to constrain the subclass of elements that are being retrieved from the ontology. The following query performs a search for the words "color" and "schema", but only analyses elements that are instances of the class `Function` (remember the ontology definition in Sec. 3.2, and that the Program Ontology is the default ontology for selecting elements):

```
[ word=color word=schema class=Function ]
```

This means that the elements of the resulting rank are only instances of functions, because the query constrains the search domain of the locate function. Another example, can be searching for variables, by selecting the class `Variable`, this includes all the members that are instances of the class `Variable` and also instances of all sub-classes of `Variable` (e.g. `Parameter`, `LocalVariable`), i.e. the resulting rank includes all kinds of variables in the original program:

```
[ word=color class=Variable ]
```

Particular types of variables can be selected, for example searching only local variables:

```
[ word=color class=LocalVariable ]
```

Another important propriety that can be defined is the scoring function, i.e. the function that will compute the semantic relatedness score between the keywords searched and each of the selected elements, this is done using the `score` propriety. For example, the query:

```
[ word=color class=Variable score=levenshtein ]
```

uses the `levenshtein` word distance algorithm [25], to compute the score. By default, the scoring function based on kPSS is used.

So far, the illustrated queries have been computing scores between elements (e.g. functions, variables) and a set of words. But more complex comparisons may provide more accurate rankings. The `aggr` propriety allows a query to define the name of a relation (defined in the ontology) to compute a score not only between each selected elements, but also including a set of elements that are related to them. The query:

[`word=color class=Function score=levenshtein aggr=inFunction`]

for example, analyses all the functions, and for each function also considers all the elements that are related with that function by the relation `inFunction` (defined in the ontology). This relation is used to link all the local variables and parameters to all the functions (or methods depending on programming language) where they are defined and used. In practice, the score for each element (function) is the average between computing the score for the element itself, and the score for every local variable and parameter defined in that function.

The scores used by the *locate* and *mapping* functions are calculated by the function defined in the query, and are discussed in the next section.

4.2   Scoring Functions

The score between two elements (or an element and a word) quantify how close they are semantically related. This score is used to sort the ranks computed by the *locate* function by relevance, or to highlight the cells that express close relatedness between elements in the matrixes computed with the *mapping* function.

The main scoring function available in the CONCLAVE system is the *kpss* function (used by default), and is based on kPSS, which defines a formalism to describe synonymous sets based on Probabilist Synonymous Sets (PSS) [6, 42]. These define synonymous sets based on statistical analysis of parallel corpora.

A kPSS of order n is formally defined as a set of orders; a synonymous set corresponds to each order n (where $n \in N$), which is defined as a list of *Term*. Each *Term* contains a word and a probability:

    type *kPSS* $= [Order]$
    data *Order* $= Order \{ n :: Int, synsen :: [Term] \}$
    data *Term* $= Term \{ word :: String, prob :: Float \}$

The sum of all the probabilities in every synonymous set, for every order is always equal to 1. A kPSS can be built for any given word, and usually only up to order 3. Every time a new order is added, probabilities need to be normalized so that the invariant is kept valid. The first order contains the original word itself only, the second order *synset* contains only words directly derived from the original word (e.g. lemma, inflection), and the third order contains terms extracted from the PSS for the same word. Table 1 illustrates the kPSS for the word *"inserting"*.

Once a kPSS is available for a pair of words, the relatedness score between these words can be calculated. The *kpss* function is used to compute this score (as a *Float*) and is defined as:

    kpss $:: kPSS \rightarrow kPSS \rightarrow Float$
    kpss $k1 \; k2 = \sum [\, \mathbf{min} \, (prob \; x) \, (prob \; y) \; |$
                                   $x \leftarrow flatten \; k1, \; y \leftarrow flatten \; k2, \; word \; x \; == word \; y \,]$

This function iterates over the flattened version of the kPSS, and sums the minimum probabilities for terms that are common. The flattened version of the kPSS is simply a single list of terms. The *flatten* function is defined as:

Table 1. kPSS of order 3 for the word *"inserting"*

Order (n)	$n = 1$	$n = 2$	$n = 3$
SynSet $(word \rightarrow prob)$	$inserting \rightarrow 0, (3)$	$insert \rightarrow 0, 1(6)$ $insertings \rightarrow 0, (3)$	$inserted \rightarrow 0, 023$ $enter \rightarrow 0, 039$ $insertion \rightarrow 0, 063$ $inserts \rightarrow 0, 014$ (...)

$$flatten \quad :: kPSS \rightarrow [\,Term\,]$$
$$flatten\ kpss = \mathbf{concat}\ [\,synset\ order\ |\ order \leftarrow kpss\,]$$

Other scoring functions can be used to produced different ranks and mappings. The *levenshtein* function is another example, this calculates the score as the word distance between terms. Another function implemented in the system is the *match* function (this helps simulating techniques based on grep[10]), that simply returns 1 if the words match, or 0 otherwise. The next section describes the experimental evaluation done to empirically verify the advantages of the use of kPSS.

5   Experimental Validation

The previous sections describe the underlying technique used in the CONCLAVE system for feature location, based on kPSS. This section describes the preliminary evaluation done, to verify if this technique introduces benefits over other common techniques. In current available IDEs, common search facilities provided to the users, are still grep-like approaches, so the following research question was formulated:

RQ1: How does the *kpss* scoring function performs, when compared to the *match* scoring function, for finding relevant elements of the code given a search query?

To help answering this question the following experience was performed:

Step 1: in order to ease the process or replicating this experience the benchmark provided by Dit *et al*[11] for the jEdit[12] editor (version 4.3) was used, instead the devising a new data set. The benchmark contains a set of 150 bug reports, including the function set that was changed to resolve the bug (refered as the gold set) – more details about the benchmark in [12];

Step 2: the title for each bug report was extracted, stop words[13] were removed, and the resulting set was archived as keywords;

---

[10] http://www.gnu.org/software/grep/ (Last accessed: 29-01-2014).
[11] http://www.cs.wm.edu/semeru/data/benchmarks/ (Last accessed: 29-01-2014).
[12] http://www.jedit.org/ (Last accessed: 29-01-2014).
[13] Common words that tend to express poor semantics (e.g. *"the"*, *"a"*, *"too"*) [29].

Step 3: for each bug report, the *locate* function to compute a rank was called, using the *match* scoring function, the keyword set computed in *Step 2*, and setting as range the `Function` program element;

Step 4: replicate *Step 3* but using the *kpss* scoring function;

Step 5: calculate the effectiveness measure for each resulting rank.

The effectiveness measure is calculated by analyzing the computed rank in order, and its value is the first position of the rank that is a relevant function. Functions that are part of the set of functions changed to resolve the bug (the gold set) are considered relevant. The rank position can be compared for different scoring functions to measure which rank produced the best results. This approach was also used in [33] and [38] for comparing feature location techniques performance.

The results of this experience are presented in Table 2. They show that for this software package the kPSS based scoring approach produced a better result 55 times, outperforming the 22 better results achieved by the simple *match* function. The remaining times either both approaches scored the same, or none of the relevant functions were found in the resulting rank.

Table 2. Results of the experimental validation

Scoring Function	Analysed Bugs	Better Eff. Measure
match	150	22
kPSS	150	51

Although these results are satisfactory, they do not provide enough empirical data to generalize the performance of kPSS based techniques. Also, the keywords used to build the queries and the functions gold sets are a threat to validity because: (a) the keywords set was built automatically from reports titles that sometimes lack relevant terms, or use only ambiguous words (e.g. "*bug*"), a human would be more prone to devise a set of terms (after reading the report) that would create a more accurate rank; (b) sometimes, when fixing bugs, the actual defect is really not related to the concepts functions are addressing, which translates in changing code unrelated to search queries.

6   Conclusion

Many PC techniques benefit from mappings between the source code and problem domain concepts. These relations help the programmer quicker understand the source code, and discover which areas of the code need changing to address a specific feature or bug fix.

Many tools and techniques can be used to gather information about the program and the problem domain. Using ontologies allows the combination of heterogenous results and data in a single representation format. Applications can take advantage of a panoply of tools available (e.g. inference engines, descriptive logics, OTK-*like* frameworks), to perform data analysis and relate elements in different domains. kPSS based feature location is a sound example of such applications. The OTK framework for abstracting ontology operations from the underlying technology has proven a valuable asset during applications implementation.

The main trends for future work include devising new scoring functions, as well as combinations of approaches to improve results. And also, design practical experiments to compare CONCLAVE-MAPPER with other state-of-the-art feature location techniques.

References

1. Antoniol, G., Guéhéneuc, Y.-G.: Feature identification: An epidemiological metaphor. IEEE Transactions on Software Engineering 32(9), 627–641 (2006)
2. Bechhofer, S., Van Harmelen, F., Hendler, J., Horrocks, I., McGuinness, D.L., Patel-Schneider, P.F., Stein, L.A., et al.: Owl web ontology language reference. W3C Recommendation 10, 2006–01 (2004)
3. Biggerstaff, T.J., Mitbander, B.G., Webster, D.: The concept assignment problem in program understanding. In: Proceedings of the 15th International Conference on Software Engineering, pp. 482–498. IEEE Computer Society Press (1994)
4. Binkley, D., Lawrie, D.: Information retrieval applications in software maintenance and evolution. In: Encyclopedia of Software Engineering (2009)
5. Binkley, D., Lawrie, D.: Information retrieval applications in software development. In: Encyclopedia of Software Engineering (2010)
6. Carvalho, N.R., Almeida, J.J., Pereira, M.J.V., Henriques, P.R.: Probabilistic synset based concept location. In: SLATE 2012 — Symposium on Languages, Applications and Technologies (June 2012)
7. Chen, K., Rajlich, V.: Case study of feature location using dependence graph. In: 8th International Workshop on Program Comprehension. IEEE (2000)
8. Chikofsky, E.J., Cross II, J.H.: Reverse engineering and design recovery: A taxonomy. IEEE Software, 13–17 (1990)
9. Corbi, T.A.: Program understanding: Challenge for the 1990s. IBM Systems Journal 28(2), 294–306 (1989)
10. Deissenboeck, F., Pizka, M.: Concise and consistent naming. Software Quality Journal 14(3), 261–282 (2006)
11. Dit, B., Guerrouj, L., Poshyvanyk, D., Antoniol, G.: Can better identifier splitting techniques help feature location? In: IEEE 19th International Conference on Program Comprehension (2011)
12. Dit, B., Revelle, M., Gethers, M., Poshyvanyk, D.: Feature location in source code: a taxonomy and survey. Journal of Software: Evolution and Process 25(1), 53–95 (2013)
13. Eisenbarth, T., Koschke, R., Simon, D.: Locating features in source code. IEEE Transactions on Software Engineering 29(3), 210–224 (2003)
14. Furnas, G.W., Landauer, T.K., Gomez, L.M., Dumais, S.T.: The vocabulary problem in human-system communication. Communications of the ACM 30(11), 964–971 (1987)

15. Happel, H.-J., Seedorf, S.: Applications of ontologies in software engineering. In: Proc. of Workshop on Sematic Web Enabled Software Engineering (SWESE) on the ISWC, pp. 5–9. Citeseer (2006)
16. Hayashi, S., Yoshikawa, T., Saeki, M.: Sentence-to-code traceability recovery with domain ontologies. In: 2010 17th Asia Pacific Software Engineering Conference (APSEC), pp. 385–394. IEEE (2010)
17. Hill, E., Pollock, L., Vijay-Shanker, K.: Exploring the neighborhood with dora to expedite software maintenance. In: Proceedings of 22nd IEEE/ACM International Conference on Automated Software Engineering, pp. 14–23 (2007)
18. Hill, E., Pollock, L., Vijay-Shanker, K.: Automatically capturing source code context of nl-queries for software maintenance and reuse. In: Proceedings of the 31st International Conference on Software Engineering. IEEE (2009)
19. Horrocks, I., Patel-Schneider, P.F., van Harmelen, F.: From SHIQ and RDF to OWL: the making of a Web Ontology Language. Web Semantics: Science, Services and Agents on the World Wide Web 1(1), 7–26 (2003)
20. Keller, W.: Mapping objects to tables. In: Proc. of European Conference on Pattern Languages of Programming and Computing, Kloster Irsee, Germany, vol. 206, p. 207. Citeseer (1997)
21. Klyne, G., Carroll, J.J., McBride, B.: Resource description framework (rdf): Concepts and abstract syntax. W3C Recommendation, 10 (2004)
22. Lattner, C.: Llvm and clang: Next generation compiler technology. In: The BSD Conference, pp. 1–2 (2008)
23. Lawrie, D., Binkley, D.: Expanding identifiers to normalize source code vocabulary. In: 2011 27th IEEE International Conference on Software Maintenance (ICSM), pp. 113–122 (2011)
24. Lawrie, D., Morrell, C., Feild, H., Binkley, D.: What's in a name? a study of identifiers. In: 14th International Conference on Program Comprehension (2006)
25. Levenshtein, V.I.: Binary codes capable of correcting deletions, insertions, and reversals. Soviet Physics Doklady 10, 707–710 (1966)
26. Marcus, A., Sergeyev, A., Rajlich, V., Maletic, J.I.: An information retrieval approach to concept location in source code. In: Proceedings of the 11th Working Conference on Reverse Engineering, pp. 214–223. IEEE (2004)
27. Marcus, A., Rajlich, V., Buchta, J., Petrenko, M., Sergeyev, A.: Static techniques for concept location in object-oriented code. In: Proceedings of the 13th International Workshop on Program Comprehension, IWPC 2005, pp. 33–42. IEEE (2005)
28. Marcus, A., Rajlich, V.: Identification of concepts, features, and concerns in source code. In: Panel Discussion at the International Conference on Software Maintenance (2005)
29. Martin, J.H., Jurafsky, D.: Speech and language processing (2000)
30. Nelson, M.L.: A survey of reverse engineering and program comprehension. Arxiv preprint cs/0503068 (2005)
31. Parr, T.: The Definitive ANTLR 4 Reference. Pragmatic Bookshelf (2013)
32. Pérez, J., Arenas, M., Gutierrez, C.: Semantics and complexity of SPARQL. In: Cruz, I., Decker, S., Allemang, D., Preist, C., Schwabe, D., Mika, P., Uschold, M., Aroyo, L.M. (eds.) ISWC 2006. LNCS, vol. 4273, pp. 30–43. Springer, Heidelberg (2006)
33. Poshyvanyk, D., Guéhéneuc, Y.-G., Marcus, A., Antoniol, G., Rajlich, V.: Feature location using probabilistic ranking of methods based on execution scenarios and information retrieval. IEEE Transactions on Software Engineering 33(6), 420–432 (2007)

34. Prud'Hommeaux, E., Seaborne, A., et al.: Sparql query language for rdf. W3C Recommendation, 15 (2008)
35. Rajlich, V., Wilde, N.: The role of concepts in program comprehension. In: Proceedings of the 10th International Workshop on Program Comprehension, pp. 271–278. IEEE (2002)
36. Ratiu, D., Deissenboeck, F.: How programs represent reality (and how they don't). In: 13th Working Conference on Reverse Engineering, WCRE 2006, pp. 83–92. IEEE (2006)
37. Ratiu, D., Deissenboeck, F.: From reality to programs and (not quite) back again. In: 15th IEEE International Conference on Program Comprehension, ICPC 2007, pp. 91–102. IEEE (2007)
38. Revelle, M., Dit, B., Poshyvanyk, D.: Using data fusion and web mining to support feature location in software. In: 2010 IEEE 18th International Conference on Program Comprehension (ICPC), pp. 14–23. IEEE (2010)
39. Robillard, M.P.: Topology analysis of software dependencies. ACM Transactions on Software Engineering and Methodology (TOSEM) 17(4), 18 (2008)
40. Safyallah, H., Sartipi, K.: Dynamic analysis of software systems using execution pattern mining. In: 14th IEEE International Conference on Program Comprehension (2006)
41. Shepherd, D., Fry, Z.P., Hill, E., Pollock, L., Vijay-Shanker, K.: Using natural language program analysis to locate and understand action-oriented concerns. In: Proceedings of the 6th International Conference on Aspect-Oriented Software Development, pp. 212–224. ACM (2007)
42. Simões, A., Almeida, J.J., Carvalho, N.R.: Defining a probabilistic translation dictionaries algebra. In: XVI Portuguese Conference on Artificial Inteligence - EPIA, pp. 444–455 (September 2013)
43. Von Mayrhauser, A., Vans, A.M.: Program comprehension during software maintenance and evolution. Computer 28(8), 44–55 (1995)
44. Wilde, N., Buckellew, M., Page, H., Rajlich, V., Pounds, L.: A comparison of methods for locating features in legacy software. Journal of Systems and Software (2003)
45. Würsch, M., Ghezzi, G., Reif, G., Gall, H.C.: Supporting developers with natural language queries. In: Proceedings of the 32nd ACM/IEEE International Conference on Software Engineering, vol. 1 (2010)
46. Zhang, Y.: An Ontology-based Program Comprehension Model. PhD thesis (2007)
47. Zhao, W., Zhang, L., Liu, Y., Sun, J., Yang, F.: Sniafl: Towards a static noninteractive approach to feature location. ACM Trans. Softw. Eng. Methodol. 15(2), 195–226 (2006)

The Dominated Coloring Problem
and Its Application

Yen Hung Chen

Department of Computer Science, University of Taipei,
Taipei 10048, Taiwan, R.O.C.
yhchen@utaipei.edu.tw

Abstract. Given a graph $G = (V, E)$, a set $\mathcal{Z}$ of V is dominated by a vertex $v \in V$ if v is *adjacent* to all vertices of $\mathcal{Z}$. A proper coloring in G is an assignment of colors to each vertex of V such that any two adjacent vertices have different colors. A dominated coloring in G is a proper coloring such that each color class is dominated by a vertex in V. The dominated coloring problem is to find a dominated coloring $\mathcal{D}$ in G such that the number of color classes in $\mathcal{D}$ is minimized. In this paper, we provide the first possible application of the dominated coloring problem that is the development of interpersonal relationships in the social network. Since the dominated coloring problem has been listed to be NP-hard, we prove that the dominated coloring problem has a lower bound of $\frac{|V|^{(1-\epsilon)}+1}{1.5}$ on the best possible approximation ratio in general graphs, for any $\epsilon > 0$, and is NP-complete even when the graphs are bipartite graphs. Then we design the first non-trivial brute force exact algorithm to solve the dominated coloring problem in general graphs. Since the dominated coloring problem has a lower bound of $\frac{|V|^{(1-\epsilon)}+1}{1.5}$ on the best possible approximation ratio in general graphs, we present an $H(d)$-approximation algorithm that is faster but still exponential-time, where d is the maximum degree of G and $H(d)$ is the d-th harmonic number. The $H(d)$-approximation algorithm can run in polynomial time in bipartite graphs.

Keywords: computational complexity, approximation algorithm, non-trivial brute force exact algorithm, dominated coloring problem, development of interpersonal relationships in the social network.

1 Introduction

The *dominating set problem* [1–3] and *graph coloring problem* [4–6] are fundamental problems in graph theory, algorithms and combinatorial optimization. Given a connected, undirected graph $G = (V, E)$ with vertex set V and edge set E, a vertex $v \in V$ is said to be *adjacent* to a vertex $v' \in V$ if vertices v' and v share a common edge. A set $\mathcal{Z}$ of V is *dominated* by a vertex $v \in V$ if v is adjacent to all vertices of $\mathcal{Z}$. We also say that the vertex v is a dominating vertex of $\mathcal{Z}$. A dominating set Υ in G is defined to be a subset of V such that

B. Murgante et al. (Eds.): ICCSA 2014, Part VI, LNCS 8584, pp. 132–145, 2014.

every vertex in $V \setminus \Upsilon$ is adjacent to at least one vertex in Υ. The *dominating set problem* (DSP) is to find a dominating set with minimum cardinality [1–3]. We use γ to denote the cardinality of the optimal solution for the DSP. A partition of V is to divide all vertices in V into two or more disjoint classes that cover all vertices in V, where a class contains a non-empty subset of V. A *proper coloring* in G is an assignment of colors to each vertex of V such that any two adjacent vertices have different colors. In other words, a *proper coloring* in G is a partition of V into color classes so that no two vertices in the same class are adjacent. The *graph coloring problem* (GCP) is to find a proper coloring $\mathcal{C}$ in G such that the number of color classes in $\mathcal{C}$ is minimized [4–6]. We use χ (chromatic number of G) to denote the number of color classes of the optimal solution for the GCP. Both problems had been shown to be NP-hard [7] and many approximation algorithms [8–12] and exact algorithms [2, 4–6, 13–22] had been proposed. It had been shown that both problems have many important applications in facility location, social network, wireless and telephone switching network routing, scheduling, register allocation, map coloring, and so on [5, 18, 23–31].

Boumediene Merouane et al. [32] studied a combination of the DSP and GCP, called as the *dominated coloring problem*. A *dominated coloring* in a graph G is a proper coloring such that each color class is dominated by a vertex in V. The *dominated coloring problem* (DCP) is to find a dominated coloring $\mathcal{D}$ in G such that the number of color classes in $\mathcal{D}$ is minimized. Let $\chi_{\mathcal{D}}$ denote the number of color classes of the optimal solution for the DCP. Boumediene Merouane et al. [32] listed the DCP is NP-hard even when the number of color classes larger than 3, a bound of $\chi_{\mathcal{D}}$ (i.e., $max\{\chi, \gamma\} \leq \chi_{\mathcal{D}} \leq \chi \cdot \gamma$) in general graphs, and some bounds of $\chi_{\mathcal{D}}$ in special graphs. However, they did not provide the motivations or applications for the DCP. Hence, we describe a possible application (or motivation) for the DCP in the following scenario. In a social network [33, 34], a set of social actors (such as individuals) and a set of the dyadic ties between these actors, each actor (stranger) wants to develop interpersonal relationships in the social network by some intermediaries. We can model the social network to a friendship graph: each vertex as representing an actor (stranger or intermediary) and each edge as representing friendship between any two actors in social network (i.e., two actors are adjacent if they are friends). Two strangers can become friends by their mutual friend (intermediary). The DCP is to find the minimum stranger groups (or intermediaries) in the social network such that each stranger in the same group can become friends later by an intermediary in the social network.

Since only one research paper studies the DCP at present [32], the computational complexity, exact algorithms, and approximation algorithms for the DCP in general graphs or some special graphs are still unclear. In this paper, we focus on the computational complexity, exact algorithms, and approximation algorithms for the DCP in general graphs and bipartite graphs. It is not hard to see the DCP is NP-hard. In general graphs, we provide the first proof that the DCP is NP-hard to approximate within any ratio less than $\frac{|V|^{(1-\epsilon)}+1}{1.5}$ by *L-reduction* from the GCP, for any $\epsilon > 0$, since the GCP has been shown to be NP-hard to approximate within any ratio less than $|V|^{(1-\epsilon)}$ [35]. Then we design the first non-trivial brute force exact algorithm to solve the DCP in $O^*(2.4423^{|V|})$ time [1], and the first approximation algorithm with approximation ratio of $H(d)$ in $O^*(1.2114^d)$ time using polynomial space or $O^*(1.2108^d)$ time using exponential space, where d is the maximum degree of G and $H(d) = 1 + \frac{1}{2} + \frac{1}{d}$ is the d-th harmonic number. Note that $d \leq |V| - 1$ and $\ln(d) \leq H(d) \leq \ln(d) + 1$. In bipartite graphs, we show that DCP is also NP-complete, and the $H(d)$-approximation algorithm runs in polynomial time.

The rest of this paper is organized as follows. In Section 2, some definitions and notations are given. In Section 3, we prove the NP-completeness in bipartite graphs and a lower bound of $\frac{|V|^{(1-\epsilon)}+1}{1.5}$ on the best possible approximation ratio in general graphs for the DCP. In Section 4, we propose the first non-trivial brute force exact algorithm for the DCP in $O^*(2.4423^{|V|})$ time. In Section 5, we present an $H(d)$-approximation algorithm running in exponential-time (respectively, polynomial-time) for the DCP in general graphs (respectively, bipartite graphs). Finally, we make a conclusion in Section 6.

2   Preliminaries

In this paper, a graph is simple, connected and undirected. By $G = (V, E)$, we denote a graph G with vertex set V and edge set E. We use $|V|$ to denote the number of vertices of G. Let (v, v') denote an edge connecting two vertices v and v'. Foe any vertex $v \in V$ is said to be *adjacent* to a vertex $v' \in V$ if vertices v' and v share a common edge. The neighborhood $N(v)$ of $v \in V$ is defined to be the set of vertices adjacent to v. The *degree* of a vertex $v \in V$ is the number of vertices in $N(v)$ and the maximum degree d of G is the largest degree among all vertices in V. A set $\mathcal{Z}$ of V is *dominated* by a vertex $v \in V$ if $N(v) = \mathcal{Z}$. We also say that the vertex v is a dominating vertex of the vertex set $\mathcal{Z}$. An *independent set* I is a subset of V such that any pair of vertices $v, v' \in I$, $(v, v') \notin E$. A partition of V is to divide all vertices in V into two or more disjoint classes that cover all vertices in V, where a class contains a non-empty subset of V. A *proper coloring* in G is a partition of V into some color classes such that each color class is an independent set. A *dominated coloring* in G is a proper coloring such that each color class is dominated by one vertex in V. A bipartite graph $G_B = (L, R, E)$ is a graph whose vertex set can be partitioned into two independent sets L and R. A subgraph $G'' = (V'', E'')$ of $G = (V, E)$ is an *induced subgraph* by a vertex set V'', if $V'' \subseteq V$ and for all $v, v' \in V''$, $(v, v') \in E$ iff $(v, v') \in E''$. We use $IG(G, V'')$ to denote an induced subgraph of G by the vertex set V''. Now, we define some problems as follows.

---

[1] Throughout this paper we use a modified O notation that suppresses all polynomially bounded factors. For two functions f and g, we write $f(n) = O^*(g(n))$ if $f(n) = O(g(n) \cdot poly(n))$, where $poly(n)$ is a polynomial function in n, introduced by Woeginger [36].

GCP (Graph Coloring Problem)
Instance: A connected, undirected graph $G = (V, E)$.
Question: Find a proper coloring $\mathcal{C}$ such that the number of color classes in $\mathcal{C}$
   is minimized.

DCP (Dominated Coloring Problem)
Instance: A connected, undirected graph $G = (V, E)$.
Question: Find a dominated coloring $\mathcal{D}$ such that the number of color classes
   in $\mathcal{D}$ is minimized.

In this paper, we use the *L-reduction* to find a lower bound of $\frac{|V|^{(1-\epsilon)}+1}{1.5}$ on
the best possible approximation ratio for the DCP in general graphs , for any
$\epsilon > 0$. We review the definition of the *L-reduction* as follows.

Definition 1. *[37, 38] Given two optimization problems π and π', we say that
π L-reduces to π' if there exist polynomial time computable algorithms f and g
and positive constants α and β such that for every instance $\mathcal{I}$ of π, the following
properties are satisfied:*

1. *Algorithm f produces an instance $\mathcal{I}' = f(\mathcal{I})$ of π' such that $OPT(\mathcal{I}') \leq
   \alpha \cdot OPT(\mathcal{I})$, where $OPT(\mathcal{I})$ and $OPT(\mathcal{I}')$ are the costs of optimal solutions
   of $\mathcal{I}$ and $\mathcal{I}'$, respectively.*
2. *Given any solution of $\mathcal{I}'$ with cost c', algorithm g produces a solution of $\mathcal{I}$
   with cost c such that $|c - OPT(\mathcal{I})| \leq \beta \cdot |c' - OPT(\mathcal{I}')|$.*

The tuple (f, g, α, β) is called an *L-reduction* from π to π'.

Theorem 1. *[37, 38] If π L-reduces to π' with positive constants α and β, there
is a λ-approximation algorithm for π' if and only if there is a $(1 + \alpha \cdot \beta \cdot (\lambda - 1))$-
approximation algorithm for π.*

Since the GCP has been shown to be NP-hard to approximate within any
ratio less than $|V|^{(1-\epsilon)}$ [35], for any $\epsilon > 0$, we will prove the lower bound of
approximation ratio for the DCP in general graphs by an *L-reduction* from the
GCP and Theorem 1 in Section 3.1.

In order to show the NP-complete for the DCP in bipartite graphs, we *reduce*
the *set cover problem* (SCP) to the DCP in bipartite graphs in Section 3.2. The
SCP is defined as follows. Given a finite set $\mathcal{U}$ of elements and a collection $\mathcal{S}$ of
(non-empty) subsets of $\mathcal{U}$. A set cover is to find a subset $\mathcal{S}' \subseteq \mathcal{S}$ such that every
element in U belongs to at least one element of $\mathcal{S}'$.

SCP (Set Cover Problem)
Instance: A finite set $\mathcal{U}$ of elements and a collection $\mathcal{S}$ of (non-empty) subsets
   of $\mathcal{U}$.
Question: Find a set cover $\mathcal{S}'$ such that the number of sets in $\mathcal{S}'$ (i.e., $|\mathcal{S}'|$) is
   minimized.

In Section 4, we need to generate all *maximal independent sets* in these induced subgraphs $IG(G, N(v))$, for all $v \in V$. A *maximal independent set* in G is an independent set that is not a proper subset of any other independent set. There are at most $3^{\frac{|V|}{3}}$ maximal independent sets in a graph G and can be generated in $O(3^{\frac{|V|}{3}})$ time [39]. In Section 5, for each $v \in V$, we need to solve the *independent set problem* (ISP) in the induced subgraph $IG(G, N(v))$ (i.e., find the *maximum independent set* in the induced subgraph $IG(G, N(v))$). The ISP is defined as follows.

ISP (Independent Set Problem)
Instance: A connected, undirected graph $G = (V, E)$.
Question: Find an independent set I such that the number of vertices in I is maximized.

For the ISP, the time-complexity of the best-known exact algorithms are $O^*(1.2108^{|V|})$ [40] using exponential space and $O^*(1.2114^{|V|})$ [41] using polynomial space, respectively.

3   Hardness Results for the DCP

In this Section, we prove that the DCP has a lower bound of $\frac{|V|^{(1-\epsilon)}+1}{1.5}$ on the best possible approximation ratio in general graphs, for any $\epsilon > 0$. Then we show that the DCP is NP-complete even when the instances are bipartite graphs. In Subsection 3.1, we transform the GCP to the DCP by the *L-reduction* in general graphs. In subsection 3.2, we transform the SCP to the DCP by the *reduction* in bipartite graphs.

3.1   In General Graphs

Let f be the polynomial-time algorithm to *reduce* an instance $\mathcal{I}$ of the GCP to the instance $\mathcal{I}'$ of the DCP (i.e., $\mathcal{I}' = f(\mathcal{I})$). Let a graph $G = (V, E)$ be an instance $\mathcal{I}$ of the GCP with $V = \{v_1, v_2, \ldots, v_\ell\}$. Algorithm f transforms $\mathcal{I}$ into an instance $\mathcal{I}'$ of the DCP, say $G' = (V', E')$, as follows.

$V' = \{V\} \cup \{v_a\}$, where v_a is an auxiliary vertex.
$E' = \{E\} \cup \{(v_a, v_i) | 1 \le i \le \ell\}$.

Next, let g be another polynomial-time algorithm to *reduce* any solution of G' for the DCP to a solution of G for the GCP. Let $\mathcal{D}$ be a dominated coloring in G'. Algorithm g transforms $\mathcal{D}$ to a proper coloring $\mathcal{C}$ in G as follows.

Let $\mathcal{D}_1 \leftarrow \mathcal{D}$.
Delete the color class which contains v_a from $\mathcal{D}_1$.
Let $\mathcal{C} \leftarrow \mathcal{D}_1$.

Lemma 1. *For any dominated coloring $\mathcal{D}$ in G', Algorithm g transforms $\mathcal{D}$ into a proper coloring $\mathcal{C}$ in G and $|\mathcal{C}| = |\mathcal{D}| - 1$.*

Proof. Since $\mathcal{D}$ is proper coloring in G' and $N(v_a) = \{V' \setminus v_a\}$ in G', the vertex $\{v_a\}$ is a color class in $\mathcal{D}$. Hence, $\mathcal{C}$ is a proper coloring in G and $|\mathcal{C}| = |\mathcal{D}| - 1$ since $v_a \notin V$.

Next theorem shows that the DCP cannot be approximated within any ratio less than $\frac{|V|^{(1-\epsilon)}+1}{1.5}$ unless the GCP can be approximated within a ratio of $|V|^{(1-\epsilon)}$, for any $\epsilon > 0$.

Theorem 2. *The DCP is NP-hard to approximate within any ratio less than $\frac{|V|^{(1-\epsilon)}+1}{1.5}$ in general graphs, for any $\epsilon > 0$.*

Proof. We prove that the tuple $(f, g, 1.5, 1)$ constitutes an *L-reduction* from the GCP to the DCP by the following two inequalities.

(1)$OPT(\mathcal{I}') \leq \alpha \cdot OPT(\mathcal{I})$, where $\alpha = 1.5$. By Definition 1, $OPT(\mathcal{I})$ and $OPT(\mathcal{I}')$ are the cardinalities (the numbers of color classes) of the optimal solutions of $\mathcal{I}$ and $f(\mathcal{I})$ for the GCP and DCP, respectively. Since the vertex v_a is adjacent to all vertices in $\{V' \setminus \{v_a\}\}$, all color classes of the optimal solution of G for the GCP can be dominated by the vertex v_a. Hence, the union of the optimal solution of G for the GCP and adding a new color class $\{v_a\}$ which be dominated by any vertex in V is a feasible solution (dominated coloring) in G' for the DCP. Hence, the cardinality of the optimal solution in G' for the DCP is less than or equal to the cardinality of the optimal solution in G for the GCP plus one (i.e., $OPT(\mathcal{I}') \leq OPT(\mathcal{I}) + 1$). Since G is connected, $OPT(\mathcal{I}) \geq 2$. Hence, we have $OPT(\mathcal{I}') \leq 1.5 \cdot OPT(\mathcal{I})$.

(2) $(c - OPT(\mathcal{I})) \leq \beta \cdot (c' - OPT(\mathcal{I}'))$, where $\beta = 1$. Note that $\mathcal{D}$ denotes any dominated coloring in G' and $\mathcal{C}$ denotes a proper coloring in G by the output of the Algorithm g. By Definition 1, we have $c = |\mathcal{C}|$ and $c' = |\mathcal{D}|$, respectively. By Lemma 1, we have

$$|\mathcal{C}| - OPT(\mathcal{I}) - |\mathcal{D}| - 1 - OPT(\mathcal{I}) \leq |\mathcal{D}| - 1 - (OPT(\mathcal{I}') - 1)$$
$$= |\mathcal{D}| - OPT(\mathcal{I}'),$$

since $OPT(\mathcal{I}) \geq OPT(\mathcal{I}') - 1$. Therefore, we have $(c - OPT(\mathcal{I})) \leq (c' - OPT(\mathcal{I}'))$.

Since it is NP-hard to approximate the GCP within any ratio less than $|V|^{(1-\epsilon)}$ [35], the DCP has a lower bound of $\frac{|V|^{(1-\epsilon)}+1}{1.5}$ on the best possible approximation ratio by the above *L-reduction* and Theorem 1. $\square$

3.2 In Bipartite Graphs

In this Section, we show that the DCP is NP-complete even when the graphs are bipartite graphs.

Dominated Coloring Decision Problem
Instance: A connected, undirected bipartite graph $G_B = (L, R, E)$, and a positive integer κ.

Question: Does there exist a dominated coloring $\mathcal{D}$ such that the number of color classes in $\mathcal{D}$ is less than or equal to κ?

The dominated coloring decision problem is also called the κ-*dominated coloring problem* [32].

Now, we show that the dominated coloring decision problem is NP-complete by the *reduction* from the SCP, which is a common NP-complete problem [7]. We define the set cover decision problem as follows.

Set Cover Decision Problem

Instance: A finite set $\mathcal{U}$ of elements and a collection $\mathcal{S}$ of (non-empty) subsets of $\mathcal{U}$, and a positive integer κ'.

Question: Does there exist a set cover $\mathcal{S}'$ such that the cardinality of $\mathcal{S}'$ is less than or equal to κ'?

Theorem 3. *The dominated coloring decision problem is NP-complete even when the instance is a* bipartite graph *.*

Proof. First, it is easy to see that the dominated coloring decision problem is in NP. Then, we show the *reduction*: the transformation from the set cover decision problem to the dominated coloring decision problem. Let a finite set $\mathcal{U} = \{u_1, \ldots, u_n\}$ of n elements and a collection $\mathcal{S} = \{s_1, \ldots, s_m\}$ of (non-empty) subsets of $\mathcal{U}$, and a positive integer κ' be an instance of the set cover decision problem. Now, we construct an instance of the dominated coloring decision problem, say a bipartite graph $G_B = (L, R, E)$, and a positive integer κ, as follows.

$\kappa = \kappa' + 1$.
$L = \{v_{s_1}, \ldots, v_{s_m} | s_i \in \mathcal{S}, 1 \leq i \leq m\}$.
$R = \{v_{u_1}, \ldots, v_{u_n} | u_i \in \mathcal{U}, 1 \leq i \leq n\} \cup \{v_{r_a}\}$, where v_{r_a} is an auxiliary vertex.
$E = \{(v_{r_a}, v_{s_i}) | 1 \leq i \leq m\} \cup \{(v_{s_i}, v_{u_j}) | \text{ if } u_j \in s_i, 1 \leq j \leq n \text{ and } 1 \leq i \leq m\}$.

Now, we show that there is a set cover $\mathcal{S}'$ such that $|\mathcal{S}'|$ is κ' if and only if there is a dominated coloring in G_B such that the number of color classes is κ. Let a color class $dc(v)$ denote an independent set in G_B which is dominated by the vertex v.

(Only if) Assume that there is a set cover $\mathcal{S}' = \{s'_1, \ldots, s'_{\kappa'}\} \subseteq \mathcal{S}$, first we assign all vertices in L to a color class $dc(v_{r_a})$ which is dominated by v_{r_a} for the dominated coloring decision problem. Then we consider all vertices in R. For each $s' \in \mathcal{S}'$ and each element $u \in s'$ for the set cover decision problem, we assign the vertex $v_u \in R$ to the color class $dc(v_{s'})$ for the dominated coloring decision problem. Note that if two sets $\{s'_i, s'_j\} \in \mathcal{S}'$, $u \in s'_i$ and $u \in s'_j$, randomly assign v_u to one of color classes $dc(v_{s'_i})$ and $dc(v_{s'_j})$. Finally, assign the v_{r_a} to any color class of $dc(v_{s'})$ for $s' \in \mathcal{S}'$. Clearly, the union of the $dc(v_{s'})$, for all $s' \in \mathcal{S}'$, is a partition of R since $\mathcal{S}'$ is a set cover of $\mathcal{U}$. Hence, the union of the $dc(v_{s'})$ for all $s' \in \mathcal{S}'$ and $dc(v_{r_a})$ is a dominated coloring in G_B. Hence, we have a dominated coloring in G_B which the number of color classes is $|\mathcal{S}'|$ plus 1.

(If) Suppose that there exists a dominated coloring $\mathcal{D} = \{dc(v_1), \ldots, dc(v_\kappa)\}$ in G_B. Let $\tilde{\mathcal{D}}$ be a subset of $\mathcal{D}$ such that each vertex in R is contained in one of color classes in $\tilde{\mathcal{D}}$. For any pair of vertices $v_l \in L$ and $v_r \in R$, v_l and v_r must be assigned to different color classes in $\mathcal{D}$ by the definitions of the bipartite graph and the dominated coloring. Hence, we have $|\tilde{\mathcal{D}}| < |\mathcal{D}|$. By definition of the dominated coloring, each color class in $\tilde{\mathcal{D}}$ must be dominated by a vertex in L. Let $\Upsilon_R = \{v_{s'_1}, \ldots, v_{s'_\tau}\}$ be the set of dominating vertices of all color classes in $\tilde{\mathcal{D}}$. We have $|\Upsilon_R| \leq |\tilde{\mathcal{D}}| < |\mathcal{D}|$. Clearly, $\mathcal{S}' = \{s'_1, \ldots, s'_\tau\}$ is a set cover for the set cover decision problem and $|\mathcal{S}'| = |\Upsilon_R|$. If $|\Upsilon_R| < \kappa'$, we have a set cover $\mathcal{S}'$ whose cardinality is less than κ'. If $|\Upsilon_R| \geq \kappa'$, $\kappa - 1 = \kappa' \leq |\Upsilon_R| < |\mathcal{D}| = \kappa$. Hence, we have $|\mathcal{S}'| = \kappa'$. □

4   A Non-brute Force Exact Algorithm for the DCP

In this Section, we propose two exact algorithms for the DCP in general graphs. These algorithms are the first known exact algorithms for the DCP. First, a brute force exact algorithm can be designed as follows. For a graph $G = (V, E)$ and each vertex $v \in V$, we generate all maximal independent sets of the induced graph $IG(G, \{N(v) \cap V\})$ which can be dominated by v. Let the $AMI(V)$ denote the collection of all maximal independent sets of the induced graphs $IG(G, \{N(v) \cap V\})$ for all vertices $v \in V$. We transform the DCP to the SCP. An instance of SCP contains a finite set $\mathcal{U}$ of elements and a collection $\mathcal{S}$ of (non-empty) subsets of $\mathcal{U}$. We let each vertex $v \in V$ correspond to each element in $\mathcal{U}$, and each maximal independent set in $AMI(V)$ correspond to each set in $\mathcal{S}$. For each maximal independent set $I' \in AMI(V)$, if the vertex v is contained in I' for G, then the corresponding element of v in $\mathcal{U}$ also belongs to the corresponding set of I' in $\mathcal{S}$. Then we use the best-known exact algorithm for the SCP running in $O^*(1.2267^{|\mathcal{S}|+|\mathcal{U}|})$ time [20] to find a minimum set cover $\mathcal{S}' \subseteq \mathcal{S}$. Then for each corresponding set of $s' \in \mathcal{S}'$ in $AMI(V)$, we assign it to the optimal solution as a color class for the DCP. Note that if two sets $\{s'_i, s'_j\} \in \mathcal{S}'$, the element $u \in s'_i$ and $u \in s'_j$, randomly assign the corresponding vertex of u in V to one of corresponding sets of s'_i, s'_j in $AMI(V)$. Clearly, these corresponding sets of $s' \in \mathcal{S}'$ partition V into some independent sets which are dominated by the vertices in V. Moreover, the number of independent sets of this optimal solution is minimized for the DCP since $\mathcal{S}'$ is the minimum set cover for the SCP. However, for each vertex v, the number of all maximal independent sets of $IG(G, \{N(v) \cap V\})$ has at most $3^{\frac{|V|-1}{3}}$ since there are at most $3^{\frac{|V|}{3}}$ maximal independent sets in any graph $G = (V, E)$ [39]. Totally, we have at most $|\mathcal{S}| = |AMI(V)| = |V| \cdot 3^{\frac{|V|-1}{3}}$ sets for the SCP.

Next, we design a better and non-brute force exact algorithm for the DCP as follows. For a graph $G = (V, E)$ with a vertex set $\mathcal{Z} \subseteq V$, let $AMI(\mathcal{Z})$ denote the collection of all maximal independent sets of the induced graphs $IG(G, \{N(v) \cap \mathcal{Z}\})$ for all vertices $v \in V$. Clearly, each maximal independent set in $AMI(\mathcal{Z})$ can be dominated by a vertex in V. For a dominated coloring in G, one of the color classes can be assumed to a maximal independent set of $IG(G, \{N(v) \cap V\})$ for all vertices $v \in V$ by the above brute force exact algorithm. For a vertex set $Z \subseteq V$, a *dominated coloring of Z* is a partition of Z into color classes such that each color class is an independent set which is dominated by a vertex in V. For a vertex set $Z \subseteq V$, let $\mathcal{D}(Z)$ be a *dominated coloring of Z* such the number of color classes is minimize and $\chi_\mathcal{D}(Z)$ be the cardinality of the $\mathcal{D}(Z)$ (i.e., the number of color classes in $\mathcal{D}(Z)$). We can find the $\chi_\mathcal{D}(Z)$ in a graph G by the following recursion:

$$\chi_\mathcal{D}(Z) = \begin{cases} 1 + \min\{\chi_\mathcal{D}(Z \setminus I)|I \in AMI(Z)\} & \text{if } Z \neq \emptyset \\ 0 & \text{if } Z = \emptyset \end{cases} \tag{1}$$

For clarification, we describe the recursive procedure $\chi_\mathcal{D}(Z)$ as follows.

Procedure $\chi_\mathcal{D}(Z)$
1. Let $AMI(Z) \leftarrow \emptyset$ and $\chi_\mathcal{D}(Z) = \infty$.
2. **For** each each vertex $v \in V$ **do**
   Find all maximal independent sets from $IG(G, N(v) \cap Z)$ and then put these sets to $AMI(Z)$.
   end for
3. **For** each independent set $I \in AMI(Z)$ **do**
   If $\chi_\mathcal{D}(Z \setminus I) + 1 < \chi_\mathcal{D}(Z)$ **then** $\chi_\mathcal{D}(Z) \leftarrow \chi_\mathcal{D}(Z \setminus I) + 1$.
   end for

If we have all $\chi_\mathcal{D}(Z')$, $Z' \subset Z$, Procedure $\chi_\mathcal{D}(Z)$ runs in $O^*(3^{\frac{|Z|}{3}})$ time since this time bound to the number of all maximal independent sets in $AMI(Z)$ and there are at most $(|V| \cdot 3^{\frac{|Z|}{3}})$ sets in $AMI(Z)$. Let $\mathcal{D}_{OPT}$ be the optimal solution for the DCP and the power set $\mathcal{P}(Z)$ of any set Z be the set of all subsets of Z. Initially, $\mathcal{D}_{OPT}$ is empty. We describe our non-brute force exact algorithm for the DCP as follows.

Algorithm OPT-DCP
Input: A connected, undirected graph $G = (V, E)$.
Output: A dominated coloring $\mathcal{D}_{OPT}$ such that the number of color classes in $\mathcal{D}_{OPT}$ is minimized.
1. Let $\mathcal{D}_{OPT} \leftarrow \emptyset$.
2. **For** each vertex set Z in $\mathcal{P}(V)$ by non-decreasing order of the cardinality, resolving ties arbitrarily **do**
   2.1. Use Procedure $\chi_\mathcal{D}(Z)$ to compute $\chi_\mathcal{D}(Z)$.
   end for
3. Backtrack to find the optimal solution $\mathcal{D}_{OPT}$.

The backtracking in Step 3 is that when $\chi_\mathcal{D}(V) \leftarrow \chi_\mathcal{D}(V \setminus I) + 1$, I is assigned to a color class in $\mathcal{D}_{OPT}$. Hence, Step 3 takes at most $O(|V|)$ time. Clearly, each color class in $\mathcal{D}_{OPT}$ can be dominated by a vertex in V. Hence, the time-complexity of Algorithm OPT-DCP is dominated by the cost of Step 2. Finally, we have that Algorithm OPT-DCP runs in

$$O^*(\sum_{\mathcal{Z} \subseteq V} \chi_D(\mathcal{Z})) = O^*(\sum_{i=0}^{|V|} \binom{|V|}{i} \cdot |V| \cdot 3^{\frac{i}{3}}) = O^*((1 + 3^{\frac{1}{3}})^{|V|})$$

time, which is $O^*(2.4423^{|V|})$ time. Note that it takes $O(2^{|V|})$ space to store all $\chi_D(\mathcal{Z})$, $\mathcal{Z} \in \mathcal{P}(V)$.

Corollary 1. *Given a general graph* $G = (V, E)$, *there is an* $O^*(2.4423^{|V|})$ *time exact algorithm for the DCP.*

5   An Approximation Algorithm for the DCP

In this Section, we present an $H(d)$-approximation algorithm for the DCP, where d is the maximum degree of G and $H(d) = 1 + \frac{1}{2} + \frac{1}{d}$ is the d-th harmonic number. Clearly, $d \leq |V| - 1$ and $\ln(d) \leq H(d) \leq \ln(d) + 1$. Since the DCP in general graphs cannot be approximated within any ratio less than $\frac{|V|^{(1-\epsilon)}+1}{1.5}$ in polynomial time unless $NP = P$ by Theorem 2, the $H(d)$-approximation algorithm is faster but still exponential-time in general graphs. This approximation algorithm runs in $O^*(1.2108^d)$ time using exponential space and $O^*(1.2114^d)$ time using polynomial space for general graphs, and runs in $O(|V|^3)$ time for bipartite graphs. Let the vertex set $\widehat{V}$ be all vertices in V which are un-colored. Initially, $\widehat{V} \leftarrow V$. Our approximation algorithm is based on the greedy strategy: choose the maximum independent set among all induced subgraphs $IG(G, \{N(v) \cap \widehat{V}\})$ for $v \in V$ and assign all vertices in this independent set to a color class and then delete these vertices from $\widehat{V}$ until each vertex $v \in \widehat{V}$ is assigned. Let EA_{ISP} be the best-known exact algorithm for the ISP (i.e., find the maximum independent set). Given a graph $G = (V, E)$, let $\mathcal{D}_{APX}$ be a dominated coloring in G. Initially, $\mathcal{D}_{APX}$ is empty. Now, for clarification, we describe the $H(d)$-approximation algorithm for the DCP as follows.

Algorithm APX-DCP
Input: A connected, undirected graph $G = (V, E)$.
Output: A dominated coloring $\mathcal{D}_{APX}$.
1. $\mathcal{D}_{APX} \leftarrow \emptyset$ and a vertex set $\widehat{V} \leftarrow V$.
2. While $\widehat{V} \neq \emptyset$
    2.1. For each $v \in V$ **do**
        Use EA_{ISP} to find a maximum independent set I_v for the induced
        subgraph $IG(G, \{N(v) \cap \widehat{V}\})$.
    end for
    2.2. Select an independent set I_M with maximum cardinality among all I_v,
        $v \in V$ (i.e., $I_M = \max\{|I_v| | v \in V\}$.
    2.3. Put the independent set I_M to $\mathcal{D}_{APX}$ (i.e., I_M is assigned to a color
        class of $\mathcal{D}_{APX}$).
    2.4. $\widehat{V} \leftarrow \widehat{V} \setminus I_M$.

Clearly, each independent set I_M selected by Step 2.2 is dominated by a vertex in V and assigned to a color class in $\mathcal{D}_{APX}$ by Step 2.3. After performing Algorithm APX-DCP, all color classes in $\mathcal{D}_{APX}$ are a partition of V. Hence, $\mathcal{D}_{APX}$ is a dominated coloring in G for the DCP. Now, we also let $\mathcal{D}_{OPT}$ be the optimal solution for the DCP. Note that $|\mathcal{D}_{APX}|$ and $|\mathcal{D}_{OPT}|$ are the numbers of color classes (independent sets) in $\mathcal{D}_{APX}$ and $\mathcal{D}_{APX}$, respectively. For showing the approximation ratio of Algorithm APX-DCP, we assign a cost of 1 to each independent set selected by Step 2.2, and then split the cost evenly among all vertices in this set. The analysis is similar to the greedy algorithm for the SCP [42]. For each color class (independent set) I in $\mathcal{D}_{APX}$ and a vertex $v \in I$, we define that $p(v) = \frac{1}{|I|}$ is the *price* of v. Clearly, $|\mathcal{D}_{APX}| = \sum_{v \in V} p(v)$.

Lemma 2. *For each independent set I of G which is dominated by a vertex in V, $\sum_{v \in I} p(v) \leq H(|I|)$.*

Proof. Let $I_M(k)$ be the independent set which is selected by k-th iteration of Step 2 of Algorithm APX-DCP. Let a vertex set $V_A(k)$ be all vertices in $\mathcal{D}_{APX}$ after running the k-th iteration of Step 2. For any independent set I of G which is dominated by a vertex in V, we use $V_{\mathcal{R}}(k) = \{I \setminus V_A(k)\}$ to denote the set of remaining vertices in I after running the k-th iteration of Step 2. Initially, $V_{\mathcal{R}}(0) = I$. Let ρ be the least iteration such that $V_{\mathcal{R}}(\rho) = \emptyset$. Clearly, $V_{\mathcal{R}}(k) \subseteq V_{\mathcal{R}}(k-1)$ and the vertices in $V_{\mathcal{R}}(k-1) \setminus V_{\mathcal{R}}(k)$ are colored when $I_M(k)$ is selected. Hence, we have

$$\sum_{v \in I} p(v) = \sum_{k=1}^{\rho} \{(|V_{\mathcal{R}}(k-1)| - |V_{\mathcal{R}}(k)|) \cdot \frac{1}{|I_M(k)|}\}.$$

Since $I_M(k)$ is the maximum independent set which is dominated by a vertex in V after running the $(k-1)$-th iteration of Step 2, $|I_M(k)| \geq |I \setminus V_A(k-1)| = |V_{\mathcal{R}}(k-1)|$. Otherwise, the independent set I will be selected in the k-th iteration of Step 2. Hence, we have

$$\sum_{v \in I} p(v) \leq \sum_{k=1}^{\rho} \{(|V_{\mathcal{R}}(k-1)| - |V_{\mathcal{R}}(k)|) \cdot \frac{1}{|V_{\mathcal{R}}(k-1)|}\}.$$

Note that for two integers a, b and $a \leq b$, we have $H(b) - H(a) \geq \frac{b-a}{b}$ [42]. Hence, we have

$$\sum_{v \in I} p(v) \leq \sum_{k=1}^{\rho} \{(|V_{\mathcal{R}}(k-1)| - |V_{\mathcal{R}}(k)|) \cdot \frac{1}{|V_{\mathcal{R}}(k-1)|}\}$$
$$\leq \sum_{k=1}^{\rho} \{(H(|V_{\mathcal{R}}(k-1)|) - H(|V_{\mathcal{R}}(k)|)\}$$
$$\leq H(|V_{\mathcal{R}}(0)|) - H(|V_{\mathcal{R}}(\rho)|) = H(|I|).$$

$\square$

Theorem 4. *Algorithm APX-DCP is an $H(d)$-approximation algorithm for the DCP, where d is the maximum degree of G and $H(d)$ is the d-th harmonic number.*

Proof. We first analyze the time-complexity of Algorithm APX-DCP as follows. For each $v \in V$, Algorithm APX-DCP uses EA_{ISP} to find a maximum independent set which is dominated by v in Step 2.1. For each $v \in V$, EA_{ISP} can be done by Robson's Algorithm [40] in $O^*(1.2108^d)$ time using exponential space and by Bourgeois' Algorithm [41] in $O^*(1.2114^d)$ time using polynomial space for the induced subgraph $IG(G, \{N(v) \cap V\})$, $d \leq |V| - 1$. It takes $O(|V|^2)$ time for bipartite graphs. Then Step 2.2 (respectively, Step 2.3 and 2.4) takes $O(|V|)$ (respectively, $O(|I_M|)$) time. As mentioned above, the time-complexity of Algorithm APX-DCP is $O^*(1.2108^d)$ (respectively, $O^*(1.2114^d)$) using exponential space (respectively, polynomial space) for general graphs, and $O(|V|^3)$ for bipartite graphs.

Next, we prove the approximation ratio for Algorithm APX-DCP. For each independent set (color class) $I_o \in \mathcal{D}_{OPT}$, $\sum_{v \in V} p(v) = \sum_{I_o \in \mathcal{D}_{OPT}} \sum_{v \in I_o} p(v)$, since $\mathcal{D}_{OPT}$ is also a partition of V into independent sets of $\mathcal{D}_{OPT}$ such that each independent set is dominated by a vertex in V. Since $|\mathcal{D}_{APX}| = \sum_{v \in V} p(v)$, we have $|\mathcal{D}_{APX}| = \sum_{I_o \in \mathcal{D}_{OPT}} \sum_{v \in I_o} p(v)$. For any independent set I of G which is dominated by a vertex in V, we have $\sum_{v \in I} p(v) \leq H(|I|)$ by Lemma 2. As a result, we have

$$|\mathcal{D}_{APX}| = \sum_{I_o \in \mathcal{D}_{OPT}} \sum_{v \in I_o} p(v) = \sum_{I_o \in \mathcal{D}_{OPT}} H(|I_o|)$$
$$= |\mathcal{D}_{OPT}| \cdot H(max|I_o| \,|\, I_o \in \mathcal{D}_{OPT}) = H(d) \cdot |\mathcal{D}_{OPT}|,$$

since the cardinality of each independent set of G which is dominated by a vertex $v \in V$ is at most the degree of the vertex v.   □

6   Conclusion

In this paper, we have investigated the DCP and provided one possible application. For the computational complexity, we have proved that the DCP has a lower bound on best possible approximation ratio in general graphs and is NP-complete in bipartite graphs, respectively. For algorithms, we have proposed a non-brute force exact algorithm in $O^*(2.4423^{|V|})$ time. Finally, we have presented an $H(d)$-approximation algorithm running in exponential-time (respectively, polynomial-time) in general graphs (respectively, bipartite graphs), where d is the maximum degree of G and $H(d)$ is the d-th harmonic number. An Immediate direction for future research could involve improving the time complexity of the exact algorithm for the DCP. Another direction for future research is whether the DCP has a polynomial time approximation algorithm in general graphs.

References

1. Berge, C.: Theory and Graphs and its Applications. Methuen, London (1962)
2. Cockayne, E.J., Hedetniemi, S.T.: Towards a theory of domination in graph. Networks 7, 247–261 (1977)
3. Ore, O.: Theory of Graphs. American Mathematical Society Colloquium Publications (1962)
4. Biggs, N.J., Lloyd, E.K., Wilson, R.J.: Graph Theory 1736–1936. Oxford University Press (1999)
5. Chartrand, G., Zhang, P.: Introduction to Graph Theory. McGraw-Hill, Boston (2005)
6. Harary, F.: Graph Theory. Addison-Wesley, MA (1969)
7. Garey, M.R., Johnson, D.S.: Computers and Intractability: A Guide to the Theory of NP-Completeness. W.H. Freeman and Company, San Francisco (1979)
8. Bar-Yehuda, R., Moran, S.: On approximation problems related to the independent set and vertex cover problems. Discrete Applied Mathematics 9, 1–10 (1984)
9. Halldorsson, M.M.: A still better performance guarantee for approximate graph coloring. Information Processing Letters 45, 19–23 (1993)
10. Johnson, D.S.: Approximation algorithms for combinatorial problems. Journal of Computer and System 9, 256–278 (1974)
11. Slavik, P.: A tight analysis of the greedy algorithm for set cover. Journal of Algorithms 25, 237–254 (1997)
12. Wigderson, A.: Improving the performance guarantee for approximate graph coloring. Journal of the ACM 30, 729–735 (1983)
13. Bjorklund, A., Husfeldt, T., Koivisto, M.: Set partitioning via inclusion-exclusion. SIAM Journal on Computing 39, 546–563 (2009)
14. Byskov, J.M.: Enumerating maximal independent sets with applications to graph colouring. Operations Research Letters 32, 547–556 (2004)
15. Chang, G.J.: Algorithmic aspects of domination in graphs. In: Handbook of Combinatorial Optimization, vol. 3. Kluwer Academic Publishers, Boston (1998)
16. Eppstein, D.: Small maximal independent sets and faster exact graph coloring. Journal of Graph Algorithms and Applications 7, 131–140 (2003)
17. Fomin, F., Grandoni, F., Kratsch, D.: A measure & conquer approach for the analysis of exact algorithms. Journal of the ACM 56, 1–32 (2009)
18. Haynes, T.W., Hedetniemi, S.T., Slater, P.J.: Fundamentals of Domination in Graphs. Marcel Dekker, New York (1998)
19. Lawler, E.L.: A note on the complexity of the chromatic number problem. Information Processing Letters 5, 66–67 (1976)
20. van Rooij, J.M.M., Nederlof, J., van Dijk, T.C.: Inclusion/Exclusion meets measure and conquer: Exact algorithms for counting dominating sets. In: Fiat, A., Sanders, P. (eds.) ESA 2009. LNCS, vol. 5757, pp. 554–565. Springer, Heidelberg (2009)
21. van Rooij, J.M.M., Bodlaender, H.L.: Exact algorithms for dominating set. Discrete Applied Mathematics 159, 2147–2164 (2011)
22. West, D.B.: Introduction to Graph Theory. Prentice Hall, USA (2001)
23. Eubank, S., Kumar, V.S.A., Madhav, M.V., Srinivasan, A., Wang, N.: Structural and algorithmic aspects of massive social networks. In: Proceedings of the 15th Annual ACM-SIAM symposium on Discrete Algorithms (SODA 2004), New Orleans, LA, USA, pp. 718–727 (2004)
24. Friedman, R., Kogan, A.: Efficient power utilization in multi-radio wireless ad hoc networks. In: Proceedings of the International Conference on Principles of Distributed Systems (OPODIS 2009), Nimes, France, pp. 159–173 (2009)

25. Fritsch, R., Fritsch, G.: The Four Color Theorem: History, Topological Foundations and Idea of Proof. Springer, New York (1998)
26. Hooker, J.N., Garfinkel, R.S., Chen, C.K.: Finite dominating sets for network location problems. Operations Research 39, 100–118 (1991)
27. Kelleher, L.L., Cozzens, M.B.: Dominating sets in social networks. Mathematical Social Sciences 16, 267–279 (1988)
28. Mihelic, J., Robic, B.: Facility location and covering problems. In: Proceedings of the 7th International Multiconference Information Society, vol. D–Theoretical Computer Science, Ljubljana, Slovenia (2004)
29. Shamizi, S., Lotfi, S.: Register allocation via graph coloring using an evolutionary algorithm. In: Panigrahi, B.K., Suganthan, P.N., Das, S., Satapathy, S.C. (eds.) SEMCCO 2011, Part II. LNCS, vol. 7077, pp. 1–8. Springer, Heidelberg (2011)
30. Stecke, K.: Design planning, scheduling and control problems of flexible manufacturing. Annals of Operations Research 3, 3–12 (1985)
31. Wang, F., Camacho, E., Xu, K.: Positive influence dominating set in online social networks. In: Du, D.-Z., Hu, X., Pardalos, P.M. (eds.) COCOA 2009. LNCS, vol. 5573, pp. 313–321. Springer, Heidelberg (2009)
32. Boumediene Merouane, H., Haddad, M., Chellali, M., Kheddouci, H.: Dominated coloring of graphs. In: 11th Cologne-Twente Workshop on Graphs and Combinatorial Optimization (CTW 2012), Munich, Germany, pp. 189–192 (2012)
33. Borgatti, S.P., Mehra, A., Brass, D.J., Labianca, G.: Network analysis in the social sciences. Science 323, 892–895 (2009)
34. Freeman, L.C.: The Development of Social Network Analysis: A Study in the Sociology of Science. Empirical Press (2004)
35. Zuckerman, D.: Linear degree extractors and the inapproximability of max clique and chromatic number. Theory of Computing 3, 103–128 (2007)
36. Woeginger, G.J.: Exact algorithms for NP-hard problems: A survey. In: Jünger, M., Reinelt, G., Rinaldi, G. (eds.) Combinatorial Optimization - Eureka, You Shrink! LNCS, vol. 2570, pp. 185–207. Springer, Heidelberg (2003)
37. Kann, V.: On the Approximability of NP-complete Optimization Problems. PhD thesis, Department of Numerical Analysis and Computing Science, Royal Institute of Technology, Stockholm (1992)
38. Papadimitriou, C.H., Yannakakis, M.: Optimization, approximation, and complexity classes. Journal of Computer and System Sciences 43, 425–440 (1991)
39. Moon, J.W., Moser, L.: On cliques in graphs. Israel Journal of Mathematics 3, 23–28 (1965)
40. Robson, J.M.: Algorithms for maximum independent sets. Journal of Algorithms 7, 425–440 (1986)
41. Bourgeois, N., Escoffier, B., Paschos, V.T., van Rooij, J.M.M.: Fast algorithms for max independent set. Algorithmica 62, 382–415 (2012)
42. Cormen, T.H., Leiserson, C.E., Rivest, R.L., Stein, C.: Introduction to Algorithms, 3rd edn. MIT Press, Cambridge (2009)

Modularity Based Hierarchical Community Detection in Networks

Vinícius da F. Vieira[1,2], Carolina R. Xavier[1,2], Nelson F.F. Ebecken[1], and Alexandre G. Evsukoff[1,3]

[1] COPPE/UFRJ - Federal University of Rio de Janeiro, Rio de Janeiro, Brazil
[2] UFSJ - Federal University of São João del Rei, São João del Rei-MG, Brazil
[3] EMAp/FGV - Getúlio Vargas Foundation, Rio de Janeiro, Brazil

Abstract. The organization of nodes in communities, i.e., groups of nodes with many internal connections and few external connections, is one of the main structural features of networks and community detection is one of the most challenging tasks in networks. The communities in networks can be observed in different levels and a great number of methods can be found in the literature in order to identify the hierarchical organization of the communities. This work proposes a methodology for the representation of the hierarchical organization of communities in complex networks based on the spectral method of Newman. The proposed methodology, in contrast to other traditional approaches found in the literature, use the modularity, one of the most adopted measures for the quality of communities, in order to define the distances between the communities in the network. The methodology provides, as output, a dendrogram in order to illustrate the hierarchical organization of communities in networks. The application of the methodology to large scale networks show that the hierarchical visualization enhances the understanding of the complex systems modelled by networks, providing a broader view of the community structures.

1 Introduction

Many complex systems can be represented by networks, where the elements are "nodes" and the connections between them are "edges". Human societies are an example of complex systems, where persons (nodes) are related through affinities or some kind of social interaction (edges). Understanding the way that the relationships among the elements occur helps to reveal the behaviour of the entire system. In a network of social interactions, for example, the way that the relationships among people occur can directly determine how news are spread or how opinions are formed [22].

In many networks, the nodes can be organized as communities, i.e. a division of the network in groups that show a great density of internal connections and a low density of external connections [24]. Moreover a large number of methods to detect such communities can be found in literature. Community detection in large scale complex networks is a challenge task and many studies focusing this subject can be found in literature [24,30,3,20,8,10].

B. Murgante et al. (Eds.): ICCSA 2014, Part VI, LNCS 8584, pp. 146–160, 2014.
© Springer International Publishing Switzerland 2014

One of the most important properties of a network is the organization of its communities in hierarchy, i.e., the way that its nodes are organized in communities and, the communities are nested in greater communities (and so on). The analysis of the levels of organization in a network allows the exploration of one of the central organization principles in complex networks and offers a better understanding about the phenomenon modelled by the network [5].

Traditional methods for community detection are able to extract the organization of the vertices in groups (regardless of the adopted criteria for the definition of a group of vertices), however, they miss the multi-scale representation of the complex systems. The study of the communities hierarchy can be considered beyond the traditional study of communities and allows the analysis in all the modular levels - and not only one level - what enhances the concept of community structure.

Although it is not trivial concept, many complex systems are naturally organized in hierarchical levels. Many works can be found in the literature with the purpose of analyzing the hierarchical structure of many systems modelled by networks, such as collaboration networks [13,5], social networks [25,8,4,12], food web networks [13,5] and metabolic networks [14,26,18,5].

The concept of analyzing networks structures hierarchically is not new, and have been considered for decades [6]. Classical methods for data analysis based on hierarchical clustering [15] rely on the idea of successive divisions (or unions) of elements.

The analysis of the hierarchical organization of a network is frequently performed with a dendrogram, a graphical representation that is often used for classical data analysis [16,15,22]. The dendrogram is a tree in which each branch represents a level of agglomeration (or division) of elements. The lowest level (the leaves) represent the isolated elements and the higher level (the root) represents the entire group of elements. The elements are placed over the horizontal axis and the vertical axis represents the distance between each pair of elements (or group of elements). Each level of the tree indicates one level of elements union.

This works aims on presenting a methodology for the identification of the hierarchical organization of communities in complex networks. The methodology is based on Newman's spectral method [21], one of the most important methods for community detection reported in the literature. The proposed methodology identifies, for each network, the hierarchical structure, which is illustrated as a dendrogram, what enables a visual analysis of how the communities are formed in different levels.

The spectral method proposed by Newman, performs successive bisection of communities aiming on the maximization of modularity [24], one of the most adopted measures for assessing the quality of a partition of a network in communities. Modularity is based on the idea that the quality of a community increases as its density is distinguished from the observed density for its random version, keeping the same degree distribution.

The methodology proposed in this work takes advantage of some quantities obtained by Newman's spectral method at each performed bisection and uses the modularity variation caused for each operation in order to identify the hierarchy of communities. Thus, in this work, the modularity variation caused by the separation of two communities is interpreted as the distance among them. This distance is represented in the dendrogram, enabling a visual analysis to be performed in respect of the hierarchy of communities in the network.

The results obtained by the methodology proposed in this work, represented by the dendrograms, allow a more sophisticated analysis of the community structure in real world networks. This work uses a high performance implementation of Newman's spectral method [28], which allows it to be applied to networks in a scale above a million nodes.

2   Background and Related Work

2.1   Community Detection in Networks

It is a consensus that a community structure in a network can be identified when their is a division of the network in groups in which there is a high internal connections density and a low external connections density. One can argue that the sense of community becomes more evident when the difference between internal and external connection density increases, being necessary to define the criteria for measuring the quality of the partitioning of a network into communities. It is fundamental, therefore, to define a criteria to quantify the quality of a network division into communities. From the analysis of such criteria, it is possible to define methods that, without any previous information on the structure of the network, are able to differentiate networks with a well defined community structure from those that present a structure formed by random connections [24].

Consider a graph $G(\mathcal{V}, \mathcal{E})$ where $\mathcal{V}$ represents the set of n nodes and $\mathcal{E}$ represents the set of m edges. The degree of a node v_i, i.e., the sum of all of its connections, is denoted as k_i. Also consider the set $\mathcal{C}$ as a community structure (of size $|\mathcal{C}| = nc$), i.e, a partition of $\mathcal{V}$ in nc communities. In other words, $\mathcal{C}$ denotes a set of communities, in which each element (each community) is a set of nodes. In the present study, a community structure $\mathcal{C}$ is considered so that $\mathcal{C} = \{\mathcal{C}_a, a = 1...nc\}$, where $\mathcal{V} = \bigcup_{a=1}^{nc} \mathcal{C}_a$ and $\bigcap_{a=1}^{nc} \mathcal{C}_a = \emptyset$. In other words, this work considers each node belonging to one and only one community.

Among many measures for the quality of communities, the modularity proposed by Newman [24] is the most adopted and it is used in this work. The main idea behind modularity is that a vertex subset may be considered a community if the number of internal connections is greater than expected in a random formation with the same degree distribution.

The modularity Q of the division of a network in communities is defined as

$$Q = \frac{1}{2m} \sum_{ij} \left(A_{ij} - \frac{k_i k_j}{2m} \right) \delta(c_i, c_j), \tag{1}$$

where A is the adjacency matrix, $\delta(\cdot, \cdot)$ is the Kronecker delta (a function that return 1 if the operands are identical and 0 otherwise) and k_i is the degree of a node v_i. The term $\frac{1}{2}$ is used to avoid the double counting of edges.

2.2   Hierarchical Methods for Community Detection

The hierarchical organization in complex systems modelled by networks and the analysis of such structure has being receiving increasing interest by researchers and several works in the literature address this topic [23,19,12,5,27,14].

A major concern that arises when studying methods for the visualization of hierarchy in networks is the clear notion of how biased can be such methods. In other words, it is difficult to understand if the hierarchy is embedded in the community structure or if it is a result of the method itself [16]. Sales-Pardo describes some conditions to which methods for hierarchy identification should satisfy [27]:(i) it should be applicable to different kinds of networks; (ii) it should identify the different levels of hierarchy, as well as the modules at each level.

In one of the first works of community detection in networks, Girvan and Newman [12] present a methodology in which the hierarchical organization in which the communities are formed can be naturally obtained. The work of Girvan and Newman is based on the well known betweenness centrality [11], generalizing the concept for edges, instead of nodes. At each step, the method identifies and removes the edge with the largest betweenness centrality, until the network is divided in two disjoint subsets, interpreted as communities. The same strategy is, then, applied to each of the subsets, generating sub-communities. Thinking of the hierarchical organization, each step of the method identifies the edges placed at the boundary of two communities at a given level.

Following this idea, the authors graphically represent the successive divisions of a network by a dendrogram [12]. The higher level of the dendrogram (the root of the dendrogram) represents the undivided network. The network divisions are represented as nodes in the dendrogram. The leaves of the dendrogram represent the unary communities.

Despite the importance of the method proposed by Girvan and Newman in the field of community detection, some drawbacks of the approach proposed by Girvan and Newman [12] can be highlighted. First, the method needs to be executed until no more edges are left in the network and the output does not provide any information about at which level the community structure is better characterized. Moreover, the high number of leaves (the same as the number of nodes in the network) harms the visualization and the interpretation of the dendrogram in large scale networks. Furthermore, the execution of the method with large scale networks is impracticable due to the high execution time of the algorithm [12].

More recently, Clauset et al. [5] proposed an hierarchical random network model, similar to trees based network models. Based on a dendrogram and a set of probabilities p, the authors generate random networks with a predetermined structural properties. Then, the method creates a probabilistic model that exhibit structural properties similar to those observed in the studied networks. A discussion about the fitting of networks to probabilistic models is made by Newman [23], who makes an analogy between such methods and fitting numerical points to a function.

Other approaches for the identification of the hierarchical organization of networks can be defined from traditional methods for community detection. In the work in which it was proposed, the fast greedy method of Clauset, Newman and Moore [4] was adjusted exhibiting, as output, the community structure of the network. Illustrating the hierarchy as a dendrogram, the root (highest level) represent a single community, corresponding to the entire network and the leaves (lowest level) correspond to the unary communities. The same strategy is used by Danon et al. [8], in a work in which the authors propose a modification on the fast greedy method.

This work proposes a methodology for the identification of hierarchy in networks based in another traditional method for community detection: the spectral method of Newman. Therefore, the next section presents the basic principle of the method and, thus, the methodology for the hierarchy identification is presented.

2.3   Newman's Spectral Method for Community Detection

The spectral approach for the community detection problem, proposed by Newman [21], is based on the definition of a modularity matrix B, where each element B_{ij} can be described as

$$B_{ij} = A_{ij} - \frac{k_i k_j}{2m}. \tag{2}$$

Considering the division of the network in just two generic communities ($\mathcal{C}_a$ e $\mathcal{C}_b$), one can represent the communities in which each node belongs to a vector $s \in \{-1, 1\}^n$, such that a generic term is $s_i = +1$ if $v_i \in \mathcal{C}_a$ and $s_i = -1$ if $v_i \in \mathcal{C}_b$. Redefining the modularity Q (Equation 1) in terms of B as

$$Q = \frac{1}{4m} s^T B s, \tag{3}$$

and relaxing the vector s in a vector u which allows any real number, the solution of the modularity maximization problem can be obtaining by solving the eigenproblem

$$Bu = \beta u, \tag{4}$$

where β is the largest eigenvalue of B and u is its corresponding eigenvector.

The nodes of the network are assigned to each community according to the signs of the elements of u: the nodes corresponding to the positive elements of u are associated to a group and the nodes corresponding to the negative elements of u are associated to another group.

The same approach can be extended to the detection of several communities, by performing a successive bisection process. This idea is described by Newman in [21].

In addition, as suggested by Newman [21], the spectral method can be combined to a variation of the Kernighan-Lin method [17], as a post-processing stage after each bisection, improving significantly the quality of the resulting partition.

The method takes, as input, an initial division of a set of nodes in C_a and C_b and, among all elements of the input set of nodes, the method seeks for that one that, when moved (from C_a to C_b or from C_b to C_a) causes the largest increment (or the lowest decrement) in the modularity. This operations are performed repeatedly, with the constraint that each node can be only moved once. The method seeks, among all the intermediate states, the one that results in the largest modularity value.

This work incorporates a reduction in the number of nodes swapped among the communities. In the fine tuning proposed by Newman, the method searches the vertex which permutation causes the largest increase (or lowest decrease) in the modularity and this operation is repeated until all the vertices in the nodes subset have been moved. However, as proposed in a previous work [28], after a number of such operations, the modularity variation does not increase. In other words, the FT firstly moves the nodes that are less attached to their communities and, after some exchanges, just the more attached nodes remain. At this stage, there is a low probability that a node permutation will increase the modularity. By moving just a portion of the nodes, the method avoids unnecessary operations and reduces the execution time. An empirical result stated in [28] is that, after moving a number between 10% and 20% of the nodes, the modularity does not change. In this work, the fine tuning stage uses a threshold of 20% of nodes swaps.

3 Hierarchy Identification Based on Newman's Spectral Method

A feature that can be observed in many methods for community detection is the hierarchical order of operations, which can be agglomerative (bottom-up), like the method of Clauset, Newman and Moore [4] and the method of Blondel et al. [3], or divisive (top-down), like the method of Girvan and Newman [12] discussed in Section 2.2 and the spectral method of Newman, discussed in the Section 2.3.

A common approach explored in the literature, is to use the order of operations performed by the methods to illustrate the hierarchy of communities in networks. In other words, the only information depicted by the hierarchical representation of the communities, which can be represented as a dendrogram, is the order in which the operations occur. Some examples of this approach are the methods of Girvan and Newman [12] and method of Clauset, Newman and Moore [4].

On the other hand, divisive and agglomerative methods that aim on modularity optimization can reveal, at each step, important information concerning the obtained communities when the modularity variation (ΔQ) caused by each operation is taken into account.

In this way, the modularity gain obtained in the division of a generic community C_a in two other C_b and C_c can be interpreted as a distance measure between them. That is, the larger is the modularity gain obtained when dividing C_a in C_b and C_c, the larger is the distance between C_b and C_c in the network structure.

This section presents a methodology for the representation of the hierarchical organization of communities in complex networks based on the spectral method of Newman. The proposed methodology, in contrast to other traditional approaches found in the literature, use the modularity in order to define the distances between the communities in the network.

The hierarchy of communities in a network, according to the proposed methodology, is identified from information about the order of bisection operations performed by the spectral method of Newman, but it also considers the modularity variation caused by such divisions.

The different levels of organization of the communities are depicted as dendrograms. Thus, in order to be consistent to the traditional approach for hierarchical representation by means of dendrograms, the horizontal axis represent the communities and the vertical axis represent the distance between the communities, which is based on the modularity. Thus, the quality of the communities division in the dendrogram can be quantified, enhancing the visualization of the hierarchy of a network. The height of a division in the dendrogram is, thus, proportional to the modularity gain caused by this operation, which reflects the intuitive idea that a pair of communities with low affinity should be placed with a large distance in the dendrogram.

The root of the dendrogram represents the undivided network (at the beginning of the execution). When a division occurs, the modularity variation obtained by this operation is used as the height of a division in the dendrogram. Each leaf in the dendrogram represents a resulting community (at the end of the execution) and this level reflects the point of modularity peak.

This idea is explained in an intuitive way by Figure 1 present the dendrogram generated by the Zachary Karate Club network [31], which is frequently used in the literature as benchmark for community detection methods. The spectral method of Newman was implemented following the approach described in Section 2.3.

The spectral method of Newman performs in a divisive way, by successively bi-sectioning the communities. Figure 1 shows that, for the Zachary Karate network, three divisions (branches) were performed: Branch 1, Branch 2 and Branch 3. As illustrated by Figure 1, the Branch 2 causes a modularity variation $\Delta Q_2 = 0.016765$, and the Branch 3 causes a modularity variation $\Delta Q_3 = 0.030243$. This results suggest that the communities C_3 and C_4 have more affinity than communities C_1 and C_2. This intuitive notion is properly illustrated by the dendrogram, in which the height of the division of the communities C_1 and C_2 is greater than the height of division of the communities C_3 and C_4. On the other hand, the height of the Branch 1 (between $C_1 \cup C_2$ and $C_3 \cup C_4$) is visually greater than the others, which reflects the modularity gain caused by this operation ($\Delta Q_1 = 0.371795$).

The Zachary Karate network is known to have two ground-truth communities (as described in [31]). However, the modularity peak calculated by the spectral

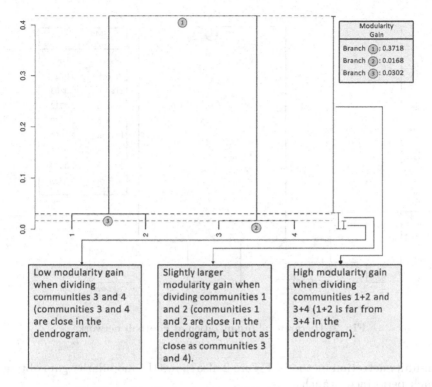

Fig. 1. Hierarchical organization of Zachary Karate network

method leads to the consideration of four communities. The hierarchical visualization allows the data analyst to notice that the Branches 2 and 3 are much less important than the Branch 1, and the division of the network in just two communities should be considered.

The previous example is quite simple, and a deeper analysis on the construction of the dendrogram should be done for situations in which the hierarchical organization shows a higher number of levels. For this end, it is important to highlight that the spectral method follows a top-down approach, while the dendrogram is built bottom-up. Taking the bottom-up way, the modularity gain caused by a branch in a certain level "pushes up" the higher levels in the same branch of the tree, i.e., the vertical axis value in which a division occurs does not reflect the modularity obtained by this isolated operation, but the sum of modularity gains obtained at each lower level. From that, the height in which each division occurs is equal to the sum of the modularity gains between this division and the leaves in its branch.

In order to better illustrate the dendrogram construction, Figure 2 presents the hierarchical organization obtained by the spectral method (as described in Section 2.3) for the Football network, also frequently used as benchmark for

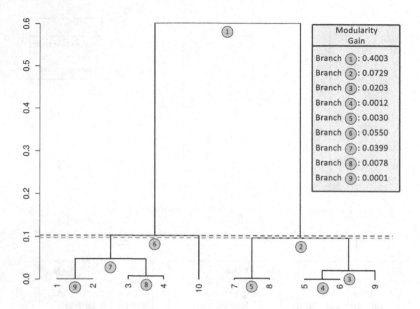

Fig. 2. Hierarchical organization of Football network

community detection methods. Figure 2 also shows the modularity gain obtained by each performed branch.

The analysis of the Branches 6 and 2, observed in Figure 2 allows a clear comprehension of the methodology proposed for the construction of the dendrogram. The height in which the Branch 6 occurs is 0.102924 and it is result of the sum of the modularity gains $\Delta Q_6 + \Delta Q_7 + \Delta Q_8 + \Delta Q_9$ (0.05503 + 0.039876 + 0.0.007856 + 0.000162). The height in which the Branch 2 occurs is 0.097492 and it is result of the sum of the modularity gains $\Delta Q_2 + \Delta Q_3 + \Delta Q_4 + \Delta Q_5$ (0.0729 + 0.0203 + 0.001245 + 0.003047).

In order to allow the construction of the dendrogram using the methodology described in this section, during the execution of the spectral method, a binary tree stores the information about the communities generated by each branch, as well as the modularity gain caused by each operation. By the end of the execution, the method performs a post-order depth-first search [7] that, from the leaves to the root, calculates the height in which each height must be represented (from the sum of modularity gains found in the lower levels of its branch). After that, a breadth-first search is performed (from the lowest to the highest level), in order to build the dendrogram from the union of the communities bottom-up.

It is worth to say that the execution of the dendrogram construction is much lower than the execution time of the spectral method itself, and it is reasonable to ignore it.

A positive feature of this methodology is that the data used for the construction of the hierarchical structure are obtained automatically during the execution of a traditional method for community detection. Thus, the proposed methodology for the identification of the hierarchical organization of communities in networks inherits various features of the spectral method of Newman in which the methodology is based, specially the quality of the obtained partitions and the execution time.

Another aspect that distinguishes the proposed methodology from other methodologies described in the literature comes from the top-down approach of the spectral method. In bottom-up methods, the leaves of the dendrogram represent the nodes of the network, harming the visualization for large scale networks. In the proposed methodology, the leaves of the dendrogram represent communities, allowing its application to large scale networks, since, normally, the number of communities is much lower than the number of nodes.

4   Experiments and Discussion

4.1   Computational Environment

The computational implementation of the spectral method of Newman was performed with C programming language and the mathematical library PETSc (Portable, Extensible Toolkit for Scientific Computation) [2]. The dendrograms were generated using R statistical analysis tool. All the experiments were executed in a PC laptop with a 64-bit Intel Core i7 2.3GHz CPU and 8Gb memory, using the Ubuntu 12.10 OS.

The methodology was test with a set of networks of different scales and contexts, which are frequently used as benchmark for community detection methods. Table 1 presents some of the benchmark networks, describing the number of nodes, the number of edges and a brief description of them.

The spectral method of Newman implemented in this work with the Fine Tuning stage modified to perform just 20% of nodes swap in each division (as described in Section 2.3), was applied to the networks presented in Table 1. The results for these executions are presented in Table 2, which describes the quality of the obtained partitions, expressed by the modularity Q and the execution time in seconds.

The results presented in Table 2, when compared to other results found in the literature by other authors and in previous works [1,9,14,29,28], show that the implemented method is able to detect communities with high modularity values in a reasonable execution time, considering a relaxed optimization method.

However, many aspects regarding the division of the networks in communities can not be captured only by the observation of the partition obtained at the modularity peak. Although the wide use of the spectral method of Newman, the hierarchical organization of the communities is not explored in the literature, what distinguishes the proposed methodology from other methodologies with the same purpose. In order to enhance the understanding about how the communities are hierarchically divided, providing a broader view of complex systems modelled

Table 1. Networks summary: number of nodes n, number of edges m and a brief description

Network	n	m	Brief Description
Dolphins[2]	62	159	Interaction network of dolphins.
Adjnoun[2]	112	425	Adjectives and nouns in a novel.
Lesmis[2]	77	254	Co-appearance in Les Miserables.
Email[3]	1133	5452	Email communication network.
C. elegans[3]	453	2040	Metabolic network.
Web-Berkeley-Stanford[4]	685230	6649470	Web network.
CA-AstroPh[4]	18772	198110	Collaboration network.
CA-HepPh[4]	12008	118521	Collaboration network.
Cit-HepTh[4]	27770	352324	Citation network.
Email-Enron[4]	36692	183831	Email communication network.
Amazon[4]	334836	925872	Amazon co-purchasing network.
Youtube[4]	1134890	2987624	Social network.

Table 2. Modularity Q and execution time (in seconds) with the spectral method

Network	Q	Time
Dolphins	0.5143	1.00×10^{-4}
Adjnoun	0.2915	1.75×10^{-2}
Lesmis	0.5443	1.00×10^{-2}
Email	0.5526	5.50×10^{-2}
C. elegans	0.4233	2.05×10^{-2}
Web-Berkeley-Stanford	0.9060	1.98×10^{4}
CA-AstroPh	0.5921	9.22×10^{1}
CA-HepPh	0.6403	5.30×10^{1}
Cit-HepTh	0.6244	2.16×10^{2}
Email-Enron	0.5957	2.40×10^{2}
Amazon	0.8694	2.56×10^{3}
Youtube	0.6930	2.42×10^{4}

as networks in different contexts, Figure 3 presents the dendrograms for each of the networks presented in Table 1, following the proposed methodology.

After the partition of the networks in communities, calculated by the spectral method, and the application of the proposed methodology, an analysis of the hierarchical organization can be performed based on the generated dendrograms (as depicted by Figure 3). Considering the Email network, the spectral method of Newman stops (i.e., refuses to keep dividing the sub communities), when the network is partitioned in ten communities (the point of modularity peak), which can be observed in Figure 3(a). However, a top-down analysis of the

---

[2] Downloaded from: http://www-personal.umich.edu/~mejn/netdata/

[3] Downloaded from: http://deim.urv.cat/~aarenas/data/welcome.htm

[4] Downloaded from: http://snap.stanford.edu/data/

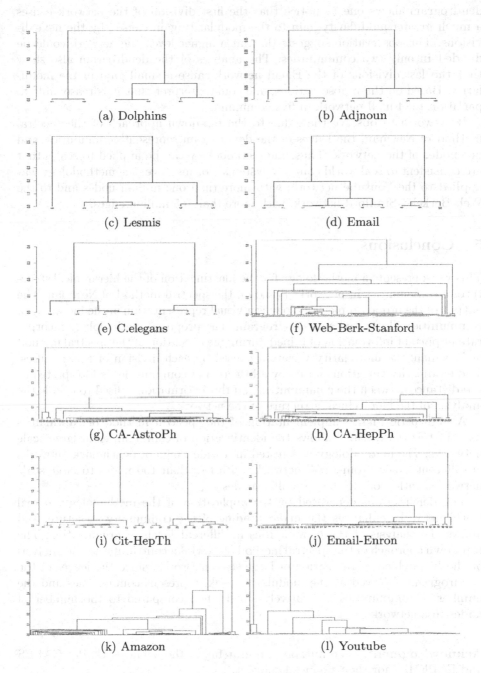

Fig. 3. Hierarchical visualization of networks: (a)Dolphins; (b)Adjnoun; (c)Lesmis; (d)Email; (e)C.elegans; (f)Web-Berkeley-Stanford; (g)CA-AstroPh; (h)CA-HepPh; (i)Cit-HepTh; (j)Email-Enron; (k)Amazon; (l)Youtube

dendrogram allows one to notice that the first division of the network causes a much greater modularity gain to the modularity gain caused by the next divisions. This observation suggests that, at a higher level, the network could be divided in only two communities. The analysis of the dendrogram also show that the last divisions of the Email network cause a small gain in the modularity. Based on these observations, one could interpret that it is reasonable to partition the Email network in five communities.

It is worth to observe that, due to the top-down approach of the spectral method of Newman, the leaves of the dendrogram represent communities, and not nodes of the network. Thus, the methodology can be applied to scales that are consistent to real world complex systems. For instance, the methodology was applied to the Youtube network, with more than one million nodes and to the Web-Berkeley-Stanford network, with more than 6.5 million edges.

5   Conclusions

This work presents a methodology for the identification of the hierarchical structure of communities in networks based on the spectral method of Newman. The output of the proposed methodology is a visual representation of the hierarchy of communities, illustrated as a dendrogram. The proposed methodology incorporate important information obtained during the execution of the spectral method of Newman: the modularity variation caused by each division of communities. The modularity variation caused by a bisection of communities is interpreted as the distance between the communities and this information is used to enable the analysis of the hierarchical structure of the network.

A high performance implementation of the spectral method of Newman is used in this work, which allows the identification of communities in large scale networks. The methodology was tested in a wide range of benchmark networks of different scales: from small networks with less than 100 nodes to large scale networks, with more than one million nodes.

The dendrograms generated by the application of the methodology to real world networks enhances the understanding of the community structure, and allows the analysis of the communities in different levels of organization. The top-down approach of the spectral method of Newman contributes to the analysis of the hierarchical organization of large scale networks, since the leaves of the dendrogram, observed at the modularity peak, represent communities and the number of communities is relatively small when compared to the number of nodes in a network.

Acknowledgment. The authors are grateful to the agencies CNPq, CAPES and FAPEMIG for their financial support.

References

1. Agarwal, G., Kempe, D.: Modularity-maximizing graph communities via mathematical programming. European Physical Journal B 66(3), 409–418 (2008)
2. Balay, S., Brown, J., Buschelman, K., Eijkhout, V., Gropp, W.D., Kaushik, D., Knepley, M.G., McInnes, L.C., Smith, B.F., Zhang, H.: PETSc users manual. Technical Report ANL-95/11 - Revision 3.3, Argonne National Laboratory (2012)
3. Vincent, D.: Blondel, Jean-Loup Guillaume, Renaud Lambiotte, and Etienne Lefebvre. Fast unfolding of communities in large networks. Journal of Statistical Mechanics: Theory and Experiment 2008(10) (2008)
4. Clauset, A., Newman, M.E.J., Moore, C.: Finding community structure in very large networks. Phys. Rev. E. 70(6), 066111 (2004)
5. Clauset, A., Moore, C., Newman, M.E.J.: Hierarchical structure and the prediction of missing links in networks. Nature 453(7191), 98–101 (2008)
6. Clauset, A., Moore, C., Newman, M.E.J.: Structural inference of hierarchies in networks. In: Airoldi, E.M., Blei, D.M., Fienberg, S.E., Goldenberg, A., Xing, E.P., Zheng, A.X. (eds.) ICML 2006. LNCS, vol. 4503, pp. 1–13. Springer, Heidelberg (2007)
7. Cormen, T.H., Leiserson, C.E., Rivest, R.L., Stein, C.: Introduction to Algorithms, 2nd edn. The MIT Press (2001)
8. Danon, L., Diaz-Guilera, A., Arenas, A.: Effect of size heterogeneity on community identification in complex networks. Journal of Stat. Mech.: Theory and Experiment 2006(11), 6 (2006)
9. Duch, J., Arenas, A.: Community detection in complex networks using extremal optimization. Physical Review E: Statistical, Nonlinear and Soft Matter Physics 72(2), 027104+ (2005)
10. Fortunato, S.: Community detection in graphs. Physics Reports 486, 75–174 (2010)
11. Freeman, L.: Centrality in social networks: Conceptual clarification. Social Networks 1(3), 215–239 (1979)
12. Girvan, M., Newman, M.E.J.: Community structure in social and biological networks. Proceedings of the National Academy of Sciences 99(12), 7821–7826 (2002)
13. Girvan, M., Newman, M.E.J.: Community structure in social and biological networks (December 2001)
14. Guimerà, R., Amaral, L.: Functional cartography of complex metabolic networks. Nature 433, 895–900 (2005)
15. Han, J., Kamber, M.: Data Mining: Concepts and Techniques, 1st edn. Morgan Kaufmann (2005)
16. Hastie, T., Tibshirani, R., Friedman, J.H.: The Elements of Statistical Learning. Springer (July 2003)
17. Kernighan, B.W., Lin, S.: An Efficient Heuristic Procedure for Partitioning Graphs. The Bell System Technical Journal 49(1), 291–307 (1970)
18. Cosentino, M., Lagomarsino, P., Jona, B.: Bassetti, and H. Isambert. Hierarchy and feedback in the evolution of the Escherichia coli transcription network. Proc. Natl. Acad. Sci. U S A 104(13), 5516–5520 (2007)
19. Lancichinetti, A., Fortunato, S., Kertesz, J.: Detecting the overlapping and hierarchical community structure of complex networks. New Journal of Physics (February 2009)
20. Leon-Suematsu, Y.I., Yuta, K.: Framework for fast identification of community structures in large-scale social networks. In: Data Mining for Social Network Data. Annals of Information Systems, vol. 12, pp. 149–175. Springer US (2010)

21. Newman, M.E.J.: Modularity and community structure in networks. Proceedings of the National Academy of Sciences of the United States of America 103(23), 8577–8582 (2006)
22. Newman, M.E.J.: Networks: An Introduction, 1st edn. Oxford University Press, USA (2010)
23. Newman, M.E.J.: Communities, modules and large-scale structure in networks. Nature Physics 8(1), 25–31 (2012)
24. Newman, M.E.J., Girvan, M.: Finding and evaluating community structure in networks. Physical Review E: Statistical, Nonlinear and Soft Matter Physics 69(2) (February 2004)
25. Radicchi, F., Castellano, C., Cecconi, F., Loreto, V., Parisi, D.: Defining and identifying communities in networks. Proceedings of the National Academy of Sciences of the United States of America 101(9), 2658–2663 (2004)
26. Ravasz, E., Somera, A.L., Mongru, D.A., Oltvai, Z.N., Barabasi, A.L.: Hierarchical organization of modularity in metabolic networks. Science 297(5586), 1551–1555 (2002)
27. Sales-Pardo, M., Guimerà, R., Moreira, A.A., Amaral, L.A.N.: Extracting the hierarchical organization of complex systems. Proceedings of the National Academy of Sciences 104(39), 15224–15229 (2007)
28. Vieira, V.F.: Detecção de Comunidades em Redes Complexas de Larga Escala. PhD thesis, Rio de Janeiro, RJ, Brazil (2013)
29. da Fonseca Vieira, V., Evsukoff, A.G.: A comparison of methods for community detection in large scale networks. In: Menezes, R., Evsukoff, A., González, M.C. (eds.) Complex Networks. SCI, vol. 424, pp. 75–86. Springer, Heidelberg (2013)
30. Wakita, K., Tsurumi, T.: Finding community structure in mega-scale social networks. Analysis 105(2), 9 (2007)
31. Zachary, W.W.: An information flow model for conflict and fission in small groups. J. Anthropol. Res. 33, 452–473 (1977)

Double Pointer Shifting Window C++ Algorithm for the Matrix Multiplication

Jerzy S. Respondek

Silesian University of Technology, Faculty of Automatic Control, Electronics and
Computer Science, Institute of Computer Science, Poland
jerzy.respondek@polsl.pl

Abstract. In this article we show how to utylize the pointer effectiveness in the
matrix multiplication algorithm. To achieve this we proposed an advanced pointer-oriented matrix multiplication numerical recipe. We involved the double
pointer matrix representation, with separate allocation of each row enabling this
way convenient element acces without multiplication. Finally, we performed
time tests which proved the high efficiency of the proposed pointer-oriented
paradigm.

Keywords: Numerical recipes, C++, Linear algebra, Pointers.

1 Introduction

The C++ programming language is based on the C programming language and
enables to create the object-oriented software with the speed of the C programs. The
C++ was designed by Bjarne Stroustrup [3,4]. The C programming language was
designed by Kernighan, Ritchie [2] as a highly effective tool for operating systems
designing. Apart from the operating systems, the majority of the severe contemporary
software is still programmed in the C++ language. The high efficiency of the software
coded in C++ follows, to a large extent, from C++ pointers. The pointers appear also
in other general purpose languages but in the C++ programs they are usually the sole
part of the algorithms code. The pointer-oriented C++ code closes the programming
style to the direct assembler programming causing its unmatched effectiveness.

The main objective of this article is to make use of the pointer effectiveness in one
of the fundamental numerical algorithms, i.e. in the matrix multiplication. The main
obstacle to this aim is iterating through a matrix column, because it requires jumping
over a separate tables. As a solution to this trouble I proposed a shifting window in a
form of a table of auxiliary double pointers.

Neither of the available monographs on numerical recipes makes use of the pointers, even if the algorithms they present are coded in C++. This probably follows from
the fact that in the past the numerical algorithms were coded with the use of such
languages as Fortran, Algol and Pascal. In those languages the pointers are used in a
very limited range. Thus in this article we want to show how we can improve the

B. Murgante et al. (Eds.): ICCSA 2014, Part VI, LNCS 8584, pp. 161–170, 2014.
© Springer International Publishing Switzerland 2014

execution time thanks to involving the C++ pointers in the numerical algorithms by the example of the matrix multiplication. We proposed the pointer-oriented matrix multiplication algorithm 4.3 that turned out to be significantly more efficient than the classic one.

The paper is organized as follows: chapter 2 provides the necessary theoretical background for the matrix multiplication problem and storing the matrices as a series of dynamic tables, in chapter 3 we present the general paradigm of the pointer programming, in chapter 4 we present the new fast C++ matrix multiplication algorithm, in chapter 5 we carry out its thorough performance tests, in chapter 6 we give some conclusions.

2   Theoretical Background

2.1   Matrices from the Mathematical Point of View

The sole matrix definition can be found e.g. in the book Bellman [1] p.37. The $m \times n$ -dimensional matrix A we will denote as $A = \left[a_{ij} \right]_{i=1,\dots,m,\ j=1,\dots,n}$. The objective of the article is to present a fast C++ algorithm for matrix multiplication. Thus let us first refer to the formal definition of the matrix multiplication.

Matrix Multiplication Definition (Bellman [1] p. 39)

Let us be given two matrices $A = \left[a_{ij} \right]_{i=1,\dots,m,\ j=1,\dots,n}$, $B = \left[b_{ij} \right]_{i=1,\dots,n,\ j=1,\dots,l}$. *As a result of the multiplication of the* A *matrix by the* B *matrix we define the matrix* $C = \left[c_{ij} \right]_{i=1,\dots,m,\ j=1,\dots,l}$ *with entries expressed by* (1):

$$c_{ij} = \sum_{k=1}^{n} a_{ik} b_{kj}, \quad i = 1,\dots,m, \quad j = 1,\dots,l \tag{1}$$

2.2   Storing the Matrices in the Memory as a Series of Dynamic Tables

The most frequently used and flexible way to store the matrix in the memory is the so-called double pointer structure. This topic is presented in the most general case, for the arbitrary order of the pointers, in the monograph [5]. Let us assume that we want to store an $m \times n$ -dimensional A matrix of the general data type T in a memory. The appropriate solution is:

- declare the double pointer of the type `T**`,
- assign to it the address of the m -element table of the `T*` type pointers,
- assign to each of the above single pointers the addresses of the n-element T-type tables.

We illustrated the idea of storing the matrices as a series of dynamic tables below.

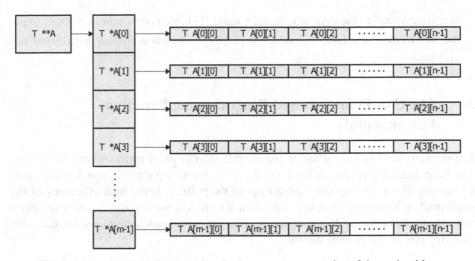

Fig. 1. Representation of the matrices in the memory as a series of dynamic tables

The proper C++ type definition code has the following form:

```
template <class T> struct TMatrix
{
  T **data;
  int rows,cols;
};
```

where *data* is the double pointer indicating the matrix data while *rows* and *cols* are the number of matrix rows and columns, respectively. The task of storing in a memory a $m \times n$-dimensional matrix A of the data type T can be performed by the following C++ code:

```
TMatrix A;

A.rows = m, A.cols = n;
A.data = new T* [m];
for(int i=0;i<m;i++)
  A.data[i] = new T [n];
```

Such a way of a computer matrix representation is used by all professional mathematical packages, like Matlab® and similar. It has two important advantages:

- The dimensions of the allocated matrix can be dynamically determined during the program execution. Thus there is no need to reserve an additional amount of memory during the code compilation; we use exactly as much memory as we need.
- We gain fast random access to each element of the allocated matrix. In order to get the value of the a_{ij} element of the A matrix the only code we have to write is A[i][j]. It is both concise and efficient. It does not need to perform any index

multiplication. On the contrary, in the simple 1D table matrix representation, to get the a_{ij} element we have to code: `A[i*n+j]` which is neither convenient nor efficient.

3   The C++ Pointer-Based Programming Paradigm Fundamentals

The pointers in C++ play a much greater role than in other programming languages. The fundamental C reference book [2] devotes a broad separate chapter 5 to the pointer notion. The most important advantage of the pointers is the high efficiency of the sequential table operations coded with their use. Below we showed, by the exemplary task of summing the table elements, its two opposite implementations: by the table indexing and by the pointer iteration.

A)
```
   float tab[1000],s=0.0;

   for(int i=0;i<100;i++)
     s += tab[i] ;
```
B)
```
   float tab[1000],*p=tab,s=0.0;

   for(int i=0;i<100;i++)
     s += *p++ ;
```

We can conclude the following reasons of the pointer-based programming execution time boost:

- In the table-based implementation (cell A) we have to determine the memory address for each summed table element in each loop iteration again and again in compliance with the expression: `&tab[i] = tab + i*`**`sizeof(float)`**. It is worth to notice that in the table implementation it is necessary to perform an additional integer multiplication in each loop iteration.
- In the pointer-based implementation (cell B), in order to sequentially sum up the table elements the only task we have to perform in each loop iteration (apart from the sole element addition) is to shift the pointer value to the next table element by a constant equal to **`sizeof(float)`**.

We illustrated the idea of pointer-based table access compared with the classic one in Fig. 2

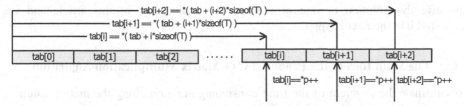

Fig. 2. The classic table access vs. pointer based table access

The main objective of this article is to apply the C++ pointer programming paradigm in order to obtain highly efficient matrix multiplication.

4 The Fast C++ Matrix Multiplication Algorithm

4.1 Problem Analysis

In item 3 we showed how to use the pointers in the sequential iterating through one dimensional table. The problem of the pointer iteration through two dimensional matrices is a more sophisticated one. We illustrated the necessary pointer paths in the matrix A by B multiplication in Fig. 3.

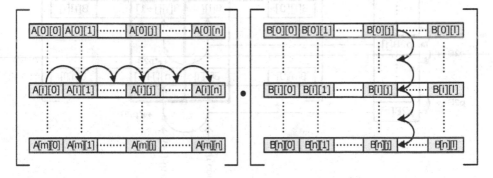

Fig. 3. The pointer iteration paths in the matrix multiplication process

We can observe the following:

- The pointer iteration along a fixed matrix row is a relatively simple task. This task is in fact equivalent to the iteration through the ordinary dimensional table explained in item 3. The simplicity of this task arises from the fact that a single matrix row occupies a coherent block in the operating memory.
- The problem significantly complicates when it comes to the pointer iteration along a matrix column. In this task we have to iterate over a series of the row tables. Each row is usually placed in a different memory location, as we presented in item 2.2.

The effective solution of that problem is the crucial part of this article and we presented it in the next chapter.

4.2   The Main Idea of the Proposed C++ Matrix Multiplication Algorithm

To eliminate the problem of the time consuming iteration along the matrix columns we proposed the following solutions:

- Introduction of an auxiliary pointer table p1B. In that table we store the addresses of the current column elements, initializing them by the addresses of the first column cells of the B matrix (code line 11 in item 4.3). Next, we sequentially increment each pointer value in the pointer table p1B (26th code line). This way we get a shifting memory window, enveloping a single matrix column. This idea is clearly illustrated in Fig. 4.

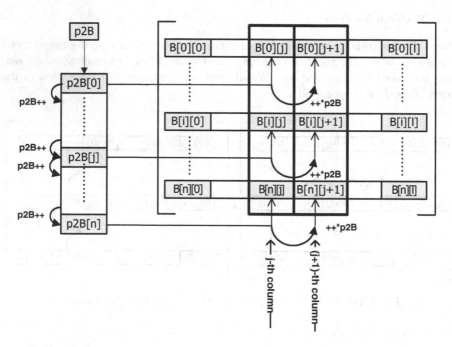

Fig. 4. Shifting the memory window

The most nested loop code (line 22) of the proposed algorithm deserves a closer look. It adds the subsequent scalar terms of the form $a_{ik}b_{kj}$, due to the definition (1). It has the following form:

```
temp += *p2A++ * **p2B++;
```
(2)

The sub-expression: `* **p2B++` performs 3 tasks at once:

— gets the value of the current B matrix cell (sub-expression `**p2B`),
— multiplies it by the value of the proper cell of the A matrix,
— finally post-increments the value of the `p2B` pointer, shifting it to the next matrix cell.

- Changing the order of calculations of the matrix product. The classic matrix multiplication algorithm calculates the elements of the $C = AB$ matrix in the natural row-by-row order. In the advanced multiplication algorithm we changed the calculation order of the $C = AB$ matrix to the column-by-column order. It enabled to minimize the number of necessary memory window shifting moves (26th code line) thanks to moving this operation to the least nested loop.

It is worth to notice that the code (2) requires no integer multiplication to determine the addresses of the a_{ik}, b_{kj} matrix cells. Instead we use the highly C++-oriented fast code involving advanced double pointer iteration. That is just the main advantage of the proposed C++ algorithm and the reason why it is so efficient.

4.3   The Final C++ Matrix Multiplication Algorithm Code

The innovations proposed in the previous item lead to the following, final C++ code for the fast matrix multiplication:

```
1 : T **p1B,**p1C;
2 :
3 : template <class T> // C=AB
4 : void Fast_Matrix_Multiply(Matrix<T> &A,Matrix<T>
&B,Matrix<T> &C)
5 : {
6 :     int rows1=A.rows,cols1=A.cols,cols2=B.cols;
7 :     int i,j,k;
8 :     T temp;
9 :     T**p1A,*p2A,**p2B=p1B,
        **p3B=B.data,**p2C=p1C,**p3C=C.data;
10:
11:     for(k=0;k<cols1;k++) *p2B++=*p3B++;
12:     for(k=0;k<rows1;k++) *p2C++=*p3C++;
13:
14:     for(j=0;j<cols2;j++)
15:     {
16:       p1A=A.data,p2C=Cp1;
17:       for(i=0;i<rows1;i++)
18:       {
19:         p2A=*p1A++,p2B=p1B;
20:         temp=0.0;
```

```
21:          for(k=0;k<cols1;k++)
22:            temp += *p2A++ * **p2B++;
23:          **p2C++ = temp;
24:        }
25:        p2B=p1B,p2C=p1C;
26:        for(k=0;k<cols1;k++)  ++*p2B++;
27:        for(k=0;k<rows1;k++)  ++*p2C++;
28:      }
29: }
```

The meaning of the variables is as follows:

— i, j, k	– loop working variables,
— temp	– the variable storing the temporary value of the scalar product,
— rows1, cols1, cols2	– number of the respective matrix rows and columns,
— A, B	– matrices to be multiplied,
— C	– result of the matrix multiplication, with respect to the formula $C = AB$,
— pxA ,pxB, pxC	– auxiliary pointers.

5   The Performance Tests

In this chapter we present the results of the performance tests we performed to verify the robustness of the proposed fast C++ matrix multiplication algorithm 4.3. The test was performed on the Intel™ Core i7 workstation under the Visual Studio™ Ultimate 2010 suite with the control of the Windows 7 64bit operating system. We stored the following:

• The time gain of the fast matrix multiplication algorithm (item 4.3) with respect to the classic one – the results are presented in Fig. 5 for the float matrix entries data type, for both types of the compiler speed optimizations: mild and high.
• The respective performance tests for the double data type matrix entries we presented on the Fig. 6.

We performed the benchmarks for the square matrices of the dimensions ranging from 1×1 up to 512×512.

We can observe that for both small-size, medium-size and large matrices the proposed fast C++ matrix multiplication algorithm 4.3 allows to shorten significantly the necessary calculation time.

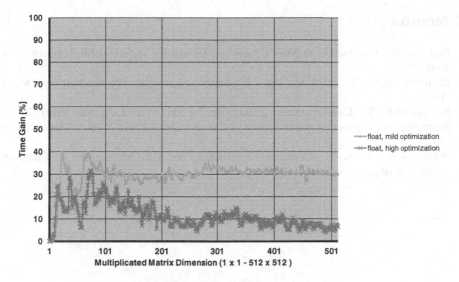

Fig. 5. Execution time gain for the float matrix entries data type

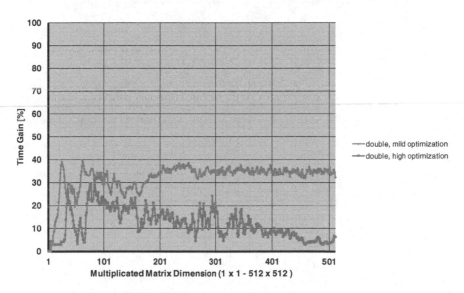

Fig. 6. Execution time gain for the double matrix entries data type

6   Summary

The main novelty of this article is that we proposed the fast C++ algorithm for the matrix multiplication. The C++ programming language appeared to be a powerful tool in the matrix multiplication task and generally in the numerical recipes field.

To sum up, we hope that many researchers and engineers will find the proposed ready-to-use algorithms useful in their work.

References

1. Bellman, R.: Introduction to Matrix Analysis. Society for Industrial Mathematics, New York (1987)
2. Kernighan, B.W., Ritchie, D.M.: The C Programming Language. Prentice Hall, New Jersey (1978)
3. Stroustrup, B.: The C++ Programming Language, 3rd edn. AT&T Labs, New Jersey (2000)
4. Stroustrup, B.: The Design and Evolution of C++, 9th edn. Addison-Wesley, Massachusetts (1994)
5. Waite, W.M., Goos, G.: Compiler Construction, 2nd edn. Monographs in Computer Science. Springer, New York (1983)

Dynamic Data Structures in the Incremental Algorithms Operating on a Certain Class of Special Matrices

Jerzy S. Respondek

Silesian University of Technology, Faculty of Automatic Control,
Electronics and Computer Science, Institute of Computer Science, Poland
jerzy.respondek@polsl.pl

Abstract. The article gives an incremental algorithm for calculating the inversion of the confluent Vandermonde matrix (CVM). We implemented all the incremental operations, i.e. adding, deleting and changing the single matrix parameter to avoid repeating the same calculations again and again. We obtained the quadratic logarithmic time complexity in the general case and the linear logarithmic complexity for the special case of small matrix parameters. Up to now, the fastest algorithms were of the cubic and quadratic class, respectively.

Keywords: Binary Tree, Data structures, Numerical recipes, Computational complexity.

1 Introduction

This paper gives two results in the CV special matrices topic:

— We derive a system of linear recursive equations for the inversion of the CVM.
— We construct a dynamical data structure to implement incremental algorithms for adding, removing and changing the value and/or multiplicity of a matrix parameter.

The traditional data structures used by the numerical algorithms, including algorithms operating on matrices, are the 1D table and the matrix. This topic is presented for the arbitrary number of dimensions in [19]. The exception to that is the algorithm for the N-body system simulation, classical problem in physics, whose implementation without the use of dynamic data structures gives weak $O(n^2)$ time complexity. The famous article *"An efficient program for many-body simulation"* ([2, 4 ch. 5.1]) proposed the adaptically changing tree which stored in its nodes the whole groups of bodies. Such an implementation of the N-body simulation algorithm with the use of tree data structure improved the time complexity to $O(n \log n)$. Afterwards it was shown in [6] that the obtained efficiency is even better i.e. $O(n)$. This problem still remains a subject of research, e.g. [3].

The objective of this work is to use a binary tree of the specially designed structure, together with a series of doubly-linked lists, to construct a set of incremental algorithms operating on the CVM. The implementation of all the incremental operations

B. Murgante et al. (Eds.): ICCSA 2014, Part VI, LNCS 8584, pp. 171–185, 2014.

in a manner that avoids repeating all the calculations anew results in a better time complexity than the one obtained by the ordinary algorithms. That was possible thanks to the flexibility of the binary tree and doubly-linked lists, which are traditional in computer science but rarely used by the numerical algorithms specialists.

2   Importance of the Confluent Vandermonde Matrices

The CVM arise in a broad range of both theoretical and practical problems. Below we surveyed the problems which need to operate on the CVM.

- Control problems: investigating the so-called controllability [17] of the higher order systems with multiple characteristic polynomial zeroes leads to the problem of inverting the CVM [13].
- Interpolation: apart from the ordinary polynomial interpolation with single nodes we consider the Hermite interpolation, allowing the multiple interpolation nodes. This augmented interpolation enables to predefine not only the sole data points values but also the derivatives of the interpolation polynomial in the data points. This way we gain better control over the interpolation curve, e.g. over its sloping and convexity in the data points. That problem leads to the system of linear equations, with the CVM ([9] pp. 363-373).
- Information coding: the CVM is used in coding and decoding information in the Hermitian code [10].

3   Comparison with Other Methods

In the literature the problem of inverting the CVM has been tackled several times:

- The article [8] propose a method which requires hand-held calculation of the fractional expansion of the characteristic polynomial inversion. The numerical part of the proposed methods is of the weak $O(n^4)$ class.
- The article [15] proposes algorithms of the $O(nr)$ order in general case, but it also requires the knowledge of all the matrix parameters at once.
- A step forward is made in the article [16], which enables to add new parameters, however it is still impossible to change or delete them.

To sum up, there are known algorithms operating on the CVM, but none of them enables to perform the incremental operations freely and efficiently. Moreover, some of the previous results cannot be even performed numerically as they require sophisticated symbolic calculations.

4   The Confluent Vandermonde Matrix Notion

Let $\lambda_1, \lambda_2, ..., \lambda_r$ be given real pair wise distinct zeroes of the polynomial $p(s) = (s - \lambda_1)^{n_1} ... (s - \lambda_r)^{n_r}$ with $n_1 + ... + n_r = n$. The CVM related to the zeroes of $p(s)$ is defined as the $n \times n$ matrix $V = [V_1\ V_2 ... V_r]$, where the block matrix $V_k = V(\lambda_k, n_k)$ is of order $n \times n_k$ having elements [8]:

$$[V(\lambda_k, n_k)]_{ij} = \begin{cases} \binom{i-1}{j-1} \lambda_k^{i-j}, & for\ i \geq j \\ \\ 0, & otherwise \end{cases} \qquad \lambda_k \in R \qquad (1)$$

for $k = 1, 2, ..., r$, $i = 1, 2, ..., n$ and $j = 1, 2, ..., n_k$. Examples of concrete matrices are given in [8] p. 1544, [15] p. 2050, [16] p. 725. If $n_1 = ... = n_r = 1$, we have the classic Vandermonde matrix. The article [8] presents theorem for inverting the CVM:

4.1   Theorem – Inverting the Confluent Vandermonde Matrices

The inverse of the CVM has the form $V^{-1} = \begin{bmatrix} W_1^T & W_2^T & \cdots & W_r^T \end{bmatrix}^T$. *The column vectors* h_{km} *of the block matrix* $W_k = [h_{kn}, ..., h_{k1}]$ *in the inverse* V^{-1} *may be computed by:*

$$\begin{cases} h_{k1} = \begin{bmatrix} K_{k,1} & \cdots & K_{k,n_k} \end{bmatrix}^T \\ h_{k(m+1)} = J_k(\lambda_k, n_k) h_{km} + a_m h_{k1}, & m = 1, 2, ..., n-1 \end{cases} \quad , k = 1, 2, ..., r \qquad (2)$$

where a_m *are the coefficients of the polynomial* $p(s)$, $J_k(\lambda_k, n_k)$ *is the elementary Jordan block, and* $K_{k,j}$ *are the auxiliary coefficients expressed by (3):*

$$K_{k,j}(s)\big|_{s=\lambda_k} = \frac{1}{(n_k - j)!} \frac{d^{(n_k - j)}}{ds^{(n_k - j)}} \left[\frac{1}{p(s)} (s - \lambda_k)^{n_k} \right]_{s=\lambda_k} , \quad \begin{matrix} j = n_k, ..., 1 \\ k = 1, 2, ..., r \end{matrix} \qquad (3)$$

5   Inversion by the Recursive Equation

The severe drawback of the theorem 4.1 operating on the CVM are symbolic derivatives necessary to calculate in order to determine the coefficients $K_{k,j}$ (3). It requires laborious algebraic transformations and thus cannot be performed numerically. The description of those auxiliary coefficients, without the necessity to calculate the derivatives, is given by the theorem 5.1 in the form of the recursive equations system.

5.1   Theorem – The Auxiliary Coefficients in the Linear Recursive Equations Form

The $K_{k,1},...,K_{k,n_k-1}$ *coefficients may be computed by the system of recursive, linear equations (4):*

$$qK_{k,n_k-q} + \left[\sum_{i=1,i\neq k}^{r}\frac{n_i}{\lambda_k-\lambda_i}\right]K_{k,n_k-q+1} - \left[\sum_{i=1,i\neq k}^{r}\frac{n_i}{(\lambda_k-\lambda_i)^2}\right]K_{k,n_k-q+2} +...+(-1)^{q+1}\left[\sum_{i=1,i\neq k}^{r}\frac{n_i}{(\lambda_k-\lambda_i)^q}\right]K_{k,n_k}=0 \quad (4)$$

for $k=1,2,...,r$, $q=1,2,...,n_k-1$ *and* λ_i,n_i *are the* $p(s)$ *polynomial zeroes with respective multiplicities. The* K_{k,n_k} *coefficients may be computed directly from the formula (3).*

Proof
The formula (4) can be rewritten in a more compact form (5):

$$qK_{k,n_k-q} = -\sum_{j=1}^{q}\left[(-1)^{j+1}K_{k,n_k-q+j}\sum_{i=1,i\neq k}^{r}\frac{n_i}{(\lambda_k-\lambda_i)^j}\right], \quad k=1,2,...,r, \ q=1,2,...,n_k-1 \quad (5)$$

With the use of the falling factorial[1] from (5) we can obtain:

$$q!K_{k,n_k-q} = -(q-1)!\sum_{j=1}^{q}\left[(-1)^{j+1}K_{k,n_k-q+j}\sum_{i=1,i\neq k}^{r}\frac{n_i}{(\lambda_k-\lambda_i)^j}\right] =$$
$$= -\sum_{j=1}^{q}\left[(-1)^{j+1}(q-1)^{\underline{j-1}}(q-j)!K_{k,n_k-q+j}\sum_{i=1,i\neq k}^{r}\frac{n_i}{(\lambda_k-\lambda_i)^j}\right] \qquad \begin{array}{l} k=1,2,...,r \\ q=1,2,...,n_k-1 \end{array} \quad (6)$$

By the auxiliary equality $K_{k,n_k-q+j} = \Big/_{(q-j)!} H_{k,n_k-q+j}$ the formula (6) can be transformed to the following form:

$$H_{k,n_k-q} = -\sum_{j=1}^{q}(-1)^{j+1}(q-1)^{\underline{j-1}}H_{k,n_k-q+j}\left[\sum_{i=1,i\neq k}^{r}\frac{n_i}{(\lambda_k-\lambda_i)^j}\right] =$$
$$= -\sum_{j=1}^{q}\sum_{i=1,i\neq k}^{r}\frac{n_i(-1)^{j+1}(q-1)^{\underline{j-1}}}{(\lambda_k-\lambda_i)^j}H_{k,n_k-q+j} \qquad \begin{array}{l} k=1,2,...,r \\ q=1,2,...,n_k-1 \end{array} \quad (7)$$

The upper limit of the internal sum is independent of the external sum variable, thus the summation order may be interchanged with one another. Observe:

---
[1] $x^{\underline{j}} = x(x-1)\cdot...\cdot[x-(j-1)]$, $\quad j\in N,\ x\in R$.

$$H_{k,n_k-q} = -\sum_{i=1,i\neq k}^{r}\sum_{j=1}^{q}\frac{n_i(-1)^{j+1}(q-1)^{\underline{j-1}}}{\left(\lambda_k-\lambda_i\right)^j}H_{k,n_k-q+j} =$$

$$= -\sum_{i=1,i\neq k}^{r}n_i\sum_{j=1}^{q}\frac{(-1)^{j+1}(q-1)^{\underline{j-1}}}{\left(\lambda_k-\lambda_i\right)^j}H_{k,n_k-q+j}, \quad k=1,2,...,r, \quad q=1,2,...,n_k-1$$

(8)

The final formula (8) is equivalent to the formula proved by the theorem 5.1 in the work [16] p. 720. Summing up, the theorem 5.1 enables to calculate the auxiliary coefficients $K_{k,j}$ (3) with no need to perform symbolic calculations $\square$.

6   Incremental Algorithm Operating on the CVM

The $O(n^2)$ order algorithms compared in section 3, in general, consist of the following two main steps:

- determine the last column of the target matrix,
- determine the remaining columns in the backward order, i.e. from $(n-1)^{th}$ to 1st.

By the theorem 4.1 the second step in the considered algorithm consists of a series of simple bi-diagonal matrix by vectors multiplications, so determining the $(n-1)^{th} \div 1^{th}$ columns requires at most two scalar multiplications and two scalar additions per element. The most time-consuming is the first step. The theorem 4.1 show that the last column to be calculated is given by the vector $\begin{bmatrix} K_{k,1} \cdots K_{k,n_k} \end{bmatrix}^T$ where $K_{k,j}$ are the coefficients (3). Let us rewrite the formula (3) in the following, more readable form:

$$K_{k,j}(s)\Big|_{s=\lambda_k} = \frac{1}{(n_k-j)!}\frac{d^{(n_k-j)}}{ds^{(n_k-j)}}\left[\frac{1}{(s-\lambda_1)^{n_1}}\cdots\frac{1}{(s-\lambda_{k-1})^{n_{k-1}}}\frac{1}{(s-\lambda_{k+1})^{n_{k+1}}}\cdots\frac{1}{(s-\lambda_r)^{n_r}}\right]_{s=\lambda_k}$$

(9)

We should keep in mind that:

- we have the arbitrary number r of parameters λ_k,
- each parameter has an arbitrary multiplicity n_k, for $k=1,...,r$,
- we need the arbitrary j^{th} order derivative in each λ_k point, for $k=1,...,r$.

Our objective is to construct efficient incremental algorithms for inverting the CVM on the basis of the derivation formula (9). To achieve this, we have to note that the operations of adding, deleting or changing the matrix parameter correspond to adding, deleting or changing the respective term of the general form $1/(s-\lambda_i)^n$ in the formula (9). The work [16] proposes the following recursive formula:

$$\frac{d^{j}}{ds^{j}}\left[\frac{1}{(s-\lambda_{r+1})^{n_{r+1}}}\cdot\frac{1}{\prod_{i=1,i\neq k}^{r}(s-\lambda_{i})^{n_{i}}}\right]_{s=\lambda_{k}}=\sum_{q=0}^{j}\binom{j}{q}\frac{(-1)^{q}(n_{r+1}+q-1)^{q}}{(\lambda_{k}-\lambda_{r+1})^{n_{r+1}+q}}\cdot\frac{d^{j-q}}{ds^{j-q}}\left[\frac{1}{\prod_{i=1,i\neq k}^{r}(s-\lambda_{i})^{n_{i}}}\right]_{s=\lambda_{k}} \tag{10}$$

In the subsequent sections we use the recursive formula (10) in combination with a properly designed binary tree to implement the incremental operations.

6.1   Data Structure

As the data structure to calculate the formula (9) we use a binary tree of a special construction. The tree must fulfil the following assumptions:

- In the leafs we store the data corresponding to a single λ_i matrix parameter i.e. the $0 \div (n_i - 1)$ order derivatives of the term $1/(s-\lambda_i)^{n_i}\big|_{s=\lambda_k}$, in compliance with the equality:

$$d^{j}\left[1/(s-\lambda_{i})^{n_{i}}\right]/ds^{j}\big|_{s=\lambda_{k}}=(-1)^{j}(n_{i}+j-1)^{j}/(s-\lambda_{i})^{n_{i}+j}\big|_{s=\lambda_{k}},\ \lambda_{k}\neq\lambda_{i},\ k=1,..,r \tag{11}$$

- In the internal nodes we combine the derivatives of the left and right child, using the formula (10).
- The tree root contains derivatives (9) which cover all the parameters $\lambda_1,...,\lambda_r$.

The tree should be balanced to make incremental operations efficient. However, the popular versions of balanced binary trees, i.e. red-black [7] or AVL [1], cannot be used because they allow to place the added element in an internal node, which violates the above presented assumptions. We use the nearly complete binary tree[2] (in the outside shape of a heap [20]), but different from the heap in terms of the organization of the data field contents, i.e. the derivatives (11) are stored in the leafs. Then an exemplary tree structure, in the case for $r = 6$[3] matrix parameters and $k = 6$, is presented in section 7, in Fig. 3. The formula (9) must be applied separately for each matrix parameter, so we have, in overall, r trees. By using the tree form data structure it is easy to implement the incremental operations on the matrix in a time-optimal way:

- To modify a given parameter λ_i, i.e. its value and/or multiplicity, we need to recalculate the nodes on the single path from the respective leaf to the root.

---

[2] A binary tree is *nearly complete*, if its leafs exist only on the last and last but one level. We treat the *complete* binary tree as the special case of the nearly complete binary tree, with its leafs only on the last level.

[3] In compliance with the equality (9), for a given k the kth parameter λ is omitted in the multiple product formula, so we have 5 leafs.

- To add a new parameter, it is required to restore the nearly complete binary tree shape. To achieve this we search for the first left leaf on the $h-1$ level, where h is the tree height (with the exception of the complete tree ([5] B.5.3) where we add the new node as the first left leaf), replacing that leaf by a new internal node. We attach the existing leaf to that new node as the left child, and the new parameter as the right child. Finally, we have to recalculate the path from the new leaf to the root.
- Deleting a matrix parameter must be designed in a way which preserves the tree in a proper form, so it deserves a closer look. For this purpose we define an auxiliary leaf-reduction operation, used to restore the tree to the nearly complete binary form (Fig. 1):

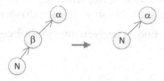

Fig. 1. The leaf reduction operation

The leaf reduction is valid for leafs which are the only children of their parent nodes. We perform it in two steps: we assign the parent field of its current parent node (β) to the parent pointer of the leaf to be reduced (N), then we remove the useless parent node (β). There are three cases how to delete the matrix parameter. Below, in Fig. 2, there is an example for 6 leafs:

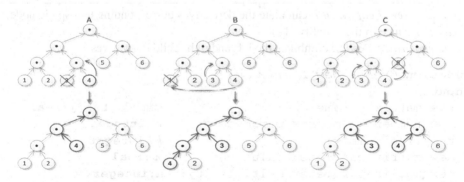

Fig. 2. Distinct cases of modifying the tree to delete the matrix parameters

(a) Deleting the last or one before last parameter on the deepest level h: delete the parameter node which belongs to the given parameter (here 3), reduce the second leaf of the common parent (here 4).
(b) Deleting other than last or one before last parameter on the deepest level h: assign to the parent field of the h level last parameter (here 4) the address of the parent of the leaf which is deleted (here 1), next reduce the new h level last leaf (now 3), finally remove the useless node.

(c) Deleting on the $h-1$ level: assign to the parent field of the h level last parameter the address of the parent of the leaf being deleted, delete the proper leaf and reduce the respective branch in a way presented in Fig. 2.

We bolded the paths which have to be recalculated. As we can see, in the cases B and C first we have to recalculate two paths from the leafs to the level where they meet, further there is a single path to the root. Please find below the key observation that allows us to use the simple data structures in tree nodes and avoid the use of dynamic data structures with random element access (e.g. dynamic tables or the lists with jumps [12]):

Observation. The formula (10) can be implemented in an incremental way by iterating all the parameters (i.e. the sum variable q, the derivative order, the falling factorial and the power) one after one. The Newton symbol can be iterated thanks to its property (12):

$$\binom{j}{q+1} = \frac{j-q}{q+1}\binom{j}{q}, \quad j, q \in N \tag{12}$$

Thus we choose to keep each tree node (both external and internal) as a doubly-linked list (e.g. [5] ch. 10.2). Each operation on the list work in $O(1)$ time. Each of the incremental operations on the matrix requires recalculating the respective path(s) of the tree, from the leaf to the root. A single step of that operation consists of respective one of the two subroutines:

- *Create_New_Leaf (λ , n):* calculate the derivatives corresponding to a single node, in compliance to the equality (11).
- *Internal_Node (Node):* combines the left and right child derivatives.

Function *Create_New_Leaf(λ , n)*
Input:
- the matrix parameters: lambda_tab[]:**real**
- the number of matrix parameters: r :**integer**
- the tree number: k :**integer**
- new matrix parameter value: λ :**real**
- new matrix parameter multiplicity: n:**integer**
Output:
- the list L containing the leaf data for a new node
```
0: begin
1:   aux ← 1 / (lambda_tab[k] - λ )^n;
2:   for j ← 0 to n-1 do
3:     L.push(aux);
4:     aux ← - aux * (n + j) / (lambda_tab[k] - λ );
5:   next j;
6:   return L;
```

```
Procedure Internal_Node(Node)
Input:
- the modified tree node: Node
- Node.left, Node.right - the left and the right child of
                             a given node, respectively
0: begin
1:   p←Node.data.front();
2:   pR_aux←Node.right.data.front();
3:   for j←0 to n-1 do
4:     pL←Node.left.data.front();
5:     pR←pR_aux;
6:     Node.data(p)←0;
7:     k←1;
8:     for q←0 to j do
9:       Node.data(p)←Node.data(p)+
                 +k*Node.left.data(pL)*Node.right.data(pR);
10:      pL←Node.left.data.next(pL);
11:      pR←Node.right.data.prev(pR);
12:      k←(j-q)/(q+1)*k ;
13:    next q;
14:    p←Node.data.next(p);
15:    pR_aux←Node.right.data.next(pR_aux);
16: next j;
```

In case of modifying the matrix parameter value, we use slightly modified procedure *Create_New_Leaf*, adding to the code the initial step 0: { p←L.front() } and replacing step 3 by the code: { L.data(p)←aux; p←L.Next(p); }.

We can summarize the above results by the following final theorems, determining the computational complexities of the proposed algorithm for the incremental operations:

6.2   Theorem – The Time Complexity

The time complexity, in general, is $O(rn_i^2 + \log r \sum_{k=1, k \neq i}^{r} n_k^2)$ for adding and modifying a matrix parameter, and $O(\log r \sum_{k=1, k \neq i}^{r} n_k^2)$ for deleting it; in the special case of all the n_k small matrix parameters, i.e. $n_k \leq n_{max} \ll n$ for $k = 1, ..., r$, the time complexity for each type of incremental operations decreases to $O(n \log r)$, where i is the number of the modified parameter, r is the overall number of the matrix parameters, n_{max} is a given small constant independent of n, n is the matrix dimension.

Proof

For each incremental operation we have to recalculate the leaf data first. That is performed by the formula (11), which can be calculated in an $O(n_k)$ time. Next we recalculate the tree path(s) by the formula (10). This operation requires to recalculate each node in the given path. Recalculating a single node requires the following number of iterations:

$$\sum_{j=0}^{n_k-1}(j+1) = \frac{(1+n_k)n_k}{2} = O(n_k^2), \quad k = 1,...,r$$

The height of the nearly complete binary tree is equal to $h = O(\log r)$ ([5] ch. 6), where r is the number of matrix parameters ($r \leq n$). Therefore, the time for a single tree path[4] can be estimated by $O\big((\log r)n_k^2\big)$. To add or modify a parameter, we also need to recalculate the whole i^{th} tree, so the time for those two operations over all trees is $O(rn_i^2 + \log r \sum_{k=1,k\neq i}^{r} n_k^2)$, and $O(\log r \sum_{k=1,k\neq i}^{r} n_k^2)$ for deleting a parameter.

This reasoning shows that in the case of all small multiplicities, often occurring in practice, the complexity reduces to $O(n \log r)$ □.

6.3   Theorem – The Space Complexity

The space complexity needed for the incremental operations on the matrix is $O(nr)$, where r is the overall number of the matrix parameters and n is the matrix dimension ($r \leq n$).

Proof

The nearly complete binary tree contains the following number of nodes:

$$O\left(r + \frac{r}{2} + .. + 1\right) = O\left(r\frac{1-(\tfrac{1}{2})^{\log_2 r+1}}{1-\tfrac{1}{2}}\right) = O(r)$$

Moreover, each node of the kth tree, $k = 1,..,r$, stores n_k derivatives of the scalar floating point type. Therefore the space needed for all trees is $O(r\sum_{k=1}^{r} n_k) = O(nr)$ □.

Table 1 features the summary of literature results on inverting of the CVM. Each time complexity is divided into two separate rows: the first one with the general case complexity (for any arbitrary parameter) and the second one for the case of small parameters. The space complexity in [15, 16] can be decreased to linear, because there is a need to keep only the intermediate results of the last iteration in the respective algorithms.

---

[4] In the cases B and C presented in Fig. 2 we have initially two paths, but the complexity remains the same.

Table 1. The complexity survey of the available algorithms for inverting of the CVM; n is the matrix dimension, r is the number of matrix parameters $1 \leq r \leq n$.

Source		Recalculating all anew	Modifying/deleting i-th parameter	Adding a parameter
		the respective operation in case $n_1, ..., n_r \leq n_{max} \ll n$		
A: [15]	Time	$O(nr)$		
		$O(nr)$		
	Space	$O(n)$		
B: [16]	Time	$\dfrac{O\left(r \sum_{k=1}^{r} n_k^2\right)}{O(nr)}$	impossible	$\dfrac{O\left(rn_r^2 + \sum_{k=1}^{r-1} n_k^2\right)}{O\left(rn_r^2 + n\right)}$
	Space	$O(n)$		$O(n)$
This paper	Time	$\dfrac{O(r \sum_{k=1}^{r} n_k^2)}{O\left(nr\right)}$	$\dfrac{O(rn_i^2 + \log r \sum_{k=1 \atop k \neq i}^{r} n_k^2)^*}{O\left(n \log r\right)}$	$\dfrac{O(rn_r^2 + \log r \sum_{k=1}^{r-1} n_k^2)}{O\left(n \log r\right)}$
	Space	$O(nr)$		

7   Numerical Example

Let us consider the CVM of the 7×7 size with the number of $p(s)$ polynomial nodes $r = 6$ of the form summarized in Table 2, with the modified 3rd node:

Table 2. Confluent Vandermonde matrix parameters

k	1	2	3 *(initial)/(modified)*	4	5	6
λ_k	1.500	−2.00	−1.00 /1.00	−0.500	3.700	2.00
n_k	1	1	1	1	1	2

7.1   Finding the Auxiliary Coefficients by the Recursive Equation

The recursive equation (4) is convenient if we know all the matrix polynomial $p(s)$ nodes at once. Let us show how it works for the sixth node:

---

[*] In the case of deleting the matrix parameter, the rn_i^2 term vanishes.

$$K_{6,2} = \left[\frac{1}{p(s)} (s-2.0)^2 \right]_{s\,=\,2.0} = -0.12; \; K_{6,1} + \left[\sum_{i=1}^{5} \frac{n_i}{(\lambda_6 - \lambda_i)^1} \right] K_{6,2} = 0.0 \Leftrightarrow \begin{cases} K_{6,2} = -0.12 \\ K_{6,1} = 0.36 \end{cases}$$

7.2 Finding the Auxiliary Coefficients by the Incremental Algorithm

Let us show how the incremental algorithm works by the example of modifying the third zero and $k = 6$. Each node of the respective tree has a doubly-linked list $D_k(\lambda_{i_1,\dots,i_a})$ which stores the derivatives. The tree has the form presented in Fig. 3:

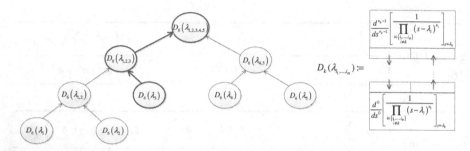

Fig. 3. The tree data structure for the exemplary matrix parameters

We bolded the path to be recalculated from the node corresponding to the modified third matrix parameter to the tree root. Table 3 presents the obtained values of the derivatives, in compliance with the formula (9). We underlined the received final values $(n_6 - j)!K_{6,j}$.

Table 3. Values of the derivatives in the nodes placed in the recalculated path

node order of the derivative	$D_6(\lambda_3)$		$D_6(\lambda_{1,2,3})$		$D_6(\lambda_{1,2,3,4,5})$	
	old	new	old	new	old	new
0	0.33	1.00	0.17	0.50	-0.04	<u>-0.12</u>
1	−0.11	-1.00	-0.43	-1.63	0.09	<u>0.36</u>

The final task is to execute the second equality in the recursive scheme (2). We can obtain the following inversion of the CVM:

$$V^{-1} = \left[W_1^T \; W_2^T \; \cdots \; W_6^T \right]^T = \begin{bmatrix} -7.69 & -1.77 & 22.18 & -12.44 & -2.99 & 3.22 & -0.52 \\ -0.01 & 0.01 & 0.02 & -0.04 & 0.02 & -0.01 & 0.001 \\ 3.65 & 2.06 & -9.05 & 3.90 & 1.37 & -1.10 & 0.16 \\ 0.38 & -0.92 & 0.69 & -0.05 & -0.16 & 0.07 & -0.01 \\ 0.01 & -0.002 & -0.05 & 0.04 & 0.001 & -0.01 & 0.002 \\ 4.65 & 0.62 & -13.80 & 8.59 & 1.75 & -2.17 & 0.36 \\ -1.31 & -0.08 & 3.93 & -2.66 & -0.43 & 0.67 & -0.12 \end{bmatrix}$$

8   The Performance Tests

In this section we present the results of the performance tests we carried out to compare the robustness of the proposed incremental algorithm, operating on the CVM, to other algorithms available in the literature, surveyed in tab. 1. The tests were performed on the Intel™ Core i7 workstation under the Visual Studio™ Ultimate 2010 suite with the control of the Windows 7 64bit. We inverted the special matrix of the dimension in the range $1 \div 350$ in the two opposite cases of the parameters: small multiplicities $n_i = i$ for $i = 1, 2, 3$, $n_i = n_i - 3$ ($n_{max} = 3$) for $i > 3$ (Fig. 4), and three zeroes with approximately equal large multiplicities $n_i = n / 3$ (Fig. 5).

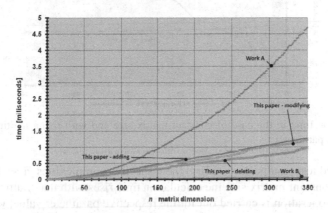

Fig. 4. Execution time in the case of small multiplicities

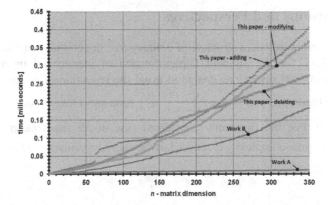

Fig. 5. Execution time in the case of small number of zeroes, each with a large multiplicity

The received test results confirm the complexities determined in a theoretical way, summarized in tab. 1. The algorithms proposed in this article are unmatched in the case of small zeroes multiplicities, which often occurs in practice, since in this case the time complexity is reduced to the linear logarithmic level. In Fig. 6 we present the efficiency comparison of the two incremental operations performed by the algorithms presented in this work, i.e. modifying and deleting a matrix parameter, in the case of constant matrix size and with the rising modified or deleted matrix parameter multiplicity $n_i = i$ for $i = 1, 2, .., r$, $r = 50$.

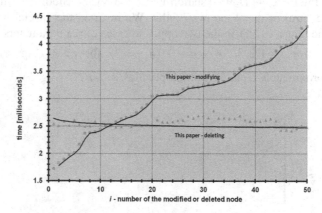

Fig. 6. Execution time in the case of constant matrix size and with rising multiplicity of the modified matrix parameter

The received test results also confirm the respective complexities given in tab. 1. In the case of a constant matrix size the calculation time rises with that multiplicity when the modifying operation is carried out for the respective parameter value, while such a rise of the calculation time cannot be observed when a parameter is deleted.

9   Conclusions and Perspectives

In this work we propose a data structure of a binary tree in combination with a series of doubly-linked lists, for the efficient implementation of incremental inverting of the CVM. The incremental operations enable to avoid the repetition of all the calculations anew, thus we obtained $O\left(rn_i^2 + \log r \sum_{k=1, k \neq i}^{r} n_k^2 \right)$ time and $O(nr)$ space complexity in the general case. For the frequently considered case of all small matrix parameters multiplicities, the time complexity is reduced to the linear-logarithmic $O(n \log r)$ level.

References

1. Adel'son-Vel'skii, G.M., Landis, E.M.: An algorithm for the organization of information. Soviet Mathematics Doklady 3, 1259–1263 (1962)
2. Appel, A.W.: An efficient program for many body simulation. SIAM J. Sci. Stat. Comput. 6(1), 85–103 (1985)
3. Augustyn, D.R., Warchal, L.: Cloud service solving N body problem based on Windows Azure platform. Comm. Comput. and Inf. Sci. 79, 84–95 (2010)

4. Bentley, J.: Programming Pearls. AT & T Laboratories, New Jersey (1986)
5. Cormen, T.H., Leiserson, C.E., Rivest, R.L., Stein, C.: Introduction to Algorithms, 2nd edn. McGraw Hill, Massachusetts (2001)
6. Esselink, K.: The order of Appel's algorithm. Inf. Proc. Lett. 41, 141–147 (1992)
7. Guibas, L.J., Sedgewick, R.: A dichromatic framework for balanced trees. In: 19th IEEE Annual Symposium on Foundations of Computer Science, pp. 8–21. IEEE Press (1978)
8. Hou, S., Pang, W.: Inversion of confluent Vandermonde matrices. Comput. and Math. with Appl. 43, 1539–1547 (2002)
9. Kincaid, D.R., Cheney, E.W.: Numerical Analysis: Mathematics of Scientific Computing, 3rd edn. Brooks Cole, California (2001)
10. Lee, K., O'Sullivan, M.E.: Algebraic soft-decision decoding of Hermitian codes. IEEE Trans. on Inf. Theory 56(6), 2587–2600 (2010)
11. Press, W.H., Teukolsky, S.A., Vetterling, W.T., Flannery, B.P.: Numerical Recipes in C, 2nd edn. Cambridge University Press, Cambridge (1992)
12. Pugh, W.: Skip lists: A probabilistic alternative to balanced trees. Comm. of ACM 33(6), 668–676 (1990)
13. Respondek, J.S.: Approximate controllability of the n-th order infinite dimensional systems with controls delayed by the control devices. Int. J. Syst. Sci. 39(8), 765–782 (2008)
14. Respondek, J.S.: On the confluent Vandermonde matrix calculation algorithm. Appl. Math. Lett. 24, 103–106 (2011)
15. Respondek, J.S.: Numerical recipes for the high efficient inverse of the confluent Vandermonde matrices. Appl. Math. and Comp. 218, 2044–2054 (2011)
16. Respondek, J.S.: Recursive numerical recipes for the high efficient inverse of the confluent Vandermonde matrices. Appl. Math. and Comp. 225, 718–730 (2013)
17. Sakthivel, R., Ganesh, R., Anthoni, S.M.: Approximate controllability of fractional nonlinear differential inclusions. Appl. Math. and Comp. 225, 708–717 (2013)
18. Spitzbart, A.: A Generalization of Hermite's interpolation formula. Am. Math. Mon. 67(1), 42–46 (1960)
19. Waite, W.M., Goos, G.: Compiler Construction, 2nd edn. Monographs in Computer Science. Springer, New York (1983)
20. Wiliams, J.W.: Algorithm 232 (Heapsort). Comm. of ACM 7, 347–348 (1964)

Interactive Data Structure Learning Platform

Estevan B. Costa[1], Armando M. Toda[1], Marcell A.A. Mesquita[2],
Fábio T. Matsunaga[1], and Jacques D. Brancher[1]

[1] Universidade Estadual de Londrina
[2] Centro Universitário do Pará

Abstract. The advent in technology in the past few years allowed an
improvement in the educational area, as the increasing in the development of educational system. One of the techniques that emerged in this
lapse is called Gamification, defined as the utilization of video game mechanics outside its bounds. Researchers in this area found positive results
in the application of these concepts in several areas from marketing to
education. In education, there are researches that covers from elementary
to higher education, with many variations to adequate to the educators
methodologies. Among higher education, focusing on IT courses, Data
Structures can be considered an important subject to be taught, as they
are base for many systems. Based on the exposed this paper describes
the development and implementation of an interactive web learning environment, called DSLEP (Data Structure Learning Platform), to support
students in higher education IT courses. The system includes basic concepts taught on this discipline as stacks, queues, lists, arrays, trees and
was implemented to receive new ones. The system is also implemented
with gamification concepts, as points, levels, and leader boards, to motivate students in the learning process and stimulate self-learning.

1 Introduction

The advent of technology in the latest years stimulated the development of
various techniques to improve and analyse human behaviour. This can be seen
on the emergence of a diversity of applications to monitor, analyse and manage
several social aspects in many areas, from health and business to educational.

In education, computers are able to assist teachers and students in several
ways from junior high to higher education. In the last one, there are many
researchers that aims to minimize the evasion and failure rate, mainly focused
on IT courses where these statistics are higher.

In order to prevent that, studies were made based on the development and
analysis of tools to aid students, focusing on Programming lessons [2]. As for
Data Structures lessons, which are also an important subject to IT courses, and
sometimes considered an extension of Programming lessons.

Data Structures is considered an important subject in many IT syllabus, as it
involves concepts of non-trivial algorithms that are utilized within many systems.
Improving the learning of these concepts is crucial to enhance the formation of
professionals in the area [12].

B. Murgante et al. (Eds.): ICCSA 2014, Part VI, LNCS 8584, pp. 186–196, 2014.
© Springer International Publishing Switzerland 2014

In order to improve the educational process, it is necessary to motivate and engage the student. This can be done with the aid of Gamification concepts, defined as the utilization of game mechanics outside its bounds to improve and modify user behaviour [3], [7], [14].

The utilization of those concepts is achieving positive results in the past few years, and studies comproved that those mechanics can be used within several areas [6]. The outcomes of the Gamification tied with educational methods proved to be efficient in many studies ranged in junior to higher education [11],[5] [4].

The observations exposed above stimulated the main objective of this work, which aims to develop an Interactive Data Structures Learning Platform. This tool aims to improve the learning process for the lessons, addressing some main concepts as stacks, queues, lists, trees and arrays.

Among those are also the Gamification concepts that aims to improve the interactions between users and the platform, in order to engage and motivate them to study by themselves. Also, it is important to point that this tool is not made to replace the Professor but to aid in the learning processes and methodologies of their disciplines.

This paper is divided as follows: section 2 describes some concepts involving gamification in education through motivation theories. Section 3 follows describing the project, architecture and tools utilized in the system and section 4 describes the results obtained by implementing it. Finally on section 5 there are the conclusions, discussions and future works of this project.

2   Gamification

The concept of gamification consists in utilizing game mechanics outside of its bounds with the perspective to improve user engagement [6] [7] [14]. It is possible to find some work reporting successful cases in areas as economy, marketing and health [6],[7].

According to [6] the utilization of these concepts must be well planned in order to obtain a positive effect and to not let the users get addicted to it. The author also focus on three major areas that are directly affected by Gamification in education: cognition, emotional and social. This happens because these areas influence the success or failures in a students academical progress [10].

The cognition area comprises the interactions of the user with the systems. As the games have a set of rules, the user is stimulated to learn these through experimentation, trial and error. During classes, the student may become reticent since the chances of failure are generally higher [6] [9].

These methods of experimentation generates the emotional transformation, as the user is compelled to overcome his challenges in a game. The transformation occurs by changing a positive emotion into a negative one and vice versa. For example, when frustration becomes pride after overcoming an obstacle inside the world of the game.

The social area is explored through the interactions which are made inside the game. As the player progresses their virtual avatar need to make choices that will influence in their actions and the others around them. [9] and [6] also states that the user may identify himself as a scholar while "playing" the school game which most of the doesn't feel like it.

These theories can be reinforced through the studies of motivation by [8]. Those describes that in order to increase engagement, three major areas must be discussed: Autonomy, Mastery and Purpose.

Autonomy is when the person has control over their decisions and actions. Mastery is the intrinsic motivation which occurs when this person learns something new. As for Purpose, it is the necessity to create bounds with something externally to fulfil an internal need.

Another theory to enforce the ones described previously is the Flow proposed by [1]. This creates a relation between a person's skills and the size of the problem. The author argues that it must have a balance between these two metrics for the person be engaged with the task.

In this theory a good learning method stays in the "Flow", which means If the problem is bigger than the skills, the person becomes frustrated however if the skills are higher, then this person may feel bored.

3 Development

The development of the system was divided in three distinct modules, defined as the System Core, User Interaction and Gamification. The system core consists in the implementation of the activities and its mechanics, while the User Interaction module can be identified as the graphical interfaces that connects the users with the system core. Finally the Gamification core is the implementation of the game concepts and mechanics to improve user engagement.

The system core englobes the main Data Structure topics that are implemented within the system, which are divided by tasks. Each one of them have two general activities. The first one is the Tutorial level which explains the basic concepts, and the second one is the exclusive task, these exemplifies the concepts in contextualized real sutuations.

As for the Gamification module, it cointains the game concepts that are used within the system. Those were choosen to improve user interactions and engagement and were based on the following mechanics: Experience points, Levels, Leader Boards, Achievements, User Profile.

After the definition of the mechanics that would be used, we started the process to implement the gamification core of the system. To achieve the desired results, there was a necessity to plain the use of those concepts to persuade the students to study and also have fun while doing it.

Based on the above, we created a table (Table 1) of the points gained per difficulty of the task. This score was planned to gain more points as the difficulty increases, so the student may feel encouraged to try harder levels of the same task without much hesitation.

Table 1. Points and their relations with each Difficulty level

Difficulty	Points per activity
Easy	1
Medium	3
Hard	5
Very Hard	7

As can be seen on Table 1, easy activities guarantee 1 point while hard activities 7 points.

Finally the User Interface Module consists in the web interface generated to integrate the System Core and Gamification Modules with the user. This graphical interface was developed using a PhP system to manage the users profiles and statistics.

4 Results

The system is designed to be divided according each Data Structure within it (Figure 1). These are unlocked through player progression. Each time the user complete a task, get more points that will be used to unlock new activities. Besides, there have some special achievements like the faster task finished, most task solved.

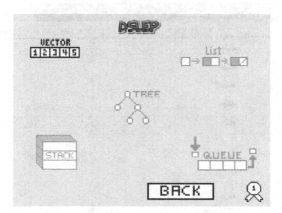

Fig. 1. Structure Menu

As can be seen figure 1 there are 5 different structures to be explored. Each one of them has their own activities and achievements. To unlock new structures the player must complete previous activities with some predefined roles that may be number of mistakes, tasks completed, etc.

However in order to unlock new structures, the player must finish a certain number of tasks or the overall structure. By unlocking new structures, the student also activate new tasks to be done.

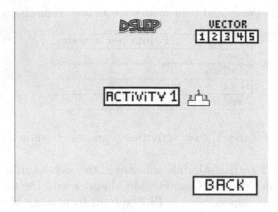

Fig. 2. Array Menu

By choosing the structure, the user is directed to the sub-menu, where he is able to choose the desired activity, see their global statistics or return to the main menu, as can be seen on Figure 2. By succesfully doing the presented tasks, more points are accumulated for the global score (Figure 3) and more tasks will be unlocked.

Fig. 3. Leader board example of the Array Structure

On figure 3 there has an example of a leader board implemented within the system. So the user may be able to see and compare his results with other students. This was made to stimulate a healthy competition among them.

Those game concepts were choosen to be agreeable with [6] major areas affected by Games (and consequently Gamification), which are cognition, emotional and social. The cognition is affected through the interactions with an intuitive environment and it set of rules.

The progressive and continuous use of the system allow a quick feedback for the student, which may trigger many emotional transformations from anger and frustration, for not being able to complete a task, to happiness and joy, by finishing it.

As for the social area, the system is implemented with a user profile (Figure 4), where he may see his overall statistics, points and achievements to measure his progress. They may also post those statistics in social networks to improve communication with others.

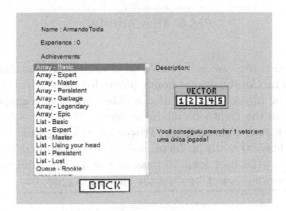

Fig. 4. Profile Screen

Figure 4 represents the profile Screen, where the user may see his statistics. Overall, 33 achievements were created and divided in the 7 implemented tasks, which will be described in the following sections.

4.1   Tasks

The 7 activities implemented within the system are divided as follow: 1 for array structure, 2 for stack structure, 1 for queue, 1 for lists and 2 for trees, along with the designed tutorials for each one.

The Array activity is called **"Piperray"**, it utilize the main property of this structure, which is to storage a group of variables of a single type. The student must connect a shape (that will fall from the top of the screen) and insert into a pipe.

By allocating the shape to that specific pipe, he locks the entry of other different shapes. This is the contextualization of the main array property as explained before. The user also gain bonus points by making the sequence presented in the screen (Figure 5).

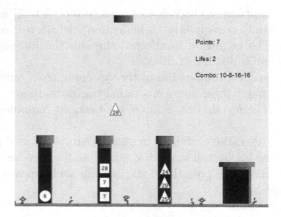

Fig. 5. Array Activity

Figure 5 demonstrates the array task, with each pipe containning values of the same shape. The Combo grants bonus points for the student, if the sequence is true in any of the pipes. The user also is able to discard a shape by throwing in the trash pipe on the right side of the screen.

As for the stack tasks, there is the **Tower of Hanoi**, which consists in the user solving the classical math problem, however using limited moviments. This problem is based on the user logic to pass all the 3 discs from pin A to pin C, therefore the bigger ones can't be over the smaller (Figure 6).

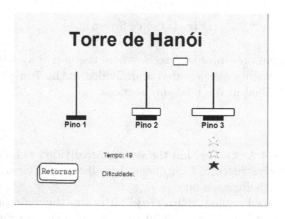

Fig. 6. Tower of Hanoi Stack Activity

On figure 6 is a screen of the task. The user must pass the 3 discs from pin A to C without overlap the smaller ones with the bigger. The student gain points by finishing this activity with the lowest pin changes and average time.

The second stack task is called **"Asterostacks"**, that is a side-scrolling shooter where the user must form a Combo, by collecting the right values. The player is able to shoot the last captured value (which is the First In First Out concept of Stack structure) to score points by destroying the upcoming ones 7.

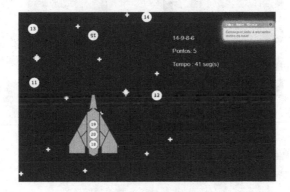

Fig. 7. Second Stack Activity

On figure 7 is a representation of the task. The user is able to control the ship through mouse buttons and is able to shoot the first value in order to destroy and collect others. By performing a Combo, he also gains a bonus score.

The queue activity is called **Queue Race**, this one explain the concept of First In Last Out of the structure. The student controls a race car, which gain points based on the distance travelled and the values collected through it (Figure 8).

Figure 8 demonstrate the functional task, the mechanics are very similar to the **Asterostacks** activity, however by discarding a value from the queue, the car gains more velocity and so it increases the travelled distance and so the user points.

The lists structure are represented through the **Snake** task. It is based on the classic snake game, however the user may eat a value through its head or tail. Which represent an auxilliary node to manipulate the Linked List structure. By colliding the extremities, the user creates a circular list, ending the task (Figure 9).

In figure 9 there is the demonstration of the activity, with the player controlling the head extremity wich also influences the tail. He gain points by colliding the head with the random scattered values in the screen.

Finally, the Tree structures are represented through 2 tasks. The first one is the **Balanced Tree** task, which aims to describes the binary tree balancing operation. It randomizes a numerical value for the root and the user must specify where the next value fits in Left or Right subtrees (Figure 10).

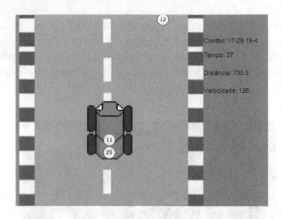

Fig. 8. Queue Race Activity

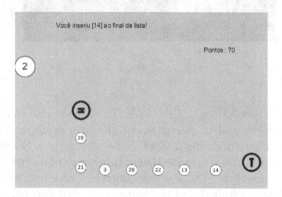

Fig. 9. Snake Activity

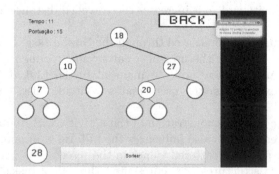

Fig. 10. Binary Balanced Tree Activity

On figure 10 there is a demonstration of the task, where the user may click where he may insert the randomized value. The student gain points by putting the values in the correct position and the activity is finished when the timer runs out.

The second tree activity is the **Transversal Methods**, which describes the concepts. The main objective is select the right nodes in order to achieve the randomized Transversal Method (Figure 11).

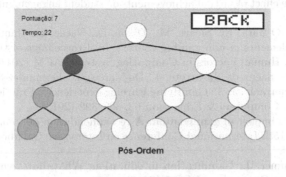

Fig. 11. Transversal Tree Activity

Figure 11 is a screen of the task. The user gain points by doing the right tranversal method and loses if selecting the wrong nodes. The method is randomized each time the user select the activity.

Also, as explained above, each activity has it set of achievements that are unlocked by doing specific tasks. The system accumulates the points acquired through the activities and increases the level of the user.

The system is currently under tests with students of Data Structure lessons in a local College. The evaluation of the system will be evaluated through a questionnaire applied at the end of the bimester, and aims to analyse the pedagogical effectiveness.

5   Conclusions

This work presents the development and implementation of an Interactive Learninv Environment for Data Structures which was also proposed by other authors, as [13]. To improve the gamification core of the system, 33 achievements and 9 leader boards were made to be used within the system and its 7 activities.

The objective of this work was achieved since the platform is finished and already being tested with students from local colleges. Also it is implemented with the gamification concepts to engage and stimulate their motivation.

To secure and guarantee the information stored within the system, we used a tool (Clay.IO) to manage the gamification concept. This also allowed the user to exchange personal information with others, as a social network system.

As future works, new activities and new concepts will be implemented. Also it is in development a profile analyser to work within the system so the teachers may create a 'class' to monitor the students development and progress within the lessons. Those analysis also may determine wich activities may be improved or balanced.

References

1. Csikszentmihalyi, M.: Flow: The Psychology of Optimal Experience. Harper and Row, New York (1990)
2. Denny, P.: The effect of virtual achievements on student engagement. In: CHI 2013. ACM (2013)
3. Deterding, S., O'Hara, K., Sicart, M., Dixon, D., Nacke, L.: Gamification: Using game-design elements in non-gaming context. In: Proccedings of the 2011 Annual Conference on Human Factors in Computing Systems. ACM (2011)
4. Domínguez, A., Saenz-de Navarrete, J., De-Marcos, L., Fernández-Sanz, L., Pagés, C., Martínez-Herráaiz, J.-J.: Gamifying learning experiences: Practical implications and outcomes. Computers & Education 63, 380–392 (2013)
5. Erenli, K.: The impact of gamification: A recommendation of scenarios for education. In: 2012 15th International Conference on Interactive Collaborative Learning (ICL) (2012)
6. Lee, J.J., Hammer, J.: Gamification in education: What, how, why bother? Academic Exchange Quarterly, AEQ 15 (2011)
7. McGonigal, J.: Reality is Broken: Why Games Make us better and How they can Change the World, 1st edn. Penguin Press, New York (2011)
8. Pink, D.: Drive: The Surprising truth about what motivate us. Canongate, New York (2010)
9. Pope, D.C.: Doing School: How We Are Creating a Generation of Stressed-Out, Materialistic, and Miseducated Students. Yale University Press (2003)
10. Rock, M.: Transfiguring it out: Converting disengaged learners to active participants. Teaching Exceptional Children, CEC 36(5) (2004)
11. Simões, J., Redondo, R.R.D., Vilas, A.A.F.: A social gamification framework for a K-6 learning platform. Computers in Human Behavior 29(2), 345–353 (2012)
12. Tan, B., Seng, J.L.K.: Game-based learning for data structures: A case study. In: 2nd International Conference on Computer Engineering and Technology. IEEE (2010)
13. Toda, A., Almeida, M., Moraes, C.R., Freires, A.F., Brancher, J.: Interactive learning enviroment for data structures with gamification concepts. In: WWW/Internet. IADIS (2013)
14. Zichermann, G., Cunningham, C.: Gamification by Design: Implementing Game Mechanics in Web and Mobile Apps, 1st edn. O'Reilly (2011)

Intelligent Systems for Monitoring and Prevention in Healthcare Information Systems

Fernando Marins[1], Luciana Cardoso[1], Filipe Portela[2], António Abelha[1], and José Machado[1,*]

[1] University of Minho, CCTC, Department of Informatics, Braga, Portugal
[2] University of Minho, Algoritmi,
Department of Information Systems, Guimarães, Portugal
{a55561,a55524}@alunos.uminho.pt, cfp@dsi.uminho.pt,
{abelha,jmac}@di.uminho.pt

Abstract. Nowadays the interoperability in Healthcare Information Systems (HIS) is a fundamental requirement. The Agency for Integration, Diffusion and Archive of Medical Information (AIDA) is an interoperability healthcare platform that ensures these demands and it is implemented in *Centro Hospitalar do Porto* (CHP), a major healthcare unit in Portugal. Therefore, the overall performance of CHP HIS depends on the success of AIDA functioning.

This paper presents monitoring and prevention systems implemented in the CHP, which aim to improve the system integrity and high availability. These systems allow the monitoring and the detection of situations conducive to failure in the AIDA main components: database, machines and intelligent agents. Through the monitoring systems, it was found that the database most critical period is between 11:00 and 12:00 and the resources are well balanced. The prevention systems detected abnormal situations that were reported to the administrators that took preventive actions, avoiding damage to AIDA workflow.

Keywords: Healthcare Information Systems, Interoperability, Availability, Monitoring Systems, Preventing Systems.

1 Introduction

In healthcare area, information systems are in fast evolution, the main goal is convert paper-based practices into computerized processes in order to ensure the healthcare delivery and improve services quality. After dematerialization, Healthcare Information Systems (HIS) can provide a better coordination between medical professionals and applications, allowing the reduction of the number and the incidence of medical errors. Furthermore it enables the reduction of healthcare costs and may provide a means to improve the hospital management [1, 2].

---

* Corresponding author.

B. Murgante et al. (Eds.): ICCSA 2014, Part VI, LNCS 8584, pp. 197–211, 2014.

Nowadays a hospital constitutes an environment with many and complex information systems. They are heterogeneous, distributed and they speak in different languages. Consequently, the impact of these systems in a hospital environment is low, once they manipulate the information in an individualized way, which does not allow a proper interaction among the different systems [1, 3].

In this way, it emerges the necessity of to establish a communication link among the systems that constitutes the HIS. This lack is filled through the interoperability, which is the ability of independent systems to exchange meaningful information and initiate actions from each other, in order to operate together to mutual benefit. The main goal of interoperability in healthcare is to connect applications and data can be shared and exchanged across the healthcare environment and distributed to medical staff or patients whenever and wherever they need it [4].

The interoperability implementation in HIS must hold mechanisms that ensure the information security. More specifically mechanisms that ensure the confidentiality, the integrity and the availability of the information. This issue will be presented in the Section 2 together with the presentation of the AIDA platform, a solution to achieve interoperability in HIS. The Section 3 presents the MEWS (Modified Early Warning Score) model, which is a fault forecasting model used in the Intensive Care Units that monitors the patient vital signs and it predicts organ failures. The work presented in this paper aims to increase the AIDA availability through monitoring and prevention systems, which will be presented in the Section 4. The systems developed and presented are based on the MEWS model. In the Section 5 the results of their implementation in the AIDA (Agency for Integration, Diffusion and Archive of Medical Information) installed in the *Centro Hospitalar do Porto* (CHP) will be discussed, critical situations conducive to failure were detected and fixed. It was possible to study the workload of the AIDA main components too. Finally, in the Section 6 the conclusions and future work are exposed.

2   Interoperability in HIS

There are a variety of methodologies and architectures through which it is possible to implement interoperability between HIS. These methodologies are based on common communication architectures and standards. The multi-agent system (MAS) technology is being to stand out in the area of interoperability, including interoperability in healthcare, addressing the concerns mentioned above. The autonomy and pro-activeness features of an agent allow it to plan and perform tasks defined to accomplish the design objectives. The social abilities enable an agent to interact in MAS and cooperate or complete to fulfil its goals. The agents in a healthcare facility configure applications or utilities that collect information in the organization. Once collected, this information can be provided directly to other entities, e.g. a doctor or to a server, stored in a file or sent by e-mail to someone for it to be treated at a later date [5].

2.1   AIDA

In this context, it arises an intelligent and dynamic platform with the purpose of making the HIS interoperable: the Agency for Integration, Diffusion and Archive of Medical Information (AIDA). This platform is developed by researchers at the University of Minho and it is implemented in several large health organizations in Portugal, including the CHP. AIDA is responsible for the process of integration of different information sources. To achieve that, it uses the Service Oriented Architecture (SOA) and MAS to implement interoperability in a distributed and specific environment in accordance with standards. This platform provides intelligent electronic agents, best known as software agents that besides they have a pro-active behaviour, they provide many services such as [4, 6]:

– The communication among different sub-systems, sending and receiving information (e.g. clinical or medical reports, images, data collections, prescriptions);
– Managing and archiving the information;
– Responding to requests, with the necessary resources to carry them out correctly and timely.

2.2   Information Security

Once AIDA stores and manipulates human related information, it is crucial to ensure its protection. The information security is based on three attributes that should be on the AIDA characteristics [7–9]:

– **Confidentiality:** provided by mechanisms that prevent unauthorized persons access and publicized private information;
– **Integrity:** this is the property that ensures that the information retains all the original features established by the owner even after handling, in other words the information should be inviolable, incorrupt and consistent;
– **Availability:** provided by mechanisms to access the required information in time by those who are authorized. In addition, it should have mechanisms for fault prevention and tolerance, so that the system be able to continue operating despite the failure of any component.

2.3   AIDA Availability

The AIDA should have a high level of availability and a proper functioning twenty-four hours a day. A short stop period or the system slowness may bring serious consequences for the services quality delivered. The AIDA platform has already implemented fault tolerance mechanisms such as: the Oracle Real Application Clusters (RAC) system, which improves the availability and the scalability of AIDA database; and a data guard solution, which consists in a replica of the original database, situated in a different place. This database is accessed when the original database is unavailable. In spite of all advantages, the fault tolerance tools do not allow the focus on the faults themselves, but only on the faults effects, trying to minimize them. Thus, it is necessary to perform a proactive monitoring on AIDA platform, taking actions before the fault occurs [10].

The AIDA availability is directly connected to its main components workflow. These main components are: its database, its agents and the machines wherein the agents perform their tasks. AIDA is a platform that ensures the interoperability in a healthcare environment and consequently, it strongly improves the quality of the healthcare services delivered to the patients. In spite of the high level of maturity of the AIDA platform, it has some flaws that influence this quality. The high availability is a strong demand in this platform and without a proper functioning of one of AIDA main components, its availability is compromised. In this context, the monitoring and prevention systems presented in this paper include these three components and they aim increase the AIDA high availability.

3   MEWS

In medicine there is already a model for the prediction, in advance, of serious health problems. This model is called Modified Early Warning Score (MEWS) and it is used in several healthcare units in the world. Particularly in the CHP, this model has been applied in the Intensive Care Unit [11].

MEWS assumes that a serious problem of health is often preceded by physiological deterioration. The use of MEWS implies a strict monitoring of the patient's vital signs. Then, using the decision table (Table 1), the scores are calculated to determine the level of risk of each patient, trying to understand when a serious problem will occur. The monitoring of patient's vital signs should be continuous and all values must be archived so that it becomes possible to understand the behaviour of the vital signs over the time [12,13].

In order to classify the health state of the patient, the seven scores are extracted from Table 1 and then they are summed, obtaining the total score. The patient's state is characterized according to the following guidelines [12,13]:

Table 1. MEWS Scores (adapted) [12]

MEWS Score	3	2	1	0	1	2	3
Temperature ($^\circ C$)		< 35.0	35.1-36.0	36.1-38.0	38.1-38.5	> 38.6	
Heart rate (min^{-1})		< 40	41-50	51-100	101-110	111-130	> 131
Systolic BP ($mmHg$)	< 70	71-80	81 - 100	101 - 199		> 200	
Respiratory rate (min^{-1})		< 8		8-14	15-20	21-29	> 30
Blood oxygen (%)	< 85	85-89	90-93	> 94			
Urine output ($ml/kg/h$)	Nil	< 0.5					
Neurological		New confusion		Alert	Reacting to voice	Reacting to pain	Unresponsive

- When one of the parameters has the score two, the patient should be observed frequently;
- If the patient's total score is four or if there is an increase of two values of the total score, the patient requires urgent medical attention;
- If the total score is more than four, the patient is at risk of life.

This model is endowed with several advantages such as: enabling to set priorities for the interventions to be carried out; knowing better the physiological tendencies of the patient's organism through the monitoring process; assisting in making medical decisions, once it uses a quantitative criteria; predicting situations that the patient need internment in the Intensive Care Unit [11–13].

4   Monitoring and Preventing Systems

This section presents the monitoring and prevention systems developed for the database, machines and agents of AIDA. Once all of them are based on MEWS, it was created three score tables, one for each AIDA main component. To develop prevention system, it is indispensable to realize a monitoring process. A high knowledge about the system is demanded in order to select the performance indicators properly. These indicators should be based on the system workload and other parameters that are important to prevent faults [14].

4.1   Database

Through the performance views, a tool provided by Oracle to collect information about database state, it was possible to collect the AIDA database performance indicators, in addition Unix scripts were responsible to collect information about the processor and memory of database server. In this way, the twelve performance indicators selected to accomplish the monitoring and prevention faults were [15]:

- **Processor utilization:** high values may seriously compromise the database workflow;
- **Memory utilization:** it is also a fundamental key to a database proper functioning;
- **DB time:** this is the time elapsing between the instant of placing of the query by the user to the reception of all results, this time should be the lowest possible;
- **Number of transactions:** It is the sum of the number of commits and the number of rollbacks effectuated by users;
- **Number of operations:** one transaction consists in a set of operations, depending on the query it can have more or less operations per transaction;
- **Calls ratio:** it is the ratio between recursive calls and total calls (sum of recursive calls and user calls). A recursive call occurs when a user request needs one query SQL that needs another SQL query. Ideally, this ratio should be lower as possible;

- **Number of current sessions:** a high number of sessions opened may cause an excessive use of memory;
- **Size of redo file:** the amount (KBytes) of redo files;
- **Buffer cache ratio:** the percentage of data that is in memory cache, if this ratio decreases, this may indicate memory problems;
- **Amount of I/O requests:** a high value of this parameter may indicate memory problems and frequent access to the disc;
- **Amount of redo space requests:** it indicates if there is enough space to write in the buffer;
- **Volume of network traffic:** it is the network that interconnects all the components of the database. If this value is very high, the database becomes slower and it compromises the users requests.

In the Table 2 it is presented the score table for AIDA database. The scores attribution is made through percentiles, inspired in the 95th percentile method used for billing in Internet Service Providers and websites [16]. In the Table 2, all performance indicators have the same limits to establish the score to each one.

Table 2. Score table for the AIDA database based on the performance indicators selected

Score	0	1	2	3
Values	$\leq p80$	$]p80, p85]$	$]p85, p95]$	$> p95$

Once this score table is based on percentiles, it was necessary collect a reasonable amount of data related to each performance indicator during a period of time. It is important to refer that there is a specific percentile for each score of each performance indicator for each hour of the day. The system developed is upgradeable, new percentiles are calculated at the end of the day based on new measurements.

After a discussion among the system administrators and Information Systems (IS) specialists, it was decided that the total score is calculated 15 in 15 minutes using the averages of all performance indicators collected during this period. In this way false positive situations are avoided and the effectiveness of this prevention system is increased. Similarly to the MEWS, if the total score is more than four, it is a critical situation that compromises the database availability. This situation is conducive to fault occurrence, so the system developed send an email to system administrators warning the database state in order to they take preventive actions, avoiding damage in the HIS. For less serious situations (total score less than 5 or an increase of 2 values), visual warnings appear in the monitoring dashboards developed with the objective to monitor the activity of the database, machines and agents.

4.2   Machines

With the purpose of to know the AIDA machines behaviour and then to prevent possible faults, three performance indicators were selected, all of them related to their workload (free percentage of processor, memory and disk space). These parameters were collected through Windows Management Instrumentation (WMI) technology, which enables a simple exchange of information through a powerful set of tools based on standards. With a simple query-based language named Windows Query Language (WQL), this technology allows the user to obtain a wide range of information about the hardware and software of a specific machine. Obviously the user must know the required credentials.

In the case of the machines, initially there was an attempt to create a score table based on percentiles as the Table 2 previously presented, but it did not succeed. The application sent several warnings false positives per day. The computer performance limits for a good operation is an issue that varies a lot. Those limits depend of the objectives that the system administrators want for a specific machine. So, the score table was created with default fixed limits based on the administrators experience, their knowledge about the system and opinions from IS specialists. Through a management page, the administrators can change these limits for each parameter either generally or specifically for one machine, anytime. The default score table for all machines, based on MEWS, it is presented in the Table 3.

Table 3. Score table for the AIDA machines

Score	0	1	2	3
Free processor (%)	≥ 50	]50, 25]	]25, 10]	< 10
Free memory (%)	≥ 15	]15, 10]	]10, 5]	< 5
Free disk space (%)	≥ 15	]15, 10]	]10, 5]	< 5

As in MEWS, the following situations can occur:

- If the sum of all parameters score is more than four, critical situations are detected and a warning (email) is sent to the administrators in order to they take preventive actions;
- If there is an increase of two values, it is considered that the situation is very grave, such as a situation wherein the total score is four. In these cases a warning will appear in the monitoring dashboards.

Another adaptation of this fault prevention system was considered that when one of the parameters showed in the Table 3 has the score three it is a critical situation, which triggers an email warning. It was important to implement this adaptation because any of the machines performance indicators is fundamental for a good performance. If one of these resources is overly consumed, the machine will have serious problems of accomplishing its tasks.

4.3   Agents

In the two systems of monitoring and prevention presented above, the performance indicators intervene in the process of prevention faults and in the monitoring process. In the case of agents, the prevention and monitoring systems are separated. The monitoring system collects information related to the workload of each agent so that they are analyzed in the monitoring dashboards. For the prevention of the faults, the system is based on the frequency of agent activity. In other words, it is the frequency that the agent is executed. It also can be interpreted by the interval of time that the agent takes to refresh the log file with its newest activity. To collect the values of this indicator, it is used the Directory class of .NET framework and a batch script that creates the mapped drives for the AIDA machines that execute the agents and it is where the log files are.

The agents monitoring system collects and disseminates the values of various performance indicators such as: processor and memory utilization by agent, I/O of data per second and execution time. The first three indicators are also collected through WQL queries. In this way, the system administrators can access the monitoring dashboards and to consult the agents behaviour anytime.

In the Table 4, it is presented the score table for AIDA agents based on its activity frequency (in minutes) and percentiles (inspiration in the 95th percentile method again [16]). Once there is one variable in the score table, it was added the score four, becoming the system more accurate. After obtaining the results from the first tests of this prevention system, the Table 4 was adapted. The 95th percentile was replaced for the 97.5^{th} percentile and consequently the other limits were modified, with also the purpose of increase the system accuracy.

Table 4. Score table for the AIDA agents based on its activity frequency

Score	0	1	2	3	4
Activity frequency (min)	$\leq p85$	$]p85, p90]$	$]p90, p95]$	$]p95, p97, 5]$	$> p97, 5$

Before beginning the prediction process, it is indispensable to collect several data about agents' activity frequency during a reasonable period of time. In this way it is possible to evaluate the normal behaviour of agents and start the prevention process based on the interpretation of MEWS scores:

- If the score obtained was less than four then a visual warning will be issued on the monitoring dashboards;
- If the score was equal to four an email is sent to the system administrators in order to they take speedy action to restore the normal workflow and to prevent future damages.

New limits are constantly calculated for each agent improving the system efficacy. Besides that, this application is endowed with persistency in relation to the database state. If the database is down, all SQL statements are recorded in a file and the administrators are warned. When the database returns back to normal all registers are inserted and the limits are refreshed. During the database down time, scores do not stop of being calculated and abnormal situations are detected, however the limits are not refreshed and the limits in the score table are the last ones calculated. It is important to refer that this application does not use the same database that the AIDA uses, avoiding an overload in the AIDA database and allowing a dependency on the application operation.

5   Results

The systems presented in the Section 4 were implemented in the AIDA installed in the CHP. Besides prevention damage to AIDA workflow, detecting critical situations through the prevention systems, these systems also allowed to realize a study about the workload of the AIDA main components. This study increased the administrators' knowledge about the AIDA platform, enabling them to detect situations that need to be fixed and fragile situations that require special attention, to establish priorities for taking action and still to manage the platform in order to obtain the best balance of resources.

All the graphs presented in this section were extracted from the monitoring dashboards developed through an open source Business Intelligence (BI) tool named Pentaho Community, which revealed to be an easy handling tool, it improved the consolidation of the data and it provided a greater support for decision making [17].

5.1   Workload Results

Database
In the Table 5 is possible to verify that the AIDA database has a high utilization, an average number of sessions in the order of 681 sessions. Furthermore, it can observe that are executed on average about 214 transactions per second, resulting 742 operations per second in the database, which proves that this is a database with a very high workload.

The Figure 1 shows the values of four performance indicators in a specific regular day: number of sessions, processor utilization, DB time and network traffic. On this day, no abnormal situations were detected in AIDA components. The graphs are composed by two limits, the 25th and 75th percentiles (p25 and p75 in the Figure 1). Consequently, 50% of the data collected is in the range defined by these limits and it is expected that the next measurements occur in the same range. All four graphs in the Figure 1 present a significant peak at the end of the morning. The other performance indicators (not included in the Figure 1) also demonstrated this behaviour on this day, although not in such an obvious way. It may be concluded that, in a regular day, the period between 11:00 and 12:00 is the most critical for AIDA database.

Table 5. Average values of AIDA database performance indicators during the study period

Performance indicators	Average values
Processor utilization (%)	18
Memory utilization (%)	98
DB time per second	6
Transactions per second	214
Operations per second	742
Calls ratio	0.14
Number of sessions	681
Redo file size (KBytes per second)	152
Buffer cache ratio	0.998
I/O requests per second	632
Redo space requests per second	0.55
Network Traffic Volume (KBytes per second)	671

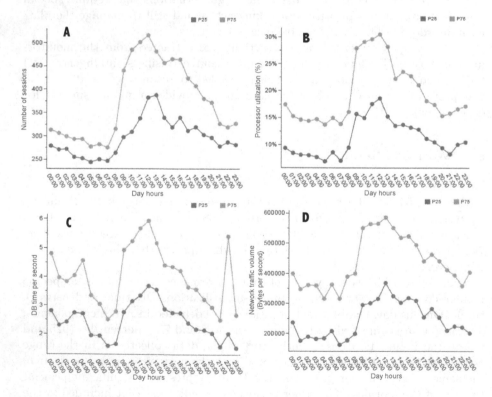

Fig. 1. Values of four performance indicators of AIDA database during a regular day. **A**: Number of sessions; **B**: Processor utilization; **C**: DB time; **D**: Network traffic.

Machines and Agents

During a month it was collected the machines performance indicators. As the Figure 2 presents, among the five machines that execute AIDA agents, the machine 08 is the one that consumes more CPU (an average of 14.09%) and the machine 01 is the one that consumes more memory RAM (an average of 42,38%). On the other hand, machine 01 is the one that consumes less CPU (an average of 5.5%) and machine 08 and machine 04 are the ones that consume less memory RAM (an average of 14.23% and 12.93%, respectively). It was also possible to confirm in the monitoring dashboards that the CPU consume was constant only varying from 5 to 10% in maximum. The consumption of memory RAM was very constant.

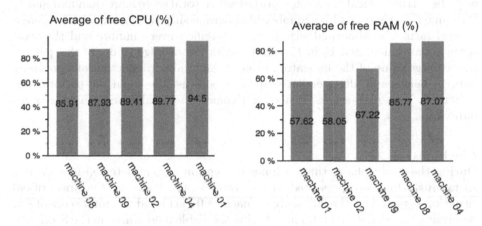

Fig. 2. AIDA machines workload (processor and memory utilization)

Relative to agents workload, it was possible to uncover that agents that are continually being executed and are responsible for archives transfer and provide web services, consume more RAM memory. That is the case of the machine 01 that only have agents with these functions. As it is possible see through Figure 2, the machine 01 is the one that consumes more RAM memory. In this case, talking about the machines prevention system, the administrators should lower the memory limits of this machine in the score table for machines (Table 3) in order to avoid being warned in regular situations.

The high number of agents that are executed in machine 08 justifies the high consumption of processor. In this machine are installed agents that are responsible for archives transfer, billing, requests processing and verifications. Besides the number of agents, most of these are often performed, which also justifies the elevated use of CPU in machine 08.

The machine 04 is the machine that has more resources available as it can be seen in the Figure 2. So, it may be concluded that when a new agent is created, it should be installed in this machine. This conclusion can be taken from the monitoring dashboards, which acts as a decision support system. These dashboards also enable to the system administrators to monitor the AIDA components in real time and to study their behaviours in the past, in a period selected by the user.

5.2   Preventing Systems Results

Database

During the period of this study, the database prevention system did not detect any situation with the total score more than four. However, abnormal situations for each performance indicator were detected. The memory utilization, the calls ratio, the amount of I/O requests and the number of transactions were the indicators with more occurrences: 377, 369, 362 and 352 respectively. The reason why there is no critical situation reported with a total score more than four and at the same time there is a high number of abnormal situations is explained through several facts: the abnormal situations are verified every minute and the total score is only calculated 15 in 15 minutes with the average values of this period; the average value of the indicators rarely is high; abnormal situations normally happen before and after regular situations; and usually when the performance indicators have high scores, it does not coincide with high scores of the other indicators.

Machines

During the study phase, the machines prevention system detected four critical situations, which are presented in the Table 6. For instance, the first critical situation detected on the 30th July in machine 02 achieved the total score of six. Analysing this situation and remembering the Table 3 presented in the Section 4, it is verified that the high utilization of processor and memory implied the total score obtained. On the other hand, the occurrences of the machine 04 achieved a total score of four, because in both situations there was an increase of two values relative to the previous score. Although the total scores are lower than five, it was considered as a critical situation because the processor utilization had the maximum score and according the mentioned in the Section 4, these are situations propitious to faults. All four situations presented in the Table 6 were reported successfully to a system administrator who quickly identified and repaired the problem. In this way, he prevented bigger damages in the normal workflow of the AIDA.

Agents

The study phase for the agents prevention system lasted almost two months and it was realized after collecting data about the agents activity frequency during two weeks. Three maintenance situations were realized in different days during this phase. The process of maintenance, naturally, caused a stop in several agents activities. In all of these three situations, the agents prevention system detected

Table 6. Critical occurrences per machine

Machine	Date-time	Free processor (%)	Free memory (%)	Free disk space (%)	Score
machine 02	30th July 16:00-16:30	3	9	50	6
	24th-25th September 22:40-00:15	1	4	50	6
machine 04	27th August 12:40	0	81	34	4
	5th September 16:15-17:30	6	88	34	4

that those agents stopped. This fact proves that the system is capable of quickly detect an irregular situation of the agents, preventing bigger damages.

Ignoring the occurrences from maintenance situations, it was detected abnormal activities of some AIDA agents. In the Figure 3, it is visible the occurrences where certain agents reached the score four, triggering the sending of warning emails.

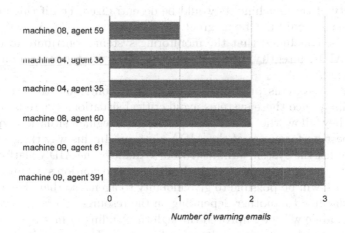

Fig. 3. Number of warning emails (critical situations) per agent during the study phase for agents prevention system. The occurrences relative to the maintenance situations are ignored in this graph.

In all occurrences shown in the Figure 3, the agents presented an abnormal behaviour. They remained too long without performing their tasks. These situations must be carefully analyzed to determine if is necessary to modify these agents or other components.

6   Conclusions and Future Work

The adaptation of the model MEWS in medicine for the AIDA main components (database, machines and intelligent agents) was conceivable, detecting critical situations conducive to failures. However, it is very important to refer that is essential have a high knowledge of the system to make possible this adaptation and the development of the monitoring and prevention systems. Through the process of monitoring, these systems should be accompanied and refined continually. Furthermore, the monitoring of these components assists in making decisions about the system and it enables a better resource management of each of these components.

Relative to workload results, it may be concluded that the AIDA database of the CHP has a high workload, mainly between 11:00 and 12:00. According to the results obtained in the Section 5, the machine 08 and the agents that it executes, deserve a special attention from system administrators. This is due to the fact that a great number of agents perform their tasks in the machine 08. The machine 02 should be observed frequently, once critical situations were detected either for the machine or for the agents that run there. The workload of the AIDA agents and machines demonstrated that the system resources are well balanced. However, minor repairs could be made such as the transfer of one or more agents of the machine 08 to the machine 04. Thus the high importance of high availability of the machine 08 would be decentralized, i.e., if this machine failed, the damage would not be as great.

It also may be concluded that the monitoring systems contribute to the improvement of AIDA integrity, because they provide a greater control on AIDA main components.

The prevention systems presented in this paper, contribute to increase the AIDA availability, once these systems avoid critical situations where failure can arise easily. They allow the administrators repair irregular situations quickly, ensuring the best performance of the AIDA platform. In this way these systems contribute to assist the system administrators to manage the AIDA platform and consequently, to the improvement of the healthcare information systems quality.

As future work will be possible to give mobility to agents so that they can migrate from a machine to another depending on the resources there are available. The warning module will be improved through the sending of messages via SMS in extreme cases or when a critical situation lasts long. The data mining process should be applied to the data that concerns about these three main components. In order to recognize correlations and standards in the components behaviour.

Acknowledgement. This work is financed with the support of the Portuguese Foundation for Science and Technology (FCT), within project PEst-OE/EEI/UI0752/2014.

References

1. Palazzo, L., Sernani, P., Claudi, A., Dolcini, G., Biancucci, G., Dragoni, A.F.: A multi-agent architecture for health information systems. In: International Workshop on Artificial Intelligence and NetMedicine, p. 41 (2013)
2. Reichertz, P.L.: Hospital information systems-past, present, future. International Journal of Medical Informatics 75(3-4), 282–299 (2006); Electronic Health Record, Healthcare Registers and Telemedicine and European Potential for Building Information Society in Healthcare
3. Miranda, M., Salazar, M., Portela, F., Santos, M., Abelha, A., Neves, J., Machado, J.: Multi-agent systems for hl7 interoperability services. Procedia Technology 5, 725–733 (2012)
4. Miranda, M., Duarte, J., Abelha, A., Machado, J., Neves, J.: Interoperability in healthcare. In: Proceedings of the European Simulation and Modelling Conference (ESM 2010), Belgium (2010)
5. Machado, J., Abelha, A., Novais, P., Neves, J., Neves, J.: Quality of service in healthcare units. International Journal of Computer Aided Engineering and Technology 2(4), 436–449 (2010)
6. Duarte, J., Salazar, M., Quintas, C., Santos, M., Neves, J., Abelha, A., Machado, J.: Data quality evaluation of electronic health records in the hospital admission process. In: 2010 IEEE/ACIS 9th International Conference on Computer and Information Science (ICIS), pp. 201–206 (August 2010)
7. Rodrigues, A.: Oracle 10g e 9i: Fundamentos Para Profissionais. FCA, Lisboa (2005)
8. Bertino, E., Sandhu, R.: Database security - concepts, approaches, and challenges. IEEE Transactions on Dependable and Secure Computing 2(1), 2–19 (2005)
9. Kim, S., Cho, N., Lee, Y., Kang, S.H., Kim, T., Hwang, H., Mun, D.: Application of density-based outlier detection to database activity monitoring. Information Systems Frontiers, 1–11 (2010)
10. Drake, S., et al.: Architecture of Highly Available Databases. In: Malek, M., Reitenspiess, M., Kaiser, J. (eds.) ISAS 2004. LNCS, vol. 3335, pp. 1–16. Springer, Heidelberg (2005)
11. Portela, C.F., Santos, M.F., Silva, Á., Machado, J., Abelha, A.: Enabling a pervasive approach for intelligent decision support in critical health care. In: Cruz-Cunha, M.M., Varajão, J., Powell, P., Martinho, R. (eds.) CENTERIS 2011, Part III. CCIS, vol. 221, pp. 233–243. Springer, Heidelberg (2011)
12. Devaney, G., Lead, W.: Guideline for the use of the modified early warning score (MEWS). Technical report, Outer North East London Community Services (2011)
13. Albino, A., Jacinto, V.: Implementação da escala de alerta precoce - EWS. Technical report, Centro Hospitalar do Barlavento Algarvio, EPE, Portimão (2010)
14. Boudec, J.: Performance Evaluation of Computer and Communication Systems. Computer and Communication Sciences. EPFL Press (2010)
15. Chan, I., et al.: Oracle database–performance tuning guide, 10g release 2(10.2). Redwood City, Oracle Corporation 474, B14211–01 (2005)
16. Dimitropoulos, X., Hurley, P., Kind, A., Stoecklin, M.P.: On the 95-percentile billing method. In: Moon, S.B., Teixeira, R., Uhlig, S. (eds.) PAM 2009. LNCS, vol. 5448, pp. 207–216. Springer, Heidelberg (2009)
17. Tereso, M., Bernardino, J.: Open source business intelligence tools for smes. In: 2011 6th Iberian Conference on Information Systems and Technologies (CISTI), pp. 1–4. IEEE (2011)

A Strength Pareto Approach to Solve
the Reporting Cells Planning Problem

Víctor Berrocal-Plaza, Miguel A. Vega-Rodríguez, and Juan M. Sánchez-Pérez

Dept. of Computers & Communications Technologies, University of Extremadura
Escuela Politécnica, Campus Universitario S/N, 10003, Cáceres, Spain
{vicberpla,mavega,sanperez}@unex.es

Abstract. This work addresses a multiobjective approach of the Reporting Cells Planning Problem. This optimization problem models a well-known strategy to manage the subscribers' movement in mobile networks. Our approach can be considered as a novel contribution because, to the best of the authors' knowledge, there are no other works in the literature that tackle this problem with multiobjective optimization techniques. Two are the main reasons for using a multiobjective approach. Firstly, we avoid the drawbacks associated with the linear aggregation of objective functions. And secondly, a multiobjective optimization algorithm provides (in a single run) a wide range of solutions, among which the network operator could select the one that best adjusts to the real state of the signaling network. In this work, we propose our version of the Strength Pareto Evolutionary Algorithm 2 (SPEA2), a well-known multiobjective evolutionary algorithm. We have checked the quality of our algorithm by means of an experimental study. This study clarifies that our proposal is very interesting because it achieves good Pareto Fronts and, at the same time, it outperforms the results obtained by other algorithms published in the literature.

Keywords: Reporting Cells Planning Problem, Mobile Location Management, Multiobjective Optimization, Strength Pareto Evolutionary Algorithm 2.

1 Introduction

In the last decade, due to the exponential increase that has occurred in the number of mobile subscribers, many research works have been focused on optimizing the different management tasks involved in the proper operation of mobile networks. This manuscript addresses one of the most important management tasks in this kind of networks: the mobile location management.

Every mobile network divides the desired coverage area into several smaller regions, called network cells. In this way, the available radioelectric resources can be distributed and reused among these cells, and hence the network operator could provide service to a huge number of mobile subscribers with few radioelectric resources [1]. However, the cell division of the coverage area has

B. Murgante et al. (Eds.): ICCSA 2014, Part VI, LNCS 8584, pp. 212–223, 2014.

a direct consequence: the subscribers' movement among the network cells must be controlled to properly redirect the incoming calls. There are several strategies to manage the subscribers' movement (e.g. Never Update, Always Update, Location Areas, and Reporting Cells), and all of them consist of two main procedures: the subscriber location update (LU) and the paging (PA) [2]. The first procedure is initiated by the mobile stations (i.e. the subscribers' terminals) to notify the network that they have changed their location. And the second one is initiated by the network to know the exact cell in which every callee subscriber is located.

This manuscript addresses the Reporting Cells Planning Problem (RCPP), a popular mobile location management strategy [3]. This planning problem defines a multiobjective optimization problem in which some network cells are selected as Reporting Cells (RCs), and the remaining ones as non-Reporting Cells (nRCs). In this way, a mobile station only updates its location when it moves to a new RC and the paging procedure is only performed in a few network cells. In previously published works, this multiobjective optimization problem was tackled by using different techniques of the single-objective optimization field [4–7]. For it, the involved objective functions were linearly combined into a weighted sum. In this work, with the aim of avoiding the drawbacks associated with the linear aggregation of objective functions, we propose our version of the Strength Pareto Evolutionary Algorithm 2 (SPEA2 [8], a multiobjective optimization metaheuristic). Furthermore, to the best of our knowledge, a multiobjective approach of the RCPP is a novel contribution of this manuscript.

The paper is organized as follows. Section 2 presents a brief discussion of the related works. The Reporting Cells Planning Problem is described in Section 3. Section 4 shows a formal definition of a multiobjective optimization problem, and presents our proposal. Experimental results are analyzed in Section 5. And finally, our conclusion and future work are discussed in Section 6

2   Related Work

The Reporting Cells Planning Problem (RCPP) was firstly proposed in [9], where A. Bar-Noy and I. Kessler presented the mathematical model and shown that this problem is in general an NP-complete optimization problem. Later, the RCPP was studied with very different optimization techniques. A. Hac and X. Zhou proposed a heuristic method to optimize a simplified version of the RCPP in which the paging cost was considered as a constraint [10]. R. Subrata and A. Y. Zomaya implemented three well-known metaheuristics [4]: Genetic Algorithm (GA), Tabu Search (TS), and Ant Colony Optimization (ACO). E. Alba et al. studied the RCPP with the Geometric Particle Swarm Optimization (GPSO) and a hybridized Hopfield Neural Network with the Ball Dropping mechanism (HNN-BD) [5]. Almeida-Luz et al. proposed the use of Differential Evolution (DE [6]) and Scatter Search (SS [7]).

In all of these previous works [4–7], the objective functions of the RCPP were linearly combined into a weighted sum. This strategy allows the use of single-objective optimization techniques, but has several drawbacks (see Section 3).

That is why we propose the use of multiobjective optimization with our version of the Strength Pareto Evolutionary Algorithm 2. As said in Section 1, this is a novel contribution of our research because, to the best of our knowledge, there are no other works in the literature that optimize the RCPP in a multiobjective way.

3   Reporting Cells Planning Problem

The mobile location management based on Reporting Cells (RCs) is a static strategy in which the mobile stations only update their location whenever they enter in a new RC. In this way, the network can partially track the subscribers' movement and the paging is only performed in a subset of network cells [9]. Therefore, the main challenge of the RCPP consists in finding the configuration of RCs such that the number of location updates (f_1) and the number of paging messages needed to find a callee subscriber (f_2) are reduced to a minimum. Formally, these two objective functions can be formulated as:

$$f_1 = \min \left\{ \sum_{i=0}^{N-1} \rho_i \cdot N_{LU}(i) \right\}, \tag{1}$$

$$f_2 = \min \left\{ \sum_{i=0}^{N-1} N_P(i) \cdot V(i) \right\}. \tag{2}$$

Where N is the number of network cells. ρ_i is equal to 1 only when the network cell i is a RC, otherwise ρ_i is equal to 0. N_{LU} is a vector that stores the number of location updates of every network cell. N_P is a vector that stores the number of incoming calls of every network cell. V is a vector that stores the vicinity factor of every network cell. The vicinity factor V(i) represents the number of network cells that have to be paged to locate a subscriber in the cell i. For an RC, V(i) corresponds to the number of nRC reachable from this RC without passing over other RC, and including the RC in question (see Fig. 1(a)). For a nRC, V(i) corresponds to the maximum vicinity factor of the RCs that have this nRC within its vicinity [4] (see Fig. 1(b)). And finally, $V(i) = N, \forall i \in [0, N-1]$ when all the network cells are nRC.

It should be noted that these two objective functions are conflicting: if we want to reduce the number of location updates, the number of RCs should also be reduced, which leads to an increase in the number of paging messages due to the increase in the vicinity factor of the remaining RCs. On the other hand, the paging overload is reduced to a minimum when every network cell is a RC, configuration in which the number of location updates achieves its maximum value. Therefore, we can conclude that the RCPP is a multiobjective optimization problem.

In previously published works, the RCPP was tackled with different techniques of the single objective optimization field by means of the linear aggregation of the objective functions [4–7], see Equation 3 where β is a weight coefficient

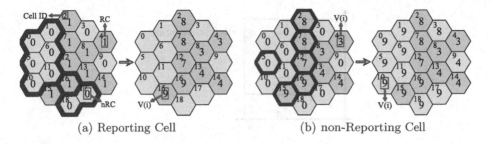

(a) Reporting Cell   (b) non-Reporting Cell

Fig. 1. Vicinity factor

used to assign more priority to one of the two objectives. However, the linear aggregation of the involved objective functions has several drawbacks. Firstly, we must have a very accurate knowledge of the problem when configuring the weight coefficient ($\beta \in \Re$). Secondly, different states of the signaling network could require different values of this coefficient. And finally, a single-objective optimizer has to perform an independent run for every value of β. These drawbacks are one of the reasons why we propose a multiobjective approach. The other reason is that, with a multiobjective optimization algorithm, we obtain in a single run a set of solutions (commonly known as non-dominated solutions) among which the network operator could select the one that best adjusts to the real state of the signaling network.

$$f_3^{SO}(\beta) = \beta f_1 + f_2. \tag{3}$$

4   Multiobjective Optimization

As mentioned in Section 3, the optimization problem addressed in this manuscript is a multiobjective optimization problem, i.e. an optimization problem in which two or more conflicting objective functions have to be optimized [11]. In the following and without loss of generality, we assume a bi-objective problem in which the objective functions have to be minimized (as the RCPP). In this kind of problems, the main challenge consists in finding the best possible set of non-dominated solutions, where a solution $\mathbf{x}^i$ is said to dominate the solution $\mathbf{x}^j$ ($\mathbf{x}^i \prec \mathbf{x}^j$) when $\forall k \in [1,2], f_k(\mathbf{x}^i) \leq f_k(\mathbf{x}^j) \wedge \exists k \in [1,2] : f_k(\mathbf{x}^i) < f_k(\mathbf{x}^j)$. Note that every non-dominated solution corresponds to a specific trade-off between objectives (or a specific value of β in Equation 3).

One of the most popular optimization techniques to solve a multiobjective problem is the use of evolutionary algorithms. These algorithms are population-based metaheuristics in which the evolutionary operators of biological systems (recombination of parents, genetic mutation, and natural selection) are iteratively applied with the aim of improving a set of initial solutions. Due to the fact that evolutionary algorithms manage populations of individuals (where an individual is an encoded solution of the problem $\mathbf{x}^i$), they can obtain a set of

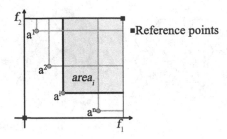

Fig. 2. Hypervolume for a minimization problem with two objectives

non-dominated solutions in a single run. In this work we propose our version (in terms of our evolutionary operators specific to solve the RCPP) of the Strength Pareto Evolutionary Algorithm 2 (SPEA2), a well-known multiobjective evolutionary algorithm, see Section 4.2.

On the other hand, there are several indicators to measure the quality of a set of non-dominated solutions (whose representation in the objective space is commonly known as Pareto Front) [11]. In this work, we use one of the most accepted indicators: the Hypervolume ($I_H(A)$). This indicator is explained in Section 4.1.

4.1   Hypervolume: $I_H(A)$

The Hypervolume ($I_H(A)$) is an indicator that associates the quality of a Pareto Front (i.e. a set of non-dominated solutions) with the area of the objective space that is dominated by this set of solutions, and is bounded by the reference points [11]. Fig. 2 shows an example of the $I_H(A)$ calculation for a minimization problem with two objectives (as the RCPP), and Equation 4 presents its formal definition. In this way, the Pareto Front A will be better than the Pareto Front B when $I_H(A) > I_H(B)$.

$$I_H(A) = \left\{ \bigcup_i \text{area}_i \mid \mathbf{a}^i \in A \right\}. \tag{4}$$

4.2   The Strength Pareto Evolutionary Algorithm 2

This algorithm is the multiobjective evolutionary algorithm that was proposed by E. Zitzler et al. in [8] as an improved version of the SPEA published in [12]. Two were the main contributions of SPEA2 with respect to its predecessor: the inclusion of an estimation of the density of solutions in the fitness function, and an enhanced archive truncation method. Basically, SPEA2 is an elitist genetic algorithm with a fitness function that estimates the quality of a solution in the multiobjective context, and with an archive of configurable size in which the best solutions found so far are stored. This fitness function is shown in

---

Algorithm 1. Pseudo-code of SPEA2

---

1   *% Initialize the parent population*
2   Ind ← Initialization (N_{pop});
3   *% Evaluate the parent population*
4   Ind ← ObjectiveFunctionsEvaluation (Ind);
5   *% Create the archive*
6   Arch ← EmptyArchive (N_{arch});
7   *% Main loop*
8   **while** *stopping condition $\neq$ TRUE* **do**
9 | *% Evaluate all the individuals*
10 | [Ind, Arch] ← FitnessEvaluation (Ind, Arch);
11 | *% Copy the fittest individuals in the archive*
12 | Arch ← NaturalSelection (Ind, Arch);
13 | *% Crossover operation*
14 | Ind ← Crossover (Arch, P_C, N_{pop});
15 | *% Mutation operation*
16 | Ind ← Mutation (Ind, P_M);
17 | *% Evaluate the offspring*
18 | Ind ← ObjectiveFunctionsEvaluation (Ind);
19 **end**

Equation 5, where $\mathbf{z}^i = \left[f_1\left(\mathbf{x}^i\right), f_2\left(\mathbf{x}^i\right) \right]$, d represents the euclidean distance between two vectors, $\mathbf{z}^k$ is the k-nearest solution of $\mathbf{z}^i$ in the objective space, where $k = \sqrt{N_{pop} + N_{arch}}$ (N_{pop} is the population size, and N_{arch} is the size of the archive). P_t is the set of solutions in the population in the time t. P_t^{arch} is the set of solutions stored in the archive in the time t. $\mathbf{S}(\mathbf{z}^j)$ represents the number of solutions dominated by $\mathbf{z}^j$. Due to the fact that the main target of a multiobjective optimization algorithm consists in finding a wide range of non-dominated solutions evenly distributed in the feasible objective space, the best solutions will be those that minimize this fitness function [8].

On the other hand, the pseudo-code of SPEA2 is presented in Algorithm 1. In this pseudo-code, P_C is the crossover probability (i.e. the probability of recombination among parents), P_M is the mutation probability (i.e. the probability of changing the genetic information of an individual), and we have selected the maximum number of generations as our stopping condition (we use the same stopping condition as in [6, 7]).

$$f_{\text{fitness}}\left(\mathbf{z}^i\right) = \frac{1}{2 + d\left(\mathbf{z}^i, \mathbf{z}^k\right)} + \sum_{j \in P_t + P_t^{arch}, \mathbf{z}^j \prec \mathbf{z}^i} \mathbf{S}\left(\mathbf{z}^j\right). \tag{5}$$

Individual Representation. Every individual ($\mathbf{x}^i$) of the population and the archive is a vector in which we store the state of each network cell: $x_j^i = 1$ if the cell j is a Reporting Cell, otherwise $x_j^i = 0$. The initial population of parents is randomly generated following a discrete uniform distribution.

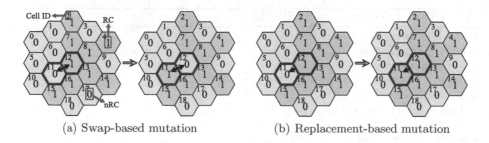

(a) Swap-based mutation   (b) Replacement-based mutation

Fig. 3. Mutation operations

Crossover Operation. This evolutionary operator is performed with probability P_C to generate a new population of N_{pop} offspring by recombining the parent population [11]. In this operation, we firstly select two pairs of individuals stored in the archive. Then, each parent will be the best solution from each pair (according to the fitness function defined in Equation 5). After that, each parent is cut in pieces and recombined with the pieces of the other parent in order to generate two new individuals. In this work, we use a maximum number of crossover points equal to 4. Finally, these two new solutions are evaluated, and only the best one is stored in the population of offspring.

Mutation Operations. The mutation operation is performed with probability P_M [11]. This evolutionary operator consists in modifying the genetic information of the offspring with the aim of finding new solutions of greater or equal quality. We propose two mutation operations specific to solve the RCPP. In the first mutation operation, we swap the value of two neighboring cells that belong to different states (i.e. RC and nRC, see Fig. 3(a)). On the other hand, in the second mutation operation, we change the value of a randomly selected network cell by the value of one of its neighboring cells that belongs to the other state. Fig. 3(b) shows an example of this mutation operation.

Natural Selection. This evolutionary operator is performed before the crossover operation with the goal of selecting the fittest individuals (i.e. the best solutions found so far according to the fitness function, Equation 5) as the parent population. Therefore, this operator is the basis of the continuous improvement by the algorithm.

5   Experimental Results

With the aim of knowing the quality of our proposal, we have tested it in 12 test networks of different complexity: TN1-TN3 (networks of 4x4 cells), TN4-TN6 (networks of 6x6 cells), TN7-TN9 (networks of 8x8 cells), and TN10-TN12 (networks of 10x10 cells). The main appeal of these test networks is that they

Table 1. Statistics of Hypervolume (I_H) for our approach

Ref. points	Test Network											
	TN1	TN2	TN3	TN4	TN5	TN6	TN7	TN8	TN9	TN10	TN11	TN12
LU_{max}	11480	11428	11867	30861	30237	29864	47854	46184	42970	54428	49336	49775
LU_{min}	0	0	0	0	0	0	0	0	0	0	0	0
PA_{max}	125184	124576	125248	256500	256788	255636	691008	680000	690112	1691300	1666400	1676400
PA_{min}	7824	7786	7828	7125	7133	7101	10797	10625	10783	16913	16664	16764
Statistics of I_H												
Aver.(%)	60.57	61.44	62.58	71.75	71.89	72.63	75.78	76.53	76.69	78.47	79.54	79.48
Dev.(%)	0.01	0.00	0.00	0.02	0.01	0.06	0.11	0.10	0.16	0.13	0.22	0.17

present a subscriber's call and mobility pattern close to the one that we can find in real mobile networks [5]. Furthermore, we have compared our results with those published in other works [5–7], where different single-objective meta-heuristics were proposed: Geometric Particle Swarm Optimization (GPSO) [5], a hybridized Hopfield Neural Network with the Ball Dropping technique (HNN-BD) [5], Differential Evolution (DE) [6], and Scatter Search (SS) [7]. Additionally, we have modeled and solved the RCPP described in Section 3 with the IBM ILOG CPLEX Optimizer (a well-known high-performance solver [13]) and with the NOMAD algorithm (Non-linear Optimization with the Mesh Adaptive Direct search [14, 15]). The NOMAD algorithm is a recent heuristic that treats the objective function as a black-box, which is very useful for non-differentiable problems [14]. These last two programs have been limited with an execution time 10 times higher than the execution time of our proposal (our proposal has an execution time of 10 minutes). The comparative study with other optimization techniques published in the literature is shown in Section 5.1.

On the other hand, in order to perform a fair comparison with the algorithms published in [5–7], we have configured our proposal with the same population size (Npop = 175) and the same stopping condition (1000 generations). A run-time comparison cannot be conducted because the execution time of GPSO, HNN-BD, DE, and SS is not available. The other parameters of SPEA2 have been configured by means of a parametric study of 30 independent runs per experiment. We have chosen the parameters configuration that maximizes the value of Hypervolume (I_H, see Section 4.1): $P_C = 0.75$, $P_M = 0.25$, $N_{arch} = 175$.

Table 1 gathers statistical data of the Hypervolume (mean and standard deviation) of 30 independent runs. In this table, we also present the value of reference points, which are obtained by means of the two extreme configurations of Reporting Cells: Never Update (when all the network cells are nRC) and Always Update (when all the network cells are RC). Fig. 4(a) - Fig. 4(l) show a graphical representation of the Pareto Fronts associated with the mean Hypervolume. In this figure, we can observe that our proposal achieves good Pareto Fronts in all the test networks, which extend from the Never Update to the Always Update strategy. However, it should be noted the appearance of some gaps in the Pareto Front of the following test networks: TN4, TN9, TN10, and TN12. The analysis of such gaps could be an interesting future work.

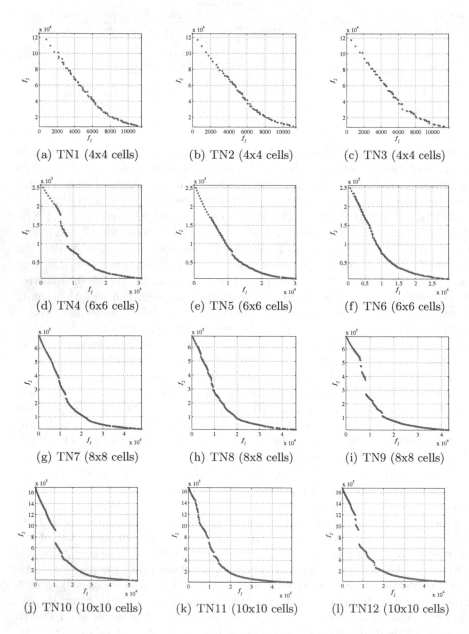

Fig. 4. Pareto Fronts associated with the mean Hypervolume

Table 2. Comparison with other works: f_3^{SO} (10). We indicate with "-" the information that is not available in the corresponding reference

Algorithm		TN1	TN2	TN3	TN4	TN5	Test Network TN6	TN7	TN8	TN9	TN10	TN11	TN12
	Min.	98,535	97,156	95,038	173,701	182,331	174,519	308,702	287,149	264,204	386,721	358,392	370,868
SPEA2	Aver.	98,535	97,156	95,038	173,701	182,331	174,711	308,822	287,149	264,279	387,764	359,077	371,331
	Dev.(%)	0.00	0.00	0.00	0.00	0.00	0.19	0.06	0.00	0.07	0.22	0.12	0.10
	Min.	98,535	97,156	95,038	181,677	200,990	186,481	375,103	351,505	407,457	514,504	468,118	514,514
CPLEX	Aver.	98,535	97,156	95,038	181,677	200,990	186,481	375,103	351,505	407,457	514,504	468,118	514,514
	Dev.(%)	0.00	0.00	0.00	0.00	0.00	0.00	0.00	0.00	0.00	0.00	0.00	0.00
	Min.	98,535	97,156	95,038	177,647	200,069	175,620	316,328	301,833	271,637	404,447	371,091	382,180
NOMAD	Aver.	100,366	100,065	100,131	189,263	221,889	182,289	326,008	312,497	289,309	415,740	379,725	391,627
	Dev.(%)	2.54	2.92	4.30	4.44	6.27	2.66	2.19	2.59	3.09	1.82	1.31	0.81
	Min	98,535	97,156	95,038	173,701	182,331	174,519	307,695	287,149	264,204	385,927	357,714	370,868
SS[7]	Aver.	-	-	-	-	-	-	-	-	-	-	-	-
	Dev.(%)	-	-	-	-	-	-	-	-	-	-	-	-
	Min.	98,535	97,156	95,038	173,701	182,331	174,519	308,401	287,149	264,204	386,681	358,167	371,829
DE[6]	Aver.	-	-	-	-	-	-	-	-	-	-	-	-
	Dev.(%)	-	-	-	-	-	-	-	-	-	-	-	-
	Min.	98,535	97,156	95,038	173,701	182,331	174,519	308,929	287,149	264,204	386,351	358,167	370,868
HNN-BD[5]	Aver.	98,627	97,655	95,751	174,690	182,430	176,050	311,351	287,149	264,695	387,820	359,036	374,205
	Dev.(%)	0.09	0.51	0.75	0.56	0.05	0.87	0.78	0.00	0.18	0.38	0.24	0.89
	Min.	98,535	97,156	95,038	173,701	182,331	174,519	308,401	287,149	264,204	385,972	359,191	370,868
GPSO[5]	Aver.	98,535	97,156	95,038	174,090	182,331	175,080	310,062	287,805	264,475	387,825	359,928	373,722
	Dev.(%)	0.00	0.00	0.00	0.22	0.00	0.32	0.53	0.22	0.10	0.48	0.20	0.76

5.1 Comparison with Other Optimization Techniques

In this section, we present a comparative study with other algorithms (by other authors) published in the literature: Geometric Particle Swarm Optimization (GPSO) [5], a hybridized Hopfield Neural Network with the Ball Dropping technique (HNN-BD) [5], Differential Evolution (DE) [6], and Scatter Search [7]. Furthermore, we have modeled and solved the RCPP with the IBM ILOG CPLEX Optimizer [13] and the NOMAD algorithm [14, 15].

It should be noted that all of these algorithms belong to the single-objective optimization field, but our proposal is a multiobjective evolutionary algorithm. Therefore, in order to perform a fair comparison with these optimization techniques, we must search in our Pareto Fronts (see Fig. 4(a) - Fig. 4(l)) the solutions that best fit the objective function used by these single-objective optimization algorithms (Equation 3 with $\beta = 10 : f_3^{SO}$ (10)).

The results of this comparison (of 30 independent runs per experiment) are gathered in Table 2. In this table, we present the following information for each test network and for each algorithm: the minimum value (Min.), the average value (Aver.), and the deviation percentage from the minimum value (Dev.(%)) of f_3^{SO} (10). Regrettably, the authors of DE [6] and SS [7] do not provide the average value (Aver.) and the deviation percentage (Dev.(%)) in their works. This table reveals that our version of SPEA2 is very interesting because it achieves the best average values of f_3^{SO} (10) in all the test networks (with the exception of the test network TN11), and it also obtains the minimum known value of f_3^{SO} (10) in most of the test networks (except in TN7, TN10, and TN11). This last is far from trivial, because we are comparing with algorithms specialized in finding only one solution. On the other hand, it is noteworthy that the two heuristic methods studied in this work (the IBM ILOG CPLEX Optimizer and the NOMAD algorithm) are only competitive in the less complex test networks (TN1 - TN3 for the IBM ILOG CPLEX Optimizer, and TN1 - TN4 and TN6 for NOMAD).

6   Conclusion and Future Work

In this work, we propose a multiobjective approach to optimize the signaling load of one of the most important management systems in mobile networks: the mobile location management system. The problem is defined and addressed in terms of Reporting Cells. This strategy consists in selecting a subset of network cells as Reporting Cells with the aim of partially controlling the subscribers' movement. In this way, a mobile station only updates its location when it enters in a new Reporting Cell, and the paging is only performed in a few network cells. Our proposal can be considered as a novel contribution because, to the best of the authors' knowledge, there are no other works in the literature that tackle the Reporting Cell Planning Problem with a multiobjective optimization algorithm. In this manuscript, we have adapted (in terms of our evolutionary operators specific to solve the RCPP) a well-known multiobjective evolutionary algorithm: the Strength Pareto Evolutionary Algorithm 2 (SPEA2). In order to known the quality of our proposal, we have tested it in 12 test networks of different complexity, and then, we have compared our results with those obtained by other optimization techniques (all of them belonging to the single-objective optimization field). This experimental study clarified that our version of SPEA2 is very interesting because it achieves good Pareto Fronts and, at the same time, it outperforms (in average) the results of these single-objective optimization algorithms. This is far from trivial, because we have compared with algorithms specialized in finding only one solution.

As a future work, it would be interesting to study other multiobjective optimization techniques, other location management strategies, and other test networks.

Acknowledgements. This work was partially funded by the Spanish Ministry of Economy and Competitiveness and the ERDF (European Regional Development Fund), under the contract TIN2012-30685 (BIO project). The work of Víctor Berrocal-Plaza has been developed under the Grant FPU-AP2010-5841 from the Spanish Government.

References

1. Agrawal, D., Zeng, Q.: Introduction to Wireless and Mobile Systems. Cengage Learning (2010)
2. Kyamakya, K., Jobmann, K.: Location management in cellular networks: classification of the most important paradigms, realistic simulation framework, and relative performance analysis. IEEE Transactions on Vehicular Technology 54(2), 687–708 (2005)

3. Mukherjee, A., Bandyopadhyay, S., Saha, D.: Location Management and Routing in Mobile Wireless Networks. Artech House mobile communications series. Artech House (2003)
4. Subrata, R., Zomaya, A.Y.: A comparison of three artificial life techniques for Reporting Cell planning in mobile computing. IEEE Trans. Parallel Distrib. Syst. 14(2), 142–153 (2003)
5. Alba, E., García-Nieto, J., Taheri, J., Zomaya, A.Y.: New research in nature inspired algorithms for mobility management in GSM networks. In: Giacobini, M., Brabazon, A., Cagnoni, S., Di Caro, G.A., Drechsler, R., Ekárt, A., Esparcia-Alcázar, A.I., Farooq, M., Fink, A., McCormack, J., O'Neill, M., Romero, J., Rothlauf, F., Squillero, G., Uyar, A.Ş., Yang, S. (eds.) EvoWorkshops 2008. LNCS, vol. 4974, pp. 1–10. Springer, Heidelberg (2008)
6. Almeida-Luz, S.M., Vega-Rodríguez, M.A., Gómez-Pulido, J.A., Sánchez-Pérez, J.M.: Applying differential evolution to the Reporting Cells problem. In: International Multiconference on Computer Science and Information Technology, pp. 65–71 (2008)
7. Almeida-Luz, S.M., Vega-Rodríguez, M.A., Gómez-Pulido, J.A., Sánchez-Pérez, J.M.: Solving the reporting cells problem using a scatter search based algorithm. In: Szczuka, M., Kryszkiewicz, M., Ramanna, S., Jensen, R., Hu, Q. (eds.) RSCTC 2010. LNCS, vol. 6086, pp. 534–543. Springer, Heidelberg (2010)
8. Zitzler, E., Laumanns, M., Thiele, L.: SPEA2: Improving the strength pareto evolutionary algorithm for multiobjective optimization. In: Giannakoglou, K.C., Tsahalis, D.T., Périaux, J., Papailiou, K.D., Fogarty, T. (eds.) Evolutionary Methods for Design Optimization and Control with Applications to Industrial Problems, Athens, Greece, pp. 95–100. International Center for Numerical Methods in Engineering (2001)
9. Bar-Noy, A., Kessler, I.: Tracking mobile users in wireless communications networks. IEEE Transactions on Information Theory 39(6), 1877–1886 (1993)
10. Hac, A., Zhou, X.: Locating strategies for Personal Communication Networks: A novel tracking strategy. IEEE Journal on Selected Areas in Communications 15(8), 1425–1436 (1997)
11. Coello, C.A.C., Lamont, G.B., Veldhuizen, D.A.V.: Evolutionary Algorithms for Solving Multi-Objective Problems (Genetic and Evolutionary Computation). Springer-Verlag New York, Inc., Secaucus (2006)
12. Zitzler, E., Thiele, L.: An evolutionary algorithm for multiobjective optimization: The strength Pareto approach. Technical Report 43, Computer Engineering and Communication Networks Lab (TIK), Swiss Federal Institute of Technology (ETH), Zürich, Switzerland (1998)
13. ILOG, Inc.: ILOG CPLEX: High-performance software for mathematical programming and optimization (2006), http://www.ilog.com/products/cplex/
14. Le Digabel, S.: Algorithm 909: NOMAD: Nonlinear Optimization with the MADS Algorithm. ACM Transactions on Mathematical Software 37(4), 44:1–44:15 (2011)
15. Currie, J., Wilson, D.I.: OPTI: Lowering the Barrier Between Open Source Optimizers and the Industrial MATLAB User. In: Foundations of Computer-Aided Process Operations, Georgia, USA (2012)

Human Emotion Estimation Using Wavelet Transform and t-ROIs for Fusion of Visible Images and Thermal Image Sequences

Hung Nguyen[1], Fan Chen[1], Kazunori Kotani[1], and Bac Le[2]

[1] Japan Advanced Institute of Science and Technology,
1-1 Asahidai, Nomi, Ishikawa, Japan
{nvhung,chen-fan,ikko}@jaist.ac.jp
[2] University of Science, Ho Chi Minh city,
227 Nguyen Van Cu, Ho Chi Minh city, Vietnam
lhbac@hcmuns.edu.vn

Abstract. Most studies in human emotion estimation focus on visible image-based analysis which is sensitive to illumination changes. Under uncontrolled operating conditions, estimation accuracy degrades significantly. In this paper, we integrate both visible images and thermal image sequences. First, to address limitations of thermal infrared (IR) images, such as being opaque to eyeglasses, we apply thermal Regions of Interest (t-ROIs) to sequences of thermal images. Then, wavelet transform is applied to visible images. Second, features are selected and fused from visible features and thermal features. Third, fusion decision using Principal Component Analysis (PCA), Eigen-space Method based on class-features (EMC), PCA-EMC is applied. Experiments on the Kotani Thermal Facial Emotion (KTFE) database show the effectiveness of proposed methods.

Keywords: Human emotions, thermal images, emotion estimation, feature fusion, decision fusion, thermal image sequences, KTFE database.

1 Introduction

In the last decade, automated estimation of human emotions has attracted the interest of many researchers, because such systems will have numerous applications in security, medicine, and especially human-computer interaction. Many previous works [1] proposed have been inclined towards developing facial expression estimation. Nevertheless, there is a lack of accurate and robust facial expression estimation methods to be deployed in uncontrolled environments. When the lighting is dim or when it does not uniformly illuminate the face, the accuracy decreases considerably. Moreover, human emotions estimation based on only the visible spectrum has proved to be difficult in cases where there are emotion changes that expressions do not show. Using thermal infrared (IR) imagery, which is not sensitive to light conditions, is a new and innovative way to fill the gap in the human emotions estimation field. Besides, human emotions could

B. Murgante et al. (Eds.): ICCSA 2014, Part VI, LNCS 8584, pp. 224–235, 2014.

be manifested by changing temperature of face skin which is obtained by an IR camera. Consequently, thermal infrared imagery gives us more information to help us robustly estimate human emotions. Although there are many significant advantages when we use IR imagery, it has several drawbacks. Firstly, thermal data are subjected to change together with body temperature caused by variable ambient temperatures. Secondly, presence of eyeglasses may result in loss of useful information around the eyes. Glass is opaque to IR, and object made of glass act as temperature screen, completely occluding the parts located behind them. Hence, the sensitivity of IR imagery is decreased by facial occlusions. Thirdly, there are some facial regions not receptive to the emotion changes. To eliminate the effects of these challenging problems above, we propose fusion of visible images and sequence of thermal images. To estimate five emotions, we use the fusion of conventional methods Principal Component Analysis (PCA), Eigenspace Method based on class-features (EMC), and PCA-EMC over obtained the fusion features.

2   Related Work

In the recent years, a number of studies have demonstrated that thermal infrared imagery offers a promising alternative to visible imagery in facial emotion estimation problems by better handling the visible illumination changes. Sophie Jarlier et al. [2] extracted the features as representative temperature maps of nine action units (AUs) and used K-nearest neighbor to classify seven expressions. The database for testing has four persons and the accuracy rate is 56.4%. M.M.Khan et al. [3] suggested using Facial Thermal Feature Points (FTFPs), which are defined as facial points that undergo significant thermal changes in presenting an expression, and used Linear Discriminant Analysis (LDA) to classify intentional facial expressions based on Thermal Intensity Values (TIVs) recorded at the Facial Thermal Feature Points (FTFPs). The database has sixteen persons with five expressions and the accuracy rate ranges from 66.3% to 83.8%. L.Trujillo et al. [4] proposed using a local and global automatic feature localization procedure to perform facial expression in thermal images. They used PCA to reduce the dimension and interest point clustering to estimate facial feature localization and Support Vector Machine (SVM) to classify three expressions. B.Hernandez et al. [5] used SVM to classify the expressions surprise, happy, neutral from two inputs. The first input consists of selections of a set of suitable regions where the feature extraction is performed, second input is the Gray Level Co-occurrence Matrix used to compute region descriptors of the IR images. B.R.Nhan et al. [6] extracted time, frequency and time-frequency features from thermal infrared data to classify the natural responses in terms of subject-indicated levels of arousal and valence stimulated by the International Affective Picture System. Y.Yoshitomi et al. [7] used two dimensional detection of temperature distribution on the face using infrared rays. Based on studies in the field of psychology, several blocks on the face are chosen for measuring the local temperature difference. With Back Propagation Neutral Network, the facial expression is recognized. The recognition accuracy reaches 90% with neutral, happy, surprising and sad expressions. However, the testing database is obtained from only one female frontal view. Y. Yoshimomi generated feature vectors by using a two-dimensional Discrete Cosine Transformation (2D-DCT) to transform the grayscale values of each block in the facial area of an image into their frequency components, and used them to recognize five expressions, including angry, happy, neutral, sad, and surprise. The mean expression accuracy is 80% with four test subjects [8]. Y.Koda et al. used the idea from [8] and added a proposed method for efficiently updating of training data, by only updating the training data with happy and neutral facial expression after an interval [9]. The expression accuracy increased from 80% to 87% with this new approach. All these studies with thermal infrared imagery have shown that the facial temperature changing is useful for estimating the human emotions.

Recently, a little attention has been paid to facial emotion estimation by using fusion information from visible images and thermal information. Wang et al. [10] proposed both decision-level and feature-level fusion methods using visible and IR imagery. In feature-level, they used tools for the Active Appearance Model (AAM) to extract features and extracted three features of head motion for visible feature and calculated several statistical parameters including mean, standard deviation, minimum and maximum as IR features. To select the feature, they used F-test statistic. They also used Bayesians networks (BNs) and SVMs to obtain the feature fusion. In decision-level, BNs and SVMs are used to classify three emotions, happiness, fear and disgust. The results show that their methods improved about 1.35% accuracy compare with only using visible features. Yoshitomi et al. [11] proposed decision-level fusion of voices, visual and IR imagery to recognize the affective states. DCT is used to extract the visible and IR features, then two neutral networks are trained for obtained visible and IR features, respectively. For voice recognition, Hidden Markov Models (HMMs) are used. To decide the results, simple weighted voting is used. Following the related work, there are a few researches using fusion of visible and thermal imagery or these approaches that use the extracted features from a single infrared thermal image may lose some useful information which could be contained in the sequences. Therefore, we consider two methods of human emotion estimation by fusing visible images and sequence of thermal imagery at decision-level and feature-level respectively.

3   Methods

In this section, we propose a feature fusion method to integrate visible images and sequence of thermal images by delicate selection of representative features (i.e. t-ROI) in Section 3.1 and a decision-level fusion which explores the best fusion weights of features in Section 3.2.

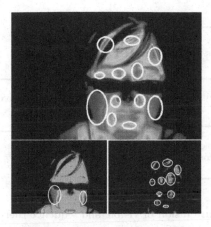

Fig. 1. An example of t-ROIs

3.1   Feature-Level Fusion

Before selecting features, we perform some preprocessings. First, with sequences of thermal images, we find the regions of interest based on t-ROIs.

In our definition, interest regions are regions in which temperature increases or decreases significantly when human emotions change. We use the two regions which are the hottest and coldest regions of the face, except the eyeglasses, usually the forehead, eyeholes, and cheek-bone regions, as our interest regions. Before finding the t-ROIs, to avoid any ambient temperature change from frame to frame, we update the temperature of each point of each frame based on the difference between mean of ambient temperature and mean of the first m frame ambient temperature.

Let f be a map from face ($F \subset R^2$) to temperature ($T \subset R$) space

$$f : F \to T$$

$$(i, j) \mapsto f(i, j)$$

We obtain the t-ROIs by using the following equations:

$$\Delta T_F = T^F_{Max} - T^F_{Min}; \delta T_F = \Delta T_F/5$$

$$L^F_{k,idx} = \{(i,j) \in F | T^F_{Min} + \delta T_F * (idx-1) \leq f(i,j) < T^F_{Max} - \delta T_F * (5-idx)\} \quad (1)$$

where T^F_{Max}, T^F_{Min} are maximum and minimum of temperature of each human face at frame k, respectively; $idx \in \{2, 5\}$.

Second, with visible images, to eliminate of effects of non-uniform illumination and to omit unnecessary details, we use wavelet transform with Antonini filter bank [12].

To select the feature between visible feature and thermal feature, we perform feature-level fusion of visible and thermal image by using t-ROIs and PCA.

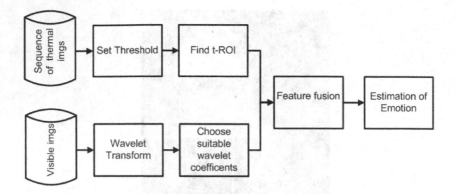

Fig. 2. Feature fusion of visible and sequence thermal image

Step 1. Find t-ROIs over sequence of thermal images.
Step 2. Apply Wavelet transform over visible facial images and keep LL.
Step 3. Apply PCA over each t-ROI.
Step 4. Build matrix from feature vectors obtained from step 2 and 3.
Step 5. Using PCA, EMC to classify emotions

3.2 Decision-Level Fusion

To estimate human emotions, we use decision fusion method of PCA, EMC and PCA-EMC.

With PCA, the aim is to build a face space, including the basis vectors called principal components, which better describes the face images [13]. The difference between PCA and EMC is that PCA finds the eigenvector to maximize the total variance of the projection to line, while EMC [14] obtains eigenvectors to maximize the difference between the within-class and between-class variance.

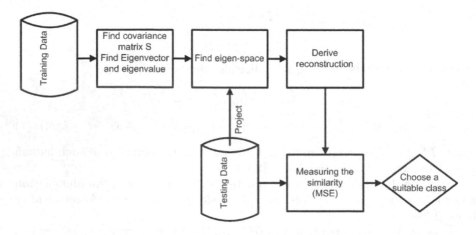

Fig. 3. Estimation of emotion using PCA

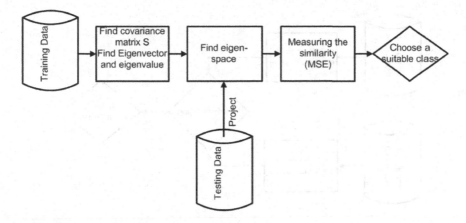

Fig. 4. Estimation of emotion using EMC

The difference between the within-class and between-class variance is calculated as following:

$$S = S_B - S_W. \tag{2}$$

$$S_B = \frac{1}{M} \sum_{f \in F} M_f (\overline{x}_f - \overline{x})(\overline{x}_f - \overline{x})^\tau. \tag{3}$$

$$S_W = \frac{1}{M} \sum_{f \in F} \sum_{f \in F}^{M_f} M_f (\overline{x}_{fm} - \overline{x}_f)(\overline{x}_{fm} - \overline{x}_f)^\tau. \tag{4}$$

$$\overline{x}_f = \frac{1}{M} \sum_{m=1}^{M_f} x_{fm}; \overline{x} = \frac{1}{M} \sum_{f \in F} \sum_{m=1}^{M_f} x_{fm}. \tag{5}$$

where F is a set of expression classes, M_f facial-patterns are given for each class $f \in F$ and x_{fm} is an N-dimension vector of the $m-th$ facial patterns, $m = \overline{1, M_f}$

Figure 3 shows the procedure of estimating human emotions using PCA. Figure 4 shows the procedure of estimating human emotions using EMC.

To estimate human emotions using PCA-EMC, first, we use PCA to reduce the dimension and apply EMC to the obtained eigentspace.

Figure 5 shows the general procedure to estimate human emotions using decision fusion. When using decision fusion of PCA, we used the estimation of emotion module as described in figure 3.

To determine the best class of emotions, after using PCA, the voting method with weights is used. The weights, 2/3 and 4/3 are set to fusion data and visible image, respectively. We determine the emotion class f of input image by choosing j satisfied minimum of following equation:

$$f = argmin \left(w_1 * MSE_j^{VI} + w_2 * MSE_j^{FU} \right) \tag{6}$$

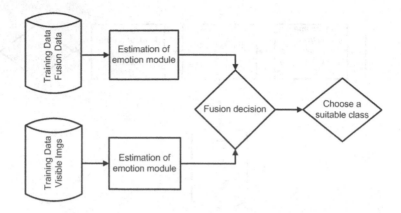

Fig. 5. Estimation of emotion using decision fusion

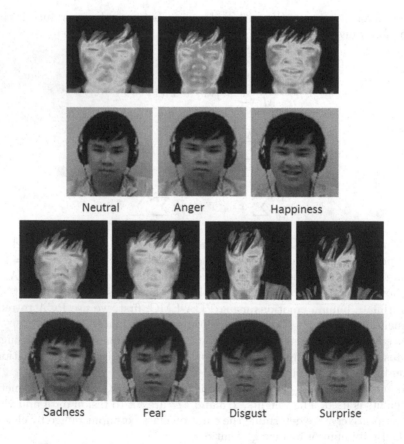

Fig. 6. Sample thermal and visible images of seven emotions

where MSE_j^{VI} and MSE_j^{FU} are mean square errors calculated at class j of visible image and fusion data. $w_1 = \frac{4}{3}$ and $w_2 = \frac{2}{3}$

To estimate human emotion using decision fusion of EMC, we used the estimation of emotion module as described in figure 4. Figure 5 shows the procedure to estimate human emotions using decision fusion of EMC.

To determine the best class of emotions, after using EMC, the voting method with weights is used. The weights, 4/3 and 2/3 are set to fusion data and visible image, respectively. We determine the emotion class f of input image by choosing j satisfied maximum of following equation:

$$k = max_i \frac{f_h^{VI} * F_i^{VI}}{\|f_h^{VI}\| * \|F_i^{VI}\|}, i = \overline{1, n} \tag{7}$$

$$g = max_i \frac{f_h^{FU} * F_i^{FU}}{\|f_h^{FU}\| * \|F_i^{FU}\|}, i = \overline{1, n} \tag{8}$$

$$f = argmax\left(w_1 * k + w_2 * g\right), \tag{9}$$

where n is a number of the training images of class j; f_h^{VI} and f_h^{FU} are testing image h of visible image and fusion data, respectively; F_i^{VI} and F_i^{FU} are vector i of eigenface of visible image and fusion data, respectively; $w_1 = \frac{2}{3}$ and $w_2 = \frac{4}{3}$

4   Database

The KTFE database [15] includes 131GB visible and thermal facial emotion videos, visible facial expression image database and thermal facial expression image data-base. This database contains 26 subjects who are Vietnamese, Japanese, Thai from 11 year-old to 32 year-old with seven emotions. The example of visible and thermal images is shown in Fig.6.

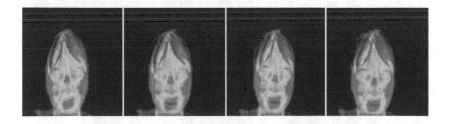

Fig. 7. Sample sequence of thermal images

From draw data of KTFE database, we extract manually visible images and sequences of thermal images based on self-reports of participants, expressions and changing of facial temperatures. Causing the time-lag phenomenon, the sequence of thermal images are designed from a frame which we extracted the visible image to a frame which is after the participant emotion is neutral.

5   Experimental Results

In our experiments, we separate the training and testing data as 70% and 30% of total visible images, thermal images, and fusion of visible and thermal image sequence. Fig.8 shows the results of emotion estimation of EMC with visible images (vi_EMC), thermal images (ther_EMC) and fusion of visible images and sequence of thermal images (fu_EMC). Accuracy of estimating human emotion using thermal images is lower than using visible images. Emotions of thermal images are always not clearer than emotions of visible images. Therefore, with EMC methods, good for classification, the results using visible images are better than results using thermal images. In general, average accuracy of each emotion increases when we use fusion information. The results prove the necessary of fusion information.

Fig.9 shows the results of emotion estimation of PCA with visible images (vi_PCA), thermal images (ther_PCA) and fusion of visible images and sequence of thermal images (fu_PCA). With PCA, accuracy using thermal images is better than accuracy using visible images. Although, emotions of thermal images are not clearer than emotion of visible image, PCA works better than EMC, which is good to classify each emotion. In general, with PCA, using fusion data gives the best results comparing using thermal and visible images.

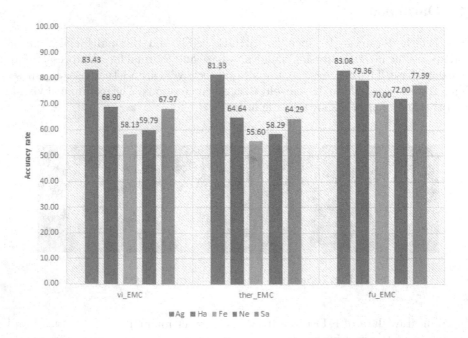

Fig. 8. Human emotion estimation results using EMC

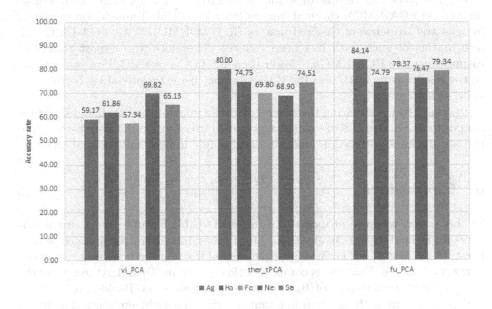

Fig. 9. Human emotion estimation results using PCA

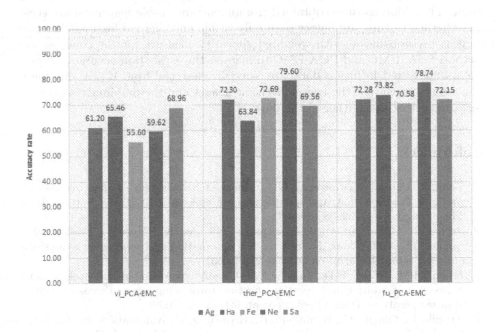

Fig. 10. Human emotion estimation results using PCA-EMC

Fig.10 shows the results of emotion estimation of PCA-EMC with visible images (vi_PCA-EMC), thermal images (ther_PCA-EMC) and fusion of visible images and sequence of thermal images (fu_PCA-EMC). With PCA-EMC, the information of each data has been reduced. Therefore, the almost results of estimation using PCA-EMC are lower than using PCA or EMC. However, similar to the results using EMC and PCA, the accuracy using fusion data is better than using other data.

In conclusion, comparing the results of visible images, thermal images, and fusion data, the accuracies of estimating emotion using fusion data are better than the accuracies of estimation emotion using visible images and thermal images.

6   Conclusions

In this paper, we have proposed the fusion of visible features and thermal features for estimating human emotions. Our method has several advantaged points. First, to the best of our knowledge, this is one of the first methods using sequence of thermal images. Emotion is complex action of human. To understand it clearly, using a single image can not figure out the exact emotion. Besides, using thermal information with single frame can not give the right emotion. Therefore, it is necessary to use sequence of thermal images. Second, with t-ROIs, we fill the gaps of thermal image, eyeglass problem. Third, using wavelet transform for visible image gives several advantages such as to reduce the unnecessary coarse, so on. The fusion features, obtained from important visible features and necessary thermal feature, are better than only visible and thermal features. We also suggest decision fusion with weighted similarity measure for the conventional method PCA, EMC and PCA-EMC to increase the estimation accuracy. Experiments are tested in fusion database, specially designed from KTFE database. The results prove that the fusion of visible images and thermal image sequences performs better than either of the data.

References

1. Zeng, Z., Pantic, M., Roisman, G.T., Huang, T.S.: A survey of affect recognition methods: Audio, visual, and spontaneous expressions. IEEE Trans. Pattern Anal. Mach. Intell. 31(1), 39–58 (2009)
2. Jarlier, S., Grandjean, D., Delplanque, S., N'Diaye, K., Cayeux, I., Velazco, M., Sander, D., Vuilleumier, P., Schere, K.: Automatic facial expression analysis: a survey. Pattern Recognition 36, 259–275 (2003)
3. Khan, M.M., Ward, R.D., Ingleby, M.: Classifying pretended and evoked facial expression of positive and negative affective states using infrared measurement of skin temperature. Trans. Appl. Percept. 6(1), 1–22 (2009)
4. Trujillo, L., Olague, G., Hammoud, R., Hernandez, B.: Automatic feature localization in thermal images for facial expression recognition. In: IEEE Computer Society Conference on Computer Vision and Pattern Recognition-Workshops, CVPR Workshops, p. 14 (2005)

5. Hernández, B., Olague, G., Hammoud, R., Trujillo, L., Romero, E.: Visual learning of texture descriptors for facial expression recognition in thermal imagery. Computer Vision and Image Understanding 106, 258–269 (2007)
6. Nhan, B.R., Chau, T.: Classifying affective states using thermal infrared imaging of the human face. IEEE Transactions on Biomedical Engineering 57, 979–987 (2010)
7. Yoshitomi, Y., Miyawaki, N., Tomita, S., Kimura, S.: Facial expression recognition using thermal image processing and neural network. In: 6th IEEE International Workshop Robot and Human Communication, ROMAN 1997 Proceedings, pp. 380–385 (1997)
8. Yoshitomi, Y.: Facial expression recognition for speaker using thermal image processing and speech recognition system. In: Proceedings of the 10th WSEAS International Conference on Applied Computer Science, pp. 182–186 (2010)
9. Koda, Y., Yoshitomi, Y., Nakano, M., Tabuse, M.: A facial expression recognition for a speaker of a phoneme of vowel using thermal image processing and a speech recognition system. In: The 18th IEEE International Symposium on Robot and Human Interactive Communication, ROMAN 2009, pp. 955–960 (2009)
10. Wang, S., He, S., Wu, Y., He, M., Ji, Q.: Fusion of visible and thermal images for facial expression recognition. J Frontiers of Computer Science (2014)
11. Yoshitomi, Y., Kim, S., Kawano, T., Kilazoe, T.: Effect of sensor fusion for recognition of emotional states using voice, face image and thermal image of face. In: Proceedings of the 9th IEEE International Workshop on Robot and Human Interactive Communication, pp. 178–183 (2000)
12. Antonini, M., Barlaud, M., Mathieu, P., Daubechies, I.: Image coding using wavelet transform. IEEE Trans. Image Processing 1, 205–220 (1992)
13. Lin, D.T.: Facial Expression Classification Using PCA and Hierarchical Radial Basis Function Network. Journal of Information Science and Engineering 22, 1033–1046 (2006)
14. Kurozumi, T., Shinza, Y., Kenmochi, Y., Kotani, K.: Facial Individuality and Expression Analysis by Eigenspace Method Based on Class Features or Multiple Discriminant Analysis. In: ICIP (1999)
15. Nguyen, H., Kotani, K., Chen, F., Le, B.: A thermal facial emotion database and its analysis. In: Klette, R., Rivera, M., Satoh, S. (eds.) PSIVT 2013. LNCS, vol. 8333, pp. 397–408. Springer, Heidelberg (2014)

Efficient Biometric Palm-Print Matching
on Smart-Cards

Rafael Soares Wyant[1], Nadia Nedjah[1], and Luiza de Macedo Mourelle[2]

[1] Department of Electronics Engineering and Telecommunication
[2] Department of System Engineering and Computation,
Engineering Faculty, State University of Rio de Janeiro
{wyant,nadia,ldmm}@eng.uerj.br

Abstract. Biometrics have been used as a solution for system access control, for many years. However, the simple use of biometrics can not be considered as final and perfect solution. Most problems are related to the data transmission way between where the users require access and the servers where the biometric data, captured upon registration, are stored. In this paper, the use smart-cards is adopted as a possible solution to this problem. we propose an efficient implementation of palm-print verification for smart-cards. In this implementation, the matching is done on-card. Thus, the biometric characteristics are always kept in the owner card.

1 Introduction

Biometrics are studies of certain human physical or behavioral characteristics that are capable of distinguishing two different persons. There are certain properties that must be present in the features so that they can be used as a biometrics. Every person or at least the vast majority of people must posses this characteristic. This property is called *universality*. Two persons should differ considering this characteristic. This property is called *distinguishibility*. The characteristic should be invariant with respect to time. This property is called it perdurability. Finally, the characteristic should be measurable. This property is called *collectability* [6]. Currently, there are many characteristics or organs used in biometrics, such as DNA, face, hand veins, fingerprint hand geometry, iris, palm print, voice, among many others. Biometrics are highly secure and convenient for identification and/or verification of individual identity, as they can not be stolen or forgotten besides the extremely high difficulty to be forged [7].

The main application of biometrics is related to access control, i.e. through the verification of biometrics, a person can be granted or denied access to the service in question. In most cases, biometry have a big advantage when compared to other kind of identity certification because only them can really guarantee the authenticity of the claimant. Banking systems, for example, often resort to the use of passwords and alphanumeric codes for letters besides the use of a card to grant an account access. The implementation of a biometric system considerably improves security against misuse. Despite the fact that the use of biometrics is a

B. Murgante et al. (Eds.): ICCSA 2014, Part VI, LNCS 8584, pp. 236–247, 2014.

solution that aims to increase security, the risk of fraud can not be ignored. Many developers believe that the use of biometrics is the final and perfect solution for all identification problems [8].

Besides the problems regarding security, there is also the acceptance problem by users. The use of biometrics has been spreading rapidly in the world and people are starting to think about their own safety when they are asked to register indiscriminately their biometrics in various institutions. After all, their biometric details would be stored in many databases, which are susceptible to attacks. Thus, biometrics, which is unique and invariable in time, would be forever compromised.

In this paper, we study the approach that makes use of smart-cards together with biometrics aiming at increasing the security of access control systems. The objective is to evaluate the possibility of using a multi-application smart-card to perform biometric comparisons. Thus, it would be possible to use a single card for several institutions and biometrics would always be stored in a single card in the possession of the owner. The biometrics details would be stored only in a unique smart-card and the matching is processed on-card.

Palm-prints have been used as a human identifier for over 100 years and is still considered one of the most reliable ways to distinguish a person from another, because of its stability and uniqueness [9]. However, only recently, studies about using it as biometrics have emerged. Comparison of palm-prints, as described in [10], was chosen to be implemented on smart-cards because it requires small amount of memory and low computing effort to obtain the comparison results.

The rest of this paper is organized in 5 sections. First, in Section 2, we define the internal representation of palm-print. Then, in Section 3, we give some details on the actual implementation on smart-cards. After that, we present and discuss the performance and effectiveness of the proposed implementation. Last but not least, in Section 5, we draw some conclusions and point out some future work.

2   Palm-Print

In the work reported in this paper, we used the method proposed in [10]. The extraction of palm-code of a region of interest of the palm-print image using a 2D Gabor filter. The result of the convolution is a matrix of complex numbers and the palm-code consists of two binary matrices representing the real and imaginary parts.

The comparison between two palm-codes is performed by computing the Hamming distance between them. The Hamming distance between two palm-codes $P : (P_R, P_I)$ and $Q : (Q_R, Q_I)$ is defined in (1), wherein $\oplus$ is the binary XOR operator and N^2 the size of the real or imaginary part of $Palm\text{-}codes$. It provides the percentage of different bits between two palm-codes.

$$H_{P,Q} = \frac{\sum\limits_{i=1}^{N} \sum\limits_{j=1}^{N} P_R(i,j) \oplus Q_R(i,j) + P_I(i,j) \oplus Q_I(i,j)}{2N^2} \tag{1}$$

This result can be improved using a relative displacement between the two compared matrices. This due the fact that often the error occurs because of a small phase shift between the stored and input templates. In order to improve the effectiveness of the palm-code matching process, we implemented both horizontal and vertical displacements as well as in both directions, i.e. left and right.

3   Implementation on Smart-Cards

The aim of this work is to use the smart-card to process the matching operation and thereby increase the security level. During the card configuration, the palm-code of the owner is transmitted to the card so that it is stored for future matching upon access requirement. To access the service being protected by the biometry, the card-holder must provide its palm-code, as an input, to confirm his/her identity through the computation of the Hamming distance to the stored template. For implementation purposes, we used the Java Card platform.

3.1   2D Gabor's Filter

Direct comparison of the extracted image with another image is very susceptible to brightness and quality of that image. Inspired by Daugman's work on Iris Biometrics [1], Zhang [10] proposed using 2D Gabor filter to extract the main features of the palm print. This filter allows to neutralize the difference in brightness and quality, bringing forth the possibility of direct comparison.

The 2D Gabor function was proposed by Daugman [3], [2] as a simple model of the visual cortex cells. It is based on the discovery of the crystalline organization of the principal cells of the cortex in the brains of mammals [5]. The 2D Gabor function, as proposed by Daugman, is a spacial bandpass filter that achieves the theoretical limit for a resolution the associated to information in the 2D spatial and 2D Fourier domains.

Gabor [4] showed that there is a "quantum principle" for information: an association of time-frequency domain for 1D signals must necessarily be quantified so that no signal or filter can fit in an area that is less than a certain minimum area. This minimum area, which reflects the inevitable trade-off between time resolution and frequency, has a lower limit of the product, analogous to the Heisenberg uncertainty principle in physics. He found that complex exponential modulated by a Gaussian provides a better result. Equation (2) presents a general form of the 2D Gabor filter, used in [10], to extract the characteristics of a palm-print.

$$G(x, y, \theta, u, \sigma) = \frac{1}{2\pi\sigma^2} \exp^{-\frac{x^2+y^2}{2\sigma^2}} \exp^{2\pi i(uxcos\theta+uysen\theta)}, \tag{2}$$

wherein $i = \sqrt{-1}$, u i a the frequency of the senoidal wave, θ that controls the orientation of the function and σ represents the standadrd deviation of the Gaussian function.

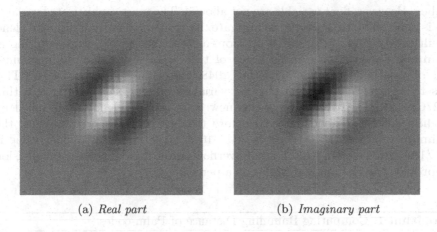

(a) *Real part*   (b) *Imaginary part*

Fig. 1. Response of the 2D Gabor's filter to a pulse

Figure 1a shows the real part of the 2D Gabor filter and Fig 1b shows the imaginary part. To make the filter more robust against brightness, it is transformed into zero CD (direct current), applying (3).

$$G'[s, y, \theta, u, \sigma] = G[x, y, \theta, u, \sigma] - \frac{\sum\limits_{i=-n}^{n} \sum\limits_{j=-n}^{n} G[i, j, \theta, u, \sigma]}{(2n+1)^2}, \tag{3}$$

wherein $(2n+1)^2$ is the filter size. Due of the symmetry, the imaginary part of the filter already has zero DC. Certainly, the success of the extraction of code depends on the choice of parameters θ, u and σ. In [10], a process of refinement was applied to optimize these parameters and found out that $\theta = \pi/4$, $u = 0.0916$ and $\sigma = 5.6179$. These values were used in the implementation of 2D Gabor filter, used in this paper.

3.2   Palm-Code Comparison

The palm-code is represented by two matrices of 32×32 bits: one for the real part and the other for the imaginary part. Thus, the total size to store the template is $2 \times 32 \times 32 = 2048$ bits or 256 bytes. Because of the limit of a single transmission within Java Card, which is of 128 bytes, the transfer of the code is be done in two transmission steps. It is noteworthy that there is no problem in terms of memory storage of the code because current smart-cards offer over 100kb EEPROM and 1 to 3Kb RAM.

The JavaCard not allow the use of 2-dimensional arrays. Therefore, the palm-code is allocated in vector for future comparisons. Bytes are compared using the XOR operator using 128 iterations to go through all the bytes that compose the palm-code. The number of bits with value 1 is counted and the accumulated result is divided by the total number of bits. This division results in a decimal number. However, float variable are not allowed. The most sophisticated variable type is *short* (2 bytes). In order to mitigate the problem of accuracy, the dividend is multiplied by 100 before final division, as described in (1), thus resulting in a Hammig distance with an accuracy of two digits. Note that the maximum value of the dividend is 2048 and thus 204800 after multiplication by 100. This leads to an overflow problem because variables of type *short* can vary within $[-32768, +32767]$. To remedy to this new issue, the dividend is multiplied by 10 while divider is divided by 10 before the final division occurs. Thus, the maximum value of the dividend is $2048/10 = 20.480$ and the divider value is $2048/10 = 204$, avoiding any kind of overflow. After this maneuver, the division will provide the Hamming distance as a percentage.

Algorithm 1. Computing Hamming Distance of Palm-codes

Require: Input Palm-code: $\mathbb{I}$ and template palm-code: $\mathbb{T}$
Ensure: Hamming distance between $\mathbb{I}$ and $\mathbb{T}$: H
1. $Nbits = 0$
2. **for** $i := 1 \rightarrow N^2$ **do**
3. $xored = \mathbb{T}_R(i) \oplus \mathbb{I}_R(i)$
4. $Nbits = Nbits + countBits(xored)$
5. $xored = \mathbb{T}_I(i) \oplus \mathbb{I}_I(i)$
6. $Nbits = Nbits + countBits(xored)$
7. **end for;**
8. $Nbits = 10 \times Nbits$
9. $H = Nbits/204$

Algorithm 1 computes the Hamming distance of the proposed modifications, where N^2 is the size of the matrix that is stored as a vector of *shorts* with 64 positions, $\mathbb{T}$ is the binary code stored in the card and $\mathbb{I}$ is the binary code of the individual who is requesting authentication. The palm-code has a real part and an imaginary that are represented as $(\mathbb{T}_R, \mathbb{T}_I)$ and $(\mathbb{I}_R, \mathbb{I}_I)$, respectively.

4   Performance Results

The images used in to evaluate this work were taken from the database of Polytechnic University of Hong Kong and is available in [11]. The database contains 8000 samples of palm-print images of 400 different hands. Figure 2 shows two examples of samples of palm-prints for different hands contained in the database. All images are in a grayscale and of 128×128 pixels.

The results are presented in terms of the *proximity* defined as $100-$ Hamming distance, which indicates the percentage of similarity between two palm-codes. To test and validate the proposed implementation on the smart-card, we compared 10 samples of 20 different hands randomly selected from the database used.

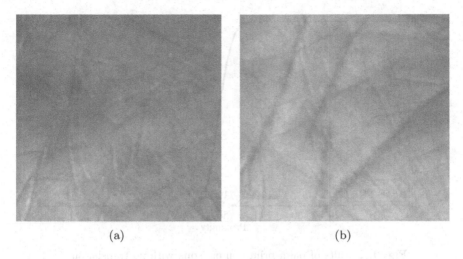

(a)   (b)

Fig. 2. Samples from plamprint images database POLYU

Fig. 3 shows the test results relating the FRR (False Rejection Rate) and FAR (False acceptance Rate) for different distances. The definition of the proximity limit which separates the most approved comparisons of the rest can also be made from the data in the chart. To improve the security of a system, it is fundamental that different individuals are not confounded. So, the FAR should be as small as possible, but without great impact on the FRR. In the chart, these two important points: the point where the percentages are equal (EER - Equal Error Rate) and the point that FAR is less than 0.1% (secure FRR) are highlighted. The chart also shows that the ERR is about 5% and secure FRR is close to 10%. The latter is the most interesting for the system security and means that, if the selection of the proximity is of 68% as threshold, the result would be that for each set of 1000 comparisons of different hands, only 1 would be considered as authentic whereas an individual would have 90% chance to get his/her access granted on each trial.

Fig. 4 shows the chart of FAR and FRR of the results for a translation of 1 bit. The percentage of equal error rate dropped to 0.17% and the secure FRR to 0.39%, thereby, validating the possibility of making use of the translation of bits it to improve the result of the comparison.

Figure 5 shows the chart that relates FAR and FRR for different distances. This time, both important points are equal to 0. This indicates that the biometric is approaching perfection. That is, there would be errors in the comparison of different individuals, and also there would be no failed attempts for authentic cardholders.

Although the comparison result is important for validating the implementation performance, it is not the only result to consider. The runtime is another important aspect as it will indicate the waiting time of the individual at every authentication attempt.

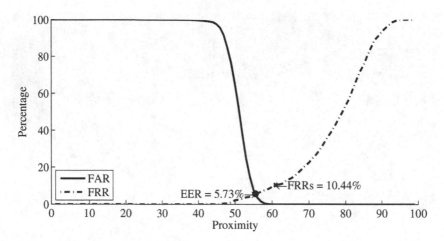

Fig. 3. Results of palm-print comparisons with no translation

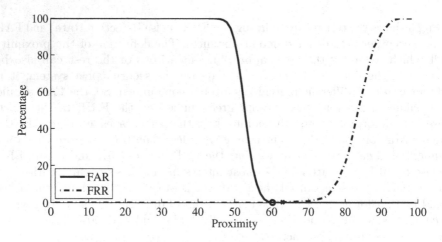

Fig. 4. Results of palm-print comparisons with 1-bit translation

The transmission time and storage in the EEPROM of each part of the palm-code is about 850ms. Two transmissions must be made to store the whole code. Thus, the total storage time is 1700ms. This time is relatively high compared to the time of transmission and storage in the RAM of part of the palm-code to be compared, which about 185ms. Although relatively slow, the time required during the storage in the EEPROM is not significant since it is a process that is performed only once during the lifetime of the card.

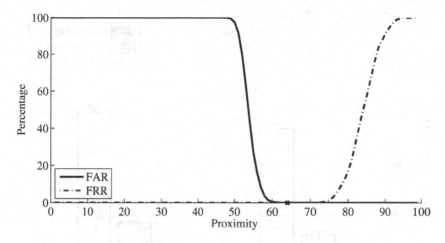

Fig. 5. Results of palm-print comparisons with 2-bit translation

The total time for a comparison takes into account the complete transmission of the palm-codes and the processing time required to compute the Hamming distance. Comparisons with translation of bits can be done to achieve optimal error rates but certainly the corresponding processing time has a bad impact on the execution time. Table 1 shows the means and standard deviations of the execution times of the comparisons without translation and translations of 1 and 2 bits.

Table 1. Execution time of comparisons varying the number of translated bits

Translation(*bits*)	Authentic comparison		False comparison	
	Mean (ms)	Std. Dev.	Mean (ms)	Std. Dev.
0	489	22	538	8
1	1520	74	1783	37
2	3725	109	4127	79

Fig. 6 presents a comparison chart the average execution times. The average execution time of the comparisons with translations of 2 bits is nearly 7× larger than the average execution time without translation while with the translation of 1 bit, the time increases approximately 3×. As biometric systems using smart-cards require a real-time response, the translation of 2 bit presents a negative factor in spite of the fact that it causes reduced error rates.

The results presented indicate that a smart-card is able to execute a biometric-based method of the palm-print, and also providing high performance and high reliability. Nonetheless, it can be weighted whether the given system should prioritize the speed of comparison or the low error rates. From the analysis of the

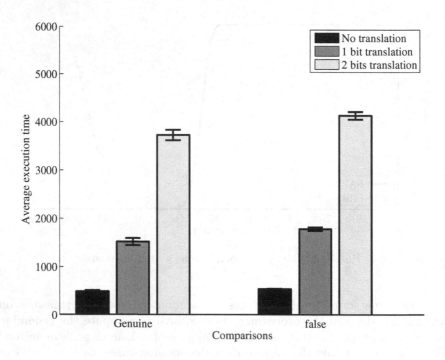

Fig. 6. Execution time of comparisons varying the number of translated bits

results and in order to improve the execution time, it is possible to use comparisons with translation, which achieved the lowest error rates, but imposing an acceptance threshold so that the processing of various displacements of bits can be halted in case of a positive authentication and thereby reducing the overall runtime.

The acceptance threshold is a specific value of proximity, from which the comparisons are considered to be correct, i.e. after reaching a predefined proximity, the smart card will have to compare the comparison outcome and may terminate the execution, returning the available result. The definition of the acceptance threshold can only reduce the execution time of the authentic comparisons, but does not affect the execution time of false comparisons since, in this case, the execution is not interrupted. Comparisons with no translation perform the Hamming distance computation only once, so cannot be interrupted. The acceptance threshold was implemented for comparisons with translation of 1 and 2 bits. Here, only the execution times are shown since no improvement is made in the comparison results.

Fig. 7 presents the charts that show the distribution of the execution times during comparisons with translation 1 bit with acceptance threshold. It is easy to observe in the chart that are 3 clusters of time to indicate the accepted values and each comparison. As the false comparisons are all denied they are in the cluster with longer execution time while accepted authentic comparisons are clustered in two groups with lower execution times. The average execution time for authentic comparisons is of 868ms with a standard deviation of 249 while the average execution time for false comparisons is of 1744ms with a standard deviation of 37. The latter is similar to the time of the comparisons with translation 1 bit, as expected.

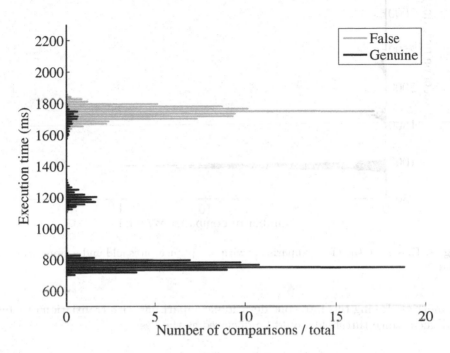

Fig. 7. Execution time for comparisons with acceptance threshold and a translation of 1 bit

Fig. 8 shows the chart related to the runtime comparisons with a translation of 2 bits and allowing an acceptance threshold. There are 5 clusters representing the 5 iterations that actually occur during comparisons. It is possible to notice that few comparisons performed the last two iterations. The average execution time of authentic comparisons is of 1217ms. When compared to the average execution time of comparisons with translation of 1 bit and acceptance threshold, which was 868ms, it can be concluded that the benefit is too small compared to the processing cost.

For the authentic comparisons, the execution times of the comparisons with translation of 1 or 2 bits have shown considerable improvement when using the threshold of acceptance. In the case of an actual utilization, it is expected that only the card-holder attempts an authentication. Therefore, the execution time for authentic comparisons has a greater relevance in relation to false comparisons. However, false comparisons cannot impose a very high runtime because it has a negative impact on the acceptance of the technology of smart-card with

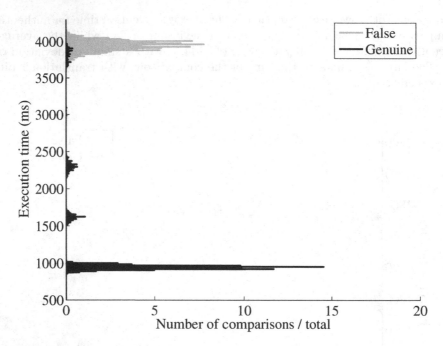

Fig. 8. Execution time for comparisons with acceptance threshold and a translation of 2 bit

biometrics. Taking this into consideration, comparisons with translation of 1 bit and acceptance threshold would be declared as the best.

5   Conclusions

Comparison of palm-print implemented in a smart-card achieved good results both in terms of execution time and effectiveness. Using the translations of the *Palm-code*, it is possible to achieve an EER 0%, i. e. using a specific acceptance threshold, there will be no error in comparison results. In other words, all false comparisons will be detected and all true comparisons will be accepted. However, the use of translations causes an increase in execution time. To mitigate this problem, a threshold of acceptance is proposed. The result is returned once the limit is reached. The results showed that the use of 1-bit translation is sufficient to achieve reduced execution times.

As a future work, other types of biometrics may be studied in order to verify the possibility of the implementation in a smart-card. As a multi-application card, it is possible that the same card offers more than one type of biometric verification. The implementation of merging of different biometrics can also improve system security.

References

1. Daugman, J.G.: High confidence visual recognition of persons by a test of statistical independence. IEEE Transactions on Pattern Analysis and Machine Intelligence 15(11), 1148–1161 (1993)
2. Daugman, J.G.: Uncertainty relation for resolution in space, spatial frequency, and orientation optimized by two-dimensional visual cortical filters. Journal of the Optical Society of America A: Optics, Image Science, and Vision 2(7), 1160–1169 (1985)
3. Daugman, J.G.: Two-dimentional spectral analysis of cortical receptive field profiles. Vision Research 20(10), 847–856 (1980)
4. Gabor, D.: The Theory of communication Part I: the analysis of information. Journal of the Institution of Electrical Engineers-Part III–Radio and Communication Engineering, IET 93(26), 429–441 (1946)
5. Hubel, D.H., Wiesel, T.N.: Ferrier Lecture: Functional architecture of macaque monkey visual cortex. Proc. of the Royal Society of London. Series B, Biological Sciences 198(1130), 1–59 (1977)
6. Jain, A.K., Ross, A., Prabhakar, S.: An introduction to biometric recognition. IEEE Transactions on Circuits and Systems for Video Technology 14(1), 4–20 (2004)
7. Healthcare Council Smart Cards and Biometrics in Healthcare Identity Applications (2012), http://www.smartcardalliance.org
8. Hachez, G., Quisquater, J.J., Koeune, F.: Biometrics, access control, smart cards: a not so simple combination. In: Smart Card Research and Advanced Applications, pp. 273–288. Springer (2000)
9. Shu, W., Zhang, D.: Automated personal identification by palmprint. Optical Engineering. International Society for Optics and Photonics 37(8), 2359–2362 (1998)
10. Zhang, D., et al.: Online palmprint identification. IEEE Transactions on Pattern Analysis and Machine Intelligence 25(9), 1041–1050 (2003)
11. The Hong Kong Polytechnic University PolyU 3D Palmprint database (2008), http://www.comp.polyu.edu.hk/biometrics/2D3DPalmprint.htm

Model-Based Test Case Generation
for Web Applications

Miguel Nabuco[1] and Ana C.R. Paiva[1,2]

[1] Department of Informatics Engineering
[2] INESC TEC
Faculty of Engineering of University of Porto
Porto, Portugal

Abstract. This paper presents a tool to filter/configure the test cases generated within the Model-Based Testing project PBGT. The models are written in a Domain Specific Language called PARADIGM and are composed by User Interface Test Patterns (UITP) describing the testing goals. To generate test cases, the tester has to provide test input data for each UITP in the model. After that, it is possible to generate test cases. However, without a filter/configuration of the test case generation algorithm, the number of test cases can be so huge that becomes unfeasible. So, this paper presents an approach to define parameters for the test case generation in order to generate a feasible number of test cases. The approach is evaluated by comparing the different test strategies and measuring the performance of the modeling tool against a capture-replay tool used for web testing.

1 Introduction

Web applications are getting more and more important. Due to the stability and security against losing data, the cloud and cloud-based web applications are progressively gaining a bigger userbase. Web applications can now handle tasks that before could only be performed by desktop applications [22], like editing images or creating spreadsheet documents.

Despite the relevance that Web applications have in the community, they still suffer from a lack of standards and conventions [23], unlike desktop and mobile applications. This means that the same task can be implemented in many different ways. Pure HTML, Javascript, Java and Flash are just some of the technologies that can be used to implement a certain feature in a web application. This makes automatic testing for Web harder, because it inhibits the reuse of test code.

However, users like to have a sense of comfort and easiness, even some familiarity, when they use their applications. For this purpose, developers use some common elements, such as User Interface (UI) Patterns [25]. In this context, UI Patterns are recurring solutions that solve common design problems. When a user sees an instance of a UI Pattern, he can easily infer how the UI is supposed to be used. A good example is the Login Pattern. It is usually composed of two

B. Murgante et al. (Eds.): ICCSA 2014, Part VI, LNCS 8584, pp. 248–262, 2014.

text boxes and a button to send the username and password data, and its function is to allow a user to access its private data.

Despite a common behavior, UI Patterns may have slightly different implementations along the Web. However, it is possible to define generic and reusable test strategies to test them. A configuration process is then necessary to adapt the test strategies to those different possible applications. That is the main idea behind the Pattern-Based GUI Testing [24] (PBGT) project in which context this research work is developed. In the PBGT approach, the user builds a test model containing instances of the before mentioned test strategies (corresponding to UI Test Patterns) for testing occurrences of UI Patterns on Web applications.

The goal of this research work is to provide different test case generation techniques from PBGT models in order to test Web applications.

The rest of the paper is structured as follows. Section 2 provides a state of the art of test generation techniques, addressing related works. Section 3 presents an overview of the PBGT approach, setting the context for this work. Section 4 presents the approach to generate and filter test cases. Section 5 presents some results and Section 6 takes some conclusions, summing up the positive points and limitations of this approach.

2   State of the Art

A good number of different test case generation techniques have been developed and researched in the past years. These techniques rely on different types of artifacts: some use software models, like Finite State Machines (FSM), others use information about the input data space, others use information from the software specifications. None of these techniques is exclusive. They can be combined to provide better results.

This paper focuses on model-based testing, which is a technique for generating a suite of test cases from models of software systems [1]. The tests are generated from these models so if they are not consistent with the software they represent, the tests will not be able to fully validate the application under test (AUT).

This testing methodology is very broad. Some approaches use axioms to build state based models using pre- and post-conditions [3]. Dick [2] developed a method to generate test cases from VDM models based on pre- and post-conditions. He created transition models where each pre- and post- condition is represented as a state. If applying a function $f()$ with an argument x on a state $S1$ reaches state $S2$ a transition is created between states $S1$ and $S2$ ($S1 \rightarrow pre(f(x))$ and $post(f(x)) \rightarrow S2$). After this state machine was created, test selection techniques (such as random testing or state coverage) was allied to filter the number of tests to be performed on the AUT. Paiva [30] also uses VDM models to test user interfaces based on their specifications, represented in state transition models.

Using a slightly different approach, Claessen developed QuickCheck [4], a combinator library that generates test cases. In QuickCheck the programmer writes assertions about logical properties that a function should fulfill; these tests are specifically generated to test and attempt to falsify these assertions.

A GUI mapping tool was developed in [14], where the GUI model was written in Spec# with state variables to model the state of the GUI and methods to model the user actions on the GUI. However, the effort required for the construction of Spec# GUI models was too high. An attempt to reduce the time spent with GUI model construction was described in [15] where a visual notation (VAN4GUIM) is designed and translated to Spec# automatically. The aim was to have a visual front-end that could hide Spec# formalism details from the testers.

Algebraic specification languages have been used for the formal specification of abstract data types (ADTs) and software components, and there are several approaches to automatically derive test cases that check the conformity between the implementation and the algebraic specification of a software component [28] [29].

There are also some approaches that use FSMs. In FSM approaches, the model is formalized by a *Mealy machine*. A *Mealy machine* is a FSM whose output values are determined both by its current state and current inputs [5]. For each state and input, at most one transition is possible.

The main goal of FSM based testing tools is complete coverage. However, this can consume a lot of computational and time resources. To filter the tests to be performed, they use structural coverage criteria, such as transition coverage, state coverage, path coverage, amongst others.

Rayadurgam [8] presents a method that uses model checkers to generate test sequences that provide structural coverage of any software artifact that can be represented as a FSM.

Since most of the real software systems cannot be fully modelled with a simple FSM, most approaches use extended finite state machines (EFSM) [6]. A typical implementation on an EFSM is a state chart.

Frohlich [7] automatically creates test cases (which he defines as sequences of messages plus test data) from UML state charts. These state charts are originated from use cases which contain all relevant information, including pre- and postconditions, variations and extensions. To generate the test cases, state chart elements are mapped to a planning language.

Similar to FSMs, labeled transition systems (LTS) are also being used to generate test cases. The difference between LTS and FSM is that LTS is non-deterministic: the same trace can lead to different states. Also, the number of states and transitions is not necessarily finite and it may not have a "start" or "end" states.

There is a particular testing theory for LTS systems, called IOCO (input/output conformance) [9]. The theory assumes the AUT to be an input enabled LTS which accepts every input in every state. There are some test generation tools that implement the IOCO theory, such as TVEDA [10], TGV [11], the Agedis Tool Set [12], and TestGen [13].

An alternative approach to IOCO is *alternating simluation* in the framework of *interface automata* (IA) [16]. IA does not require input completeness of the

AUT, as opposite to IOCO. The IA approach was deeply refined and composes the foundation of Microsoft Spec Explorer Tool [17].

Some works were developed to provide test selection strategies on top of these frameworks. Feijs [18] implemented test selection based on metrics for IOCO and Nachmanson [19] implemented test selection based on graph traversal and coverage for IA.

However, for the testing of web applications, none of these approaches seemed adequate enough. The approaches mentioned require a significant effort to model the system and do not take advantage of the existence of recurrent behaviors in the application under test.

Pattern Based GUI Testing (PBGT) project aims to promote reuse by defining generic test strategies for testing common recurrent behavior on the web [26]. A big number of Web applications are built on top of UI patterns that may have slightly different implementations. PBGT defines User Interface Testing Patterns (UITP) that allows testing UI patterns and their different implementations promoting reuse and diminishing the effort of the modeling activity within a model based testing process. The description of this approach can be seen in Section 3.

3   PBGT Overview

Pattern Based GUI Testing (PBGT) is a testing approach that aims to increase the level of abstraction of test models and to promote reuse. PBGT [20] has the following components:

- **PARADIGM** — A domain specific language (DSL) for building GUI test models based on UI patterns;
- **PARADIGM-ME** — A modelling and testing environment to support the building of test models;
- **PARADIGM-TG** — A test case generation tool that builds test cases from PARADIGM models;
- **PARADIGM-TE** — A test case execution tool to execute the tests, analyze the coverage and create reports;

Currently, PARADIGM is used to test web and mobile applications, but it can be expanded to test desktop applications. By using PARADIGM-ME, can build the test model written in PARADIGM language with UI Test patterns [21]. Each UI pattern instance may have multiple configurations (for example, for a login test pattern this includes the set of usernames and passwords used, as seen in Figure 1). There are two types of configurations: Valid and Invalid. An Invalid configuration is one that simulates an erratic behavior of the user (for example, inserting wrong username and password for a Login UI test pattern, or insert letters in a text box that only accepts numbers).

Field/Value Entries	Validity	Check	Message
[password/pass; username/john_doe]	Valid	ChangePage	
[username/test; password/test]	Invalid	PresentInPage	Welcome to...

Fig. 1. Login UI test pattern configurations

A UI Test Pattern describes a generic test strategy, formally defined by a set of test goals, denoted as $<$ Goal, V, A, C, P $>$ where:

- **Goal** is the ID of the test;
- **V** is a set of pairs variable, inputData relating test input data (different for each configuration)with the variables involved in the test;
- **A** is the sequence of actions to be performed during test execution;
- **C** is the set of checks to be performed during test execution;
- **P** is the precondition defining the set of states in which is possible to perform the test.

In PARADIGM-ME, the tester build the web application test model, by creating the respective UITPs and joining them with connectors. These connectors define the order that the UITPs will be performed. Then, the tester has to configure each UITP with the necessary data (input data, pre-conditions, and checks). The list of supported UITPs can be seen in Table 1.

In Figure 2, a model of Tudu (an online application for managing a todo list) is shown.

Table 1. List of UITPs supported

Name	Description
Login	The Login UITP is used to verify user authentication. The goal is to check if it is possible to authenticate with a valid username/password and check if it is not possible to authenticate otherwise.
Find	Find UITP is used to test if the result of a search is as expected (if it finds the right set of values).
Sort	The Sort UITP is used to check if the result of a sort action is ordered accordingly to the chosen sort criterion (such as sort by name, by price and ascending or descending).
Master Detail	Master Detail UITP is applied to elements with two related objects (master and detail) in order to verify if changing the master's value correctly updates the contents of the detail.
Input	The Input UITP is used to test the behavior of input fields for valid and invalid input data.
Call	The Call UITP is used to check the functionality of the corresponding invocation. It is usually a link that may lead to a different web page.

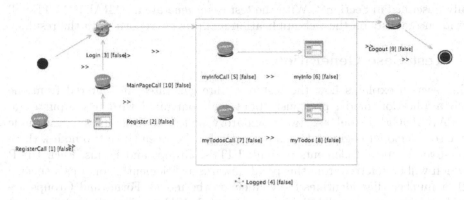

Fig. 2. Model of the web application Tudu

To perform the tests, it is necessary to establish a mapping between the model UITPs and the UI patterns of the web application to be tested. This will allow the test execution module to know which web elements to interact with that corresponds to a certain UITP described in the model. For mapping the Login UI Pattern of the Tudu model, the user must relate the login (and password) of the model with the login (and password) text box within the application under test (Figure 3) by clicking on it. PARADIGM-ME will save the following information from the user clicks:

- **Text boxes ID's** — the ID property of text boxes.
- **Images** — The image of the authentication form. This is saved through Sikuli [27] and is used when the tool is not able to identify the object by its ID.
- **Area coordinates** — The coordinates of the object. When the two previous methods fail, the tool uses the coordinates in order to interact with the web object.

After the mapping is made, PARADIGM-TG will then generate test cases from the model. The test case generation is the focus on this paper, and it is

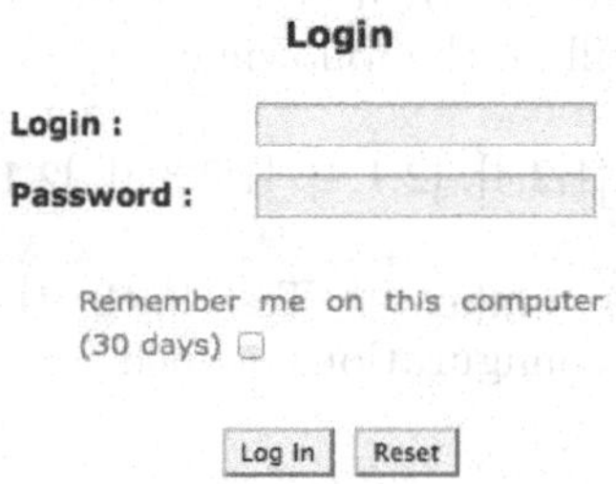

Login

Login :

Password :

Remember me on this computer
(30 days) ☐

Log In   Reset

Fig. 3. Login form web application Tudu

fully described in Section 4. With the test cases generated, PARADIGM-TE will perform the tests on the web application and provide reports with the results.

4   Test Case Generation

This section explains how the test cases are generated and filtered from the web application model previously built and configured with test input data. A PARADIGM model has two mandatory nodes: an Init and an End node (where the model begin and ends respectively). Between these two nodes, there are several types of elements present: UITPs, Groups and Forms. Each UITP (which will be referred from this point onwards as "element") contains a specific ID, a number that identifies the element in the model. Forms and Groups are used to structure a model. A Group (Figure 2, Node 4, *Logged*) contains a set of UITPs that can be performed in any order and also be optional. A Form (Figure 2, Node 2, 6 and 8) contains a PARADIGM model inside it. It is used for structural reasons, to simplify the view and modeling of different sections of a web application.

4.1   Path Generator

The first step for test case generation is to flatten the model. A PARADIGM model, after being flattened, can be seen as a graph that starts in Init and finished in End and has nodes in between that are only UITP (there are no groups nor forms). So, the next step required for the test case generation is to calculate all possible paths within that graph. A path is a sequence of elements that define a possible execution trace. A model can generate many paths, based on the connectors between elements.

Considering that a model contains **x** different branches (considering the graph), **y** elements that can be performed in any order and **z** optional elements, the number of total paths generated will be:

$$TestPaths = 2^z xy!$$
(1)

As an example, consider a web form where the user has two mandatory inputs (ID=1 and ID=2) in any order followed by other 2 optional inputs (ID=3 and ID=4). The number of paths will then be $2^2 2! = 8$.
The sequences generated will be the following:

$\{[1,2],\ [2,1],\ [1,2,3],\ [2,1,3],\ [1,2,4],\ [2,1,4],\ [1,2,3,4],\ [2,1,3,4]\}$

Each of these sequences compose a Test Path, that can contains multiple test cases according to the configurations defined for each UITP.

4.2   Path Filter

In case the model generates too many test paths for the available resources or if the testing goal is to test a single component and not the whole web application, a path filter component was implemented to reduce the number of Test Paths generated.

The path filter will restrict the number of test paths generated based on three simple conditions:

- **Cycles** — A Test Path can have cycles, i.e., a sequence of elements that can repeat itself multiple times. It is possible to reduce the number of Test Paths created by limiting the number of times each cycle is performed. In Figure 4, a cycle can be seen in the elements {[1,2]}.
- **Mandatory and Exclusion Elements** — If the tester wants to test a certain feature or a certain workflow that can only be reached following a certain path, he can explicitly define the mandatory and exclusion elements. By excluding one element (or one sequence of elements), every Test Path that contains that element will not be generated. Also, by excluding an element other elements may be automatically excluded.
- **Random** — The tester can also define a maximum number of Test Paths to be randomly generated. These Test Paths may already be filtered by the two conditions described above, so this is the last condition to be applied.

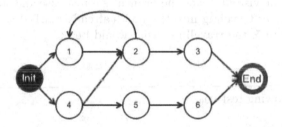

Fig. 4. Sample Test Model

4.3   Path Tree

After the initial Test Path Filtering, each Test Path must be decomposed in multiple test cases.

One element can contain multiple configurations. A Login element can contain valid and invalid configurations and each one of these can lead to different outcomes. Therefore, each configuration must generate a different test case.

Considering a Test Path with $\mathbf{x}$ elements, each with $\mathbf{y_x}$ configurations, the total number of Test Cases generated will be:

$$Tc = y_x \times y_{x-1} \times ... \times y_1 \qquad (2)$$

To make the test case generation easier, a tree structure was created, as seen in Figure 5.

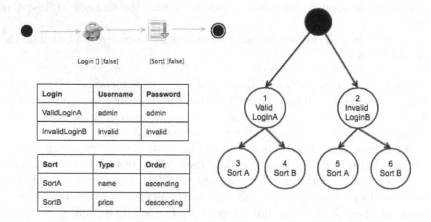

Fig. 5. PARADIGM model, configurations and the corresponding Test Tree

The tree is then traveled using a depth-first search algorithm. An iterator starts from the tree root node and follows one branch until the end. Then, it marks the last node visited, saves the branch as a test case and starts again from the beginning, never traveling into the nodes already marked as visited.

From the Figure 5, the travelling order would be:

[1,3,1,4,2,5,2,6]

Creating the following test cases:

{[1,3], [1,4], [2,5], [2,6]}

4.4   Test Case Filtering

From the example above, it can be seen that from two simple Login and Sort UI Test Patterns with two different configurations, 4 test cases are generated. If this is scaled to real-life applications, there can be a test case explosion, with too many test cases to perform regarding the time and computational resources available. Therefore, just like a filter was applied to the Test Paths, a different filter may be applied to reduce the number of test cases. The metrics applied in the filter are the following:

- **Specific configurations** — If the tree is filtered so that only test cases with specific configurations are considered, the number of test cases to be performed can drastically reduce. In Figure 5, if it is specified that only valid logins are of interest, the total number of test cases drops from 4 to 2, decreasing 50%. The modelling tool provides enough flexibility so that the tester can generate test cases for the specific components he wants to test.

– **Random** — As with the Path Filtering explained in Section 4.2, the tester can also specify a number of test cases to be randomly chosen from the total set.

4.5   Test Case Generation Modes

To apply the filters, the user can choose to do it manually (specifying for each option described previously which are the parameters), provide a text file with a set configuration or choose one of the automatic test strategies options available. The test strategies implemented are the following:

– **Default Strategy** – This strategy generates all test paths, but only performs N test cases per each test path (N is defined by the user).
– **All Invalids Strategy** – The goal of this test strategy is to test all invalid configurations of all the elements, to find erratic behaviors in the application under testing. So, this strategy generates one test case for each invalid configuration defined.
– **Random Strategy** – This strategy picks a random test case from a random test path and runs it. Then it picks a random test case again. This cycle will continue, until the user stops the program execution.

In Figure 6, the option menu to select the test strategy can be seen.

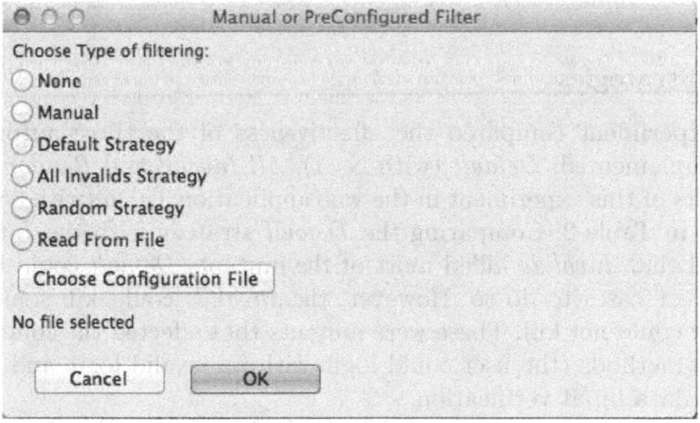

Fig. 6. Test Strategies Menu

4.6   Exporting to PARADIGM-TE

After all the filters are applied, and the test cases are selected, the tool will then export the test cases into XML files, so that the test execution tool (PARADIGM-TE) can run them and generate reports. Each Test Path (with several test cases) is stored in a different XML file. This allows an easier test execution and debug, as it is easier to know if there is any Path not correctly defined. An example of a XML file of a simple two nodes test case can be seen in Figure 7.

```
<?xml version="1.0" encoding="UTF-8" standalone="no"?>
<Script>
<Path value="[1.0, 1.1, 1.2, 1.3, 1.4, 1.5, 1.6]">
<Config/><Init/>
<Call check="presentInPage" flag="true" name="LinkNormal" number="1.1" optional="false"
result="yes"><Field name="call_1.1"/></Call>
<Login check="changePage" flag="false" message="" name="LoginNormal" number="1.2"
optional="false" validity="Valid"><Value field="username_1.2" fieldName="username"
value="user"/><Value field="password_1.2" fieldName="password" value="pass"/></Login>
</Path></Script>
```

Fig. 7. Two nodes XML file

5   Evaluation

To evaluate this approach, two different experiments were made: one to see the results of using different test strategies and the other to see how efficient this approach could be against another web testing tool.

The experiment was performed over three different open-source web applications: **Tudu** (Figure 2), a web application for managing todo lists; **iAdress-Book**, a PHP address book application; and **TaskFreak**, a task management web application. Each model contained several configurations for each element.

To measure the effectiveness of the approach, each web application source code was injected with mutants. Each mutant would change boolean and arithmetic operations and conditions (i.e., change a *true* to *false*, or a *greater than* for a *smaller or equal than*).

5.1   Test Strategies

The first experiment compared the effectiveness of the three automatic test strategies implemented: *Default* (with N=1), *All Invalid* and *Random*.

The results of this experiment in the web application TuDu (with 35 mutants) can be seen in Table 2. Comparing the *Default* strategy with the *Invalids* one, it was found that *Invalids* killed most of the mutants *Default* could kill, but it took more test cases to do so. However, the *Invalids* could kill some mutants that *Default* could not kill. These were mutants that affected the code of the authentication methods (the user could login with an invalid login and password) or incorrect data input verification.

Comparing the *Default* strategy with the *Random* one, it was possible to see that Random could kill more mutants than the default strategy (it killed some that were only killed by the *Invalids* strategy), but it would take more time and

Table 2. Comparison between different test strategies

	Default	Random	Invalids
Number of mutants killed	24	32	27
Test Cases to kill one mutant (avg)	1.3	5.2	3.7
Time (min) per mutant (avg)	9.1	26.5	18.5

test cases to kill them. Also, since it is totally Random, it repeated some test cases, therefore spending more time to kill different mutants.

In conclusion, the *Default* strategy can kill a significant number of mutants in a short time, while the *Random* strategy kills more mutants, but it takes more time.

5.2   Comparison against a Capture-Replay Tool

The goal of the second experiment was to compare how many mutants PARADIGM and a capture-replay tool would kill. The capture-replay tool chosen for this experiment was Selenium-IDE. Selenium was chosen because PARADIGM uses Selenium libraries in its core, and so it makes sense to see what advantages could this approach bring over regular Selenium testing. Both of the tools are black-box testing methods, as none of them has access to the application source code. They both perform the tests by following through a sequence of actions previously defined (in the case of PARADIGM, the model). The test strategy chosen was the default with N=1 (one test case per test path, as explained in Section 4.5).

To perform the experiment, a set of test goals were delivered to two different testing teams (one for PARADIGM and other for Selenium), that had no interaction with each other. Both had to configure the application, run the tests, and measure the results.

The effort required to build and configure the application scripts (for Selenium) and the models (for PARADIGM) can be seen in Table 3. The effort to build the application models in PARADIGM is greater than to create test scripts in Selenium. However, in case the web application layout changes, it is easier to change the model in PARADIGM than to change the scripts in Selenium.

The applications were then tested with both tools and the results can be seen on Table 4.

PARADIGM tool managed to successfully kill 82.7% of the mutants, when Selenium killed 72.3%.

The extra effort needed to model the application (as opposed to the *crawling* nature of Selenium, in which the testers simply interact with the application and their steps are recorded), compensates in the number of mutants killed.

Selenium uses scripts to run the tests, which limits the range of inputs and checks. If a tester wants to add a different set of input and check data, he has to create different script. When the tester creates the model in PARADIGM, it is

Table 3. Effort (in minutes) required to build and configure models/scrips

	PARADIGM	SELENIUM	Difference
TuDu	73	60	+13
TaskFreak	164	101	+63
iAdressBook	137	75	+62

Table 4. Evaluation results

	Mutants	PARADIGM	SELENIUM
TuDu	28	24	16
TaskFreak	76	59	53
iAdressBook	69	60	56
Total	**173**	**143(82.7%)**	**125(72.3%)**

easy to continuously add different input data and checks, since the model stays the same. With more configurations, the number of test cases increases and the application under test is more tested.

Also, while the Selenium team created an average of 5 to 7 scripts to test each web application according to the test goals defined, the PARADIGM tool could generate hundreds of different test cases from each model (for TuDu alone, 245 test cases were generated).

6   Conclusions and Future Work

This paper proposed a test generation and filtering technique for model-based testing of Web applications. The models contain information in the form of UI Test Patterns linked with connectors. Each UI Test Pattern contains specific configurations with the data needed for test execution.

Considering that test cases cannot run forever, this paper presented several filters that were applied to provide flexibility and reduce the number of test cases generated. Firstly the filters were applied to the Test Paths and then to each Test Path element (an UI Test Pattern) configurations. Some test strategies were then created, each with its own broadness and coverage.

This approach was tested on three different web applications. The several test strategies were compared between themselves. Also, this approach was compared with the same set of tests performed with a capture-replay tool. PARADIGM proved to be effective, as the filtering can provide better coverage, finding more bugs and killing more mutants.

In general, it is possible to see clear advantages of this approach:

- **Distributing Test Suites** — Since the test process can be configured in order to generate different test suites, it is possible to spread different test suites through different workstations and run tests concurrently and in less time.
- **Easier debugging** — If a critical error occurs that may stop the execution of the testing program altogether, it is easy to see which element or configuration caused that error. All the information is stored so that the user can replay the steps even manually. Also, as the results are updated after each Test Path, there is no need to run all the selected tests again in case of error.

The features planned for the near future include a broader set of filtering options, test strategies and test coverage statistics.

Acknowledgments. This work is financed by the ERDF — European Regional Development Fund through the COMPETE Programme (operational programme for competitiveness) and by National Funds through the FCT — Fundação para a Ciência e a Tecnologia (Portuguese Foundation for Science and Technology) within project FCOMP-01-0124-FEDER-020554.

References

1. Dalal, S.R., Jain, A., Karunanithi, N., Leaton, J.M., Lott, C.M., Patton, G.C., Horowitz, B.M.: Model-based testing in practice. In: Proceedings of the 21st International Conference on Software Engineering (ICSE 1999), pp. 285–294. ACM, New York (1999)
2. Dick, J., Faivre, A.: Automating the generation and sequencing of test cases from model-based specifications. In: Larsen, P.G., Wing, J.M. (eds.) FME 1993. LNCS, vol. 670, pp. 268–284. Springer, Heidelberg (1993)
3. Katsiri, E., Mycroft, A.: Model checking for sentient computing: an axiomatic approach. In: Proceedings of the First International Workshop on Managing Context Information in Mobile and Pervasive Environments (SME 2005), CEUR-WS, Ayia Napa (May 2005)
4. Claessen, K., Hughes, J.: QuickCheck: a lightweight tool for random testing of haskell programs. In: Odersky, M., Wadler, P. (eds.) Proc. of the 5th ACM SIGPLAN International Conference on Functional Programming (ICFP 2000), pp. 268–279. ACM (2000)
5. Mealy, G.H.: A Method for Synthesising Sequential Circuits. Bell Systems Technical Journal 34, 1045–1079 (1955)
6. Cheng, K.-T., Krishnakumar, A.S.: Automatic Functional Test Generation Using The Extended Finite State Machine Model. In: 1993 30th Conference on Design Automation, June 14-18, pp. 86–91 (1993)
7. Fröhlich, P., Link, J.: Automated Test Case Generation from Dynamic Models. In: Bertino, E. (ed.) ECOOP 2000. LNCS, vol. 1850, p. 472. Springer, Heidelberg (2000)
8. Rayadurgam, S., Heimdahl, M.P.E.: Coverage based test-case generation using model checkers. In: Proceedings. Eighth Annual IEEE International Conference and Workshop on the Engineering of Computer Based Systems-ECBS, pp. 83–91 (2001)
9. Tretmans, J.: Test Generation with Inputs, Outputs, and Repetitive Quiescence. Software-Concepts and Tools 17, 103–120 (1996)
10. Phalippou, M.: Relations d Implantation et Hypotheses de Test sur des Automates a Entrees et Sorties. PhD thesis, L Universite de Bordeaux I, France (1994)
11. Jard, C., Jeron, T.: TGV: Theory, Principles and Algorithms: A Tool for the Automatic Synthesis of Conformance Test Cases for Non-Deterministic Reactive Systems. Software Tools for Technology Transfer 7(4), 297–315 (2005)
12. Hartman, A., Nagin, K.: The AGEDIS Tools for Model Based Testing. In: Int. Symposium on Software Testing and Analysis - ISSTA 2004, pp. 129–132. ACM Press, New York (2004)
13. He, J., Turner, K.: Protocol-Inspired Hardware Testing. In: Csopaki, G., Dibuz, S., Tarnay, K. (eds.) Int. Workshop on Testing of Communicating Systems, vol. 12, pp. 131–147. Kluwer Academic Publishers (1999)

14. Paiva, A.C.R., Faria, J.C.P., Tillmann, N., Vidal, R.A.M.: A Model-to-Implementation Mapping Tool for Automated Model-Based GUI Testing. In: Lau, K.-K., Banach, R. (eds.) ICFEM 2005. LNCS, vol. 3785, pp. 450–464. Springer, Heidelberg (2005)

15. Moreira, R.M.L.M., Paiva, A.C.R.: Visual Abstract Notation for Gui Modelling and Testing - VAN4GUIM. In: ICSOFT 2008, pp. 104–111 (March 4, 2008)

16. de Alfaro, L., Henzinger, T.A.: Interface automata. In: ESEC / SIGSOFT FSE, pp. 109–120 (2001)

17. Campbell, C., Grieskamp, W., Nachmanson, L., Schulte, W., Tillmann, N., Veanes, M.: Testing concurrent object-oriented systems with spec explorer. In: Fitzgerald, J.S., Hayes, I.J., Tarlecki, A. (eds.) FM 2005. LNCS, vol. 3582, pp. 542–547. Springer, Heidelberg (2005)

18. Feijs, L.M.G., Goga, N., Mauw, S., Tretmans, J.: Test selection, trace distance and heuristics. In: Proceedings of the IFIP 14th International Conference on Testing Communicating Systems (TestCom 2002), pp. 267–282. Kluwer (2002)

19. Nachmanson, L., Veanes, M., Schulte, W., Tillmann, N., Grieskamp, W.: Optimal strategies for testing nondeterministic systems. In: Avrunin, G.S., Rothermel, G. (eds.) Proc. of the 2004 ACM SIGSOFT International Sym- posium on Software Testing and Analysis (ISSTA 2004), pp. 55–64 (2004)

20. Moreira, R., Paiva, A., Memon, A.: A Pattern-Based Approach for GUI Modeling and Testing. In: Proceedings of the 24th annual International Symposium on Software Reliability Engineering, ISSRE 2013 (2013)

21. Welie, M., Gerrit, C., Eliens, A.: Patterns as tools for user interface design. In: Workshop on Tools for Working With Guidelines, Biarritz, France (2000)

22. Garrett, J.J.: Ajax: a new approach to Web applications (2006), http://www.adaptivepath.com/publications/essays/archives/000385.php

23. Constantine, L.L., Lockwood, L.A.D.: Usage-centered engineering for Web applications. IEEE Software Journal 19(2), 42–50 (2002)

24. Monteiro, T., Paiva, A.: Pattern Based GUI Testing Modeling Environment. In: 4th International Workshop on Testing Techniques & Experimentation Benchmarks for Event-Driven Software, TESTBEDS 2013 (2013)

25. Nabuco, M., Paiva, A., Camacho, R., Faria, J., Inferring, U.I.: Patterns with Inductive Logic Programming. In: 8th Iberian Conference on Information Systems and Technologies (2013)

26. Cunha, M., Paiva, A., Ferreira, H., Abreu, R.,, P.: A Pattern-Based GUI Testing Tool. In: 2nd International Conference on Software Technology and Engineering (ICSTE 2010), pp. 202–206 (2010)

27. Sikuli API (last acessed February 2014), https://code.google.com/p/sikuli-api/

28. Andrade, F.R., Faria, J.P., Paiva, A.: Test generation from bounded algebraic specifications using alloy. In: ICSOFT 2011, 6th International Conference on Software and Data Technology (January 2011)

29. Rebello de Andrade, F., Faria, J.P., Lopes, A., Paiva, A.C.R.: Specification-driven unit test generation for java generic classes. In: Derrick, J., Gnesi, S., Latella, D., Treharne, H. (eds.) IFM 2012. LNCS, vol. 7321, pp. 296–311. Springer, Heidelberg (2012)

30. Paiva, A.C.R., Faria, J.P., Vidal, R.M.: Specification-based Testing of User Interfaces. In: Proceedings of the 10th DSV-IS Workshop - Design, Specification and Verification of Interactive Systems, Funchal, Madeira, de Junho 4-6 (2003)

Characterizing the Control Logic of Web Applications' User Interfaces

Carlos Eduardo Silva and José Creissac Campos

Departamento de Informática, Universidade do Minho
& HASLab/INESC TEC
Braga, Portugal
{cems,jose.campos}@di.uminho.pt

Abstract. In order to develop an hybrid approach to the Reverse Engineer of Web applications, we need first to understand how much of the control logic of the user interface can be obtained from the analysis of event listeners. To that end, we have developed a tool that enables us to perform such analysis, and applied it to the implementation of the one thousand most widely used Websites (according to Alexa Top Sites). This paper describes our approach for analyzing the user interface layer of those Websites, and the results we got from the analysis. The conclusions drawn from the exercise will be used to guide the development of the proposed hybrid reverse engineering tool.

Keywords: Web applications, user interfaces, reverse engineering.

1 Introduction

Reverse Engineering [1,2] is the process of moving from more concrete representations of a system, typically its source or binary code, to more abstract representations of the same system. Doing this we gain knowledge about the design of the system, avoiding details that are relevant only to its implementation. Reverse engineering can be useful in a number of different situations, be it during development, testing or maintenance. Two basic types of techniques can be used. Static techniques start from the code (the source code or some form of byte-code) analyzing it statically. Dynamic techniques take a black box approach and analyze the behavior of the system while executing.

We are particularly interested in applying reverse engineering to Web applications. While development for the Web has changed how the software industry approached software development and distribution, it is recognized that applying adequate standards and methodologies is still a challenge [3]. Through reverse engineering we aim to provide support to the process of engineering and re-engineering Web applications. By helping the process of gaining an understanding of an existing system, it becomes possible to better support its testing and validation, or its maintenance and evolution. For example, by enabling the comparison of different versions of the same software.

B. Murgante et al. (Eds.): ICCSA 2014, Part VI, LNCS 8584, pp. 263–276, 2014.
© Springer International Publishing Switzerland 2014

More specifically, we are interested in the user interface layer of Web applications. This is both because they are typically more susceptible to changes (due to changes in user requirements or the technology), and are particularly hard to develop and maintain (due, not only to the event based nature of the code, but also the the wide variety of technologies available for that development).

Early attempts at applying static analysis techniques to analyze the user interface layer of Web applications [4], have shown that this type of technique, while feasible for native applications (see, for example, [5]), suffers from a number of shortcomings when applied to the Web. The two main issues are related to the existence of many alternative implementation technologies, on the one side, and to the on the fly generation of the user interfaces, on the other.

The fact that many different languages, toolkits and frameworks exist, which can be used to implement Web applications, and their user interfaces in particular, means that a static analysis approach will either have to be able to parse and interpret the different languages and/or toolkits/frameworks, or commit to a particular technology. The first solution is not viable, in practice, as it means that developing such an approach would be prohibitively expensive. The second, means that the utility of the approach becomes rather restricted. This is specially the case when one considers the fast pace of technological development in the field.

Dynamic analysis avoids the above problems by taking a black box view of the system under test, and exploring its behavior at run time. Here what is needed is some adapter layer to programmatically interact with the user interface. Frameworks such as Selenium [6] provide this layer for the case of Web applications, and have already been used in tools such as Crawljax [7].

The fact that dynamic analysis takes a black box view of the system under test, however, also poses some limitations. Most notably, apart from very simple user interfaces, it is not possible to be sure that all possible behaviors/states of the system have been explored. Additionally, when in presence of alternative system behaviors, it is not easy to determine what are the conditions that triggers each alternative behavior [8]. The danger, then, is that models obtained through a dynamic approach to reverse engineering will be incomplete, both regarding the full behavior of the system under test, and regarding the semantics of the observed behaviors.

To avoid these issues, we propose to use an hybrid approach to reverse engineering [9]. The approach combines dynamic analyzes of the application, with static analyzes of the source code of the event handlers found during interaction. Information derived from the source code is both directly added to the generated models, adding semantics to the model, and used to guide the dynamic analysis, thus enabling it to be more complete.

In order to avoid the pitfalls of static analysis, the goal is to keep the analysis of event handlers as simple as possible. This, however, begs the question of whether enough control logic is present in the event handlers of Web applications, and whether we can adequately identify and process the event handlers. To answer those questions, we have carried out a study of the top one thousand most used

Websites according to Alexa Top Sites[1]. In this paper we describe how the study was setup, and the results that were achieved.

The paper is organized as follows: Section 2 defines the criteria used in the analysis; Section 3 explains how information about event handlers was extracted from the Web pages, and Section 4 describes the tool which was developed; Section 5 describes the analysis of the Websites, and the paper concludes with Section 6.

2   Criteria for Analysis

Since the interest is in identifying possible alternative behaviors of the system, and the conditions under which these alternative behaviors occur, the focus of the static analysis are the conditions in if statements and loops in the event handlers.

The analysis of the Websites was thus made according to a set of criteria defined to help understand how much information could be obtained from the event handlers. The following criteria were defined:

- **Number of hrefs** – The analysis of how many hrefs are used in the page. A high number of hrefs and the failure to detect events handlers might lead us to infer the usage of Web 2.0 (i.e. static) techniques in the page.
- **Number of Events** – This criterion is the total number of events we were able to find on the Website.
- **Number of Click Events** - It is important to discern, from all the events in the page, those which are triggered by clicks, since they are easier for us to simulate. Moreover, only the visible click events are being considered in this criterion.
- **Number of Ifs** – This criterion is important since we need to measure how much used these constructs are, in comparison to others that affect the control flow of the application.
- **Number of other Control Flow constructs** – The other type of constructs related to the control flow we are analyzing, these include while/for loops and ternary conditional operators.
- **Number of Element Variables** – This criterion counts the number of variables we found, whose value is obtained from elements in the page. We are currently not distinguishing between input and non-input elements. Moreover, this analysis was performed by detecting usage of classic JavaScript *getElementBy* function calls only.
- **Number of Synthesized Variables** – Variables related to function calls that we could not ascertain were used to retrieve elements from the page.
- **Number of Object Variables** – Variables associated with objects or object properties.
- **Number of Global Variables** – Variables whose declaration is made outside the scope of the handler/function being analyzed.

---

[1] http://www.alexa.com/topsites (last accessed: May 8, 2014).

- **Number of Control Flow Variables** – Variables that might be assigned different types depending on the control flow structures.
- **Number of Hybrid Variables** – Variables that can be classified using more than one of the previously defined types, at different points in the code.
- **Event Delegation** – If our analysis infers the Website uses the event delegation approach for event handlers (see Section 3).

3   Extracting Web Page Behavior Information

In order to combine dynamic analysis with the static analysis of event handlers, we must be able to identify those event handlers at run time. Identifying the event handlers in a page, depends on how those event handlers were added to the page in the first place.

There are two main approaches to add events to a Web page. The classic approach is to add an event handler to an element. However, even using this approach, there are several different ways of adding event handlers to the elements. A simple example is:

```
element.onclick = event_handler;
```

An event handler added in this way is retrieved simply by querying *element.onclick* in JavaScript. However, in our analysis, we soon discovered that most Websites use other methods for adding event handlers. For example:

```
element.addEventListener('click', event_handler, false)
```

The above source code is another option of adding an *onclick* event to an element. Using *element.onclick* in this case will not retrieve any results. Moreover, we also have to consider all the different JavaScript frameworks, for instance, in jQuery [10] the event handler is added as follows:

```
$(element).click(handler);
```

To address this we resorted to Visual Event, an open source framework[2] which is able to parse several JavaScript libraries and retrieve the event handlers. It currently works with a number of different libraries, and can be extended by developing new parsers for those not supported and adding them to the framework.

The other approach to adding event handling code to a Web page is called event delegation[3]. The idea is to take benefit of the browsers' event bubbling features. That is, when we have nested elements in HTML, triggering an event handler in an element also implies triggering the handlers for that event in the parent elements. For example, in Figure 1 on the left side we have the

---

[2] http://www.sprymedia.co.uk/article/Visual+Event+2 (last accessed: May 8, 2014).

[3] http://icant.co.uk/sandbox/eventdelegation/ (last accessed: May 8, 2014).

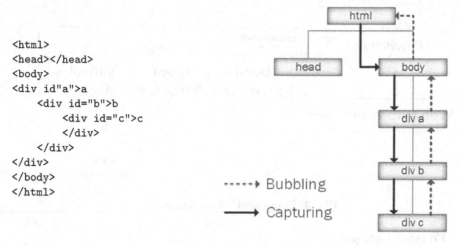

```
<html>
<head></head>
<body>
<div id"a">a
    <div id="b">b
        <div id="c">c
        </div>
    </div>
</div>
</body>
</html>
```

Fig. 1. Bubbling and Capturing Example

source code of a simple Web page with three *divs*. Triggering an event on the innermost element, that is, *div c* , causes the browser to traverse the Document Object Model (DOM) from the root node (*html*) to the element, this is called the capturing phase. Afterwards the browser also needs to traverse the DOM from the element to the root node, this is called the bubbling phase. Depending on the browser and how event handling is configured, relevant event handlers will be triggered in each element in either the capturing or bubbling phase.

This behavior of the Web browsers led to the usage of event delegation. That is, instead of assigning event handlers to each element, we assign a single handler to a parent container (the extreme case is to add the handler to *body*) and that container then controls the events depending on which child element has triggered the event.

Discerning which technique is being used is important since the approach to reverse engineering the events is completely different. After all, on the classic approach we can simply analyze the handlers for each element whereas in the event delegation approach we must use additional logic to identify which element triggers which behavior.

4   Approach

In order to gather the data of the Websites we have developed a framework to analyze pages according to the criteria defined in Section 2. The architecture of our tool is depicted in Figure 2. The framework is composed by three components: DOM Analyzer, Event Detection and Variables Analyzer. Selenium WebDriver[4], a tool to automate the Web browser, was used to perform the communication with the Web browser.

---

[4] http://docs.seleniumhq.org/projects/webdriver/ (last accessed: May 8, 2014).

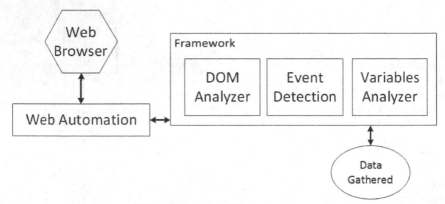

Fig. 2. Framework's architecture

4.1 DOM Analyzer

This component is responsible for parsing the HTML and gathering data from the DOM. The DOM provides an Application Programming Interface (API) for valid HTML and well-formed XML documents[5]. Whenever a Web page is loaded, the Web browser creates a DOM of that page, which in terms of HTML is the standard on how to manipulate HTML elements.

We parse the DOM with an HTML cleaner to overcome malformed HTML tags that may exist. Afterwards we inject JavaScript to discover which elements are visible and which elements are invisible. This is important since our analysis of the elements that are responsible for the behavior of the page will focus on the visible elements only.

In order to do this, the JavaScript we are injecting in the page is using the jQuery *find(':hidden')* function. Obviously, this also means that we have to add the jQuery framework to all the sites we are analyzing. In order to avoid problems with sites using other JavaScript frameworks, since many other frameworks use the $ character as a shortcut, we are using the jQuery noconflict() method.

4.2 Event Detection

The Event Detection component analyzes the page and searches for the event handlers that are present therein. As we discussed previously in Section 3, we must be able to search for the event handlers independently of what approach is being used in the Web page. Thus, Visual Event is first used to retrieve all the event handlers assigned to the Web page using the classic approach. However, the event delegation approach is significantly harder to analyze.

Since it can be implemented in several different ways, we are currently only identifying if that approach is being used. Even to perform that identification, we require an analysis of the entire JavaScript source code, in search of usages

---

[5] http://www.w3.org/TR/2004/REC-DOM-Level-3-Core-20040407/
introduction.html (last accessed: May 8, 2014).

```
1   document.body.onclick = function(e) {
2       e = ( e ) ? e : event
3       var el = e.target || e.srcElement;
4       if (el.id == "c") {
5           //Handler code
6           if (e.preventDefault) {
7               e.preventDefault();
8           }
9           else {
10              e.returnValue = false;
11          }
12      }
13  }
```

Fig. 3. An example of using Event Delegation

of JavaScript tokens similar to the ones present on Figure 3, which depicts the source code of an example of assigning an event handler to an HTML element using the event delegation approach.

We are currently using a combination of the Firebug[6] and NetExport[7] extensions of the Firefox Web browser in order to retrieve all the JavaScript files that are requested from the server when a page is loaded. Then we analyze all the JavaScript source code in search for patterns similar to the ones presented on Figure 3.

4.3   Variables Analyzer

After extracting the relevant JavaScript code from the event handlers, we create an Abstract Syntax Tree (AST). In order to do this we use Mozilla's Rhino[8] to parse the JavaScript source code and generate the AST.

The code is analyzed to identify the statements that affect control flow. The relevant constructs are: ifs/elses, ternary operators, and while/for loops. Moreover, it is also necessary to analyze the variables used in those constructs.

Initially, we statically analyzed and classified those variables, based on how they are used on the source code, into the following four categories:

- **Constants** - are variables that remain unaffected by any type of function call. Our analysis will ignore these variables since we consider them of no interest, as they cannot cause changes in the logic of the control flow.
- **Element variables** - are variables that use function calls like getElement-ById(...) to retrieve elements from the page. Moreover, these variables are further divided into:

---

[6] http://getfirebug.com/ (last accessed: May 8, 2014).

[7] http://www.softwareishard.com/blog/netexport/ (last accessed: May 8, 2014).

[8] https://developer.mozilla.org/en-US/docs/Mozilla/Projects/Rhino (last accessed: May 8, 2014).

- **Input variables** - elements of the page that a user can manipulate (e.g textboxes).
- **Static variables** - elements of the page that a user cannot manipulate (e.g labels).
- **Synthesized variables** - are variables that use other types of function calls, typically these will be calls to auxiliary functions at the user interface level, or to the business logic of the application.
- **Object variables** - are variables which are associated with an object or one object property.

An analysis of more complex pieces of source code led us to the need of also identifying three more types of variables:

- **Global variables** - If the variable declaration or assignment is outside the scope of our function we consider that variable to be global.
- **Control Flow variables** - are variables that have simultaneous assignments in different parts of the source code under different control constructs. For example, lets imagine we were analyzing the source code in Figure 3. If we added a new if condition after line 11 using the *e* variable, our analysis would conclude that, at that stage, *e* would have either been affected by the previous invocation of the *preventDefault()* method, or by the assignment of *false* to its *returnValue* property. In these situations we call that variable a Control Flow variable since its origin may differ according to the flow of the application.
- **Hybrid variables** - are variables for whose classification we identified more than one type of the previously discussed types of variables (except constants since those can be ignored). For instance, in the following source code variable *b* would be obtained from both an input variable and a synthesized variable:

```
var b = document.getElementById('c');
b = b + getServerData();
if(b>0){...}
```

Another aspect we can see from the source code above is that our analysis must include all previous assignments of that variable in the source code. Otherwise, in the previous code we would define the variable as Synthesized instead of Hybrid, which would be inaccurate.

It is also important to notice that our analysis is focusing only on events associated with visible elements. While the depth of analysis is customizable (that is, as long as they are available in the browser, it is possible to configure the tool to analyze the source code of the functions called from the event handlers, and the ones called from those, etc.), in what follows we will only be analyzing the event handlers up to a depth of one. If an event handler has several function calls, we are also going to analyze those functions (if those functions are available), but not the functions called by them. This happens because the goal of this analysis was to evaluate how much information about control flow we could access, analyzing as few JavaScript as possible.

Our approach to gather the data regarding variables is the following: we start by gathering the control flow constructs present in the portion of source code we are analyzing. This source code is either an event handler function or a single function in the code we retrieved by analyzing a function call. For each construct we gather which variables are used.

The following source code shows an example of an if construct:

```
if(b==document.getElementById('c'){...};
```

We differentiate our analysis according to the variable. If we have a single name token followed by any type of operator we have to analyze the previous code in search for that variable assignment, such as, b in the source code above. In that case we extract all the previous assignments of that variable on the source code and analyze each one for the types of variables being used. Afterwards, according to the number of different types found we identify the variable under analysis conforming to the previously explained catalog.

However, if we have more complex constructs before or after the operators, e.g., document.getElementById('c') in the above source code, we process them as a variable. Therefore, we need to identify which type of variable that part of the source code is according to our catalog, in this example we would identify it as an Element variable.

5 Top Sites Analysis

In this section we describe how the tool just described was used to investigate how much control logic information might be possible to extract from event handlers. As already explained, the goal is to assess the viability of developing an hybrid approach to the reverse engineering of Web applications.

5.1 Scope of the Analysis

Since the approach we have developed is fully automated, we could define any number of Websites for analysis. We decided to focus our analysis on the most popular Websites globally, thus we used the Alex top Websites list. Our analysis covered the first one thousand sites on that list. The analysis was performed on the 26th of February, 2014.

It is important to note that in order for the analysis to be automated we had to bypass several errors that could occur analyzing these applications. For example, one important problem we had was that some sites could never finish the page loading. This problem was unrelated to our tool, since opening those sites on different Web browsers, no matter how much time we waited the page would never finish loading. When this happens we cannot extract any information from the page. In order to overcome the application being set on a loop waiting for the loading we set a timeout of one minute for each analysis. The final result showed that forty five of the one thousand sites could not finish loading in the

Table 1. Events and control-flow constructs

	hrefs	Events	Click Events	Ifs	Other Control Flow Constructs	Function Calls
Mean	214.09	26.57	18.53	40.26	40.34	116.96
Standard Deviation	301.28	74.78	68.19	153.71	153.71	412.49
Median	104	5	1	3	3	9
Maximum value	3349	902	887	2109	2109	5121
Percentage of zeros	6.4	21.8	46.5	34.8	34.6	31
Mean excluding zeros	228.73	33.98	34.63	61.74	61.68	169.51

minute we set. This means that almost five percent of our analysis scope has no data gathered because of this problem.

Another problem we had in many sites was that while analyzing the event handlers source code, we got JavaScript parsing errors. In these cases we skipped only those handlers analysis. At this point we could not identify if the malformed JavaScript was coded on purpose, to prevent third party analysis such as this one, or were simply coding errors.

5.2　Data Analysis

We decided to group the criteria into two groups, one with the data on events and constructs, and the other with the data on variables. The event delegation criterion is going to be treated independently of the other criteria since is the only one with a boolean as a result and not a number.

Table 1 shows a summary of the results we retrieved from the analysis of the one thousand sites. Something we can immediately gather, from the table by analyzing the maximum value in comparison with the mean and the median, is that there clearly exist outliers in the data. Moreover, we can see that the data is quite disperse, by comparing the standard deviation values with the mean values.

In terms of hrefs, and discounting the 4.5% of sites that were not computed, numbers show that only around 2% of the analyzed pages that did not use any type of hrefs. This means that the wide majority of Websites still uses hrefs for navigation between pages. This was, of course, to be expected, but shows that the reverse engineering tool should not be restricted to single page Web applications, and consider also navigation between pages.

The analysis of the event handlers shows that only approximately 22% of sites did not have any event we could find. Moreover, when rstricting the events to only clicks we get a 46.5% in total. That means that there were approximately 25% of sites that had events we could detect but none of those events were clicks. Also, having a mean of around 26.57 events and 18.53 clickable events shows that

Table 2. Variables comparison

	Element	Synthesized	Object	Global	Control Flow	Hybrid
Mean	0.98	15.09	16.96	29.66	4.98	0.37
Standard Deviation	4.83	79.40	84.28	142.37	45.48	2.84
Median	0	0	1	1	0	0
Maximum value	49	1008	1830	2520	1155	64
Percentage of zeros	89.2	65	44.4	48.6	86.7	93.1
Mean excluding zeros	9.1	43.11	30.49	57.7	37.41	5.27

an analysis based on this type of events would have an important impact on the Web overall.

In terms of control flow structures the data we gathered showed an interesting result. The usage of if constructs is almost identical to use of the other control flows constructs. We were expecting a lot more usage of ifs than the other constructs but that was not the case in this analysis. One hypothesis is that since these are the most popular sites globally, most source code is done with performance and space constraints and a significant part of those other constructs are ternary conditional operators.

Regarding these criteria global values, finding an average of 80 control flow constructs per site with only an analysis of a depth of one function call is quite significant for our approach. Obviously, we must also take into consideration that around 35% of sites had no construct at all. We can only assume that either no logic was present on the client side, or that the event delegation approach was being used.

Concerning function calls, we found an average of 111.96 function calls per site, only on the subset of source code we were analyzing. This shows that there is a significant amount of other source code that is not analyzed in our approach. Moreover, considering those function calls might have other function calls and so on, that is even more significant. Also important is that most JavaScript code obfuscation techniques increase the number of function calls significantly to hide the logic behind the code [11]. Thus, even an increase in our analysis to a depth of 5 or more function calls might not retrieve interesting results, despite the added computational load.

Table 2 depicts a summary of the data we gathered about our analysis of variables on the source code. Element variables were not found in 89.2% of the sites. This was something we were expecting since not only there are a lot of different frameworks in JavaScript, but also there are several ways to shorten the usage of *document.getElementBy*. For instance, by wrapping those calls inside auxiliary functions:

```
function getId(id){ return document.getElementById(id); }
```

Nevertheless, about 10% the most popular sites use this construct unaltered to get elements on their event handlers. Also interesting is that excluding the sites with no variables of this type found, we got an average of 9 element variables found.

In terms of synthesized variables, they were found in 35% of sites and excluding zeros we got an average of 43.11 variables found. It is interesting, however, that the number of sites where these variables were found is quite lower than what we were expecting, particularly when comparing with object and global variables whose results were higher. Thus, most sites are not currently using these variables, which means that they might be using functional references on variables, which we are currently interpreting as object variables.

This conclusion is important because it means that we have to analyze each object variable of our analysis to see if the type of that variable is or not a function. Although this might lead to a lot more computation, this statistical analysis showed us that it is important for us to do add this feature.

Both object and global variables were found in most sites. In fact, if we exclude the 35% of sites where no control flow constructs were found, thus no analysis of variables was performed, only around 10% of sites were analyzed and got none of these variables. It is also interesting that the type of variable that clearly got more matches in our analysis was the global.

Both control flow and hybrid variables were found in a significantly smaller number than other types of variables excluding element variables. Since handling these variables is quite more complex than the others, these results were promising to our approach. Moreover, the hybrid variables were clearly the ones that were identified less in our analysis.

In terms of event delegation we identified 30.6% of sites that were using this approach for adding event handlers. It was also interesting and something unexpected, that most of these sites also had click events that we were able to identify. Thus, there are a significant number of sites that use both approaches for adding event handlers.

6   Conclusions

This paper has presented two main contributions. First a tool for analyzing Web applications, based only on the event handlers' source code, was proposed. The tool works by extracting information about the control flow of the application, based on the types of variables identified. Second that approach was used on the top thousand most popular Web sites globally. The analysis of those sites, in terms of events and variables used in control flow constructs, has been presented and discussed.

That analysis enabled us to retrieve useful information towards our goal of developing an hybrid tool for the reverse engineering of Web applications. For instance, an analysis of the two approaches of adding event handlers to a page as discussed in Section 3 shows that the classic approach is widely used, since we got results in approximately 78% of sites while the event delegation just appeared on approximately 30% of sites. Therefore, a tool that reverse engineers sites based on the classic approach would work on the majority of Websites.

Another important result is that the amount of if constructs used in the source code we analyzed is similar to the amount of other control flow constructs. This means that if our analysis focus only on ifs we would be analyzing only half the constructs that affect the control flow of the application.

In terms of our variable's analysis we infer that both control flow and hybrid variables are used significantly less than other types of variables, thus the added computational logic we would need to handle these variables might not compensate. Furthermore, we were expecting more synthesized variables than what we found, this mean we must further inspect object variables to identify if they are functions.

Future work will comprise the usage of this data into developing a Reverse Engineering tool that will be able to extract information on most sites. Moreover, there were a few shortcomings in our analysis that could be improved such as extending the element variables identification to other frameworks or techniques, this could mean an analysis of the entire JavaScript files similar to the one we are doing to identify event delegation approaches. Furthermore, trying to find correlations between the several criteria could also yield interesting results.

Acknowledgments. This work was partly funded by project LATiCES (Ref. NORTE-07-0124-FEDER-000062) financed by the North Portugal Regional Operational Programme (ON.2 – O Novo Norte), under the National Strategic Reference Framework (NSRF), through the European Regional Development Fund (ERDF), and by national funds, through the Portuguese funding agency, Fundação para a Ciência e a Tecnologia (FCT). Carlos Eduardo Silva is further funded by the Portuguese Government through FCT, grant SFRH/BD/71136/2010.

References

1. Eilam, E.: Reversing: Secrets of Reverse Engineering. Wiley (2005)
2. Telea, A.C.: Reverse Engineering – Recent Advances and Applications. InTech (2012)
3. Mikkonen, T., Taivalsaari, A.: Web applications – spaghetti code for the 21st century. Technical Report SMLI TR-2007-166, Sun Microsystems (2007)
4. Silva, C.E.: Reverse engineering of rich internet applications. Master's thesis, Universidade do Minho (2009)
5. Campos, J.C., Saraiva, J., Silva, C., Silva, J.C.: GUIsurfer: A reverse engineering framework for user interface software. In: Telea [2], ch.2, pp. 31–54
6. de Kleijn, R.: Learning Selenium: Hands-on tutorials to create a robust and maintainable test automation framework. Leanpub (2014)
7. Mesbah, A., van Deursen, A., Lenselink, S.: Crawling Ajax-based web applications through dynamic analysis of user interface state changes. ACM Transactions on the Web (TWEB) 6(1), 3:1–3:30 (2012)
8. Morgado, I.C., Paiva, A.C.R., Faria, J.P., Camacho, R.: GUI reverse engineering with machine learning. In: 2012 First International Workshop on Realizing AI Synergies in Software Engineering (RAISE), pp. 27–31. IEEE (June 2012)

9. Silva, C.E., Campos, J.C.: Combining static and dynamic analysis for the reverse engineering of web applications. In: Forbrig, P., Dewan, P., Harrison, M., Luyten, K., Santoro, C., Barbosa, S.D.J. (eds.) Proceedings of the 5th ACM SIGCHI Symposium on Engineering Interactive Computing Systems (EICS 2013), pp. 107–112. ACM (2013)
10. Jakob, J.: jQuery Compressed. Jenkov Aps (2011)
11. Schrittwieser, S., Katzenbeisser, S.: Code obfuscation against static and dynamic reverse engineering. In: Filler, T., Pevný, T., Craver, S., Ker, A. (eds.) IH 2011. LNCS, vol. 6958, pp. 270–284. Springer, Heidelberg (2011)

Monitoring Recommender Systems:
A Business Intelligence Approach

Catarina Félix[1,2], Carlos Soares[1,3], Alípio Jorge[2,4], and João Vinagre[2,4]

[1] INESC TEC, Portugal
[2] Faculdade de Ciências da Universidade do Porto, Portugal
[3] Faculdade de Engenharia da Universidade do Porto, Portugal
[4] LIAAD-INESC TEC, Portugal
{cfo,joao.m.silva}@inescporto.pt, csoares@fe.up.pt, amjorge@fc.up.pt

Abstract. Recommender systems (RS) are increasingly adopted by e-business, social networks and many other user-centric websites. Based on the user's previous choices or interests, a RS suggests new items in which the user might be interested. With constant changes in user behavior, the quality of a RS may decrease over time. Therefore, we need to monitor the performance of the RS, giving timely information to management, who can than manage the RS to maximize results. Our work consists in creating a monitoring platform - based on Business Intelligence (BI) and On-line Analytical Processing (OLAP) tools - that provides information about the recommender system, in order to assess its quality, the impact it has on users and their adherence to the recommendations. We present a case study with Palco Principal[1], a social network for music.

1 Introduction

Websites like Amazon[2] or eBay[3] incorporate Recommender Systems (RS) that can use, for example, history from users purchases to train a recommendation model able to predict new products that they may be interested in.

Because the behavior of users may change [1, 2], it is possible that a RS degrades its quality over time. Also, as it is possible to have many different recommendation models (by using different algorithms or parameterization) for a given problem. Therefore, even if the performance of the current RS does not degrade, it is possible that a different model will provide better recommendations. All these factors generate the need for methods and systems to monitor or support the management of RS.

The goal of this work is to develop a Business Intelligence-based system to monitor RS. Our approach consists of implementing BI tools such as dashboards, reports and OLAP that provide information on the behavior of the RS used on

---

[1] http://www.palcoprincipal.com
[2] http://www.amazon.com/
[3] http://www.ebay.com/

B. Murgante et al. (Eds.): ICCSA 2014, Part VI, LNCS 8584, pp. 277–288, 2014.
© Springer International Publishing Switzerland 2014

a site. These tools are based on a data warehouse (DW), which is the central component of our proposal. We also illustrate the use of the proposed approach with the system that was developed for Palco Principal, a social network for music.

2   Business Intelligence

Business Intelligence is based on tools and techniques to store, manage and analyze large amounts of business data to support the decision making process. The data is stored in data warehouses (DW) and some of the tools that are commonly used for analysis include OnLine Analytical Processing (OLAP) and data mining.

A DW is a database used for storing the data needed for decision making processes in an organization. It is usually separated from the organization's operational database because the two databases are organized in a different way, as they serve different purposes.

The DW is used for information processing, allowing the analysis of consolidated and historical data. The data is stored in the DW by ETL (Extract, Transform and Load) processes: they extract data from the operational system, transform it in order to suit the needs of the decision support system and load the data into the DW. The data warehouse consists of dimension and fact tables. A fact table is the primary table in the dimensional model and also stores numerical measurements (measures) that represent the indicators that are relevant to support decisions, and the foreign keys that connect to the dimension table's primary keys [3]. The dimension attributes serve as the source for query constraints, grouping and report labels, making the dimension tables the entry point to the fact table. A generic example is: to answer the question "What was the store's revenue, by user age?", the fact table would contain the measure "revenue" and one of the dimension tables would have a "user" attribute.

2.1   Pentaho

Pentaho [4] is a Business Intelligence suite that includes a wide range of analytics tools, including data integration and On-line Analytical Processing (OLAP) analysis. It is based on open source projects and there are two versions available: Pentaho Enterprise and Pentaho Community.

Pentaho's Business Analytics Platform is Web-based, which enables it to be accessible anywhere and using any platform. Its interface allows the user to create, edit or view analysis: charts that illustrate the DW's information. Here, the administrator can also manage users, database connections and scheduled tasks.

The data integration tool of Pentaho is Spoon [5], from the Kettle [5] project. It is used to create the DW and also the ETL processes used to populate it with data from the operational databases. The ETL processes transform the data according to the requirements of the monitoring system. We can use the

Schema Workbench [5] to design the cube, with its dimensions and measures, to be used by the OLAP and other analytics tools. For data analysis, Pentaho uses the Saiku plugin [6], which is a web based analytics solution that allows users to perform OLAP analysis. Pentaho's workflow can be viewed in Fig. 1.

3 Recommender Systems

Recommender Systems [7] allow websites to dynamically display up-to-date content that suit the preferences of the user, also satisfying the users' complex and diverse needs and behaviors. This is usually accomplished by collecting site information and storing it into a database. The data typically concerns the characteristics of the content in the site (content data: e.g. the products sold in the web site), user (user data: e.g. age, gender, location) and the interactions between the user and the site (usage data: e.g. products browsed and bought, alongside with the users that performed those actions). The RS analyses information from this database and uses it to generate recommendations. Examples of services using Recommender Systems are Amazon.com and eBay [8], which recommend new products to the user, based on his browsing and purchasing history.

To generate the recommendation, as we can see in Fig. 2, the system can use different algorithms and parameter configurations to originate the models. The next phase consists in selecting the model to be used, and it can be performed manually or automatically. The selected model will generate the recommendations that will be sent to de web site.

An example of a recommendation technique is Collaborative Filtering. It suggests new items to a user based on the items the user likes and the opinion of users with similar tastes. This technique can predict the likelihood of a user being interested in one item and, then, recommend him the items he will probably like most, provided that the items haven't been yet rated (or bought, accessed, etc.) by the user [9].

Monitoring recommender systems has been recognized as an important part of the system itself. This can done obtrusively by conducting surveys or focused studies [10, 11] or by seamlessly collecting and analyzing user behavior data [12, 13]. Our work follows the later approach. We propose a business intelligence system to provide the manager of the recommender system to explore user behavior data in response to recommendations made by the system. This approach has the advantage of not having to explicitly ask questions to the users and also not making strong assumptions concerning the questions that the site manager might ask.

4 BI Architecture for Monitoring RS

Monitoring Recommender Systems enables the assessment of the behavior of recommendation models. Although, the most important aspect is the quality (e.g. whether a recommendation is followed by the user), other aspects may be

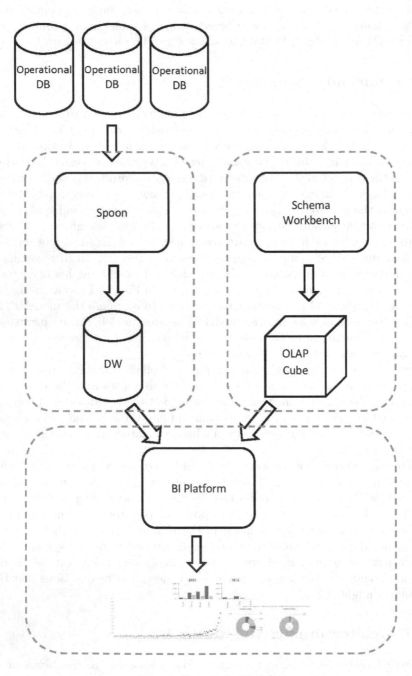

Fig. 1. Pentaho's Workflow

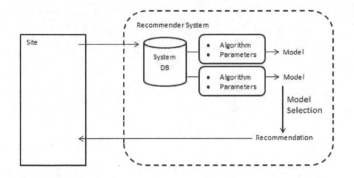

Fig. 2. Recommendation Generation Workflow

interesting as well (e.g. variety of the recommendations). Additionally, monitoring may focus on the evaluation of a single model over time or it may compare different models. This way, the website administrator can switch between recommenders in order to achieve the business goals. He/she may also take measures to improve the ones that are not behaving the way they are intended to. To do this a tool is required to support the analyses of the recommendations and help to find the circumstances under which they are performing better or worse.

Monitoring a RS enables:

- Evaluating recommendation performance: we can measure the acceptance rate of the recommendations made by this system and compare it to other values from previous points in time;
- Evaluating the recommender operation mode: this leads to being able to adapt the recommenders to the users;
- Managing recommenders: if the monitor shows that recommender A is performing worse than usual, the site administrator can switch to recommender B and see how it behaves;
- If integrated with the the BI system for the whole business, the RS monitoring system may help understand how it is contributing (or not) the the business goals.

For example, Fig. 3 shows a comparison of the number of positive actions (accepted recommendations) and negative actions (rejected recommendations) for each of the recommendation models used. The models here are represented by the date in which the system has started to use them. As we can see in the chart, some models originate more positive actions, while the negative ones are nearly the same for every model. Using this information, the site administrator can select the model to be used to generate the recommendations.

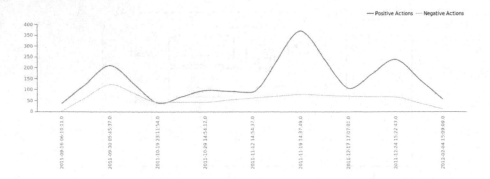

Fig. 3. Comparison of positive and negative actions for recommendation model

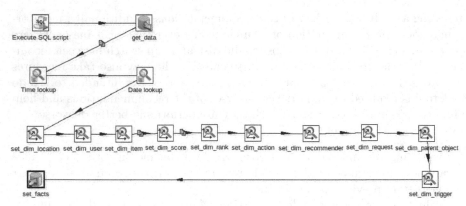

Fig. 4. Spoon Transformation

5   Case Study: Music Portal

Palco Principal (PP) is a social network based on music. Its members are Internet users with interests on that kind of content: artists and listeners. The artists upload their music to the website, along with metadata characterizing the uploaded tracks, such as genre and pace.

The listeners, upon registration, are invited to choose their preferred music genre and, then, the Recommender System can start recommending content based on the user's preferences. After listening to the recommended items, the user can either add them to a playlist or to a blacklist. These factors will hereafter be considered by the RS to generate new recommendations.

To make the most effective use of recommendations, the manager of the web site needs to know how the recommender system is behaving and how users are responding to it. Therefore, every recommendation that is shown is stored in a database. This data is processed in Spoon, using the process shown in Fig. 4, and loaded to the data warehouse represented in Fig. 5.

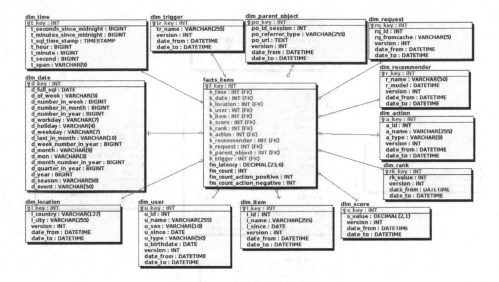

Fig. 5. PP's data warehouse schema

The data warehouse used for this job (Fig. 5) can be represented by a star schema: it is composed by a fact table surrounded by several dimension tables. This design style was chosen because it provides better query performance. This is accomplished due to the faster aggregation capabilities, when compared to the snowflake schema (that is composed by various stars that are connected to each other). This performance increase is achieved because queries can be written with simple inner joins between the fact table and the dimension tables.

Here we have the fact table surrounded by the dimension tables: *dim_time*, *dim_date*, *dim_location*, *dim_user*, *dim_item*, *dim_score*, *dim_rank*, *dim_action*, *dim_recommender*, *dim_request*, *dim_parent_object* and *dim_trigger*. The dimension tables are then linked by the foreign keys in the fact table, which also stores some measurements: *latency* (time interval between the recommendation and the user reaction) and indicators to the type of the action (positive or negative). These attributes are the ones we want to measure for the analysis. For each item recommended, a row in the fact table is created, and it is either linked to the corresponding rows on the dimension tables (if they already exist) or the dimension tables rows are created and then the fact row is linked to them.

For the proper functioning of the monitoring system we expect certain data to be imported from the database of the operational system of the website: the content of the site, the users and the offered recommendations. This way we can relate the recommendations with the characteristics of the items and of the users. We can relate any of the attributes from the recommendations with any of the attributes of the users and/or the items, as shown in Fig. 6, which shows a fragment of the schema in Figure 5.

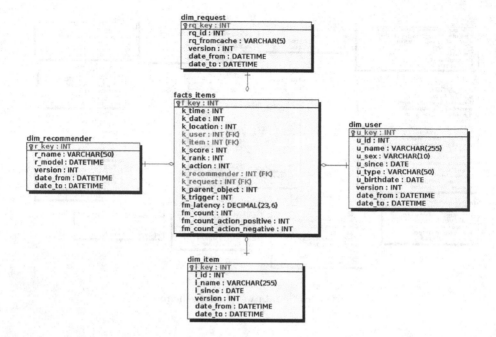

Fig. 6. Relation between Users, Items and Recommendations

5.1   Recommendation Monitoring Functionalities

With this DW we can, for example, analyze the difference between positive (adding an item to a playlist) and negative (adding the item to the blacklist) actions, taking into consideration the date when the recommendations were made, as shown in Fig. 7. In this figure, we can see that the number of positive actions is always bigger than the negative ones, but also that the system registered more actions (positive or negative) during 2011 than during 2012. With these results the administrator could decide to switch back to the recommender used in 2011, if it was a different one, in order to obtain better results.

Another possible analysis can be in terms of the time of the day when the recommendations are made and the actions on those recommendations are performed, as depicted in Fig. 8. In the chart, we can see a peak of actions in the afternoon. This can be a consequence of the global usage of the site. However, if during this time of the day the system was using a different recommender, the administrator can decide to use that recommender for the rest of the day, expecting to increase user interaction with the website. On the other hand, if the recommendation model is the same, this could mean that different models should be used at different times of the day. He could also take precautions (in terms of system resources) so that during that period there will not occur problems in the website due to an overload of requests.

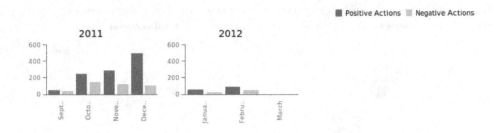

Fig. 7. Positive and Negative Actions by Date (Year and Month)

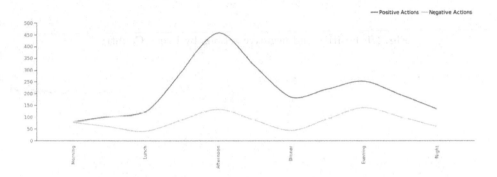

Fig. 8. Positive and Negative Actions by time of day

The site administrator can also analyze reactions to recommendations according to the gender of the user Fig. 9 shows that, while the gender has nearny no impact in negative actions, male users tend to perform more positive actions than females.

The administrator can also view the distribution of actions in terms of the user countries. Fig. 10 shows that most actions are made by users from Portugal (70.4% of positive and 88.8% of negative actions), followed by Brazil, Angola and Mozambique. This is expected because the site was first launched in Portugal and the other portuguese-speaking countries.

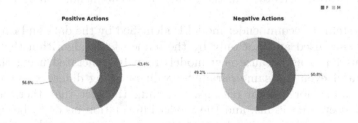

Fig. 9. Positive and Negative Actions by user gender

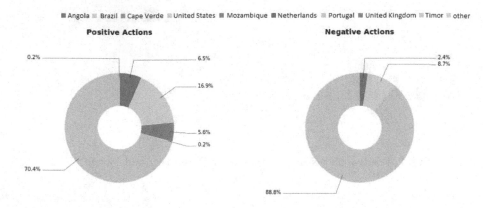

Fig. 10. Positive and negative actions by User's Country

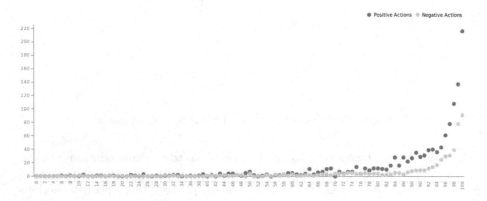

Fig. 11. Positive and negative actions by Item's Rank in Previous Recommendations

We may also want to know if the rank of the item in the recommendation affects the likelihood of it being accepted by users. Fig. 11 shows that when the rank is high (i.e., the recommendation is stronger), there are more actions on the item. The administrator can then tune the recommendation system based on this fact.

In this system, a recommender model is identified by the date and time when it started being used and not only by the name of the algorithm that generated it. This happens because new models may be generated using the same algorithm (and even the same parameter values), using different sets of data. This is necessary for several reasons, including the fact that there are constantly new users and items, and the model information needs to be updated. Also, the recommendations can be made from cache memory, when similar requests have been made. In Fig. 12 we can see the difference between the newly

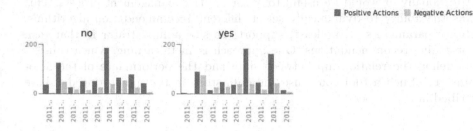

Fig. 12. Positive and negative actions by Recommender Model and whether the Recommendation was (chart with title "yes") or not (chart with title "no") made from the cache

calculated recommendations (left chart) and the ones made from cache memory (right chart).

5.2 Summary

Using Business Intelligence tools, the site administrator can monitor and assess the performance of the recommendation systems, simply by analyzing charts containing data from the acceptance or rejection of the recommendations. These analyses can help the administrator in tasks such as switching on and off the recommenders, or tuning their parameters differently, in order to improve the system's performance.

The system can also be used to analyze other aspects that were not illustrated here. Information such as, for instance, what are the most recommended items, what type of music is most often recommended, may also be useful for the site administrator.

6   Conclusions and Future Work

With the introduction of recommender systems in e-business and social websites, and since the performance of the recommendation models may vary over time (due to, for example, changes in user behavior), there is a need to develop tools for monitoring them.

In this paper we have proposed the use of Business Intelligence Tools in order to complete that task. Using these tools, system administrators may assess the online performance of the system and use this information to improve the overall system's quality, either by switching between algorithms, triggering model updates or fine-tuning parameters.

Despite the difficulty in evaluating our proposal, it would be worthwhile considering a process that allows the site manager to score the utility of each feature at each moment.

It would also be useful to have a tool or process that would grant the administrator the capacity of measuring the impact of the monitoring architecture and tools on a production system.

Additionally, it would be useful to automate the management process. This means the ability to dynamically select different recommendation algorithms, calibrate parameters or, at least, support the site administrator in that task by providing recommendations. One approach is metalearning, which consists of modeling the relationship between data and the performance of the algorithms [14]. Such a tool could use the data store in the monitoring database described in this paper.

Acknowledgments. This work is partially funded by FCT/MEC through PID-DAC and ERDF/ON2 within project NORTE-07-0124-FEDER-000059.

References

1. Billsus, D., Pazzani, M.: User modeling for adaptive news access. User Modeling and User-Adapted Interaction 10(2-3), 147–180 (2000)
2. Koychev, I., Schwab, I.: Adaptation to drifting user's interests. In: Proceedings of ECML2000 Workshop: Machine Learning in New Information Age, pp. 39–46 (2000)
3. Kimball, R.: The data warehouse toolkit: The Complete Guide to Dimensional Modeling. Wiley (2006)
4. Pentaho: Pentaho website, http://www.pentaho.com/ (accessed: January 17, 2014)
5. Pentaho: Pentaho community website, http://community.pentaho.com/ (accessed: January 17, 2014)
6. Saiku: Saiku website, http://meteorite.bi/saiku (accessed: January 17, 2014)
7. Resnick, P., Varian, H.R.: Recommender systems. Commun. ACM 40(3), 56–58 (1997)
8. Schafer, J.B., Konstan, J., Riedl, J.: Recommender systems in e-commerce. In: Proceedings of the 1st ACM Conference on Electronic Commerce. EC 1999, pp. 158–166. ACM, New York (1999)
9. Sarwar, B., Karypis, G., Konstan, J., Riedl, J.: Item-based collaborative filtering recommendation algorithms. In: Proceedings of the 10th International Conference on World Wide Web. WWW 2001, pp. 285–295. ACM, New York (2001)
10. Nowak, M., Nass, C.: Effects of behavior monitoring and perceived system benefit in online recommender systems. In: Proceedings of the SIGCHI Conference on Human Factors in Computing Systems. CHI 2012, pp. 2243–2246. ACM, New York (2012)
11. Pu, P., Chen, L., Hu, R.: A user-centric evaluation framework for recommender systems. In: Proceedings of the Fifth ACM Conference on Recommender Systems. RecSys 2011, pp. 157–164. ACM, New York (2011)
12. Domingues, M.A., Gouyon, F., Jorge, A.M., Leal, J.P., Vinagre, J., Lemos, L., Sordo, M.: Combining usage and content in an online recommendation system for music in the long tail. IJMIR 2(1), 3–13 (2013)
13. Kohavi, R., Longbotham, R., Walker, T.: Online experiments: Practical lessons. IEEE Computer 43(9), 82–85 (2010)
14. Brazdil, P., Giraud-carrier, C., Soares, C., Vilalta, R.: Metalearning: Applications to Data Mining. In: Cognitive Technologies. Springer (2009)

Two Parallel Algorithms for Effective Calculation of the Precipitation Particle Spectra in Elaborated Numerical Models of Convective Clouds

Nikita O. Raba[1] and Elena N. Stankova[1,2]

[1] Saint-Petersburg State University
198504 St.-Petersburg, Russia, Peterhof, Universitetsky pr., 35
[2] Saint-Petersburg Electrotechnical University "LETI",
197376, St.-Petersburg, Russia, ul.Professora Popova 5
no13@inbox.ru, lena@csa.ru

Abstract. Effective calculation of the spectra of precipitation particles, i.e. the spectra of water drops, ice and snow crystals, graupel and hail is one of the most challenging problems in 2-D and 3-D numerical models of natural convective clouds. Algorithms for spectrum calculation are usually proportional to the cubic degree of spectral bin number and therefore are computationally very expensive. The problem becomes even more complicated taking into account the fact that the spectrum of each precipitation particle should be calculated in each spatial grid point of 2-D and 3-D model. The algorithm of Kovetz and Olund and the algorithm of Bott have been chosen as two of the most popular algorithms intended for calculation of the evolution of cloud particle spectra for subsequent optimization and parallelization. Kovetz and Olund algorithm has been optimized and parallelized using both CPU and GPU. Its optimal version is quadratic in time and allows using more than 1000 threads for effective parallelization. Our results show that speed-up of the optimized algorithm is equal to 2.6–13 depending upon the number of spectrum grid points and the use of GPU can accelerate calculations 15-20 times. Bott's algorithm has been parallelized using only CPU and provides speed up equal to 5 on 8 threads. The developed algorithms are universal: they can be applied to models of different dimensions and to different types of cloud particles. They can be effectively used in elaborated numerical cloud models for operational forecast of dangerous weather phenomena, such as thunderstorm, heavy rain and hail.

Keywords: parallel algorithm, GPU, CUDA technology, stochastic collection equation, numerical modeling.

1 Introduction

Clouds play the key role in the formation of weather and climate on the Earth. Power and intensity of precipitation have a great influence on people's life. Radiation characteristics of clouds significantly affect the transfer of radiant energy in the

B. Murgante et al. (Eds.): ICCSA 2014, Part VI, LNCS 8584, pp. 289–299, 2014.

atmosphere. The latent heat of phase transitions, released in the process of precipitation formation, is the source of energy for atmospheric effects of different temporal and spatial scales, such as thunderstorms, squalls, powerful tropical cyclones and hurricanes.

Investigation of dangerous convective phenomena such as thunderstorms, hail and rain storms requires consideration of various processes having different nature and scale and presents an extremely complex problem for numerical modeling [1,2]. Cloud model should reproduce both thermodynamical and microphysical processes. The former describe interaction of updraft and downdraft convective flows, turbulence vortexes and temperature variations. The latter describe transformations and interactions of small cloud particles – water drops, aerosols and various kinds if ice particles (ice crystals, snowflakes, graupel and frozen drops). Calculations of the whole set of the processes require a large number of computational resources and time, especially in case of using 2 - D and 3 - D cloud models. Numerical simulation of the microphysical processes is the most computationally expensive part in such models especially in case of the so-called "detailed" description of the microphysical process, which demands calculation of cloud particle spectra.

Droplet or ice particle spectrum evolves mainly due to the collection process, numerical description of which is conducted by means of numerical solution of stochastic collection equation (SCE). SCE is the complicated integro-differential equation describing the evolution of the spectrum (distribution function) of particles in dispersed medium through collision and subsequent coalescence. Analytical solution of such equation is possible only in case of the special kernel type; in other cases it should be solved numerically [3-5].

Sequential algorithms which are used for SCE numerical solution, as a rule, have the computational complexity of $O(N^3)$ order, where N is the total number of spectrum bins or intervals. These algorithms are quite computationally expensive, even for calculation of one type of cloud particles. Computations of the spectra of several particle types in each spatial grid point of 2-D and 3-D model demand tremendous computational resources and time. This timing does not allow the use of the models in the operational practice for prediction of dangerous convective phenomena such as thunderstorms, hail and heavy rain. The way out is the optimization and parallelization of the sequential algorithms for SCE numerical solution.

We use two of such algorithms – the algorithm of Kovetz and Olund [6,7] and the algorithm of Bott [8]. The approach of Kovetz and Olund is known to be numerically very efficient. The approach of Bott provides both numerical efficiency and the absence of strong numerical diffusivity.

Kovetz and Olund algorithm have been parallelized using both CPU and GPU. Its optimal version is quadratic in time and allows using more than 1000 threads for effective parallelization. Our results show that the use of GPU can accelerate calculations in 15-20 times. Bott's algorithm has been parallelized using only CPU and provides speed up equal to 5 on 8 threads.

The developed algorithms are universal: they can be applied to models of different dimensions and to different types of cloud particles. They can be effectively used in elaborated numerical cloud models for operational forecast of dangerous weather phenomena, such as thunderstorm, heavy rain and hail.

2   Optimization of Kovetz and Olund Algorithm

Algorithms of Kovetz and Olund provide very efficient and mass conservative numerical solution of SCE, which can be written in the following form:

$$\frac{\partial f(m)}{\partial t} = \frac{1}{2} \int_0^\infty \int_0^\infty K(m',m'')\delta(m - m' - m'')f(m')f(m'')dm'dm'' - f(m)\int_0^\infty K(m,m')f(m')dm', \quad (1)$$

where $f(m, t)$ is drop mass distribution function at time t and $K(m, m')$ is the collection kernel describing the rate at which a drop of mass $(m-m')$ is collected by a drop of mass m' thus forming a drop of mass m. The first integral on the right hand side of (1) describes the gain rate of the drops of mass m by collision and coalescence of two smaller drops, while the second integral denotes the loss of drops with mass m due to collection by other drops. $\delta(m - m' - m'')$ is the delta function. This expression is the continuous analogue of the formulas proposed by Kovetz and Olund [6].

We use the modified algorithm [7,9] for the numerical solution of SCE, where M_i is the mass of the droplets in the interval $[m_{i-1/2}, m_{i+1/2}]$. M_i is used as a variable instead of the mass distribution function $f(m,t)$. [6] can been written in the following form:

$$\frac{dM_i}{dt} = S_i^+ - S_i^-; \quad S_i^- = \sum_{j=1}^{N-1} S_{ij}; \quad S_{ij} = \frac{K(m_i, m_j)M_i M_j}{m_j}; \quad S_i^+ = \sum_{j=1}^{i-1} \sum_{k=j+1}^{i} B_{jki} S_{jk}$$

$$B_{jki} = \begin{cases} (m_j + m_k - m_{i-1})/(m_i - m_{i-1}), \text{if } m_{i-1} < m_j + m_k < m_i, \text{and } i < N \\ (m_{i+1} - m_j - m_k)/(m_{i+1} - m_i), \text{if } m_i \leq m_j + m_k < m_{i+1}, \text{and } i < N-1 \\ 0 \quad \text{in other cases} \end{cases} \quad (2)$$

where S_i^+, S_i^- describe respectively the gain and loss of the particles from the interval i due to collection of the particles from other intervals. Coefficients B_{jki} provide redistribution of the drops formed by the collision of the particles from the intervals j and k, mass of which appeared to be between the intervals i and $i+1$.

Computational complexity of initial (not optimized) Kovetz and Olund algorithm is of the order of $O(N^3)$, where N is the total number of grid intervals (bins) of the mass grid [11].

The algorithm has been optimized before parallelization in order to decrease its computational complexity. Optimization is based on the fact that the matrix B_{ijk} is very sparse. There are at most only two nonzero elements of B_{jki} for any j and k, the rest of their elements are equal to zero. So we can introduce two 2-D arrays A_{jk} and I_{jk} instead of 3-D array B_{jki}. A_{jk} presents matrix of weight coefficients and I_{jk} presents matrix of indices.

$$A_{jk} = (m_{i+1} - m_j - m_k)/(m_{i+1} - m_i) = B_{jki} \quad (3)$$

$$I_{jk} = i, \text{ if } m_i \leq m_j + m_k < m_{i+1} \text{ and } I_{jk} = 0 \text{ if there is no such } i. \quad (4)$$

Let us note, that

$$B_{jki} = A_{jk}, \text{ if } i = I_{jk}; \quad B_{jki} = 1 - A_{jk} \text{ if } i = I_{jk} + 1; \quad B_{jki} = 0 \text{ in other cases} \tag{5}$$

Collision of the particles from j and k intervals results in the decrease of particle concentration in j and k intervals and increase of particle concentration in the intervals i and $i+1$ if $m_i \le m_j + m_k < m_{i+1}$ ($i = I_{jk}$). The newly formed particles are redistributed between the intervals i and $i + 1$ with the weight coefficients Ajk and $1-A_{jk}$. This process can be formulated in terms of time evolution of particle concentration in the following way:

$$\partial f_j/\partial t|_{jk} = -K_{jk} f_j f_k; \quad \partial f_k/\partial t|_{jk} = -K_{jk} f_j f_k,$$

$$\partial f_{i+1}/\partial t|_{jk} = K_{jk} f_j f_k (1 - A_{jk}),$$

$$\partial f_x/\partial t|_{jk} = 0, \text{ if } y \text{ do not equal } j, k, i \text{ and } i + 1 \tag{6}$$

where $\partial f_y/\partial t|_{jk}$ is variation in particle concentration in y interval due to collisions of the drops of j and k intervals. The total mass of particles in the collision does not change.

Summing up all the collisions in all mass intervals we obtain the total change of particle concentration in y interval due to the process of collection (collision + coalescence):

$$\partial f_y / \partial t = \sum_{j=1}^{N-1} \sum_{k=j+1}^{N} \partial f_y / \partial t|_{jk} , \tag{7}$$

Microphysical block of modern cloud models is usually not limited by the description of one type of colliding particles. At least 7 types of hydrometeors should be taken into account: water drops, columnar ice crystals, plate ice crystals, ice dendrites, snowflakes, graupel and frozen drops. So the system of seven equations of type (7) should be numerically solved describing variation of seven functions f_i due to the collision of different types of hydrometeors with each other. This task is much more complex as a collision may occur between different types of particles, and the type of the resulting particles depends on the type and mass of the colliding particles as well as on the ambient temperature. The equation for the change in the concentration f_i of the particles of i-th type with mass m is as follows:

$$\partial f_i(m)/\partial t = \sum_{j=1}^{N_{PT}} \sum_{k=1}^{N_{PT}} C_{ijk}(T) \int_0^{m/2} K_{jk}(m-m',m') f_j(m-m') f_k(m') dm' -$$

$$- f_i(m) \sum_{j=1}^{N_{PT}} \int_0^{\infty} K_{ij}(m,m') f_j(m') dm', \tag{8}$$

where N_{PT} is the number of particle types, $C_{ijk}(T)$ is the coefficient, which is equal to 1, if a particle of i-th type is formed due to collision of particle of j-th type with the particle of k-th type at temperature T (mass of k-th type particle is less than mass of j - th type particle). $C_{ijk}(T)$ is equal to zero in other cases. $K_{ij}(m, m')$ is collection kernel (probability of collision and merging of i-th type particle with mass m and j -th type particle with mass m').

The following algorithm [10] has been developed for calculation of change of concentrations of all particles type due to the process of collection. (Please note, that arrays S^- and S^+ have dimension $(N_{PT} \times N)$; f_{ni} is the concentration of particles of type n in the interval i; grid intervals are equal for all types of particles).

1. Set initial values for S^- and S^+: $S^-_{ni} := 0$, $S^+_{ni} := 0$, for n from 1 to N_{PT}, i from 1 to N.
2. Sort out all pairs of intervals (j, k) of colliding particles, and sort out types of colliding particles m (type of particles from j interval) and n (type of particles from k interval) $m \le n$ for each pair when $I_{jk} \ne 0$. Let us define p as the type of a particle, formed due collision for each four indices (m, n, j, k) if $j < k$, or $m \ne n$. Implement $i := I_{jk}$, $F := K_{mnjk} f_{mj} f_{nk}$, $S^-_{mj} := S^-_{mj} + F$, $S^-_{nk} := S^-_{nk} + F$, $S^+_{pi} := S^+_{pi} + F A_{jk}$, $S^+_{p\,i+1} := S^+_{p\,i+1} + F(1 - A_{jk})$.
3. Calculate $\partial f_{ni} / \partial t := S^+_{ni} - S^-_{ni}$, for n from 1 до N_{PT}, i from 1 до N.

Such algorithm is universal: it can be applied to models of different dimensions and with different types of colliding particles (N_{PT}).

The developed algorithm has computational complexity of $O(N^2)$ order if we consider one type of colliding particles and of order of $O(N^2_{PT} N^2)$ if we consider several types of colliding particles.

Optimization results are presented in Table 1.

Table 1. Calculation time (in seconds) obtained with the help of the CPU (E6400) and the GPU (GTX460 and GTX470). N is the number of grid points

Algorithm	N				
	50	70	100	150	250
not optimized	29.3	68.1	177.3	547.2	2349.9
optimized	11.3	18.8	32.5	71.6	181.2

As it can be seen from the table, speed-up of the optimized algorithm is equal to 2.6–13.0 depending upon the number of mass grid points N. The results have been obtained using Core 2 Quad Q8200 computer.

3   Parallel Version of Kovetz and Olund Algorithm

On the one hand, parallelization of the Kovets and Olund algorithm is rather simple, as all the collisions are independent. It is only necessary to sum up all the changes caused by all collisions and the final particle concentrations will be obtained.

Particle spectrum can be divided into several different parts which evolution can be calculated in parallel.

On the other hand, parallelization of calculation of collision process comes across the famous problem of data race, when multiple threads try to change the values in the same memory location. The problem appears when collision of the particles from different parallel threads results in the appearance of the new particle from the same mass interval. In that case several threads will try to change the same cell of the array S^+ and it is not possible to predict what particular value appears there as a result.

To avoid the problem the optimized algorithm was modified by adding the array MP (instead of arrays S^- and S^+) where four memory cells have been allocated separately for each pair of intervals (j, k) of the colliding particles of types m and n. The first memory cell is responsible for the reduction of the concentration of particles of type m in j interval. The second cell is responsible for the reduction of the concentration of particles of type n in k interval. The third and the forth cells are responsible for the increase of the concentration of particles of type p in the intervals i and $i + 1$ respectively, provided that $i = I_{jk}$

Four matching indices for the array MP are stored for every four values (m, n, j, k): $ind0mnjk$, $ind1mnjk$, $ind2mnjk$, $ind3mnjk$. The indices are calculated in such a way that the memory cells related to the increase (or decrease) in the concentration of particles of the same type should be next to each other (i.e. form a coherent field in the array MP). Moreover, the memory cells related to an increase (or decrease) in the concentration of particles of the same type in the same mass range, too, must be close to each other. Some cells may be left unfilled, they may not have an appropriate index ind. Indices of all these successive areas: $pos0ni$ is the index of the last element in MP array, related to the decrease in the concentration of particles of type n in the interval i, $pos1ni$ - the index of the last element in the array MP, relating to the increase in the concentration of particles of type n in the interval i. The method of calculating index ind , pos and dimension of the array MP is presented in [11].

Calculation of the concentration variation is performed in 3 stages.

At the first stage, each thread calculates $F = K_{mnjk} f_{mj} f_{nk}$ for each corresponding collision and fills the four cells of the array MP: $MP [ind0mnjk] = -F$, $MP [ind1mnjk] = -F$, $MP [ind2mnjk] = F Ajk$, $MP [ind3mnjk] = F (1 - Ajk)$. The array MP is completely filled after the calculation of all collisions. Initially array MP is filled with zeros, so those cells of MP for which index ind is not defined remain equal to zero.

Summation of the separate array cells is provided at the second and the third stages of calculation. Please note, that cells of the array under summation should belong to the same grid intervals.

The performance of such algorithms is limited mainly to the bandwidth of memory bus. Since bandwidth of memory bus of modern video cards is ten times greater than that of the computer's memory, the graphics processors (GPU) have been used for parallel algorithm realization.

The second reason for using GPU is the possibility for dividing calculation of SCE into thousands of parallel threads. The maximum number of threads can be equal to 1750 ($N_{PT} \times N$). This fact is very important taking into account that modern graphics processors (GPU) have much more processing cores (hundreds) compared to CPU

(2-8). To take full advantage of GPU, special "substantially parallel" algorithms are needed (their implementation should be divided into thousands of threads which perform the same operations on different data). There are additional restrictions: operations must be as simple as possible, the number of possible branching should be minimal as well as thread synchronization and address to the global memory.

Currently, there are several techniques to use GPUs for general-purpose computation (GPGPU - General-Purpose computing on Graphics Processing Units): CUDA, ATI Stream, OpenCL, DirectCompute [12-16]. One of the most popular technologies is CUDA presented by NVidia company. It allows to make full use of the GPU, and it is based on the language C / C $^{+\,+}$ with some extensions and restrictions, making it easier to write programs. The program for CUDA uses both CPU and GPU. CPU executes sequential calculations, and GPU executes massive parallel calculations. These parallel calculations are presented in the form of concurrent threads.

We have investigated several schemes for thread allocation and chosen the most efficient one, which allows creating the largest number of parallel threads and to take into account some specific features of CUDA. It is the following: each thread processes the collision of particles of type m from the mass interval j with particles of different types ($n \geq m$) from the mass interval k, i.e. one thread handles with N_{PT} collisions maximum (or less, for example, when $k \leq j$). The values of j, m and k depend on the index of thread (*threadIdx*) and block (*blockIdx*) (in CUDA threads are combined in blocks, the blocks are combined in a grid, threads and blocks have their multidimensional indices, threads within a block may be synchronized and have access to the fast shared memory).

The final parallel algorithm consists of two steps: initialization and collision calculation. Initialization step includes the preliminary calculation of the indices I,*ind*, *pos* and the coefficients A; loading of I, A and coagulation kernels K' to the video card memory; allocation of the video card memory for the resulting concentration of mass distribution function f and the auxiliary array MP. Collision calculation step includes loading the values of mass distribution function f to the video card memory; calculation of the values of MP array using video cards; summation of the separate MP array cells and calculation of the new particle concentration f; copying the new values of f to the main memory of the computer.

Test results have been performed for the same initial conditions. We compared the periods of time necessary for calculating the process of collection with the help of CPU (Intel Core 2 Duo E6400) and the GPU (NVidia GeForce GTX470 - 448 cores) and assessed the benefits of using the latter. The results are presented in Fig. 1 and 2.

As we can see from the figures the use of GPU allows to achieve the substantial decrease of computation time and corresponding increase of speed-up values up to 15 – 20 times depending upon the number of grid intervals.

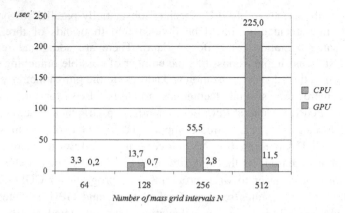

Fig. 1. Calculation time (in seconds) obtained with the help of the CPU (E6400) and the GPU (GTX470

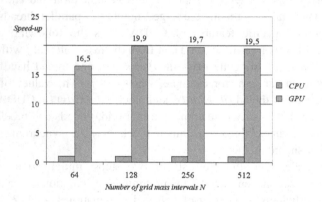

Fig. 2. Speed-up obtained on GPU (GTX470) versus CPU (E6400)

4   Optimization and Parallelization of Bott's Algorithm

The approaches of Kovetz and Olund are known to be numerically very efficient. However, due to its strong numerical diffusivity this method produces artificial broadening of the raindrop spectrum. The approach of Bott docs not have this drawback. Furthermore, this method is also numerically effective and mass conservative.

The method consists of the two-step procedure. At the first step the mass distribution of drops with mass m' that has been newly formed in the collision process is added to grid box k with $m_k \leq m' \leq m_{k+1}$. By solving an advection equation, at the second step the certain fraction of the cloud water mass is transported from k to $k+ +$. The detailed description of the method is given in [8].

Unlike algorithm of Kovets and Olund, Bott's algorithm is essentially sequential. It implies iterative procedure for treating all collisions of drops during the time of step Δt. At the first step the collision of the smallest drops from the first grid interval with drops of the second mass interval is calculated yielding new values of the mass distribution function in the mass intervals of colliding particles. Then the collision of the remaining drops in the first mass interval having now the new mass distribution function, with drops from the third interval is calculated. The procedure is continued until all collisions of the drops in the first grid box with drops of the rest of the grid boxes are accounted for. At the next iteration step the collisions of drops in the second grid box with all larger drops are treated in the same way. This procedure is repeated for the drops in the rest grid boxes until at the last step the collision of drops of the two last grid boxes has been calculated. Due to this iteration approach, the drop distribution is updated before the next collision process is calculated. And this is the main difficulty for parallelization, because collision process in different parts of particle spectrum is not independent.

At first we provide optimization of Bott's algorithm. Optimization is based on the explicit calculation of the grid box numbers k and $k+1$, in which the newly formed drops with mass x' has to be split. Let us introduce the mass distribution function $g(y,t)$ as

$$g(y, t)dy = mf(m, t)dm,$$
(9)

where $y = ln\ r$ and r is the radius of the drops with mass m, and choose the logarithmically equidistant mass grid, that is,

$$m_{k+1} = \alpha m_k, \quad k = 1,...N$$
(10)

Then the number of the grid box k, for the drops with mass m' which have been newly formed by the collision of the drops from the grid boxes i and j can be calculated as follows:

$$k = int(i + \log_\alpha (1 + \alpha^{j-i}))$$
(11)

Knowledge of the exact k value allows us to avoid additional calculations of k and $k+1$ by sorting all values of k index for which the condition $m_k \leq m' \leq m_{k+1}$ is implemented.

The parallel algorithm is based on the calculation of the changes of the mass distribution function in each of the intervals i, j, k, and subsequent summation of these changes. The loss of drops in each of the intervals is written in the two-dimensional array $dg(i, j)$ and the gain is written in the array $dgk(i, j)$. Each cell of the arrays $dg(i,j)$ and $dgk(i\ j)$. is responsible for the results of individual collision of the particles from the intervals i and j and thus allows to avoid the problem of data race. This approach is similar to the one used for parallelization of Kovets and Olund algorithm which has been described above.

We provide parallel calculations using CPU. Each thread has been responsible for the calculation of evolution of the mass distribution function from the grid box i due to collision with the particles from the drop boxes j (j= 1,...N). Preliminary test results of the parallel Bott's algorithm are presented in Fig. 3.

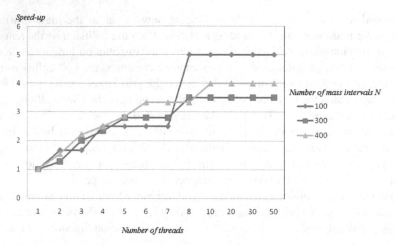

Fig. 3. Speed-up obtained by parallelization of Bott's algorithm

As it can be seen from Fig. 3, parallelization allows to reduce the computation time essentially. Maximum speed-up is equal to 5 and is achieved when 8 threads are launched.

Further investigations will deal with the elaboration of the parallel version of Bott's algorithm for GPU calculations.

5   Conclusions

Two of the most popular algorithms for numerical solution of stochastic collection equation have been optimized and parallelized.

Optimized version of Kovetz and Olund algorithm has the computational complexity on the order of magnitude smaller than non-optimized algorithm and decreases the time of calculations 2.6–13 times depending upon the number of spectrum grid points.

The multi-threaded parallel algorithms are developed for numerical solution of stochastic collection equation on CPU and graphical processors using CUDA technology. The way of splitting of computations on several stages is discussed; the method of redistribution of calculations on different threads is presented and analyzed. We also describe some methods for preventing record collisions when multiple threads try to change the values in the same memory cell.

The results of the numerical tests show that parallel Kovetz and Olund algorithm developed for GPU allows the reduction of computation time 15-20 times depending upon the number of grid intervals. Maximum speed –up achieved by the parallel version of Bott's algorithm is equal to 5 when 8 threads are launched.

The developed algorithms are universal: they can be applied to cloud models of different dimensions and to different types of colliding particles. So the algorithms are the prompt instruments for their use in the hardware and software systems intended for operational forecast of dangerous weather phenomena.

Acknowledgment. This research was sponsored by the Saint-Petersburg State University under the project 0.37.155.2014 "Research in the field of designing and implementing effective computational simulation for hydrophisical and hydro-meteorological processes of Baltic Sea (and the open Ocean and offshores of Russia)"

References

1. Khain, A., Ovtchinnikov, M., Pinsky, M., Pokrovsky, A., Krugliak, H.: Notes on the state-of-the-art numerical modeling of cloud microphysics Review. Atmospheric Research 55, 159–224 (2000)

2. Kogan, Y., Mazin, I.P., Sergeev, B.N., Khvorostyanov, V.I.: Numerical cloud modeling, p. 183. Gidrometeoizdat, Moscow (1984)

3. Pruppacher, H.R., Klett, J.D.: Microphysics of Clouds and Precipitation, p. 954. Kluwer Academic (1997)

4. Tzivion, S., Feingold, G., Levin, Z.: An efficient numerical solution to the stochastic collection equation. J. Atmos. Sci. 44, 3139–3149 (1987)

5. Prat, O.P.: A Robust Numerical Solution of the Stochastic Collection–Breakup Equation for Warm Rain. J. of Applied Meteorology and Climatology 46, 1480–2014 (2007)

6. Kovetz, A., Olund, B.: The effect of coalescence and condensation on rain formation in a cloud of finite vertical extent. J. Atm. Sci. 26, 1060–1065 (1969)

7. Stankova, E.N., Zatevakhin, M.A.: Modified Kovetz and Olund method for the numerical solution of stochastic coalescence equation. In: Proceedings 12th International Conference on Clouds and Precipitation, Zurich, August 19-23, pp. 921–923 (1996)

8. Bott, A.: A flux method for the numerical solution of the stochastic collection equation. J. Atmos. Sci. 55, 2284–2293 (1997)

9. Raba, N.O., Stankova, E.N.: On the effectiveness of using the GPU for numerical solution of stochastic collection equation. In: Murgante, B., Misra, S., Carlini, M., Torre, C.M., Nguyen, H.-Q., Taniar, D., Apduhan, B.O., Gervasi, O. (eds.) ICCSA 2013, Part V. LNCS, vol. 7975, pp. 248–258. Springer, Heidelberg (2013)

10. Raba, N.: Optimization algorithms for the calculation of physical processes in the cloud model with detailed microphysics. In: Raba, N. (ed.) Applied Mathematics. Informatics. Control Processes, West. SPSU. Ser. 10, vol. 3, pp. 121–126 (2010) (In Russian)

11. Raba, N.: Development and implementation of the algorithm for calculating the coagulation in a cloud model with mixed phase using CUDA technology. In: Applied Mathematics. Informatics. Control Processes, Bulletin of St. Petersburg State University. Series 10, vol. 4, pp. 94–104 (2011) (In Russian)

12. Sanders, J., Kandrot, E.: CUDA by Example: An Introduction to General-Purpose GPU Programming, p. 312. Addison-Wesley Professional (2010) ISBN-13: 978-0131387683

13. Luebke, D., Harris, M., Krüger, J., Purcell, T., Govindaraju, N., Buck, I., Woolley, C., Lefohn, A.: GPGPU: general purpose computation on graphics hardware. In: ACM SIGGRAPH 2004 Course Notes, Los Angeles, CA, August 08-12, p. 33 (2004), doi:10.1145/1103900.1103933

14. Buck, I., Fatahalian, K., Hanrahan, P.: GPUbench: evaluating GPU performance for numerical and scientific applications. In: Poster Session at GP2 Workshop on General Purpose Computing on Graphics Processors (2004), http://gpubench.sourceforge.net

15. Gaster, B., et al.: Heterogeneous Computing with OpenCL, p. 296. Morgan Kaufmann, Waltham (2011) ISBN 978-0-12-387766-6

16. Banger, R., Bhattacharyya, K.: OpenCL Programming by Example, p. 304. Packt Publishing (2013) ISBN: 1849692343, ISBN 13: 9781849692342

Iterative Remeshing for Edge Length
Interval Constraining

João Vitor de Sá Hauck, Ramon Nogueira da Silva, Marcelo Bernardes Vieira,
and Rodrigo Luis de Souza da Silva

Universidade Federal de Juiz de Fora, Departamento de Ciência da Computação,
36036-900, Juiz de Fora-MG, Brazil
{jhauck,ramon.nogueira,marcelo.bernardes,rodrigoluis}@ice.ufjf.br

Abstract. This paper presents an iterative method to remesh an arbitrary surface into a mesh with all edge lengths within an interval. The process starts with a triangular 2-manifold mesh. It uses stellar operations to achieve the necessary amount of vertices and triangles. Subsequently, it applies a constrained version of the Laplacian filter in order achieve a more uniform distribution of the vertices over the surface. In order to prevent the natural shrink caused by the Laplacian filter, we perform a projection over the original surface. We also apply a post processing step to correct the lengths of troubling edges. Our method results in a regular mesh, with vertices uniformly distributed. The dual mesh obtained can be useful for several applications. The main contribution of this work is a new approach for edge length equalization, with explicit constraints definition, lower global geometry losses and lower memory cost if compared to previous works.

Keywords: iterative remeshing, edge length equalization, interval constraining.

1 Introduction

Computational models of real objects are currently used for several applications. The growing need of geometric models conducted to the development of many technologies for mesh generation, such as computer vision algorithms with 3D scanners [1,2] or direct modeling softwares [3,4]. However, these technologies not always lead to an optimal mesh representation for a specific application. Therefore, the improvement of the quality of these representations became a primal research area in computer graphics.

The precise quality criteria for an arbitrary mesh depends on its usage. For real time applications, for instance, it is usually required a simplification of the model, in order to achieve high performance. In physics and chemistry simulations [5,6], some constraints may be necessary to guarantee the fidelity of the results, e.g. constraints on edge lengths, valid vertices valency, proper distributions of vertices, etc. Bommes *et al* [7] enumerate the quality aspects most commonly required.

B. Murgante et al. (Eds.): ICCSA 2014, Part VI, LNCS 8584, pp. 300–312, 2014.

This work is interested in the regularization of the edge lengths of a triangular 2-manifold mesh. Specifically, our goal is to impose a constraining interval for the lengths. So, we iteratively remesh the model until all the edge lengths satisfy the defined constraints. Since the average of the edge lengths in a region impact on the amount of the edges and faces found there, the method applies stellar operations to adjust the amount of edges locally. It also applies an approximation of the Laplacian filter to relax the mesh on each iteration. Then, it projects the vertices over the original surface, in order to preserve the geometry. Finally, after the execution of the iterations, it performs a post processing step that eliminates the remaining problems. Although the process is designed to maintain the original geometry of the model, some local geometry losses may occur, specially in regions with high curvature.

As our results indicate, the method generates models which satisfy the input constraining for most cases. Moreover, in the final mesh the standard-deviation of edge lengths tends to be low. The resulting mesh can also be used to generate a very regular trivalent mesh, by computing its dual. This kind of mesh can be greatly useful for engineering and physics applications, such as nano carbon simulations, which firstly motivated this work.

2   Related Works

There are many remeshing processes focused on obtaining more regular meshes for applications. N-Symmetric fields [8] can be used for generating highly regular polygonal meshes. The work of [7] applies the method for remeshing an arbitrary triangular mesh into a high quality quadrangular mesh. The method is computationally expensive, since the formalism proposed by [8] results in a mixed-integer system, which is a well known NP-Hard problem.

Using a similar approach, Huang J. et al [9] aims to obtain a mesh where the angle between two arbitrary edges of a triangle is $60°$. Computing the N-Symmetric field for this technique requires a set of feature lines of the model, which can be either automatically estimated or defined by the user. That estimation has a high computational cost.

The work presented in [10,11] proposes a method to obtain a regular trivalent mesh. At first it estimates a quadrangular mesh, as proposed by [7]. The following step computes the rhomboid mapping of that quadrangular mesh, resulting in the desired trivalent mesh. Although is not the main goal of that work, a very regular triangular mesh could be obtained by computing the dual from that resulting trivalent mesh.

Pietroni et al [12] use a global parametrization to obtain an almost isometric mesh. This method leads to a very regular triangular mesh. Their results are really competitive as showed in the results, however the global parametrization is complex and computationally expensive, since it requires a set of feature lines of the model. As well, they are not explicitly concerned with edges lengths.

In the approach of [13], a regular triangulation is not the final objective, but a necessary step for achieving a self-supporting surface, i.e., a surface that stands in static equilibrium without external support. The method used for obtaining the regular triangulation is the power diagram [14].

The method proposed by [15] processes a mesh to obtain a more regular version. The first step is to apply the edge split operation in edges that are longer than $\frac{4}{3}l$ and edge collapse in edges that are shorter than $\frac{4}{5}l$. Next, they perform edges flips to correct the valency. Finally, they equalize the triangle area with an area based tangential vertex smoothing. This work is not focused in edge length so it did not constrain the edge length into an interval.

Surazhsky *et al* [16] proposed a remeshing method based on area equalization and angle smoothing. Their method aims a mesh with triangle areas almost uniform and maximizes the minimum angle of the triangle. This interesting approach build a high regular surface. They also propose a new method to smooth the surface based on angles. However, this method does not remove elements and is not suitable for simplifications.

For the problem of edge length equalization [17], the goal is to obtain an average edge length that is close to a user defined value, with low standard deviation. Our work uses a similar approach. That method solves a large linear system to apply the Laplacian filter. In our work, that process is no longer necessary due to a new explicit approximation. This allows the method to process larger models with the same memory cost.

3   Proposed Method

Our method is an extension of the work of [17]. However, while [17] aims to obtain a mesh where the average edge length is as close as possible to a target value, our goal is to generate a mesh without any *long* or *short* edges a_j, classified as:

$$\text{long, if } |a_j| > e_{max}$$
$$\text{short, if } |a_j| < e_{min} \text{ ,}$$

where e_{min} is the shortest edge length allowed and e_{max} is the longest edge length allowed.

The input for this algorithm is a tuple $(\mathcal{M}, e_{min}, e_{max}, n, k, p)$, where $\mathcal{M}$ is the triangular mesh, n the number of iterations, k the number of rings used at the Laplacian optimization step and p is the number of iterations before the original mesh is replaced by the current mesh in order to relax next projections. At the end of each iteration, we save the resulting mesh if it has a lower amount of *long* and *short* edges than the current saved mesh. The Algorithm 1 is an overview of the proposed method.

Detailed information about the *CorrectValency* and *Projection* procedures can be found in [17]. The other steps will be explained ahead.

3.1   Stellar Operations

The amount of edges necessary to achieve the constraining interval is directly related to the average length m of the interval. Thus, in this step we modify

$$\mathcal{M}' = \text{Copy}(\mathcal{M})$$
$$m = \frac{e_{min}+e_{max}}{2}$$
for $i = 1$ **to** n **do**
> **if** $p > 0$ *and $(i \bmod p) = 0$* **then**
> > | $\mathcal{M} = \text{Copy}(\mathcal{M}')$
>
> **end**
> StellarOperations($\mathcal{M}'$)
> CorrectValency($\mathcal{M}'$)
> LowPassFiltering($\mathcal{M}'$, k)
> Projection($\mathcal{M}$, $\mathcal{M}'$)

end
PostProcess($\mathcal{M}'$)
return $\mathcal{M}'$

Algorithm 1. UniformRemeshing($\mathcal{M}, e_{min}, e_{max}, n, k, p$)

the number of edges in the model to reach a feasible amount. However, if m is much greater than the current average edge length, a strong simplification of the mesh is required. In order to prevent the degeneration of the mesh, we calculate intermediate values for e_{min} and e_{max} that only allow smooth transformations. These values are defined as:

$$ei_{min} = MIN(2 \cdot m_i, m) - \frac{e_{max}-e_{min}}{2}$$
$$ei_{max} = MIN(2 \cdot m_i, m) + \frac{e_{max}-e_{min}}{2} \; ,$$

where m_i is the current average edge length of the model.

The order of appliance of the stellar operations is important. Therefore, we create a priority list of edges, as presented in [17]. Once the list is set, the algorithm traverses it processing each edge.

If the edge length is shorter than ei_{min}, it is collapsed. Otherwise, if the edge length is longer than ei_{max}, then it is split. This modifies the amount of edges locally, since in an arbitrary mesh some regions need to be refined while others need to be simplified. If the edge is neither shorter or longer nothing is done. This may occur when one of the vertices were modified by another stellar operation. After an edge is processed, all vertices affected are marked as processed. When an edge has its two vertices marked, it is removed from the list.

Originally, both the remaining vertex of the edge collapse operation and the new vertex of the edge split operation can be placed in an arbitrary position over the operated edge. Hence, in order to optimize the convergence of our algorithm we compute the position that minimizes the equation:

$$\sum_{V_j} (|V_i - V_j| - m)^2,$$

where V_i is the vertex we want to position and V_j the vertices connected to V_i.

3.2 Low-pass Filter

To achieve equalized edge lengths, we globally distribute the vertices uniformly over the surface. To do so, after the valency correction step, as described in [17], we proceed to the low-pass filtering. In this work, we use a modified version of the Laplacian filter.

The classic Laplacian filter is defined as:

$$\nabla^2 f = \frac{\partial^2 f}{\partial^2 x_1} + ... + \frac{\partial^2 f}{\partial^2 x_n}.$$

It is a measurement of the dispersion in $\mathbb{R}^n$ of a function f. Taubin [18] propose a discrete approach to the Laplacian operator. The approach is:

$$L(V_i) = \sum_{V_j} w_{ij}(V_i - V_j),$$

with V_j in the neighborhood of V_i. In the literature, many weights were proposed for w_{ij}. There are schemes based on cotangent [19] and neighborhood [17].

The discrete Laplacian is largely used due to its simplicity. Basically, its appliance moves each vertex to the average of its neighbors. This procedure tends to equalize edge lengths, minimizing the standard deviation. The Laplacian must be zero to achieve these properties and the system to be solved is given by:

$$\sum_{v_j} w_{ij}(v_i - v_j) = 0.$$

The technique employed in this work does not solve the system. Instead, we run an iterative approximation that gives us almost the same results, significantly reducing the memory cost.

In the classical Laplacian filter, we add some additional constraints to reduce the geometry loss:

$$N_i \cdot D_i = 0, \ \forall D_i \in \mathcal{M}',$$
$$|D_i| = 0 \ \forall D_i \in \mathcal{B},$$

where N_i is the normal of the current mesh in the vertex V_i and D_i is the unknown displacement of the vertex V_i. These constraints were imposed in the approximation in a extremely restrict way. It means that all restrictions were exactly matched.

The iterative Algorithm 2 approximates the constrained Laplacian filtering described above. It calculates the new vertex position based on the k-neighborhood as proposed in [17]. The first step is to compute for each vertex the new position without the application of the new constraints. This position is defined by the center of mass of all neighbors vertices weighted by their ring number in such a way that distant vertices contribute less than near vertices.

The second step is to impose the constraint $N_i \cdot D_i = 0$ by removing the vector projection of D_i in N_i. When all displacements are computed, the vertices V_i are updated, except those on the borders.

foreach $V_i \in \mathcal{M}'$ **do**
    $kStar$=getKStar (V_i,k)
    fat=0
    foreach $V_j \in kStar$ **do**
        $V_i' = V_i' + \frac{V_j}{star}$
        $fat = fat + \frac{1}{star}$
    end
    $V_i' = \frac{V_i'}{fat}$
    $D_i = V_i' - V_i$
    $D_i = D_i$ -projection (D_i, N_i)
end
foreach $V_i \in \mathcal{M}'$ **do**
    if $V_i \notin \mathcal{B}$ **then**
        $V_i = V_i + D_i$
    end
end

Algorithm 2. LowPassFiltering($\mathcal{M}'$, k)

3.3   Post Processing

After n iterations, we proceed to the last step of the algorithm. This is step is run over the saved mesh with best results from the previous iterations. It distributes the vertices locally, only in the problematic regions.

First, we propose an error measurement for a region around an edge:

$$\sum_{V_i} \sum_{V_j} (|V_i - V_j|^2 - m^2)^2, \tag{1}$$

where V_i are the vertices in the forth star of the edge and V_j the vertices in the first star of V_i.

If we want to approximate the edge lengths to m, it is enough to minimize the Equation 1. Nonetheless, it may greatly modify the local geometry, once there are no constraints for the vertices displacement. Thence, we restrict those displacements to the tangent plane. For each vertex V_i, we obtain an orthonormal base with its normal vector.

This local base is $< N_i, T_{i1}, T_{i2} >$, where N_i is the normal direction, and T_{i1}, T_{i2} are the directions over the tangent plane. Using this local base we impose a restriction in the Equation 1, and the final error measurement to minimize is:

$$\sum_{V_i} \sum_{V_j} (|V_i + \alpha_i \cdot T_{i1} + \beta_i \cdot T_{i2} - (V_j + \alpha_j \cdot T_{j1} + \beta_j \cdot T_{j2})|^2 - m^2)^2, \tag{2}$$

where α_i and β_i are the variables in the function and represent the displacement over the tangent plane. If the index i does not exist both α_i and β_i are set zero.

In this work we use a conjugate gradient method [20] for minimizing Equation 2. Due to performance and numerical problems, we do not minimize it for all the vertices in the model. Instead, we process the model per edge.

As we expect to minimize the geometry losses, we seek to apply the transformations firstly in the regions with the greatest amount of *long* or *short* edges. To do so, we first create a new priority list of edges. For this list, the priority assigned to each edge is proportional to the number of *long* and *short* edges in its neighborhood. Next, we perform the minimization of Equation 2 for each edge on the list, in order.

In the cases in which even this technique does not solve the problem for all the edges, we perform the minimization without constraints, allowing three degrees of freedom to each vertex. This is enough for most of remaining cases, at the cost of geometry losses.

4   Experimental Results

In this section we discuss the generated results of the proposed method. The algorithm was implemented using C++ programming language and compiled using GCC 4.6.3. All tests were performed in a Intel Xeon(R) CPU E31220 @ 3.10GHz x 4 computer with 8 GBs of RAM. The graphics card was an AMD Radeon HD 5700 series.

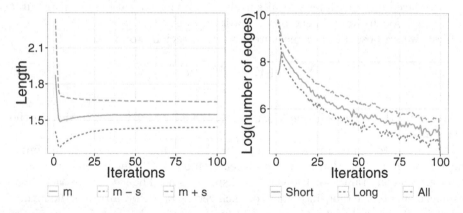

Fig. 1. Progression of Egea model through time

First we analyze the progression of the method over the time. Figure 1 exhibits the graphics for the progression of the *Egea* model, and Figure 2 shows the graphics for *Fertility* model. The graphics on the left illustrates the average edge length of the model through the iterations, and the deviation around it, where m is the average length and s is the standard-deviation. The graphic on the right shows the decrease of the amount of *long* and *short* edges over the time. For visualization purposes, the y-axis is presented in logarithmic scale. In both

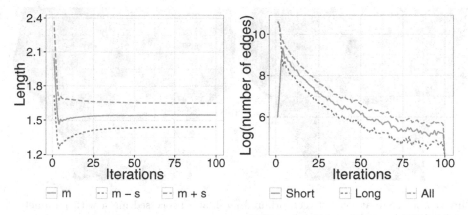

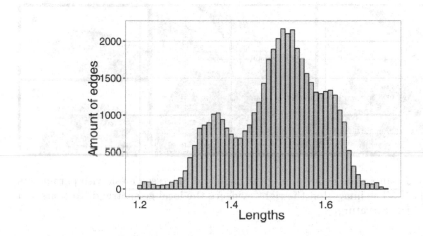

Fig. 2. Progression of Fertility model through time

Fig. 3. Edge lengths for the processed *Egea* model, with $e_{min} = 1.2$ and $e_{max} = 1.8$.

graphics the hundredth iteration corresponds to the post processing step of the algorithm.

One may notice that the average edge length quickly converges for a value very close to the mean of the endpoints of the interval constraining. Furthermore, the amount of *long* and *short* edges for both models is clearly reduced as the number of iterations increase. The final results for *Egea* mesh is depicted on Figure 4, and the Figure 3 illustrates the distribution of edge lengths for the final result.

We also illustrates the final results for the *Bunny* model in the Figure 5, pointing out the method behavior in regions with high curvature. As we can notice, some local geometry distortions may occur in those cases.

The experimental data from Egea model can be found in Table 1, where $\bar{x}$ is the average edge length, S the standard-deviation; O_{iter}, O_{cpp}, O_{final} are the total number of *short* and *long* edges after the iterations, after the constrained

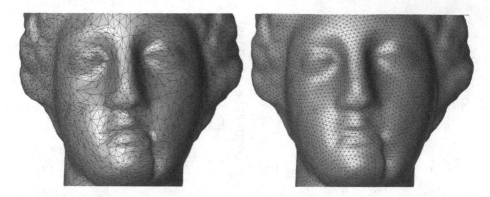

Fig. 4. *Egea* comparison between original mesh and processed mesh with parameters (1.2,1.8,100,2,0)

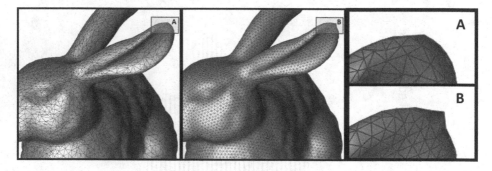

Fig. 5. *Bunny* comparison between original mesh and processed mesh with parameters (1.2,1.8,100,2,0). The images on the right evidence the local geometric distortions in a region with high curvature.

post processing step and after the free post processing, respectively, and *Reg. Vert.* is the percentage of vertices with valency 6.

As the data reveals, the best results are usually achieved when $k = 2$. It also reveals that parameter p can accelerate the convergence of the algorithm, once the number of final *long* and *short* edges is lower as the frequency in which the mesh is replaced is higher. However, it is important to note that it can also slightly modify the geometry, as the projection is performed in a mesh gradually more distinct from the original, and the geometry distortions caused by the Laplacian filter become permanent.

Figure 6 presents the surface *rockerarm* with variant values for e_{min} and e_{max}. The refined version greatly represents the original surface. When the process aims to larger edges, however, details of the model are lost. This is a natural consequence when a strong simplification is performed.

Table 1. *Egea* experiments with different k and p values

(e_{min},e_{max},n,k,p)	$\overline{x}$	S	O_{iter}	O_{cpp}	O_{final}	Reg. Vert.	Time(s)
Model	2.25	0.702222	-	-	33673	76.87	-
(1.2,1.8,100,1,0)	1.528363	0.150698	5568	790	3	79.482452	469.759777
(1.2,1.8,100,2,0)	1.549898	0.104768	321	13	0	89.262466	**229.957724**
(1.2,1.8,100,3,0)	1.558817	0.111454	803	47	0	88.976165	239.299956
(1.2,1.8,100,4,0)	1.56931	0.121848	1229	110	2	88.402027	270.382254
(1.2,1.8,100,1,10)	1.525323	0.158969	6119	887	4	77.620942	592.869659
(1.2,1.8,100,2,10)	1.550390	0.101163	**66**	**0**	0	90.076699	302.429551
(1.2,1.8,100,3,10)	1.558510	0.10615	199	**0**	0	89.833185	299.922871
(1.2,1.8,100,4,10)	1.569758	0.105946	243	10	1	90.880503	312.314491
(1.2,1.8,100,1,25)	**1.524605**	0.155988	5856	753	2	78.148691	561.070489
(1.2,1.8,100,2,25)	1.550454	**0.100802**	91	5	0	90.104096	282.562190
(1.2,1.8,100,3,25)	1.556311	0.107920	372	18	0	89.789667	288.441856
(1.2,1.8,100,4,25)	1.573154	0.108984	595	70	0	**90.528588**	301.354567
(1.2,1.8,100,1,50)	1.526876	0.155968	5966	684	0	78.074253	517.338868
(1.2,1.8,100,2,50)	1.549427	0.102719	177	2	0	89.747730	266.794273
(1.2,1.8,100,3,50)	1.559335	0.109731	508	46	0	89.532036	273.067145
(1.2,1.8,100,4,50)	1.569257	0.116656	907	102	0	89.014995	298.019677

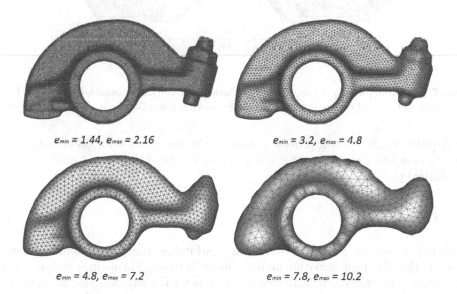

e_{min} = 1.44, e_{max} = 2.16

e_{min} = 3.2, e_{max} = 4.8

e_{min} = 4.8, e_{max} = 7.2

e_{min} = 7.8, e_{max} = 10.2

Fig. 6. Final result for *rockerarm* model with variant intervals for edge lengths. Each picture illustrates a specific length constraining.

4.1   Applications

A direct application of regular meshes is computational simulation. Uniform hexagonal meshes can be used for the simulation of nano carbon-structures, in which each vertex represents a carbon. This kind of mesh can be generated by regular triangular meshes. In this work, we compute the dual mesh of our final triangular mesh to obtain an highly uniform hexagonal mesh. One may observe that if the primal mesh is not regular, there will be non hexagonal polygons on the dual.

To analyze the quality of the resulting dual mesh we calculate the Lennard-Jones potential with $eps = 10.1$ and $\sigma = 0.9$. As depicted in Figure 7 the processed model is much more uniform, avoiding great energy variations. This will lead to a much more stable structure.

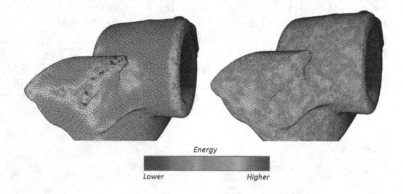

Fig. 7. The first picture is the original *rockerarm* model with potential from -33.4 to 1.21 and the second one is the processed model with potential from -16.94 to -2.99.

Another application is process a model to be more stable for other numerical methods as finite elements. Due the great regularity of the output mesh, several numerical problems are avoid.

5   Conclusion

This paper presents a method to remesh an arbitrary triangular 2-manifold mesh with all the edge lengths within an user defined interval. The main contribution of this work is an explicit definition of constraints for the edge lengths. In addition, we achieve a lower memory cost than previous approaches. We apply stellar operations to adjust the number of edges of the model, and we make use of a new approximation of the Laplacian filter for mesh relaxation, including constraints that prevent geometry losses. We also introduce a post processing step for the problems unsolved at the standard iterations.

Our results indicate that the method achieves the goal for a wide range of lengths. Moreover, the resulting mesh fairly represent the original surface in most of cases, and can be useful for several applications.

The parameters have important roles on the algorithm. For higher number of iterations n, the algorithm is more likely to reduce the number of *long* and *short* edges before the post processing step. Nonetheless, if the parameter p is low, but not zero, the geometry is more likely to be softened, and the effect is aggravated as the number of iterations increase. Lower values of p are useful for accelerating the convergence, but it should be balanced with a low n value if one wants to preserve the geometry of the surface. We can also observe that the number of neighbors k for the Laplacian transformation can also accelerate the convergence, accelerating the smoothing effect as well. For the minimum and maximum values allowed for edge lengths, as the values e_{min} and e_{max} are greater and the difference e_{min} - e_{max} is smaller, the final geometry losses are greater and the convergence of the method is slower.

The major problem faced by the method is to maintain the local geometry in regions where the curvature is high. As a future work, a new approach that avoids local geometry distortions could be proposed.

Acknowledgments. Authors thank to FAPEMIG and CAPES for financial support.

References

1. Rocchini, C., Cignoni, P., Montani, C., Pingi, P., Scopigno, R.: A low cost 3d scanner based on structured light. Computer Graphics Forum 20(3), 299–308 (2001)
2. Vieira, M.B., Velho, L., Sa, A., Carvalho, P.C.: A camera-projector system for real-time 3d video. In: IEEE Computer Society Conference on Computer Vision and Pattern Recognition-Workshops, CVPR Workshops, pp. 96–96. IEEE (2005)
3. Dembogurski, R., Dembogurski, B., de Souza da Silva, R.L., Vieira, M.B.: Interactive mesh generation with local deformations in multiresolution. In: Murgante, B., Misra, S., Carlini, M., Torre, C.M., Nguyen, H.-Q., Taniar, D., Apduhan, B.O., Gervasi, O. (eds.) ICCSA 2013, Part I. LNCS, vol. 7971, pp. 646–661. Springer, Heidelberg (2013)
4. Schöberl, J.: Netgen an advancing front 2d/3d-mesh generator based on abstract rules. Computing and Visualization in Science 1(1), 41–52 (1997)
5. Iijima, S., et al.: Helical microtubules of graphitic carbon. Nature 354(6348), 56–58 (1991)
6. Rapaport, D.C.: The Art of Molecular Dynamics Simulation. Cambridge University Press, New York (1996)
7. Bommes, D., Zimmer, H., Kobbelt, L.: Mixed-integer quadrangulation. ACM Trans. Graph. 77, 1–77 (2009)
8. Ray, N., Vallet, B., Li, W.C., Lévy, B.: N-symmetry direction field design. ACM Transactions on Graphics (2008); Presented at SIGGRAPH
9. Huang, J., Zhang, M., Pei, W., Hua, W., Bao, H.: Controllable highly regular triangulation. Science China Information Sciences 54(6), 1172–1183 (2011)

10. Pampanelli, P.C.P.: Mesh generation through the mapping of triangular models into rhomboid space. M.sc. dissertation, Universidade Federal de Juiz de Fora, Advisor: Marcelo Bernardes Vieira, Co-advisors: Marcelo Lobosco and Sócrates de Oliveira Dantas (2011)
11. Pampanelli, P.P., Peanha, J., Campos, A.M., Vieira, M.B., Lobosco, M., de Oliveira Dantas, S.: Rectangular hexagonal mesh generation for parametric modeling. In: 2009 XXII Brazilian Symposium on Computer Graphics and Image Processing (SIBGRAPI), pp. 120–125. IEEE (2009)
12. Pietroni, N., Tarini, M., Cignoni, P.: Almost isometric mesh parameterization through abstract domains. IEEE Transactions on Visualization and Computer Graphics 16(4), 621–635 (2010)
13. Liu, Y., Pan, H., Snyder, J., Wang, W., Guo, B.: Computing self-supporting surfaces by regular triangulation. ACM Trans. Graph. 32(4), 92:1–92:10 (2013)
14. Aurenhammer, F.: Power diagrams: properties, algorithms and applications. SIAM J. Comput. 16(1), 78–96 (1987)
15. Botsch, M., Kobbelt, L.: A remeshing approach to multiresolution modeling. In: Proceedings of the 2004 Eurographics/ACM SIGGRAPH Symposium on Geometry Processing, pp. 185–192. ACM (2004)
16. Surazhsky, V., Gotsman, C.: High quality compatible triangulations. Engineering with Computers 20(2), 147–156 (2004)
17. de Oliveira, J.P.P.N.: Iterative method for edge length equalization. In: International Conference on Computational Science, Barcelona, Spain, pp. 481–490 (2013)
18. Taubin, G.: A signal processing approach to fair surface design. In: Proceedings of the 22nd Annual Conference on Computer Graphics and Interactive Techniques, SIGGRAPH 1995, pp. 351–358. ACM, New York (1995)
19. Alliez, P., Meyer, M., Desbrun, M.: Interactive geometry remeshing. ACM Trans. Graph. 21(3), 347–354 (2002)
20. Press, W.H., Teukolsky, S.A., Vetterling, W.T., Flannery, B.P.: Numerical Recipes in C: The Art of Scientific Computing, 2nd edn. Cambridge University Press, New York (1992)

Topology Preserving Mapping for Maritime Anomaly Detection

Ying Wu[1], Anthony Patterson[1], Rafael D.C. Santos[2],
and Nandamudi L. Vijaykumar[2]

[1] Coastal and Marine Research Centre, ERI, University College Cork
Glucksman Marine Facility, Naval Base, Haulbowline, Ireland
{y.wu,a.patterson}@ucc.ie
[2] Brazilian National Institute for Space Research,
Av dos Astronautas, 1.758, Jd. Granja - CEP 12227-010,
São José dos Campos, São Paulo, Brazil
{vijay.nl,rafael.santos}@inpe.br

Abstract. In this paper, we present the topology preserving mapping for maritime anomaly detection. Specifically, the topology preserving mapping is applied as an unsupervised learning method, which captures the vessel behaviors and visualizes the extracted underlying data structure. At the same time, the topology preserving mapping is used as the probability estimator, where the data likelihood can be evaluated and the anomalies can be detected. Real satellite AIS data, used by the Next Generation Recognized Maritime Picture project (NG-RMP) funded by the European Space Agency, is used in this paper as the main data source. We demonstrate that the topology preserving mapping can classify the vessel observations and detect the anomalies reasonably and with high accuracy.

1 Introduction

The maritime anomaly detection has been recognized in recent years as the crucial component in order to achieve the situation awareness in maritime surveillance. The typical sensors used for maritime surveillance include radars, video cameras, aircraft, etc. Recently, however, a number of self-reporting maritime systems, such as the Satellite Automated Identification System (S-AIS), have been introduced. The S-AIS messages are transmitted from vessels in the maritime domain, reporting their position, speed, heading, and other detailed information, such as their destination, vessel type and ship identifier. Although the main purpose of the self-reporting AIS system is safety in navigation and collision avoidance, it also provides a wealth of information for maritime surveillance. The major challenge of processing such an abundant source of data is that the amount of information could be very large and difficult to monitor with the limited operator sources and the large number of vessels on the seas makes observing the vessel behavior time consuming and error prone for the human analysts. Therefore, it is necessary and highly demanded to explore the S-AIS data

B. Murgante et al. (Eds.): ICCSA 2014, Part VI, LNCS 8584, pp. 313–326, 2014.

in an automated manner in order to improve the situation awareness and draw the operators' attention to those vessels with suspected anomalous behaviors.

A wide range of research has been carried out for the anomaly detection, which can be classified by the top-down approaches and the bottom-up approaches. The top-down approaches usually require explicitly defined models, rules and templates for the particular events and the anomalous situations can thus be identified through the goal-driven pattern matching and reasoning. Roy [17,16] has developed a rule-based expert system, where the domain expert knowledge is represented by using the *if...then...* type of statement or the ontology. The vessel situations are transformed into the facts that can be accepted by the expert system. The inference engine will match all the rules in the IF part against the facts and execute the THEN part when one rule is matched. Another example is the Bayesian Belief Network [4] where the challenging tasks are represented by the variables and edges in the Bayesian belief network. The probability of vessel situation is evaluated and when it falls down to the pre-defined threshold, the vessel is flagged as anomalies.

The fact that there is a large amount of AIS data available motivates the use of bottom-up approaches, where the data-driven methods are applied in the sense that the normal vessel behaviors are determined by analyzing the collected large set of historical data. A clear advantage of these approaches is that no *a priori* information is required to model the normalcy. The machine learning techniques have been widely considered to detect the anomalies as the bottom-up approaches. For instance, the adaptive Kernel Density Estimation is used in [11,6] to model the normal vessel tracks. In [12], the Gaussian Mixture Model (GMM) has been presented as the probability estimation model to capture the normal vessel behaviors. The Self-Organized Mapping (SOM) has been studied to cluster the motion trajectory of moving objects [13], to visually and interactively monitor and control the clustering process of trajectory data [2] and to determine the parameters of the statistical model of normal behaviors [1]. In [15], a neural network has been derived for exploiting track data to learn normal patterns of motion behavior and detect the deviations from normalcy. In [9], Gaussian process method, as a non-parameter Bayesian model, is applied to built the model of normal vessel behaviors. The Active Learning paradigm is integrated to select an informative subsample of the data to reduce the computational complexity of training.

The Topology Preserving Mapping (TPM), a topology preserving model that is closely related to the generative topographic mapping [3], is presented in this paper to capture the underlying structure in the data set, in which points which are mapped close to one another have some common features. The topology preserving mapping was applied for the data visualization [7,14] and data clustering [19]. In this paper, the topology preserving mapping is applied for anomaly detection, where the normal vessel behavior is modelled by the TPM in order to detect the anomaly.

The paper is organized as follows. In section 2, we first describe the data processing given the S-AIS data to build the proposed feature model that will be used to represent the vessel behaviors and the topology preserving mapping method is then presented for the anomaly detection. Real S-AIS data will be used in section 3 to demonstrate that the underlying structure of the S-AIS data can be extracted and visualized and the vessels in anomalous situation can be detected reasonably. We discuss anomaly detection with topology preserving mapping and compare it with other widely applied anomaly detection methods, Gaussian mixture model and self-organized mapping, in section 4.

2   Topology Preserving Mapping in Anomaly Detection

For the maritime anomaly detection, a critical aspect is how the vessel behaviors can be parameterized in the right way that the anomalous situations can be identified, i.e. an appropriate feature model is defined, based on which the topology preserving mapping can be applied to capture those vessels in anomalous situations. In this section, the data processing is described to build the feature model of vessel behavior firstly. Then the topology preserving mapping is applied to the anomaly detection.

2.1   Feature Model and Data Processing

In this paper, we will use the identical feature model proposed in [6,12,11], where the positions in latitude and longitude coordinates and the velocity in latitude and longitude directions of individual vessels at a time point are used to characterize the vessel situation, which forms a four-dimensional feature space for a single vessel observation.

The S-AIS data is considered as the important data source in maritime anomaly detection, as it provides information on positioning and movements of maritime vessels directly. In this paper, real S-AIS data, used in the Next Generation Recognized Maritime Picture project (NG-RMP) funded by the European Space Agency, covering the area from the north-west point (at latitude $63°$, longitude $-24°99'$) to the south-east point (at latitude $48°$, longitude $0°99'$) is used. From the satellite transmitted messages, we have extracted the Maritime Mobile Service Identity (MMSI), the momentary latitude and longitude position, course, speed and the absolute timestamp. To fit the S-AIS data into the proposed feature model, the vessel speed and course are processed to the latitudinal and longitudinal velocities. Therefore, the transformed dataset as input to the topology preserving mapping contains six attributes: the MMSI, timestamp, position and the latitudinal and longitudinal velocities. The vessel observations are grouped into trajectories in order to investigate those trajectory segments being classified as anomalous situations easily. The "grouping" of the vessels is based on the MMSI which is assumed to uniquely identify each vessel in the data set.

2.2   Topology Preserving Mapping

A topographic mapping captures the underlying structure in the data set, in which points which are mapped close to one another have some common feature while points that are mapped far from one another do not share this feature. The most common topographic mapping is Kohonen's self-organizing map (SOM) [10]. The Generative Topographic Mapping (GTM) [3] is a mixture of experts model which treats the data as having been generated by a set of latent points where the mapping is *non-linear*. Fyfe [7] has derived an alternative topology preserving model, called the *Topographic Products of Experts* (ToPoE), based on the analysis of products of experts [8], which is closely related to the generative topographic mapping.

Specifically, the base model is defined as

$$p(\mathbf{t}_i|\theta) \propto \prod_{k=1}^{K} \left(\frac{\beta}{2\pi}\right)^{\frac{D}{2}} \exp\left(-\frac{\beta}{2}\|\mathbf{m}_k - \mathbf{t}_i\|^2\right). \tag{1}$$

To fit this model to the data we can define a cost function as the negative logarithm of the probabilities of the data so that

$$J = \sum_{i=1}^{N} \sum_{k=1}^{K} \frac{\beta}{2}\|\mathbf{m}_k - \mathbf{t}_i\|^2. \tag{2}$$

In [8], the Product of Gaussian model was extended by allowing latent points to have different responsibilities depending on the data point, where we have:

$$p(\mathbf{t}_i|\theta) \propto \prod_{k=1}^{K} \left(\frac{\beta}{2\pi}\right)^{\frac{D}{2}} \exp\left(-\frac{\beta}{2}\|\mathbf{m}_k - \mathbf{t}_i\|^2 r_{ik}\right). \tag{3}$$

Given a set of data points $\mathbf{t}_1, ..., \mathbf{t}_N$, we follow [3,7] to create a latent space of points $\mathbf{x}_1, ..., \mathbf{x}_K$ which lie equidistantly on a line or at the corners of a grid. To allow non-linear modeling, we define a set of M basis functions, $\phi_1(), ..., \phi_M()$, with centres μ_j in latent space. Thus we have a matrix Φ where $\phi_{kj} = \phi_j(\mathbf{x}_k)$, each row of which is the response of the basis functions to one latent point, or, alternatively each column of which is the response of one of the basis functions to the set of latent points. Typically, the basis function is a squared exponential. These latent points are then mapped to a set of points $\mathbf{m}_1, ..., \mathbf{m}_K$ in data space where $\mathbf{m}_j = (\Phi_j \mathbf{W})^T$, through a set of weights, $\mathbf{W}$. The matrix $\mathbf{W}$ is $M \times D$ and is the sole parameter which we update during training,

$$\Delta_n \mathbf{w} = \sum_{k=1}^{K} \eta(\mathbf{t}_n - \mathbf{m}_k)\Phi(\mathbf{x}_k)r_{nk}. \tag{4}$$

where the r_{nk} is the responsibility of the kth latent point for the data point, $\mathbf{t}_n$,

$$r_{nk} = \frac{\exp(-\gamma d_{nk}^2)}{\sum_j \exp(-\gamma d_{ij}^2)} \tag{5}$$

where $d_{nk} = \|\mathbf{t}_n - \mathbf{m}_k\|$, the Euclidean Distance between the nth data point and the projection of the kth latent point through the basis functions.

2.3   Anomaly Detection

The topology preserving mapping defines a constrained mixture of Gaussian in the data space. The latent points are nonlinearly mapped to the prototype points, $(\mathbf{m}_1, ..., \mathbf{m}_K)$, in the data space and these prototypes form the centres of the Gaussian distributions in the data space, where the centres of the mixture components, however, can not move independently of each other. Moreover, in practice, all components of the mixture share the same variance, and the mixing coefficients are defined by $\frac{1}{K}$. We can thus represent the distribution of the data points in terms of a smaller number of latent points in a low dimensional nonlinear manifold:

$$p(\mathbf{t}) = \frac{1}{K} \sum_{k=1}^{K} \left(\frac{\beta}{2\pi} \right)^{\frac{D}{2}} \exp \left(-\frac{\beta}{2} ||\mathbf{t} - \mathbf{m}_k||^2 \right) \tag{6}$$

In this paper, the S-AIS data is used as the input to the topology preserving mapping model for the anomaly detection. With the assumption that the given vessel observation is in normal situation and is independent of its previous observations, all the vessel observations can be modelled statistically by a joint probability density function (PDF) through the TPM method, which represents the normal vessel behaviors. Thus the data likelihood for a new vessel observation can be estimated by (6) as the indicator that the observed vessel is in normal situation. For the anomaly detection, a threshold is set for the likelihood value, which allows the operator to control the false alarm rate in a feasible way. When the estimated likelihood is below the threshold, it is very unlikely that the data point was generated from the normal PDF and thus it is considered to be an anomaly.

It is worth noting that with the trained TPM model, the underlying structure of the data set can be extracted and visualized by projecting the data points from the data space to the latent space, where we have

$$\mathbf{t}_n^{latent} = \sum_{k=1}^{K} r_{nk} \mathbf{x}_k. \tag{7}$$

3   Results

To demonstrate the topology preserving mapping for the anomaly detection, the S-AIS data described in section 2.1 is used as the input data. Specifically, the satellite AIS message 1, 2, 3 collected between 09/10/2012 and 08/11/2012 is used as the training data. In total, there are 28,667 vessel observations, including the information of MMSI, timestamp, latitudinal position, longitudinal position, velocity in the latitudinal direction and velocity in the longitudinal direction.

It has been pointed out in [6] that the anomaly detection becomes infeasible due to the complexity of the statistical model with the growth of the size of data set. We follow [6] and [12] that the surveillance area is discretized into a 3×5

grid with 15 uniformly sized cells, where each cell corresponds to a sub-area of the surveillance area. The vessel observations located in one cell form the local training data and the topology preserving mapping method is performed in each cell individually. We set the threshold to be 100, where the vessel observation is reported as anomalous directly when the number of samples in the local training data set is below this threshold.

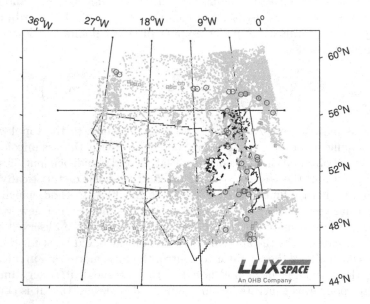

Fig. 1. The anomaly detection with the S-AIS data by topology preserving mapping

To demonstrate the TPM method for anomaly detection, 100 vessel observations collected between 09/11/2012 and 13/11/2012 have been selected sequentially as the testing data set. The reason to select the testing data in sequence but not randomly is that we presume the vessel observations in a time interval are more likely from the same trajectories, which makes it easier for us to determine whether the vessel observations have been classified reasonably.

Figure 1 has shown the output of topology preserving mapping, where the training data is plotted by the grey points. The testing data is plotted by the green squares and the red circles, where the green squares correspond to the vessel observations classified as in normal situations and the red circles correspond to the vessels with anomalous situations. By examining the results qualitatively, it has been found that the detected anomalies mainly correspond to the vessels that are traveling across or in the opposite direction to the sea lanes formed by the training data, and/or at the anomalous speed. Such results illustrate that the topology preserving mapping captures the normal vessel behaviors and detects the anomalies in terms of the traveling direction and velocity. We consider this is reasonable in that the feature model described in subsection 2.1 determines the TPM method extracts the vessel behaviors from aspects of the momentary position and speed jointly.

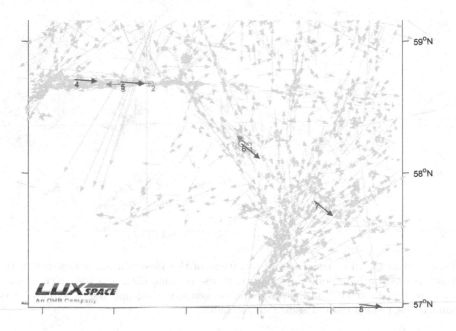

Fig. 2. The vessel observations in the 1rd row and the 5nd column sub-area. The grey arrows show the vessel observations as training data and the green arrows correspond to the observations as testing data classified as in normal situations and the red arrows mean anomalous situations.

To investigate those vessels with suspected anomalous situations in detail, a sub-area in the 1st row and the 5th column in Figure 1, as an example, is zoomed in as shown in Figure 2, where the grey arrows show the vessel observations as training data and the green arrows correspond to the observations as testing data classified as in normal situations and the red arrows mean anomalous situations. The length of the arrows in Figure 2 shows the speed of the traveling vessel. Among the eight vessel observations as the testing data in that sub-area, we can see that the observation 1, 2, 3 are reasonably marked as normal situations as they follow the common sea lanes formed by the training data and at the normal speed. However, although the observation 4, 5 are close to the observation 2, 3 in position and observation 6 is close to observation 1, we can see that the vessels in the observation 4, 5, 6 were traveling in the opposite direction of the sea lane at high speed and thus classified as anomalies. The anomaly is detected in observation 7, as the vessel was obviously traveling in an anomalous direction as it crosses the common sea lane formed by the training data. After further investigation, we find that the observation 4, 5, 6, 7, 8 are from the same vessel within a short period, which means the reported anomalies can be considered to be from the same trajectory, which is particularly interesting to the operator.

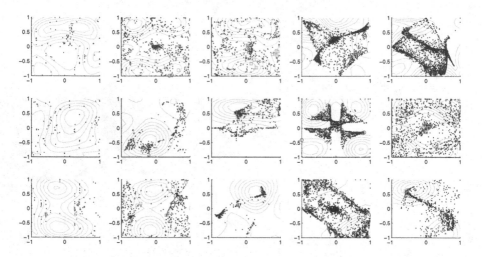

Fig. 3. Visualization of the vessel observations of the 15 sub-areas in Figure 1 in the latent space. The blue points correspond to the training data and the green squares correspond to the testing data classified as normal situations and the red circles correspond to the testing data classified as anomalies.

Furthermore, we have pointed out in [19] that we can have the high-level representation of the data points by the projection of the data points in the latent space, where the data points in the original data space are represented by a lower q-dimensional nonlinear manifold. We plot the projection of the vessel observations in the 2-dimensional latent space in Figure 3, where each subplot corresponds to one sub-area as in Figure 1. We can see that the TPM method clusters the vessel observations accurately. For example, the vessels, in the sub-area in 3rd row and the 2nd column in Figure 1, usually travel from the west to the east or in the opposite direction as shown in Figure 4, which forms two clusters. The corresponding subplot in the 3rd row and the 2nd column in Figure 3 has clearly identified these two separated clusters. It is worth noting since the data points with similar features will be mapped closely in the latent space, the vessel observations following similar sea lanes will tend to have similar behaviors and thus form a high-density area in the latent space. In Figure 5, we visualize the vessel observations in the same sub-area as in Figure 2 in a 2-dimensional latent space, where the blue points correspond to the training data and the green squares correspond to the testing data classified as normal situations and the red circles correspond to the testing data classified as anomalies. It is clear to see that the vessel observations, which are marked by green squares and classified as in normal situations, are in or very close to the area of high density, which means the vessel has followed the common sea lanes. However, the vessel observations, which are marked by the red circles and reported as anomalies, have been pushed to the edge of data cloud and in the area with low density, which means these vessel observations are different from the training data and are very likely to happen rarely and thus considered as anomalies. This illustrates that the vessel observations are classified reasonably by the topology preserving mapping.

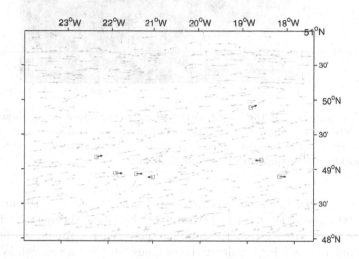

Fig. 4. Visualization of the vessel observations in the 3rd row and the 2nd column sub-area as shown in Figure 1 . We can see that the vessels in this subarea usually travel from the west to the east or in the opposite direction, which thus can be clustered into two groups.

4   Analysis and Discussion

Therefore, we can see that the topology preserving mapping can learn the vessel behaviors accurately and detect the anomalies reasonably. At the same time, the learned vessel behaviors can be visualized, which allows the operator to have more control in the learning process. It is interesting to compare the topology preserving mapping with other widely used methods, Gaussian Mixture Model and Self-Organized Mapping, for anomaly detection. The topology preserving mapping defines a constrained mixture of Gaussian, where the mapped latent points form the nonlinear manifold in the data space in which the distribution of the data points can thus be represented by maximizing the log-likelihood function. Therefore the topology preserving mapping can be applied as the probability estimator directly as the Gaussian mixture model does. By comparing the results of anomaly detection by the topology preserving mapping presented in

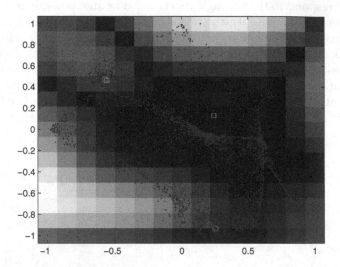

Fig. 5. Visualization of the vessel observations of the sub-area as shown in Figure 2 in the latent space. The blue points correspond to the training data and the green squares correspond to the testing data classified as normal situations and the red circles correspond to the testing data classified as anomalies.

this paper and the Gaussian mixture model method, the same training data and testing data have been used to perform the Gaussian mixture model. As shown in Figure 6, we can see that the topology preserving mapping method has produced similar classifying results with the Gaussian mixture model. We follow the method used in [12], by which the degree of similarity of the results by the two methods is calculated in Formula (8),

$$S(A_1, A_2) = \frac{|A_1 \cap A_2|}{|A_1 \cup A_2|} = 56.75\%, \tag{8}$$

where A_1 is the anomalies detected by the topology preserving mapping and A_2 is the anomalies detected by the Gaussian mixture model. This indicates that there is a considerable overlap in the anomalies detected by each method. However, apart from the performance, the Gaussian mixture model is lack of the capacity of visualizing the learned vessel behaviors in a lower-dimensional space, which means the operator has less control in the vessel behavior modeling. The self-organized mapping can extract the underlying structure of the data set and map the data into a lower dimensional space as a visualization method. However, it has been pointed out in [1] that although the SOM is a useful method for clustering and visualizing the data, it may not have strong anomaly detection capability by itself since the large amount of vessel observation may not clearly fall into well-defined clusters. This motivates the SOM is performed as the parameter initialization of the GMM model in [1], where the learning process involves two stages: the SOM method is performed first and then the output is

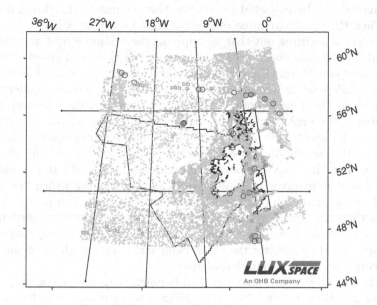

Fig. 6. The anomaly detection with the S-AIS data by Gaussian mixture model

transformed to build the GMM model. We find such a learning process becomes more time-consuming when the data set becomes huge and at the same time, it becomes impractical in the runtime monitoring where new vessel observations are streamed into the monitoring system and the model is required to be trained by the online learning.

To apply the topology preserving mapping for the anomaly detection in the NG-RMP project, we find that the likelihood threshold, as an important parameter, should be adjusted carefully by the operator. The likelihood threshold corresponds to the anomaly threshold, which determines the system level of sensitivity. Given a vessel observation, when the estimated likelihood is below this threshold, it is very likely that the data point is not generated from the normal PDF and thus it is considered to be an anomaly. Therefore, when the threshold is raised, there are more vessel observations considered to be anomalous. However, this means the false alarm rate will also increase as there could be more anomalies reported which the operator may not have interest. On the other hand, if the threshold is set too low, some true anomalies that are of interest for the operator could be dismissed by the system, which affects the accuracy of anomaly detection. Therefore, in real practice, the operator should be able to adjust the threshold according to their experience during the system running. Furthermore, as one of the future work, we aim to set the likelihood threshold automatically by the system by applying some self-adaptive methods, such as reinforcement learning [18] and cross entropy method [5,20]. The system keeps monitoring the interaction between the operator and the user interface and the feedback information can be collected to identify how accurately the threshold has been set. Then the threshold value can be optimized automatically.

It is worth pointing out that by applying the feature model described in subsection 2.1, we only consider the momentary positions and speed of the vessels, i.e. without considering the previous vessel states. Thus we consider vessel observation is independent and identically distributed (i.i.d.) from a underlying PDF, which forms the fundamental assumption to perform the topology preserving mapping as the probability estimator. On the other hand, this fact implies that the anomaly detection is not performed based on the trajectory analysis and the states of the surrounding vessels are not considered either. Therefore, certain types of anomalies may not be identified by the TPM method. For example, the anomalies that involves multiple vessels, such smuggling which two vessels are usually close with each other in an anomalous distance, can not be detected by our method. However, we consider that the feature model applied in this paper is still usable, based on which the topology preserving mapping can be the generic method to monitor the vessel behaviors in one maritime domain without requiring particular domain knowledge.

Although we presume that the vessel observations as the training data are in normal situation, we find it is very difficult to guarantee this fact in practice. For example, it is obviously to see that some vessels in Figure 2 as the training data plotted by the grey arrows are traveling across the sea lanes at extremely high speed. Actually, given the training data, we find it is impractical to make a judgement for each training vessel observation, considering the complicated marine environment and the large number of training S-AIS data generated even in a short time. We regard those vessel observations that are assumed to be in normal situation but actually are anomalies as noise observations. Although the noise observations affect the performance of anomaly detection, we find the topology preserving mapping still works well enough. This is reasonable because the topology preserving mapping is applied to model the vessel behaviors and it can capture those vessel behaviors that happen with high probability. Since the noise observations are regarded as the rare events, which happen with low probability, the topology preserving mapping can evaluate the derivation of anomalies from those normal vessel behaviors with high probability accurately.

5   Conclusion

In this paper, we have presented the topology preserving mapping for maritime anomaly detection by capturing, visualizing and classifying the vessel behaviors. The satellite AIS data, used in the Next Generation Recognized Maritime Picture project (NG-RMP), is used in this paper as the main data source for the anomaly detection and we considered the feature model where the vessel behaviors are represented in terms of momentary position and velocity. We presented the anomaly detection by the topology preserving mapping where the TPM method is performed as the probability estimator to model the vessel behaviors so that the anomalies can be detected. The results have shown that the

TPM method can capture the vessel behaviors accurately and thus the vessel observations in the testing data have been classified reasonably, which means the anomalies can be detected with high performance. We demonstrated that the vessel behaviors can be visualized by the projection of the data points in the latent space. It has been shown that the vessels having similar behaviors in the same sea lanes tend to gather together to form the area with high density in the latent space and the anomalous observations are pushed far away from the normal observations, which are usually in low-density area of the latent space. We have compared the topology preserving mapping with other widely applied methods, the Gaussian mixture model and self-organized mapping, for the anomaly detection. We point out the topology preserving mapping has similar performance with the Gaussian mixture model in anomaly detection. However, one advantage of the topology preserving mapping is that the vessel observations can be classified and visualized in one model, which not only achieves high performance , but also the operator can have more control in the learning process.

Acknowledgement. The European Space Agency (ESA) Integrated Application Promotion program under contract AO/1-6124/09/NL/US has funded this work. For more information, please visit the NG-RMP IAP web site: http://iap.esa.int/project/security/ng-rmp. Rafael Santos would like to thank FAPESP, the São Paulo Research Foundation, for its support (grant 2014/05453-6). We would like to thank LuxSpace company for providing the Satellite AIS data from VesselSat2 and terrestrial AIS data.

References

1. Arouh, S.L., Webb, M.L., Kraiman, J.B.: Automated anomaly detection processor. In: Proceedings of SPIE: Enabling Technologies, for Simulation Science (2002)
2. Bernard, J., Tekusova, T., Kohlhammer, J., Schreck, T.: Visual cluster analysis of trajectory data with interactive kohonen maps. Information Visualization 8 (2009)
3. Bishop, C.M., Svensen, M., Williams, C.K.I.: Gtm: The generative topographic mapping. Neural Computation 10, 215–234 (1998)
4. Das, S., Grey, R., Gonsalves, P.: Situation assessment via bayesian belief networks. In: Proceedings of the 5th International Conference on Information Fusion, Maryland (July 2002)
5. de Boer, P.-T., Kroese, D.P., Mannor, S., Rubenstein, R.Y.: A tutorial on the cross-entropy method. Annals of Operations Research 134(1), 19–67 (2004)
6. Falkman, G., Sviestins, E., Laxhammar, R.: Anomaly detection in sea traffic - a comparison of the gaussian mixture model and the kernel density estimator. In: The 12th International Conference on Information Fusion, Seattle, WA, USA (2009)
7. Fyfe, C.: Two topographic maps for data visualization. Data Mining and Kownledge Discovery 14, 207–224 (2007)
8. Hinton, G.E.: Training products of experts by minimizing contrastive divergence. Technical Report 2000-004, Gatsby Computational Neuroscience Unit, University College, London (2000)
9. Kowalska, K., Peel, L.: Maritime anomaly detection using gaussian process active learning. In: The 15th International Conference on Information Fusion (2012)

10. Kohonen, T.: Self-organising maps. Springer (1995)
11. La Scala, B., Morelande, M., Gordon, N., Ristic, B.: Statistical analysis of motion patterns in ais data: Anomaly detection and motion prediction. In: The 11th International Conference on Information Fusion (2008)
12. Laxhammar, R.: Anomaly detection for sea surveillance. In: The 11th International Conference on Information Fusion, Cologne, German (2008)
13. Manohara, P., Rajpurohit, V.: Using self organizing networks for moving object trajectory prediction. International Journal on Artificial Intelligence and Machine Learning 9 (2009)
14. Pena, M., Barbakh, W., Fyfe, C.: Topology-Preserving Mappings for Data Visualisation. In: Principal Manifolds for Data Visualization and Dimension Reduction, pp. 131–150. Springer, Heidelberg (2008)
15. Rhodes, B.J., Bomberger, N.A., Zandipour, M., Garagic, D.: Adaptive mixture-based neural network approach for higher-level fusion and automated behavior monitoring. In: International Conference on Communications (2009)
16. Roy, J.: Anomaly detection in the maritime domain. In: Proceedings of the SPIE (2008)
17. Roy, J.: Rule-based expert system for maritime anomaly detection. In: Proceedings of the SPIE (2010)
18. Sutton, R.S., Barto, A.G.: Reinforcement Learning: An Introduction. The MIT Press (1998)
19. Wu, Y., Doyle, T.K., Fyfe, C.: Multi-layer topology preserving mapping for K-means clustering. In: Yin, H., Wang, W., Rayward-Smith, V. (eds.) IDEAL 2011. LNCS, vol. 6936, pp. 84–91. Springer, Heidelberg (2011)
20. Wu, Y., Fyfe, C.: The on-line cross entropy method for unsupervised data exploration. WSEAS Transactions on Mathematics 6(12), 865–877 (2007)

Heuristics for Semantic Path Search in Wikipedia

Valentina Franzoni[1], Marco Mencacci[1], Paolo Mengoni[1], and Alfredo Milani[1,2]

[1] Department of Mathematics and Computer Science, University of Perugia, Perugia, Italy
[2] Department of Computer Science, Hong Kong Baptist University, Hong Kong
{valentina.franzoni,milani}@dmi.unipg.it
{marco.mencacci,paolo.mengoni}@studenti.unipg.it

Abstract. In this paper an approach based on Heuristic Semantic Walk (HSW) is presented, where semantic proximity measures among concepts are used as heuristics in order to guide the concept chain search in the collaborative network of Wikipedia, encoding problem-specific knowledge in a problem-independent way. Collaborative information and multimedia repositories over the Web represent a domain of increasing relevance, since users cooperatively add to the objects tags, label, comments and hyperlinks, which reflect their semantic relationships, with or without an underlying structure. As in the case of the so called Big Data, methods for path finding in collaborative web repositories require solving major issues such as large dimensions, high connectivity degree and dynamical evolution of online networks, which make the classical approach ineffective. Experiments held on a range of different semantic measures show that HSW lead to better results than state of the art search methods, and points out the relevant features of suitable proximity measures for the Wikipedia concept network. The extracted semantic paths have many relevant applications such as query expansion, synthesis of explanatory arguments, and simulation of user navigation.

Keywords: heuristics search, semantic networks, collaborative networks, semantic similarity measures, random walk, information retrieval.

1 Introduction

In the era of Big Data, searching, browsing and collaboratively building online semantic networks are the typical tasks that absorb a lot of users activity e.g., in social networks, collaborative encyclopaedias, media sharing repositories et cetera. Collaboratively updating/adding a reference to a Wikipedia entry, or labelling a multimedia object in a repository in order to make it clearer, are examples of those *collaborative* actions. The action is called collaborative since the user, in order to share, inform or facilitate other users, purposely adds the content/relationship to the network.

The connections of a semantic collaborative network can provide important information about the semantic relationship among the network entities i.e., the nodes. In this work, we focus on the problem of finding a chain between two entries of the Wikipedia online encyclopedia i.e., a path of articles between a source and a target, by

B. Murgante et al. (Eds.): ICCSA 2014, Part VI, LNCS 8584, pp. 327–340, 2014.

following hyperlinks among articles. This problem is of great importance, since it can provide useful information concerning the subject entities, such as relationships and explanations. In particular, considering the notion as a *context* consisting of two concepts with corresponding Wikipedia articles, the intermediate nodes in the Wiki path or *semantic chain*, as well as the *surrounding nodes*, can represent *hidden concepts* or *underlying concepts*, relevant for a number of applications, such as natural language understanding, query expansion and automatic explanation.

Let consider for instance two Wikipedia entries like *Mars* and *Scientist* and let the path `Mars->planet->science->scientist`. The intermediate concepts in the chain can be used to generate an explanation of the relationship between *Mars* and *Scientist*, by focusing on meanings that are consistent with the underlying context, which will most probably link the concept of *Mars* as a *planet in the Solar System* rather than to *the ancient Greek god Mars*. The problem of the semantic chain search can be reduced to a problem of search in a graph. To establish which concepts a pair of terms implies in a dialogue in concept explanation, we can consider the path between them, where the starting and ending nodes in the path form the context.

The basic idea of our approach is to apply the *Heuristic Semantic Walk (HSW)* [17][18] framework, where a proximity measure m, defined between pair of concepts, and derived from the statistical results of a query [1][2][4] to a search engine S, is used as *heuristic*, and is applied to guide a path search the Wikipedia concept network. The *HSW* approach is a general framework which can be instantiated with different heuristics h_m based on a different proximity measure m (e.g. *confidence*, *Pointwise Mutual Information*, *Normalized Google Distance*) [2][3][4], used by different informed search algorithms. In general, the proximity measure m between two term t_1 and t_2 is computed by submitting simple queries to a search engine S, and using the statistics about frequency and co-occurrence of terms in the indexed resources. In other words, the measure $m(t_1, t_2)$ reflects the collective knowledge embedded in the search engine S, with respect to answer the question about how far away are concepts t_1 and t_2. An informed search algorithm A can then calculate the value of the heuristics for each candidate successors, and decide the direction where to expand the search.

The experiments, held on Wikipedia on a range of different semantic proximity measures, show that the proposed approach outperforms classical uninformed search methods. In particular, HSW with heuristic randomized search returns the path that connects two concept nodes in much faster times than an uninformed blind random search; moreover HSW returns a higher quality path, in a semantic point of view, than an uninformed blind search. This latter result is particularly important when the HSW is used for semantic applications e.g., in *query expansion*, where the nodes of the path are used as candidates for the query expansion.

This paper is organised as follows. In the second section, the main features of the proposed heuristic walk approach are described, and semantic walk strategies are considered in section 3, where the experimented proximity measures are also exposed. Conclusions are drawn and future directions of the research are finally discussed.

2 The Heuristic Semantic Walk Model

In the Heuristic Semantic Walk model, we consider is to browse a semantic network in order to connect a pair of concepts, formulated as the problem of searching paths between two nodes over an oriented graph.

Definition: a *semantic network graph* $\Sigma = (V, E)$, is defined by a pair where V is a set of vertices/concepts (e.g., the entries in Wikipedia), and $E \subseteq (V \times V)$ is a set of oriented edges, representing the links between concepts in the network (e.g., the anchor links in the text of a Wikipedia article toward a referenced article).

Definition: the *semantic path finding problem* or Path(s,g), given a semantic network $\Sigma = (V, E)$ and two nodes s, $g \in V$, consists in finding if a sequence of vertices $(v_0, v_1, ..., v_n)$ exists, such that $v_0 = s$, $v_n = g$ and for each $i \in [0, n-1]$ *the edge* $(v_i, v_{i+1}) \in E$. Similarly.

Shortest Path search problem SPath(s,g) can be defined straightforwardly.

2.1 Shortest Path and Plan Quality

In the following of this paper, we will consider the *Semantic Path Extraction* problem as broadly equivalent to the *Shortest Path* extraction on Wikipedia, although it is not. Intuitively Wikipedia is a *network of concept definitions and explanations*, then the *shortest the path* between to concepts, should be also the more "meaningful" i.e., the shortest the path the more *direct* is the concepts relationships. In the real case Wikipedia linked entries can also contain *user introduced noise, personal* and *structural biases*. Although policies, guidelines and form of controls are in place, the users are completely free to arbitrarily modify a Wikipedia article, thus introducing unwanted *errors*, placing links on irrelevant concepts or on common terms thus influencing the *semantic quality* of a possible path. This and other problems, such structural biases and hub terms will be further discussed in this paper.

2.2 Semantic Proximity Measures as Search Heuristics

It is well known how path search in state space, could greatly benefit from an informed search strategy, but unfortunately there are not inherent properties of the problem at hand, which can be used to define heuristics using classical technique like problem constraints relaxation. In fact the Wikipedia network, or other collaborative concept networks, cannot straightforwardly be seen as a state space. In the case of a semantic network, the relaxation technique cannot be applied to define heuristics, since the node is not a state generated by an action. On the other hand, it can be observed that links among concepts are added by the collective collaborative effort of the users, with the purpose of providing further explanations and insight knowledge e.g., the Wikipedia linked articles, or providing explanations about the content e.g., labels and tags in a multimedia repository. In term of state space problem, we can say that it is possible move from a state A (or Wiki article) to another state B if a *is_related_to* action is applicable from state *A* to state *B* i.e., if the community of users have decided it. The basic idea is to use a measure of *relatedness* and a heuristic. If we broadly assume that *relatedness* is monotonically non-decreasing, and we look at it transitively, then we can support the intuition that there is a high chance that following a path with higher relatedness to the goal will likely lead to the target goal. As candidate for such heuristics, we focus on *proximity measure* in the literature which can be calculated from statistical data extracted from collaborative collective sources of information, such as any search engine, that can be both a generalized one (e.g., Google, Bing) or an embedded one in specialized media repositories (e.g. YouTube, Flickr) or social networks (e.g. Facebook, Twitter). This approach will guarantee that the source of data reflects the dynamical judgment of the community of the engine/repository, and at the same time, it can be easily calculated by querying the engine. Semantic proximity measures have been widely used for semantic extraction and automatic clustering of terms [12] but, as far as we know, in the HSW model for the first time they were used as heuristics for semantic search.

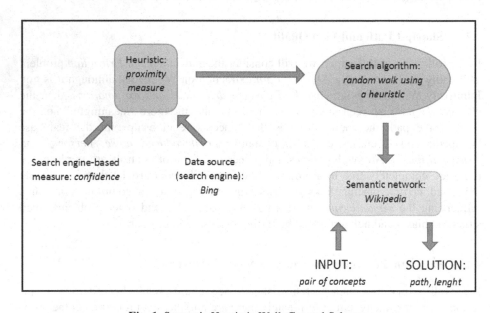

Fig. 1. Semantic Heuristic Walk General Scheme

2.3   Heuristic Semantic Walk (HSW): The General Scheme

The goal of a HSW is to return the path between a pair of terms *(s, g)* following the semantic links (e.g., the anchor links from the text of a starting Wikipedia article *s* to a goal article *g*), and using a semantic proximity measure to drive the search towards the best successor candidate (c_i) toward the goal node (g). A general scheme can be defined in order to characterize the class of HSW solutions to the semantic SPath problem.

Definition. A *Heuristic Semantic Walk instance*, or HSW, is an algorithm for SPath problems characterized by a *semantic collaborative network* graph $\Sigma=(V,E)$, a search engine S, which can return statistics about the occurrences of elements in $\wp(V)$, a proximity measure m defined in terms of those statistics, and a *search strategy* or *walk strategy A* using heuristics derived from m.

Note that $\wp(V)$ indicates the *part-of* set i.e., the set of all subsets of V; in other words, the engine/repository S should be able to return statistics for co/occurrences of objects. Most engines/repositories are in fact able to return statistics for queries like $Q= (a \wedge b \vee \neg c)$. Since the actual returned objects are not relevant, for our purposes the engine S can be formally defined.

Definition: a search engine/repository S is a function $S: \wp(V) \rightarrow N$, where $S(Q)$ returns the number of occurrences of Q according to the engine/repository i.e., the statistics will reflect the frequency contextual use of terms in Q, according to the community related to the domain that feeds objects to S.

Given the previous general scheme, we will use the following notation, where *HSW(S,m,A)* will denote an instance of HSW for a semantic network Σ where search strategy A uses heuristics h_m computed from source S, and *HSW(A)* denotes a non-informed algorithm with a walk strategy A. For example, *HSW(Bing, NGD, A*)* for Wikipedia denotes an A^* search algorithm operating on Wikipedia, which uses as a heuristic the Normalized Google Distance computed using Bing statistics (see Fig.1).

3   Semantic Walk Strategies

The size and morphological features of online collaborative semantic networks pose important challenges and factors to be considered when choosing a walk strategy for a HSW scheme. In particular, the *high degree of nodes* i.e., the number of outgoing edges from semantic objects; the *high connectivity degree of the graph* i.e., the high number of *alternative paths* existing between a pair of nodes; and the nearly *unlimited size* of the network, point out drawbacks such as memory requirements, branching factors, endless loops and deadlock on optimality plateau, which can heavily affect the performance of the classical "best" graph search algorithms.

3.1   Non Informed Walks

Since the branching factor in Wikipedia can be very high, the BFS approaches are deeply penalized even in the case in which the target terms are only moderately distant from the source, by the exponential growth of the memory requirements. On the other hand, Depth First (DF) approaches cannot avoid to fall into loops, when re-encountering a node previously visited on a discarded branch, founding either loops or eventually very suboptimal paths [7][9][13] Iterative Deepening Search (IDS), which has low memory requirements and guarantees completeness, is extremely inefficient with high branching factors, and it is further penalized from the high time cost of the repeated online access to the candidate nodes. We can conclude that non informed walk strategies cannot guarantee an efficient SPath search in the online collaborative semantic network scenario; on the other hand, complete algorithms based on IDS or BFS [5] still represent an important reference point, allowing to find benchmark optimal values for comparison purposes.

3.2   Heuristic Walks

Heuristic Walks are in general informed search strategies, which make use of a heuristic to estimate a score of each candidate node, and to sort them for the expansion. The evaluation is performed in terms of closeness/relatedness to the goal, where relatedness is computed by measuring the semantic proximity of the current node to the goal node. In the area of heuristic algorithms, we certainly have to consider A^* and algorithms derived from it, where its optimality properties would make it an ideal candidate for semantic heuristic walk strategies. The core of the A^* class of algorithms is the selection function $f(n)=c(n)+h(n)$ where $c(n)$ represents the cost to reach n from the source node (the path distance, in our case), while $h(n)$ is the heuristic estimate of the distance of n from the goal.

In A^*, a node n_e in the fringe F of an unexpanded node is selected for expansion if $n_e=min(n\in F,\ f(n))$ i.e., if the value of evaluation function is minimal for n_e. Unfortunately, as preliminary experiments have shown, when the target is not close to the source the A^* fringe F will grow too large with real online semantic networks, making the approach unpractical.

A *Greedy best-first* (GBF) approach has then been considered, since it avoids maintaining different nodes' path histories. In typical GBF, the evaluation $f(n)=h(n)$ does not consider previous costs, then only the nodes of the fringe that are the closest to the goal are chosen.

The problem of a growing large fringe have led to a particular implementation, *Local GBF* (LGBF), where no fringe memory is maintained except for the current path, and only the current node descendants are considered in the fringe. LGBF implements a local search with loop avoidance control.

3.3   Heuristic Weighted Random Walk (WRW)

In order to eliminate the problems with LGBF, we have decided to add a degree of randomization to the original LGBF. The basic idea of the semantic heuristic *Weighted Random Walk (WRW)* is to choose among a set of successors for the current node, making a random tournament among the candidate successors, weighted by the probability distribution induced on the candidates by the values of the proximity measure between the candidate concepts and the target concept. No memory of unexpanded nodes is maintained, except for the current solution path. In WRW, the proximity measure values need to be normalized to [0,1] in order to build the probability distribution. The probability of the candidate c_i in the random tournament is defined as

$$P(c_i) = \frac{m\ (c_i, g)}{\sum_j\ m\ (c_j, g)} \qquad \forall j \in Succ(curr)$$

$$(1)$$

where g is the target node, c_j are the successors of the current node *curr*, and $m(c_j, g)$ is the proximity measure. The computation of the value of $m(c_j, g)$ requires to query a given search engine in order to obtain the appropriate data. The hypothesis underlying this approach is that Wikipedia pages of close concepts are linked by a short path.

3.4   Web-Based Semantic Proximity Measures

Search engines are based on documents that are dynamically updated by a great number of users. Therefore using information on indexed terms provided by a search engine is then a valid approach to evaluate proximity semantics of pairs of terms, or groups of terms.

The general idea is to use a search engine or an object repository search tool S as a black box, to which submit queries and from which to extract useful statistics about the occurrence of a term or a set of terms in the indexed objects, as simple as counting the number of results for terms/objects queries. A proximity measure m is then used as heuristic $h(n)=m(n,g)$ to evaluate the most promising node n to browse, with the aim of reaching the goal node.[1] Since we normalize all the measures, distance and proximity measures can be compared as complementary.

- **Confidence** is an asymmetric statistical measure, used in rule mining for measuring trust in a rule $X \rightarrow Y$ i.e., given the number of transactions that contain X, then the *Confidence* in the rule indicates the percentage of transactions that contain also Y.

$$confidence(x \to y) = \frac{P(x,y)}{P(x)} = \frac{f(x,y)}{f(x)} \tag{2}$$

From a probabilistic point of view, confidence approximates the conditional probability, such that

$$confidence\ (x \to y) = P(y/x) \tag{3}$$

- **Pointwise Mutual Information (PMI)** [4] is a point-to-point measure of association used in statistics and information theory. Mutual information between two particular events w_1 and w_2, in this case two words w_1 and w_2 in webpages indexed by a search engine, is defined as:

$$PMI(w_1, w_2) = log_2 \frac{P(w_1, w_2)}{P(w_1)P(w_2)} \tag{4}$$

$PMI(w_1, w_2)$ is an approximate measure of how much a word gives information on the other word of the pair, in particular the quantity of information provided by the

---

[1] Let denote by $f(x)=S(x)$ and $f(x,y)=S(x \wedge y)$ the cardinalities of the results of a query to S, with search term x and $x \wedge y$ respectively, and we denote by N the number of documents/objects which are indexed by the search engine S. The probability estimate can be directly computed from the frequency, as $P(x)=f(x)/N$ (2), summarizing the frequency-based approach to probability, whenever the total N is known or can be realistically approximated with a value that will be greater than $f(x)$ for each possible x in the considered domain or context.

occurrence of the event w_2 about the occurrence of the event w_1. A high value of PMI represents a decrease of uncertainty. PMI has been successfully used in [15] to recognize synonyms, using only word counts, despite that on low frequency data, PMI does not provide reliable results and is not a good measure of independence, since values near zero indicate frequency, but at the same time is a bad measure of dependence, since the dependency score is related to the frequency of individual words.

- χ^2 **(Chi-squared or Chi-square)** [8] measures the significance of a relation between two categorical variables, checking if the values, observed by measuring frequency, differ significantly from the frequencies obtained by the theoretical distribution. The intuition behind Chi-square is that two events are associated only when they are more related than by pure chance. χ^2 coefficient has also been used in algorithms for community discovering.

$$\chi^2 = \frac{(ad-bc)^2 n}{(a+b)(a+c)(b+d)(c+d)} \qquad (5)$$

where a,b,c,d represent the number of document in which w_1 and w_2 occurs according to $a= w_1 \wedge w_2$ $b= w_1 \wedge \neg w_2$ $c= w_2 \wedge \neg w_1$ $d=\neg w_1 \wedge \neg w_2$ and $n=N=a+b+c+d$.

- **Normalized Google Distance (NGD)** has been introduced in [3] as a measure of semantic relation, based on the assumption that similar concepts occur together in a large number of documents in the Web i.e., the frequency of documents returned by a query on a search engine S approximates the distance between related semantic concepts. The NGD between two terms x and y is formally defined as follows:

$$NGD(x,y) = \frac{max\{\log f(x), \log f(y)\} - \log f(x,y)}{\log M - min\{\log f(x), \log f(y)\}} \dots \dots \dots \dots \qquad (6)$$

where $f(x)$, $f(y)$ and $f(x,y)$ are the cardinalities of results returned by S for the query on x, y, $x \wedge y$ respectively, and M is the number of pages indexed by S, or a value which is reasonably greater than $f(x)$.

- **PMING Distance** [5][7][11] consists of NGD and PMI locally normalized, with a correction factor of weight ρ, which depends on the differential of NGD and PMI.

    More formally, the PMING distance of two terms x e y in a context W is defined, for $f(x) \geq f(y)$, as a function $PMING:W \times W \rightarrow [0,1]$:[2]

$$PMING(x,y) = \rho \left(1 - \log \frac{f(x,y)M}{f(x)f(y)\mu_1}\right) + (1-\rho)\left(\frac{\log f(x) - \log f(x,y)}{(\log M - \log f(y))\mu_2}\right) \dots \qquad \dots (7)$$

---

[2] μ_1 and μ_2 are constant values which depend on the context of evaluation, defined as: $\mu_1=\max PMI(x,y)$; $\mu_1=\max NGD(x,y)$, with $x,y \in W$.

4   Experiments and Comparison

4.1   Experimental Setting

After a preliminary phase in order to evaluate general-purpose walk strategies and to tune the *WRW* strategy, systematic experiments on pairs of terms had been held on Bing as a search engine, with NGD and Confidence (CF) as heuristics, in order to evaluate the suitability of the different proximity measures as heuristics. Comparisons have been made among *HSW(BFS)*, *HSW(Bing, CF, WRW)* and *HSW(Bing, NGD, WRW)*, and ongoing work among *HSW(Wikipedia, CF, WRW)* and *HSW(Wikipedia, CF, WRW)* and *HSW(Wikipedia, NGD, WRW)* using in any case *Wikipedia* as a network, where a shortest path between the initial and target goal is searched, and a list of candidate terms for expansion was generated using the anchor links to other Wikipedia articles present in the textual content of the wiki page.

4.2   Pre-filtering and Bounds

Since the network is large and highly connected with non-relevant links, a pre-processing information filtering phase has been necessary, in order to filter out users' biases and non-semantic outliers:

Div Content. Only the anchor links in the content of the article are evaluated. [6][11].The parsing of the page can be furthermore limited by considering only the main content HTML *div* element of the article and not any other Wikipedia div or box.

Maximum number of candidate links (sub-optimality). In order to prune the graph and reduce computing time, a threshold can be stated on the number of linked candidate node [7]. In fact in Wikipedia the first lines of text are more related to the essential definition of a concept, while the longer a page is, the less significant links are provided at the end of the page.

Hub Links filtering. Filtering links that may lead to a hub in the Wikipedia network, with loss of semantic quality of the path: e.g., categories such as years, centuries and millenniums, first names of person, et cetera, nearly connect all the Wikipedia pages, without carrying a relevant semantic value.

Blank pages elimination. Pruning of pages without anchor links in the main text of the article i.e., dead ends.

Depth limitation. it has been proven useful for search speed-up to establish a limitation of the depth of the search, i.e. a maximum number of steps, although it is compromising completeness, it is in practice is not introducing a limitation.

4.3   Results Comparisons

Experimental result analysis shows that HSW has better performances than non-informed search and Pure Random Walk. In particular, all the considered proximity measures used as a heuristics leads on the average to better results than a pure random walk, while non-informed search algorithms quickly exhausted memory resources or do not terminated for non-trivial searches.

For the sake of clarity, the results for non-informed search algorithms and the other experimented proximity measures are not shown in table 1 and 2: we can summarize the omitted result by noting that PMI generally performed worse than NGD and in some cases even worse than the PR pure randomness. Furthermore, the PMING distance, tested in preliminary experiments, was better performing than PMI itself, but required excessive computational cost, due to the many Wiki queries needed for context calculation, and was then discarded for the systematic experiments.

Chi-squared (CHI) was the third performing measure on the average although far behind NGD and Confidence.

The Bing search engine has been chosen as source of statistics because the more popular Google was found in [4] to lead to poorer results from a semantic point of view, due the bias and inconsistencies observed in the returned, probably for user modelling and commercial purposes, which no longer represent the real statistics of the engine. Analysing the results, in fact, it was observed that Google's results, both with and without API, greatly differ, while in and among the other search engines the difference is much smaller. This issue is explained by Google support as the lack of some additive services with the use of the API, where the lacking services are the ones which Google provides with the use of personal information about browsing, through cookies and/or accounting.

E.g. submitting the query "franzoni" to Google and Bing, the following results were obtained the following occurrences:

	With API	Without API
Google	1840000	5460000
Bing	2220000	2750000

And for the concurrent query "franzoni milani":

	With API	Without API
Google	28100	321000
Bing	44200	44300

We then decided to create a script to submit the queries and read the results through page scraping, instead of using the API, to automate the process and at the same time to keep the same results that would be obtained with a manual submission. Page scraping was implemented simply putting the HTML content of the page with query results in a string variable and extracting the data about the number of results matching regular expressions.

Notice that the cardinality of the results of a query in a search engine can vary and give different results in different times, so manually submitting a bunch of data may not return the best results, because of the time gap between pairs of terms or sets of terms which results have to be compared. Both with and without API is therefore suggested to submit queries with an automated program, to shorten the time lapse.

4.3.1 Performance Comparison

The performance comparison has been held by considering the average *path length* found from source to the target node, and the number of cases in which the algorithm does not converge at all on the given bounds. The best results were obtained with *HSW(Bing,CF,WRW)* and second best *HSW(Bing,NGD,WRW)*, resulting generally much more performing of PRW, see Fig.2(a).

Using the *CF+Bing* heuristics, the average number of steps needed to converge to the target node and the number of non-convergence cases is on average 50% of those required for the PRW, while for NGD+Bing the reduction is up to 25%, where comparison are made on 100 runs for each pair. Table 2 shows the performance for 100 runs on the pairs ("computer", "software"), ("planet", "galaxy"), ("drug", "abuse") and ("student", "professor"). In the first and second columns, the starting and goal terms of the pair are shown. In columns from third to fifth, the average number of steps needed for RW+Bing, NGD+Bing and CF+Bing to converge to the solution is shown. It is worth to mention that the NGD on worst case performed even better than the RW on average. Ongoing experiments with *HSW(Wikipedia, CF/NGD,WRW)* confirm the trend.

Table 1. experiment on h(n)={NGD, confidence} for the pair ("arithmetic", "counting")

Step 1: arithmetic=>{"mathematics", "science", "business"}

	t1	f(t1)	t2	f(t2)	f(t1,t2)	NGD(t1,t2)	confidence(t1->t2)
n1	mathematics	101000000	counting	38800000	2230000	0,5495821639	0,0220792079
n2	science	435000000	counting	38800000	7030000	0,5945563276	0,0161609195
n3	business	864000000	counting	38800000	8780000	0,6614232480	0,0101620370

Step 2: mathematics=>{"quantity", "structure", "space"}

	t1	f(t1)	t2	f(t2)	f(t1,t2)	NGD(t1,t2)	confidence(t1->t2)
n1	quantity	21600000	counting	38800000	1730000	0,4133869026	0,0800925926
n2	structure	80100000	counting	38800000	2560000	0,4962758938	0,0319600499
n3	space	383000000	counting	38800000	6190000	0,5945477630	0,0161618799

Step3: quantity=>{"property (philosophy)", "magnitude (mathematics)", "counting"}

	t1
n1	property (philosphy)
n2	magnitude (mathematics)
n3	counting

HSW("arithmetic", "counting")={"arithmetic"→"mathematics" →"quantity" →"counting"}

Table 2. experiments on h(n)={NGD, PMI} on 100 runs for each pair

Start term	Goal term	Avg RW	Avg NGD	Avg CF
Computer	Software	159,48	91,1	19,42
Planet	Galaxy	94,66	48,52	17,94
Drug	Abuse	418,7	326,7	152,82
Student	Professor	225,16	184,6	153,21

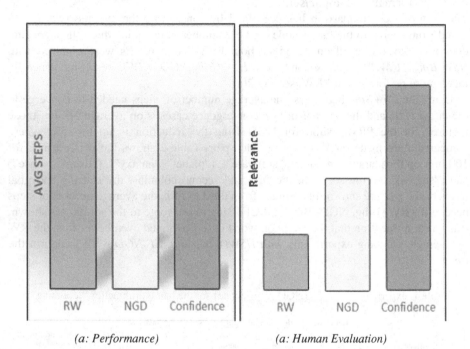

(a: Performance)   *(a: Human Evaluation)*

Fig. 2. Walk trends for Pure Random RW, NGD and Confidence)

4.3.2 Semantic Path Quality

Although a more systematic quality evaluation is an ongoing work, a preliminary evaluation confirms the best results of NGD and Confidence also with respect to the quality of the semantic paths returned, assessed by a blind human evaluation (see Fig.2(b) and Table 3) on a group of 15 users and 97 pairs (100 runs for Rt-HSW each pair). All the HSW variants returns a higher quality path, from a semantic point of view, than the PRW. The best quality resulted in HSW(Bing, NGD, WRW) as the best quality evaluation and HSW(Bing, Confidence, WRW) as the second best while the others navigate through less relevant intermediate nodes. This preliminary evaluation allows to draw the conclusion that HSW path is suitable to be used as a semantic explanation chain, for natural language contexts. The best quality and performance of Confidence over the other measures can be explained with the nature of the Wikipedia network, which contains articles explaining the meaning of terms, and links, which usually point to related explanatory information i.e., a directed graph, where links means explanation.

We argue that Confidence is more effective because is measuring more a casual relationships than a mere co-occurrence, like for example with NGD. The point is that Confidence is expected to be more suitable effective in order to navigate Wikipedia, since casual explanation is intuitively bringing higher quality to the semantic path. Furthermore, we have to consider, for instance, that more general concepts tend to have more inbound links, which reflects exactly the notion underlying Confidence; this pushes the search to go through more general concepts with respect to just co-occurring ones, then a semantic chain that is more easy to follow.

Table 3. Human evaluation (Avg) on heuristic-driven randomized strategies

Rt-HSW(Random Tournament)	Relevance human evaluation (points from 0 to 5)
PRW	1
RT-Confidence	4
RT-NGD	2

5   Conclusions

In order to solve the problem of extracting semantic paths between concepts from the Wikipedia collaborative network, we have proposed the Heuristic Semantic Walk approach, which uses search engine-based proximity measures as search heuristics. A remarkable feature of HSW is than it can be used for online search on large size semantic collaborative networks like Wikipedia, and that the HSW framework can be parameterized with respect to proximity measures and to the sources of information used for computing it. A proximity measure reflects some relationships between terms embedded in the indexed corpora of documents. The statistics extracted from search engines or from social networks reflect the relationships between terms and semantics as seen by the members of the specific community or social network. Experiments have shown that HSW with *Confidence* and *NGD* heuristics using *Bing* as a statistic source for semantic path searching in Wikipedia, outperforms classical search and other proximity measures, both with respect to *path length* and *quality*.

Ongoing research regards experimenting different proximity measures and variants of other informed search algorithms for semantic networks exploration, as well as systematic evaluation of semantic path quality. Future research will focus on further applications of the basic principle behind the HSW i.e., *using a semantic based heuristic to drive semantic search* to other contexts, such as, modelling user navigation in information repositories, modelling user associative reasoning in the area of natural language understanding and brain informatics.

References

1. Bollegala, D., Matsuo, Y., Ishizukain, M.: A Web Search Engine-Based Approach to Measure Semantic Similarity between Words. IEEE Transactions on Knowledge and Data Engineering (2011)
2. Cilibrasi, R., Vitanyi, P.: The Google Similarity Distance. ArXiv.org (2004)
3. Church, K.W., Hanks, P.: Word association norms, mutual information and lexicography. In: ACL, vol. 27 (1989)

4. Franzoni, V., Milani, A.: PMING Distance: A Collaborative Semantic Proximity Measure. In: WI-IAT, 2012 IEEE/WIC/ACM International Conferences on Web Intelligence and Intelligent Agent Technology, vol. 2, pp. 442–449 (2012)
5. Kurant, M., Markopoulou, A., Thiran, P.: On the bias of BSF. ITC (2010)
6. Milne, D., Witten, I.H.: An effective, low-cost measure of semantic relatedness obtained from Wikipedia links. WIKIAI (2008)
7. Yeh, E., Ramage, D., Manning, C.D., Agirre, E., Soroa, A.: WikiWalk: Random walks on Wikipedia for Semantic Relatedness. In: Proc. Graph-based Methods for Natural Language Processing (2009)
8. Newman, M.E.J.: Fast algorithm for detecting community structure in networks. University of Michigan, MI (2003)
9. Cao, G., Gao, J., Nie, J.Y., Bai, J.: Extending query translation to cross-language query expansion with markov chain models. CIKM, ATM (2007)
10. Turney, P.D.: Mining the web for synonyms: PMI-IR versus LSA on TOEFL. In: Flach, P.A., De Raedt, L. (eds.) ECML 2001. LNCS (LNAI), vol. 2167, pp. 491–502. Springer, Heidelberg (2001)
11. Xu, Z., Luo, X., Yu, J., Xu, W.: Measuring semantic similarity between words by removing noise and redundancy in web snippets. Concurrency Computat: PE 23 (2011)
12. Wu, L., Hua, X.S., Yu, N., Ma, W.Y., Li, S.: Flickr Distance. Microsoft Research Asia (2008)
13. Leung, C.H.C., Li, Y., Milani, A., Franzoni, V.: Collective Evolutionary Concept Distance Based Query Expansion for Effective Web Document Retrieval. In: Murgante, B., Misra, S., Carlini, M., Torre, C.M., Nguyen, H.-Q., Taniar, D., Apduhan, B.O., Gervasi, O. (eds.) ICCSA 2013, Part IV. LNCS, vol. 7974, pp. 657–672. Springer, Heidelberg (2013)
14. Gori, M.,, P.: A random-walk based scoring algorithm with application to recommender systems for large-scale e-commerce. In: 12th ACM SIGKDD International Conference on Knowledge Discovery and Data Mining (2006)
15. Franzoni, V., Milani, A.: Heuristic Semantic Walk. In: Murgante, B., Misra, S., Carlini, M., Torre, C.M., Nguyen, H.-Q., Taniar, D., Apduhan, B.O., Gervasi, O. (eds.) ICCSA 2013, Part IV. LNCS, vol. 7974, pp. 643–656. Springer, Heidelberg (2013)
16. Lew, M.S., Sebe, N., Djeraba, C., Jain, R.: Content-based multimedia information retrieval: State of the art and challenges. ACM Trans. Multimedia Comp. Com. App (2006)
17. Franzoni, V., Milani, A.: Heuristic semantic walk for concept chaining in collaborative networks. International Journal of Web Information Systems 10(1), 85–103 (2014), doi:10.1108/IJWIS-11-2013-0031
18. Franzoni, V., Milani, A., Mengoni, P., Mencacci, M.: Semantic Heuristic Search in Collaborative Networks: Measures and Contexts. In: WI-IAT, 2014 IEEE/WIC/ACM International Conferences on Web Intelligence and Intelligent Agent Technology (2014) (accepted for)
19. Cheng, V.C., Leung, C.H.C., Liu, J., Milani, A.: Probabilistic Aspect Mining Model for Drug Reviews. IEEE Transactions on Knowledge and Data Engineering 99, 1 (preprint, 2014), doi:10.1109/TKDE.2013.175
20. Milani, A., Santucci, V.: Community of scientist optimization: An autonomy oriented approach to distributed optimization. AI Commun. 25(2), 157–172 (2012), doi:10.3233/AIC-2012-0526
21. Leung, C.H.C., Chan, A.W.S., Milani, A., Liu, J., Li, Y.: Intelligent Social Media Indexing and Sharing Using an Adaptive Indexing Search Engine. ACM TIST 3(3), 47 (2012), doi:10.1145/2168752.2168761
22. Baioletti, M., Milani, A., Poggioni, V., Rossi, F.: Experimental evaluation of pheromone models in ACOPlan. Ann. Math. Artif. Intell. 62(3-4), 187–217 (2011), doi:10.1007/s10472-011-9265-7

Constructing Virtual Private Supercomputer Using Virtualization and Cloud Technologies

Ivan Gankevich, Vladimir Korkhov, Serob Balyan, Vladimir Gaiduchok,
Dmitry Gushchanskiy, Yuri Tipikin, Alexander Degtyarev,
and Alexander Bogdanov

St. Petersburg State University,
Universitetskii 35, Petergof, 198504, St. Petersburg, Russia
igankevich@ya.ru, vkorkhov@apmath.spbu.ru, serob.balyan@gmail.com,
{gajduchok,dgushchanskii,iutipikin}@cc.spbu.ru,
{deg,bogdanov}@csa.ru

Abstract. One of efficient ways to conduct experiments on HPC platforms is to create custom virtual computing environments tailored to the requirements of users and their applications. In this paper we investigate virtual private supercomputer, an approach based on virtualization, data consolidation, and cloud technologies. Virtualization is used to abstract applications from underlying hardware and operating system while data consolidation is applied to store data in a distributed storage system. Both virtualization and data consolidation layers offer APIs for distributed computations and data processing. Combined, these APIs shift the focus from supercomputing technologies to problems being solved. Based on these concepts, we propose an approach to construct virtual clusters with help of cloud computing technologies to be used as on-demand private supercomputers and evaluate performance of this solution.

Keywords: virtualization, supercomputer, virtual cluster, cloud computing.

1 Introduction

Virtual supercomputer can be seen as a collection of virtual machines working together to solve a computational problem much like a team of people working together on a single task. There is a known definition of personal supercomputer as a kind of metacomputer which was given in [1], however, in our approach virtual supercomputer is not only a personal supercomputer but it also offers a way of creating virtual clusters that are adapted to problem being solved and to manage processes running on these clusters (figure 1). This is the case where virtual shared memory cannot be used directly because of high latency and low performance caused by complex data transfer patterns. Migration of processes to data as well as methods of workload balancing [2] can solve this problem and form a basis of load balancing technique for a virtual supercomputer.

B. Murgante et al. (Eds.): ICCSA 2014, Part VI, LNCS 8584, pp. 341–354, 2014.

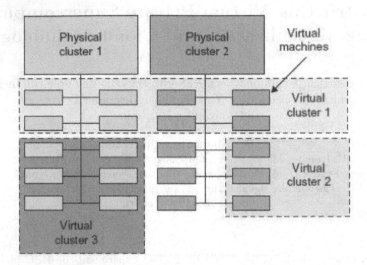

Fig. 1. A cloud platform example with three virtual clusters over two physical clusters

Computers like people need some sort of collective board to share results of their work and advance problem solution one step further. In a distributed computing environment distributed file systems and distributed databases act as such a board, storing intermediate and final results of computation. Apart from a shared desk people in a team need some sort of management to solve a problem in time and computers need a way of combining them into hierarchy helping efficiently distribute tasks among available computing nodes. Finally, from a technical point of view, problem solution should be decoupled from actual execution of tasks by a virtualization layer as not every problem has efficient mapping on physical architecture of a distributed system. So, virtual supercomputer is not only a cluster of machines but also virtualization and middleware layers on top of it.

There are many ways to construct such a supercomputer and it is time-consuming to compare and assess benefits of all technology combinations, however, it is convenient to tailor technologies to the needs of chosen problems and to show advantages of virtual supercomputer approach in these particular cases. Section 2 gives an overview of related work in this area. The chosen problems should be general enough to cover a wide area of potential applications. We selected two examples of such generic problems to be solved in personal supercomputer environment: ontology storage, retrieval and analysis involving use of a distributed database; fluid dynamics simulations involving execution of highly parallel code. These problems are discussed in Section 3. Corresponding virtual supercomputer configuration, its key principles are discussed in Section 4. Experimental evaluation of our solution is presented in Section 5. Section 6 concludes the paper and shows the directions of future work.

2   Related Work

One of the first approaches to construct clusters from virtual machines was proposed in [3] and partially realized in In-VIGO [4], VMPlants [5], and Virtual Clusters on the Fly [6] projects. In-VIGO focused on the end-to-end design of a Web service that could employ VMs as part of a cluster computing system, while VMPlants and Virtual Clusters on the Fly focused on rapid construction of virtual clusters. All three systems were particularly concerned with the issue of specifying and adapting to requirements and constraints imposed by the user.

Dynamic Virtual Clustering (DVC) [7] implemented the scheduling of VMs on existing physical cluster nodes within a campus setting. The motivation for this work was to improve the usage of disparate cluster computing systems located within a single entity.

The idea of an adaptive virtual cluster changing its size based on the workload was presented in [8] describing a cluster management software called COD (Cluster On Demand), which dynamically allocates servers from a common pool to multiple virtual clusters.

Grid architecture that allows to dynamically adapt the underlying hardware infrastructure to changing Virtual Organization (VO) demands is presented in [27]. The backend of the system is able to provide on-demand virtual worker nodes to existing clusters and integrate them in any Globus-based Grid.

The goal of current work is to investigate possibilities provided by modern cloud and virtualization technologies to enable personal supercomputing meaning creation of dedicated virtual clusters based on user's and application's requirements. Unlike many of other projects in this area, our intention is to use created clusters of VM as a single resource provided to a single parallel application but not generating sets of worker nodes provided to different applications separately.

3   Large-Scale Supercomputer Problems

3.1   Ontology Storage and Retrieval

One way of applying virtual supercomputer is graphs storage and processing. Transport logistics, articles citation or social networks are common examples of such tasks. In some cases graphs may have thousands or millions vertices and edges, as in Web graph related tasks [9]. That kind of structures can be handled by various types of algorithms such as shortest path computations, a special subgraphs allocation, different varieties of clustering etc. It is challenging to efficiently process large graphs since some graph's properties work poorly with high performance techniques [10].

- Parallelism based on partitioning of computation can be difficult to express because the structure of computations in the algorithm is not known a priori.
- The irregular structure of graph data makes it difficult to extract parallelism by partitioning the problem data. Scalability can be quite limited by unbalanced computational loads resulting from poorly partitioned data.

- Graphs can represent complex irregular relationships between entities thus it may provide the lack of locality for computations and data access patterns.
- Runtime can be dominated by the wait for memory fetches because usually graph algorithms are based on exploring the structure of a graph in preference to performing large numbers of computations on the data.

For the sake of effective handling large graphs require particular storage and processing tools: graph databases such as Pregel [11] or hypergraph oriented HyperGraphDB (http://www.hypergraphdb.org) [12]. They permit direct operation with a graph without any intermediate relational data representations. The tools support replication and distributed transactions hence make work with a graph size independent.

A special case of graphs is semantic network, which is considered a widespread method for knowledge representation. Due to this fact creation and processing of knowledge bases and ontologies constructed upon them may be seen as a another resource consuming task [13,14]. Such networks may not be as big as graphs related to Web graph problems in terms of numbers of elements, but they often have complex hierarchical relations between vertices and compound nodes and edges structure. This complicates the methods of their processing as the graph structure and graph data are interconnected. Knowledge extraction and ontology-based reasoning can be used as examples of such complex tasks.

Growing interest in ontologies development and processing generates demand for tools which are capable of handling complex operational problems on their own. Such tools have been created and already mentioned HyperGraphDB is one of them. HyperGraphDB implements OWL 2.0 standard of ontology representation with operating multiple ontologies in one database as subgraphs. Usage of subgraphs as the base allows representations of ontologies to use all benefits of distributive graph database. HyperGraphDB has an integration with Protege Editor, the most popular ontology editor, and permits using popular reasoners such as Hermit, Fact++ and Pellet. Thereby the database hides all the internal work and allows users to work with familiar tools.

3.2 Fluid Dynamics Simulations

Another way of applying virtual supercomputer is fluid dynamics simulations. This class of applications demands highly scalable architecture. In particular, experiments in a virtual testbed can be carried out on a single multiprocessor machine [15] only in the most simple cases involving small simulation region and time interval. However, large-scale simulations with multiple atmospheric and ship motion models involved require use of multiple machines comprising distributed computing system. Moreover, hierarchy of mathematical models and high number of dimensions of these models demand a way of organizing computations into a single distributed workflow [16], for example, WRF, Wavewatch3 and wind wave model. So, a capability of a virtual supercomputer to dynamically compose distributed pipelines can accelerate execution of experiments in a virtual testbed.

4   Principles of Virtual Supercomputer

Although virtual supercomputer can be implemented in many ways and using different combinations of technologies, there are some principles that such implementation is considered to obey. On one hand these principles arise from similarity of different technologies and their implementations, on the other hand the purpose of some principles is to solve problems inherent to existing general-purpose distributed systems. In any case, the principles are useful for solving large-scale problems on virtual supercomputer and some of them can be neglected for problems of small sizes. So, the principles follow.

- Virtual supercomputer is completely determined by its application programming interface (API) and this API should be platform-independent. The use of API as the only interface in distributed processing systems is common, but its dependency on operating system or programming language leads to problems in the long run. For example, the first API for portable batch systems (PBS) was implemented in low-level C language and only for UNIX-like platforms which led to inability or inefficiency of its usage in other programming languages and in exposing it as a web service. Moreover, the API does not cover all the functions of underlying PBS [17]. So, using platform-independent API is one of the ways to avoid such integration and connectivity problems. In other words, API is a programming language of a virtual supercomputer and the only way of interacting with it.
- Virtual supercomputer API provides functions to connect with other virtual supercomputers and such interaction is seamless. Interaction of different distributed systems is the way of solving large-scale problems [18] and seamless interaction helps compose hybrid distributed systems dynamically: to extend capacity when needed [21]. So it is the way of scaling virtual supercomputer to solve problems that are too complex for one virtual supercomputer.
- Virtual supercomputer processes data stored in a single distributed database and this processing is done using virtual shared memory. Efficient data processing is achieved by distributing data among available nodes and by running small programs (queries) on each host where corresponding data resides; this approach helps not only run query concurrently on each host but also minimizes data transfers [11,19]. However, in existing implementations these programs are not general-purpose: they are parts of algorithm and they are specific to data model this algorithm was developed for. For example, in MapReduce framework programs represent map and reduce functions that are run on each row of table (or line of file) and it is difficult to compose general-purpose program to process any data within this framework [19]. On the other hand, virtual shared memory interface allows processing of data located on any host [20] and does it in efficient way. So, distributed database is a way of storing large data sets and virtual shared memory is a way of writing general-purpose program to process it.
- Experiments show that using paravirtualization instead of full virtualization is advantageous in terms of performance [22] and virtual computing nodes

should be created using paravirtualization technologies only. However, not every operating system can be paravirtualized and it should be possible to access virtual supercomputer facilities through fully-virtualized hosts. So, paravirtualization is inevitable in achieving balance between good performance and ease of system administration in distributed environment and as a consequence operating system should be UNIX-like for paravirtualization to work.

- Load balance is achieved using virtual processors with controlled clock rate and process migration. The first technique allows balancing coarse-granularity tasks and the second is suitable for fine-grained parallelism.
- Virtual supercomputer uses complex gridlike security mechanisms. One of the cloud problems is security issues [23] but we feel that proper combination of GRID security tools with cloud computing technologies is possible.

To summarize, virtual supercomputer is an API offering functions to run programs, to work with data stored in a distributed database and to work with virtual shared memory and this API is the only programming language of a virtual supercomputer.

5   Experimental Evaluation

5.1   Experimental Setup and Evaluation of Virtualization Impact

Implementation of a virtual supercomputer will not be possible without use of server virtualization technologies: virtual machine migration provides load-balancing and fault-tolerance capabilities and it is necessary to evaluate their performance relative to physical machines.

We conducted most of our experiments at Resource Centre Computer Centre of SPbSU [25]. This centre offers a specific approach to manage resources. Each user is given a virtual machine with necessary characteristics. Such a machine can be flexibly customized since user is granted administrative rights. When resources of a single virtual machine become insufficient to meet all user requirements, they can be easily extended, or even additional VMs can be created in order to form a virtual cluster. This is how dynamic allocation of computational resources is carried out. Further on, access to HPC resources is provided via this personal virtual machine.

In the same way virtual clusters can be created automatically when more resources are needed and processes are dynamically migrated to newly constructed virtual machines. Automated creation of virtual machines can be achieved by rewriting PBS prologue script where desired properties of virtual machines can be described in the same way as they are described when submitting PBS job to cluster of physical machines, however in case of virtual machines operating system can also be specified. So, there is no difference from a user point of view whether a job is submitted to real or virtual cluster.

Alternatively, user can run jobs on dedicated HPC clusters. In case of our resource centre they are T-Platforms cluster and HP cluster. User home directory

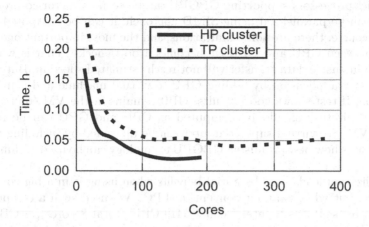

Fig. 2. Performance of clusters with different interconnect bandwidth based on GRO-MACS workload

is mounted via NFS on clusters. It provides universal access to computational data: raw data and results are stored in a single place.

We chose GROMACS as an example of real application running on clusters as it is commonly used by the users of St.Petersburg University Computer Center. GROMACS is used for efficient molecular simulations [26]. Figure 2 illustrates the GROMACS runs (2 different tasks) on T-Platforms (maximum 376 CPU cores were used) and HP (maximum 192 cores were used) clusters. Picture shows that these tasks have different scalability on different clusters. Without going into details, we can say that the root causes of this behavior is network bandwidth (HP has twice as much better network), memory size (swapping to disk substantially increase run time; HP cluster has 96 GB RAM per node while T-Platforms has only 16 GB) and intensive communication between worker processes.

But what can user do when network communication prevents scalability? The right way is using multicore SMP machine with large amount of memory. Computer centre has 3 machines of this type. In usual case in order to harness such a machine user has to migrate his applications, environment and data. In our case virtual machine is migrated to SMP node. It can be done with ease and it solves many problems: user does not need to do any actions, even get accustomed to new environment because his tuned virtual machine is completely migrated to powerful physical machine, and all applications, libraries and user settings remain unchanged. So, virtual machine migration is another way of extending dynamic computational resource pool.

Moreover, the resource pool can contain different accelerators. Contemporary hypervisors have means to harness available GPUs in a system. One can dedicate up-to-date GPUs supporting GPGPU to virtual machines, that is, to provide users with GPGPU computations. Physical GPUs can be shared between several virtual machines. Several GPUs can be utilized by hypervisor.

Graphics processors supporting GPGPU can be seen as vector co-processors. If a task can be parallelized using SIMD approach, it can be computed fast on GPU. Of course, there are some limitations, and the most important one is data transfer between GPU and CPU (it becomes a bottleneck). That is why some tasks with intensive data transfer will not reach estimated speedup. But one can test necessary application by adding GPU to virtual machine and running the application. If test shows good results, GPU remains in the VM configuration. Otherwise this task should be computed on CPU and GPU can be removed from the VM. So, virtual supercomputer can be upgraded by including modern GPUs. That is how new promising GPGPU technology can be used within virtual systems.

Virtualization leads to substantial benefits when using it in a big computing center [28], but what is about conventional PC? We used such a computer for additional tests. It has 2 Intel Xeon E5410 CPU (total 8 cores), 8 GB RAM, 250 GB HDD. Such systems become ubiquitous today. Xen technology was used for virtualization. We created paravirtualized guests. Both the host and the guest systems run Debian 7.0. We were interested in testing such a PC with practical workloads. We tested a GROMACS workload that puts heavy load on CPUs. We tried to run this job on the host system without virtualization and on the guest paravirtualized OS. After series of tests we can say that in our case (paravirtualized guest using Xen) virtualization led to 5% time overheads only (Figure 3), so virtualization offers benefits even on conventional PC.

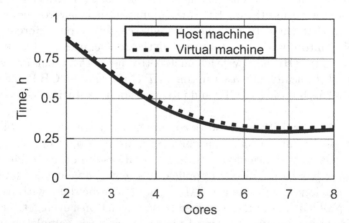

Fig. 3. Performance comparison for host and virtual machine based on GROMACS workload

5.2   Cloud Management Tools for Virtual Clusters

For creation and management of virtual clusters we tested the FishDirector software system developed by Sardina Systems [24]. FishDirector software supports heterogeneous platforms and is hypervisor agnostic. It is architected to scale from a few physical servers to well beyond 100,000, on-premise and in-Cloud.

It is based on OpenStack and allows to create, monitor and manage virtual clusters. Software lets users to create and manage clusters by using web interface or by command line scripts.

FishDirector's decision engines provide power-down/power-up automation which can massively help reduce the carbon emissions of physical servers across the data centre and lowers total power draw. When running HPC jobs on the cluster there is a possibility of conflicts between big and small jobs within a computing estate. To avoid the small job and big job biases FishDirector's Operations Manager predetermines the policies upon how big and small HPC jobs are to be run and adds those constraints to the system, jobs are squeezed into the resource pool, freeing up the remaining resources for other jobs to run and resources are returned to the overall pool once any job is completed. In addition, HPC users with jobs that are set to run for lengthy periods of time will benefit from not having to create time consuming job snapshots for checkpoint restarts.

FishDirector monitors the stability and performance of HPC nodes in the system. Should one of the nodes become unstable, FishDirector can transfer the VM to a different node with no loss of data and without having to roll-back to a checkpoint restart. This reduces the overall time taken to process the HPC job without having to take regular time-consuming job snapshots with the added confidence of full data integrity. Overall structure of FishDirector architecture is shown in Figure 4.

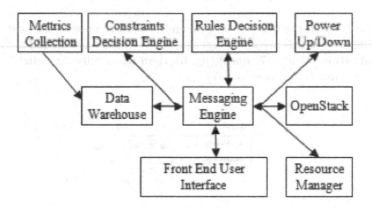

Fig. 4. FishDirector reference architecture

As it was mentioned above, not all hardware and operating systems support paravirtualization, and in that case full virtualization should be used. But not every machine can be fully virtualized considering underlying hardware capabilities. To show the importance of efficient full virtualization we conducted tests on two different machines: the first one supported full virtualization and the second one did not. For this purpose we created virtual guests using FishDirector package on each host and then ran the same test script on physical and virtual machines. The script found an inverse of a matrix with given sizes (100x100,

500x500 and 1000x1000) and then checked whether these matrices are really inverse of each other or not. Besides showing the significance of full virtualization, this test could show efficiency of FishDirector and mathematical correctness of virtual guests (created by FishDirector) work. We ran this script twenty times for each size of matrix on every machine using only RAM and twenty times with writing data on HDD and collected average values of the test time (fig. 5).

Memory		Size Machine	100x100	500x500	1000x1000
Without full virtualization	HDD	VM	0.18	4.8	20.16
		Physical	0.06	1.02	3.7
	RAM	VM	0.1	2.2	10.5
		Physical	0.025	0.32	1.5
Supports full virtualization	HDD	VM	0.11	0.92	5.11
		Physical	0.083	0.7	3.93
	RAM	VM	0.09	0.35	1.25
		Physical	0.067	0.3	1.05

Fig. 5. Average values of test results

As it can be seen the difference between the performance of the host and the guest (using only RAM) are much bigger when the used machine did not support full virtualization (around 7 times) (fig. 6), than when fully-virtualized machine was used (around 1.15 times) (fig. 7).

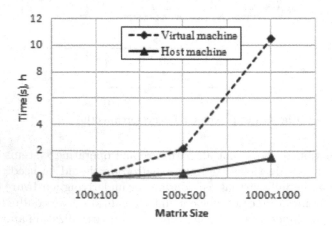

Fig. 6. Performance of host and virtual machines using only RAM, when full virtualization is not supported

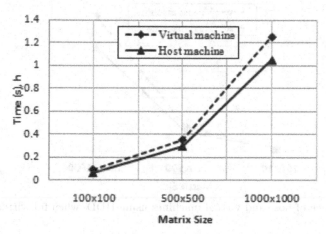

Fig. 7. Performance of host and virtual machines using only RAM, when full virtualization is supported

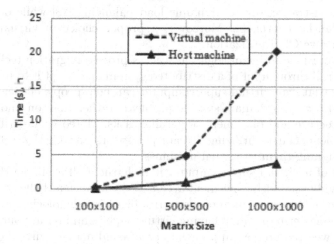

Fig. 8. Performance of host and virtual machines using HDD, when full virtualization is not supported

The situation was almost the same when HDD was used: the performance of virtual machine, which worked on a host with support of full virtualization, was lower of physical host's performance for about 7 times (fig. 8), and in case of full virtualization the difference was about 1.3 times (fig. 9).

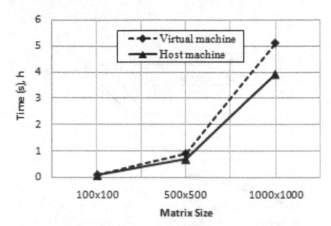

Fig. 9. Performance of host and virtual machines using HDD, when full virtualization is supported

6   Conclusions and Future Work

It is known that virtualization improves security, resilience to failures, substantially eases administration due to dynamic load balancing [28] while does not introduce substantial overheads. Moreover, a proper choice of virtualization package can improve CPU utilization.

Usage of standard cloud technologies as well as process migration techniques can improve overall throughput of a distributed system and adapt it to problems being solved. In that way virtual supercomputer can help people efficiently run applications and focus on domain-specific problems rather than on underlying computer architecture and placement of parallel tasks. Moreover, described approach can be beneficial in utilizing stream processors and GPU accelerators dynamically assigning them to virtual machines.

The key idea of a virtual supercomputer is to harness all available HPC resources and provide user with convenient access to them. Such a challenge can be effectively solved only using contemporary virtualization technologies. They can materialize the long-term dream of having virtual supercomputer at your desk.

In this paper we presented an approach to create and manage virtual clusters based on virtualization and cloud technologies. We presented generic experimental evaluation of the virtualized hardware used as building blocks for the virtual cluster. In the future we plan to perform specific experiments with popular software packages to estimate performance and usability of our solution in real-life problems provided by end-users.

Acknowledgment. The research was carried out using computational resources of Resource Centre Computer Centre of Saint Petersburg State University (T-EDGE96 HPC-0011828-001) and supported by Russian Foundation for Basic Research (project N 13-07-00747) and Saint Petersburg State University (projects N 9.38.674.2013, 0.37.155.2014).

References

1. Smarr, L., Catlett, C.E.: Metacomputing. Communications of the ACM 35(6), 44–52 (1992)
2. Korkhov, V.V., Moscicki, J.T., Krzhizhanovskaya, V.V.: The user-level scheduling of divisible load parallel applications with resource selection and adaptive workload balancing on the grid. IEEE Systems Journal 3(1), 121–130 (2009)
3. Figueiredo, R.J., Dinda, P.A., Fortes, J.A.B.: A case for grid computing on virtual machines. In: Proceedings of the 23rd International Conference on Distributed Computing Systems (2003)
4. Matsunaga, A.M., Tsugawa, M.O., Adabala, S., Figueiredo, R.J., Lam, H., Fortes, J.A.B.: Science gate- ways made easy: the In-VIGO approach. Concurrency and Computation: Practice and Experience 19(6), 905–919 (2007)
5. Krsul, I., Ganguly, A., Zhang, J., Fortes, J.A.B., Figueiredo, R.J.: VMPlants: Providing and manag- ing virtual machine execution environments for grid computing. In: Proceedings of the 2004 ACM/IEEE Conference on Supercomputing (2004)
6. Nishimura, H., Maruyama, N., Matsuoka, S.: Virtual clusters on the fly - fast, scalable, and flexible installation. In: CCGRID 2007: Seventh IEEE International Symposium on Cluster Computing and the Grid (May 2007)
7. Emeneker, W., Stanzione, D.: Dynamic virtual clustering. In: IEEE Cluster 2007, Austin, TX (September 2007)
8. Chase, J.S., Irwin, D.E., Grit, L.E., Moore, J.D., Sprenkle, S.E.: Dynamic virtual clusters in a grid site manager. In: HPDC 2003: Proceedings of the 12th IEEE International Symposium on High Performance Distributed Computing, p. 90. IEEE Computer Society, Washington, DC (2003)
9. Shuming, S., Zhang, H., Yuan, X., Wen, J.-R.: Corpus-based semantic class mining: distributional vs. pattern-based approaches. In: Proceedings of the 23rd International Conference on Computational Linguistics, pp. 993–1001 (2010)
10. Andrew, L., Gregor, D., Hendrickson, B., Berry, J.: Challenges in parallel graph processing, Parallel Processing Letters, vol. Parallel Processing Letters 17(01), 5–20 (2007)
11. Grzegorz, M., Austern, M.H., Bik, A.J., Dehnert, J.C., Horn, I., Leiser, N., Czajkowski, G.: Pregel: a system for large-scale graph processing. In: Proceedings of the 2010 ACM SIGMOD International Conference on Management of Data, pp. 135–146 (2010)
12. Iordanov, B.: HyperGraphDB: A generalized graph database. In: Shen, H.T., Pei, J., Özsu, M.T., Zou, L., Lu, J., Ling, T.-W., Yu, G., Zhuang, Y., Shao, J. (eds.) WAIM 2010. LNCS, vol. 6185, pp. 25–36. Springer, Heidelberg (2010)
13. Kevin, K., Luk, S.K.: Building a large-scale knowledge base for machine translation. In: Proceedings of the National Conference on Artificial Intelligence, p. 773 (1994)
14. Ravi, K., Raghavan, P., Rajagopalan, S., Tomkins, A.: Extracting large-scale knowledge bases from the web. In: Proceeding of the International Conference on Very Large Data Bases, pp. 639–650 (1990)
15. Degtyarev, A., Gankevich, I.: Efficiency comparison of wave surface generation using OpenCL, OpenMP and MPI. In: Proceedings of 8th International Conference Computer Science & Information Technologies, Yerevan, Armenia, pp. 248–251 (2011)
16. Wibisono, A., Vasyunin, D., Korkhov, V.V., Zhao, Z., Belloum, A., de Laat, C., Adriaans, P.W., Hertzberger, B.: WS-VLAM: A GT4 based workflow management system. In: Shi, Y., van Albada, G.D., Dongarra, J., Sloot, P.M.A. (eds.) ICCS 2007, Part III. LNCS, vol. 4489, pp. 191–198. Springer, Heidelberg (2007)

17. Peter, T., et al.: Standardization of an API for distributed resource management systems. In: Seventh IEEE International Symposium on Cluster Computing and the Grid, CCGRID 2007. IEEE (2007)
18. Douglas, T., Tannenbaum, T., Livny, M.: Distributed computing in practice: The Condor experience. Concurrency and Computation: Practice and Experience 17(2-4), 323–356 (2005)
19. Jeffrey, D., Ghemawat, S.: MapReduce: simplified data processing on large clusters. Communications of the ACM 51(1), 107–113 (2008)
20. Ping, A., et al.: STAPL: An adaptive, generic parallel C++ library. In: Dietz, H.G. (ed.) LCPC 2001. LNCS, vol. 2624, pp. 193–208. Springer, Heidelberg (2003)
21. Bogdanov, A., Dmitriev, M.: Creation of hybrid clouds. In: Proceedings of 8th International Conference Computer Science & Information Technologies, Yerevan, Armenia, pp. 235–237 (2011)
22. Paul, B., et al.: Xen and the art of virtualization. ACM SIGOPS Operating Systems Review 37(5), 164–177 (2003)
23. Hamlen, K., Kantarcioglu, M., Khan, L., Thuraisingham, B.: Security Issues for Cloud Computing
24. Sardina Systems, http://www.sardinasystems.com
25. Resource Center Computer Center of St.Petersburg State University, http://cc.spbu.ru
26. Berendsen, H.J.C., van der Spoel, D., van Drunen, R.: GROMACS: A message-passing parallel molecular dynamics implementation. Computer Physics Communications 91(1-3), 43–56 (1995) ISSN 0010-4655
27. Rodríguez, M., Tapiador, D., Fontán, J., Huedo, E., Montero, R.S., Llorente, I.M.: Dynamic Provisioning of Virtual Clusters for Grid Computing. In: César, E., Alexander, M., Streit, A., Träff, J.L., Cérin, C., Knüpfer, A., Kranzlmüller, D., Jha, S. (eds.) Euro-Par 2008. LNCS, vol. 5415, pp. 23–32. Springer, Heidelberg (2009)
28. Bogdanov, A.V., Degtyarev, A.B., Gankevich, I.G., Gayduchok, V.Y., Zolotarev, V.I.: Virtual workspace as a basis of supercomputer center. In: Proceedings of the 5th International Conference on Distributed Computing and Grid-Technologies in Science and Education, Dubna, Russia, pp. 60–66 (2012)

Multi-Objective Evolutionary Algorithm for Oil Spill Detection from COSMO-SkeyMed Satellite

Maged Marghany

Institute of Geospatial Science and Technology (INSTeG)
Universiti Teknologi Malaysia
81310 UTM, Skudai, Johor Bahru, Malaysia
maged@utm.my, magedupm@hotmail.com

Abstract. This study has demonstrated a design tool for oil spill detection in COSMO-SkyMed satellite data using Multi-Objective Evolutionary Algorithm which based on Pareto optimal solutions. The COSMO-SkyMed along the Gulf of Thailand is involved in this study. The study also shows that Multi-Objective Evolutionary Algorithm provides an accurate pattern of oil slick in COSMO-SkyMed data. This shown by 96% for oil spill, 1% look–alike and 3% for sea roughness using the receiver –operational characteristics (ROC) curve. The MOGA also shows excellent performance in COSMO-SkyMed data. In conclusion, Multi-Objective Evolutionary Algorithm can be used as an automatic detection tool for oil spill in COSMO-SkyMed satellite data.

Keywords: Multi-Objective Evolutionary Algorithm, COSMO-SkyMed, oil spill, Pareto optimal solutions, Automatic detection.

1   Introduction

Oil spill pollution has a substantial role in damaging the marine ecosystem. Oil spill that floats on top of water, as well as decreasing the fauna populations, affects the food chain in the ecosystem [1-5]. In fact, oil spill is reducing the sunlight penetrates the water, limiting the photosynthesis of marine plants and phytoplankton. Moreover, marine mammals for instance, disclosed to oil spills their insulating capacities are reducing, and so making them more vulnerable to temperature variations and much less buoyant in the seawater. Therefore, oil coats the fur of sea otters and seals, reducing its insulation abilities and leading to body temperature fluctuations and hypothermia. Ingestion of the oil causes dehydration and impaired digestions [5].

Over recent years, there has been an explosive increments in marine oil spill pollutions. Deepwater Horizon oil spill in 2010, for instance, is the most serious marine pollution disaster has occurred in the history of the petroleum industry (Fig. 1). This disaster has dominated by three months of oil flows in coastal waters of the Gulf of Mexico. Incidentally, the Deepwater Hoizon oil spill has serious effects on feeble maritime, wildlife habitats, Gulf's fishing, coastal ecologies and tourism industries [30][31]. Therefore, the resulting oil slicks are difficult to control, as their

B. Murgante et al. (Eds.): ICCSA 2014, Part VI, LNCS 8584, pp. 355–371, 2014.

evolution be influenced by weather, currents, tides, and many chemical and physical factors [1,10,31]. Further, oil sources are challenging to verify and be subject to the type of oil, its volume , location and duration of the seepage, and surrounding environmental conditions [1].

Fig. 1. Oil spill disaster in Gulf of Mexico

There are several SAR sensors which are involved in the oil spill detection and monitoring. These data are from ERS-1/2, [1-5], ENVISAT [30], ALOS, [32] [33], RADARSAT-1/2,[18,19] and TerraSAR-X [25] which have been globally used to identify and monitor the oil-spill. Further, Airborne SAR sensors like Uninhabited Aerial Vehicle Synthetic Aperture Radar (UAVSAR, by JPL, L-band) with a 22-km-wide ground swath at 22° to 65° incidence angles[33]and E-SAR, (by DLR, multi-land) have also proven their excellent potential for monitoring coastal zone oil pollutions. Recently, the multi-polarimetric SAR high resolution data have become a very hot research area for oil spill detection [19] [34].

1.1   Principle of Oil Spill Detection in SAR Data

The main theory of oil spill imaging using SAR images is based on the concept of resonance fundamental theory. Very short ocean waves travel on the larger ocean waves, or swells. As a superposition, radar backscatter at incident angles of 20° to 70° is principally produced by Bragg resonance [3].

1.1.1   General Concept of Bragg Scattering

According to According to Topouzelis [22] and Trivero [24], backscatter can model in two ways: specular reflection and Bragg scattering. Specular reflection occurs when the water surface is tilted which creates a small mirror pointing to the radar. For the perfect specular reflector, radar returns (backscatter) exist only near vertical incidence. This is because of a 90°-depression angle of the slope of the surface. In addition, the reflected energy is localized to the small angular region around the angle of reflection. Even for non-vertical incidence, however, backscatter can exist for a rough subsurface. This can occur if the radar is penetrating deep enough [25]. Nevertheless, because of the highest dielectric of water ocean, the radar ' s signal cannot penetrate sea surface.

Bragg or resonant scattering is scattered from a regular surface pattern. Resonant backscattering occurs when phase differences between rays backscattered from subsurface pattern interfere constructively. The resonance condition is $2\lambda_w \sin\theta = \lambda_r$, where λ_w and λ_r are water wavelength and the radar wavelength, respectively. θ is the local angle of incidence (Fig. 2). According to Topouzelis et al., [20], the short Bragg-scale waves form in response to wind stress, if the sea surface is rippled by a light breeze and no long waves are present. The radar backscatter is due to the component of the wave spectrum. This resonates with a radar wavelength.

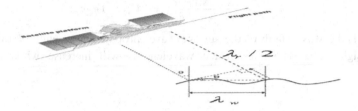

Fig. 2. Bragg Scatter concept

Swell waves being imaged have much longer wavelengths than the short gravity waves, which cause the Bragg resonance. Further, Bragg-resonance from the ocean might be considered as coming from facets. The term facet refers to a relatively flat portion of the long wave structure with cover of ripples containing Bragg-resonant facets. This behaves like specular points. The beam-width and scattering gain of each facet are determined by its length in the appropriate direction [4][10][22].

1.1.2 Mathematical Expression of Bragg Scattering

As the incidence angle of the SAR is oblique to the local mean angle of the ocean surface, there is almost no direct specular reflection except at very high sea states. It is therefore assumed that at first approximation Bragg resonance is the primary mechanism for backscattering radar pulses. The Bragg equation defines the ocean wavelengths λ_w for Bragg scattering as a function of radar wavelength λ_r and incidence angle θ:

$$\lambda_w = \frac{\lambda_r}{2 \sin \theta} \tag{1}$$

The short Bragg-scale waves are formed in response to wind stress. If the sea surface is rippled by a light breeze with no long waves present, the radar backscatter is due to the component of the wave spectrum which resonates with the radar wavelength. The threshold wind speed value for the C-band waves is estimated to be at about 3.25 m/s at 10 meters above the surface. The Bragg resonant wave has its crest nominally at right angles to the range direction [3]. Under the circumstance of Bragg scattering, largest incidence angle reduces the SAR backscatter. Indeed, the SAR backscattered power is proportional to the spectral energy density of the Bragg and the spectral distribution decays at shorter wavelength (Fig.3). SAR images tend to become darker with increasing range. Backscatter is related to the local incident angle (i.e. as the local incident angle increases, backscatter decreases), which is in turn related to the distance in the range direction. Backscatter is also related to wind speed [3].

According to Brekke and Solberg [3] ,For RADARSAT-1 and ENVISAT ASAR with C-band frequency, a radar wavelength of 5.7 cm and incidence angles in the range of 20^0-50^0 will this model give Bragg resonant sea wavelengths λ_w in the range of 8.3-3.7 cm. For surface waves with crests at an angle ϕ to the radar line-of-sight (Fig. 3) the Bragg scattering criterion is

$$\lambda'_w = \frac{\lambda_r . \sin \phi}{2 \sin \theta} = \lambda_w . \sin \phi \tag{2}$$

where λ'_w is the wavelength of the surface waves propagating at angle ϕ to the radar line- of sight. The resonant surface wavelengths will increase when frequency increases.

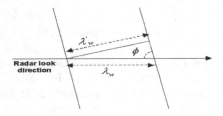

Fig. 3. Crests at an angle ϕ to the look direction of the SAR [3]

The SAR directly images the spatial distribution of the Bragg-scale waves. The spatial distribution may be affected by longer gravity waves, through tilt modulation, hydrodynamic modulation and velocity bunching. Moreover, variable wind speed, changes in stratification in the atmospheric boundary layer, and variable currents associated with upper ocean circulation features such as fronts, eddies, internal waves and the bottom topography effect the Bragg waves [3] [22].

1.1.3   Oil spill and Surface Films Impact on Bragg Scattering

Bragg scattering is a significant concept to understand the radar signal interaction with the ocean surface. In this regard, the presence of capillary wave will produce backscatter that assists the reader in imagining sea surface. Short gravity waves and capillary waves by the are damped by dynamic elasticity of the water surface, that is by changes in surface tension which occur when the surface is stretched or compressed [3]. This has the effect of the extracting of this energy from those waves which depends wholly or partly on surface tension to provide the restoring force necessary for wave propagation. If a surface film is present, the surface tension is lower than it would be in the absence of the film, and stretching and compression of the film because of the presence of waves provides the dynamic elasticity which enhances the wave damping. Thus capillary waves and short gravity waves are always dumped in the presence of surface films [4]. If the surface film is spatially patchy,varying in thickness, or lined up in slicks because of surface convergence , it is to expect that the capillary/gravity wave energy will reflect that patchiness, being greatest where is no surface film. Oil films on the sea surface damp the capillary waves of the surface height spectrum. This hydrodynamic damping influences the normalized radar cross section (NRCS) of contaminated seas, comparatively to clean seas [22].

Therefore, oil slicks dampen the Bragg waves (wavelength of a few cm) on the ocean surface and reduce the radar backscatter coefficient. This results in dark regions or spots in satellite SAR images. Topouzelis et al., [21], emphasizes the importance of weathering processes, as they influence the oil spills physicochemical properties and detect-ability in SAR images. The processes that play the most important role for oil spill detection are evaporation, emulsification and dispersion. Lighter components of the oil will evaporate to the atmosphere. The rate of evaporation is dependent on oil type, thickness of the spill, wind speed and sea temperature. Emulsification is estimated based on water uptake as a function of the wind exposure of the actual oil type. Dispersion is an important factor in deciding the lifetime of an oil spill and it is strongly dependent on the sea state [3].

1.2 Oil Spill Detection in SAR Data

Scientists which have agreed with that oil spill detection with SAR data is based on: (i) all dark patches present in the SAR images are isolated [3][16]; (ii) features for each dark patch are then extracted [1][12]; (iii) these dark patches are tested against predefined values [22]; and (v) the probability for every dark patch is calculated to determine whether it is an oil spill or a look-alike phenomenon [1][14][20][22][25]. Therefore, Topouzelis et al., [21] and Topouzelis [22] reported that most studies use the low resolution SAR data, such as quick-looks, with the nominal spatial resolution of 100 m x 100 m, to detect oil spills. In this regard, quick looks' data are sufficient for monitoring large scale area of 300 km x 300 km. On the contrary, they cannot efficiently detect small and fresh spills [22].

Further, SAR data have distinctive features as equated to optical satellite sensors which makes SAR extremely valuable for spill watching and detection [5] [7] [9] [13]. These features are involved with several parameters: operating frequency, band resolution, incidence angle and polarization [10] [12]. Marghany and Hashim [7] develop comparative automatic detection procedures for oil spill pixels in Multimode (Standard beam S2, Wide beam W1 and fine beam F1) RADARSAT-1 SAR satellite data post the supervised classification (Mahalanobis), and neural network (NN) for oil spill detection. They found that NN shows a higher performance in automatic detection of oil spill in RADARSAT-1 SAR data as compared to the Mahalanobis classification with a standard deviation of 0.12. In addition, they W1 beam mode is appropriate for oil spill and look-alikes discrimination and detection [10] [11] [12]. Recently, Skrunes et al., [9], nevertheless, reported that there are several disadvantages are associated with Current SAR based oil spill detection and monitoring. They stated that, SAR sensors are not able to detect thickness distribution, volume, the oil-water emulsion ratio and chemical properties of the SAR data. In this regard, they recommended to utilize multi-polarization acquisition data such as RADARSAT-2 and TerraSAR-X satellites. They concluded that the multi-polarization data show a prospective for prejudice between mineral oil slicks and biogenic slicks. Finally Marghany, [30] used Genetic algorithm for oil spill detection in ENVISAT ASAR data along Singapore Straits. Marghany [30] found that crossover process, and the fitness function generated accurate pattern of oil slick in SAR data. He used receiver –operational characteristics (ROC) curve to verify oil spill detection in SAR data. He stated that ROC verified 85% for oil spill, 5% look–alike and 10% for sea roughness.

1.3   Problem Statement

The main challenge with SAR satellite data is the difficulty to drawbacks, which makes it difficult to develop a fully automated detection of oil spill. Due to the inherent difficulty of discriminating between oil spills and look-likes, an automatic algorithm with a reliable confidence estimator of oil spill would be highly desirable. The needs for automatic algorithms rely on the number of images to be analyzed, but for monitoring large ocean areas, it is a cost-effective alternative to manual inspection. Automatic detection algorithms of oil spill are normally divided into three steps: (i) dark spot detection; (ii) dark spot feature extraction; and (iii) dark spot classification [3,10, 22,31].

One of the main problems in oil slick combat and management are forecasting the behavior (movement and spreading) of oil slicks. Commonly, the objective of predicting the behavior of oil slicks is to determine the time-evolving shape of the slick under different weather patterns in waters where currents exist [11-14]. Wind direction and speed are the most important climate parameters can impact the oil spill imagine in SAR data. Although great progress made in detecting and surveying oil slicks, a general model for oil slick movement and spreading has not yet been devised [1]. Models for oil slick behavior are important in environmental engineering and used as a decision support tool in environmental emergency responses. These models also use to help ships avoid oil slicks [1].

Modeling the progress and extent of oil in the ocean are not always tested against authentic spills, and models are not regularly developed with real databases, but rely instead on theoretical scenarios. In fact, some spill-threat scenarios have not been based on real oil movement data at all [15]. Yet there are frequent demands to provide just such models with credible precision. Consequently, it is important to study the behavior and movement of spilled oil in the sea in order to describe a suitable management plan for mitigating adverse impacts arising from such accidents. Simulation of oil spills using mathematical models form an important basis for subsequent study, according to Marghany [9]. Collectively, with the information about position of weak resources in time and space, the simulation outcome may develop a basis for evaluating the damage potential from an eventual oil spills. This may help the regulatory authorities to take direct preventive measures [9].

In addition, most of the studies that have been conducted in the coastal waters of Malaysia used a single radar image, which is inadequate to ensure a high degree of accurate detection of oil spills using SAR data. Some of the work involved the implementation of non-appropriate techniques for oil spill automatic detection. For instance, Mohamed et al., [16] have used data fusion techniques in a single RADARSAT-1 SAR image, with different co-occurrence texture algorithm 'results. However, the data fusion technique must apply with two or more different sensors, for example, ERS-1, LANDSAT, and SPOT. In addition, according to Brekke and Solberg [3], using such PCA analysis does not consider an appropriate method for data fusion. Data fusion technique involves several methods such as high pass filtering technique, IHS transformation method, Brovey method, and à Trous wavelet method [3].

1.4   Hypotheses and Objective

Concern to above prospective, we address the question of Multi-Objective Evolutionary Algorithm (MOEA) ability for oil spill detection in COSMO-SkyMed satellite. This work has hypothesized that the dark spot areas (oil slick or look-alike pixels) and its surrounding backscattered environmental signal complex looks in the COSMO-SkyMed data can detect using MOEA Algorithm. However, previous work has implemented post classification techniques [4][9][16][18] or artificial neural network [19],[21],[24] which are considered as semi-automatic techniques. The objective of this work can divide into two sub-objectives: (i) To examine MOEA [27][31] for oil spill automatic detection in COSMO-SkyMed satellite data; and (ii) To design the multi-objective optimization algorithm based on Pareto optimal solutions for oil spill automatic detection.

2   Study Area and Data Acquisition

The study area is located in Royang, Thailand. About 50 tons of crude oil spilled into the sea of Rayong province, Thailand, on Saturday morning, July 27, 2013 after a leak occurred in a pipeline (Fig.4a). The oil spill has moved away from the mainland and has started to disperse to an extent. However, what is worrying now is that it seems to have reached a group of islands dominated by Koh Kudee. The stag-horn and giant clam coral reef is dominated natural features of Koh Samet island (Fig.4b) with water depth less than 20 m depth.

(a)   (b)

Fig. 4. Oil spill covers beach of (a) Koh Samet Island and (b) Google map of Koh Samet Island

COSMO-SkyMed-2 imagery was acquired on August 1, 2013 at 06:09 a.m shows oils pill at Rayong. COSMO-SkyMed is a 4-spacecraft constellation for Earth observation, funded by ASI and the Italian Ministry of Defense. Each of the four satellites is equipped with SAR instruments [22]. All the spacecrafts of the SAR constellation are to be positioned in the same orbital plane with a defined phasing. The nominal repeat cycle is 16 days. However, each single satellite is believed to have a near revisit time of 5 days. The nominal full constellation should have a revisit time of a few hours on a global scale. [35]. The first constellation satellite is designed to observe in X-band (9.6 GHz with a wavelength of 3.1 cm). Multi-band scenarios (X-, C- L- and P bands) are planned. The X-band instrument has a swath between 10 and 200 km (depending on support mode); a total FOR (Field of Regard) of 1300 km in the cross-track direction.

3   Multi-Objective Evolutionary Algorithm (MOEA)

3.1   Data Organization

Following Marghany [30], entire backscatter of dark patches in COSMO-SkyMed data are $[\beta_1, \beta_2, \beta_3, ..., \beta_K]$ where K is the total number backscatter of dark patches in the SAR data. Therefore, K is made up from genes which is representing the backscatter β of dark patches and its surrounding environment and genetic algorithms is started with the population initializing step.

3.2   Problem Formulation

Coello et al., [36] stated that the Multi-Objective Evolutionary Algorithm (MOEA) deliver a set of compromised solutions called Pareto optimal solution since no single solution can optimize each of the dark objectives separately in SAR data. The Pareto optimal solution can deliver the exact solution based on the input of different pixels backscatter in SAR data. The decision maker is provided with the set of Pareto optimal solutions in order to choose Following Marghany [30] a constrained multi-objective problem for oil spill discrimination in SAR data deals with more than one objective and constraint namely look-alikes, for instance, currents, eddies, upwelling or downwelling zones, fronts and rain cells). The general form of the problem is adapted from Sivanandam and Deepa [29] and described as

$$\text{Minimize } f(\beta) = [f_1(\beta), f_2(\beta), ..., f_k(\beta)]^T \tag{4}$$

Subject to the constraints:

$$g_i(\beta) \leq 0, \quad i = 1, 2, 3, ..I \tag{5}$$

$$h_j(\beta) \leq 0, \quad j = 1, 2, 3, ...J \tag{6}$$

$$\beta_s \leq \beta \leq \beta_U \tag{7}$$

where, $f_i(\beta)$ is the *i-th* pixel backscatters β in SAR data, $g_i(\beta)$ and $h_j(\beta)$ represents the *i-th* and *j-th* constraints of backscatter in raw direction and column direction, respectively. β_L and β_U are the lower and upper limit of values of the backscatter. The transition rules for the cellular automata oil spill detection is designed using the input of different backscatter values β to identify the slick conditions required in the neighbourhood pixels of kernel window size of 7x7 pixels and lines for a β pixel to become oil slick. These rules can be summarized as follows:

1. IF test pixel is sea surface, OR current boundary features THEN $\beta \geq 0$ not oil spill.

2. IF test pixel is dark patches (low wind zone, OR biogenic slicks OR shear zones) $\beta \leq 0$ THEN It becomes oil slick if its.

3.3   Pareto Optimal Solution

In this research, two objectives are considered. One is oil spill backscatter and the other is sea surface, ship, look-alikes, and land backscatters. The definitions of oil spill pixels and non- oil spill pixels are given as follows:

1. Oil Spill pixels (P_{max}): the variation of maximum pixels which contain oil spill pixels i.e. $P_{max} - \max\{P_1, P_2, ..., P_k\}$. Where P_j denotes the oil spill pixel j, $\forall j = 1,2,...,k$.

2. total SAR pixels ($\sum P_i$): the sum of pixels surrounding oil spill pixels of each row and column in SAR data.

   Pareto optimal solutions are applied to retain the discrimination of oil spill backscatter diversity and surrounding backscatter environment.

3.3.1   Definition: Pareto Optimal Solutions

Let $\beta_0, \beta_1, \beta_2 \in \beta_{SAR}$, and β_{SAR} is a feasible backscatter in whole SAR image. And β_0 is called the Pareto optimal solution in the minimization problem for identification of oil spill pixels. if the following conditions are satisfied.

1. If $f(\beta_1)$ is said to be partially greater than $f(\beta_2)$, i.e. $f_i(\beta_1) \geq f_i(\beta_2), \forall i = 1,2,...,n$ and $f_i(\beta_1) > f_i(\beta_2), \exists i = 1,2,...,n$, Then β_1 is said to be dominated by β_2 .

2. If there is no $\beta \in \beta_{SAR}$ s.t. β dominates β_0, then β_0 is the Pareto optimal solutions for identifying oil spill pixels P_{max} .

To optimize the oil spill automatic detection from SAR data using MOEA, the oil spill backscatter must be coded into a Genetic Algorithm syntax form. In this context, the oil spill backsacatter is coded into the chromosome form. In this problem, the chromosome consists of a number of genes where every gene corresponds to a coefficient in the nth-order surface fitting polynomial as given by.

$$f(i,j) = \beta_0 + \beta_1 i + \beta_2 j + \beta_3 i^2 + \beta_4 ij + \beta_5 j^2 + \ldots + \beta_m j^n \tag{8}$$

where $\beta [0,1.....m]$ are the backscatter parameter coefficients that will be estimated by the genetic algorithm to approximate the minimum error for oil spill backscatter discrimination from surrounding environment. i and j are indices of the pixel location in the image respectively, m is the number of coefficients (Fig.5).

Fig. 5. Coding scheme of the coefficients of the nth-order surface fitting polynomial into the chromosome syntax form

Then the weighted sum to combine multiple objectives into single objective is given by

$$f(\beta) = w_1 f_1(\beta) + w_2 f_2(\beta) + \cdots + w_n f_n(\beta) \tag{9}$$

where $f_1(\beta), f_2(\beta), \ldots, f_n(\beta)$ are the objective functions and $w_1, w_2, \ldots, w_n$ are the weights of corresponding objectives that satisfy the following conditions.

$$w_i \geq 0 \quad \forall i = 1,2,\ldots,n$$
$$w_1 + w_2 + \ldots\ldots + w_n = 1 \tag{10}$$

Once the weights are determined, the searching direction is fixed. To search Pareto optimal solutions as much as possible, the searching directions should be changed again and again to sweep over the whole solution space. Therefore the weights have to be changed again and again. The weights consist of random numbers and they are generated as the following way:

$$w_i = \frac{r_i}{r_1 + r_2 + \cdots + r_n}, \quad \forall i = 1,2,\ldots, n \tag{11}$$

where $r_1, r_2, \ldots, r_n$ are random numbers within (0, 1). Solutions searched through changing directions are collected in a set. Then the definition of Pareto optimal solution is applied to determine which solutions in the set are Pareto optimal. The step repeats in every generation in MOGA.

3.4   The Fitness Function

Following Kahlouche et al., [27] and Marghany [30] a fitness function is selected to determine the similarity of each individual backscatter of dark patches in COSMO-SkyMed data. Then the backscatter of dark patches in COSMO-SkyMed data be symbolized by β_i where $i=1,2,3, \ldots, K$ and the initial population P_i^j where $j=1,2,3,$ $\ldots, N$ and $i=1,2,3\ldots, K$. Formally, the fitness value $f(P^j)$ of each individual of the population is computed as follows [27]:

$$f(P^j) = [\sum_{i=1}^{K} |P_i^j - \beta_i|]^{-1} \qquad j=1, \ldots, N. \qquad (12)$$

where, N and K are the number of individuals of the population used in fitness process. Generally, Equation 11 used to determine the level of similarities of dark patches that are belong to oil spill in COSMO-SkyMed data.

3.5   Selection Step

The key parameter in the selection step of genetic algorithm which is chosen the fittest individuals $f(P^j)$ from the population P_i^j. The threshold value τ is determined by the maximum values of fitness of the population $Max\, f(P^j)$ and the minimum values of fitness of the population of $Min\, f(P^j)$. Indeed, in the next generations, this step serves the populations P. Therefore, the values of the fittest individuals dark patches in COSMO-SkyMed data data are greater identifies threshold τ which is given by

$$\tau = 0.5\,[Max\, f(P^j) + Min\, f(P^j)] \qquad (13)$$

Equation 13 used as selection step to determine the maximum and minimum values of fitness of the population, respectively. This is considered as a dark patches' population generation step in GA algorithms[30].

3.6   The Reproduction Step

On the word of Sivanandam and Deepa [29], Genetic algorithm is mainly a function of the reproducing step which involves the crossover and mutation processes on the backscatter population P_i^j in SAR data. In this regard, the crossover operator constructs the P_i^j to converge around solutions with high fitness. Thus, the closer the crossover probability is to 1 and the faster is the convergence [27][40]. In crossover step the chromosomes interchange genes. Following Marghany [30], a local fitness value effects each gene as

$$f(P_i^j) = |\beta_i - P_i^j| \qquad (14)$$

Then the crossfire between two individuals consists to keep all individual populations of the first parent, which have a local fitness greater than the average local fitness $f(P_{av}^j)$ and substitutes the remained genes by the corresponding ones from the second parent. Hence, the average local fitness is defined by:

$$f(P_{av}^j) = \frac{1}{K} \sum_{i=1}^{K} f(P_i^j) \qquad (15)$$

Therefore, the mutation operator denotes the phenomena of extraordinary chance in the evolution process. Truly, some useful genetic information regarding the selected population could be lost during reproducing step. As a result, mutation operator introduces a new genetic information to the gene pool [27][30].

4   Results and Discussion

In this study, COSMO-SkyMed image is acquired on July 29, 2013 at 11:23:33 UTC which is implemented for oil spill detection in the Koh Samet island, Thailand. This data covered 12° 31′48″ N to 12° 37′ 48″ N latitude and 101° 2′24″ E to 101 33 37″ E longitude (Fig.6). The Satellite has a Synthetic Aperture Radar (SAR) with multiple polarization modes, including a fully polarimetric mode in which HH, HV, VV and VH polarized data are acquired. Its highest resolution is 1 m in Spotlight mode 2 with the maximum coverage is ≤20 x 20 km², geometric resolution is 1.0 m², pixel spacing is 0.5 m x 0.5 m, and the incident angle is between 20° to 59°. Fig. 6 shows the COSMO-SkyMed data where the oil spill is heading by 16.5° towards inland within 6.59 km length of the island to inland (Fig.6).

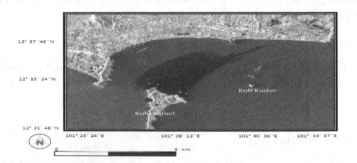

Fig. 6. COSMO-SkyMed data along Koh Samet island, Thailand

COSMO-SkyMed data shows the evolution of the spill, which has a darker tone than the surrounding water, as well as some boats are existed with bright dots (Fig. 8). Clearly, there are various of dark patches which are scattered over a large area of coastal waters. The lowest backscatter of -20 dB is noticed close to island. However, the surrounding environment has highest backscatter of -9 dB.   According to Marghany [31], oil spills change the roughness of the ocean surface to smoothness surface which appear as dark pixels as compared to the surrounding ocean [1-22].

Therefore, the speckle caused difficulties in dark patch identifications in SAR data [14][16]. Further, wind speed was recorded during July 29 2013 was ranged between 1 to 7 m/s. In addition, the measured reductions of backscattered radar power at X-band could be impacted by instrumental limitations, i.e. by the fact that the backscattered radar power reaches the noise floor [35].

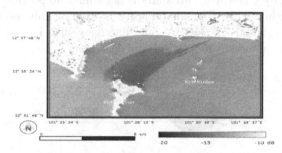

Fig. 7. Backscatter variations in COSMO-SkyMed

However, the result of backscatter values is different as compared to previous studies of Marghany et al., [10] [11] and Marghany and Hashim [12]. This is because of previous studies used different radar sensor of the RADARSAT-1 SAR and these studies have done under different weather and ocean conditions compared to recent work. Figure 8 shows the result of Pareto optimal solutions for oil spill discrimination in COSMO-SkyMed data. It is clear that the Pareto optimal provides excellent decision of oil spill pixels in COSMO-SkyMed data.

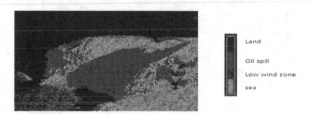

Fig. 8. Pareto optimal solution for oil spill discrimination in COSMO-SkyMed

In fact, the Multi-Objective Evolutionary Algorithm (MOEA) provide a set of compromised solutions called Pareto optimal solution since no single solution can optimize each of the objectives separately. The decision maker is provided with the set of Pareto optimal solutions in order to choose solution based on the decision maker's criteria. This sort of MOEA solution technique is called a posteriori method since decision is taken after searching is finished. This confirms the work done by Coello et al., [36]. In this context, the Pareto-optimization approach does not require any a priori preference decisions between the conflicting of oil spill pixel, look-alike pixels, land pixels, and surrounding sea pixels. Further, Pareto-optimal points have form Pareto-front as shown in Fig.8 in the multi pixel objectives function the COSMO-SkyMed data space.

Clearly the Multi-Objectives Evaluation Algorithm (MOEA) which based on Pareto optimal solutions able to isolate oil spill dark pixels from the surrounding environment. In other words, look-alike, low wind zone, sea surface roughness, and land are marked by white colour while oil spill pixels are marked all black (Figures 10). Further, Fig. 9 show the results of the MOEA, where 100% of the oil spills in the test set were correctly classified. This study is not similar to previous work done by Marghany and Hashim [12]. The dissimilarity is because this work provides the automatic classifier based on MOEA but Marghany and Hashim work [12] is considered as a semi-automatic tool for oil spill detection.

Fig. 9. Oil spill automatic detection by Multi-Objectives Evaluation Algorithm (MOEA)

Multi-Objectives Evaluation Algorithm (MOEA) which based on the Pareto optimal solutions provides excellent discrimination of oil spill pixels. This can be confirmed by the receiver–operator characteristics (ROC) curve (Fig.10). In this regard, the existing of weight sum of objective function converts a conflicting multiobjective problem of oil spill and surrounding sea features objective one.

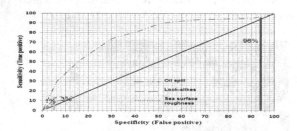

Fig. 10. ROC for oil spill discrimination using MOEA

This can be seen in ROC curve where oil spill has an area difference of 96% which is larger than look-alike and sea surface areas. Further, p probability of 0.0005 another proof for excellent of MOEA for oil spill detection. This study shows a great performance as compared to previous work done by Marghany [30]. This because of Pareto-front contains the Pareto-optimal solutions and in case of continuous front, it divides the pixels objective function space into two parts, which are non-optimal solutions and infeasible solutions. In this regard, it improved the robustness of pattern search and improved the convergence speed of MOEA. This confirms the work of Yudong et al., [37].

5   Conclusions

This work has utilized Multi-Objective Evaluation Algorithm (MOEA) for oil spill automatic detection. In doing so, Pareto optimal solution is implemented with MOEA to minimize the conflict of oil spill automatic detection in coherence data. This is applied to along the coastal water of along Koh Samet island, Thailand. The study shows that the implementation of Pareto optimal solution and weight sum in MOEA generated accurate pattern of oil slick. This is confirmed by using the receiver – operational characteristics (ROC) curve. The ROC curve approves the existing of oil spill along Koh Samet island, Thailand with 96 % which is larger than other surrounding environment features. In conclusion, MOEA can be which based on Pareto optimal solution can be used as an automatic detection tool for oil spill. It can be said that COSMO-SkyMed data are an excellent sensor for oil spill detection and monitoring.

Acknowledgement. The author would like to thank Geo-informatics and Space Technology Development Agency (GISTDA) of Thailand for providing COSMO-SkyMed data.

References

1. Adam, J.A.: Specialties: Solar Wings, Oil Spill Avoidance, On-Line Patterns. IEEE Spect. 32, 87–95 (1995)
2. Aggounc, M.E., Atlas, L.E., Cohn, D.A., El-Sharkawi, M.A., Marks, R.J.: Artificial Neural Networks For Power System Static Security Assessment. IEEE Int. Sym. on Cir. and Syst. Portland, Oregon, pp. 490–494 (1989)
3. Brekke, C., Solberg, A.: Oil Spill Detection by Satellite Remote Sensing. Rem. Sens. of Env. 95, 1–13 (2005)
4. Fiscella, B., Giancaspro, A., Nirchio, F., Pavcsc, P., Trivero, P.: Oil Spill Detection Using Marine SAR Images. Int. J. of Rem. Sens. 21, 3561–3566 (2000)
5. Frate, F.D., Petrocchi, A., Lichtenegger, J., Calabresi, G.: Neural Networks for Oil Spill Detection Using ERS-SAR Data. IEEE Tran. on Geos. and Rem. Sens. 38, 2282–2287 (2000)
6. Hect-Nielsen, R.: Theory of the Back Propagation Neural Network. In: Proc. of the Int. Joint Conf. on Neu. Net., pp. 593–611. IEEE Press (1989)
7. Marghany, M., Hashim, M.: Comparative algorithms for oil spill detection from multi mode RADARSAT-1 SAR satellite data. In: Murgante, B., Gervasi, O., Iglesias, A., Taniar, D., Apduhan, B.O. (eds.) ICCSA 2011, Part II. LNCS, vol. 6783, pp. 318–329. Springer, Heidelberg (2011)
8. Marghany, M.: RADARSAT Automatic Algorithms for Detecting Coastal Oil Spill Pollution. Int. J. of App. Ear. Obs. and Geo. 3, 191–196 (2001)
9. Marghany, M.: RADARSAT for Oil spill Trajectory Model. Env. Mod. and Sof. 19, 473–483 (2004)
10. Marghany, M., Cracknell, A.P., Hashim, M.: Modification of Fractal Algorithm for Oil Spill Detection from RADARSAT-1 SAR Data. Int. J. of App. Ear. Obs. and Geo. 11, 96–102 (2009)

11. Marghany, M., Cracknell, A.P., Hashim, M.: Comparison between Radarsat-1 SAR Different Data Modes for Oil Spill Detection by a Fractal Box Counting Algorithm. Int. J. of Dig. Ear. 2, 237–256 (2009)

12. Marghany, M., Hashim, M., Cracknell, A.P.: Fractal Dimension Algorithm for Detecting Oil Spills Using RADARSAT-1 SAR. In: Gervasi, O., Gavrilova, M.L. (eds.) ICCSA 2007, Part I. LNCS, vol. 4705, pp. 1054–1062. Springer, Heidelberg (2007)

13. Marghany, M., Hashim, M.: Texture Entropy Algorithm for Automatic Detection of Oil Spill from RADARSAT-1 SAR data. Int. J. of the Phy. Sci. 5, 1475–1480 (2010)

14. Michael, N.: Artificial Intelligence: A guide to Intelligent Systems, 2nd edn. Addison Wesley, Harlow (2005)

15. Migliaccio, M., Gambardella, A., Tranfaglia, M.: SAR Polarimetry to Observe Oil Spills. IEEE Tran. on Geos. and Rem. Sen. 45, 506–511 (2007)

16. Mohamed, I.S., Salleh, A.M., Tze, L.C.: Detection of Oil Spills in Malaysian Waters from RADARSAT Synthetic Aperture Radar Data and Prediction of Oil Spill Movement. In: Proc. of 19th Asi. Conf. on Rem. Sen., Hong Kong, China, November 23-27, vol. 2, pp. 980–987. Asian Remote Sensing Society, Japan (1999)

17. Provost, F., Fawcett, T.: Robust classification for imprecise environments. Mach. Lear. 42, 203–231 (2001)

18. Samad, R., Mansor, S.B.: Detection of Oil Spill Pollution Using RADARSAT SAR Imagery. In: CD Proc. of 23rd Asi. Conf. on Rem. Sens., Birendra International Convention Centre in Kathmandu, Nepal, November 25-29, Asian Remote Sensing (2002)

19. Skrunes, S., Brekke, C., Eltoft, T.: An Experimental Study on Oil Spill Characterization by Multi-Polarization SAR. In: Proc. European Conference on Synthetic Aperture Radar, Nuremberg, Germany, pp. 139–142 (2012)

20. Topouzelis, K., Karathanassi, V., Pavlakis, P., Rokos, D.: Potentiality of Feed-Forward Neural Networks for Classifying Dark Formations to Oil Spills and Look-alikes. Geo. Int. 24, 179–191 (2009)

21. Topouzelis, K., Karathanassi, V., Pavlakis, P., Rokos, D.: Detection and Discrimination between Oil Spills and Look-alike Phenomena through Neural Networks. ISPRS J. Photo. Rem. Sens. 62, 264–270 (2007)

22. Topouzelis, K.N.: Oil Spill Detection by SAR Images: Dark Formation detection, Feature Extraction and Classification Algorithms. Sens. 8, 6642–6659 (2008)

23. Trivero, P., Fiscella, B., Pavese, P.: Sea Surface Slicks Measured by SAR. Nuo. Cim. 24C, 99–111 (2001)

24. Trivero, P., Fiscella, B., Gomez, F., Pavese, P.: SAR Detection and Characterization of Sea Surface Slicks. Int. J. Rem. Sen. 19, 543–548 (1998)

25. Velotto, D., Migliaccio, M., Nunziata, F., Lehner, S.: Dual-Polarized TerraSAR-X Data for Oil-Spill Observation. IEEE Trans. Geosci. Remote Sens. 49, 4751–4762 (2011)

26. Chaiyaratana, N., Zalzala, A.M.S.: Recent developments in evolutionary and genetic algorithms: theory and applications. In: Second International Conference on Genetic Algorithms in Engineering Systems: Innovations and Applications, GALESIA 1997, Glasgow, September 2-4, pp. 270–277 (1997)

27. Kahlouche, S., Achour, K., Benkhelif, M.: Proceedings of the 2002 WSEAS International Conferences, Cadiz, Spain, June 12-16, pp. 1–5 (2002), http://www.wseas.us/e-library/conferences/spain2002/papers/443-164.pdf

28. Gautam, G., Chaudhuri, B.B.: A distributed hierarchical genetic algorithm for efficient optimization and pattern matching. Pattern Recognition Journal 40, 212–228 (2007)

29. Sivanandam, S.N., Deepa, S.N.: Introduction to Genetic Algorithms. Springer, Heidelberg (2008)

30. Marghany, M.: Genetic Algorithm for Oil Spill Automatic Detection from Envisat Satellite Data. In: Murgante, B., Misra, S., Carlini, M., Torre, C.M., Nguyen, H.-Q., Taniar, D., Apduhan, B.O., Gervasi, O. (eds.) ICCSA 2013, Part II. LNCS, vol. 7972, pp. 587–598. Springer, Heidelberg (2013)
31. Marghany, M.: Genetic Algorithm for Oil Spill Automatic Detection from Multisar Satellite Data. In: Proceedings of the 34th Asian Conference on Remote Sensing 2013, Bali, Indonesia, October 20-24, pp. SC03-671-SC0-3677 (2013)
32. Zhang, B., Perrie, W., Li, X., Pichel, W.: Mapping sea surface oil slicks using RADARSAT-2 quad-polarization SAR image. Geophys. Res. Lett. 38, L10602 (2011)
33. Zhang, Y., Lin, H., Liu, Q., Hu, J., Li, X., Yeung, K.: Oil-spill monitoring in the coastal waters of Hong Kong and vicinity. Marine Geodesy 35, 93–106 (2012)
34. Shirvany, R., Chabert, M., Tourneret, J.-Y.: Tourneret: Ship and Oil-Spill Detection Using the Degree of Polarization in Linear and Hybrid/Compact Dual-Pol SAR. IEEE Journal of Selected Topics in Applied Earth Observations and Remote Sensing 5, 885–892 (2012)
35. Trivero, P., Biamino, W., Nirchio, F.: High resolution COSMO - SkyMed SAR images for oil spills automatic detection. In: IEEE International Geoscience and Remote Sensing Symposium, IGARSS 2007, pp. 2–5 (2007)
36. Coello, C.A., Lamont, G.B., Van Veldhuizen, D.A.: Evolutionary algorithms for solving multi-objective problems, 2nd edn. Springer, Berlin (2007)
37. Yudong, Z., Shuihua, W., Genlin, J., Zhengchao, D.: Genetic Pattern Search and Its Application to Brain Image Classification. Math. Prob. in Eng., 1–8 (2013)

Cubic B-Spline Collocation Method for Pricing Path Dependent Options

Geraldine Tour[1] and Désiré Yannick Tangman[2]

Department of Mathematics, University of Mauritius,
Réduit, Mauritius
marie.tour1@umail.uom.ac.mu,
y.tangman@uom.ac.mu
http://www.uom.ac.mu/

Abstract. We consider an efficient pricing method based on the cubic B-spline collocation method in which the approximation to the exact solution can be expressed as a linear combination of cubic B-spline basis functions. An earlier work has proposed to use such technique to solve the generalised Black–Scholes PDE to price European options only. In this work, we extend the application of the B-spline collocation method to price path-dependent options such as American, Barrier and Asian options under the Black–Scholes model. Our numerical results show that the new scheme is a method of choice for option pricing. Indeed the scheme has the same computational complexity as the standard second order finite difference scheme since they both require the solution of tridiagonal linear systems. However our numerical experiments reveal that the B-spline collocation method is more accurate in the infinity norm when solving convectively dominated financial problems and yields second order convergent prices and hedging parameters for both continuous and discretely sampled path-dependent options.

Keywords: collocation method, cubic B-splines, basis functions, path-dependent option pricing.

1 Introduction

A financial derivative derives its value from an underlying financial security, whose price can be modeled by some stochastic process. A derivative is path-dependent if its payoff depends on the entire path traversed by the underlying security. Path-dependent options are defined using either discrete or continuous price sampling and due to regulatory and practical issues, most of path-dependent options traded in financial markets are discrete path-dependent options. Most research has focused on either partial differential equation (PDE), Monte Carlo, lattice based methods (i.e. binomial and trinomial trees) or Fourier transform techniques for pricing these options. The latter is highly accurate and is among the best method for option pricing [18,19]. Various other methodologies used for solving American options include the front-fixing transformations of Wu and Kwok [1], operator

B. Murgante et al. (Eds.): ICCSA 2014, Part VI, LNCS 8584, pp. 372–385, 2014.

splitting techniques of Ikonen and Toivanen [2] and penalty methods by Forsyth and Vetzal [3]. For barrier options, a review of various analytical approximation methods and numerical schemes for pricing discrete barrier options can be found in [4]. Asian option pricing relies on closed-form approximations [5], computed Laplace transforms [6] or methods based on PDEs [7,16]. By their nature, path-dependent options are very easily implemented using the PDE approach.

Currently, there are a number of numerical methods for finding numerical solutions of differential equations. Indeed, the theory of spline functions is a very active field of approximation theory and boundary-value problems (BVPs) when numerical aspects are considered. The basic idea of splines [8] is to construct piecewise polynomial curves that not only provide a better curve fit than other interpolation methods but they are also smooth, i.e. they must be continuously differentiable to some degree. The collocation method was developed to seek numerical solutions in the form of linear combination of coordinate function with linear coefficients. Recurrence relations for the purpose of computing coefficients can also be found in [9]. The cubic B-spline collocation method was proposed for Burgers's equation [10] and used for the numerical solution of the differential equations in [11,12] where the method was based on redefined cubic B-splines basis functions. Kadalbajoo, Tripathi and Kumar [13] recently focused on the implementation of the cubic B-spline collocation in option pricing to find the numerical solution of the generalised Black–Scholes equation but they priced only European style options.

Our contribution is to extend the use of the cubic B-spline collocation method [13] to price discrete and continuous American, Barrier and Asian options under the standard Black–Scholes model [14] by using the Crank-Nicolson scheme. This leads to tridiagonal linear systems which can be solved easily with a linear computational effort.

The paper is organised as follows: In §2, we describe the partial differential equations approach to the pricing of the European, American, Bermudan, Barrier and Asian options in the Black–Scholes framework. In §3, the description of the cubic B-spline collocation method is given and §4 presents the procedure for the implementation of the present method to price the different path-dependent options mentioned earlier. §5 contains the numerical experiments and we present our conclusions in §6.

2   Option Pricing in the Black–Scholes Framework

It is well-known that under the risk-neutral measure $\mathbb{Q}$, in the Black–Scholes model, the asset prices can be modeled by the geometric Brownian motion

$$dS = (r - \delta)Sdt + \sigma SdW_t \, ,$$

where S is the asset value and dW_t is the standard Wiener process at time t. The interest rate r, the continuous dividend yield δ and the volatility σ are allowed to depend on both S and t under the generalised Black–Scholes model but are taken as constant in [14]. Kadalbajoo, Tripathi and Kumar [13] implemented the uniform cubic B-Spline collocation method to find the numerical solution of the generalised Black–Scholes partial differential equation

$$\frac{\partial V}{\partial \tau} = \mathcal{L}_S V = \left(\frac{1}{2}\sigma^2 S^2 \frac{\partial^2}{\partial S^2} + (r-\delta)S\frac{\partial}{\partial S} - r\right)V \tag{1}$$

to find the value of a European option with strike K and maturity date T. Here $\tau = T-t$ and $\mathcal{L}_S$ represents the spatial operator obtained after using Itō's lemma under the no-arbitrage assumption.

For simplicity, we assume σ, r and δ to be constant but it is easy to extend the discretisation techniques presented here to the generalised Black–Scholes. The price of any particular derivative security is obtained by solving the forward PDE (1) together with the appropriate auxiliary conditions for the corresponding financial derivatives. We describe these conditions next.

2.1 European Options

For European-style options, the contract can be exercised with a strike price K only on the expiry date T, so that the conditions for such an option are given by

$$V(S,0) = \max(\phi(S-K),0) ,$$
$$V_{SS} = 0 \quad \text{as} \quad S \to 0 \quad \text{or} \quad S \to \infty , \tag{2}$$

where $\phi = 1$ gives a call and $\phi = -1$ defines a put payoff.

2.2 American Options

American options give their holders the right to exercise the option at any time up to and including the expiration date. Thus their value is at least as much as the value of the corresponding European option. Usually the American option pricing problem can be posed as a linear complementary problem (LCP) [15] of the form

$$V_\tau = \mathcal{L}_S ,$$
$$V(S,0) = \max(\phi(S-K),0) ,$$
$$V(S,\tau) \geq V(S,0) ,$$
$$(V_\tau = \mathcal{L}_S) \wedge (V(S,\tau) = V(S,0)) , \tag{3}$$

where $\mathcal{L}_S$ represents the spatial operator in (1).

2.3 Bermudan Options

Bermudan options are discretely early exercisable American options meaning that they can only be exercised at a predetermined set of dates $\tau_k \in (0, T]$,

$k = 1, 2, ..., n$ in contrast with an American option which is continuously exercisable. Consequently they have a value between that of an American and a European option. If $C(S, \tau)$ denotes the continuation value obtained by solving (1) and $g(S, \tau)$ the payoff at time τ, then at each observation dates τ_k, we simply need to enforce the condition,

$$V(S, \tau_k) = \max(g(S, \tau_k), C(S, \tau_k)) ,$$

to obtain the value of a Bermudan option.

2.4   Barrier Options

Options with the barrier feature are considered to be one of the simplest type of path-dependent options. For instance, the "up-and-out" call barrier option ceases to exist if S hits the barrier level b. In fact, the problem is identical to that of a European option where they have similar payoff and lower boundary condition. The valuation problem differs in that the upper boundary condition is applied at $S = b$ rather than as $S \to \infty$, so that

$$V(b, \tau) = 0 . \tag{4}$$

For a discretely sampled "up-and-out" barrier option, it is sufficient to enforce the condition

$$V(S \geq b, \tau_k) = 0$$

at the observation dates τ_k. The attractiveness of barrier options is that they are cheaper than their corresponding vanilla options.

2.5   Asian Options

An Asian option is an option whose payoff depends on the arithmetic or geometric average price of the underlying asset over a certain period of time. We consider here the continuously sampled arithmetic average of S_t over $[0, t]$ defined by

$$a_t = \frac{1}{t} \int_0^t S_u \, du \Rightarrow da_t = \frac{S - a_t}{t} dt .$$

Let $V(S, a, t)$ be the value of an Asian option, then the two-dimensional PDE satisfied by the option is determined as [16]

$$\frac{\partial V}{\partial \tau} = \mathcal{L}_S V + \mathcal{L}_a V , \tag{5}$$

where we have the spatial operators

$$\mathcal{L}_a = \left(\frac{S - a}{T - \tau} \right) \frac{\partial}{\partial a}$$

and $\mathcal{L}_S$ is as in (1) .

The conditions for a fixed strike Asian option are given as

$$V(S, a, 0) = \max(\phi(a - K), 0) ,$$

with similar boundary conditions as for a European option.

For discretely sampled Asian options, we let

$$a_k = \frac{1}{k} \sum_{j=1}^{k} S_j \Rightarrow a_k = a_{k-1} + \frac{S_k - a_{k-1}}{k}$$

represents the discrete average over observation times $t_1, t_2, \ldots t_N$. Away from monitoring dates, (5) loses its dependence on a so that we only need to solve a one dimensional PDE (1) for each value of a. At observation dates τ_k, we simply need to apply the jump conditions

$$v(S, a_k, \tau_k) = v\left(s, a_{k-1} + \frac{S_k - a_{k-1}}{N - k}, \tau - \varepsilon\right) \tag{6}$$

where $\varepsilon > 0$, $\varepsilon \to 0$ and N is the total number of observation dates.

For continuously sampled Asian options we note that the solution to the split problem

$$V_\tau = \mathcal{L}_a V = \left(\frac{S - a}{T - \tau}\right) \frac{\partial V}{\partial a} , \tag{7}$$

is

$$V(S, a, \tau + \Delta\tau) = V\left(S, a + \Delta\tau \left(\frac{S - a}{T - \tau}\right), \tau\right) .$$

If we let the solution operators $\mathcal{L}_1$ and $\mathcal{L}_2$ be the solution operators to problems (1) and (7) respectively, we can use a dimensional splitting technique to decouple (5) into a set of one dimensional problems:

$$v(\tau + \Delta\tau) = \mathcal{L}_1(\Delta\tau)\mathcal{L}_2(\Delta\tau)v(\tau) + \mathcal{O}(\Delta\tau) .$$

The temporal accuracy can then be improved using the Strang splitting technique [17]

$$v(\tau + \Delta\tau) = \mathcal{L}_2\left(\frac{1}{2}\Delta\tau\right) \mathcal{L}_1(\Delta\tau)v(\tau)\mathcal{L}_2\left(\frac{1}{2}\Delta\tau\right) v(\tau) + \mathcal{O}(\Delta\tau^2) .$$

We refer to [16] for more details about the pricing of Asian options using the PDE approach.

3   Description of the Collocation Method

In the cubic B-spline collocation method, the approximation $v(S,t)$ to the exact solution $V(S,t)$ can be expressed as a linear combination of cubic B-spline basis functions in the form [8]

$$v(S,t) = \sum_{j=-1}^{m+1} \alpha_j(t)B_j(S) \tag{8}$$

where $\alpha_j(t)$ are time dependent quantities to be determined.
Consider a closed interval $[S_{\min}, S_{\max}]$ with equally spaced knots as

$$S_{\min} = S_0 < S_1 < ... < S_m = S_{\max}$$

with $h = S_{j+1} - S_j = (S_{\max} - S_{\min})/m$. Since each cubic B-spline covers four elements, the construction of the cubic B-splines is established by introducing four additional knots S_{-2}, S_{-1}, S_{m+1} and S_{m+2} such that $S_{-2} < S_{-1} < S_0$ and $S_m < S_{m+1} < S_{m+2}$.
The cubic B-splines $B_j(S)$, for $j = -1, ..., m+1$ are defined as

$$B_j(S) = \begin{cases} (S - S_{j-2})^3, & S \in [S_{j-2}, S_{j-1}) \, ; \\ (S - S_{j-2})^3 - 4(S - S_{j-1})^3, & S \in [S_{j-1}, S_j) \, ; \\ (S_{j+2} - S)^3 - 4(S_{j+1} - S)^3, & S \in [S_j, S_{j+1}) \, ; \\ (S_{j+2} - S)^3, & S \in [S_{j+1}, S_{j+2}) \, ; \\ 0, & \text{otherwise} \, , \end{cases} \tag{9}$$

where $\{B_{-1}(S), B_0(S), ..., B_m(S), B_{m+1}(S)\}$ forms a basis over $[S_{\min}, S_{\max}]$. Since each function $B_j(S)$ is twice continuously differentiable on the entire real line, using (9), the values of $B_j(S)$ and its derivatives $B_j'(S)$ and $B_j''(S)$ can be evaluated. Their coefficients are summarised in Table 1 for a uniform grid.

Table 1. Coefficient of cubic B-splines and its derivatives at knots S_j

S	S_{j-2}	S_{j-1}	S_j	S_{j+1}	S_{j+2}
$B_j(S)$	0	1	4	1	0
$B_j'(S)$	0	$3/h$	0	$-3/h$	0
$B_j''(S)$	0	$6/h^2$	$-12/h^2$	$6/h^2$	0

4   Application of the collocation Method

In order to apply the collocation method to (1) we use the trial function (8) and cubic splines (9) to approximate the values of $v(S), v'(S)$ and $v''(S)$ at the knots S_j in terms of the element parameters α_j by

$$v(S) = \alpha_{j-1} + 4\alpha_j + \alpha_{j+1} , \tag{10}$$

$$v'(S) = \left(\frac{3}{h}\right)(\alpha_{j+1} - \alpha_{j-1}) ,$$

$$v''(S) = \left(\frac{6}{h^2}\right)(\alpha_{j-1} - 2\alpha_j + \alpha_{j+1}) . \tag{11}$$

We can then approximate the Black–Scholes operator $\mathcal{L}_S$ in (1) in the matrix form

$$L = \frac{1}{2}\sigma^2 I_S^2 D_{SS} + (r - \delta)I_S D_S - rD ,$$

where

$$D = \text{tridiag}[1, 4, 1] ,$$
$$D_S = \text{tridiag}[-3/h, 0, 3/h] ,$$
$$D_{SS} = \text{tridiag}[6/h^2, -12/h^2, 6/h^2]$$

and I_S is a diagonal matrix containing S_j.

In order to implement boundary conditions (2), we note from (11) that

$$\alpha_{-1} = 2\alpha_0 - \alpha_1 \quad \text{and} \quad \alpha_{m+1} = 2\alpha_m - \alpha_{m-1} .$$

The same boundary implementation is used for discretely sampled barrier and Asian option except conditions (4) and (6) are enforced respectively at observation date. The continuous barrier boundary condition (4) can also be easily implemented by using (10) to obtain

$$\alpha_{m+1} = -4\alpha_m - \alpha_{m-1} ,$$

where we assume here that $S_m = S_{\max} = b$.

Once the spatial derivatives have been discretised, the time derivative can be approximated by the Crank-Nicolson scheme and we end up with the linear system,

$$Av^{k+1} = Bv^k ,$$

where

$$A = \left(I - \frac{1}{2}\Delta_T L\right) \quad \text{and} \quad B = \left(I + \frac{1}{2}\Delta_T L\right)$$

are $(m + 1) \times (m + 1)$ tridiagonal matrices.

In the case of American options, we apply the operator splitting method of Ikonen and Toivanen [2] to solve the linear complementarity problem (3). This technique adds a Lagrangian multiplier λ^k as follows

$$\frac{v^{k+1} - v^k}{\Delta_T} - \frac{1}{2}Lv^{k+1} - \frac{1}{2}Lv^k - \lambda^{k+1} = 0$$

and since λ^{k+1} is unknown at time t_k, we solve the above linear system using the following two fractional time steps

$$A\tilde{v}^{k+1} = Bv^k + \Delta\tau\lambda^k$$

and

$$\lambda^{k+1} = \lambda^k + \frac{v^{k+1} - \tilde{v}^{k+1}}{\Delta\tau} \, ,$$

where λ^{k+1} is the zero vector in $\mathcal{R}^{m+1}$. It then follows that

$$v^{k+1} = \max\left((K - S), \tilde{v}^{k+1} - \Delta\tau\lambda^k\right) \, .$$

The complete MATLAB® code for pricing American option using the collocation method under the Black–Scholes is given in Figure 1.

```
m = 2^10; n = m; S0 = 100; E = 100; sigma = 0.2; r = 0.05; delta = 0.00;
smin = 0; smax = 200; ds = (smax-smin)/m; vecs = [smin: ds: smax]';
T = 0.5; dt = T/n; vect = [0: dt: T]';

e = ones(m+1,1); I = speye(m+1);
I = spdiags([e 4*e e], -1:1, m+1, m+1); I(1,1) = I(1,1) + 2;
I(1,2) = I(1,2) - 1; I(end,end) = I(end,end) + 2;
I(end,end-1) = I(end,end-1) - 1;
Ds = spdiags([-3*e 0*e 3*e], -1:1, m+1, m+1)/ds;
Ds(1,1) = Ds(1,1) + 2*(-3/ds); Ds(1,2) = Ds(1,2) - 1*(-3/ds);
Ds(end,end) = Ds(end,end) + 2*(3/ds);
Ds(end,end-1) = Ds(end,end-1) - 1*(3/ds);
Dss = spdiags([6*e -12*e 6*e], -1:1, m+1, m+1)/(ds^2);
Dss(1,1) = Dss(1,1) + 2*(6/(ds^2)); Dss(1,2) = Dss(1,2) - 1*(6/(ds^2));
Dss(end,end) = Dss(end,end) + 2*(6/(ds^2));
Dss(end,end-1) = Dss(end,end-1) - 1*(6/(ds^2));
L = 0.5*sigma^2*spdiags(vecs.^2, 0, m+1, m+1)...
*Dss + (r-delta)*spdiags(vecs, 0, m+1, m+1)*Ds...
- r*I;

V = max(E-vecs, 0); Payoff = I\V;
V = Payoff; A = I - 0.5*dt*L; B = I + 0.5*dt*L;
[L,U]=lu(A); lambda = zeros(m+1,1);

for j = 1: n,
    Vprime = U\( L\(B*V + dt*lambda) );
    V = max( Vprime - dt*lambda , Payoff );
    lambda = lambda + (1/dt)*( V - Vprime );
end

Delta = Ds*V; Gamma = Dss*V; V = I*V;
```

Fig. 1. MATLAB® code for pricing an American put option under the Black–Scholes model using the collocation method with operator splitting

5 Numerical Results

In this section, numerical tests are performed to assess the efficiency and accuracy of the cubic collocation method for the European, American, Bermudan, Barrier and Asian options under the Black–Scholes model. The parameters to be used for our numerical experiments are presented in Table 2. All computations have been performed using MATLAB® on a Core i5 laptop with 8GB RAM and speed 2.50 GHz.

It is well known that the standard finite difference method is very popular for the valuation of option pricing problems. However when the Black–Scholes PDE becomes convection-dominated, the finite difference method is less accurate than the cubic B-spline collocation method as we can see from Figure 2 which depicts a log plot of the infinity norm error against the number of steps. Noting that we have the same level of computational complexity for both methods since we only need to solve tridiagonal linear systems, we can clearly see the superior performance of the cubic B-spline collocation method. Computed option prices, errors, convergence rates, cpu timings are shown for the cubic collocation method to price the European, American, "up-and-out" barrier and the Asian options under the Black–Scholes model in Table 3 for the continuously monitored and in Table 4 for the discretely monitored options. We observe that second order convergence is achieved as expected and the algorithms run within a few CPU seconds only. The benchmark for all option prices were obtained by using the highly accurate COS method by Fang, Zhang and Oosterlee [18,19] with 2^{14} grid nodes.

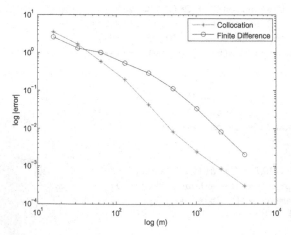

Fig. 2. Plot of log error against log m for the cubic B-spline collocation and finite difference schemes for pricing an European call option using the convectively dominated Black–Scholes PDE with $\sigma = 0.01$ and $r = 0.2$

Table 2. Test parameters for pricing the different options

Options	Parameters
European	$S_0 = 100, K = 100, \sigma = 0.2, r = 0.1, \delta = 0.00, T = 1, S_{min} = 0, S_{max} = 200$
American	$S_0 = 100, K = 100, \sigma = 0.2, r = 0.05, \delta = 0.00, T = 0.5, S_{min} = 0, S_{max} = 200$
Bermudan	$S_0 = 100, K = 110, \sigma = 0.2, r = 0.1, \delta = 0.00, T = 1, S_{min} = 0, S_{max} = 200$
Barrier	$S_0 = 100, K = 100, \sigma = 0.2, r = 0.05, \delta = 0.00, T = 0.5, S_{min} = 0, S_{max} = 200, UB = 120$
Asian	$S_0 = 2.0, K = 2.0, \sigma = 0.5, r = 0.05, \delta = 0.00, T = 1, S_{min} = 0, S_{max} = 2K$

Table 3. Error and order of convergence for the European, American, Barrier and Asian options

European Option

m	Price	Error	Order	cpu(s)
2^4	13.10740	0.1623	-	0.0004
2^5	13.22947	0.0402	2.0129	0.0006
2^6	13.25965	0.0100	2.0033	0.0007
2^7	13.26717	0.0025	2.0008	0.0012
2^8	13.26905	6.2637e–4	2.0002	0.0036
2^9	13.26952	1.5659e–4	2.0001	0.0093
2^{10}	13.26964	3.9147e–5	2.0000	0.0328
2^{11}	13.26967	9.7867e–6	2.0000	0.1296

Reference price: 13.269677

American Option

m	Price	Error	Order	cpu(s)
2^4	5.53481	0.8791	-	0.0009
2^5	4.83334	0.1776	2.3070	0.0013
2^6	4.69635	0.0406	2.1278	0.0021
2^7	4.66316	0.0075	2.4454	0.0044
2^8	4.65681	0.0011	2.7440	0.0105
2^9	4.65582	1.1920e–4	3.2242	0.0303
2^{10}	4.65567	2.5195e–5	2.2422	0.0904

Reference price: 4.655675

Barrier Option

m	Price	Error	Order	cpu(s)
2^4	1.93317	0.1945	-	0.0003
2^5	2.16582	0.0506	1.9432	0.0005
2^6	2.19965	0.0123	2.0417	0.0008
2^7	2.20845	0.0030	2.0238	0.0012
2^8	2.21052	7.5646e–4	1.9977	0.0029
2^9	2.21109	1.8900e–4	2.0009	0.0093
2^{10}	2.21123	4.7264e–5	1.9995	0.0348
2^{11}	2.21127	1.1814e–5	2.0002	0.1301
2^{12}	2.21128	2.9538e–6	1.9999	0.5177

Reference price: 2.211281

Asian Option

m	Price	Error	Order	cpu(s)
2^4	0.25184	0.0054	-	0.0768
2^5	0.24783	0.0014	1.9378	0.1898
2^6	0.24679	3.6922e–4	1.9394	0.5724
2^7	0.24651	9.6153e–5	1.9411	2.2976
2^8	0.24644	2.4817e–5	1.9540	10.3477
2^9	0.24642	6.0711e–6	2.0313	53.6703

Reference price: 0.246416

Another important aspect of option pricing is the measure of risk. We calculate the Delta (Δ) and Gamma (Γ) hedging parameters for the European and discretely monitored barrier options and the convergence rates are shown in Table 5 and Table 6 respectively. The results indicate that the Greeks are highly accurate and also possess second order convergence. We further observe from Figure 3 and Figure 4 that there are no spurious oscillations in the greeks calculations either.

Table 4. Error and order of convergence for the discrete path dependent options

	Discrete Barrier Option					Bermudan Option			
m	Price	Error	Order	cpu(s)	m	Price	Error	Order	cpu(s)
20×2^4	1.84541	0.0039	-	0.0153	2^4	11.19241	0.7129	-	0.0011
20×2^5	1.84837	9.8697e–4	1.9990	0.0471	2^5	10.71074	0.2312	1.6244	0.0015
20×2^6	1.84911	2.4679e–4	1.9997	0.1683	2^6	10.54905	0.0695	1.7336	0.0022
20×2^7	1.84929	6.1699e–5	1.9999	0.6389	2^7	10.49408	0.0146	2.2552	0.0047
20×2^8	1.84934	1.5425e–5	2.0000	3.0037	2^8	10.48315	0.0036	2.0041	0.0122
20×2^9	1.84935	3.8562e–6	2.0000	11.7159	2^9	10.48041	8.9013e–4	2.0282	0.0374
					2^{10}	10.47977	2.53225e–4	1.8136	0.1103
					2^{11}	10.47959	7.2112e–5	1.8121	0.4169
					2^{12}	10.47953	1.4438e–5	2.3204	1.6023

Reference price: 1.849353   Reference price: 10.479520

Discrete Asian Option

m	Price	Error	Order	cpu(s)
2^4	12.29264	0.3877	-	0.0943
2^5	11.99346	0.0885	2.1312	0.1426
2^6	11.92525	0.0203	2.1251	0.2259
2^7	11.90978	0.0048	2.0765	0.3959
2^8	11.90604	0.0011	2.1724	0.7465
2^9	11.90510	1.3016e–4	3.0350	1.4538

Reference price: 11.90497

Table 5. Convergence rates of the European Delta and Gamma Greeks

m	Δ error	Order	Γ error	Order
2^4	0.0053	-	4.6808e–4	-
2^5	0.0012	2.0869	1.1044e–4	2.0835
2^6	3.0730e–4	2.0204	2.7237e–5	2.0197
2^7	7.6557e–5	2.0050	6.7863e–6	2.0049
2^8	1.9123e–5	2.0013	1.6951e–6	2.0012
2^9	4.7796e–6	2.0030	4.2370e–7	2.0003
2^{10}	1.1948e–6	2.0001	1.0592e–7	2.0001
2^{11}	2.9871e–7	2.0000	2.6479e–8	2.0000
2^{12}	7.4663e–8	2.0003	6.6184e–9	2.0003
$\Delta = 0.725746882$		$\Gamma = 0.166612301$		

Table 6. Convergence rates of the discrete "up-and-out" Barrier Delta and Gamma Greeks

m	Δ error	Order	Γ error	Order
20×2^4	5.7284e–5	-	1.4472e–5	-
20×2^5	1.4369e–5	1.9952	3.6209e–6	1.9989
20×2^6	3.5952e–6	1.9988	9.0540e–7	1.9997
20×2^7	8.9900e–7	1.9997	2.2636e–7	1.9999
20×2^8	2.2470e–7	1.9999	5.6592e–8	2.0000
20×2^9	5.6191e–8	2.0000	1.4151e–8	1.9997
$\Delta = -0.010058016$			$\Gamma = -0.007935846$	

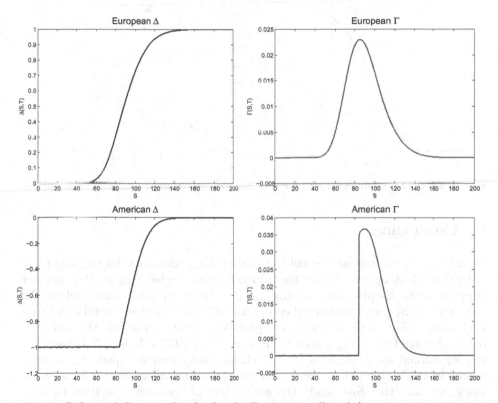

Fig. 3. Delta and Gamma Greeks for the European call and American put options

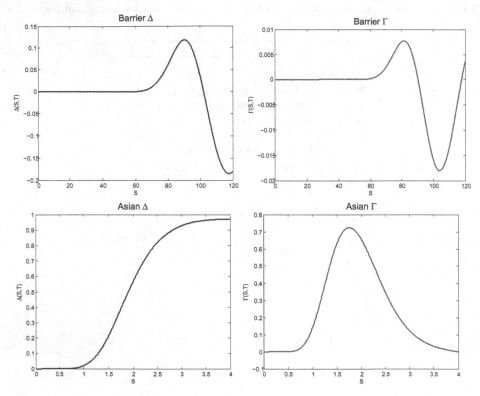

Fig. 4. Delta and Gamma Greeks for the "up-and-out" Barrier and Asian call options

6   Conclusion

In this paper, we extend the cubic B-spline collocation method to price some path-dependent options under the Black–Scholes model. Due to the compact support of the B-spline, the arising linear systems are tridiagonal and can be solved with linear computational complexity. We observe that second order convergence is achieved when pricing the path-dependent options and the scheme is more accurate for solving convectively dominated PDE's. It is well known that the log-normal assumption of Black–Scholes model cannot explain the sudden jumps in stock prices that occur infrequently in the stock market. In a future work, we shall therefore study the application of collocation method to price these derivatives under the jump-diffusion Lévy processes which can fully capture the empirical leptokurtic distributions that have been observed in econometric financial time series.

Acknowledgments. The research of Geraldine Tour was supported by a post-graduate research scholarship from the Tertiary Education Commission.

References

1. Wu, L., Kwok, Y.K.: A front-fixing difference method for the valuation of American options. Journal of Financial Engineering 6, 83–97 (1997)
2. Ikonen, S., Toivanen, J.: Operator splitting methods for American option pricing. Applied Mathematics Letters 17, 809–814 (2004)
3. Forsyth, P.A., Vetzal, K.R.: Quadratic convergence of a penalty method for valuing American options. SIAM J. Sci. Comput. 23, 2096–2123 (2002)
4. Kou, S.G.: Discrete barrier and lookback options. Handbooks in Operations Research and Management Science 15, 343–373 (2008)
5. Turnbull, S., Wakerman, L.: A quick algorithm for pricing European average options. Journal of Financial and Quantitative Analysis 26, 377–389 (1992)
6. Linetsky, V.: Spectral expansions for Asian (average price) options. Operations Research 52, 856–867 (2004)
7. Věceř, J.: A new PDE approach for pricing arithmetic average Asian options. The Journal of Computational Finance 4, 105–113 (2001)
8. Prenter, P.M.: Splines and Variational Methods. Wiley (1975)
9. de Boor, C.: A Practical guide to splines. Springer (1972)
10. Mittal, R.C., Arora, G.: Numerical solution of the coupled viscous Burger's equation. Communications in Nonlinear Science and Numerical Simulation 16, 1304–1313 (2011)
11. Mittal, R.C., Jain, R.K.: Redefined cubic B-splines collocation method for solving convection-diffusion equations. Applied Mathematical Modelling 36, 5555–5573 (2012)
12. Mittal, R.C., Jain, R.K.: Numerical solutions of nonlinear Fisher's reaction-diffusion equation with modified cubic B-spline collocation method. Mathematical Sciences (2013)
13. Kadalbajoo, M.K., Tripathi, L.P., Kumar, A.: A cubic B-spline collocation method for a numerical solution of the generalized Black–Scholes equation. Mathematical and Computer Modelling 55, 1483–1505 (2012)
14. Black, F., Scholes, M.: The pricing of options and other corporate liabilities. Journal of Political Economy 81, 637–654 (1973)
15. Zhu, Y.-L., Wu, X., Chern, I.-L., Sun, Z.-Z.: Derivatives Securities and Different Methods. Springer, New York (2004)
16. Tangman, D.Y., Peer, A.A.I., Rambeerich, N., Bhuruth, M.: Fast simplified approaches to Asian option pricing. The Journal of Computational Finance 14, 3–36 (2011)
17. Strang, G.: On the construction and comparison of difference schemes. SIAM Journal on Numerical Analysis 29, 209–228 (1968)
18. Fang, F., Oosterlee, C.W.: Pricing Early-Exercise and Discrete Barrier Options by Fourier-Cosine Series Expansions. Numerische Mathematik 114, 27–62 (2009)
19. Zhang, B., Oosterlee, C.W.: Efficient Pricing of Asian Options under Lévy Processes Based on Fourier Cosine Expansions, Report submitted to TU Delft University of Technology (2011)

Accelerating Band Linear Algebra Operations on GPUs with Application in Model Reduction

Peter Benner[1], Ernesto Dufrechou[2], Pablo Ezzatti[2], Pablo Igounet[2], Enrique S. Quintana-Ortí[3], and Alfredo Remón[1]

[1] Max Planck Institute for Dynamics of Complex Technical Systems, Magdeburg, Germany
{benner,remon}@mpi-magdeburg.mpg.de
[2] Instituto de Computación, Universidad de la República, 11.300–Montevideo, Uruguay
{edufrechou,pezzatti,pigounet}@fing.edu.uy
[3] Dep. de Ingeniería y Ciencia de la Computación, Universidad Jaime I, Castellón, Spain
quintana@icc.uji.es

Abstract. In this paper we present new hybrid CPU-GPU routines to accelerate the solution of linear systems, with band coefficient matrix, by off-loading the major part of the computations to the GPU and leveraging highly tuned implementations of the BLAS for the graphics processor. Our experiments with an NVIDIA S2070 GPU report speed-ups up to 6× for the hybrid band solver based on the LU factorization over analogous CPU-only routines in Intel's MKL. As a practical demonstration of these benefits, we plug the new CPU-GPU codes into a sparse matrix Lyapunov equation solver, showing a 3× acceleration on the solution of a large-scale benchmark arising in model reduction.

Keywords: Band linear systems, linear algebra, graphics processors (GPUs), high performance, control theory.

1 Introduction

Linear systems with band coefficient matrix appear in a large variety of applications, including finite element analysis in structural mechanics, domain decomposition methods for partial differential equations in civil engineering, and as part of matrix equations solvers in control and systems theory. Exploiting the structure of the matrix in these problems yields huge savings, both in number of computations and storage space. This is recognized by LAPACK [1, 2] that, when linked to a (multi-threaded) implementation of BLAS, provides an efficient means to solve band linear systems on general-purpose (multicore) processors.

In the last few years, hybrid computer platforms consisting of multicore processors and GPUs (graphics processing units) have evolved from being present only in a reduced niche, to become common in many application areas with high computational requirements [3]. A variety of reasons have contributed to the progressive adoption of GPUs, including the introduction of NVIDIA's programming framework CUDA [4, 5] and the OpenACC application programming

B. Murgante et al. (Eds.): ICCSA 2014, Part VI, LNCS 8584, pp. 386–400, 2014.
© Springer International Publishing Switzerland 2014

interface (API), combined with impressive raw performance, affordable price, and an appealing power-performance ratio. In particular, in the area of dense linear algebra many studies have now demonstrated remarkable performance improvements by using GPUs; see e.g., among many others, [6–8].

In this paper we present new LAPACK-style routines that leverage the large-scale hardware parallelism of hybrid CPU-GPU platforms to accelerate the solution of band linear systems. In particular, the experimental results collected on a hardware platform equipped with an Intel i7-2600 processor and an NVIDIA S2070 ("Fermi") GPU with the accelerator-enabled codes demonstrate superior performance and scalability over the highly-tuned multithreaded band solver in Intel MKL (Math Kernel Library). Furthermore, the integration of the GPU solvers with Lyapack [9], a library for the solution of linear and quadratic matrix equations, reveals that these benefits carry over to the solution of model reduction problems arising in control theory.

The rest of the paper is structured as follows. In Section 2 we review the LAPACK routines for the solution of band linear systems, while in Section 3 we introduce our new hybrid CPU-GPU routines for band matrix factorization and the solution of triangular band systems in detail. Section 4 summarizes the experimental evaluation of our new solvers and Section 5 analyzes the application of the developed kernels to the solution of linear matrix equations arising in model reduction (specifically, sparse Lyapunov equations). Finally, a few concluding remarks and a discussion of open questions close the paper in Section 6.

2   Solution of Band Linear Systems with LAPACK

The solution of a linear system of the form

$$AX = B, \tag{1}$$

where $A \in \mathbb{R}^{n \times n}$ is a band matrix with upper and lower bandwidth k_u and k_l respectively, $B \in \mathbb{R}^{n \times m}$ contains a collection of m right-hand side vectors (usually with $m \ll n$), and $X \in \mathbb{R}^{n \times m}$ is the sought-after solution can be performed in two steps using LAPACK. First, the coefficient matrix A is decomposed into two triangular band factors $L, U \in \mathbb{R}^{n \times n}$ (LU factorization) using the routine GBTRF. Then, X is obtained by solving two triangular band systems with coefficients L and U using the routine GBTRS. In this section we describe the process implemented in LAPACK and point out some of the drawbacks of the corresponding routines.

2.1   Factorization of Band Matrices

LAPACK includes two routines for the computation of the LU factorization of a band matrix: GBTF2 and GBTRF, which encode, respectively, unblocked and blocked algorithmic variants of the operation. The former performs most of the computations in terms of BLAS-2 operations, while GBTRF is rich in BLAS-3 operations. As a consequence, GBTRF is more efficient for large matrices, like

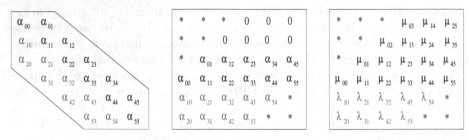

Fig. 1. 6×6 band matrix with upper and lower bandwidths $k_l = 2$ and $k_u = 1$, respectively (left); packed storage scheme used in LAPACK (center); result of the LU factorization where $\mu_{i,j}$ and $\lambda_{i,j}$ stand, respectively, for the entries of the upper triangular factor U and the multipliers of the Gauss transforms.

those appearing in our control applications. Therefore we will focus hereafter on that particular algorithmic variant.

Routine GBTRF computes the *LINPACK-style LU factorization with partial pivoting*

$$L_{n-2}^{-1} \cdot P_{n-2} \cdots L_1^{-1} \cdot P_1 \cdot L_0^{-1} \cdot P_0 \cdot A = U \qquad (2)$$

where $P_0, P_1, \ldots, P_{n-2} \in \mathbb{R}^{n \times n}$ are permutation matrices, $L_0, L_1, \ldots, L_{n-2} \in \mathbb{R}^{n \times n}$ represent Gauss transforms, and $U \in \mathbb{R}^{n \times n}$ is upper triangular with upper bandwidth $k_l + k_u$. Figure 1 illustrates the packed storage scheme used for band matrices in LAPACK and how this layout accommodates the result of the LU factorization with pivoting. We note there that A is stored with k_l additional superdiagonals initially set to zero to make space for fill-in due to pivoting during the factorization. Upon completion, the entries of the upper triangular factor U overwrite the upper triangular entries of A plus these k_l additional rows in the packed array, while the strictly lower triangular entries of A are overwritten with the multipliers which define the Gauss transforms.

Assume that the algorithmic block size, b, internally employed in the blocked routine GBTRF, is an integer multiple of both k_l and k_u, and consider the partitioning

$$A = \left(\begin{array}{c|c|c} A_{TL} & A_{TM} & \\ \hline A_{ML} & A_{MM} & A_{MR} \\ \hline & A_{BM} & A_{BR} \end{array} \right) \rightarrow \left(\begin{array}{c|c|c|c|c} A_{00} & A_{01} & A_{02} & & \\ \hline A_{10} & A_{11} & A_{12} & A_{13} & \\ \hline A_{20} & A_{21} & A_{22} & A_{23} & A_{24} \\ \hline & A_{31} & A_{32} & A_{33} & A_{34} \\ \hline & & A_{42} & A_{43} & A_{44} \end{array} \right), \qquad (3)$$

where $A_{TL}, A_{00} \in \mathbb{R}^{k \times k}$, (with k an integer multiple of b,) $A_{11}, A_{33} \in \mathbb{R}^{b \times b}$, and $A_{22} \in \mathbb{R}^{l \times u}$, with $l = k_l - b$ and $u = k_u + k_l - b$.

Routine GBTRF encodes a right-looking factorization procedure; that is, an algorithm where, before iteration k/b commences, A_{TL} has been already factorized; A_{ML} and A_{TM} have been overwritten, respectively, by the multipliers and the corresponding block of U; A_{MM} has been correspondingly updated; and the rest of the blocks remain untouched. We note that, with this partitioning, A_{31} is upper triangular while A_{33} is lower triangular. Furthermore, at this point, A_{13}, A_{23}, A_{24}, and A_{34} contain only zeros.

In order to move the computation forward by b rows/columns, during the current iteration of the routine the following operations are performed (the annotations to the right of some of these operations correspond to the name of the BLAS routine that is used):

1. Obtain $W_{31} := \text{TRIU}(A_{31})$, a copy of the upper triangular part of A_{31}; and compute the LAPACK-style LU factorization with partial pivoting

$$P_1 \begin{pmatrix} A_{11} \\ A_{21} \\ W_{31} \end{pmatrix} = \begin{pmatrix} L_{11} \\ L_{21} \\ L_{31} \end{pmatrix} U_{11}. \tag{4}$$

   The blocks of L and U overwrite the corresponding blocks of A and W_{31}. (In the actual implementation, the copy W_{31} is obtained while this factorization is being computed.)

2. Apply the permutations in P_1 to the remaining columns of the matrix:

$$\begin{pmatrix} A_{12} \\ A_{22} \\ A_{32} \end{pmatrix} := P_1 \begin{pmatrix} A_{12} \\ A_{22} \\ A_{32} \end{pmatrix} \quad \text{and} \quad \text{(LASWP)} \tag{5}$$

$$\begin{pmatrix} A_{13} \\ A_{23} \\ A_{33} \end{pmatrix} := P_1 \begin{pmatrix} A_{13} \\ A_{23} \\ A_{33} \end{pmatrix}. \tag{6}$$

   A careful application of permutations is needed in (6) as only the lower triangular part of A_{13} is physically stored. As a result of the application of permutations, A_{13}, which initially equals zero, may become lower triangular. No fill-in occurs in the strictly upper part of this block.

3. Compute the updates:

$$A_{12}(= U_{12}) := L_{11}^{-1} A_{12}, \qquad \text{(TRSM)} \tag{7}$$
$$A_{22} := A_{22} - L_{21} U_{12}, \qquad \text{(GEMM)} \tag{8}$$
$$A_{32} := A_{32} - L_{31} U_{12}. \qquad \text{(GEMM)} \tag{9}$$

4. Obtain the copy of the lower triangular part of A_{13}, $W_{13} := \text{TRIL}(A_{13})$; compute the updates

$$W_{13}(= U_{13}) := L_{11}^{-1} W_{13}, \qquad \text{(TRSM)} \tag{10}$$
$$A_{23} := A_{23} - L_{21} W_{13}, \qquad \text{(GEMM)} \tag{11}$$
$$A_{33} := A_{33} - L_{31} W_{13}; \qquad \text{(GEMM)} \tag{12}$$

and copy back $A_{13} := \text{TRIL}(W_{13})$.

5. Undo the permutations on $\left[L_{11}^T, L_{21}^T, W_{31}^T\right]^T$ so that these blocks store the multipliers used in the LU factorization in (6) (as corresponds to a LINPACK-style LU factorization) and W_{31} is upper triangular; copy back $A_{31} := $ TRIU(W_{31}).

In our notation, after these operations are carried out, A_{TL} (the part that has been already factorized) grows by b rows/columns so that

$$
A = \left(\begin{array}{c|c|c}
A_{TL} & A_{TM} & \\
\hline
A_{ML} & A_{MM} & A_{MR} \\
\hline
 & A_{BM} & A_{BR}
\end{array}\right) \leftarrow \left(\begin{array}{c|c|c|c|c}
A_{00} & A_{01} & A_{02} & & \\
\hline
A_{10} & A_{11} & A_{12} & A_{13} & \\
\hline
A_{20} & A_{21} & A_{22} & A_{23} & A_{24} \\
\hline
 & A_{31} & A_{32} & A_{33} & A_{34} \\
\hline
 & & A_{42} & A_{43} & A_{44}
\end{array}\right),
\tag{13}
$$

i.e., $A_{TL} \in \mathbb{R}^{(k+b)\times(k+b)}$, in preparation for the next iteration.

Provided $b \ll k_u, k_l$ (in practice, to optimize cache usage, $b \approx 32$ or 64) the update of A_{22} involves most of the floating-point arithmetic operations (flops). This operation can be cast in terms of the matrix-matrix product, a computation that features an ample degree of parallelism and, therefore, we can expect high performance from GBTRF provided a tuned implementation of GEMM is used.

On the other hand, the algorithm presents also two important drawbacks regarding its implementation in parallel architectures:

– The triangular structure of block A_{13}/U_{13} is not exploited in computations (10)–(12) as there exists no kernel in BLAS to perform such a specialized operation. Consequently, an additional storage is required, as well as two extra copies, and a non-negligible amount of useless flops that involve null elements are performed in these operations.
– Forced by the storage scheme and the lack of specialized BLAS kernels, the updates to be performed during an iteration are split into several small operations with reduced inner parallelism.

2.2   Solution of Triangular Systems

Given the LU factorization computed by GBTRF in (2), routine GBTRS from LAPACK tackles the subsequent band triangular systems to obtain the solution of (1). For this purpose, the routine proceeds as follows:

1. For $i = 0, 1, \ldots, n-2$, (in that strict order,) apply the permutation matrix P_i to the right-hand side term B, and update this matrix with the corresponding multipliers in L_i:

$$
\begin{aligned}
B &:= P_i B, & \text{(SWAP)} \\
B &:= L_i^{-1} B. & \text{(GER)}
\end{aligned}
\tag{14}
$$

2. For $j = 1, 2, \ldots, m$ solve a triangular system with coefficient matrix U and the right-hand side vector given by the j-th column of B (denoted as B_j)

$$
B_j := U^{-1} B_j. \qquad \text{(TBSV)}
\tag{15}
$$

Despite the operation tackled by GBTRS belongs, by definition, to the Level-3 BLAS, in this implementation it is entirely cast in terms of less efficient BLAS-2 kernels. This is due to the adoption of the packed storage scheme, the decision of not forming the triangular factor L explicitly (to save storage space), and the lack of a routine in the BLAS specification to solve a triangular system with multiple right-hand sides when the coefficient matrix presents a triangular band structure. We note that routine GETRS, which performs the analogous operation in the non-banded case, does not suffer from these shortcomings.

3   New Hybrid CPU-GPU Band Solvers

The algorithm underlying the routine GBTRF invokes, at each iteration, routine TRSM twice and routine GEMM four times (steps 3 and 4). This partitioning of the work is due to the particular storage scheme adopted for band matrices in LAPACK. However, since in LAPACK, concurrency is extracted from the usage of multithreaded implementations of the BLAS kernels, the fragmentation of the computations into small operations potentially limits the performance of the codes. This feature is specially harmful when the algorithm is executed on many-core architectures, like the GPUs, where computations involving large data sets and many flops are mandatory to exploit the capabilities of this type of architectures.

Similarly, routine GBTRS casts its computations in terms of BLAS-2 kernels, e.g. solving for every column of the right-hand side independently. Again, the computations are fragmented, and consequently, their inner concurrency is re duced.

In this section we present GPU-friendly implementations for the routines GB-TRF and GBTRS. To adapt their execution to the target platforms, we perform a reordering of the computations and minimum changes in the data storage scheme that permit to merge computations, the use of BLAS-3 kernels (instead of BLAS-2), and improve the inner concurrency of some kernels. Therefore, the new method is more suitable for architectures with a medium to large number of computational units, like current multi-core processors and GPUs. The drawback of the proposal is that the new storage format implies a moderate increment in the memory requirements but, as will be demonstrated later, at the same time it also yields important gains in terms of performance.

3.1   Routine GBTRF+M

Assume the packed data structure containing A (see Figure 1-center) is padded with b extra rows at the bottom, with all the entries in this additional space initially set to zero. Then, steps 1–4 in the original implementation of GBTRF can be transformed as follows:

1. In the first step, the LU factorization with partial pivoting

$$P_1 \begin{pmatrix} A_{11} \\ A_{21} \\ A_{31} \end{pmatrix} = \begin{pmatrix} L_{11} \\ L_{21} \\ L_{31} \end{pmatrix} U_{11} \tag{16}$$

is computed and the blocks of L and U overwrite the corresponding blocks of A. There is no longer need for workspace W_{31} nor copies to/from it as the additional rows at the bottom accommodate the elements in the strictly lower triangle of L_{31}. Although GBTF2 can be used to complete this computation, it will still require to undo the permutations performed by GBTF2 to keep the upper triangular structure of the block L_{31}.

2. Apply the permutations in P_1 to the remaining columns of the matrix:

$$\begin{pmatrix} A_{12}\ A_{13} \\ A_{22}\ A_{23} \\ A_{32}\ A_{33} \end{pmatrix} := P_1 \begin{pmatrix} A_{12}\ A_{13} \\ A_{22}\ A_{23} \\ A_{32}\ A_{33} \end{pmatrix} \quad \text{(LASWP)}. \tag{17}$$

A single call to LASWP suffices now as the zeros at the bottom of the data structure and the additional k_l superdiagonal set to zero in the structure ensure that fill-in may only occur in the elements in the lower triangular part of A_{13}.

3. Compute the updates:

$$(A_{12},\ A_{13})\,(= (U_{12},\ U_{13})) := L_{11}^{-1}\,(A_{12},\ A_{13}) \quad \text{(TRSM)}, \tag{18}$$

$$\begin{pmatrix} A_{22}\ A_{23} \\ A_{32}\ A_{33} \end{pmatrix} := \begin{pmatrix} A_{22}\ A_{23} \\ A_{32}\ A_{33} \end{pmatrix} \\ - \begin{pmatrix} L_{21} \\ L_{31} \end{pmatrix} (U_{12},\ U_{13}) \quad \text{(GEMM)}. \tag{19}$$

The lower triangular system in (18) returns a lower triangular block in A_{13}.

4. Undo the permutations on $\left[L_{11}^T, L_{21}^T, L_{31}^T \right]^T$ so that these blocks store the multipliers used in the LU factorization in (6) and L_{31} is upper triangular.

This reorganization of the algorithm is rich in matrix-matrix products, and hence, it is suitable for massively parallel architectures. The implementation in GBTRF+M takes profit of this enhanced concurrency with the purpose of efficiently exploiting the capabilities of the CPU-GPU platform. In particular, the factorization of the narrow panel (16), which presents a fine-grain parallelism and a modest computational cost, is executed in the CPU. On the other hand, the application of the permutations (17) and the updates (18)–(19) are performed in the GPU, in order to reduce the CPU-GPU communications and exploit the massively parallel architecture of the graphics accelerator.

This implementation executes each operation in the most convenient device while incurring into a moderate number of data transfers between CPU and GPU. In particular, it requires an initialization phase where the matrix A is moved to the GPU before the factorization commences. Then, during the factorization, two copies must be performed per iteration:

1. The entries of $[A_{11}^T, A_{21}^T, A_{31}^T]^T$ are transferred to the GPU after the factorization of this block in (16).
2. The entries that will form $[A_{11}^T, A_{21}^T, A_{31}^T]^T$ during the next iteration are transferred to the CPU after their update as part of (19).

Note that the amount of data transferred at each iteration is moderate in relation to the number of flops, since the number of rows and columns of the block $[A_{11}^T \, A_{21}^T \, A_{31}^T]^T$ are $(k_u + k_l + k_l)$ and b, respectively. Furthermore, upon completion of the algorithm, the resulting matrix is replicated in the CPU and GPU, so it can be used during subsequent computations in both devices.

3.2   Routine GBTRF+**LA**

Routine GBTRF+LA is an improved variant of routine GBTRF+M which incorporates look-ahead [10] to further overlap the computations performed by CPU and GPU. Concretely, GBTRF+LA reorders the computations as follows: the updates in (17)–(19), which involve $k_u + k_l$ columns of the matrix, are split column-wise, such that the first b columns are computed first. Then, the updated elements that form the block $[A_{11}^T, A_{21}^T, A_{31}]^T$ of the next iteration (b columns), are sent to the CPU, where the factorization of this block can be performed in parallel with the updates of the remaining $k_u + k_l - b$ columns corresponding to (17)–(19) in the GPU.

This variant requires minimal changes to the codes. Despite it demands the execution in the GPU of kernels with a moderate number of flops, it permits that computations proceed concurrently in both devices, reporting higher performance whenever $b \ll k_u + k_l$.

3.3   Routine GBTRS+**M**

The main drawback of LAPACK routine GBTRS is the absence of BLAS-3 kernels in its implementation. This is due to the adoption of the packed storage format and the lack of the appropiate BLAS routines. Unfortunately, the modifications introduced in the storage scheme still limit the use of BLAS-3 kernels. In particular, as the matrix L is not explicitly formed, in principle the update in (14) must be performed by means of a rank-1 update operation (routine GER from BLAS). However, it is possible to employ BLAS-3 in the solution of the systems in (15). For this purpose, we developed a new routine, named TBSM following the LAPACK convention, that performs this operation via BLAS-3 kernels (mainly matrix-matrix products). Thus, GBTRS presents two parts, the first one is a loop that updates B as in (14). Afterwards, a single call to TBSM solves (15) for all j, as described in Section 3.4.

We provide two implementations for this routine, a CPU and a GPU variant. The convenience of these implementations depends on the coefficient matrix dimension (n) and in the number of columns of B (m). But as usually $m \ll n$, we can focus our analysis on the coefficient matrix dimension. The CPU variant is suitable for medium to small values of n, since it does not require any CPU-GPU data-transfer and the computational cost of the operation is moderate. On the contrary, when n is large, the GPU implementation is more suitable due to the large computational cost and the inner concurrency of the operations involved. Note that the GPU implementation only requires to transfer the matrix B to/from the GPU, as the use of the routines GBTRF+M or GBTRF+LA ensures that the factors L and U are already stored in the GPU memory. Additionally, as all the computations are performed in the GPU, no CPU-GPU synchronizations are required.

3.4 Routine TBSM

Consider the row partitioning of the right-hand side matrix B, to be overwritten with the solution X to (1),

$$
B = \left(\frac{\frac{B_T}{B_M}}{B_B} \right) \rightarrow \left(\begin{array}{c} B_0 \\ \hline B_1 \\ \hline B_2 \\ \hline B_3 \\ \hline B_4 \end{array} \right), \tag{20}
$$

where B_B, B_4 have both k rows (with k an integer multiple of b), B_1, B_3 have b rows each, and B_2 has $u = k_u + k_l - b$ rows. Here, B_B represents the part of the right-hand side which has already been overwritten with the corresponding entries of the solution X.

Consider also the following conformal partitioning for the upper triangular factor U resulting from the factorization

$$
U = \left(\begin{array}{c|c} U_{TL} & U_{TM} \\ \hline & \begin{array}{c|c} U_{MM} & U_{MR} \\ \hline & U_{BR} \end{array} \end{array} \right) \rightarrow \left(\begin{array}{c|c|c|c|c} U_{00} & U_{01} & U_{02} & & \\ \hline & U_{11} & U_{12} & U_{13} & \\ \hline & & U_{22} & U_{23} & U_{24} \\ \hline & & & U_{33} & U_{34} \\ \hline & & & & U_{44} \end{array} \right), \tag{21}
$$

where U_{BR}, $U_{44} \in \mathbb{R}^{k \times k}$; $U_{13}, U_{33} \in \mathbb{R}^{b \times b}$ are lower and upper triangular, respectively; and $U_{22} \in \mathbb{R}^{u \times u}$.

Then, in order to proceed forward, the following operations are required at this iteration:

$$
\begin{align}
B_3 &:= U_{33}^{-1} B_3, & \text{(TRSM)} \tag{22} \\
B_2 &:= B_2 - U_{23} B_3, & \text{(GEMM)} \tag{23} \\
B_1 &:= B_1 - U_{13} B_3. & \tag{24}
\end{align}
$$

The last update involves a triangular matrix (U_{13}) and can be performed by means of the BLAS routine TRMM. However, this requires an auxiliary storage,

$W_{13} \in \mathbb{R}^{b \times m}$, since this BLAS kernel only performs a product of the form $M :=$ $U_{13}\, M$. Therefore, we perform the next operations for (24):

$$W_{13} := B_3, \tag{25}$$

$$W_{13} := U_{13}W_{13}, \qquad \text{(TRMM)} \tag{26}$$

$$B_1 := B_1 - W_{13}. \tag{27}$$

After these operations are completed, in preparation for the next iteration, the boundaries in B and U are simply shifted as

$$B = \begin{pmatrix} B_T \\ B_M \\ B_B \end{pmatrix} \leftarrow \begin{pmatrix} \dfrac{B_0}{B_1} \\ \dfrac{B_2}{B_3} \\ \overline{B_4} \end{pmatrix}$$

$$U = \left(\begin{array}{c|cc} U_{TL} & U_{TM} & \\ \hline & U_{MM} & U_{MR} \\ & & U_{BR} \end{array} \right) \leftarrow \left(\begin{array}{ccc|cc} U_{00} & U_{01} & U_{02} & & \\ & U_{11} & U_{12} & U_{13} & \\ & & U_{22} & U_{23} & U_{24} \\ \hline & & & U_{33} & U_{34} \\ & & & & U_{44} \end{array} \right). \tag{28}$$

4 Experimental Evaluation

In this section we analyze the computational performance of the new routines for the band LU factorization (GBTRF+M and GBTRF+LA) as well as two implementations of the triangular band solvers that follow the algorithm for this phase described in the previous section, but differ in the target architecture: CPU or GPU (denoted as GBTRS+M$_{CPU}$ and GBTRS+M$_{GPU}$ hereafter). Their performances are compared with that of the analogous routines in release 11.1 of Intel MKL (denoted hereafter as GBTRF$_{Intel}$ and GBTRS$_{Intel}$).

Table 1. Hardware platform employed in the experimental evaluation

Platform	Processors	#cores	Frequency (GHz)	L3 cache (MB)	Memory (GB)
ENRICO	intel i7-2600	4	3.4	8	16
	NVIDIA S2070	448	1.15	–	6

The performance evaluation was carried out using a hardware platform, EN-RICO, equipped with an NVIDIA S2070 ("Fermi") GPU and an Intel four-core processor; see Table 1. All experiments reported next were performed using IEEE double-precision real arithmetic. We employed band linear systems with 6 different coefficient matrix dimensions $n = 12{,}800,\ 25{,}600,\ 38{,}400,\ 51{,}200,\ 64{,}000$

and 76,800. For each dimension, we generated 3 instances which varied in the bandwidth, $k_b = k_u = k_l =1\%$, 2% and 4% of n. We evaluated several algorithmic block sizes (b) for each kernel, but for brevity, we only include the results corresponding to the best block size tested.

Table 2 compares the three codes for the band LU factorization: the two new hybrid CPU-GPU implementations, GBTRF+M and GBTRF+LA, and MKL's GBTRF$_{\text{Intel}}$. These results demonstrate the superior performance of the new implementations when the volume of computations is large. Concretely, both hybrid codes outperform the MKL routine for large matrices while they are still competitive for relatively small matrices. This was expected, as the hybrid routines incur a communication overhead that can be compensated only when the problem is considerably large. In summary, GBTRF+M and GBTRF+LA are superior to MKL for the factorization of matrices with n >25,600 and k_b =2% of n. When n >51,200, the new variants are faster than MKL even when k_b =1% of n. For the largest problem tested, n =76,800, the acceleration factors obtained by GBTRF+LA with respect to the MKL code are 2.0, 3.9 and 5.5× for k_b =1, 2 and 4% of n respectively. The performance obtained by GBTRF+M is slightly lower, reporting acceleration factors of 1.9, 3.5 and 5.0× for the same problems.

Additionally, we compared MKL's triangular band solver GBTRS$_{\text{Intel}}$ against both alternative codes proposed in this work (GBTRS+M$_{CPU}$ and GBTRS+M$_{GPU}$), on the solution of a linear system with a single right-hand side ($m = 1$). In this sce-

Table 2. Execution time (in seconds) for the LU factorization of band matrices in ENRICO

Matrix Dimension	Bandwidth $k_b = k_u = k_l$	GBTRF$_{\text{Intel}}$	GBTRF+M	GBTRF+LA
12,800	1%	0.066	0.174	0.180
	2%	0.142	0.240	0.245
	4%	0.385	0.358	0.341
25,600	1%	0.313	0.482	0.493
	2%	0.786	0.701	0.691
	4%	3.397	1.339	1.231
38,400	1%	0.684	0.867	0.844
	2%	2.588	1.502	1.393
	4%	11.742	3.517	3.407
51,200	1%	1.898	1.537	1.399
	2%	6.989	3.131	2.496
	4%	31.745	7.217	6.627
64,000	1%	3.104	2.175	2.029
	2%	12.241	4.465	4.053
	4%	52.701	12.796	11.660
76,800	1%	5.749	3.044	2.808
	2%	24.490	6.914	6.286
	4%	103.264	20.462	18.769

nario, the execution times were comparable though, in general, the performance of the MKL implementation was slightly higher than that of our routines. It is important to note that the new optimizations should be more beneficial when several systems are solved for the same coefficient matrix (i.e., $m > 1$). This is the case in several engineering applications and, in particular, in our target control application. We also remark that MKL is not an open source library, so its implementation may differ from that described in Section 2. In particular, it is likely that MKL also uses BLAS-3 kernels for the triangular band solver, which could explain the similarities between its performance and that of the new implementations.

Considering the results and the reduced impact of GBTRS in the total runtime of the solver, we decided to use the MKL implementation for the solution of the triangular band linear systems. Figure 2 illustrates the speed-up achieved by the best CPU-GPU routine for LU factorization (left) and for the complete band system solver (right). In both cases the reference to compute the acceleration is the solver provided by the MKL library (i.e., routines GBTRF$_{\text{Intel}}$ and GBTRS$_{\text{Intel}}$). As most of the flops correspond to the computation of the LU factorization, the speed-ups obtained for the complete solver are similar to those for the LU.

5   Application to Model Reduction

In this section we evaluate the impact of the new CPU-GPU banded solvers on the solution of Lyapunov equations of the form

$$AX + XA^T + BB^T = 0, \tag{29}$$

where $A \in \mathbb{R}^{n \times n}$ is sparse, $B \in \mathbb{R}^{n \times m}$, with $m \ll n$, and $X \in \mathbb{R}^{n \times n}$ is the sought-after solution. This linear matrix equation has important applications, among others, in model reduction and linear-quadratic optimal control problems; see, e.g., [11].

The Lyapunov solver employed in our approach consists in a modified version of the *low-rank Cholesky factor - alternating directions implicit* (LRCF-ADI)

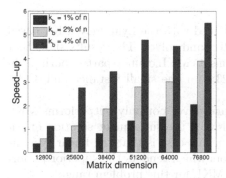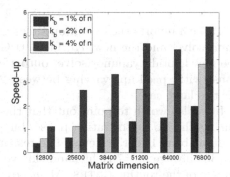

Fig. 2. Speed-ups of the new hybrid CPU-GPU solvers against their MKL counterparts for the factorization (left) and the complete solver (right)

method [12]. This iterative solver benefits from the frequently encountered low-rank property of the BB^T factor in (29) to deliver a low-rank approximation to a Cholesky or full-rank factor of the solution matrix. Specifically, given an "l–cyclic" set of complex shift parameters $\{p_1, p_2, \ldots\}$, $p_k = \alpha_k + \beta_k \imath$, with $\imath = \sqrt{-1}$ and $p_k = p_{k+l}$, the cyclic *low-rank alternating directions implicit* (LR-ADI) iteration can be formulated as follows:

$$
\begin{aligned}
V_0 &= (A + p_1 I_n)^{-1} B, & \hat{S}_0 &= \sqrt{-2\,\alpha_1}\ V_0, \\
V_{k+1} &= V_k - \delta_k (A + p_{k+1} I_n)^{-1} V_k, & \hat{S}_{k+1} &= \left[\hat{S}_k \ , \ \gamma_k V_{k+1} \right],
\end{aligned}
\tag{30}
$$

where $\gamma_k = \sqrt{\alpha_{k+1}/\alpha_k}$, $\delta_k = p_{k+1} + \overline{p_k}$, with $\overline{p_k}$ the conjugate of p_k, and I_n denotes the identity matrix of order n. On convergence, after $\hat{k}$ iterations, a low-rank matrix $\hat{S}_{\hat{k}} \in \mathbb{R}^{n \times \hat{k}m}$ is computed such that $\hat{S}_{\hat{k}} \hat{S}_{\hat{k}}^T \approx X$.

It should be observed that the main computations in (30) consist in solving linear systems with multiple (m) right-hand sides. Therefore the application of our new solver should significantly accelerate the ADI iteration.

Our approach to tackle the sparse structure of the coefficient matrix A in (29) applies a reordering based on the Reverse Cuthill-McKee algorithm [13] to transform the sparse linear systems in the expressions for V_0 and V_{k+1} in (30) into analogous problems with band coefficient matrix. In particular, we evaluated this approach using the Lyapunov equations associated with two instances of the RAIL model reduction problem from the Oberwolfach benchmark collection [14]; see Table 3.

Table 3. Instances of the RAIL example from the Oberwolfach model reduction collection employed in the evaluation

Problem	n	$k_u = k_l$	# nonzeros	m
RAIL$_S$	5,177	139	35,185	7
RAIL$_L$	20,209	276	139,233	7

Table 4 reports the execution times obtained with the Lyapunov MKL-based band solver and the new hybrid CPU-GPU band solver. The results show that the new hybrid Lyapunov solver outperforms its MKL counterpart in both problems, with speed-ups varying between 2.23× for the small instance and 3.14× for the large case.

It is also worth to point out that the new solver not only outperforms MKL during the LU factorization phase, but also for the subsequent solution of triangular band linear systems. The reason in this case is that the linear systems in (30) involve $m = 7$ right-hand side vectors which renders the superior performance of the routine GBTRS+M$_{CPU}$ over MKL for this problem range.

Table 4. Execution time (in seconds) and speed-ups obtained with the hybrid CPU-GPU Lyapunov solvers over the corresponding MKL-based counterparts in ENRICO

Problem	MKL solver	GPU-based solver	Speed-up
RAIL$_S$	2.34	1.05	2.23
RAIL$_L$	17.71	5.65	3.14

6   Concluding Remarks

We have presented new hybrid CPU-GPU routines that accelerate the LU factorization and the subsequent triangular solves for band linear systems, by off-loading the computationally expensive operations to the GPU. Our first CPU-GPU implementation for the LU factorization stage computes the BLAS-3 operation on the hardware accelerator by invoking appropriate kernels from NVIDIA CUBLAS while reducing the amount of CPU-GPU communication. The second GPU variant for this operation incorporates a look-ahead strategy to overlap the update in the GPU with the factorization of the next panel in the CPU.

The experimental results obtained using several band test cases (with dimensions between 12,800 and 76,800 and a bandwidth of 1%, 2% and 4% of the problem size), in a platform equipped with an NVIDIA 2070, reveals speed-ups for the CPU-GPU LU factorization of up to 6×, when compared with the corresponding factorization routine from Intel MKL. The advantages of the hybrid band routines carry over to the solution of sparse Lyapunov solvers, with an acceleration factor around 2-3× with respect to the analogous solver based on MKL.

As part of future work, we plan to enhance the performance of the Lyapunov solver by off-loading to the GPU other band linear algebra operations present in the LR-ADI method as, e.g., the band matrix-vector product. Furthermore, we plan to study the impact of the new GPU-accelerated algorithms on energy consumption.

Acknowledgements. Ernesto Dufrechou, Pablo Ezzatti and Pablo Igounet acknowledge support from Programa de Desarrollo de las Ciencias Básicas, and Agencia Nacional de Investigación e Innovación, Uruguay. Enrique S. Quintana-Ortí was supported by project TIN2011-23283 of the Ministry of Science and Competitiveness (MINECO) and EU FEDER, and project P1-1B2013-20 of the Fundació Caixa Castelló-Bancaixa and UJI.

References

1. Anderson, E., Bai, Z., Demmel, J., Dongarra, J.E., DuCroz, J., Greenbaum, A., Hammarling, S., McKenney, A.E., Ostrouchov, S., Sorensen, D.: LAPACK Users' Guide. SIAM, Philadelphia (1992)
2. Du Croz, J., Mayes, P., Radicati, G.: Factorization of band matrices using level 3 BLAS. LAPACK Working Note 21, Technical Report CS-90-109, University of Tennessee (1990)

3. The Top500 list (2013), http://www.top500.org
4. Kirk, D., Hwu, W.: Programming Massively Parallel Processors: A Hands-on Approach, 2nd edn. Morgan Kaufmann (2012)
5. Farber, R.: CUDA application design and development. Morgan Kaufmann (2011)
6. Volkov, V., Demmel, J.: LU, QR and Cholesky factorizations using vector capabilities of GPUs. Technical Report UCB/EECS-2008-49, EECS Department, University of California, Berkeley (2008)
7. Barrachina, S., Castillo, M., Igual, F.D., Mayo, R., Quintana-Ortí, E.S.: Solving dense linear systems on graphics processors. In: Luque, E., Margalef, T., Benítez, D. (eds.) Euro-Par 2008. LNCS, vol. 5168, pp. 739–748. Springer, Heidelberg (2008)
8. Benner, P., Ezzatti, P., Quintana-Ortí, E.S., Remón, A.: Matrix inversion on CPU–GPU platforms with applications in control theory. Concurrency and Computation: Practice and Experience 25, 1170–1182 (2013)
9. Penzl, T.: LYAPACK: A MATLAB toolbox for large Lyapunov and Riccati equations, model reduction problems, and linear-quadratic optimal control problems. User's guide, version 1.0. (2000), http://www.netlib.org/lyapack/guide.pdf
10. Strazdins, P.: A comparison of lookahead and algorithmic blocking techniques for parallel matrix factorization. Technical Report TR-CS-98-07, Department of Computer Science, The Australian National University, Canberra 0200 ACT, Australia (1998)
11. Antoulas, A.: Approximation of Large-Scale Dynamical Systems. SIAM Publications, Philadelphia (2005)
12. Penzl, T.: A cyclic low-rank Smith method for large sparse Lyapunov equations. SIAM J. Sci. Comput. 21, 1401–1418 (1999)
13. Cuthill, E., McKee, J.: Reducing the bandwidth of sparse symmetric matrices. In: Proceedings of the 1969 24th National Conference, ACM 1969, pp. 157–172. ACM, New York (1969)
14. IMTEK (Oberwolfach model reduction benchmark collection), http://www.imtek.de/simulation/benchmark/

A Video Tensor Self-descriptor
Based on Block Matching

Ana M. O. Figueiredo[1], Helena A. Maia[1], Fábio L. M. Oliveira[1],
Virgínia F. Mota[2], and Marcelo Bernardes Vieira[1]

[1] Universidade Federal de Juiz de Fora, Juiz de Fora, Brazil
{anamara,helena.maia,fabio,marcelo.bernardes}@ice.ufjf.br
[2] Universidade Federal de Minas Gerais, Belo Horizonte, Brazil
virginiaferm@dcc.ufmg.br

Abstract. In this paper, we propose a new motion descriptor which uses
only block matching vectors. This is a different and simple approach considering that most works on the field are based on the gradient of image
intensities. The block matching method returns displacements vectors as
a motion information. Our method computes this information to obtain
orientation tensors and to generate the final descriptor. It is considered
a self-descriptor, since it depends only on the input video. The global
tensor descriptor is evaluated by a classification of KTH, UCF11 and
Hollywood2 video datasets with a non-linear SVM classifier. Our results
indicate that the method runs fast and has fairly competitive results
compared to similar approaches. It is suitable when the time response
is a major application issue. It also generates compact descriptors which
are desirable to describe large datasets.

Keywords: self-descriptor, compact descriptor, block matching, human
action recognition.

1 Introduction

Human action recognition has been extensively researched over the past years
due to its application in many areas such as: video indexing, surveillance, human-
computer interfaces, among others. In order to approach this problem, many
authors have proposed video descriptors using motion representation, which is
one of the main characteristics that describes the semantic information of videos.

Among the methods to detect motion, block matching is used to find vectors
that indicate block displacements between two video frames. We chose this technique as it has not been extensively applied to human action recognition and
several works on literature use block displacement vectors for other applications,
for example [1–3]. Moreover, this method runs fast and can potentially generate
compact descriptors since it is widely used in video compression.

This work is motivated by the possibility of generating a compact and easy
to compute descriptor. Our main contribution is a new motion descriptor, based
on orientation tensor [4], which uses only block matching information. This is

B. Murgante et al. (Eds.): ICCSA 2014, Part VI, LNCS 8584, pp. 401–414, 2014.
© Springer International Publishing Switzerland 2014

a different approach, considering that most works on the field are based on the gradient of image intensities [5, 6]. The global tensor descriptor created is evaluated by a classification of KTH, UCF11 and Hollywood2 video datasets with a non-linear Support Vector Machine (SVM) classifier.

We present three variants of our method. The first, called Single Scale Single Vector (SSSV), is the simplest and fastest. It has the same elements as the block matching method: one fixed block size and one vector field generated for each pair of frames. The second, called Multiple Scales Single Vector (MSSV), yields better results than the first, at the cost of slower execution speed. Since it considers more than one block size, it requires multiple computations for each pair of frames. The third version, called Multiple Scales Multiple Vectors (MSMV), yields even better recognition rates but is also even slower. It considers multiple block sizes and also goes through more than two consecutive frames.

2   Related Works

Several works in literature use the motion intensity obtained from block matching in many applications. Hafiane et al [1] presents a method for video registration based on Prominent Feature (PF) blocks. Block matching was used to identify moving objects in a video. Structured tensors sensitive to edge and corners were used to extract a point of interest in each block. In order to find region correspondences between images, block matching was used along with Normalized Cross-Correlation (NCC) or Sum of Absolute Differences (SAD) as an error estimate. NCC is less sensitive to absolute intensity changes between the reference and target images due to normalization, but is much more expensive to compute than SAD. In this work, we employ SAD as an error function and the Four Step Search (4SS) as a fast search strategy since it is computationally more efficient than Full Search (FS). As another less costly alternative for handling intensity outliers, we apply a smoothing filter on each frame, so that SAD obtains quality results.

Similar to [1], a block matching method was used for extracting motion information in [2]. However, this information was used to generate a motion activity descriptor for shot boundary detection in video sequences. The chosen method for quickly computing the motion vectors was Adaptive Rood Pattern Search (ARPS). These vectors were used to calculate the intensity of motion and also classify among the categories presented by the authors. Vectors with higher values indicate a greater probability of being a shot.

An activity descriptor, consisting of a temporal descriptor and a spatial descriptor, is presented in [3]. The temporal descriptor is obtained through the ratios between moving blocks and all the blocks on each frame. In order to be labelled as a moving block, the error must be within a margin of tolerance. These ratios are then adjusted into quantized levels. The spatial descriptor, also used in [2], is obtained through a matrix containing all the motion vectors norms from each frame.

Other video descriptors were proposed using different methods for extracting information, such as the human action descriptor shown in [4–8]. Klaser et al [9] propose a local feature based descriptor for video sequences generalizing Histogram of Oriented Gradient (HOG) concepts to 3D. In [7], they extend the Features from Accelerated Segment Test (FAST) method for the 3D space. The information of shape and motion was obtained detecting spatial and temporal features. Following [9], they produced a descriptor based on HOG3D which describes corner features. Both use KTH and Hollywood2 databases to evaluate performance. We also use these databases to evaluate our descriptor, but we generate a global descriptor using information from Motion Estimation (ME).

Mota et al [10] presented a tensor motion descriptor for video sequences using optical flow and HOG3D information. They use an aggregation tensor based technique. This technique combines two descriptors, one includes polynomial coefficients which approximate optical flow, and the other accumulates data from HOGs. This descriptor is evaluated by a SVM classifier using KTH, UCF11 and Hollywood2 datasets. In our work, the approach of using block matching vectors reduces considerably the effort of tracking motion as compared to the use of HOG3D. Moreover, the bidimensional nature of block displacements reduces significantly the size of the histogram coded into a tensor. Compared to [10], our descriptor is more compact and easier to compute, while still yielding competitive results.

3 Block Matching Descriptor: Single Scale Single Vector

There are several block matching methods which can be used to extract motion information. The simplest is the Full Search. This method searches for each 16×16 block from the reference (current) frame in the target (next) frame. The corresponding, or matching, block is the one which minimizes a cost function such as SAD or MAD (Mean Absolute Difference), representing high similarity between blocks. The search window on the target consists of all the possible blocks differing from -7 to 7 pixels in both directions from the reference frame block. Thus, all the 225 neighbouring blocks are evaluated. Although it is a precise method, it is computationally expensive. Therefore, several fast methods were proposed such as 4SS [11], ARPS [12] and DS (Diamond Search) [13], based on steepest descent methods.

The 4SS consists of four steps with three distinct search patterns. In the first step, it checks nine points in a 5×5 window. The point referring to the block with the lowest Block Distortion Measure (BDM) becomes the center of the search window in the following step. Whenever the minimum BDM is at the center window point, the algorithm proceeds to the fourth step. In the second and third steps, it checks five or three blocks depending on whether the previous step results on a corner or a side point, respectively. The last step consists on checking eight points in a 3×3 window. In the worst case scenario, 27 blocks are evaluated.

The DS is fairly similar to 4SS. In both, the first step checks nine points and the following steps check three or five points. However, DS uses a diamond shaped search window and instead of having four steps, it repeats the second step until the lowest BDM is found at the center of the pattern or it reaches an iteration limit. It then proceeds to the third step for the final four BDM evaluations.

In this work, we use the displacement map generated by the 4SS algorithm. The input is a video, i.e., a set of frames $V = \{F_k\}$, where $k \in [1, n_f]$, n_f is the number of frames and the output is a tensor descriptor $\mathbf{T} \in \mathbb{R}^{n \times n}$ which, in fact, can be viewed as a vector in $\mathbb{R}^m$, $m = n^2$. Our descriptor is considered a self-descriptor, since it depends only on the input video. It is computed by extracting and accumulating information from each frame of the video. Basically, the frame is divided into blocks and their displacement vectors are computed. As proposed in [10], this vector field is represented by a histogram which is encoded into an orientation tensor. The tensors of each frame are accumulated to form the final tensor descriptor $\mathbf{T}$ of the video. This first method is simpler and faster than the following methods.

3.1 Computing the Motion Estimation Histogram

In the 4SS schemes, each frame k of the video is subdivided into $n_x \times n_y$ non-overlapping blocks of exactly $s \times s$ pixels. Thus, our first method (SSSV) is constrained to use square sub-images of the frame. If the frame dimension is not a multiple of s, the remaining right and bottom pixels do not form blocks. For each block, displacement vectors $v_k(i, j) = (x, y) \in \mathbb{R}^2$ are calculated, where $(i, j) \in [1, n_x] \times [1, n_y]$ are the block indexes. These vectors are converted to equivalent polar coordinates $c_k(i, j) = (\theta, r)$ with $\theta = \tan^{-1}(\frac{y}{x})$, $\theta \in [0, 2\pi]$ and $r = \| v_k(i, j) \|$.

A motion estimation histogram is used as a compact representation of the motion vector field obtained from each frame. It is defined as the column vector $h_k = (h_1, h_2, \ldots, h_{n_\theta})^T$, where n_θ is the number of cells for the θ coordinate. We use an uniform subdivision of the angle intervals. Each interval is populated as the following equation:

$$h_l = \sum_{i,j} r(i, j) \cdot \omega(i, j) \ , \tag{1}$$

where $l = 1, 2, \ldots, n_\theta$ and $\omega(i, j)$ is a vector weighting factor, which is a Gaussian function with $\sigma = 0.01$ in our experiments. The whole frame vector field is thus represented by a vector h_k with n_θ elements.

3.2 Tensor Descriptor: Coding the Motion Estimation Histogram

An orientation tensor is a representation of local orientation which takes the form of a $n \times n$ real symmetric matrix for n-dimensional signals [14]. Given a vector $v \in \mathbb{R}^n$, it can be represented by the tensor $\mathbf{T} = vv^T$. Then, we use the orientation tensor to represent the histogram $h_k \in \mathbb{R}^{n_\theta}$. The frame tensor for the size s, $\mathbf{T}_k(s) \in \mathbb{R}^{n_\theta \times n_\theta}$, is given by:

$$\mathbf{T}_k(s) = h_k \cdot h_k^T \ . \tag{2}$$

Individually, these frame tensors have the same information as h_k, but several tensors can be combined to find component covariances.

3.3 Orientation Tensor: Accumulating the Motion Estimation Tensors

The motion average of consecutive frames can be expressed using a series of tensors. The average motion is given by

$$\mathbf{T}(s) = \sum_{k=1}^{n_f} \frac{\mathbf{T}_k(s)}{\| \mathbf{T}_k(s) \|_2} \, ,$$

using all video frames. By normalizing $\mathbf{T}$ with a L_2 norm, we are able to compare different video clips or snapshots regardless their length or image resolution. Since $\mathbf{T}$ is a symmetric matrix, it can be stored with $d = \frac{n_\theta(n_\theta+1)}{2}$ elements.

If the motion captured in the histograms are too different from each other, we obtain an isotropic tensor which does not hold useful motion information. But,

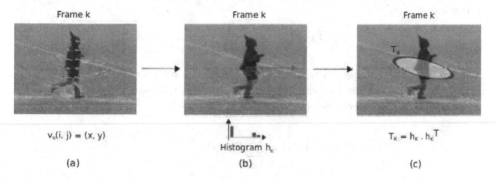

Fig. 1. Example of a tensor descriptor computed for one frame. The ellipse is merely an illustration since generally $n_\theta > 2$. (a) Extracted block displacement vectors. (b) Vectors represented by a histogram h_k. (c) Coding histogram into an orientation tensor.

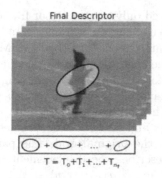

Fig. 2. Frame tensors accumulated in order to model the temporal evolution of motion

if accumulation results in an anisotropic tensor, it carries meaningful average motion information of the frame sequence [10].

Figure 1 shows an example of a video tensor descriptor. To a better understanding of the method, the tensors are represented as ellipses. However, this is only for illustrative purposes since, at this point, the tensor is totally anisotropic and generally $n_\theta > 2$. Figure 2 shows the video tensor, which is the sum of all frame tensors, and its ellipse representation.

4   Block Matching Descriptor: Multiple Scales Single Vector

In this variation, we use multiple block sizes to obtain vectors between only two consecutive frames. This allow us to extract coarse and fine movements, since some blocks will have regions with different motion vectors. We first define a block size set $S = \{s_1, s_2, ..., s_{n_s}\}$. Each block size results in a tensor as defined in 2. The final tensor for a given frame k is the combination of the multiple scales by

$$\mathbf{T}_k(S) = \sum_{s \in S} \frac{\mathbf{T}_k(s)}{\| \mathbf{T}_k(s) \|_2} \ . \tag{3}$$

Note that the histogram will have the same size for any block size, but this parameter is important to define how many vectors will be represented.

Then, the final video tensor is given by:

$$\mathbf{T}(S) = \sum_{k=1}^{n_f} \mathbf{T}_k(S) \ .$$

5   Block Matching Descriptor: Multiple Scales Multiple Vectors

Heretofore, our methods used only one successor frame to extract displacement vectors. However, a sequence of successor frames could be used in order to track the block displacement. Thus, this new method (MSMV) uses pairs of adjacent frames of this sequence. The correspondent block found for the previous pair is used as a reference block for the next matching (Fig. 3). Note that with this method it is possible to have block overlaps, which might lead to redundant information in contrast with the previous variations.

Thus, we use the frame k and its t successor frames, generating t vectors for each trajectory starting in the original grid. The parameter t is fixed for all frames. The vector that describes the displacement between a block in frame a and $a + 1$ is defined by $v_{k,a}(i,j) = (x,y) \in \mathbb{R}^2$, where $a \in [k, k + t]$. All the displacement vectors are included in the histogram of the base frame k, i.e. $h_l = \sum_{i,j} r_{k,a}(i,j) \cdot \omega(i,j)$ (analogous to 1), where $r_{k,a}(i,j)$ is the magnitude of

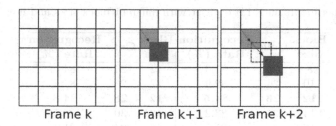

Fig. 3. Block trajectory scheme with two vectors. The first match for frame k is similar to SSSV. The vector found (frame $k + 1$) indicates the position of the new reference block (dark green block in frame $k + 1$) and this block is used to make a second match between frame $k + 1$ and frame $k + 2$, generating another displacement vector.

the displacement vector $\boldsymbol{v}_{k,a}(i,j)$. Then, the frame tensor using t frames and S scales is defined as:

$$\mathbf{T}_k^t(S) = \sum_{a=k}^{k+t} \sum_{s \in S} \frac{\mathbf{T}_{k,a}(s)}{\| \mathbf{T}_{k,a}(s) \|_2} . \tag{1}$$

and the final video tensor is given by:

$$\mathbf{T}^t(S) = \sum_{k=1}^{n_f} \mathbf{T}_k^t(S) .$$

6   Experimental Results

We chose the 4SS method to generate our descriptor because it is a fast block matching method. We apply a Gaussian filter on each frame to reduce noise. The experiments were made using the three methods shown in Sect. 3, 4 and 5. We use a SVM classifier to evaluate our descriptor on KTH [15], UCF11 [16] and Hollywood2 [17] datasets, which contains six, eleven and twelve human action classes, respectively.

For the SSSV method, we evaluate the descriptor varying its main parameters: block size and the number of histogram cells. The results for KTH dataset are shown in Tab. 1. The best result was achieved with 12×12 blocks and 28 cells. Note that other block sizes and number of cells also produce satisfactory rates. In some applications involving large datasets, for example, the size of the final descriptor play a major role. In that case, the number of cells might be reduced to obtain smaller descriptors.

We achieve 84.8% of recognition rate on KTH dataset with the previous parameters and the confusion matrix of this experiment is shown in Tab. 2. Note that the method obtains good recognition rates for walking because this motion class has many blocks moving to the same direction. This is the same reason that it has difficulty to differ clapping to boxing where the key motion occurs in

Table 1. Experiments on KTH dataset with different block sizes and number of cells

Block Size	Cells	Recognition Rate (%)	Block Size	Cells	Recognition Rate (%)
8	26	81.6	12	24	84.1
10	26	82.9	12	26	84.5
12	**26**	**84.5**	**12**	**28**	**84.8**
14	26	83.2	12	30	83.7
16	26	81.6	18	24	81.9
18	26	83.1	18	26	83.1
20	26	81.7	18	28	84.4
22	26	81.1	18	30	84.1
24	26	81.6			

Table 2. Confusion matrix of the best result on KTH dataset with SSSV method. The average recognition rate is 84.8%.

	Box	HClap	HWav	Jog	Run	Walk
Box	90.2	7.7	2.1	0.0	0.0	0.0
HClap	20.1	79.2	0.7	0.0	0.0	0.0
HWav	3.5	8.3	88.2	0.0	0.0	0.0
Jog	2.1	0.7	0.0	82.6	8.3	6.2
Run	0.7	0.7	0.0	21.5	71.5	5.6
Walk	1.4	0.0	0.0	0.0	1.4	97.2

small regions of the frame. One may see the classical problem to differ running and jogging because of their similarity.

The confusion matrix of the best UCF11 experiment with SSSV is shown in Tab. 3. We achieve 57.2% of recognition rate in this experiment with the same parameters of KTH test. Note that other objects moving in the scene causes difficulty to describe the human movement. It confuses the biking action as riding, for example. The motion direction in both actions is the same but it is hard to infer the vehicle. As in KTH dataset, the best recognition rate was in classes with many vectors having similar directions.

The best result achieved on Hollywood2 dataset was 33.9% and the average precision (AP) for each class of Hollywood2 dataset is given in the Tab. 4. Again, we achieve better recognition rates in classes with more expressive movement in one direction. As expected, this is a challenging dataset where the actions in the video are highly mixed with uncontrolled scenes and are subjected to several sudden cuts. Our result in this dataset is competitive if compared to other global descriptors [10], but with faster processing (Tab. 13).

In MSSV experiments, we combine the block sizes to obtain better results. Using KTH as a reference, the best combination was $S = \{18, 12\}$ (Tab. 5). This combination improves the previous results because bigger blocks tracks bigger objects reducing motion confusion with smaller blocks that capture more

Table 3. Confusion matrix of the best result on UCF11 dataset with SSSV. The average recognition rate is 57.2%.

	Bike	Dive	Golf	Juggle	Jump	Ride	Shoot	Spike	Swing	Tennis	WDog
Bike	58.5	1.7	1.0	0.0	0.7	18.5	1.0	1.7	6.2	4.3	6.4
Dive	0.6	72.2	2.3	2.7	0.0	3.4	8.0	4.8	1.2	1.2	3.5
Golf	0.0	2.6	74.3	5.4	0.0	2.0	1.7	4.0	0.0	9.4	0.7
Juggle	1.4	5.6	5.5	32.1	13.3	1.2	9.5	7.8	11.5	8.7	3.2
Jump	2.5	0.7	1.0	6.9	71.3	0.0	2.7	0.0	14.3	0.7	0.0
Ride	6.9	0.9	0.0	1.2	0.0	81.0	1.3	2.2	0.7	0.6	5.2
Shoot	2.4	16.8	6.3	8.9	3.0	3.0	26.8	12.3	2.9	14.8	2.9
Spike	0.7	11.9	5.0	4.0	1.8	1.1	2.7	60.4	0.0	7.3	5.0
Swing	6.4	0.8	0.5	8.9	17.2	0.0	0.0	0.0	59.4	2.6	4.1
Tennis	2.49	8.6	10.6	6.7	1.7	2.3	6.9	5.9	1.1	50.5	3.4
WDog	14.2	4.6	5.9	3.4	0.0	18.9	3.5	1.8	2.6	2.6	42.5

Table 4. Best result on Hollywood2 dataset with SSSV method. The average recognition rate is 33.9%.

Action	APhone	DCar	Eat	FPerson	GetOutCar	HShake
AP(%)	14.4	70.4	19.8	56.5	21.3	10.2
Action	HPerson	Kiss	Run	SDown	SUp	StandUp
AP(%)	21.3	41.5	48.4	49.3	8.0	45.5

Table 5. Experiments on KTH dataset using multiple scale

Block Sizes Set	Recognition Rate (%)
{24, 18, 12}	86.2
{24, 18}	85.1
{24, 12}	86.1
{14,12}	85.8
{14,18}	85.4
{18,14, 12}	85.5
{18, 12}	**86.8**

details. MSSV has increased the recognition rate for all classes (Tab. 6) resulting in 86.8% recognition rate.

We also used the block sizes $S = \{18, 12\}$ to test the UCF11 dataset. It resulted in 58.8% average recognition as shown in Tab. 7. Note that most recognition rates have increased compared to Tab. 3. The same test was performed with Hollywood2 dataset yielding exactly the same average recognition of 33.9% but with different rates for some classes.

In MSMV experiments, we used multiple vectors in two cases: with and without multiple scales. The results for KTH are shown in Tab. 8. Multiple vectors with one block size produces better results than the SSSV method. However,

Table 6. Confusion matrix of the best result on KTH dataset with MSSV method. The average recognition rate is 86.8%.

	Box	HClap	HWav	Jog	Run	Walk
Box	91.6	7.7	0.7	0.0	0.0	0.0
HClap	17.4	81.9	0.7	0.0	0.0	0.0
HWav	2.8	7.6	89.6	0.0	0.0	0.0
Jog	2.1	0.7	0.0	85.4	7.6	4.2
Run	0.7	0.0	0.0	22.9	74.3	2.1
Walk	1.4	0.0	0.0	0.0	0.7	97.9

Table 7. Confusion matrix of the best result on UCF11 dataset with MSSV. The average recognition rate is 58.8%.

	Bike	Dive	Golf	Juggle	Jump	Ride	Shoot	Spike	Swing	Tennis	WDog
Bike	56.2	1.0	0.0	0.0	1.0	21.1	1.0	1.	7.0	4.3	6.8
Dive	0.6	73.4	1.6	4.0	0.0	2.8	8.5	1.0	1.2	2.4	4.5
Golf	0.0	2.3	80.1	5.4	0.0	2.0	1.0	2.4	0.7	5.4	0.7
Juggle	0.6	4.6	7.9	36.0	11.9	1.2	10.3	4.3	9.1	10.8	3.2
Jump	1.8	0.0	1.0	8.1	74.8	0.0	2.0	0.0	12.3	0.0	0.0
Ride	6.7	1.4	0.0	0.5	0.0	80.0	0.7	2.9	1.3	0.7	5.9
Shoot	1.8	18.4	4.4	13.7	3.0	4.0	25.0	13.8	2.1	10.9	2.8
Spike	0.7	9.1	3.0	1.0	2.8	2.1	6.0	62.9	0.0	7.3	5.0
Swing	5.4	0.8	1.0	7.1	16.4	0.8	0.0	0.0	64.6	2.0	2.0
Tennis	3.0	8.0	11.1	7.8	1.7	1.7	7.5	4.1	1.1	53.4	0.6
WDog	12.7	4.7	5.6	1.6	0.0	21.2	3.5	3.8	3.6	2.6	40.7

we achieve even better results combining multiple vectors and multiple scales. Another interesting observation is that the recognition rate increases along with the number of trajectory vectors up to a certain point, and then decreases. This occurs because using more frames augments the probability of objects disappearing from the scene or to cover other blocks causing redundancies. The best recognition rate was 87.7% with $S = \{24, 18, 12\}$ and 3 vectors. Its confusion matrix is shown in Tab. 9. Note that the major benefit in this variant is the recognition gain in clapping class, reducing the confusion between clapping and boxing.

We obtain 59.5% on UCF11 dataset using MSMV and its confusion matrix is shown on Tab. 10. Note that dive and swing improved considerably, which contributed to improve the recognition rate. The Hollywood2 dataset rates are shown in Tab. 11, we also obtain a better result than previous variants of our method: 34.9%.

Table 12 shows that our results are close to, but lower than the state-of-the-art results. The best results for these databases are presented in [18, 19]. However this method is more complex than ours. It combines trajectories, HOG, Histogram of Optical Flow (HOF) and Motion Boundary Histogram (MBH), along

Table 8. Experiments on KTH dataset using multiple vectors

Block Sizes Set	Vectors	Recognition Rate(%)	Block Sizes Set	Vectors	Recognition Rate(%)
{18}	2	**85.5**	{18,12}	2	86.9
{18}	3	85.5	{18,12}	3	86.4
{18}	4	84.7	{18,12}	4	86.4
{18}	5	84.4	{18,12}	5	86.0
{18}	6	85.1	{24,12}	2	86.2
{12}	2	84.7	{24,12}	3	86.5
{12}	3	85.2	{24,12}	4	87.1
{12}	4	84.8	{24,12}	5	86.7
{12}	5	84.8	{24,18}	2	86.9
{12}	6	84.7	{24,18}	3	87.0
			{24,18}	4	86.2
			{24,18}	5	86.0
			{24,18,12}	2	87.6
			{24,18,12}	**3**	**87.7**
			{24,18,12}	4	86.5
			{24,18,12}	5	85.9

Table 9. Confusion matrix of the best result on KTH dataset with MSMV method. The average recognition rate is 87.7%.

	Box	HClap	HWav	Jog	Run	Walk
Box	91.6	7.0	1.4	0.0	0.0	0.0
HClap	13.2	86.1	0.7	0.0	0.0	0.0
HWav	2.1	4.2	93.8	0.0	0.0	0.0
Jog	1.4	0.0	0.0	84.7	6.2	7.6
Run	0.7	0.0	0.0	25.0	70.8	3.5
Walk	0.7	0.0	0.0	0.0	0.0	99.3

with a Bag-of-Features (BoF) technique, which includes a clustering overhead in order to generate the final descriptor. Yet, for KTH dataset using block matching approach, our method can achieve high execution speed as shown in Tab. 13. All experiments where generated in a machine with Intel Xeon E-4610, 2.4 GHz, 32 GB of memory using 10 threads. Using the SSSV on KTH dataset achieves around 2 ms per frame. As expected, in Hollywood2 the frame rate is lower because of its higher resolution.

There are several works on literature proposing compact image descriptors. For video descriptors, this property has not been exploited yet. One compact image descriptor is presented in [20] and each image is stored using 256 bits. With 28 histogram cells, our resulting video descriptor has only 406 elements. In KTH dataset for example, this represents an average of 138.2 bits per frame. According to them, this property is an advantage for large datasets and also enables fast search in retrieval systems.

Table 10. Confusion matrix of the best result on UCF11 dataset with MSMV. The average recognition rate is 59.5%.

	Bike	Dive	Golf	Juggle	Jump	Ride	Shoot	Spike	Swing	Tennis	WDog
Bike	56.0	1.0	0.0	0.0	2.3	20.9	2.0	0.7	5.2	4.3	7.6
Dive	0.6	76.8	1.6	4.0	0.0	1.8	7.4	1.7	1.2	2.2	2.8
Golf	0.0	3.3	78.3	4.9	0.0	1.0	1.0	2.4	0.0	7.8	1.2
Juggle	1.4	4.5	10.8	35.3	9.0	0.6	9.3	4.0	9.4	11.9	3.7
Jump	1.8	0.0	1.0	8.3	73.6	0.0	2.0	0.0	13.3	0.0	0.0
Ride	7.2	1.4	0.0	0.0	0.0	78.9	0.7	2.9	1.3	0.5	7.1
Shoot	1.0	19.2	7.1	10.9	2.3	3.0	27.5	14.7	3.4	9.8	1.2
Spike	1.7	10.6	3.7	1.0	2.8	1.1	6.0	61.5	0.0	7.7	4.0
Swing	5.4	0.8	1.0	5.1	13.0	0.8	0.0	0.0	71.3	1.3	1.3
Tennis	4.2	8.1	8.8	7.8	1.7	0.6	5.0	7.0	2.3	53.9	0.6
WDog	15.0	4.3	3.5	2.4	0.0	21.8	1.0	3.8	4.9	1.8	41.6

Table 11. Average precision for each class of Hollywood2 dataset with MSMV. The average recognition rate is 34.9%.

Action	APhone	DCar	Eat	FPerson	GetOutCar	HShake
AP(%)	16.1	68.7	24.1	60.6	23.9	8.9
Action	HPerson	Kiss	Run	SDown	SUp	StandUp
AP(%)	18.7	42.9	50.3	48.6	11.5	45.2

Table 12. Comparison with state-of-the-art for KTH, UCF11 and Hollywood2 datasets

	KTH	UCF11	Hollywood2
Klaser et al. (2008)	91.0		24.7
Liu et al. (2009)	93.8		
Mota et al. (2013)	93.2	72.7	40.3
Sad et al. (2013)	93.3	72.6	41.9
Wang et al. (2013)	**95.3**	**89.9**	59.9
Wang and Schmid (2013)			**64.3**
Our method	**87.7**	**59.5**	**34.9**

Table 13. Frame rate for each method

Method	KTH (fps)	UCF11 (fps)	Hollywood2 (fps)
SSSV	502.7	117.6	26.8
MSSV	319.3	69.2	15.5
MSMV	89.1	18.5	2.2

7   Conclusion

In this paper, we present a novel approach for motion description in videos using block matching. The resulting tensor descriptor is a simple but effective approach for video classification. It is simple because of its low complexity in terms of time and space. Our recognition rate is lower than the approaches in the literature but fairly competitive in KTH, UCF11 and Hollywood2 datasets, if compared to other self-descriptors. We obtain 84.8% recognition rate with KTH in the SSSV version with the processing rate of 502.7 frames per second.

The main advantage of our method is that it reaches good recognition rates depending uniquely on the input video. This is a different and simple approach considering that most works on the field are based on the gradient of image intensities.

Our method is fast and the descriptor is compact, making it suitable for big datasets. The addition of new videos and/or new action categories with our approach does not require any recomputation or changes to the previously computed descriptors. Finally, it might be valuable in a scenario where the application demands fast processing and a compact descriptor.

Acknowledgements. Authors thank to FAPEMIG, CAPES and UFJF for funding.

References

1. Hafiane, A., Palaniappan, K., Seetharaman, G.: Uav-video registration using block-based features. In: IEEE International Geoscience and Remote Sensing Symposium (IGARSS), vol. 2, pp. 1104–1107 (2008)
2. Amel, A.M., Abdessalem, B.A., Abdellatif, M.: Video shot boundary detection using motion activity descriptor. Journal of Telecommunications 2(1), 54–59 (2010)
3. Sun, X., Divakaran, A., Manjunath, B.S.: A motion activity descriptor and its extraction in compressed domain. In: Shum, H.-Y., Liao, M., Chang, S.-F. (eds.) PCM 2001. LNCS, vol. 2195, pp. 450–457. Springer, Heidelberg (2001)
4. Mota, V.F., Souza, J.I., de A. Araújo, A., Vieira, M.B.: Combining orientation tensors for human action recognition. In: Conference on Graphics, Patterns and Images (SIBGRAPI), pp. 328–333. IEEE (2013)
5. Sad, D., Mota, V.F., Maciel, L.M., Vieira, M.B., de A. Araújo, A.: A tensor motion descriptor based on multiple gradient estimators. In: Conference on Graphics, Patterns and Images (SIBGRAPI), pp. 70–74. IEEE (2013)
6. Perez, E.A., Mota, V.F., Maciel, L.M., Sad, D., Vieira, M.B.: Combining gradient histograms using orientation tensors for human action recognition. In: 21st International Conference on Pattern Recognition (ICPR), pp. 3460–3463. IEEE (2012)
7. Ji, Y., Shimada, A., Taniguchi, R.I.: A compact 3d descriptor in roi for human action recognition. In: IEEE TENCON, pp. 454–459 (2010)
8. Mota, V.F., Perez, E.A., Vieira, M.B., Maciel, L., Precioso, F., Gosselin, P.H.: A tensor based on optical flow for global description of motion in videos. In: Conference on Graphics, Patterns and Images (SIBGRAPI), pp. 298–301. IEEE (2012)

9. Kläser, A., Marszałek, M., Schmid, C.: A spatio-temporal descriptor based on 3d-gradients. In: British Machine Vision Conference (BMVC), pp. 995–1004 (September 2008)

10. Mota, V.F., Perez, E.A., Maciel, L.M., Vieira, M.B., Gosselin, P.H.: A tensor motion descriptor based on histograms of gradients and optical flow. Pattern Recognition Letters 31, 85–91 (2013)

11. Po, L.M., Ma, W.C.: A novel four-step search algorithm for fast block motion estimation. IEEE Transactions on Circuits and Systems for Video Technology 6(3), 313–317 (1996)

12. Nie, Y., Ma, K.K.: Adaptive rood pattern search for fast block-matching motion estimation. IEEE Transactions on Image Processing 11(12), 1442–1449 (2002)

13. Zhu, S., Ma, K.K.: A new diamond search algorithm for fast block-matching motion estimation. IEEE Transactions on Image Processing 9(2), 287–290 (2000)

14. Johansson, B., Farnebäck, G.: A theoretical comparison of different orientation tensors. In: Proceedings of the SSAB Symposium on Image Analysis, pp. 69–73 (2002)

15. Schuldt, C., Laptev, I., Caputo, B.: Recognizing human actions: a local svm approach. In: Proceedings of the 17th International Conference on Pattern Recognition (ICPR), vol. 3, pp. 32–36. IEEE (2004)

16. Liu, J., Luo, J., Shah, M.: Recognizing realistic actions from "videos in the wild". In: IEEE Conference on Computer Vision and Pattern Recognition (CVPR), pp. 1996–2003. IEEE (2009)

17. Marszalek, M., Laptev, I., Schmid, C.: Actions in context. In: IEEE Conference on Computer Vision and Pattern Recognition (CVPR), pp. 2929–2936. IEEE (2009)

18. Wang, H., Kläser, A., Schmid, C., Liu, C.L.: Dense trajectories and motion boundary descriptors for action recognition. International Journal of Computer Vision 103(1), 60–79 (2013)

19. Wang, H., Schmid, C., et al.: Action recognition with improved trajectories. In: International Conference on Computer Vision (2013)

20. Torralba, A., Fergus, R., Weiss, Y.: Small codes and large image databases for recognition. In: IEEE Conference on Computer Vision and Pattern Recognition, CVPR (2008)

Event Driven Approach for Simulating Gene Regulation Networks

Marco Berardi[1,2] and Nicoletta Del Buono[2,*]

[1] IRSA-CNR, via F. De Blasio,5 70132 Bari, Italy
[2] Dipartimento di Matematica, Università degli Studi di Bari Aldo Moro,
via E.Orabona 4, 70125 Bari, Italy
{marco.berardi,nicoletta.delbuono}@uniba.it

Abstract. Gene regulatory networks can be described by continuous models in which genes are acting directly on each other. Genes are activated or inhibited by transcription factors which are direct gene products. The action of a transcription factor on a gene is modeled as a binary on-off response function around a certain threshold concentration. Different thresholds can regulate the behaviors of genes, so that the combined effect on a gene is generally assumed to obey Boolean-like composition rules. Analyzing the behavior of such network model is a challenging task in mathematical simulation, particularly when at least one variable is close to one of its thresholds, called switching domains. In this paper, we briefly review a particular class model for gene regulation networks, namely, the piece-wise linear model and we present an event-driven method to analyze the motion in switching domains.

Keywords: Gene regulatory networks, piecewise-linear differential equation, event-driven method.

1 Introduction

Many physical phenomena are described by discontinuous time-dependent problems in which relations between the state variables are subject to irregularities or discontinuities. The study of discontinuous dynamical systems has been undertaken for several decades and has produced several theoretical results and a wide range of applications in various fields as Biological Sciences, Medicine and Engineering. Particular importance is covered by the mathematical models described by systems of differential equations characterized by discontinuities that occur when the state variable reaches a surface, said switching surface. Among the different kind of discontinuity there is the well known discontinuity of Filippov type [17,22,15], where the vector field is discontinuous on a switching region of the space. These systems have an interesting dynamic behavior, since the state of the system can be forced to remain on the discontinuity surface, in order to

* This paper has been supported by the project "Modelli Matematici Discontinui per l'Analisi delle Reti di Geni: Applicazioni al Diabete", sponsored by Fondazione Cassa di Risparmio di Puglia-FCRP (Anno 2013).

B. Murgante et al. (Eds.): ICCSA 2014, Part VI, LNCS 8584, pp. 415–425, 2014.

have a sliding motion on it [14,11,10,9]. Discontinuous systems are often used to describe physical models subjected to the action of threshold regulatory. An example is provided by the activities of the cell that are regulated by several mechanisms which provide the activation/deactivation (On/Off) of the genes. The mechanisms of gene regulation involve not only genes, but also proteins and other molecules and produce a dense network of adjustments (Gene Regulatory Network, GRN). These complex phenomena, although not yet described in detail, allow to predict the molecular mechanisms thanks to which cells respond to environmental changes, their enzymatic functions and possible diseases caused by genetic alterations [6,7]. Simple gene interactions have been described using mathematical models which are of discontinuous type and based on the assumption that the behavior of the activation/deactivation of genes is approximated by discontinuous functions which measure the expression of genes. The peculiarity of these models is the presence of various switching surfaces identified by the thresholds of activation/deactivation of involved genes.

In this paper, we are going to present a numerical mechanisms, namely the event driven approach, which can be adopted to simulate GRN. In particular, the event driven mechanism will be used to locate exactly the point at which a trajectory hits the switching surface. The paper is organized as follows. Section 2 briefly reviews the piecewise-linear differential equation models for gene networks and the concepts related to the non-smooth behavior of this kind of systems. Particularly, the Filippov approach is adopted to define the dynamics of the GRN on a switching surface. An example of a simple GRN (where two genes and two proteins are involved) is presented to illustrate the dynamics of the Filippov approach. In section 3, we describe the ideas behind event driven methods and we provide a discrete model based on the well known Explicit Euler scheme. Finally, a section of numerical simulations closes the paper.

2   Piecewise-Linear Differential Equation Models for Regulatory Networks

A question of crucial importance for describing GRN is the choice of an appropriate model structure to fit empirical observations. Examples of models for GRN include: directed graphs, Bayesian networks, "Boolean" methods that describe the states of genes simply as either on or off, "continuous" methods that use ordinary or partial differential equations to describe the variations of gene product concentrations. There are also hybrid approaches that blend elements of the Boolean and continuous methods. Discrete approaches (such as Boolean networks method) are favored for ease of formulation and computation [2,7].

On the other hand, continuous models have the advantage of greater physical accuracy and could reveal to be more adequate to study the dynamical properties of a continuous time system. Among continuous approaches, we concern with models based on piecewise-linear differential equation. A couple of motivations justify the choice of this kind of model: piecewise-linear functions are able to approximate arbitrary functions with an adjustable accuracy and they have a corresponding difference equation or discrete model [5].

The dynamics of GRN can be modeled by differential equations in which the state variables correspond to the concentrations of proteins encoded by genes in the network, while the differential equations represent the interactions arising from the regulatory influence of some proteins on the synthesis and degradation of others. Two simplifying assumption have to be adopted: the model has to abstract from the biochemical details of regulatory interactions by directly relating the expression of genes in the networks, and the switch-like behavior of genes (whose expression is regulated by continuous sigmoid curves) is approximated by discontinuous step function [7].

We will assume that $x = (x_1, \ldots, x_n)^\top \in \mathbb{R}^n$ is a vector of cellular protein concentrations and that it takes its values in a bounded hyper-rectangular region $\Omega \subset \mathbb{R}^n$, where $\Omega = \Omega_1 \times \Omega_2 \times \ldots \times \Omega_n$. Each Ω_i is given by $\Omega_i = [0, \max_i]$, where $\max_i$ indicates the maximum concentration for x_i. For each proteins i, we associate threshold concentrations $\theta_i^j \in \Omega_i$, $j \in \{1, 2, \ldots, p_i\}$ and $i = 1, \ldots, n$, such that, for $j = 1, \ldots, p_i$, θ_i^j are ordered as $\theta_i^1 < \theta_i^2 < \ldots < \theta_i^{p_i}$. When the concentration x_i of proteins crosses a threshold value, the mode of regulations of the synthesis or degradation of itself and of the other proteins change abruptly. The differential equations describing this regulatory interactions are:

$$\dot{x} = f(x) - G(x)x, \tag{1}$$

where, $f(x) = (f_1(x), \ldots, f_n(x))^\top$ expresses how the rate of synthesis of the protein encoded by gene i depends on the concentrations of proteins in the cell, $G(x) = \mathrm{diag}(g_1(x), \ldots, g_n(x))$ is the $n \times n$ strictly positive, diagonal matrix describing the regulation of protein degradation, (i.e., $g_i(x)x_i$ represents the rate of degradation of the protein). Particularly, the functions f_i are given by:

$$f_i(x) = \sum_{l \in L} k_{il} b_{il}(x) \tag{2}$$

where $b_{il}(\cdot)$ are boolean valued regulation functions expressed in terms of sums and/or multiplications of step functions, and L is a subset of indices $\{1, \ldots, n\}$. The functions b_{il} quantify the conditions under which the gene i is expressed at a rate k_{il}. These conditions are specified by step functions s^+ and s^-:

$$s^+(x_i, \theta_i^j) = \begin{cases} 1, & x_i > \theta_i^j, \\ 0, & x_i < \theta_i^j, \end{cases} \quad s^-(x_i, \theta_i^j) = 1 - s^+(x_i, \theta_i^j) = \begin{cases} 0, & x_i > \theta_i^j, \\ 1, & x_i < \theta_i^j, \end{cases} \tag{3}$$

The functions g_i can be defined analogously as $g_i(x) = \sum_{l \in L} \gamma_{il} \tilde{b}_{il}(x)$, where again $\tilde{b}_{il}$ are boolean described as combinations of the step functions (3).

2.1   Dynamics of PWL System

A key point to take into account when the behavior of the solution of (1) has to be determined is that the step functions (3) are not defined at a threshold concentrations, i.e., when a state variables $x_i = \theta_i^j$, for $j = \{1, \ldots, p_i\}$. Therefore, the hyperplanes Σ_i^j defined by the following event function

$$h_i^j(x) = x_i - \theta_i^j = 0, \qquad i = 1, \ldots, n, \tag{4}$$

represents switching surfaces identifying a change in the dynamical behavior of the solution of (1).

These surfaces partition the region Ω into regulatory domains D_r, namely regions in which none of the variables assumes a threshold value, i.e., $x_i \neq \theta_i^j$ for all i and j. Moreover, each Σ_i^j is further separated in switching domain D_s, when there exists at least one i, $i = 1, \ldots, n$, such that, there is some j ($j = 1, \ldots, p_i$) with $x_i \in \Sigma_i^j$. Figure 1 provides an example of the partition of Ω into 12 regulatory domains D_i ($i = 1, \ldots, 12$), for a GRN with two genes and three threshold values for concentration x_2 and two for x_1.

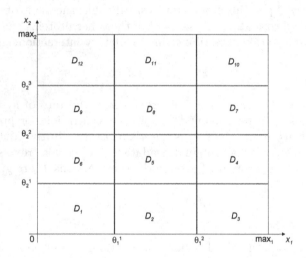

Fig. 1. Regulatory domains for a network with two genes. The system presents two threshold vslues (θ_1^1 and θ_1^2), for the concentration x_1 and three threshold values (θ_2^1, θ_2^2 and θ_2^3) for concentration x_2.

It should be observed that in a regulatory domain $D \in D_r$, the rate of synthesis $f_i(x)$ and degradation $g_i(x)$ reduces to some constant μ_i^D and ν_i^D, respectively. Particularly, μ_i^D and ν_i^D are the sums of the rate constants k_{il} and γ_{il}. Hence, inside a regulatory domain, the system (1) reduces to linear, uncoupled differential equations:

$$\dot{x} = \mu^D - \nu^D x, \qquad x \in D, \tag{5}$$

where $\mu^D = (\mu_1^D, \ldots, \mu_n^D)^\top$ and $\nu^D = \operatorname{diag}(\nu_1^D, \ldots, \nu_n^D)$. The dynamic of GRN in $D \in D_r$ is therefore simply governed by the smooth differential system (5):

starting from an initial condition x_0 in D, the solution trajectory will converge to an equilibrium point $x^D = (\mu_1^D/\nu_1^D, \ldots, \mu_n^D/\nu_n^D)^\top$, if this point belongs to D, otherwise it will leave D at some point on the boundary of D.

Hence, problems occur when a solution trajectory arrives at a switching plane from some regulatory domain. In this case, in order to describe the dynamics of the system, differential equations (1) have to be extended to differential inclusion. Hence, Filippov approach to non-smooth systems has to be considered [17].

The values where a trajectory reaches a switching surface are called *events*. Loosely speaking, there are two things which can occur when an event is reached: the trajectory can cross the switching surface, or it may stay on it. In the latter case, since the step functions (3) are not defined on Σ_i^j, a description of the motion on this surface is required. Following the Filippov approach for a general non-smooth system, we extend the system of differential equations with discontinuity right-hand sides into a system of differential inclusions. This mechanism allows to define how the system behaves on a threshold hyperplane Σ_i^j. Particularly, let D_i and D'_i be two regulatory domains for a switching variable x_i, separated by a threshold hyperplane Σ_i^j with event function $h_i^j(x) = 0$. Hence, for each $x \in D_i$, $h_i^j(x) < 0$ and for $x \in D'_i$, $h_i^j(x) > 0$. Hence, the differential system of Filippov type describing the GRN is given by:

$$
\dot{x} = \begin{cases} f_{D_i}(x) = \mu^{D_i} - \nu^{D_i} x, & x \in D_i, \\ f_F(x) = (\mu^{D_i} - \alpha(\mu^{D_i} - \mu^{D'_i})) - (\nu^{D_i} - \alpha(\nu^{D_i} - \nu^{D'_i})), & x \in \Sigma_i^j, \quad (6) \\ f_{D'_i}(x) = \mu^{D'_i} - \nu^{D'_i} x, & x \in D'_i. \end{cases}
$$

where, $f_F(x)$ is a convex combination of $f_{D_i}(x)$ and $f_{D'_i}(x)$ and $\alpha \subset [0,1]$ is a function of the vector fields $f_{D_i}(x)$ and $f_{D'_i}(x)$. The dynamics of (6) inside the regulatory domain D_i (resp. D'_i) are completely governed by the vector fields f_{D_i} (resp., $f_{D'_i}$). Now, consider a trajectory of (6), and suppose, without loss of generality, that $x_0 \in D_i$ and that the solution is going to reach a point on the threshold hyperplane Σ_i^j. At this point, loosely speaking, there are two events which can occur: (i) the trajectory may cross Σ_i^j and enter D'_i, or (ii) it may stay on Σ_i^j. Filippov theory helps to decide what to do in this situation. We briefly summarize it below. Let $t_{D_i}^{h_i^j}$, $t_{D'_i}^{h_i^j}$ be the scalar functions defined as

$$
t_{D_i}^{h_i^j}(x) = \nabla h_i^j(x)^\top f_{D_i}(x), \qquad t_{D'_i}^{h_i^j}(x) = \nabla h_i^j(x)^\top f_{D'_i}(x), \qquad (7)
$$

which provide information on the directions of the vector fields f_{D_i} and $f_{D'_i}$ at x with respect to Σ_i^j.

– **Transversal Intersection.** In the case in which, at $x \in \Sigma_i^j$, we have

$$
t_{D_i}^{h_i^j}(x) t_{D'_i}^{h_i^j}(x) > 0, \qquad (8)
$$

the trajectory will leave the threshold hyperplane Σ_i^j and continue to move towards the steady state with vector fields $f_{D'_i}$.

- **Sliding Mode.** In the case in which, at $x \in \Sigma_i^j$,

$$t_{D_i}^{h_i^j}(x) t_{D'_i}^{h_i^j}(x) < 0, \tag{9}$$

we have the so called *sliding mode* through x. In fact, the vector fields f_{D_i} and $f_{D'_i}$ are locally both pointing away from or towards the discontinuity hyperplane Σ_i^j, hence the dynamics is assumed to be locally constrained to it and the motion should follow the new vector field f_F until reaching some point on Σ_i^j where one of the two vector fields, f_{D_i} or $f_{D'_i}$, changes its direction. In this case, f_F must be tangent to the hyperplane threshold, that is $\nabla h_i^j(x)^\top f_F(x) = 0$, which yields to the parameter α

$$\alpha = \frac{\nabla h_i^j(x)^\top f_{D_i}(x)}{\nabla h_i^j(x)^\top (f_{D_i}(x) - f_{D'_i}(x))}. \tag{10}$$

Sliding mode can be further classified as attracting or repulsive depending on the directions of the vector fields f_{D_i} and $f_{D'_i}$ (more details can be found in [17]). An example of piecewise linear dynamical systems modeling a simple class of gene regulatory networks together with some discussions on the definition of the vector field on the intersection of two discontinuity hyperplanes (under assumptions of attractivity) has been recently proposed in [8].

2.2   A Simple Example

To better highlight the dynamics of Filippov system (6) let us consider the GRN with two proteins and two genes described by the following dynamical system:

$$\begin{cases} \dot{x}_1 = k_1(1 - s^+(x_1, \theta_1^1)) - \nu_1 x_1, \\ \dot{x}_2 = k_2(1 - s^+(x_2, \theta_2^1)) - \nu_2 x_2, \end{cases} \tag{11}$$

where θ_1^1 and θ_2^1 are the thresholds for the variables x_1 and x_2, respectively. The two thresholds individuate two switching lines Σ_1^1 and Σ_2^1, which divide the state space into four regulatory domains (D_i, $i = 1, \ldots, 4$). If the synthesis and degradation constants satisfy the inequalities $\frac{k_1}{\nu_1} > \theta_1^1$ and $\frac{k_2}{\nu_2} > \theta_2^1$, then sliding mode occurs on the threshold lines Σ_1^1 and Σ_2^1. Figure 2 illustrates the qualitative behavior of the system: the arrows represent the directions of the vector fields defining (11). As it should be observed, the intersection point of the two threshold lines is an attracting point for all trajectories starting from any point into regulatory domain.

3   Discrete Model and Event-Driven Method for Genetic Network

A numerical strategy often adopted to analyze non-smooth system is to regularize (or smoothing) the system [18,20]. Undoubtedly, this leads to simplifications

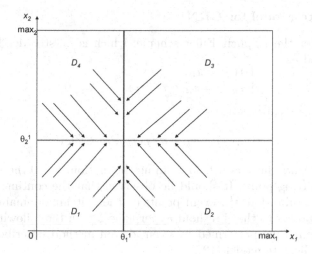

Fig. 2. Stable sliding mode occurring for GRN (2.2) when $\frac{k_1}{\nu_1} > \theta_1^1$ and $\frac{k_2}{\nu_2} > \theta_2^1$

in the theory, since existence and uniqueness of solutions may be derived from the classical theory of ODEs, together with a preliminary exploration of the problem at hand. However, the numerical simulation of the regularized system can suffer of some stiffness and, further, the regularization will lead to changing the dynamics of the original non smooth system [10].

In this section we consider a discrete approach to solve piece-wise linear systems for GRN. For the sake of clearness, we restrict our presentation to the case in which only two regulatory domains, D and D' are present and they are separated by a switching domain Σ_i^j. With this assumption, we are going to refer to the system (6) to derive the discrete event-driven method. It should be pointed out that this assumption is not restrictive, since crossing and sliding behaviors occur at a threshold hyperplane at time.

An event driven method for the numerical solution of non-smooth GRN has to accomplish the following steps:

i. Integrate in a regulatory domain starting from a point outside Σ_i^j;
ii. Check if the approximate trajectory reaches Σ_i^j;
iii. Locate the event point on Σ_i^j reached by the trajectory;
iv. Control of the transversality or sliding condition of the event point on Σ_i^j;
  v. If the sliding condition is satisfied:
    v. (i) Start to integrate on Σ_i^j with vector field f_F;
    v. (ii) Control the trajectory to find when it exits the switching hyperplane Σ_i^j;
  v. If the trasversality condition is satisfied:
    v. (i) Switch the vector field;
vi. Continue to integrate outside Σ with the proper vector field.

3.1   Discrete Model for GRN

Let us consider the Explicit Euler scheme, which is a 1st order Runge-Kutta scheme defined as

$$\begin{cases} x(0) = x_0 \\ x_{k+1} = x_k + \tau f(x_k), \qquad k > 0, \end{cases} \tag{12}$$

and it continuous extension

$$x_{k+1}(\sigma) = x_k + \sigma \tau f(x_k), \forall \sigma \in [0, 1], \tag{13}$$

where f represents the vector field defining the motion, $\tau > 0$ the step-size, and x_0 a given starting point. It should be observed that the continuous extension (13) will be useful to find the event points on the switching domains, that is the entry or exit points to the threshold hyperplane Σ_i^j. In the following, we briefly detail the main steps performed by event-driven method described above and based on the discrete model (12).

Integration in a Regulatory Domain. Let $x_0 \in D_i$ be the initial point from which we want to compute a numerical approximation of (6). Since f_{D_i} is the vector field governing the motion in D_i, we get the discrete equation

$$x_{k+1} = x_k + \tau(\mu^{D_i} - \nu^{D_i} x_k), \qquad k \geq 0. \tag{14}$$

Until $x_{k+1} \in D_i$, we continue to compute the numerical approximation using model (14).

Check Belonging to D_i . If, at some k the approximate trajectory $x_{k+1} \notin D_i$, that is $h_i^j(x_{k+1}) > 0$, then there exists a value $\bar{k} > k$ such that $x_{\bar{k}} \in \Sigma_i^j$. Hence, we need to locate the event point where the trajectory reaches the threshold hyperplane Σ_i^j.

Locate Event Points . In order to local the event point on Σ_i^j, we consider the continuous extension starting from the approximation $x_k \in D_i$, given by:

$$x_{k+1}(\sigma) = x_k + \sigma \tau(\mu^{D_i} - \nu^{D_i} x_k), \qquad \sigma \in [0, 1]. \tag{15}$$

Particularly, we consider the function $h_i^j(x_{k+1}(\sigma))$, $\sigma \in [0, 1]$. This is a continuous function, taking values of opposite sign at the endpoints $\sigma = 0$ and $\sigma = 1$. Therefore, it has a zero, say, at $\bar{\sigma}$. Particularly, solving with respect to σ the equation $h_i^j(x_{k+1}(\sigma)) = 0$, we get $[x_{k+1}(\sigma)]_i = \theta_i^j$, and

$$\bar{\sigma} = \frac{\theta_i^j - [x_k]_i}{\tau(\mu_i^D - \nu_i^D [x_k]_i)} \tag{16}$$

where $[\cdot]_i$ indicates the i-th component of the vector. Hence, the event point $x_{\bar{k}}$ on the threshold hyperplane Σ_i^j is

$$x_{\bar{k}} = x_{k-1}(\bar{\sigma}) = x_k + \frac{\theta_i^j - [x_k]_i}{(\mu_i^{D_i} - \nu_i^{D_i} [x_k]_i)}(\mu^{D_i} - \nu^{D_i} x_k). \tag{17}$$

Control Transversality or Sliding Conditions. To identify whenever the event point $x_{\bar{k}}$ defines a sliding mode, we check inequalities in (8) and (9).

(a) If $t_{D_i}^{h_i^j}(x_{\bar{k}})t_{D_i}^{h_i^j}(x_{\bar{k}}) > 0$, than $x_{\bar{k}}$ is a transversal intersection point, so that, starting from it, the motion will be defined in D_i' by the discrete model (12), with $f = f_{D_i'}$.

(b) if $t_{D_i}^{h_i^j}(x_{\bar{k}})t_{D_i}^{h_i^j}(x_{\bar{k}}) < 0$, than $x_{\bar{k}}$ is a sliding point, so that, starting from it, the motion will be defined on Σ_i^j by the discrete model (12), with $f = f_F$. While we integrate on Σ_i^j, we will monitor if the solution continues to slid on it or if it leaves Σ at some other event point.

It should be highlighted that more accurate discrete models could be obtained using an event-driven approach base on high order Runge Kutta schemes and their continuous extensions [3,4]. Examples of event-driven methods based on a second order Runge–Kutta scheme for a general non-smooth Filippov system are presented in [16,13,12].

4   Simulation with the Discrete Model

In this section, we are going to show the results of numerical simulation of the GRN system (11)), with parameter values $k_1 = 20$, $v_1 = 2$, $k_2 = 20$, $v_2 = 2$, and thresholds $\theta_1^1 = \theta_2^1 = 4$. Figure 3 illustrates numerical approximation

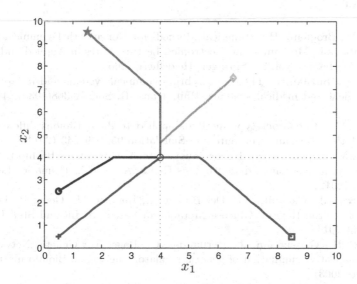

Fig. 3. Numerical simulation of GRN (2.2) obtained by the event-driven method for different initial value points ($x_0 = [0.5, 0.5]^\top$ marked by a circle, $x_0 = [0.5, 2.5]^\top$ marked by a plus, $x_0 = [8.5, 0.5]^\top$ marked by a square, $x_0 = [6.5, 7.5]^\top$ marked by a diamond, $x_0 = [0.5, 0.5]^\top$ marked by a star).

of the trajectory obtained using the discrete model, previously described, for several initial conditions in the regulatory domains. As it could be observed, the different trajectories exactly reproduce the qualitative behavior of the system. Particularly, the intersection between the two threshold lines is an attracting point: once the trajectories reach it they stay on it.

5   Conclusions

In this work, an event driven approach for simulating Gene Regulatory Network has been proposed. This allowed the construction of a discrete model which can be adopted to derive correct numerical approximations of the vector of cellular protein concentrations especially, when for some proteins i the threshold concentrations is reached. The application of the event driven method to a simple gene regulatory system showed good performance and accuracy in reproducing the behaviour of the system.

Acknowledgment. We would like to thank the Fondazione Cassa di Risparmio di Puglia (FCRP) for having supported the project "Modelli Matematici Discontinui per l'Analisi delle Reti di Geni: Applicazioni al Diabete". This work in part of the researches conducted during the afore-mentioned project.

References

1. Acary, V., Brogliato, B.: Numerical Methods for Nonsmooth Dynamical Systems. Applications in Mechanics and electronics. Lecture Notes in Applied and Computations Mechanics, vol. 35. Springer, Heidelberg (2008)
2. Aihara, K., Suzuki, H.: Theory of hybrid dynamical systems and its applications to biological and medical systems. Phil. Trans. R. Soc. A 368(1930), 4893–4914 (2010)
3. Berardi, M.: Rosenbrock-type methods applied to discontinuous differential systems. Mathematics and Computers in Simulation 95, 229–243 (2014)
4. Berardi, M., Lopez, L.: On the continuous extension of Adams-Bashforth methods and the event location in discontinuous ODEs. Applied Methematics Letters 25, 995–999 (2012)
5. D'Abbicco, M., Calamita, G., Del Buono, N., Berardi, M., Gena, P., Lopez, L.: A model for the Hepatic Glucose Metabolism based on Hill and Step Functions (Preprint 2014)
6. de Jong, H., Geiselmann, J., Hernandez, C., Page, M.: Genetic Network Analyzer: qualitative simulation of genetic regulatory networks. Bioinformatics 19(3), 336–344 (2003)
7. de Jong, H.: Modeling and Simulation of Genetic Regulatory Systems: A Literature Review. J. of Comp. Biol. 9(1), 67–103 (2002)
8. Del Buono, N., Elia, C., Lopez, L.: On the equivalence between the sigmoidal approach and Utkin's approach for piecewise-linear models of gene regulatory networks. To appear on SIAM J. Appl. Dynam. Syst. (2014)

9. Dieci, L., Elia, C., Lopez, L.: Sharp sufficient attractivity conditions for sliding on a codimension 2 discontinuity surface. To appear on: Mathematics and Computers in simulations (2014), doi:http://dx.doi.org/10.1016/j.matcom.2013.12.005, ISSN: 1872-7166

10. Dieci, L., Elia, C., Lopez, L.: A Filippov sliding vector field on an attracting co-dimension 2 discontinuity surface, and a limited loss-of-attractivity analysis. J. of Differential Equations 254, 1800–1832 (2013)

11. Dieci, L., Lopez, L.: Numerical Solution of Discontinuous Differential Systems: Approaching the Discontinuity Surface from One-Side. Applied Numerical Mathematics 67, 98–110 (2013)

12. Dieci, L., Lopez, L.: A survey of numerical methods for IVPs of ODEs with discontinuous right-hand side. Journal of Computational and Applied Mathematics 236(16), 3967–3991 (2012)

13. Dieci, L., Lopez, L.: Fundamental matrix solutions of piecewise smooth differential systems. Mathematics and Computers in Simulation 81(5), 932–953 (2011)

14. Dieci, L., Lopez, L.: Sliding motion on discontinuity surfaces of high co-dimension. A construction for selection a Filippov vector field. Numerische Mathematik 117, 779–811 (2011)

15. Dieci, L., Lopez, L.: On Filippov and Utkin Sliding Solution of Discontinuous Systems. In: DeBernardis, E., Spigler, R., Valente, V. (eds.) 9th Conference of the SIMAI, Rome, September 15-19, 2008. Applied and Industrial Mathematics in Italy III, Series on Advances in Mathematics for Applied Sciences, vol. 82, pp. 323–330 (2010)

16. Dieci, L., Lopez, L.: Sliding Motion in Filippov Differential Systems: Theoretical Results and a Computational Approach. SIAM J. Numer. Anal. 47(3), 2023–2051 (2009)

17. Filippov, A.F.: Differential Equations With Discontinuous Right hand Sides: Control Systems. Kluwer Academic Publisher (1988)

18. Ironi, L., Panzeri, L., Plahte, E., Simoncini, V.: Dynamics of actively regulated gene networks. Physica D 240, 779–794 (2011)

19. Leine, R.I.: Bifurcations in Discontinuous Mechanical Systems of Filippov's type. PhD thesis, Techn. Univ. Eindhoven, The Netherlands (2000)

20. Libre, J., da Silva, P.R., Teixeira, M.A.: Regularization of discontinuous vector fields via singular perturbation. J. Dynam. Diff. Equat. 19, 309–331 (2009)

21. Piiroinen, P.T., Kuznetsov, Y.A.: An event-driven method to simulate Filippov systems with accurate computing of sliding motions. ACM Trans. Math. Soft. 34(13), 1–24 (2008)

22. Utkin, V.: Sliding Modes in Control Optimization. Springer, New York (1992)

Implementation Aspects of the 3D Wave Propagation in Semi-infinite Domains Using the Finite Difference Method on a GPU Based Cluster

Thales Luis Sabino[1], Diego Brandão[2], Marcelo Zamith[3], Esteban Clua[1], Anselmo Montenegro[1], Mauricio Kischinhevsky[1], and André Bulcão[4]

[1] Instituto de Computação, Universidade Federal Fluminense, Niterói, RJ, BRA
tsabino,esteban,anselmo,kisch@ic.uff.br
[2] Centro Federal de Ensino Tecnológico (CEFET-RJ), Nova Iguaçu, RJ, BRA
dbrandao@ic.uff.br
[3] Universidade Federal de Viçosa, Florestal, MG, BRA
mzamith@ic.uff.br
[4] CENPES, Petrobrás, RJ, BRA

Abstract. The scattering of acoustic waves has been a matter of practical interest for the petroleum industry, mainly in the determination of new oil deposits. A family of computational models that represent this phenomenon is based on Finite Difference Methods (FDM). The simulation of these phenomena demands a high computational processing power and large amounts of memory. Furthermore, solving this problem in a high performance computing (HPC) environment requires the use of tools such as MPI (Message Passing Interface) and GPUs in order to soften the effort necessary on implementation. In this work a GPU based cluster environment is employed for the development of an efficient scalable solver for a 3D wave propagation problem using the FDM. The details related to the implementation of the FDM applied to wave propagation in GPUs are presented. A performance analysis for several simulations is also discussed. The solution discussed herein is suitable not only for a single GPU system, but for clusters of GPUs as well.

Keywords: Wave Propagation, Finite Difference Method, GPU Cluster.

1 Introduction

The scattering of acoustic waves has been of practical interest for the oil and gas industry, mainly for the determination of new oil deposits. The scattering process is commonly described with Hyperbolic Partial Differential Equations (PDEs). This kind of equation describes a large variety of physical phenomena governed by oscillatory behavior.

In order to solve such PDEs, numerical methods are employed a huge computational effort is required. When one tries to solve the scattering problem

B. Murgante et al. (Eds.): ICCSA 2014, Part VI, LNCS 8584, pp. 426–439, 2014.

with a level of precision that allows meaningful results. While solutions based on multi core CPU clusters have been widely used with successful results [1], the advent of GPU computing technology is making it possible to solve these numerical solution of PDEs from 20 to 200 times faster than traditional CPU implementation, depending on problem characteristics([2],[3], [4]).

GPU computing has become an important choice for many parallellizable computational problems. GPUs are potentially more powerful for massively parallel numerical computations than CPUs. Cluster based environments usually need to work on some communication protocol. The widely used MPI (Message Passing Interface) have been extensively used for numerical computation, to simulate large complex scattering problems. The use of GPUs on each node of MPI based clusters shall increase further the environment's computational power.

It should be emphasized that GPU architectures where originally designed for graphical purposes and exhibit some intrinsic limitations. For example, they work faster with single precision floating point numbers and double precision support only became available with recent releases, with lower performance. In order to achieve maximum performance, GPUs dedicate more transistors to arithmetic units than control units so it is important to coalesce memory reads and writes; GPUs have tightly limited global (8 GB for NVIDIA Kepler K20) and local (64 KB for NVIDIA Kepler K20) memories.

The memory limitation may be an important issue for industry scale applications, since these may require usage of hundreds of GB of memory and a single GPU may not suffice to solve such problems timely. Considering, for instance, a 3D acoustic wave propagation problem with a domain discretized in a cube with 800 grid points in each direction requires, at least, 4GB of storage capacity. Such memory requirements grow as the number of grid points per direction to the third power.

Many different non-graphical computations, simulations and numerical problems, including Protein Structure Prediction [5], Solution of Linear Equation Systems [6], Options Pricing [7], Flow Simulation [8], Wave Propagation ([1], [9]), have been solved in GPUs.

This paper discusses a parallel implementation for the scattering of 3D acoustic waves. The problem is solved with a Finite Difference Method (FDM) in a semi-infinite non-homogeneous domain. In order to solve the discretized problem, the representative variables are mapped to a GPU based cluster. Thus, the problem domain is partitioned in a way that each subdomain is handled by one GPU. An asynchronous communication scheme that avoids idle GPU computation time is also presented herein.

2   Acoustic Wave Equation

The longitudinal wave equation can be expressed as a scalar second order linear differential equation that describes the behavior of sound waves over time. The acoustic wave field is described by $P(x, y, z, t)$ and $u(x, y, z, t)$, where P is the pressure field the u is the particle's displacement. The relation between P and

u is given by $P(x, y, z, t) = -k\nabla u(x, y, z, t)$ with k representing the volumetric compression module. Thus, the 3D wave equation is given by:

$$\frac{\partial^2 P(x, y, z, t)}{\partial t^2} = c^2(x, y, z)\Phi P(x, y, z, t) + f(x, y, z, t) \tag{1}$$

where x, y and z are cartesian coordinates, t is time, c is the velocity acoustic wave and $f(x, y, z, t)$ is the source term. Moreover, ΦP represents the application of the spatial second order differential operator to P (since the problem is 3D, Φ encompasses the spatial derivatives with respect to x, y, z). The boundary condition for the wave equation is explained in the next subsection, while the initial conditions are:

$$P(x, y, z, t_0) = f(x, y, z, t_0)$$
$$\tfrac{\partial}{\partial t} P(x, y, z, t_0) = 0 \tag{2}$$

In order to solve the partial differential equation (Eq. 1) numerically, one may initially discretize it into a set of finite-difference (FD) equations by replacing partial derivatives with central differences. A central-difference approximation can be derived from the Taylor series. Thus, using a second order approximation for space and time, assuming $h = \Delta x = \Delta y = \Delta z$ and $t = n\Delta t$, Eq. 1 is rewritten as:

$$P_{(i,j,k)}^{n+1} = 2P_{(i,j,k)}^{n} - P_{(i,j,k)}^{n-1} + A\left[P_{(i-1,j,k)}^{n} + P_{(i+1,j,k)}^{n}\right] +$$
$$+A\left[-6P_{(i,j,k)}^{n} + P_{(i,j-1,k)}^{n} + P_{(i,j+1,k)}^{n} + P_{(i,j,k-1)}^{n} + P_{(i,j,k+1)}^{n}\right] \tag{3}$$

where $A = \left(\frac{c(x,y,z)\Delta t}{h}\right)^2$ and $n = 1, 2, \dots$ represent the time slice. The velocity field does not vary with time.

2.1   Semi-infinite Domains

The scattering phenomena takes place on an arbitrarily large region; however the region of interest is much smaller. This leads one to consider artificial boundaries that restrict the computational domain to the region of interest. These fictitious boundaries become part of the computational model and shall give rise to non-physical wave reflection, that have to be taken into account.

In order to dampen the non-physical wave reflections originated at the computational boundaries, a simple solution would be to increase the domain's limits in order to simulate the wave until the time slice when reflections would begin. However, the larger the domain, the higher the amount of memory required to hold all the discretized data.

Aiming to simulate semi-infinite domains, this work considers the boundary conditions proposed by [10]. Reynold's absorbing boundaries are determined by decomposing the unidimensional scalar wave equation, thus obtaining the product of two terms, each one representing the spread of the wavefront in one direction. When the wave propagation occurs to the right direction, this condition is represented as follows:

$$\frac{\partial P}{\partial \mathbf{n}} = \frac{1}{c}\frac{\partial P}{\partial t} \tag{4}$$

where $\mathbf{n}$ is the normal vector. The boundary conditions for other directions can be obtained analogously.

2.2 Source Function and Numerical Stability

The choice of an appropriate source function is essential to the FDM. In fact, any function could be employed as a source function on Equation 1. The pulse's frequency directly affects the numerical dispersion of the method [11]. Numerical dispersion is an undesired non-physical effect which is often present in finite-difference schemes. It occurs when different wavelengths propagate with different phase velocities, what causes non-realistic behavior of wave fronts. Some consequences of this phenomenon are the cumulative phase error and the non-physical refraction [12]. In order to simulate the wave behavior a gaussian-based pulse function is employed. This function has some properties that guarantee a reduction or control of numerical dispersion [11]. Thus, the following wave pulse function is adopted as the source input:

$$f(t) = \left[1 - 2\pi\left(\pi f_c t_d\right)^2\right] c^{-\pi(\pi f_c t_d)^2} \tag{5}$$

where $t_d = t - \frac{2\sqrt{t}}{f_c}$, $f_c = 3\sqrt{\pi}C_F$, C_F represents the cutoff frequency and t is the current time.

2.3 Numerical Stability

When successive steps of the integration in time of the differential equation are performed, the numerical solution may get out of control, leading to non-physical values. The stability requirement establishes, in *naive* terms, under what conditions successive time steps keep the numerical solution reasonable, namely, with errors under control. Such requirement can be computed by imposing that the amplification factor of numerical errors, for any two successive time steps, has norm smaller than one. Thus, the application (represented by multiplication) of successive amplification factors to the solution for the error equation makes the numerical error tend to zero.

The stability of the explicit FDM for a second order 3D wave propagation problem, as the one employed herein, can be obtained under the Courant-Friedrichs-Lewy (CFL) condition. This requires, in more technical grounds, that the domain of dependence of the PDE must lie within the domain of dependence of the finite difference scheme for each mesh point. Thus, the CFL that guarantees the stability in this case is [13]:

$$\sqrt{A} = \left(\frac{c(x,y,z)\Delta t}{h}\right) \leq \frac{1}{\sqrt{s}} \tag{6}$$

where s corresponds to the domain's dimension, ($s = 3$ in the present case).

3   CUDA and GPU Computing

With the release of NVIDIA's CUDA, the world of high-performance computing became more accessible. A GPU with thousand of cores is now available by a tenth of the price one needs to build a CPU cluster with the same number of processor cores and this cost tends to fall further.

A few years ago, scientific computing based on GPU architecture was developed using graphical shader languages together with some graphics API that allow the execution of such shaders. The programmer needed to map the problem as a set of vertices and fragments to generate a texture representing the final solution. One of the advantages of programming using the CUDA API, instead of the conventional Shader Language is that it allows one to work with familiar concepts while developing kernels that run on GPUs.

CUDA extends existing programming languages with a set of instructions that allows code execution on NVIDIA GPUs. Such extensions expresses the GPU hardware memory hierarchy. The GPU memory structure is divided as global, texture and shared. Shared memory is a small but extremely fast memory. This speed comes from its physical proximity to the processing core. Texture memory is a read only memory, slower than the shared one, but it is almost twice as fast as global memory.

4   Finite Difference Method on GPU

Several natural phenomena are described by evolutionary PDEs. In order to solve these equations computationally, numerical methods are often employed. A class of evolutionary problems involves a first derivative in time and a set of other derivatives with respect to the space variables. The presence of a first derivative as the sole derivative with respect to the time variable allows one to discretize the system as an Initial (and usually Boundary, as well) Value Problem (IVP). In this scheme, the unknown at a certain grid point has its numerical value computed from the previous scenario that occurred in the neighborhood of such grid point. The sequence of computed values has to begin from known values (the initial ones) at all grid points. That is, from the values available for all grid points, the scheme provides, for each of the grid points, a numerical estimate of the solution to the differential equation one time step ahead by averaging the values at the initial time. In the next time step those estimates obtained for the first computed time step provide the averaging that generates all numerical values for the second computed time step. The algorithm proceeds until the prescribed final time is reached. The overall description encompasses the so-called explicit methods. Thus, the discretization in time is performed with two time instants which we can refer to as the current (that uses the previous condition) and the set of newest (or future) values. Given the current set of grid point values, one obtains the future values.

The second order wave equation, which is the one this work deals with, exhibits a second order derivative in time. Thus, the IVP generated has to employ two time steps of numerical data in order to compute the next time steps grid point values. This is so for one can not produce a general approximation for a second derivative with only two points. Then, the procedure has to start with two sets of successive time steps data. Apart from the initial value, which must be provided, the first computed time step data shall be obtained from the first derivative with respect to time. Thus, when one discretizes the wave equation along time, the explicit procedure needs an information that is referred to as the past data and the current time step data in order to obtain the future grid points' values [14].

As described in [2], finite difference methods implemented on GPUs are more efficient when they take advantage of shared memory. In order to use a single GPU one needs to allocate enough memory to store three times the domain size. In order to compute the next time step, one needs the present and the past time steps stored. Figure 1 illustrates a 3D finite difference explicit scheme with a second order spatial FD operator.

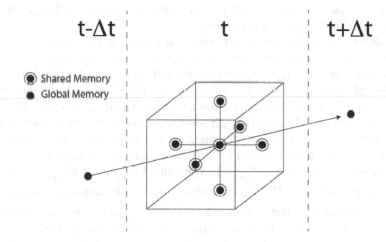

Fig. 1. Second order 3D finite difference operator. To compute the next step $(t + \Delta t)$ it is necessary to fetch the present, or current (t), and the past one $(t - \Delta t)$.

A typical CUDA application workflow consists of 4 basic steps: (1) Initialize the necessary data on the host, (2) copy the data from host to device, (3) invoke the kernel that will process the data in device and (4), read the data back from device to the host. Initially, the velocity field is allocated and copied to the device memory. Memory is allocated to contain the values for the past, present and future instants wave amplitude values. The amplitude values for $P^0_{(x,y,z)}$ and $P^1_{(x,y,z)}$ are set with a forward explicit approximation scheme. After initialization, Eq. 3 is then used for subsequent time steps.

A GPU can run millions of lightweight threads. To help the programmer manage these threads, CUDA works with the concept of grid of thread blocks. Basically the threads are organized into thread blocks, which in turn are organized into a grid of blocks. The approach herein considers that each new value $P_{(x,y,z)}^{n+1}$ is computed by only one thread. Hence, the 3D domain is divided in a set of 3D blocks, observing the constrains described in [15]. The maximum number of blocks that can be allocated is 65,535 with 512 threads each, resulting in a total of 33,553,920 points that can be processed by a single GPU.

4.1   Efficient Mapping of the Finite Difference Operator into GPU Architecture

One of the main challenges when implementing a finite difference method using CUDA kernels comes from the access pattern of the operator. Looking at Figure 1 one may see that the thread that will handle the calculation of a point (i,j,k) needs to read values corresponding to points (i,j,k), $(i-1,j,k)$, $(i+1,j,k)$, $(i,j-1,k)$, $(i,j+1,k)$, $(i,j,k+1)$, $(i,j,k-1)$ and there is a need to access the previous value of point (i,j,k). This yields 9 global memory accesses for each thread. In case of a fourth order finite difference operator, 13 global memory accesses per thread will be necessary.

As described in Section 2 and based on Figure 1, in order to solve the discretized wave equation, one needs to access amplitude values $P_{(x,y,z)}$ from times $t - \Delta t$ and t in order to compute a new amplitude value for the time $t + \Delta t$. In practice, this can be solved by storing some value $P_{(x,y,z)}^{n+1}$ in its corresponding position in an allocated memory cube which has the domain's size. This requires three cubes allocated to hold amplitude values. As described in Section 1, even a problem that may be considered a small problem could have prohibitive memory requirements. In order to avoid a third cube allocated in memory, one can store the new amplitude value $P_{(x,y,z)}^{n+1}$ in the same location of $P_{(x,y,z)}^{n-1}$. The value at $P_{(x,y,z)}^{n-1}$ overwritten by $P_{(x,y,z)}^{n+1}$ will not be required by any other thread different from the one used in the computation of $P_{(x,y,z)}^{n+1}$. Hence we can ensure that no relevante data will ever be overwritten. After all the calculations, the pointers are switched so they can be accessed by the same code in the next simulation step.

Shared memory is used to hold the values in the finite difference stencil centered in $P_{(x,y,z)}^{n}$ step. This avoids unnecessary reads from the global memory. Figure 1 shows that a thread must access seven amplitude values of the same instant $P_{(x,y,z)}^{n}$. Bringing this values to shared memory makes this accesses much faster. The values of instants $P_{(x,y,z)}^{n-1}$ are fetched directly from global memory: since all threads of the same block will access only one position, a coalesced memory read obtaining maximum performance of the GPU architecture is ensured. Shared memory's size is based on the number of threads per block plus a region called buffer border whose size is twice the size of neighborhood corresponding to the 3D operator size.

Each thread copies its data in the current time from the global memory to the shared memory. Besides, the threads at the border of a block also copies data corresponding to the data of its neighbors which belong to a neighboring block. By doing this, the shared memory structure has all data necessary to compute new values of points inside a block without any additional access to global memory. Thus, it is ensured that all threads are able to access the memory addresses corresponding to an instant t from shared memory necessary to define a new point value.

4.2   Using Asynchronous Copies to Optimize Execution Time

Section 4.1 stated the challenges when optimizing the FDTD in order to run in a single GPU. This section has the objective to explain the optimizations that can be performed in terms of tasks that may be executed asynchronously. Knowing that industry scale problems potentially may not be solved by a single GPU, one should coordinate the domain's division into various processing nodes in, for instance, a computing cluster. Taking the domain's partition that should be processed by a single GPU one may note that there is a data dependency because of the borders. Figure 2 shows this data dependency from one iteration to another in the hatched area.

The tasks that should be performed by a single process in an iteration of the simulation can be summarized as follows:

1. **Process the Borders:** As the borders should be communicated to neighbor processes in order to solve the data dependency for the next iteration, making this data available at the beginning of the iteration can save time while the center is being processed.
2. **Copy Borders to CPU Memory While Process the Center:** Taking advantage of current GPU architecture, one can copy a memory area from global memory to CPU memory at the same time a kernel is writing another area of the global memory.
3. **Transmit/Receive Border Data to/from Neighbors Processes:** MPI allows asynchronous transfers of data between processes. This capability is used to allow sending and receiving data at the same time.
4. **Copy Received Borders Data to GPU Memory:** Once the borders data are received they should be copied back to GPU memory in order to execute the simulation's next iteration.

Figure 2 shows a diagram of these tasks and how they can overlap each other. Tasks that can be executed at the same time are represented by the same number. The numbers also represent the order of task execution over time.

4.3   Work Division across Multiple GPUs

As discussed before, even a problem which is considered small for industry standards might not be treated by a single GPU. In this work we proposed an implementation for the $3D$ wave propagation problem that consists a solution

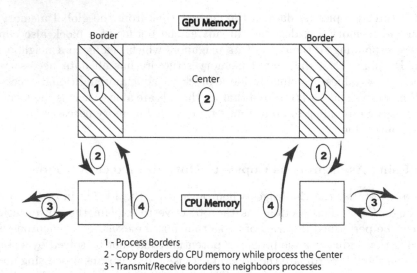

Fig. 2. Diagram of tasks that can be executed asynchronously

that consists of splitting the problem domain in a way that each partition fits in a single GPU. This allows the use of a GPU based cluster to solve industry scale problems with the power of GPUs.

A $3D$ computational domain has the shape of a parallelepiped. Considering that one GPU has not enough memory to handle the simulation, the domain is partitioned along the z-axis in order to reduce the amount of data that need to be transferred across the network.

This partition ensures that no more than two interfaces that must be communicated per node (see Figure 3). The amount of data on such interfaces depends only on the spatial order of FD operator chosen. A process must communicate its subdomains' interfaces in order to let other processes continue with the simulation. For the purpose of avoiding idle computation time, while the "data communication" stage is being performed, a process begins the computation of the bulk part of the associated subdomain while the interfaces are not received from its neighbors.

Figure 3 shows how the data that need to be communicated between time steps were organized for distribution to the devices. For 3D problems, each device handles a parallelepiped whose borders need to be sent to other processes in order to have the entire array of devices solving one single problem. The division planes are chosen in such a way that communication data are reduced to a minimum. This partitioning ensures a maximum of two communication interfaces. Taking for instance a partition scheme where an internal sub-portion of a subdomains' parallelepiped is chosen for a GPU, there is a need to communicate data to up to

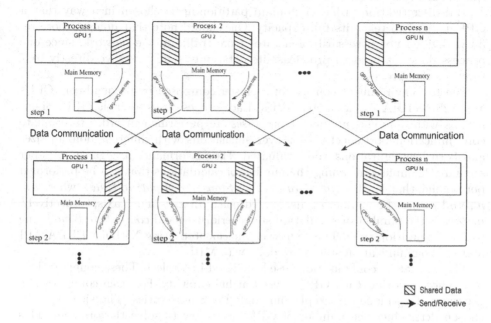

Fig. 3. Problem division across multiple devices. Each GPU is assumed to be handled by one CPU Process. Each CPU runs one or more processes depending on the number of GPUs attached and communication of necessary data is done using MPI when all GPUs finishes with their computation of associated data.

six interfaces. This may become impractical for both communication costs and by memory consumption since data near communication interfaces are redundant.

The "data communication" stage is done at the end of each simulation step. Processes need to communicate the edge values of each $P_{(x,y,z)}^{n+1}$ state, as depicted in the diagram of Figure 3. The amount of data traffic depends on the discretization order of the FD spatial operator chosen. Practical simulations use 6th up to 10th order spatial operators. In such situations, the final time of simulation may not be smaller than one that uses a pure CPU implementation.

A communication scheme for 3D simulation of scattering of acoustic wave equation using a finite differences method that is capable of dealing with industry scale problems that may not fit entirely in a single GPU has been presented.

5   Results and Analysis

Even a problem that is considered a small problem for industry standards may easily exceed a single GPU memory capacity. In order to run large finite difference simulations, we implement a communication scheme like the one shown in Figure 3. The amount of data transferred from one process to another is given mainly by the spatial discretization order of the FD method being used. More specifically, this amount of data is given by $x \times y \times d$, where d is the method spatial discretization order. A domain partitioning is chosen in a way that a GPU is filled to reach its full capacity. One should note that doing a more refined division not necessarily generates more traffic in the network, since one process does not need to broadcast its data, only send / receive directly to / from another process.

The test environment consists of a cluster composed by 64 quad-core CPUs and 128 NVIDIA Tesla C1060 GPUs. Each process uses a single GPU. Simulations with different number of processes are performed in order to evaluate communication bottlenecks. Moreover simulations with different domain sizes and kernel configurations were evaluated. The performance in giga samples per second (GSamples), meaning the number of calculations that can be performed per second, that is, $10^{-9}(nx \times ny \times nz \times nSteps)/(computing\_time)$, where nx, ny and nz are the number of discretized points in x, y and z axis, respectively, $nSteps$ is the number of simulation steps performed and $computing\_time$ is the total computation time. The code was implemented using NVIDIA CUDA API and the communication stages are done with MPI.

The simulation configurations used are listed in Table 1. These configurations were chosen so that each GPU is used at full capacity. For each configuration, the sizes of xy-planes are kept, thus only the z-axis varies. The size of z-axis chosen determines the number of GPUs necessary to solve the problem. This values are also listed in Table 1.

Figure 4 shows theoretical GSample values. For each configuration, the theoretical GSample values are obtained taking the time spent for running the simulation in a single GPU times the number of GPUs that is going to be used. Figure 5 shows the measured GSamples values for the same simulations. The discrepancy between the theoretical and measured GSample values occurs because the theoretical model does not account for the communication time. But, as one can see in Figure 5 the simulations scales linearly with the number of processes being used.

Table 1. Simulation configurations

Simulation Configurations					
z	Hosts	Processes	z	Hosts	Processes
64	1	1	1,280	10	20
128	1	2	1,408	11	22
256	2	4	1,536	12	24
384	3	6	1,664	13	26
512	4	8	1,792	14	28
640	5	10	1,920	15	30
768	6	12	2,048	16	32
896	7	14	4,096	32	64
1,024	8	16	8,192	64	128
1,152	9	18			

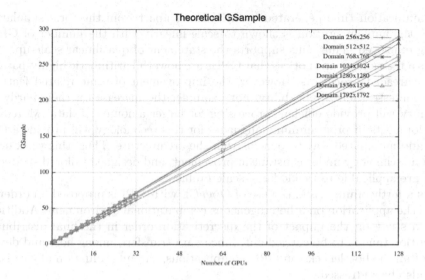

Fig. 4. Theoretical GSample result for various domain sizes

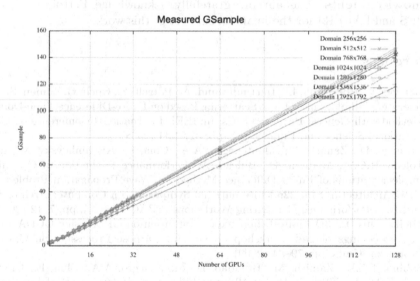

Fig. 5. Measured GSample result for various domain sizes

6   Conclusion and Future Works

A communication scheme was presented for the simulation of 3D scattering of acoustic waves through the solution of a finite difference discretization of the elastic acoustic wave. The scheme introduced herein is capable of dealing with industry scale problems that may not fit in a single GPU. Once the communication time generated by MPI is set apart from the total simulation time, the implementation is shown to scale linearly with the number of GPUs being employed. Thus, this supports the statement of quasilinear scale-up.

As a technology, one notices that GPUs are powerful arithmetic highly-parallel multithreaded processors. However, the improvement of some related features may increase their applicability. For example, the increase on their hardware memory will provide efficient processing for larger amounts of data. Moreover, developments in programming languages for this technology will provide a flexible and automated way to benefit from the architecture. Thus, this work dealt with a technology under constant improvement, and designed tailored strategies that are applicable to tackle large-scale problems.

As a forthcoming work, the use of OpenCL with MPI is proposed, in order to scale the application on a heterogeneous computational environment. Additionally, a study on the impact of the discretization order in the final distributed execution time is to be carried out. Since data transfers among host and device are often the bottlenecks on GPU computations, careful evaluation of this issue will also be addressed.

Acknowledgments The authors gratefully acknowledge Petrobras, CNPq, CAPES and FAPERJ for the financial support of this work.

References

1. Balevic, A., Rockstroh, L., Tausendfreund, A., Patzelt, S., Goch, G., Simon, S.: Accelerating Simulations of Light Scattering Based on Finite-Difference Time-Domain Method with General Purpose GPUs. In: IEEE International Conference on Computational Science and Engineering, pp. 327–334 (2008)
2. Brandao, D., Zamith, M., Montenegro, A.A., Clua, E., Kischinhevsky, M., Leal-Toledo, R.C.P., Madeira, D., Bulcao, A.: Performance Evaluation of Optimized Implementations of Finite Difference Method for Wave Propagation Problems on GPU Architecture. In: 22nd International Symposium on Computer Architecture and High Performance Computing Workshops (SBAC-PADW), pp. 7–12 (2010)
3. Micikevicius, P.: 3D Finite Difference Computation on GPUs using CUDA ACM. In: Proceedings of 2nd Workshop on General Purpose Processing on Graphics Processing Units, pp. 79–84 (2009)
4. Sabino, T.L.R., Zamith, M., Brandao, D., Montenegro, A.A., Clua, E., Kischinhevsky, M., Leal-Toledo, R.C.P., Teixeira-Filho, O.S., Bulcao, A.: Scalable Simulation of 3D Wave Propagation in Semi-Infinite Domains Using the Finite Difference Method on a GPU Based Cluster. In: Proceedings of the V e-Science Workshop, Congress of Brazilian Computer Society, pp. 110–118 (2011)
5. Langdon, W.B., Banzhaf, W.: A SIMD Interpreter for Genetic Programming on GPU Graphics Cards. In: O'Neill, M., Vanneschi, L., Gustafson, S., Esparcia Alcázar, A.I., De Falco, I., Della Cioppa, A., Tarantino, E. (eds.) EuroGP 2008. LNCS, vol. 4971, pp. 73–85. Springer, Heidelberg (2008)
6. Bolz, J., Farmer, I., Grinspun, E., Schroder, P.: Sparse matrix solvers on the GPU: conjugate gradients and multigrid. ACM Transactions on Graphics: Proceedings of ACM SIGGRAPH, 917–924 (2003)

7. Abbas-Turki, L.A., Lapeyre, B.: American Options Pricing on Multicore Graphic Cards. In: International Conference on Business Intelligence and Financial Engineering, pp. 307–311 (2009)
8. Rozen, T., Boryczko, K., Alda, W.: A GPU-based method for approximate real-time fluid flow simulation. Machine Graphics and Vision International Journal 17(3), 267–278 (2008)
9. Michea, D., Komatisch, D.: Accelerating a three-dimensional finite-difference wave propagation code using GPU graphics cards. Geophysical Journal International 182, 389–402 (2010)
10. Reynolds, A.C.: Boundary Condition for the Numerical Solution of Wave Propagation Problems. Geophysics 43, 1099–1110 (1978)
11. Bording, R.P., Lines, L.R.: Seismic Modeling and Imaging with the Complete Wave Equation. Society of Exploration Geophysicists, Tulsa (1997)
12. Juntunen, J.S., Tsiboukis, T.D.: Reduction of Numerical Dispersion in FDTD Method Through Artificial Anisotropy. Transactions on Microwave Theory and Techniques 48(4), 582–588 (2000)
13. Mitchell, A.R.: Computational Methods in Partial Differential Equations. John Wiley and Sons (1969)
14. Ames, W.F.: Numerical Methods for Partial Differential Equations. Mathematics of Computation 62 (1994)
15. NVIDIA: NVDIA - CUDA Programming Guide (2010)

Part-Based Data Analysis with Masked Non-negative Matrix Factorization

Gabriella Casalino[1], Nicoletta Del Buono[2], and Corrado Mencar[1]

[1] Department of Informatics
University of Bari, Bari 70125, Italy
[2] Department of Mathematics
University of Bari, Bari 70125, Italy
{gabriella.casalino,nicoletta.delbuono,corrado.mencar}@uniba.it

Abstract. We face the problem of interpreting parts of a dataset as small selections of features. Particularly, we propose a novel masked non-negative matrix factorization algorithm which is used either to explain data as a composition of interpretable parts (which are actually hidden in them) and to introduce knowledge in the factorization process. Numerical examples prove the effectiveness of the proposed algorithm as a useful tool for Intelligent Data Analysis.

Keywords: Nonnegative matrix factorization, mask matrix, structure retrieval.

1 Introduction

Non-negative Matrix Factorization (NMF) is a computational technique for low-rank approximation of a numerical dataset [13,14]. Differently to other low-rank approximation techniques, such as Principal Component Analysis (PCA), Singular Value Decomposition (SVD) *et similia* [8], NMF is able to explain data in terms of combination of nonnegative factors (provided that data are nonnegative too). In more formal terms, given a dataset represented as a nonnegative matrix

$$X = [\mathbf{x}_1, \mathbf{x}_2, \dots, \mathbf{x}_m] \in \mathbb{R}^{n \times m}_+$$

where $\mathbf{x}_i \in \mathbb{R}^n_+$ are column vectors representing samples [1], NMF algorithm aims to approximate X into the product of two non-negative matrices – a *base matrix* $W \in \mathbb{R}^{n \times k}_+$ and an *encoding matrix* $H \in \mathbb{R}^{k \times m}_+$ – such that

$$X \approx WH. \tag{1}$$

The value of k is user-specified and identifies the number of factors used to explain data. In fact, each sample is approximated as a nonnegative linear

---

[1] In the following, a matrix is denoted with an uppercase letter, e.g. X, its elements with the corresponding lowercase letter, e.g. x_{ij}, a column vector in lowercase bold-face, e.g. $\mathbf{x}_i$

B. Murgante et al. (Eds.): ICCSA 2014, Part VI, LNCS 8584, pp. 440–454, 2014.

combination of factors, that is:

$$\hat{\mathbf{x}}_j = \sum_{i=1}^{k} \mathbf{w}_i h_{ij}. \tag{2}$$

The non-negativity characterization of NMF makes it a useful tool for Intelligent Data Analysis (IDA). In fact, the non-negativity makes NMF capable of representing data as a additive combination of common factors. Moreover, if such factors have some physical meaning (i.e., they can be interpreted in the domain of the considered problem), NMF makes possible to explain data as a composition of parts, being each part a factor. For examples: student questionnaire results can be explained in terms of basic student skills [9], news can be cathegorized according to the arguments they refer to [17], objects can be detected and localized in still in gray-scale images [6] and so on.

The main issue of NMF is therefore related to the ability of interpreting factors in the problem domain: unfortunately, decomposition (1) is not unique; also, it may be not easy to bring out useful knowledge from the representation of W and H [11].

In order to overcome the limitations of classical NMF and to introduce knowledge in the factorization process, additional constraints to the nonnegativity of the matrices W and H can be added. Some examples of constraints are: the sparseness adopted to increase the parts-based representations of the decomposition [12], several forms of orthogonality used to improve the cluster ability of NMF [4,10,16], local manifold structure [3], label information to perform a semi-supervised learning decomposition process [15], binary structure to produce biclustering structures explicitly [18]. Obviously, any additional constraint leads to the development of different optimization algorithms for NMF factorization.

In this paper, we face the problem of interpreting parts as small selections of features. More precisely, we constrain the column vectors of W so that only a small subset of elements is non-zero. This representation of parts could be very useful for IDA, since it is able to highlight some local linear relationships existing among features that hold for a subset of data. To this purpose we introduce a new optimization problem for NMF, which constrains the columns of the base matrix W to possess a small number of non-zero elements. Then, we adopt a query-based approach, where the structure of the base matrix is defined by a user-provided mask matrix. In this way, the analyst can specify the parts she is interested to discover in data; the proposed technique, in fact, extracts the subset of data that are actually represented by the parts.

The proposed approach has been tested on a number of synthetic datasets in order to show its effectiveness in correctly selecting parts in data according to the provided queries. Moreover, a preliminary experiment on the well known Iris data is also reported in order to demonstrate the validity of the proposed approach on real dataset.

The paper is organized as follows: in the next section, the query-based approach is skeched together with the masked NMF algorithm. This latter is derived by minimizing a novel weighted penalized objective function which has been proposed to matching the query matrix and to preserve nonnegativity of data. In section 3 some numerical simulations are reported in order to illustrate the effectiveness of the proposed approach. In section 4 some conclusive remarks are outlined, along with future extensions of the proposed approach.

2   Masked NMF

Classical NMF algorithms used to compute the approximating factors W and H as in (1) are typically derived by solving the constrained least square minimization problem:

$$\min_{W \geq 0, H \geq 0} \frac{1}{2} \|X - WH\|_F^2 \tag{3}$$

where $\| \cdot \|_F$ indicates the Frobienus matrix norm on matrices.

The non-negativity constraints imposed by (3) often are not enough to provide for factors (i.e. the columns $\mathbf{w}_k$ of W) that represent useful knowledge. In fact, usually these columns are very dense; moreover, different configurations of W and H lead to the same approximation of X, thus it is difficult to associate a physical meaning to the factors. Indeed

$$X = (WC)\left(C^{-1}H\right)$$

for every $C \in \mathbb{R}^{k \times k}$ such that both C and C^{-1} are non-negative.

In order to overcome the limits of classical NMF and to inject a-priori knowledge in the factorization process we introduce the concept of part. From the vector representation of data derives that each sample is represented by a vector of n features $\{f_1, ..., f_n\}$. A part p is defined as a sparse vector in $\mathbb{R}^n$ where at least two components are non-zero. A feature belongs to a part iff its value is non-zero. In this way we constrain the factorization process to describe data as a linear correlation of different parts, whose features are linear correlated among them. The structure of the part (i.e. the features set to zero, thus excluded by the part), as well as the number of parts, constitutes the a-priori knowledge and is user-defined.

In order to obtain basis factors that are able to extract parts, we constrain the columns $\mathbf{w}_k$ in W to contain only few non-zero elements. We observe that factors possessing this type of structure enable the elicitation of local linear relationships in subsets of data.

In order to incorporate the previously explained additional constraints, we design the following minimization problem:

$$\min_{W \geq 0, H \geq 0} \frac{1}{2} \|X - (P \odot W)H\|_F^2 + \frac{1}{2}\lambda \left\|P \odot \tilde{W}\right\|_F^2 \tag{4}$$

where

$$\tilde{W}_{ij} = \exp\left(-W_{ij}\right)$$

and

$$P \in \{0, 1\}^{m \times k}$$

is binary matrix, with a fixed number of non-zero elements per column which is used as query for the NMF problem, while $\lambda \geq 0$ is a regularization parameter.

The objective function in (4) is composed by two terms: the first one represents a weigthed modification of the classical NMF problem where the mask matrix P is used to fix the structure the base matrix W has to possess. The second term is a penalty term used to enhance the elements in W that match the mask structure. For this purpose the exponential function has been chosen: in fact when the value of an entry w_{ij} of W is low it is increased by the penalty term, when it is high the penalty tends to zero. The choice of the exponential function allows us to prevent that zero values correspond to features that we want to include in the parts. The query matrix P is used to identify the parts that the analyst would like to extract from data. This is accomplished by defining P as a set of k column vectors, where each element in a column is 1 if the corresponding feature has to be selected, 0 if it has not be considered.

The objective function (4) automatically imposes the structure of the query P in the factor matrix W, minimizing the non-relevant elements in W and maximizing (when they are actually present) the relevant elements in it. It should be observed, however, that the objective function (4) is not convex in both variables W and H. So, it is thus unrealistic to find the global minima for it. However, an iterative updating algorithm to obtain the local optima of (4) can be derived. Particularly, denoted by $\Psi = [\psi_{ij}]$ and $\Phi = [\phi_{ij}]$ the Lagrangian multipliers for the constraints $W_{ij} \geq 0$ and $H_{ij} \geq 0$, the Lagrangian function associated to the minimization problem in (4) is given by:

$$\mathcal{L} = \frac{1}{2} trace\left((X - (P \otimes W)H)^{\mathrm{T}} (X - (P \otimes W)H) \right)$$
$$+ \lambda trace\left(\left(P \otimes \tilde{W}\right)^{\mathrm{T}} \left(P \otimes \tilde{W}\right) \right) + trace\left(\Psi W\right) + trace\left(\Phi H\right) \quad (5)$$

Imposing the Karush-Kuhn-Tucker conditions for the optimality, it follows that the derivative of the Lagragian with respect to W and H are

$$\frac{\partial \mathcal{L}}{\partial W} = (P \otimes W)HH^{\mathrm{T}} - P \otimes (XH^{\mathrm{T}}) + \frac{1}{2}\lambda(-2)P \otimes \tilde{W} + \Psi = 0 \quad (6)$$

$$\frac{\partial \mathcal{L}}{\partial H} = (P \otimes W)^{\mathrm{T}} (P \otimes W)H - (P \otimes W)X + \Phi = 0 \quad (7)$$

Solving the previous equations with respect to the elemements in W and H, the following updating formulas for W_{ij} and H_{ij} can be derived:

$$W_{ij} \leftarrow W_{ij} \frac{\left[P \odot (XH^{\mathrm{T}})\right]_{ij} + \lambda \left(P \odot \tilde{W}\right)_{ij}}{\left[(P \odot W)(HH^{\mathrm{T}})\right]_{ij} + \epsilon} \quad (8)$$

$$H_{ij} \leftarrow H_{ij} \frac{\left[(P \odot W)^{\mathrm{T}} X\right]_{ij}}{\left[(P \odot W)^{\mathrm{T}} (P \odot W)H\right]_{ij} + \epsilon} \quad (9)$$

where the constant $\epsilon = 10^{-12}$ has been introducted to prevent division by zero. We will refer to (8) and (9) as Masked NMF (MNMF).

Regarding these two updating rules, it is not difficult to prove that the objective function (4) is nonincreasing under the updating rules (8) and (9).

2.1   Query Based MNMF

Algorithm 1 formally describes the proposed approach to analyse data through MNMF. Particularly, the steps the proposed approach is composed by are justified and described in the following.

MNMF is an iterative updating algorithm based on the multiplicative algorithm proposed by Lee and Seung [14]. It alternatively updates the matrices W and H according to the rules (8) and (9) (lines 3 and 4 of algorithm 1) while stopping criteria is not satisfied (line 2). Since the algorithm converges to zero, the adopted stopping criteria is based on the difference between two following values of the objective function: the computation of updates stops when the difference is lower than a prescribed small value ε.

Formally, set $E = Obj(t) - Obj(t-1)$, where $Obj(t)$ indicates the value of the objective function at the $t-th$ iterate, then

$$stop = \begin{cases} true & \text{if } E \leq \varepsilon \\ false & \text{otherwise} \end{cases}$$

In the reported experiments $\varepsilon = 10^{-6}$. A maximum number of iterations (set equal to 1000) is also used as an additional stopping criteria just to avoid a high computational effort when very low convergence rate occurs.

Being the MNMF algorithm based on the gradient descent method, it is sensitive to the starting point. As stardart choice, the matrices W and H have been initialized using two random matrices W_0 and H_0 (line 1), however different initializations can lead to better results [2,5,7], further experiments will be aimed to examine this aspect.

It should be pointed out that data in the matrix X has been normalized (line 1) to lay in the unit sphere (i.e. $\|X_{:i}\|_2 = 1$ for $i = 1, \ldots, m$). This representation has been preferred because NMF works in vectorial space, where data are vectors with an own direction and not points. Normalization eliminates information related to the lenght of the vectors, preserving relationships in data.

After MNMF runs, columns of W are normalized in L_2 (line 6) together with the matrix H (line 7) in order to preserve the factorization results. This is accomplished multiplying the factor W for the diagonal matrix N, and multiplying H for its inverse:

$$\bar{W} = WN \tag{10}$$

$$\bar{H} = N^{-1}H \tag{11}$$

where $N = diag\,(norm(W_{:,1}), norm(W_{:,2}), \ldots, norm(W_{:,k}))$.

Algorithm 1. QMNMF

Require: $X \in \mathbb{R}_+^{n \times m}$ {dataset}
Require: $P \in \{0,1\}^{n \times k}$ {mask}
Require: λ {regularization parameter}
Require: $W_0 \in \mathbb{R}_+^{n \times k}$ and $H_0 \in \mathbb{R}_+^{k \times m}$ {initial matrices W and H}
Require: $t > 0$ {threshold}
Require: hardmode $\in \{true, false\}$ {hard-mode selection criterion}
  1. Normalize X
  2. **while** stopping criterion not satisfied **do**
  3.   update matrix W according to (8)
  4.   update matrix H according to (9)
  5. **end while**
  6. Normalize matrix W according to (10)
  7. Adjust matrix H according to (11)
  8. Binarize H into $\bar{H}$ according to eq. (12)
  9. **if** hardmode **then**
 10.   Compute the column index set $J = j : \bar{h}_{I,j} = 1$
 11. **else**
 12.   Compute the column index set $J = j : \bar{h}_{I,j} \neq 0$
 13. **end if**
 14. Select data samples $X' = X[1:n, J]$ {all rows and columns in J}
 15. **return** $W \in \mathbb{R}_+^{n \times k}$ {selected parts}
 16. **return** $H \in \mathbb{R}_+^{k \times m}$ {coefficents}
 17. **return** $X' \in \mathbb{R}_+^{n \times m'}$ {data subset}

When a NMF of a given data matrix X is computed, each sample is approximated in a low-rank subspace (of k dimensionality) by equation (3) . Particularly, the elements of each columns of the encoding matrix H codify the information needed to identify the factors (columns of W) used to reconstruct each sample of X in the low-rank subspace. From a geometrical point of view, the columns of W define the basis vectors of a subspace of dimension k, and each column of H defines the coefficients for each basis vector that is needed to approximate the corresponding data sample in X. Therefore, the elements in a column of H identify the importance of each basis vector in approximating the data sample: if a coefficient is very small, then the corresponding basis vector is useless in approximating the sample; as a consequence, the data sample does not contain the part represented by this basis vector. Information stored in the matrix H can be used therefore for Intelligent Data Analysis. After MNMF optimization a possible occurrence is finding samples of H which have low values corresponding to a part. This means that MNMF was not able to find the corresponding part in that subset, so the analysis could be restricted to the subset of data where parts have been recognized. This can be accomplished by "binarizing" each column of H into $\bar{H}$, i.e.

$$\bar{h}_{ij} = \begin{cases} 1, & h_{ij} \geq t \\ 0, & h_{ij} < t \end{cases} \tag{12}$$

for a user-defined threshold $t > 0$ (line 8). Samples in the matrix H that have not been reconstructed using parts which we are looking for, are then removed from the matrix X. The remaining columns after this removal procedure form a new data matrix that is denoted by X' (line 14). This approach allows the selection of the samples in data that are actually represented by the specified parts. However, the new dataset X' can be extracted according to two alternative criteria: (i) a data sample is selected if it contains at least one part in W (soft-mode) (line 12); (ii) a data sample is selected if it contains all the parts represented in W (hard-mode) (line 10). The first criterion is more conservative as it selects data samples that are represented by possibily other parts that are not included in W. On the other hand, the second criterion is more selective because it selects a data sample only if it can be well reconstructed by all the parts in W. At the end of the selection process, MNMF is re-run for the subset of the selected data samples. The objective of this last step is to re-compute the values in the base and encoding matrices without taking into account data samples that are not composed by the selected parts. This provides a more precise estimation of the parts and their contribution in the data samples.

For technical reasons related to the optimization algorithm, columns of the query matrix P have to be mutually orthogonal, i.e. $p_i^\mathrm{T} p_j = 0$ with $i, j = 1 \ldots k; i \neq j$. This constrain ensures independence of user queries in the columns of P. Moreover, the method requires extra bases. In order to avoid stretching in the factorization process we add many canonical bases to the mask as the feature that have not been selected in the query are. This ensures the factorization process to find parts that the analyst is actually interested in discovering in data. Without this precaution MNMF would attempt to force the reconstruction of data using only the specified bases. Hence if parts in the query are not enough to describe data, MNMF uses the extra bases. For this reason in the algorithm we analyze the subset of rows of the matrix H that don't refer to extra bases (named I).

3   Numerical Simulations

We illustrate the results of some numerical simulations performed on a synthetic dataset $X \in \mathbb{R}^{6 \times 350}$ that is generated in a specific way to evaluate the ability of the proposed approach in finding parts in data (fig. 1).

To generate the data samples, we made use of two random variables, $s_1 \sim \mathcal{N}(5, 1)$ and $s_2 = \alpha s_1$ with $\alpha = 3$ (we cropped to 0 negative values). Then we generated three combinations of two out of six features (mutually orthogonal), namely $(1, 2)$, $(3, 4)$, $(5, 6)$. We considered each combination of features (i_1, i_2) and defined a correponding random basis $\mathbf{c}_h$, $h = 1, 2, 3$, so that $c_{i_1} = s_1$, $c_{i_2} = s_2$ and $c_i = 0$ for $i \notin \{i_1, i_2\}$. We finally generated the dataset in blocks of 50 samples, each block being defined as a combination of the random bases $\mathbf{c}_h$ (i.e. $\mathbf{c}_1, \mathbf{c}_2, \mathbf{c}_3$, $\mathbf{c}_1 + \mathbf{c}_2$, $\mathbf{c}_1 + \mathbf{c}_3, \mathbf{c}_2 + \mathbf{c}_3, \mathbf{c}_1 + \mathbf{c}_2 + \mathbf{c}_3$).

Figure 1 shows a graphical representation of the data matrix X, which has been constructed as linear combination of the three bases c_i. It should be ob-

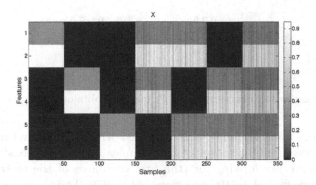

Fig. 1. Graphical illustration of the synthetic dataset X

served that the boxes represent fifty sequential data generated with the same linear combination.

Figure 2 reports the two different masks P_1 and P_2 adopted to query X. The mask P_1 is used to impose on the factors matrix W the same structure occurring in the dataset X, while the mask P_2 represents only partially the structure hidden in data (i.e., the first column in P_2 represents parts that are actually present in X, while the second represents parts that are not present in X). It has been pointed out that in both masks parts we are looking for in data are expressed by the first two columns, the remaining two have been added for techical reasons, and we will refer to them as extra bases. The mask P_1 allows us to verify if the proposed Query-based MNMF is able to recognize as relevant all the examples in dataset that have been constructed using the parts specified by the query mask; while the use of P_2 should show the behaviour of the algorithm when the analyst is looking for parts that are not actually in data.

$$P_1 = \begin{pmatrix} 1 & 0 & 0 & 0 \\ 1 & 0 & 0 & 0 \\ 0 & 1 & 0 & 0 \\ 0 & 1 & 0 & 0 \\ 0 & 0 & 1 & 0 \\ 0 & 0 & 0 & 1 \end{pmatrix} \quad P_2 = \begin{pmatrix} 1 & 0 & 0 & 0 \\ 1 & 0 & 0 & 0 \\ 0 & 1 & 0 & 0 \\ 0 & 0 & 1 & 0 \\ 0 & 0 & 0 & 1 \\ 0 & 1 & 0 & 0 \end{pmatrix}$$

Fig. 2. Mask matrices used in the MNMF algorithm

This means that in both cases the query we submitted to the algorithm does not cover all the examples in the dataset, so we expect that the procedure selects a subset of it containing only the relevant data.

Figure 3 illustrates the mask matrix P_1 together with the factor matrix W computed by MNMF with parameter $\lambda = 0.00001$ and P_1. As it can be observed, the factor W possesses the same of structure of P_1 and the extracted bases $\mathbf{w}_i$ preserve also the multiplicative factor α that has been used to generate the data, being the ratio of the two non-zero features approximately equal to 3.

$$P_1 = \begin{pmatrix} 1\,0\,0\,0 \\ 1\,0\,0\,0 \\ 0\,1\,0\,0 \\ 0\,1\,0\,0 \\ 0\,0\,1\,0 \\ 0\,0\,0\,1 \end{pmatrix} \quad W = \begin{pmatrix} 0.316 & 0 & 0\,0 \\ 0.949 & 0 & 0\,0 \\ 0 & 0.316 & 0\,0 \\ 0 & 0.949 & 0\,0 \\ 0 & 0 & 1\,0 \\ 0 & 0 & 0\,1 \end{pmatrix}$$

Fig. 3. Comparison between W and P_1 matrices obtained with $\lambda = 0.00001$

Figure 5 illustrates the matrix H obtained with MNMF and query P_2. As it can be observed only a subset of the samples has been reconstructed using only the parts we are looking for (the first two basis of the matrix W). It is made by samples from 1 to 100 and from 151 to 200. The algorithm in the hard mode returns this subset of data. The algorithm in the soft mode returns the samples that have been reconstructed using at least one of the parts in P_1. All the samples in dataset satisfy this requirement except the subset from 100 to 150.

Below a detailed analysis shows the behavior of MNMF in reconstructing samples when they contain the parts in P_1, linear relationship of these parts, and when there are not parts in the mask adequate to describe them.

Samples from 1 to 50 composed by the the first two features in X (whose linear correlation is captured by $\mathbf{w}_1$) have been correctly reconstructed in H using only the first part. Similarly, samples from 51 to 100 have been reconstructed using the part $\mathbf{w}_2$ representing the relationship between features three and four.

MNMF can recognize parts that are composed linearly to describe data. This is the case of the samples from 151 to 200 that have been generated adding data composed by the first two features and data composed by the third and fourth. These samples have been reconstructed in the matrix H using both the bases $\mathbf{w}_1$ and $\mathbf{w}_2$ capturing the linear correlation between respectively first and second features, and third and fourth features.

When the algorithm does not find parts that are able to correctly reconstruct the samples in data it uses the extra bases in the mask. This beaviour suggests to the analyst that the parts he is looking for in data are not enough to describe them. This is the case of samples from 201 to 350 that have been constructed adding data composed by feature caught in the mask and data composed by parts that are not in P_1. Matrix H correctly suggests which part in P_1 we need to reconstruct the samples and extra bases.

An extreme case are the samples from 101 to 150 that have been completely constructed using a part that is not in P_1, the algorithm returns no parts for this samples. Hence, our proposed Query Based MNMF algorithm suggests which parts correctly reconstruct data.

Figure 4 shows the reconstruction error (MSE) obtained for each sample in data. It has been obtained using the equation (13) where $i = i \ldots 350$ indicates the i−th sample and n is the total number of samples in the dataset.

$$MSE\,(i) = \frac{1}{2}\frac{\|X_{*i} - WH_{*i}\|_F^2}{n} \tag{13}$$

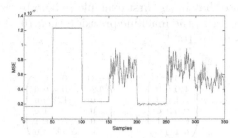

Fig. 4. MSE of each sample obtained with MNMF and P_1

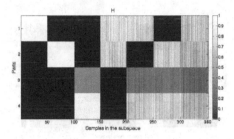

Fig. 5. Matrix H obtained with MNMF and P_1

Mask P_2 shows the behaviour of the Query Based MNMF algorithm when it is looking for parts that not correctly decribe data.

Figure 3 illustrates the basis matrix W computed by MNMF with matrix mask P_2 and $\lambda = 0.00001$. As it can be observed, whilst the first base $\mathbf{w}_1$ that catches the structure of the data has significative values, the second base $\mathbf{w}_2$ tries to describe data with parts that not completely explain them. For this reason one of the two values is very close to the maximum value 1, and the second one is low. This could suggest that the structure imposed does not allow a good reconstruction of the data.

Running the algorithm in the hard mode, it returns only the block of samples from 1 to 50. This result is correct, in fact only this subset is completely explained using the parts in P_2, particularly only the part w_1, instead the part w_2 doesn't match with the structure of the data. Soft mode algorithm returns all the samples except of the block from 51 to 100, it means that there is not in P_2 a part that describes these samples. In fact they have been constructed using the third and forth features, since the value w_{32} is negligible, there is any part in P_2 that allows to reconstruct these samples. Moreover samples in X that have been constructed using the fifth and the sixth features, are partially described by the part w_2, for this reason samples in the block from 100 to 150 have been reconstructed using both the part w_2 and an extra bases. This behaviour confirms that the part w_2 does not completely represent data, it needs extra information.

The graph of the MSE illustrated in figure 8 provides a further confirmation of the previously discussed behavior. Comparing the evolution of the MSE for both cases, we can note that in the first example (where bases well explain data) the obtained error is near to the machine precision (i.e., 10^{17}) while for the latter example, MSE grows up to 10^{-4}.

$$P_2 = \begin{pmatrix} 1 & 0 & 0 & 0 \\ 1 & 0 & 0 & 0 \\ 0 & 1 & 0 & 0 \\ 0 & 0 & 1 & 0 \\ 0 & 0 & 0 & 1 \\ 0 & 1 & 0 & 0 \end{pmatrix} \quad W = \begin{pmatrix} 0.316 & 0 & 0 & 0 \\ 0.949 & 0 & 0 & 0 \\ 0 & 0.123 & 0 & 0 \\ 0 & 0 & 1 & 0 \\ 0 & 0 & 0 & 1 \\ 0 & 0.993 & 0 & 0 \end{pmatrix}$$

Fig. 6. Comparison between W and P_2 matrices obtained with $\lambda = 0.00001$

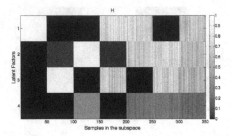

Fig. 7. Matrix H obtained with MNMF and P_2

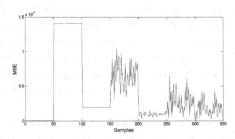

Fig. 8. MSE of each sample obtained with MNMF and P_2

A further synthetic dataset has been constructed to better explain the behaviour of the proposed method when it is forced to find parts that are not actually present in data. Points in $X \epsilon \mathbb{R}^{3x500}$ compose a circumference lying in the plane x,y on the bisector of the first quadrant, thence a linear relationship between features one and two exists in data (i.e. $x = y$). The components on the axes z are equal, or almost equal to 0. MNMF has been executed with two masks $P_3 = \begin{pmatrix} 1 & 0 \\ 0 & 1 \\ 1 & 0 \end{pmatrix}$ and $P_4 = \begin{pmatrix} 1 & 0 \\ 1 & 0 \\ 0 & 1 \end{pmatrix}$ and $\lambda = 0.0001$. The first mask P_3 tries to find a linear relationship between the features on the axes x and z. As it can be observed from the figure 9, MNMF returns a matrix W_3 whose component w_{31} has a negligible value, while the component w_{11} has the maximum value 1. This behaviour confirms that there is no part in data corresponding to the axes z. Figure 10 shows the matrices H obtained when the masks P_3 (picture on the left) and P_4 (picture on the right) are used . As it can be expected data have been reconstructed using both bases. In the second example, the mask P_4 tries to find a relationship between the components on the axes x and y. MNMF returns a base matrix W_4 (figure 9) whose components w_{11} and w_{21} are significant, moreover it catches the linear relationship between these two components.The component w_{32}, even though data do not have a z-components, has a value equals to 1. It should be noted that this behavior has been produced by the normalization process. Despite this, the matrix H (as it can be observed by the right picture in Figure 10) confirms that the base w_2 is not necessary in reconstructing data, in fact all the samples use only the first base.

$$W_3 = \begin{pmatrix} 1 & 0 \\ 0 & 1 \\ 0.0004 & 0 \end{pmatrix} W_4 = \begin{pmatrix} 0.7 & 0 \\ 0.7 & 0 \\ 0 & 1 \end{pmatrix}$$

Fig. 9. Masks matrices used in the MNMF algorithm

Fig. 10. Comparison between matrices H obtained with MNMF and masks P_3 and P_4

3.1   Iris Dataset

In this section we briefly illustrate the behaviour of the proposed approach when the well known Iris dataset is adopted [1]. The dataset is composed by 150 samples grouped in three different classes: Iris-Setosa, Iris-Veriscolor, Iris-Virginica. This example highlights the use of a specific mask to select features and extract samples which are described by these parts. Particularly, the aim is to discover if there exist any linear correlation between the features in the data samples (i.e., sepal and petal length, sepal width and petal length, sepal and petal width). MNMF has been executed with two different masks $P_5 = \begin{pmatrix} 1 & 0 \\ 0 & 1 \\ 1 & 0 \\ 0 & 1 \end{pmatrix}$

and $P_6 = \begin{pmatrix} 1 & 0 \\ 1 & 0 \\ 0 & 1 \\ 0 & 1 \end{pmatrix}$ and with parameter $\lambda = 0.0001$. Mask P_5 aims to discovery linear correlations in data between the lengths and the widths, whilst P_6 between sepal and petal feature. The use of a real dataset better highlights the semantic associated to the parts that have been selected.

Figure 11 illustrates the factor matrices W_5 and W_6 obtained respectively with masks P_5 and P_6. As it can be observed, the factor matrices preserve the structure imposed by the query masks, moreover the parts are represented by significative values; this means that they are actually present in data, i.e., there is correlation between the features selected in data.

$$W_5 = \begin{pmatrix} 0.85 & 0 \\ 0 & 0.96 \\ 0.53 & 0 \\ 0 & 0.29 \end{pmatrix} \quad W_6 = \begin{pmatrix} 0.88 & 0 \\ 0.48 & 0 \\ 0 & 0.95 \\ 0 & 0.31 \end{pmatrix}$$

Fig. 11. Comparison between W_5 and W_6 matrices obtained with masks P_5 and P_6 $\lambda = 0.0001$

Fig. 12. Comparison between matrices H obtained with MNMF and masks P_5 and P_6

Figure 12 illustrates the encoding matrices H obtained with the masks P_5 and P_6. Observing the left graph one can figures out that samples from 51 to 151 (belonging to the two classes Versicolor and Virginica) can be represented using only the first bases w_1. In fact, the elements in w_1 assume almost the maximum value, that is 1, while the elements in w_2 assume values close to zero. This means that in this subset of data there is a linear relationship between the lengths of the iris, but not between the widths. On the contrary samples from 1 to 50 (belonging to Setosa), have been reconstructed using both bases w_1 and w_2. We executed the MNMF on the modified dataset composed by the first fifty samples, with the query mask P_5. The reconstruction error obtained removing samples that have not been well reconstructed is 4.7475×10^{-4} much smaller than that obtained with the entire dataset 1.76×10^{-2}. This result confirms that data belonging to the class Setosa have linear relationships between the lengths and widths.

Similarly from the matrix H on the right, we can observe that samples from 1 to 50 are reconstructed mainly using the basis w_1. In this case it means that there is a linear relationship between the sepal features but not between the petal features. On the contrary, there is a linear relationship between both sepal and petal features in the subset of data composed by samples from 51 to 150. The reconstruction error obtained after removing samples from 1 to 50 is 8.3235×10^{-4} much lower than that obtained with the whole dataset which is 3.5×10^{-3}.

4   Final Remarks

A novel NMF algorithm, namely Masked NMF has been proposed in order to overcome the limitations of classical NMF and to introduce knowledge in the factorization process, making the proposed MNMF algorithm a useful tool for IDA. The query-based approach has been adopted to allow the analyst to specify what parts she is interested to discover. As shown in the numerical examples, the proposed approach is able to extract the subset of data that are actually represented by the parts, discarding the data in the matrix X that do not find a neat representation by the parts and returning the subset of samples that contains the selected parts.

Future work can be addressed to assess the performance of the query based MNMF approach on different real datasets as well as to further investigate its capability of selecting local features hidden in data.

Acknowledgements This work was supported by *"National Group of Computing Science (GNCS-INDAM)"*.

References

1. Bache, K., Lichman, M.: UCI Machine Learning Repository. University of California, School of Information and Computer Science, Irvine, CA (2013), http://archive.ics.uci.edu/ml
2. Boutsidis, C., Gallopoulos, E.: Svd based initialization: a head start for nonnegative matrix factorization. Pattern Recognition 41, 1350–1362 (2008)
3. Cai, D., He, X., Wu, X., Han, J.: Non-Negative Matrix Factorization on Manifold. In: Proc. Eighth IEEE Int'l Conf. Data Mining, pp. 63–72 (2008)
4. Yoo, J., Choi, S.: Orthogonal nonnegative matrix tri-factorization for co-clustering: multiplicative updates on Stiefel manifolds. Information Processing and Management 46, 559–570 (2010)
5. Casalino, G., Del Buono, N., Mencar, C.: Subtractive initialization of nonnegative matrix factorizations for document clustering. In: Petrosino, A. (ed.) WILF 2011. LNCS, vol. 6857, pp. 188–195. Springer, Heidelberg (2011)
6. Casalino, G., Del Buono, N., Minervini, M.: Nonnegative matrix factorizations performing object detection and localization. Applied Computational Intelligence and Soft Computing 15 (2012)
7. Casalino, G., Del Buono, N., Mencar, C.: Subtractive clustering for seeding nonnegative matrix factorizations. Information Sciences 257, 369–387 (2014)

8. Chu, M., Del Buono, N., Lopez, L., Politi, T.: On the low rank approximation of data on the unit sphere. SIAM Journal Matrix Analysis Appl. 27(1), 46–60 (2005)
9. Desmarais, M.C.: Conditions for effectively deriving a q-matrix from data with non-negative matrix factorization. In: Conati, C., Ventura, S., Calders, T., Pechenizkiy, M. (eds.) Proceedings of the 4th International Conference on Educational Data Mining, pp. 41–50 (2011)
10. Ding, C., Li, T., Peng, W., Park, H.: Orthogonal nonnegative matrix tri factorizations for clustering. In: Proceedings of the 12th ACM SIGKDD International Conference on Knowledge Discovery and Data Mining, pp. 126–135 (2006)
11. Donoho, D., Stodden, V.: When does non-negative matrix factorization give a correct decomposition into parts. In: Advances in Neural Information Processing Systems, vol. 16 (2003)
12. Hoyer, P.O.: Non-negative matrix factorization with sparseness constraints. J. Machine Learning Research 5, 1457–1469 (2004)
13. Lee, D.D., Seung, S.H.: Learning the parts of objects by non-negative matrix factorization. Nature 401, 788–791 (1999)
14. Lee, D.D., Seung, S.H.: Algorithms for non-negative matrix factorization. In: Proc. Adv. Neural Information Proc. Syst. Conf., vol. 13, pp. 556–562 (2000)
15. Liu, H., Wu, Z., Cai, D., Huang, T.S.: Constrained Non-negative Matrix Factorization for Image Representation. IEEE Transactions on Pattern Analysis and Machine Intelligence 34(7), 1299–1311 (2012)
16. Ma, H., Zhao, W., Tan, Q., Shi, Z.: Orthogonal nonnegative matrix tri-factorization for semi-supervised document co-clustering. In: Zaki, M.J., Yu, J.X., Ravindran, B., Pudi, V. (eds.) PAKDD 2010. LNCS, vol. 6119, pp. 189–200. Springer, Heidelberg (2010)
17. Shahnaz, F., Berry, M.W., Pauca, V.P., Plemmons, R.J.: Document clustering using nonnegative matrix factorization. Information Processing and Management 42(2), 373–386 (2006)
18. Zhang, Z.Y., Li, T., Ding, C., Ren, X.W., Zhang, X.S.: Binary matrix factorization for analyzing gene expression data. Data Min. Knowl. Discov. 20(1), 28–52 (2010)

A Strategy to Workload Division for Massively Particle-Particle N-body Simulations on GPUs[*]

Daniel Madeira[1,2], José Ricardo[2], Diego H. Stalder[3],
Leonardo Rocha[1], Reinaldo Rosa[3],
Otton T. Silveira Filho[2], and Esteban Clua[2]

[1] Universidade Federal de São João del Rei, São João del Rei, Minas Gerais, Brasil
{dmadeira,lcrocha}@ufsj.edu.br
[2] Universidade Federal Fluminense, Niterói, Rio de Janeiro, Brasil
josericardo_jr@gmail.com, {esteban,otton}@ic.uff.br
[3] Instituto Nacional de Pesquisas Espaciais, São José dos Campos, São Paulo, Brasil
diego.s.eik@gmail.com, reinaldo@lac.inpe.br

Abstract. The new programmable graphical processor units (GPUs) are now often used as high parallel mathematical co-processors, allowing many computational-intensive problems to be executed in less time. It became common and convenient to use single GPUs to implement different kinds of simulations, or to group them in a grid, so the computational power can be highly increased, while the power consumption and physical space increase are significantly lower. The N-body simulation has been successfully ported to the GPUs. This algorithm is typically applied to gravitational simulations, among many other physical problems. In cosmology, one alternative approach seeks to explain the nature of dark matter as a direct result of the non-linear space-time curvature, due to different types of deformation potentials. In order to develop a detailed study of this new approach, our group is developing the COsmic LAgrangian TUrbulence Simulator ($COLATUS\_ENVIU$). The simulator uses the direct particle-particle method to calculate the forces between the particles, so we eliminate the errors included on the hierarchical strategies. It gave robust initial results, but was limited to systems with a small amount of particles. However one limitation on the use of GPUs on N-body simulation to study the details on the mass distribution, is that these systems must have millions (or even billions) of particles, making the use of particle-particle method in one GPU unfeasible due to memory or time computational constraints. In this present work we propose a novel and efficient subdivision method to allow the workload division, allowing a single GPU to be able to solve large simulations, while allowing to watch over a particle during all the simulation.

1 Introduction

With the high improvements in the architectures of modern graphics processing units (GPU), many problems had been ported to this high parallel platform,

[*] This work is partially supported by CNPq, CAPES, Finep, Fapemig and Faperj.

B. Murgante et al. (Eds.): ICCSA 2014, Part VI, LNCS 8584, pp. 455–465, 2014.
© Springer International Publishing Switzerland 2014

drastically increasing the computational performance. It became common to use single GPUs to do all kinds of simulations, or to group them in a grid, so the computational power can be highly increased, while the power consumption and physical space increase are significantly lower. To efficiently exploit this collection of GPUs, the problem must to be modeled as a multiple thread and single instruction parallel paradigm.

The n-body algorithm is applied to a wide variety of problems in computational physics, such as gravitation, electrostatic, crowd simulation and even fluid simulation, among others. In order to solve this problems, the properties of all bodies in the system are must be updated considering the effects produced by all of the other bodies, thus presenting an $O(n^2)$ complexity. Fast algorithms based on multipole accelerated treecode exist, but requires the reconstruction of the tree during all time steps and introduce errors in calculations that have to be considered.

An alternative and relevant cosmology model for explaining the nature of dark matter uses the direct result of the non-linear space-time curvature, due to different types of deformation potentials. To evaluate these approaches, a key test is to examine the effects of deformation on the evolution of large scales structures, watching the position of a galaxy during its trajectory to the gravitational collapse of super clusters at low redshifts. In this context, each element in a gravitational N-body simulation behaves as a tracer of collapse governed by the process known as chaotic advection (or Lagrangian turbulence). In order to develop a detailed study of this new approach our group has created the COsmic LAgrangian TUrbulence Simulator ($COLATUS\_ENVIU$). The simulator uses the common particle-particle method to calculate the forces between the elements, so we eliminate the approximation errors included on the hierarchical strategies. It gave robust initial results, but was limited to a small amount of particles due hardware limitations.

In the present work, we propose a regular-grid subdivision method to allow the workload division, allowing large simulations to be feasible in a single GPU.

The remainder of this paper is divided as follows: in Section 2 we describe some related work. In Section 3 we describe the N-body problem. Section 4 describes the serial and parallel N-body algorithm and details our proposed strategy. Section 5 presents the experiments performed to evaluate our algorithm, and the obtained results. Finally, we present our conclusions in Section 6.

2   Related Work

The n-body algorithm is suitable for cosmological simulations, as presented in [1,2,3,4]. The most common algorithm is based in a simple update for each body using all properties of the elements in the system, thus presenting an $O(n^2)$ complexity. In this method, all bodies take part on the calculations, no matter how far apart they are.

Barnes and Hut, in [5], proposed a hierarchical algorithm, with lower complexity ($O(n \log(n))$). The main advantage of this algorithm was to update each body not using all other bodies. This method group a set of close bodies that are sufficiently far from a specific particle and treat this set as a single body. The set of bodies are calculated subdividing the space in cells. So, if a body is close to a cell, a single particle-particle algorithm updates the body, but if the body is far away from a cell, it will be updated using only the resulting body of the cell. Figure 1 shows an example of the Barnes-Hut method. Others hierarchical methods have also been proposed, such as [6]. With the rise of the processing power in the modern GPUs, numerous works appeared, adapting the hierarchical structure and the hierarchical methods to the GPU ([7,8,9,10,11,12]). Furthermore, some methods have been proposed to share the work in clusters of GPUs [13].

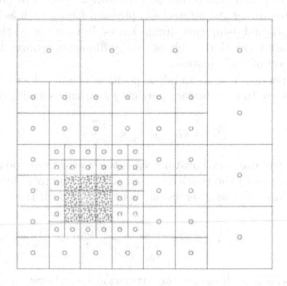

Fig. 1. Example of the Barnes-Hut hierarchical method. To evaluate the particles marked in red, we use only its neighbors cells and the centers of the other cells.

While the hierarchical method is the most common approach found in recent literature, as it allows a significant improvement in the simulation execution time, it introduces an approximation error due the clusterization of particles. This error can be tracked and controlled, but can't be eliminated. The particle-particle method gives the exact result on the calculations, but impacts on execution time.

Initially, the GPU can handle efficiently small particle-particle simulations, but two factor limit the GPU to be able to be used in larger simulations: a computational limitation, since the GPU have a maximum workload that can be given in a single step, and memory limitation. A strategy of work division can address these two limitations, as it gives smaller tasks to GPU at a time. Another important limitation, even when the GPU can handle the simulation itself, is the time constraint: one simulation is not feasible if it takes too long to finish. A work division strategy can also minimize this problem, making the simulation feasible.

Our proposed method aims workload division in a single GPU. Instead of using a hierarchical approach to calculate the iteration between the bodies, we use a subdivision scheme that allows the GPU to update only a subset of the particles at each iteration. With this method, we can run a particle-particle simulation and watch a single particle in all steps of the simulation. This particle tracer is a major contribution of this work, since it can't be implemented while using hierarchical methods, due to approximations made.

3   N-body Problem

A N-body simulation is a simulation of a dynamical system of particles. These particles are usually under the influence of physical forces, such as gravity. Many processes can be studied using this simulation as, in Cosmology, the processes of non-linear structure formation, such as galaxy filaments, galaxy halos, and the dynamical evolution of star clusters.

Given a single particle i, with an initial position x_i and velocity v_i, its gravitational attraction force to an arbitrary particle j is calculated with the gravitation equation:

$$f_{ij} = G \frac{m_i m_j}{||\vec{r}_{ij}||^2} \times \frac{\vec{r_{ij}}}{||\vec{r_{ij}}||}$$

where G is the gravitational constant, $\vec{r_{ij}}$ is the vector from body i to j and m_i and m_j are the masses of the particles i and j. Thus, the total force F_i on body i is the summation of all the interactions between i and all the other particles

$$F_i = G m_i \sum_{1 \le j \le N}^{i \ne j} \frac{m_j \vec{r_{ij}}}{||\vec{r_{ij}}||^3}$$

As bodies approach each other, the attraction force between them grows to infinite, which is undesirable. So, to contour this situation, an softening factor ϵ^2 is added, and the rewritten equation is

$$F_i \approx G m_i \sum_{1 \le j \le N} \frac{m_j \vec{r_{ij}}}{(||\vec{r_{ij}}||^2 + \epsilon^2)^{3/2}}$$

Considering $\epsilon^2 > 0$, and the force $f_{ii} = 0$, the condition $i \ne j$ is no longer needed. For the interactions between the particles, we need the acceleration, which is given by the equation $F_i = m_i a_i$. So, it is possible to cancel the m_i factor on the above equation, having as a result:

$$F_i \approx G \sum_{1 \le j \le N} \frac{m_j \vec{r_{ij}}}{(||\vec{r_{ij}}||^2 + \epsilon^2)^{3/2}}$$

For this work, we use the Leapfrog method to integrate the forces. The Leapfrog is a second-order integrator, in contrast to the Euler method, which is only first order. Yet, the number of functions evaluations needed per step by

Leapfrog is the same on Euler. Unlike Euler integration, the Leapfrog is stable for oscillatory motion. Two primary strengths of Leapfrog integration when applied to mechanics problems is the time-reversibility and its symplectic nature. So it allows that one integrate forward n steps, and then reverse the direction of integration and integrate backwards n steps to arrive at the same starting position and it conserves the (slightly modified) energy of dynamical systems. This is especially useful when computing orbital dynamics, as many other integration schemes, such as the higher order Runge-Kutta method, do not conserve energy and allow the system to drift substantially over time. [14].

4 N-body Algorithm and Proposed Method

In this section, we present the sequential and parallel N-body algorithm, as well the proposed strategy of this work.

4.1 Direct N-body

Having a system with many particles under the influence of their mutual gravitational forces, the direct gravitational N-body simulation integrates numerically all forces without any simplification or approximation. The first direct N-body simulations were done by [15].

As a simple vision of the problem, a direct N-body simulation in a system with N particles is a grid of size NxN, containing all pairwise forces between the bodies. The final force of a body i is the sum of all forces on the line i Each of those forces are completely independent, so it is possible to perform all the $O(N^2)$ calculation in parallel. The problem with this approach is that it demands a high memory space ($O(N^2)$), limiting its use on the GPU.

So, as presented in [16], the simulation could be parallelized in lines, using *tiles*, as square regions of a grid. Considering that a sub-part of the grid has a resolution of $p \times p$ particles, it is possible to evaluate all the p^2 interactions, reusing p of them on the next tile.

For optimal reuse of data, the tile is arranged so each row is evaluated in sequential, while separate rows are evaluated in parallel. Each row updates the acceleration vector. This allows the acceleration vector to use much less memory space (from $O(N^2)$ to $O(N)$). In the Figure 2, the sequential and the parallel strategies to calculate a tile are shown.

The sequential implementation, even on modern CPUs, can't achieve good time results, being discarded over the parallel GPU implementation. The main disadvantage of this parallel implementation is to require all N particles inside the GPU memory at run-time. For smalls simulations, this is not a problem, but bigger simulations can't be executed, due to memory limits.

4.2 Proposed Method

This work proposes a workload division in order to allow bigger direct N-body simulations to be run on GPUs. Systems with billions of particles cannot be

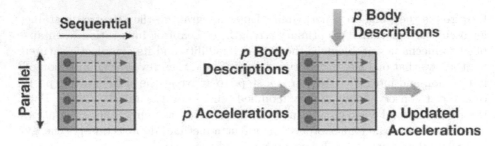

Fig. 2. Sequential and Parallel algorithms for direct N-body simulations [16]

stored in GPU, as its memory requirements are high. Besides the memory limit, there is a computational limit in the GPU. As the system increases, it is not possible to launch a single thread per particle. A solution for this limitation is to allow one thread to do multiple particles updates. This can reach other limitation of the kernel: the maximum number of instructions allowed per kernel.

Our work also allows the execution of big simulations in one GPU, avoiding the GPU limitations. The main feature for this is to parallelize the processing of each tile, setting them in a regular grid. This grid is formed by small tiles that can be easily executed by the GPU, with only some dependencies to be observed. Since the GPU can focus only in a small set at each time, the theoretical limit for the number of particles is the host memory (CPU).

Our algorithm subdivides the $N \times N$ grid of particles in total of $N/A \times N/B$ tiles. The tiles in a row must be calculated in sequence, because they depends on the same set of particles. This is the same dependency of the first algorithm proposed by [16]. As each row is completely independent, we can stream them to different GPUs. In Figure 3, we show the tasks dependencies on the grid.

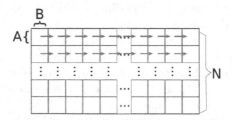

Fig. 3. Proposed workload division. Red lines indicates dependency. Each tile in a row must be evaluated in sequence to accumulate its results, while each row is completely independent and can be assigned to a different stream or GPU.

The implementation uses two sets of particles and an acceleration vector, for the result. The first set is composed by the vectors dposA and dvelA holds the particles that will be updated. The second set holds the vectors dposB and dvelB, and are the particles that will be used to update the first set.

Each row is used to calculate the acceleration of a single particle. As the acceleration of all particle are calculated, we compute the stepsize used by the Leapfrog integrator. After that, finally the particle's new position and velocity are stored in a new vector. Each step is calculated in multiple GPUs, returning their results to the CPU. After the entire system evaluation, the original and the updated position and velocity vectors are switched, passing to the next simulation step. Algorithm 1 illustrate this process.

Algorithm 1. Algorithm for the N-body simulation.

---

function CPU($*pos, *vel, *newpos, *newvel, *accell, A, B, N, maxSimSteps$)
    while ($simStep < maxSimSteps$) **do**
        for all (row $R \in N/A$ rows) **do**
            for all (tile $T \in N/B$ tiles) **do**
                CalculateInteractions(T)
            end for
            CalculateStepSize($accell$)
            IntegrateRows($R, newpos, newvel$)
        end for
        $pos = newpos$
        $vel = newvel$
        $simStep + +$
    end while
end function

The three functions mentioned in Algorithm 1 (CalculateInteractions, CalculateStepSize and IntegrateRows) runs on the GPU. The function CalculateInteractions, presented in Algorithm 2 calculates the particles interactions. One thread is launched for each particle in the vector dposA. Each thread perform a loop in all particles in the vector dposB, updating the acceleration. Shared memory optimization were omitted in the pseudocode for clearness.

The stepsize is calculated according to the biggest acceleration norm for all particles. So each GPU search all accelerations stored in its memory and return the biggest norm found. After that, the CPU can calculate the biggest of all GPUs, calculate the stepsize, and push the value to all GPUs. For this step, we use the Thrust library, as it offers optimized transform and reduce operators.

The function IntegrateRow is called after the stepsize were processed. This function is the Leapfrog integrator. Again, one thread is launched for each body in dposA and updates the body position and velocity, saving the results on a new vector.

5   Experimental Evaluation

In this section, we present the experiments conducted in order to evaluate the techniques implemented in our algorithm.

Algorithm 2. Algorithm for body-body interactions evaluation

---

function CALCULATEINTERACTIONS($*dposA, *dposB, *accell$)
    $tid = threadIdx$
    $mypos = dposA[tid]$
    $acc = \{0, 0, 0\}$
    for all (particles $b_j \in$ dposB) **do**
        $acc+ =$ bodyBodyInteraction($mypos, b_j, acc$)
    end for
    $accell[tid] = acc$
end function

In the following subsections we present our parameters and the initial condition data used in our experiments. All experiments were performed on a computer with Intel Core i7-3930K CPU at 3,20GHz, 32GB of main memory, and equipped with a GeForce K20c graphics card. All results presented are averages of 10 independent executions of the algorithms.

5.1 Multiple Streams

In this test, the objective is to evaluate the impact of stream smaller particles groups instead of pushing them all inside the GPU in one memory copy, in one GPU. Each time the kernel is launched, the GPU evaluates all the particles against the group streamed to the GPU. With this approach, each kernel do less work and uses less memory, allowing multiple kernels running side by side on the GPU. At the same time, we allow kernel to overlap memory copies, minimizing the latency.

The tests groups contains from 65536 to 1048576 particles. Initially, to test if the streams would allow improvement in our application, the group size was set to 32768, and the number of streams varies from 1 to 8. The results of this test is in Figure 4.

This tests shows the use of stream can give a boost to the execution time. Passing from one stream to two the execution time decreases significantly. After that, the time still decreases until four streams. After that, the gain is marginal. Table 1 illustrates the speedup from increasing from 1 to 4 streams.

We can see that the improvement increases as the number of particles increases, showing that our algorithm benefits from the use of data streaming. This behaviour comes from overlapping copy operations and kernel executions, minimizing latency. More than that, when using multiple GPUs, the time is strictly proportional to the number of GPUs used, since smaller copies does not overload the bus between CPU and GPUs.

5.2 Tile Size

The objective of this test is to evaluate the ideal tile size parameter. The tile size impacts in the memory copy time, as it change the group of particles sent

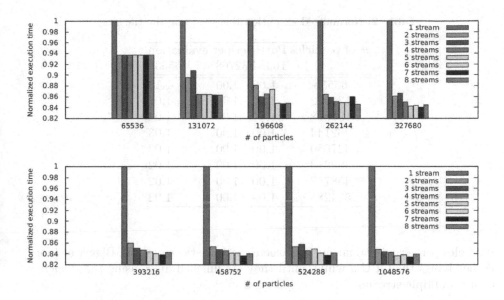

Fig. 4. Kernel and memory copy execution times for the simulation, varying the number of streams, with the group size fixed. The times were normalized to allow easier comparation between the configurations.

Table 1. Normalized execution times from one and four streams and the improvement achieved. When the number of particles increases, the improvement increases.

# of particles	1 stream	4 streams	Improvement
65536	1.00	0.937	−6.30%
131072	1.00	0.864	−13.60%
196608	1.00	0.864	−13.60%
262144	1.00	0.851	−14.90%
327680	1.00	0.850	−15.00%
393216	1.00	0.847	−15.30%
458752	1.00	0.846	−15.40%
524288	1.00	0.846	−15.40%
1048576	1.00	0.842	−15.80%

to the GPU in each stream. We fixed the number of streams in 4, as supported by the previous experiment. The group size was set to 16384, 32768 and 65536, using the same datasets from the previous experiments. So, each row of the grid contains 32768 rows, and $N/\texttt{group\_size}$ iterations are made to fully evaluate the row. The results are condensed in Table 2.

This result shows that the size of the tile has a low influency in the behaviour of the algorithm, as long as it stays small. When we increase it to 32768 × 65536

Table 2. Normalized execution times varying the tile size

# of particles	Particles per evaluation		
	16384	32768	65536
65536	1.00	1.00	1.10
131072	1.00	1.00	1.04
196608	1.00	1.00	1.04
262144	1.00	1.00	1.03
327680	1.00	1.00	1.03
393216	1.00	1.00	1.02
458752	1.00	1.00	1.02
524288	1.00	1.00	1.02

particles per kernel evaluation, the execution time starts to grow. Bigger memory copies hangs the GPU a while, until they are finished, decreasing the efficiency of use multiple streams.

6   Conclusions and Future Work

In this paper, we proposed a novel and efficient subdivision method to divide the workload, allowing a GPU to be able to solve simulations that doesn't fit on it. The results were encouraging, as the data transfers doesn't impact significantly the execution times in our scenario. After this observation we propose as future work to extend this work to multiple GPUs on a host and also a grid of GPUs, over a network. In a homogeneous grid, the algorithm can be statically instantiated for the best execution. We also aim to evaluate our algorithm in a heterogeneous grid. Since in this scenario each GPU can have different computational performance, a strategy to balance the size of each task is important, so the less powerful GPUs (or with less memory) can process tasks compatible to its architecture.

References

1. Makino, J., Taiji, M.: Scientific Simulations with Special-Purpose Computers–the GRAPE Systems (April 1998)
2. Felix Soehr, S.D.M.W.: Simulation of Galaxy Fomation and Large Scale Structure. Phd thesis, Ludwig-Maximilian Universitat Munchen (2003)
3. Springel, V.: The cosmological simulation code gadget-2. Monthly Notices of the Royal Astronomical Society 364(4), 1105–1134 (2005)
4. Stalder, D.H., Rosa, R.R., da Silva Junior, J.R., Clua, E., Ruiz, R.S.R., Velho, H.F.C., Ramos, F.M., Araújo, A.D.S., Conrado, V.G.: A new gravitational n-body simulation algorithm for investigation of cosmological chaotic advection. In: The Sixth International School on Field Theory and Gravitation-2012. AIP Conference Proceedings, vol. 1483, pp. 447–452 (2012)

5. Barnes, J., Hut, P.: A hierarchical O(N log N) force-calculation algorithm. Nature 324(6096), 446–449 (1986)
6. Warren, M., Salmon, J.: Astrophysical N-body simulations using hierarchical tree data structures. In: Proceedings of Supercomputing 1992, pp. 570–576. IEEE Comput. Soc. Press (1992)
7. Ajmera, P., Goradia, R., Chandran, S., Aluru, S.: Fast, Parallel, GPU-based. Direct (2008)
8. Grama, A.Y., Kumar, V., Sameh, A.: Scalable parallel formulations of the barnes-hut method for n -body simulations. In: Proceedings of the 1994 ACM/IEEE Conference on Supercomputing, Supercomputing 1994, p. 439. ACM Press, New York (1994)
9. Hamada, T., Narumi, T., Yokota, R., Yasuoka, K.: 42 TFlops hierarchical N-body simulations on GPUs with applications in both astrophysics and turbulence. In: Proceedings of the, p. 1 (2009)
10. Jeroen, B.: A sparse octree gravitational N-body code that runs entirely on the GPU processor. Journal of Computational Physics, 1–34 (2011)
11. Nakasato, N.: Oct-tree Method on GPU: $ 42 / Gflops Cosmological Simulation. Engineering (2009)
12. Stock, M.J., Gharakhani, A.: Toward efficient GPU-accelerated N -body simulations, 1–13 (2008)
13. Jetley, P., Wesolowski, L., Gioachin, F., Kal, L.V.: Scaling Hierarchical N -body Simulations on GPU Clusters. Science (November 2010)
14. Hairer, E., Lubich, C., Wanner, G.: Geometric numerical integration illustrated by the störmer-verlet method (2003)
15. von Hoerner, S.: Die numerische integration des n-körper-problemes für sternhaufen. i. Z. Astrophys. 50, 184–214 (1960)
16. Nyland, L., Harris, M., Prins, J.: Fast N-Body Simulation with CUDA. In: Nguyen, H. (ed.) GPU Gems 3. Addison Wesley Professional (August 2007)

A Visual Analytics System
for Supporting Rock Art Knowledge Discovery

Vincenzo Deufemia, Valentina Indelli Pisano,
Luca Paolino, and Paola de Roberto

Department of Management and Information Technology
University of Salerno
Via Giovanni Paolo II, 132, Fisciano(SA), ITALY
{deufemia,vindellipisano,lpaolino,pderoberto}@unisa.it

Abstract. This paper presents a visual analytics system, named DARK, for supporting rock art archaeologists in exploring repositories of rock art scenes each consisting of hundreds of petroglyphs carved by ancient people on rocks. With their increasing complexity, analyzing these repositories of heterogeneous information has become a major task and challenge for rock art archaeologists. DARK combines visualization techniques with fuzzy-based analysis of rock art scenes to infer information crucial for the correct interpretation of the scenes. Moreover, the DARK views allow archaeologists to validate their hypothesis against the information stored in the repository.

Keywords: visual analytics, information visualization, fuzzy analysis, cultural heritage.

1 Introduction

Rock art is a term coined in archaeology for indicating any human made markings carved on natural stone [5,22]. The most part of the symbols concerning rock art are represented by *petroglyphs*, which were created by removing part of a rock surface by incising, picking, carving, and abrading. Although it is not possible to give certain interpretations to these petroglyphs, archaeologists have proposed many theories to explain their meaning, e.g., astronomical, cultural, or religious [32]. For example, Fig. 1(a) and 1(b) show the pictures of two petroglyphs interpreted as a Christ [2] and the stellar cluster of the Pleiades [15], respectively, while Fig. 1(c) depicts a digitalized relief interpreted as a priest making water spout from the water basin [9].

In order to digitally preserve, study, and interpret these artifacts the archaeologists have created repositories containing heterogeneous informations, like pictures, 3D images, textual descriptions, GPS coordinates, black and white reliefs, and so on. The exponential growth of these repositories and the high dimensionality of the stored data have made their analysis a major task and challenge for rock art archaeologists. Even though in the recent years several

B. Murgante et al. (Eds.): ICCSA 2014, Part VI, LNCS 8584, pp. 466–480, 2014.

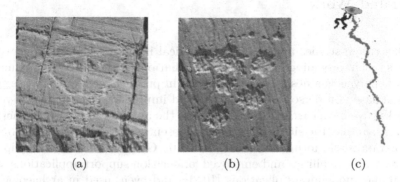

(a)   (b)   (c)

Fig. 1. The picture of a petroglyph supposed to represent a Christ (a) [2], a picture interpreted as the stellar cluster of the Pleiades (b) [15], and the relief depicting priests making water spout from the rock (c) [9].

image recognition approaches have been proposed for automating the segmentation and classification of petroglyphs [11,27,28,35], few work has been done to automate the analysis of these repositories in order to support archeologists in their investigative tasks.

In this paper we introduce novel abstraction techniques for ontology-based, interactive visual analysis of large repositories of petroglyphs. These abstraction techniques are based on structural and semantic information and allow users to easily discover new information about petroglyph shapes and relationships. The structural information includes the shape of the carved symbols and the spatial relationships between them. The latter represent a valuable information for the correct interpretation of petroglyphs since many correlations exist between the spatial relationships of the carved symbols and their interpretations. The semantic information is extracted from the interpretations given by archaeologists and organized according to an ontology. We also present a new data view that uses fuzzy-spatial relationships to visually summarize the scene topological structure. This data view can be used to study petroglyph correlations, infer new information useful to correctly interpret the scenes, and validate new hypothesis.

A prototype system, named DARK (Discovery of Ancient Rock art Knowledge), has been implemented as a component of IndianaMAS, a framework for the digital protection and conservation of rock art sites [24,31]. DARK supports archaeologists in their investigation activities by performing analytics on the dataset and providing appropriate visualization and interaction techniques. The prototype has been evaluated using data from the database containing information about the rock carving reliefs of Mount Bego.

The remainder of this paper is organized as follows. In Section 2, we present related work about visualization of archaeological information. Section 3 describes the IndianaMAS digital platform for the preservation of rock art. Section 4 introduces the proposed system and the interactive visual analysis process. Finally, conclusions are discussed in Section 5.

2   Related Work

Cultural heritage studies are basically analytical in nature and involve several disciplines, each providing a different contribution. Although the systematic study and analysis is a desirable requirement, in practice most of them are subjective and based on personal observations and impressions. In the recent years some works have been carried out to automate the analysis and reasoning activities. As an example, the site geographical representation can assist archeologists to interpret data and formulate new theories [6]. GIS tools have been equipped with increasing capabilities and employed for decision support applications [30], analytical and modeling applications [10,21], and even, used in archaeological research for analysis and reasoning activities [23]. The GIS spatial analytical application presented in [33] allows domain experts to investigate the potential extent of a habitat/environment through the analysis of a series of maps, graphs, and tabular data.

The geovisual analytics environment presented in [17] exploits the space time cube to investigate the relationships between sites and artifacts discovered at various sites in order to understand the interaction between cultures. The space time cube is a GIS based implementation of Hägerstrand's original Spacetime Aquarium, which allows to represent the three main components of spatio-temporal data, namely when, where, and what. The space time cube visualizes the space time archaeological data and provides interactive filtering and sorting functions that can be applied to clarify patterns and relationships hidden in the data. They also present the preliminary results on the development of functions for archaeological investigation within a geovisual analytics environment.

The Cyber-Archaeology project is focused on developing technologies and tools for the documentation of cultural heritage [18]. A central component of Cyber-Archaeology is a visual analytics system that supports the collaborative analysis of multispectral data spanning the broad scales of time (temporal) and space (spatial). This system can also be used to guide the knowledge discovery process and unlock the underlying meaning of an artifact, while determining if the artifact needs to be monitored or requires any type of intervention.

In [19] Inkpen *et al.* proposed a protocol based on GIS to monitor stone degradation. The proposed system integrates images from different time periods and different sites to yield useful information.

In [8] visual analytics has been used to facilitate the interpretation of multi-temporal thermographic imagery for the purpose of restoration of cultural heritage. The proposed visual environment allows to explore thermographic data from the unifying spatiotemporal perspective aiming to detect spatial and spatiotemporal patterns that could provide information about the structure and the level of decay of the material.

Vis4Heritage is a visual analytics framework for discovering wall painting degradation patterns [34]. The framework provides users with a set of analytic and hypothesis verification tools to support the multi-resolution degradation analysis.

In the context of the IndianaMAS project, in [12] we presented Indiana Finder, a visual analytic system supporting the interpretation of new archeological findings, the detection of interpretation anomalies, and the discovery of new insights. We introduced a data view based on ring charts and an interpretation summary view based on 3D maps.

To the best of our knowledge, no specific approach has been developed for supporting archaeologists to discover new relationships between petroglyphs. The most related idea is proposed in [29] where authors introduced ArchMatrix, a visual interactive system that supports archaeologists in archiving, managing and studying the findings collected during archaeological excavations. The system implements the Harris Matrix method, used to describe the position of stratigraphic units, and provides advanced information retrieval strategies through the use of a graph database.

The approach proposed in this paper differs from the previous analytics systems both in the managed information, i.e., the relationships between interpretations and spatial relationships between the petroglyphs in the scenes, and how to represented them. Indeed, the proposed data views are able to summarize this information allowing archaeologists to rapidly infer new ones.

3   Digital Preservation of Rock Art

Petroglyphs are among the oldest form of art known to humans. A petroglyph consists of a single carved symbol as in Fig. 1(a), or can be the composition of two or more symbols as in Fig. 2(a) up to several hundreds. Usually the archaeologists working on a rock art site collect petroglyphs, classify them based on their shape, and define dictionaries. Moreover, since petroglyphs were used to convey messages and ideas, archeologists try to give interpretations to the compositions. The latter are called *scenes* in the rest of this paper.

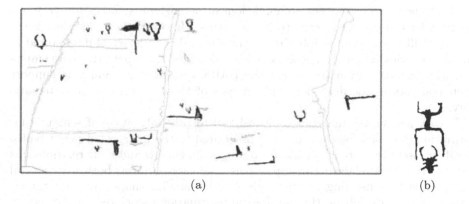

(a)   (b)

Fig. 2. A scene composed of several corniculates and halberds (a) [2], the pattern mother goddess giving birth to the bull (b) [9].

The scenes of petroglyphs can be composed of groups of animals, weapons, and so on, and they can be interpreted as religious beliefs, aesthetic concepts, warfare, and modes of life. Although in some cases the symbols composing a scene seem to be arranged without an apparent order, in many others, repetitive logical relationships can be recognized [2,9]. Such recurrent combinations are called *pattern*. As an example, Fig. 2(b) depicts a petroglyph interpreted as the mother goddess located above a bull. By analyzing the Mount Bego's petroglyphs, archaeologists have observed that this motif frequently occur in scenes and have interpreted this recurrent pattern as the *birth* [9].

To promote the awareness and the preservation of rock art and to support archaeologists in their investigation activities we are developing a platform, named *IndianaMAS* [31], that integrates and complements the techniques usually adopted to preserve cultural heritage sites. The platform exploits ontologies to provide a shared and human-readable representation of the domain [16], intelligent software agents to analyze the digital objects analysis and perform reasoning and comparison activities over them [20], and standard tools and technologies for Digital Libraries to manage and share digital objects [3]. IndianaMAS enables the digital preservation of all kinds of available data about rock carvings, such as images, geographical objects, textual descriptions of the scenes. It provides the means to organize and structure such data in a standard way and supplies domain experts with facilities for issuing complex queries on the data repositories and making assumptions about the way of life of ancient people.

The database of petroglyphs consists of textual interpretation descriptions, ontological information, geographical coordinates, pictures and drawings of the reliefs. The ontological information is a list of concepts that summarizes the interpretations given by archaeologists. Fig. 3 shows a part of the ontology we defined for the rock art archaeology domain.

4   The DARK System

As said above, the recurrent combinations of petroglyph symbols are very important for the correct interpretation of scenes. However, their identification is a very challenging task mainly due to the size and complexity of data stored in the repositories. So far, archaeologists have identified these patterns in an empirical way. In this section we present the DARK system whose aim is to support archaeologists in the detection and analysis of these patterns in a systematic way.

DARK is a visual analytics tool designed to take advantage of semantic information (i.e., ontological concepts) associated to the interpreted petroglyphs. It allows users to more easily isolate petroglyph symbols and their relationships for inferring and validating new patterns. The tool performs both a structural abstraction by clustering the petroglyphs with similar shapes and a semantic abstraction by exploiting the ontological information associated to the scenes. Moreover, to visually represent and summarize the spatial arrangement of the petroglyphs in the scenes, DARK includes a data view exploiting fuzzy theory to

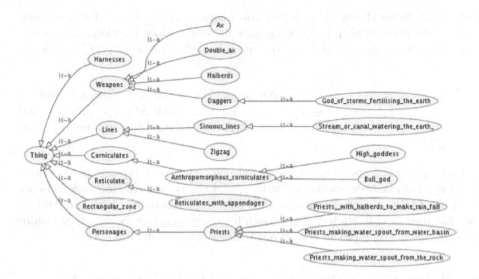

Fig. 3. Part of the ontology defined for the rock art archaeology domain

manage the uncertainty of spatial relationships. The tool also includes common visual interaction techniques, as zooming and filtering, in order to assist users in the investigation process.

4.1 Structural and Semantic Abstraction through Bubble View

The aim of the structural and semantic abstractions is to extract relevant information from the scene repository in order to simplify the detection of recurrent patterns. In the following we formalize the abstraction process and present the data view implemented in DARK that visually summarize the abstracted information.

The Abstraction Process. Let $S = \{s_1, \ldots, s_n\}$ be the set of symbol classes defined by archaeologists to classify petroglyphs and $O = \{o_1, \ldots, o_m\}$ be the set of concepts contained in the ontology. The set of scenes P stored in the repository can be formally defined as $P = \{p_1, \ldots, p_t\}$ where $p_i = \{r_{i,1}, \ldots, r_{i,q}\}$ contains the petroglyph reliefs of the i-th scene. Each relief $r_{i,j} \in p_i$ is associated to a symbol in S by using the function $Shape$, i.e., $Shape(r_{i,j}) = s_k \in S$, which applies an approximate image matching algorithm. Moreover, the function Sem provides the set of ontology concepts associated to a scene, i.e., $Sem(p_i) \subseteq O$.

Given an ontology concept o_i, DARK constructs a graph where the nodes correspond to the symbol classes obtained by applying the $Shape$ function to the reliefs of p_j such that $o_i \in Sem(p_j)$ of petroglyph symbol classes from P by including all nodes whose classes are related to o_i. The goal is to visualize the most frequent symbol classes co-occuring with a concept. In order to achieve this goal, abstraction hides the symbol classes non relevant for the selected concept and clusters the petroglyph reliefs having similar shape. More formally, an abstraction with respect to an ontology concept o_i is defined as a graph $G_{o_i} = (V, E)$ with:

- $V = \{s_x \in S \mid \exists\, r_{j,k} \in p_j$ with $Shape(r_{j,k}) = s_x$ and $o_i \in Sem(p_j)\}$ and
- $E = \{(s_x, s_y) \mid s_x, s_y \in V,\, \exists\, r_{j,k},\, r_{l,m} \in p_j$ with $Shape(r_{j,k}) = s_x$ and $Shape(r_{l,m}) = s_y$, and there exists a spatial relationship between $r_{j,k}$ and $r_{l,m}\}$.

Thus, the graph constructed for an ontology concept o_i represents the petroglyph symbol classes co-occurring with o_i and the spatial relationships between them. In DARK, the *Shape* function has been implemented by using the IDM algorithm used in [13], while the approach for computing the spatial relationships between the petroglyphs in the scenes is described in Subsection 4.2.

The Bubble View. DARK visualizes the graphs obtained from the abstraction process by using the Bubble view.

In particular, the view shows a colored bubble B_{o_i} for each graph G_{o_i}. The color of each B_{o_i} depends on the position of the concept o_i in the ontology hierarchy, namely we assign a different hue to each subtree of the nodes at level two, then we decreases the saturation on the basis of the level. As an example, Fig. 4 depicts the Bubble view constructed for the ontology concepts at level three[1]. *Fertilize* and *storm*, being children of the *Agriculture* level-two node are pink-colored, while *birth* and *sacrifice*, being children of *Earth* and *Religion*, are green and cyan-colored, respectively. It is worth to notice that the bubbles corresponding to semantically similar concepts are depicted with similar colors so as the *fertilize* and *storm* bubbles.

The size of a bubble indicates the frequency of the concept o_i in the scenes (small = low frequency, big = high frequency). In the example, we notice that the most frequent concept is *storm*.

The bubble B_{o_i} also visualizes both the most frequent symbols co-occurring with o_i and their relationship degree. In particular, the size of symbols indicates how much they frequently appear in the scenes, while the size of the links provides a suggestion about their relationship degree.

In this way, the Bubble view allows archeologists to discover candidate patterns by highlighting the prevalent key symbols and common relationships. As an example, the graph inside the *sacrifice* bubble shows a potential pattern. The *bull* and *halberd* symbols are very frequent in the scenes associated to the *sacrifice* concept, thus, their size appears bigger with respect to the other symbols. Moreover, the bold line indicates a high spatial relationship degree between them in the scene set.

---

[1] Archaeologists can select the ontology concepts either by interacting with a zooming bar, where it is possible to choose the granularity level of the concepts, or by navigating the ontology through the taxonomy relation (IS-A).

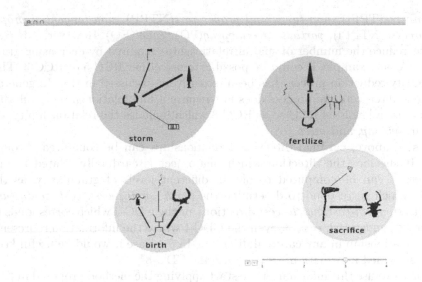

Fig. 4. A Bubble view constructed for the concepts: *fertilize, storm, birth*, and *sacrifice*

4.2   Scene Spatial Analysis Using Ring View

The information visualized in a Bubble view provides an overview of the correlations between the symbol relationships and the ontological concepts extracted
from their interpretations. When the archaeologist discovers a candidate pattern, s/he needs fine-grained information on the symbols relationships involved
in the pattern in order to validate it. In particular, s/he has to analyze the spatial relationships existing between the petroglyph symbols and their correlation
with the selected concept.

Fuzzy Spatial Analysis of Scenes. Since the petroglyph symbols were carved
in an inaccurate way, the spatial relationships between petroglyphs cannot be
computed using crisp relations. We use the fuzzy logic to model the uncertainty,
as this logic allows to express the concept of belonging where it cannot be easily
defined [35]. For example, in spatial reasoning, two entities cannot fully belong
to only one relation space so the fuzzy logic helps us to model the concept
of uncertainty. In particular, in order to obtain the topological structure of a
scene we use the region connection calculus (RCC) [25], while the directional
properties between objects have been computed using the Cardinal Direction
Calculus (CDC).

The RCC is a topological approach to the qualitative spatial representation
in which the regions are regular subsets of a topological space. This method
gives information about the topological structure of an image by associating it
to one of the topological relationships in a two-dimensional space. The RCC-8
calculus considers eight basic topological relations, which are: *externally connected*(EC), *disconnected*(DC), *tangential proper part*(TPP), *tangential proper*

part inverse(TPPi), *non-tangential proper part*(NTPP), *non-tangential proper part inverse*(NTPPi), *partially overlapping*(PO), *equal*(EQ). However, it is possible to reduce the number of spatial relationships required by decreasing granularity. As an example, it could be possible to use either RCC-5 or RCC-3. This granularity reduction process has been extensively discussed in [18]. In general, the scene interpretative process does not require a fine relationship granularity, thus we use a low level RCC so as RCC-3 which includes the relationship types, *disjoint, overlap*, and *meet*.

As said above, cardinal directional relationships can be computed through CDC. It specifies the direction which one object is cardinally related to one another. It can be computed to obtain different levels of granularity, as the CDC-8 which correspond to determine the *north, south, west, east, south-east, south-west, north-east, north-west* directions, or the CDC-4 which corresponds to the *north, south, west, east*, or even the CDC-1 where the information represents only a relationship in any cardinal direction. In our case it would be useful keep the fine-grained information, thus we use the CDC-8.

In order to use this information we start applying the method proposed in [26], that combines fuzzy Allen's relationships, specifically designed to manage topological relationships, to directional relationships in the two-dimensional space. The Allen's relationships, explained in [1], are based on the algebra of the time intervals and are labeled as follow: $A = \{<, m, o, s, f, d, eq, d_i, f_i, s_i, o_i, m_i, >\}$. These relationships assume the meaning of: *before, meet, overlap, start, finish, during, equal, during by, finish by, start by, overlap by, meet by*, and *after*. Successively, the method applies a fuzzification process to obtain the fuzzy value of the spatial information. It creates a matrix $8x8$, where the rows hold the CDC-8 topological relationships and the columns have the qualitative directional aspects of the 2D scene information. The relationships are expressed using numeric values representing the percentage area of two objects under a specific topological relationship in the qualitative direction. More precisely, the cell value for a topology relationship of a given direction is obtained using the fuzzy contributes of some Allen's relationships calculated on specific angles depending on the direction. For instance, in case of *north* direction, it is possible to calculate the fuzzy value of the *disjoint* topological relationship by means of the *after* Allen's relationship over the angles between $1/4\pi$ and $3/4\pi$ according to the following formula:

$$Disjoint\_North = \sum_{\Theta=\pi/4}^{3/4\pi} after(\Theta) * \cos^2(2 * \Theta)$$

For the sake of clarity, let us shown two examples describing how this model works. In the first example, shown in Figure 5(a), we have a scene where the *god* petroglyph is located above (*north*) the *goddess* petroglyph. The petroglyphs are disjoint to each other as well. The matrix associated with this scene contains higher values in correspondence of the cells indicated by the rows 1-*disjoint* and the columns 2-*north-east*, 3-*north*, 4-*north-west* (see Figure 6(a)).

In the latter example, shown in Figure 5(b), *bull* and *goddess* petroglyphs overlap to each other. A the same time, *goddess* is located at *north* side of *bull.* As shown in Figure 6(b), the matrix representation of this relationship involves several rows, 1-*disjoint*, 2-*meet*, 3-*partially overlap*, 4-*tangent proper part*, 5-*nontangent proper part*, and columns, 2-*north-east*, 3-*north*, 4-*north-west*. In this case, it is possible to notice that the higher value is obtained in correspondence of the cell [3-*partially overlap*, 3-*north*].

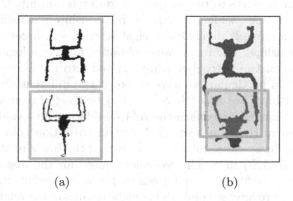

(a)   (b)

Fig. 5. A petroglyph scene containing the *god* and the *goddess* (a), and an example of interpretation of the *overlap-north* relationship between the *mother goddess* and the *bull* (b) [11]

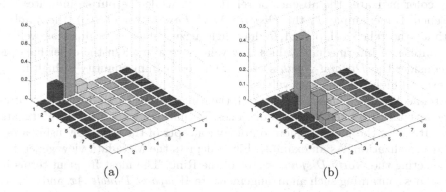

(a)   (b)

Fig. 6. The matrices representing the fuzzy relationships between the symbols in the scenes of Figure 5. The correspondence between numbers and relationships is the following: on rows, 1-disjoint, 2-meet, 3-partially overlap, 4-tangent proper part, 5-nontangent proper part, 6-tangent proper part inverse, 7-nontangent proper part inverse, 8-equal, and on columns, 1-east, 2-north-east, 3-north, 4-north-west, 5-west, 6-south-west, 7-south, and 8-south-east.

The Ring View. For assessing a topological-directional fuzzy relationship, DARK analyzes its frequency (how many times the relationship is satisfied) and its strength (how high is the fuzzy value) and visualizes this information in the Ring view. In order to visually represent this information the Ring view shows three concentric circles, which corresponds to the topological relationships of RCC-3 obtained following the reduction process from RCC-8 described in [18]. The inner circle refers to the *partially overlap* relationship, the middle one to the *meet* relationship, while the *disjoint* relationship is associated with the outer sector. Each circle is divided into eight sectors, which correspond to the directional relationships. The fuzzy values obtained for a topological-directional relationship between two petroglyphs are mapped into the corresponding cells of the Ring view. In particular, we have proposed two different visualizations of the fuzzy relationships. Let $S = \{S_1, S_2, \ldots, S_n\}$ be a set of scenes, A and B be two petroglyphs related by means of the $R(A,B)$ relationships, and $V(R)$ be the corresponding cell in the Ring view. The first visualization colors $V(R)$ using the average of the fuzzy values obtained evaluating the fuzzy relationship R for each instance of $R(A,B)$ in S. The red color is used for the *disjoint* sectors in the outer circle, purple for *meet*, and green for *partially overlap* as shown in Fig. 7(a). Notice that more intense colors imply higher values of the relationship while the black color indicates the total absence. In the second visualization, the fuzzy values of $R(A,B)$ in S are mapped using the color gradient that highlights at the same time the frequency and the strength of the relationship. Let us suppose we divide a sector into n parts $P = \{P_1, P_2, \ldots, P_n\}$ each associated with a strength range $F = \{F_1, F_2, \ldots, F_n\}$, the size of the sector part P_i is proportional to the frequency of the $R(A, B)$ fuzzy values falling into the range F_i, while the color intensity of P_i is given by calculating the average of such fuzzy values. The gray color indicates the absence of relationship, while the purple indicates the presence. For example, in the *Disjoint-North-East* sector of 7(b), the number of times the relationship $R(A,B)$ has high fuzzy values is medium as well as the number of intermediate values, low values are a few. On the other hand, if we consider the *Disjoint-North* sector, we have medium quantity of high fuzzy values and high intermediate values.

Archaeologists can also interact with the Ring sectors for analyzing the ontology concepts associated to the set of scenes. To this aim, we defined the Metadata view. It shows the most frequent ontology concepts of the scenes by using a tag cloud visualization. As an example, Fig. 8 depicts the Metadata view generated by selecting the *North-Disjoint* sector of the Ring. The most frequent terms in the scenes containing such an arrangement are *Weapons*, *Double_Ax*, and *Ax*. By using this view, archaeologists verify the correctness of the discovered pattern.

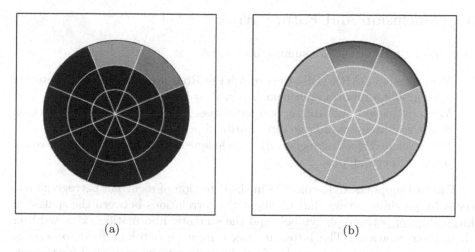

(a)   (b)

Fig. 7. Two different visualizations of the Ring view: average fuzzy values (a) and gradients (b)

Fig. 8. A Ring view with the Metadata view generated by selecting the North-Disjoint sector

5   Conclusion and Future Work

Our main contributions are summarized as follows:

– We presented DARK (Discovery of Ancient Rock art Knowledge) a prototype
  system that supports archaeologists in their investigation activities,
– We introduced novel abstraction techniques for ontology based and interactive visual analysis of large repositories of petroglyphs, and
– We developed a set of visualization techniques which helps to analyze repositories of petroglyphs.

The tool supports archaeologists in the detection of recurrent petroglyph patterns by providing views that highlight the correlations between the spatial relationships of petroglyph symbols and the semantic information extracted from
their interpretations. The latter are taken from an ontology used to organize
data associated with petroglyphs. The abstraction techniques used for the construction of the views are based on the semantic and structural information of
the petroglyphs, while fuzzy theory is used to evaluate the spatial relationships
between the symbols in the scenes.

In the future, we plan to exploit visual grammars for modeling the language
of identified patterns [7]. We will also investigate the use of query by sketch
as a technique to ease user interaction and improve retrieval effectiveness in
the repository by extending the agent-based framework presented in [4] with
a LDCRF-based sketch recognition algorithm [14]. Finally, we plan to validate
DARK with usability studies involving domain experts.

Acknowledgments. This research is supported by the "Indiana MAS and the
Digital Preservation of Rock Carvings" FIRB project funded by the Italian Ministry of University and Research (MIUR): under grant RBFR10PEIT.

References

1. Allen, J.F.: Maintaining knowledge about temporal intervals. Communication of
   ACM 26(11), 832–843 (1983)
2. Bianchi, N.: Mount bego: prehistoric rock carvings. In: Adoranten, Scandinavian
   Society for Prehistoric Art, pp. 70–80 (2010)
3. Candela, L., Athanasopoulos, G., Castelli, D., El Raheb, K., Innocenti, P., Ioan-
   nidis, Y., Katifori, A., Nika, A., Vullo, G., Ross, S.: The digital library reference
   model. Technical Report Deliverable D3.2b, DL.org: Coordination Action on Digital Library Interoperability, Best Practices and Modelling Foundations. Funded
   under the 7th Framework Programme. Project Number 231551 (2011)
4. Casella, G., Deufemia, V., Mascardi, V., Costagliola, G., Martelli, M.: An agent-
   based framework for sketched symbol interpretation. J. Vis. Lang. Comput. 19(2),
   225–257 (2008)
5. Chippindale, C., Taçon, P.S.C.: The archaeology of rock-art. Cambridge University
   Press (1998)

6. Constantinidis, D.: Seeing is believing: The visual analysis of archaeological data. In: Proceedings of the 2002 Advanced Information Visualization in Archaeology Workshop (2002)
7. Costagliola, G., Polese, G.: Extended positional grammars. In: Proc. Int'l Conf. on Visual Languages, pp. 103–110 (2000)
8. Danese, M., Demar, U., Masini, N., Charlton, M.: Investigating material decay of historic buildings using visual analytics with multi-temporal infrared thermographic data. Archaeometry 52(3), 482–501 (2010)
9. de Lumley, H., Echassoux, A.: The rock carvings of the chalcolithic and ancient bronze age from the mont bego area. the cosmogonic myths of the early metallurgic settlers in the southern alps. L'Anthropologie 113(5), 969–1004 (2009)
10. Dean, J.S., Gumerman, G.J., Epstein, J.M., Axtell, R.L., Swedlund, A.C., Parker, M.T., McCarroll, S.: Understanding Anasazi Culture Change Through Agent-based Modeling. In: Dynamics in Human and Primate Societies, pp. 179–205. Oxford University Press, Oxford (2000)
11. Deufemia, V., Paolino, L., de Lumley, H.: Petroglyph recognition using self-organizing maps and fuzzy visual language parsing. In: Proc. of IEEE Int'l Conf. on Tools with Artificial Intelligence (ICTAI 2012), pp. 852–859 (2012)
12. Deufemia, V., Paolino, L., Tortora, G., Traverso, A., Mascardi, V., Ancona, M., Martelli, M., Bianchi, N., De Lumley, H.: Investigative analysis across documents and drawings: Visual analytics for archaeologists. In: Proceedings of the International Working Conference on Advanced Visual Interfaces, AVI 2012, pp. 539–546. ACM, New York (2012)
13. Deufemia, V., Paolino, L.: Combining unsupervised clustering with a non-linear deformation model for efficient petroglyph recognition. In: Bebis, G., et al. (eds.) ISVC 2013, Part II. LNCS, vol. 8034, pp. 128–137. Springer, Heidelberg (2013)
14. Deufemia, V., Risi, M., Tortora, G.: Sketched symbol recognition using latent dynamic conditional random fields and distance-based clustering. Pattern Recognition 47(3), 1159–1171 (2014)
15. Echassoux, A., de Lumley, H., Pecker, J.C., Rocher, P.: Les gravures rupestres des pléiades de la montagne sacrée du bego, tende, alpes-maritimes, france. Comptes Rendus Palevol 8(5), 461–469 (2009)
16. Gruber, T.R.: A translation approach to portable ontology specifications. Knowledge Acquisition 5, 199–220 (1993)
17. Huisman, O., Santiago, I.F., Kraak, M.J., Retsios, B.: Developing a geovisual analytics environment for investigating archaeological events: Extending the space-time cube. Cartography and Geographic Information Science 36(3), 225–236 (2009)
18. Ibrahim, Z.M., Tawfik, A.Y.: Spatio-temporal reasoning for vague regions. In: Tawfik, A.Y., Goodwin, S.D. (eds.) Canadian AI 2004. LNCS (LNAI), vol. 3060, pp. 308–321. Springer, Heidelberg (2004)
19. Inkpen, R., Duane, B., Burdett, J., Yates, T.: Assessing stone degradation using an integrated database and geographical information system (gis). Environmental Geology 56(3-4), 789–801 (2008)
20. Jennings, N.R., Sycara, K.P., Wooldridge, M.: A roadmap of agent research and development. Autonomous Agents and Multi-Agent Systems 1(1), 7–38 (1998)
21. Kohler, T.A., Kresl, J., van West, C., Carr, E., Wilshusen, R.H.: Be There then: A Modeling Approach to Settlement Determinants and Spatial Efficiency Among Late Ancestral Pueblo Populations of the Mesa Verde Region, U.S. Southwest. In: Dynamics in Human and Primate Societies, pp. 145–178. Oxford University Press, Oxford (2000)

22. Lin, S.K.: Rock art science: The scientific study of palaeoart, 2nd edn. Arts, vol. 2(1), pp. 1–2 (2013)
23. Lock, G., Stancic, Z.: Archaeology and GIS: A European Perspective. Taylor and Francis (1995)
24. Mascardi, V., Deufemia, V., Malafronte, D., Ricciarelli, A., Bianchi, N., de Lumley, H.: Rock art interpretation within indiana MAS. In: Jezic, G., Kusek, M., Nguyen, N.-T., Howlett, R.J., Jain, L.C. (eds.) KES-AMSTA 2012. LNCS, vol. 7327, pp. 271–281. Springer, Heidelberg (2012)
25. Randell, D.A., Cui, Z., Cohn, A.: A Spatial Logic Based on Regions and Connection. In: Nebel, B., Rich, C., Swartout, W. (eds.) Proceedings of the Third International Conference on Principles of Knowledge Representation and Reasoning, pp. 165–176. Morgan Kaufmann, San Mateo (1992)
26. Salamat, N., Zahzah, E.H.: Two-dimensional fuzzy spatial relations: A new way of computing and representation. Advances in Fuzzy Systems 2012, 1–15 (2012)
27. Seidl, M., Breiteneder, C.: Detection and classification of petroglyphs in gigapixel images – preliminary results. In: VAST11: The 12th International Symposium on Virtual Reality, Archaeology and Intelligent Cultural Heritage - Short Papers, pp. 45–48 (2011)
28. Seidl, M., Breiteneder, C.: Automated petroglyph image segmentation with interactive classifier fusion. In: Proceedings of the Eighth Indian Conference on Computer Vision, Graphics and Image Processing, ICVGIP 2012, pp. 66:1–66:8. ACM, New York (2012)
29. Valtolina, S., Barricelli, B.R., Gianni, G.B., Bortolotto, S.: ArchMatrix: Knowledge management and visual analytics for archaeologists. In: Yamamoto, S. (ed.) HIMI/HCII 2013, Part III. LNCS, vol. 8018, pp. 258–266. Springer, Heidelberg (2013)
30. van Leusen, P.: GIS and archaeological resource management: a European agenda. In: Archaeology and GIS: A European Perspective, pp. 27–41. Taylor & Francis (1995)
31. Viviana, M., Briola, D., Locoro, A., Grignani, D., Deufemia, V., Paolino, L., Bianchi, N., de Lumley, H., Malafronte, D., Ricciarelli, A.: A holonic multi-agent system for sketch, image and text interpretation in the rock art domain. International Journal of Innovative Computing, Information and Control (IJICIC) 10(1), 81–100 (2014)
32. Whitley, D.: Handbook of rock art research. G - Reference, Information and Interdisciplinary Subjects Series. AltaMira Press, Beverly Hills (2001)
33. Williams, M.: Archaeology and gis: Prehistoric habitat reconstruction. In: Proceedings of ESRI User Conference (2004)
34. Zhang, J., Kang, K., Liu, D., Yuan, Y., Yanli, E.: Vis4heritage: Visual analytics approach on grotto wall painting degradations. IEEE Transactions on Visualization and Computer Graphics 19(12), 1982–1991 (2013)
35. Zhu, Q., Wang, X., Keogh, E., Lee, S.H.: An efficient and effective similarity measure to enable data mining of petroglyphs. Data Mining and Knowledge Discovery 23(1), 91–127 (2011)

On the Localization of Zeros and Poles of Chebyshev-Padé Approximants from Perturbed Functions

João Carrilho de Matos[1], José Matos[1], and Maria João Rodrigues[2,*]

[1] Instituto Superior de Engenharia do Porto, Porto, Portugal
{jem,jma}@isep.ipp.pt
[2] Faculdade de Ciências da Universidade do Porto, Portugal
mjsrodri@fc.up.pt

Abstract. We present some numerical results about the localization of zeros and poles of Chebyshev-Padé approximants from functions perturbed with random series. These results are a natural generalization of the Froissart's numerical experiments with power series. Our results suggest that the Froissart doublets of Chebyshev-Padé approximants are located, with probability one, on the Joukowski transform image of the natural boundary of the random power series.

Keywords: Froissart doublets, Padé approximation, Natural boundary.

1 Introduction

Padé approximation is a very useful tool for approximation, for localization of singularities and for analytic continuation of functions. Let f be a function with power expansion at zero $f(z) \sim \sum_{k=0}^{\infty} c_k z^k$ and let m, n be two non-negative integers. The Padé approximant $\Phi_{m,n}(f)$ is a rational function $N_{m,n}/D_{m,n}$ satisfying

$$D_{m,n}(z)f(z) - N_{m,n}(z) = O(z^{m+n+1}), \quad z \to 0,$$

with $N_{m,n}(z) = \sum_{k=0}^{m} b_k z^k$ and $D_{m,n}(z) = \sum_{k=0}^{n} a_k z^k \neq 0$, [1, Section 1].

There are uniform convergence results for some classes of functions: row sequences for meromorphic functions (Montesus' theorem), and para-diagonal sequences for some classes of function (represented by Stieltjes or by Pólya series), [1, Sections 5-6]. However the existence of uniform convergence results for others classes of functions remains an open problem. The reason for this absence of results, for general functions, is due essentially to the lack of information about the localization of poles/zeros. This knowledge is also relevant to provide useful information on the singularity structure of the function.

Frequently, in practical problems, we just have the knowledge of a finite number of coefficients of a given series and moreover these coefficients are not exact. The coefficient errors are commonly caused by the use of finite arithmetic, which implies

* Research funded by the European Regional Development Fund through the program COMPETE and by the Portuguese Government through the FCT – Fundação para a Ciência e a Tecnologia under the project PEst-C/MAT/UI0144/2013.

B. Murgante et al. (Eds.): ICCSA 2014, Part VI, LNCS 8584, pp. 481–492, 2014.

round-off errors on Taylor coefficients. Froissart had the idea to simulate this noise (due to round-off errors) using random series and to study the localization of poles and zeros. It is well known that Padé approximants from perturbed power series have the so called "artificial" Froissart doublets [4, section 6.4]. Roughly speaking, a Froissart doublet is a nearby pair zero/pole from a Padé approximant. These pairs zero/pole almost cancel, thus the existence of Froissart doublets are a drawback to produce Padé approximants with more accuracy. We must remark that some functions have Padé approximants (computed with exact arithmetic) with "genuine" Froissart doublets [2]. However, here we will only consider "artificial" Froissart doublets generated by noise (caused by the use of finite arithmetic). We also want to mention that all numerical computations were done using MATLAB®, using double precision floating point arithmetic. Our main goal is to compare our results (obtained with Chebyshev polynomials) with Froissart's results (obtained with power series).

2   Zeros and Poles of Taylor-Padé Approximants

In this section we reproduce the results of two experiments performed by Froissart.

Example 1. Consider the function $f(z) = 1/(1-z)$ with Taylor expansion around zero $T(z) = \sum_{k=0}^{\infty} z^k$. We note (as f itself is a rational function) that the Padé table of f has an infinite block. To be more precise, we have $\Phi_{m,n}(f) = f$ for all integers m and n such that $m \geq 0$ and $n \geq 1$. Now, in order to simulate round-off errors, we introduce a perturbation on the coefficients of T of the form

$$T_\varepsilon(z) = \sum_{k=0}^{\infty} (1 + \varepsilon r_k) z^k, \tag{1}$$

where ε is a small positive number and r_k are complex random numbers i.i.d. uniformly distributed on the disk $|r_k| \leq 1$. We note that the Padé table of the power series T_ε has not the infinite block present in Padé table of T, and we can analyse the behaviour of the zeros and poles of diagonal Padé approximants, $\Phi_{n,n}(T_\varepsilon)$, when we change the values of m and n. Prior to giving the numerical results obtained by Froissart, we need to make some observations: T_ε is the sum of two parts, one part being deterministic, with a pole at $z_1 = 1$ and convergent on the disk $|z| < 1$, and another part being non-deterministic with noise of order ε. The non-deterministic part of T_ε, $\sum_{k=0}^{\infty} \varepsilon r_k z^k$, has a natural boundary on the circle $|z| = 1$ with probability one [5], and consequently T_ε has a natural boundary on the circle $|z| = 1$. In the sequel, we will denote by "ε-noise" this type of perturbation.

In this experiment Froissart observed the following behaviour of the zeros and poles of $\Phi_{n,n}(T_\varepsilon)$:

1. There is a stable pole ξ_1 for $\Phi_{n,n}(T_\varepsilon)$ near the singularity of f ($z_1 = 1$), with $|z_1 - \xi_1| = O(\varepsilon)$ and it decreases as n increases;
2. There is an unstable zero η_1, usually called "ghost zero", with $|\eta_1| = O(\varepsilon^{-1})$;

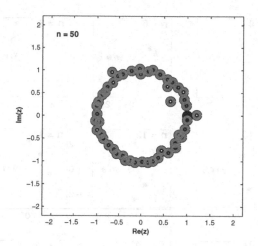

Fig. 1. Localization of zeros (*magenta circles*) and poles (*black dots*) of $\Phi_{n,n}(f_\varepsilon)$, with $n = 50$ and $\varepsilon = 10^{-4}$.

3. The other poles and zeros, ξ_k and η_k respectively, $k = 2, \ldots, n$, are Froissart doublets lying near the natural boundary $|z| = 1$, representing thus the the natural boundary of T_ε.

These remarks are represented in figure 1. We note that, in this case, we have $|\eta_1| = O(10^4)$. Thus, in order to show clearly the localization of the Froissart doublets we do not include the zero ghost on the picture.

Example 2. Froissart also considered the function $g_\omega(z) = g(z) + \sum_{k=0}^{\infty} \omega \frac{r_k}{2^k} z^k$, where $g(z) = \ln(1 - z)$. Now the deterministic part of $g_\omega(z)$ has a branch cut and Taylor series $\sum_{k=1}^{\infty} -\frac{z^k}{n}$, $|z| < 1$, while the random part of $g_\omega(z)$, $R_\omega(z) = \sum_{k=0}^{\infty} \omega \frac{r_k}{2^k} z^k$ converges on the disk $|z| < 2$ (the r_k are random numbers as in example 1 and ω is a small positive number). In the sequel we will denote by "ω-noise" this type of perturbations.

Here we can observe the following behaviour for the diagonal Padé approximants $\Phi_{n,n}(g_\omega)$:

1. There is a stable zero near z=0 (the zero of g).
2. There is a set of interlaced poles and zeros, which indicate the branch cut of g and the distance between two consecutive pairs pole/zero increases with their distance to the nearest branch point of the interval $[-1, 1]$ (z=1).
3. The remaining poles and zeros are unstable Froissart doublets, some representing the natural boundary of R_ω and the others left lying on the annular region 1<|z|<2.

We illustrate this pattern in figure 2.

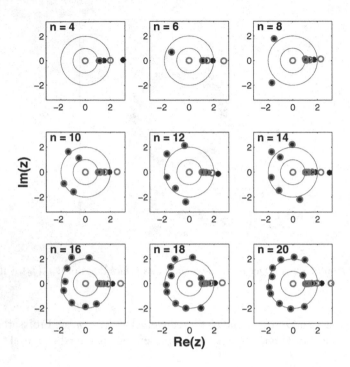

Fig. 2. Localization of poles (*black dots*) and zeros (*magenta circles*) of a sequence $\Phi_{n,n}(g_\omega)$, $n = 4, 6 \ldots 20$ with $\omega = 10^{-4}$. For the sake of clarity we did not include all the zeros/poles located in the vicinity of the branch cut.

3   Orthogonal Series and Padé Approximation

3.1   Orthogonal Series

Let $\theta(x)$ be a real function that is nondecreasing in a interval $I \subset \mathbb{R}$, with infinitely many points of increase there. Let $\{P_k(x)\}_{k \geq 0}$ be the family of orthogonal polynomials with respect to the inner product $\langle *, * \rangle$ defined by $\langle F(x), G(x) \rangle = \int_I F(x) G(x) d\theta(x)$, and associated norm $\| * \|$ (i.e., $\| F(x) \| = \langle F(x), F(x) \rangle^{1/2}$). It is well known that the polynomials $\{P_k(x)\}_{k \geq 0}$ satisfy a three term recurrence relation, [8, Chapter III, section 3.2]

$$xP_k(x) = \alpha_k P_{k+1}(x) + \beta_k P_k(x) + \gamma_k P_{k-1}(x),$$
$$P_0(x) = 1, \quad P_1(x) = (x - \beta_0)/\alpha_0. \tag{2}$$

Consider now a function $f \in L(I)$, where $L(I)$ denotes the family of real mensurable functions F, such that $\int_I F^2(x) d\theta(x) < \infty$. The orthogonal series of the function f is a series

$$f(x) = \sum_{k=0}^{\infty} f_k P_k(x) \tag{3}$$

where the coefficients f_k are given by

$$f_k = \frac{\langle f, P_k \rangle}{\|P_k\|^2}, \quad k = 0, 1, 2, \dots \tag{4}$$

We will summarize some results concerning Chebyshev polynomials of first kind (I=[-1,1]).

Chebyshev Polynomials: Let $x = \cos(\zeta)$, $0 \leq \zeta \leq \pi$. The Chebyshev polynomials are defined by $T_k(x) = \cos(k\zeta)$ $k = 0, 1, \dots$. If $f \in L([-1,1])$, then its Chebyshev series is given by

$$f(x) = \sum_{k=0}^{\infty}{}' c_k T_k(x), \quad c_k = \frac{2}{\pi} \int_{-1}^{1} \frac{f(x) T_k(x)}{\sqrt{1 - x^2}} dx$$

where the primed summation means that the first term is halved.

While power series from analytic functions converge on open disks D_ρ, Chebyshev series converge throughout Bernstein ellipses (ellipses with foci at ± 1)

$$E_\rho = \{z \in \mathbb{C} : |\varphi(z)| = \rho\},$$

where $\varphi(z) = z + \sqrt{z^2 - 1}$. To be more precise, let f be an analytic function in a open set $U \subset \mathbb{C}$ that contains the origin. Then the power series of f converges uniformly in every open disk $D_\rho = \{z \in \mathbb{C} : |z| < \rho\}$ such that $D_\rho \subset U$. Thus, if z_v is the closest singularity (of f) to the origin, the power series has radius of convergence $\rho = |z_v|$. In the Chebyshev series case, if f is analytic in a open set $U \subset \mathbb{C}$ that contains the interval $[-1, 1]$ and if z_v is the closest singularity to the origin, then the Chebyshev series of f converges uniformly on the interior of the Bernstein ellipse, $E_{r(z_v)}$, where

$$r(z_v) = \frac{\lambda_{z_v} + \sqrt{\lambda_{z_v}^2 - 4}}{2}, \quad \lambda_{z_v} = |z_v + 1| + |z_v - 1|.$$

We note that the function $\varphi(z)$ is the inverse transform of the well known Joukowski transform $w(z) = (z + z^{-1})/2$. The Joukowski transform maps the circles $|z| = \rho$ and $|z| = \rho^{-1}$, $\rho > 0$ and $\rho \neq 1$, to the Bernstein ellipse $|\varphi(z)| = \rho$ and if $\rho = 1$ the Joukowski transform maps the unit circle to the interval $[-1, 1]$.

The following results will be useful in the next section. The Chebyshev polynomials have the following generating function [6, Section 18.12]

$$\sum_{k=1}^{\infty} \frac{t^k}{k} T_k(x) = -\frac{1}{2} \ln\left(1 - 2tx + t^2\right), \quad |t| < 1. \tag{5}$$

Next we will show how to compute the Chebyshev coefficients of meromorphic functions with simple poles, derived by D. Elliot [3].

If $f(z)$ is meromorphic on the complex plane, we can use Cauchy's integral formula to represent $f(x)$ and

$$c_k = \frac{1}{i\pi^2} \int_\gamma f(z) \left(\int_{-1}^1 \frac{T_k(x)dx}{\sqrt{1-x^2}(z-x)} \right) dz, \quad k = 0, 1, 2 \dots \tag{6}$$

where γ is a suitable contour that contains the interval $[-1, 1]$. We can see, [3], that

$$c_k = \frac{1}{i\pi} \int_\gamma \frac{f(z)dz}{\sqrt{z^2-1}(z \pm \sqrt{z^2-1})^k}, \quad k = 0, 1, 2 \dots.$$

where the signal is chosen so that $|z \pm \sqrt{1-z^2}| > 1$.

If f has M simple poles, z_ℓ, with residues $\operatorname{res}(z_\ell)$, respectively, then we have

$$c_k = -2 \sum_{\ell=1}^M \frac{\operatorname{res}(z_\ell)}{\sqrt{z_\ell^2-1}\left(z_\ell \pm \sqrt{z_\ell^2-1}\right)^k}. \tag{7}$$

The notion of Padé approximant of a power series easily generalizes to orthogonal series, as we will see below.

3.2   Padé Approximants of Orthogonal Series

Let f be a function such that

$$f(x) \sim \sum_{k=0}^\infty c_k P_k(x), \quad c_k = \frac{\langle f(z), P_k(z) \rangle}{\|P_k(z)\|^2}. \tag{8}$$

Given two non-negative integers m and n, we say that the rational function

$$\Phi_{m,n}(f;x) = \frac{N_{m,n}(x)}{D_{p,q}(x)} = \frac{\sum_{i=0}^p a_i P_i(x)}{\sum_{i=0}^q b_i P_i(x)}$$

is the linear Padé approximant of f, see e.g. [7], if

$$D_{m,n}(x)f(x) - N_{m,n}(x) = \sum_{i=p+q+1}^\infty e_i P_i(x).$$

Although this definition is very similar to the Padé approximants of power series, we remark that to compute a Padé approximant, $\Phi_{m,n}$, of power series (of orthogonal polynomials) we need to know the first $m+n+1$ $(m+2n+1)$ coefficients.

In order to compute $\Phi_{m,n}$ let

$$P_i(x)f(x) = \sum_{j=0}^\infty h_{j,i} P_j(x), \quad i = 0, 1, \dots.$$

Thus, $h_{j,i}$, j=0,1,..., are the coefficients of the orthogonal series of $P_i(x)f(x)$, $i = 0, 1, \dots$. Then, the coefficients a_j and b_i of $\Phi_{m,n}(t)$ are solutions of the following homogeneous system of $m+n+1$ linear equations and $m+n+2$ unknowns.

$$\sum_{i=0}^q h_{j,i} b_i - a_j = 0, \quad j = 0, \dots, p \tag{9}$$

$$\sum_{i=0}^{q} h_{j,i} b_i = 0, \quad j = p+1, \ldots, p+q. \tag{10}$$

Thus, a non trivial solution always exist. If we set, for instance, $b_q = 1$ then equations 9 and 10 can be written in the following matricial form

$$\mathbf{H}^{[p/q]} \cdot \mathbf{b} = -\mathbf{h}^{[p/q]} \tag{11}$$

$$\mathbf{a} = \mathbf{G}^{[p/q]} \cdot \mathbf{b} + \mathbf{g}^{[p/q]} \tag{12}$$

where,

$$\mathbf{a} = \begin{bmatrix} a_0 \ldots a_p \end{bmatrix}^T, \quad \mathbf{b} = \begin{bmatrix} b_0 \ldots b_{q-1} \end{bmatrix}^T,$$

$$\mathbf{g}^{[p/q]} = \begin{bmatrix} h_{0,q} \ldots h_{p,q} \end{bmatrix}^T, \quad \mathbf{h}^{[p/q]} = \begin{bmatrix} h_{p+1,q} \ldots h_{p+q,q} \end{bmatrix}^T,$$

and

$$\mathbf{G}^{[p/q]} = \begin{bmatrix} h_{0,0} & \cdots & h_{0,q-1} \\ \vdots & & \vdots \\ h_{p,0} & \cdots & h_{p,q-1} \end{bmatrix}, \quad \text{and } \mathbf{H}^{[p/q]} = \begin{bmatrix} h_{p+1,0} & \cdots & h_{p+1,q-1} \\ \vdots & & \vdots \\ h_{p+q,0} & \cdots & h_{p+q,q-1} \end{bmatrix}.$$

The coefficients of the denominator, b_i, $i = 0, 1, \ldots, q-1$, provided that $\mathbf{H}^{[p/q]}$ is regular, are uniquely determined by solving 11. Once the coefficients b_i are determined, we use 12 to compute the numerators coefficients a_i, $i = 0, 1, \ldots, p$. The coefficients $h_{i,j}$ can be computed using the following recurrence relation, [7]

$$h_{i,j+1} = \frac{1}{\alpha_j} \left(\frac{\mu_{i+1}}{\mu_i} \alpha_i h_{i+1,j} + (\beta_i - \beta_j) h_{i,j} + \frac{\mu_{i-1}}{\mu_i} \gamma_i h_{i-1,j} - \gamma_j h_{i,j-1} \right), \ i,j \geq 1 \tag{13}$$

and,

$$h_{i,0} = c_i, \ i \geq 0, \quad h_{0,j} = \frac{\mu_j}{\mu_0} h_{j,0}, \quad j \geq 1,$$

where, α_i, β_i and γ_i are the coefficients given in relation 2 and $\mu_i = \|P_i\|^2$.

In particular, it is easy to see that for Chebyshev polynomials we have the following result:

Chebyshev Polynomials: The Chebyshev polynomials, normalized with $T_i(1) = 1$, $i \geq 0$, satisfy the recurrence relation 2 with

$$\begin{cases} \alpha_i = \gamma_i = \frac{1}{2}, \ \beta_i = 0, \ i \geq 1 \\ \alpha_0 = 1, \ \beta_0 = 0 \end{cases}$$

and $\mu_0 = \pi$, $\mu_i = \pi/2$, $i \geq 1$. Thus, using the recurrence relation 13 we get:

$$\begin{cases} h_{i,0} = c_i, \ i \geq 0 \\ h_{0,j} = \frac{1}{2} c_i, \ j \geq 1 \\ h_{1,1} = h_{0,0} + \frac{1}{2} h_{2,0} \\ h_{i,1} = \frac{1}{2} h_{i-1,0} + \frac{1}{2} h_{i+1,0}, \ i \geq 2 \\ h_{1,j} = 2 h_{0,j-1} + h_{2,j-1} - h_{1,j-2}, \ j \geq 2 \\ h_{i,j} = h_{i-1,j-1} + h_{i+1,j-1} - h_{i,j-2}, \ i,j \geq 2 \end{cases}$$

In the next section we present the results of three numerical experiments using Taylor and Chebyshev-Padé approximants.

4   Zeros and Poles of Chebyshev-Padé Approximants

Example 3. In this experiment our goal is to analyze the behaviour of poles and zeros of Chebyshev-Padé approximants from a perturbed rational function with a single
pole. We also want to compare this behaviour with the behaviour of poles and zeros of
diagonal Taylor-Padé approximants of the same function.

Consider the function $f(z) = 1/(z-2)$ whose Taylor series is $T(z) = \sum_{k=0}^{\infty} -2^{-(k+1)}z^k$,
$|z| < 2$ and whose Chebyshev series is $C(z) = \sum_{k=0}^{\infty}{}' c_k T_k(z)$, $z \in E_{r(2)}$. The Chebyshev
coefficients c_k can be computed using relation (7), with $z_1 = 2$ and $res(z_1) = 1$. If we
perturb the coefficients of $T(z)$ and of $C(z)$ with ε-noise, (such as in example 1), we
define

$$T_{\varepsilon}(z) = \sum_{k=0}^{\infty} \left(-2^{-(k+1)} + \varepsilon r_k\right) z^k, \quad \text{and} \quad C_{\varepsilon}(z) = \sum_{k=0}^{\infty}{}' (c_k + \varepsilon r_k) T_k(z),$$

where the primed summation on C_{ε} means that the deterministic part is halved. As
expected, the results concerning the zeros/poles of the Taylor series present the same
pattern of the behaviour of poles and zeros of diagonal Taylor-Padé approximants in
Froissart's experiment 1: There is a stable pole close to the singularity of the function
f ($z_1 = 2$), there is an unstable "ghost zero" and the others zeros/poles are Froissart
doublets lying in the vicinity of the natural boundary of T_{ε} (the unity circle $|z| = 1$).

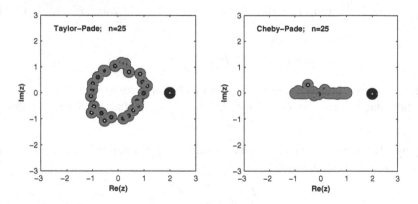

Fig. 3. Localization of Poles (*black dots*) and zeros (*magenta circles*) of $\Phi_{n,n}(T_{\varepsilon})$ (*left*) and
$\Phi_{n,n}(C_{\varepsilon})$ (*right*), with $\varepsilon = 10^{-4}$. Note that the Froissart doublets located near the unit circle
(*left figure*) were sent close to the Joukowski transform of the unit circle, i.e. the real interval
$[-1, 1]$ (*right figure*). For the sake of clarity we did not include the *ghost zero* on picture.

The poles and zeros of diagonal Chebyshev-Padé approximants, $\Phi_{n,n}(C_{\varepsilon})$, have the
following behaviour:

1. There is a stable pole ξ_1 near the singularity of f ($z_1 = 2$), where $|z_1 - \xi_1|$ is pro-
   portional to ε;

2. There is an unstable ghost zero η_1, where $|\eta_1|$ is proportional to ε^{-1};
3. The others poles ξ_k and zeros η_k, $k = 2, \ldots, n$ lie in the vicinity of the interval $[-1, 1]$.

Thus, the results concerning the Froissart doublets obtained from the Chebyshev series C_ε show a different pattern, they lie in the vicinity of the interval $[-1, 1]$ (or the degenerate Bernstein ellipse E_1).

Now if we change the perturbation with ω-noise (such as in example 2), i.e.,

$$T_\omega(z) = \sum_{k=0}^{\infty} \left(-2^{-(k+1)} + \omega \frac{r_k}{2^k}\right) z^k, \quad \text{and} \quad C_\omega(z) = \sum_{k=0}^{\infty}{}' \left(c_k + \omega \frac{r_k}{2^k}\right) T_k(z),$$

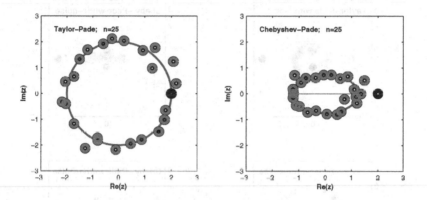

Fig. 4. Localization of poles (*black dots*) and zeros (*magenta circles*) of $\Phi_{n,n}(T_\omega)$ (*left*) and $\Phi_{n,n}(C_\omega)$ (*right*), with $\omega = 10^{-4}$. For the sake of clarity we did not include the *ghost zeros*.

the poles and zeros of the Chebyshev-Padé approximants exhibit the following behaviour:

1. There is a stable pole ξ_1 near the singularity of f ($z_1 = 2$), where $|z_1 - \xi_1|$ is proportional to ε;
2. There is an unstable ghost zero η_1, where $|\eta_1|$ is proportional to ε^{-1};
3. The others poles ξ_k and zeros η_k, $k = 2, \ldots, n$ lie in the vicinity of the Joukowski transform of the natural boundary of the noise function $\sum_{k=0}^{\infty} \omega \frac{r_k}{2^k} z^k$, i.e., they lie close to Bernstein ellipse E_2.

We illustrate this pattern in figure (4).

Example 4. The goal of this example is to test the behaviour when the convergence radius of Taylor series is less then the radius of the natural boundary caused by the noises. To this purpose, we choose the rational function f defined by

$$f(z) = \frac{4z^2 + 4z - 11}{4z^3 - 12z^2 + z - 3},$$

which has: 3 poles at points $z_1 = i/2$, $z_2 = -i/2$, $z_3 = 3$, with residues $\mathrm{res}(z_1) = -i$, $\mathrm{res}(z_2) = i$, $res(z_3) = 1$, and 2 real zeros at points $\bar{z}_1 = -(\sqrt{3}+1/2)$ and $\bar{z}_2 = \sqrt{3} - 1/2$. The Taylor series of f is given by

$$T(z) = \sum_{k=0}^{\infty} a_k z^k, \quad \text{where} \quad a_k = \begin{cases} -3^{-(k+1)}, & \text{k odd} \\ (-1)^{\frac{k}{2}} 4^{\frac{k}{2}+1} - 3^{-(k+1)}, & \text{k even} \end{cases},$$

and the Chebyshev series is $C(z) = \sum_{k=0}^{\infty} c_k T_k(z)$, where the Chebyshev coefficients are computed using relation 7. We note that: $T(z)$ converges on the disk $|z| < 1/2$, $C(z)$ converges throughout Bernstein ellipse $E_{r(z_1)}$, $T_\varepsilon(z)$ has a natural boundary on $|z| = 1$ and $T_\omega(z)$ has a natural boundary on $|z| = 2$.

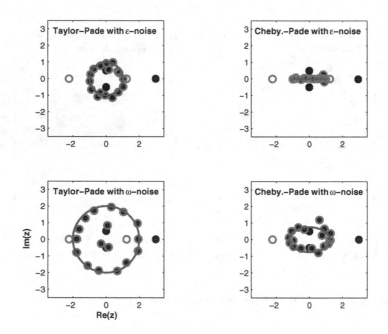

Fig. 5. Localization of poles (*black dots*) and zeros (*magenta circles*) of:$\Phi_{n,n}(T_\varepsilon)$ and $\Phi_{n,n}(T_\omega)$ (*left column*); $\Phi_{n,n}(C_\varepsilon)$ and $\Phi_{n,n}(C_\omega)$ (*right column*), $n = 20$ with $\varepsilon = \omega = 10^{-5}$. For the sake of clarity we did not include the *ghost zeros*.

In this test, the Padé approximants of the perturbed function, with both type of noises, have three stable poles in the vicinity of the three poles of f, two stable zeros in the vicinity of the two zeros of f and an unstable ghost zero. While the remainder poles/zeros of the Taylor-Padé approximants are in the vicinity of the natural boundary ($|z| = 1$ to ε-noise and $|z = 2|$ in ω-noise case), the remainder poles/zeros of the Chebyshev-Padé approximants are in the vicinity of the Joukowski transform of the boundary transform (the interval $[-1,1]$ in the ε-noise and the Bernstein ellipse E_2 in ω-noise case). We illustrate this pattern in figure 5.

Example 5. Here our goal is to analyse the behaviour of poles/zeros of Padé approximants of a function with a logarithm branch cut, as in Froissart's example 2. We choose the function f defined by $f(z) = \ln(5/4 - z)$ with branch points at $z = 5/4$ and $z = \infty$. The Taylor series and Chebyshev series are given by:

$$T(z) = \ln(5/4) - \sum_{k=1}^{\infty} \frac{4^k}{k5^k}; \quad C(z) = -\sum_{k=1}^{\infty} \frac{1}{k2^{k-1}} T_k(z),$$

where the Chebyshev series were derived from generating function (5) (setting $t = 1/2$).

The results concerning perturbed Taylor series T_ε and T_ω are similar to the Froissart's example 2 results. There is one stable pole in the vicinity of the branch point $z = 5/4$, there is a stable zero in the vicinity of the zero of f ($z=1/4$), there are two sets, B_t and F_t of poles and zeros with two distinct behaviours. The poles and zeros in B_t interlaced and represent the branch cut of function f. The poles and zeros in F_t are Froissart doublets and are located in the vicinity of the natural boundary of the non deterministic part of the perturbed Taylor series. The results concerning the poles/zeros of Chebyshev-Padé approximants are similar to the results obtained with Taylor-Padé approximants with the exception of the Froissart doublets (as in experiments 3 and 4). There is one stable pole and one stable zero close to the branch point $z = 5/4$ and the zero of f, $z = 1/4$, respectively, and there are also two sets B_c and F_c. The poles/zeros of B_c exhibit the same behaviour of the elements of B_t and the poles/zeros of F_c are Froissart doublets. Moreover, given a non-negative number n, the number of poles (zeros) of $\Phi_{n,n}(T)$ in B_t is equal to the number of poles (zeros) of $\Phi_{n,n}(C)$ in B_c and consequently the number of Froissart doublets of $\Phi_{n,n}(T)$ and of $\Phi_{n,n}(C)$ is equal. Then Froissart doublets in B_c are located, as in examples 3 and 4, on Joukowski transform of the natural boundaries of random part of Taylor series (interval $[-1, 1]$, in the ε-noise case, and in Bernstein ellipse E_2 in the ω-noise case). We illustrate these results in figures 6 and 7.

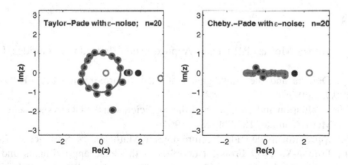

Fig. 6. Localization of poles (*black dots*) and zeros (*magenta circles*) of: $\Phi_{n,n}(T_\varepsilon)$ (*left column*); and $\Phi_{n,n}(C_\varepsilon)$ (*right column*), $n = 20$ with $\varepsilon = 10^{-6}$. For the sake of clarity we did not include all elements of B_t and B_c.

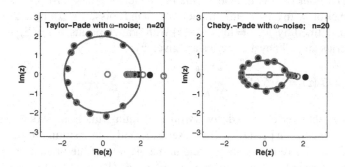

Fig. 7. Localization of Poles (*black dots*) and zeros (*magenta circles*) of: $\Phi_{n,n}(T_\omega)$ (*left column*); and $\Phi_{n,n}(C_\omega)$ (*right column*), $n = 20$ with $\varepsilon = 10^{-6}$. For the sake of clarity we did not include all elements of B_t and B_c.

5   Conclusions and Future Work

Our numerical results support the following conjectured behaviour.

Let $S(z) = \sum_{k=0}^{\infty} c_k z^k$ and $C(z) = \sum_{k=0}^{\infty} f_k T_k(z)$ be the power expansion and the Chebyshev expansion, respectively, of the same function f and let $\sum_{k=0}^{\infty} \delta_k z^k$ be a random power series with a natural boundary on $\mathcal{B} = \{z \in \mathbb{C} : |z| = \rho\}$ with probability one, then the "artificial" Froissart doublets of a diagonal Chebyshev approximant $\Phi_{n,n}(f_\delta)$ are in the vicinity of $w(\mathcal{B})$, where $f_\delta(z) \sim \sum_{k=0}^{\infty} (f_k + \delta_k) T_k(z)$ and w is the Joukowski transform.

Numerical experiments performed with Legendre-Padé approximants also show the same behaviour. It will be interesting to prove this conjecture and extend it to other orthogonal polynomial families.

References

1. Baker, G.A., Graves-Morris, P.R.: Padé Approximants, 2nd edn. Cambridge Univ. Press (1996)
2. Baker, G.A.: Defects and the convergence of Padé Approximants, LA-UR-99-1570, Los Alamos Nat. Lab. (1999)
3. Elliot, D.: The evaluation and estimation of the coefficients in the Chebyshev series expansion of a function. Math. Comput. 18, 274–284 (1964)
4. Gilewicz, J.: Approximants de Padé. Lecture notes in Mathematics. Springer (1978)
5. Gilewicz, J., Truong-Van, B.: Froissart doublets in the Padé approximants and noise. In: Sendov, B. (ed.) Construtive Theory of Function 1987, Varna, pp. 145–151. Publishing House of Bulgarian Academy of Science (1987)
6. Olver, F.W., et al.: NIST Hanbook of Mathematical Functions. Cambridge Univ. Press (2010)
7. Matos, A.C.: Recursive computation of Padé-Legendre; approximants and some acceleration properties. Numerische Mathematik 89, 535–560 (2003)
8. Szegő, G.: Orthogonal Polynomials. AMS (1967)

Elliptic Curve Cryptography on Constrained Microcontrollers Using Frequency Domain Arithmetic*

Utku Gülen and Selçuk Baktır

Bahçeşehir University
Department of Computer Engineering
Istanbul, Turkey
utku.gulen@bahcesehir.edu.tr
selcuk.baktir@bahcesehir.edu.tr

Abstract. We implemented elliptic curve cryptography in the frequency domain on the MSP430 constrained microcontroller. Our implementation of 169-bit elliptic curve cryptography (ECC) on MSP430, one of the most popular microcontrollers for wireless sensor network (WSN) nodes, performs an ECC scalar point multiplication operation, for random points, in only 1.55 ms which is similar to or faster than existing implementations. To our knowledge, this work proposes the first ever software implementation of ECC in the frequency domain on a constrained low-power microcontroller.

Keywords: Elliptic curve cryptography, ECC, finite field multiplication, discrete Fourier transform, DFT, frequency domain, wireless sensor networks, WSN, MSP430.

1 Introduction

Wireless sensor networks (WSN) have many applications [1], including:

1. Environmental monitoring, e.g. forest fires, air pollution, humidity, etc.,
2. Battlefield reconnaissance,
3. Emergency rescue operations,
4. Surveillance.

WSN nodes are cheap devices typically containing only a constrained tiny microcontroller. However, since these devices are spread around in the field, they are vulnerable against potential attacks. Since they communicate sensitive information to each other, providing security and privacy to these devices is crucial. Data confidentiality in WSNs is achieved most easily by using symmetric-key cryptographic algorithms. However, distribution of the symmetric keys remains

* This work is supported by the grant EU FP7 Marie Curie IRG 256544.

B. Murgante et al. (Eds.): ICCSA 2014, Part VI, LNCS 8584, pp. 493–506, 2014.
© Springer International Publishing Switzerland 2014

a problem, and can be overcome most efficiently by utilizing public-key cryptography [15]. RSA [27] and elliptic curve cryptography (ECC) [21,22] are the two most popular public-key cryptographic algorithms. RSA requires using at least 1024-bit long keys which necessitates computations over 1024-bit operands. For the same level of security, ECC requires 160-bit keys and computations over 160-bit numbers. WSN nodes usually run on battery and hence power efficiency is an important criterion for algorithms to be run on these devices. Furthermore, these are tiny devices with usually minimal storage available. ECC is computationally less costly and also requires less storage due to its shorter key size. Therefore, it is considered the public-key cryptographic algorithm of choice for WSNs [30,35].

Efficiency of an ECC implementation is dependent on the underlying finite field arithmetic. In ECC, assuming projective coordinates are used and inversions are avoided, multiplication is the most time-consuming arithmetic operation. Any speed-up in the multiplication operation would contribute to a significant speedup in the performance of an ECC implementation. A 160-bit multiplication, e.g. for implementing ECC, is realized by running around 100 word addition and 100 word multiplication instructions, using the classical schoolbook method, on a 16-bit microcontroller. On a constrained microcontroller, a word multiplication usually takes several times longer to execute than a word addition. This usually holds true even when the constrained microcontroller has a built-in hardware multiplier circuitry and a related multiply instruction. For instance, on the MSP430 microcontroller, a 16-bit multiply instruction takes 14 clock cycles using its hardware multiplier, whereas a 16-bit add instruction takes only a 1 clock cycle.

In this paper, for the implementation of ECC, we realize finite field multiplication in the frequency domain using a discrete Fourier transform (DFT) based method. Frequency domain multiplication is potentially more efficient for implementations on constrained microcontrollers, because they typically require performing a small number of word multiplications, in addition to a large number of simpler operations such as addition. In an earlier work, frequency domain arithmetic was shown to be efficient for constrained hardware implementations of ECC [7]. In [19], a hardware architecture for large integer multiplication in the frequency domain was proposed, however no specific implementation results were provided in terms of timing performance or circuit area. Authors presented analytical results claiming their proposed hardware multiplier architecture is more efficient than multiplication with the classical method for the operand size of 4096-bit or longer. In[10,6], analytical results for the implementation of frequency domain inversion for ECC were provided, however no implementation results were given. In [17,25,3,18], frequency domain arithmetic was proposed for lattice-based cryptography. In [28], an algorithm was proposed for performing repeated modular multiplications in the frequency domain for application to RSA. Hardware architectures for modular multiplication in the frequency domain, for RSA bitlengths, were proposed in [13] and [34], with actual implementation results. Finally in [14], a frequency domain hash function was proposed. In [8], a method for frequency domain multiplication was proposed for ECC with theoretical complexity figures but no actual implementation results. With this work, we present an efficient software implementation of frequency domain multiplication for ECC on constrained microcontrollers. To our knowledge, this is the first work that presents a practical software implementation of ECC in the frequency domain.

2 Mathematical Background

Finite Field Arithmetic

In ECC, data is represented as an element of a finite field, and cryptographic operations on this data are realized by performing finite field arithmetic operations. The choice of the finite field representation that is used influences the performance of ECC. In this work, we implement ECC over a finite field of the form $GF(p^m)$ using the Optimal Extension Field (OEF) representation [4,5]. OEFs are a special class of finite extension fields which use a field generating polynomial of the form $f(x) = x^m - w$ and have a *pseudo-Mersenne prime* field characteristic given in the form $p = 2^n \pm c$ with $\log_2 c < \lfloor \frac{n}{2} \rfloor$. Having a pseudo-Mersenne prime field characteristic and a binary field generating polynomial, the OEF representation allows very efficient reduction both in the base field $GF(p)$ and in the extension field operations.

ECC requires performing a large number of finite field arithmetic operations, such as multiplication, inversion, addition and subtraction. Inversion is the slowest arithmetic operation in finite fields, however it can be avoided in ECC if projective coordinates are made use of. Keeping the inversion operation aside, multiplication is the most complex and time-consuming finite field arithmetic operation required in ECC implementations.

Elements of $GF(p^m)$ are represented by polynomials of degree $m - 1$ with coefficients in $GF(p)$, e.g. $A \in GF(p^m)$ is represented as $A = \sum_{i=0}^{m-1} a_i x^i = a_0 + a_1 x + a_2 x^2 + \ldots + a_{m-1} x^{m-1}$ where $a_i \in GF(p)$. Multiplication of A and B, both elements of $GF(p^m)$, is performed by first computing the polynomial product $C' = A \cdot B = \sum_{i=0}^{2m-2} c_i' x^i$ and then the modular reduction $C = C' \mod f(x)$. Using OEFs, the modular reduction $C = C' \mod f(x)$ has only linear, i.e. $O(m)$, complexity. However, the polynomial multiplication $C' = A \cdot B = \sum_{i=0}^{2m-2} c_i' x^i$ has quadratic, i.e. $O(m^2)$, complexity using the typical schoolbook method. Algorithms such as the Karatsuba method [20] can achieve this polynomial multiplication operation with subquadratic complexity, however due to their irregular structure, they are considered not practical for implementations over constrained microcontrollers.

Frequency Domain Representation for Finite Field Elements

We use the DFT modular multiplication algorithm [9] for performing arithmetic in $GF(p^m)$. DFT modular multiplication achieves multiplication in $GF(p^m)$ in the frequency domain. It is based on the number theoretic transform (NTT) [24], also known as the discrete Fourier transform (DFT) [29,12] over a finite field. The algorithm takes as input two elements of $GF(p^m)$ represented in the frequency domain, and outputs their Montgomery product [23], also represented in the frequency domain. Using the NTT, the frequency domain representation of an element of $GF(p^m)$ can be computed as follows.

1. $a(x) = a_0 + a_1x + a_2x^2 + \ldots + a_{m-1}x^{m-1}$, in $GF(p^m)$, is represented as the sequence $(a) = (a_0, a_1, a_2, \ldots, a_{m-1}, 0, 0, \ldots, 0)$ after appending $d - m$ zeros to the right, where $d \geq 2m - 1$ is length of the NTT,

2. The frequency domain representation (A) of the time domain sequence (a) is obtained by applying the following NTT computation:

$$A_j = \sum_{i=0}^{d-1} a_i r^{ij} \ , \ 0 \leq j \leq d - 1 . \tag{1}$$

Likewise, the time domain representation (a) can be obtained from the frequency domain representation (A) back by applying the following inverse NTT computation:

$$a_i = \frac{1}{d} \cdot \sum_{j=0}^{d-1} A_j r^{-ij} \ , \ 0 \leq i \leq d - 1 . \tag{2}$$

The NTT computations require using a dth primitive root of unity, denoted by r. In this work we use $r = -2 \in GF(p)$.

Finite Field Multiplication in the Frequency Domain

The DFT modular multiplication algorithm [9], given in Algorithm 1, achieves multiplication in $GF(p^m)$ in the frequency domain. In this algorithm, all parameters are in their frequency domain representations. These parameters are the input operands $a(x), b(x) \in GF(p^m)$ and the Montgomery product $c(x) = a(x) \cdot b(x) \cdot x^{-(m-1)} \bmod f(x) \in GF(p^m)$. The time domain sequence representations of these parameters are $(a), (b)$ and (c) , respectively, and their frequency domain sequence representations, as used in the algorithm, are $(A), (B)$ and (C).

The DFT modular multiplication algorithm achieves multiplication in $GF(p^m)$ in the frequency domain with only a linear number of base field $GF(p)$ multiplications, in addition to a quadratic number of simple additions/subtractions and bitwise rotations in $GF(p)$.

Algorithm 1. [9] DFT modular multiplication for Montgomery multiplication in $GF(p^m)$ where $p = 2^n - 1$, m odd, $m = n$ and $f(x) = x^m - 2$

Require: $(A) \equiv a(x) \in GF(p^m)$, $(B) \equiv b(x) \in GF(p^m)$
Ensure: $(C) \equiv a(x) \cdot b(x) \cdot x^{-(m-1)} \bmod f(x) \in GF(p^m)$
 1. **for** $i = 0$ to $d - 1$ **do**
 2. $C_i \leftarrow A_i \cdot B_i$
 3. **end for**
 4. **for** $j = 0$ to $m - 2$ **do**
 5. $S \leftarrow 0$
 6.   **for** $i = 0$ to $d - 1$ **do**
 7. $S \leftarrow S + C_i$
 8.   **end for**
 9. $S \leftarrow -S/d$
10. $S_{half} \leftarrow S/2$
11. $S_{even} \leftarrow S_{half}$
12. $S_{odd} \leftarrow S + S_{half}$
13.   **for** $i = 0$ to $d - 1$ **do**
14.   **if** $i \bmod 2 = 0$ **then**
15. $C_i \leftarrow C_i + S_{even}$
16.   **else**
17. $C_i \leftarrow -(C_i + S_{odd})$
18.   **end if**
19. $C_i \leftarrow C_i/2^i$
20.   **end for**
21. **end for**
22. Return (C)

---

3   Implementation of $GF(p^m)$ Arithmetic on MSP430

We use the finite field $GF(p^m)$ with the Mersenne prime field characteristic $p = 2^{13} - 1$ and $m = 13$. Hence, we can use $d = 26$ for the NTT operations, and $r = -2$ as the 26th primitive root of unity in $GF(2^{13} - 1)$. Note that, for $r = -2$ and $p = 2^{13} - 1$, a modular multiplication in $GF(p)$ with a power of r can be achieved very efficiently with a simple bitwise rotation, in addition to a negation if the power is odd. The number theoretic transform computed modulo a Mersenne prime, as in this case, is called the *Mersenne transform* [26].

We realize our ECC implementation over the finite field $GF((2^{13} - 1)^{13})$ in the frequency domain. Addition and subtraction can easily be achieved in the frequency domain by pairwise addition/subtraction of frequency domain coefficients which are elements of $GF(2^{13} - 1)$. We perform multiplication in $GF((2^{13} - 1)^{13})$ using the DFT modular multiplication algorithm. All operations needed for implementing DFT modular multiplication are modular addition, subtraction, multiplication and rotation operations in $GF(2^{13} - 1)$. Hence efficient implementation of $GF(2^{13} - 1)$ arithmetic is critical. For modular multiplication in $GF(2^{13} - 1)$, we utilize the hardware multiplier available on the MSP430 microcontroller.

The read/write instructions from/to the memory has a significant impact on the efficient implementation of arithmetic operations. The MSP430 microcontroller has a RISC architecture with only 27 instructions and 7 addressing modes. The number of clock cycles for an instruction execution depends on the addressing mode that is used. The register addressing mode is more desirable, however there are only 12 general purpose registers available. In our implementations, we optimized our operations by using these registers for storing our operands as much as possible. Also, we stored frequently used constants in registers to avoid extra clock cycles.

3.1 Addition and Subtraction

Modular addition in $GF(2^{13} - 1)$ is the most commonly used operation in our implementation. Modular subtraction in $GF(2^{13} - 1)$ is the same as addition with the additional "xor" instruction to flip the bits of the operand which costs 1 extra clock cycle. We allocate two registers to store constant values for masking and checking the Most Significant Bit (MSB) of operands during these operations. We can achieve modular addition and subtraction in $GF(2^{13} - 1)$ in 4 and 5 clock cycles, respectively, using the Assembly routines given below.

Assembly Routine 1. Modular addition

```
add     r15,r14
bit     r11,r14 ; (R11 = 0x2000)
adc     r14
and     r13,r14 ; (R13 = 0x1FFF)
```

Modular subtraction is achieved with one extra clock cycle (for the **xor** instruction) on top of modular addition.

Assembly Routine 2. Modular subtraction

```
xor     r13,r14 ; (R13 = 0x1FFF)
add     r15,r14
bit     r11,r14
adc     r14
and     r13,r14
```

3.2 Modular Multiplication

It is crucial to perform modular multiplication in $GF(2^{13} - 1)$ efficiently, since it directly influences the efficiency of multiplication in $GF((2^{13} - 1)^{13})$. There is a 16-bit hardware multiplier available on most MSP430 microcontrollers. We made use of this hardware multiplier in order to meet the desired timing performance for multiplication in $GF(2^{13} - 1)$. We postpone the modular reduction operation until after the 26-bit integer product is computed.

There is no multiplication instruction in the MSP430 instruction set, in spite of the available 16-bit hardware multiplier. The cost of a 16-bit multiplication using the available hardware multiplier is just the cost of write/read operations to/from the special function registers for the hardware multiplier. Utilizing the hardware multiplier, a 16x16-bit multiplication can be achieved in 14 clock cycles. Our implementation of modular multiplication in $GF(2^{13} - 1)$ using the hardware multiplier takes 26 clock cycles. Modular reduction is performed after reading the 26-bit product from the hardware multiplier, as implemented with the Assembly routine given below.

Assembly Routine 3. Modular multiplication in $GF(2^{13} - 1)$

```
mov    r14,&MPY
mov    r15,&OP2
mov    &RESLO,r14
mov    &RESHI,r15
mov    r14,r10
rlc    r10
rlc    r15
rlc    r10
rlc    r15
rlc    r10
rlc    r15
and    r13,r14    ; (R13 = 0x1FFF)
add    r15,r14
bit    r11,r14    ; (R11 = 0x2000)
adc    R14
and    r13,r14
```

3.3 Bitwise Rotation

It is crucial to implement bitwise rotations over $GF(2^{13} - 1)$ elements efficiently, since a large number of such operations are performed in the **DFT modular multiplication** algorithm. Hence, we optimized this operation as much as possible. The MSP430 instruction set has 1-bit **arithmetic shift** and **shift with carry** instructions. Both instructions execute in a single clock cycle, therefore we utilized these instructions as often as possible. We also used the **set bit**, **test bit** and **swap byte** instructions to achieve rotations in $GF(2^{13} - 1)$ with the minimal number of clock cycles. For rotations by different numbers of bits, we pursued various strategies to reduce the number of required clock cycles.

1-Bit Left-Rotation

1-bit left-rotation is carried out by checking the MSB of the operand and then shifting it to the right through carry. A masking operation is required to complete the operation, thus a 1-bit left-rotation, as shown in Figure 1 and given in the below Assembly routine, is accomplished in 3 clock cycles.

Assembly Routine 4. 1-bit left-rotation

```
bit    R14,R10 ; (R14 = 0x1000)
rlc    R10
and    R13,R10 ; (R13 = 0x1FFF)
```

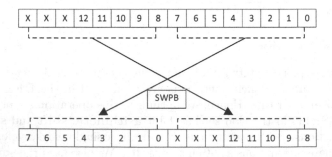

Fig. 1. 1-bit left-rotation

2,3 and 4-Bit Left-Rotations

2,3 and 4-bit left-rotations are performed by repeated 1-bit left-rotations.

For rotations by more than 4 bits, the swap byte instruction is used to simplify the operation. The swap byte instruction is used to exchange the low and high bytes of a 13-bit operand, as shown in Figure 2.

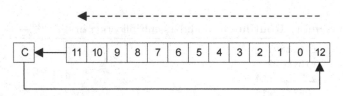

Fig. 2. Swap byte instruction

5,6,7,8,9 and 10-Bit Left-Rotations

We optimized the $5, 6, 7, 8, 9$ and 10-bit left-rotation operations through the use of the **swpb** instruction and mask/store operations. Rotations by different numbers of bits are handled slightly differently, i.e. they have different orders of codes, although a similar strategy is used to achieve the best cycle time in each case. Exemplarily, the Assembly code for the 6-bit left-rotation operation, which takes 9 clock cycles, is shown below.

Assembly Routine 5. 6-bit left-rotation

```
mov r15,r10 ; Store operand in r10
and R7,R10 ; (r7 = 0x007F)
swpb R10
rra R10
rra R10
rla R15
swpb R15
and R6,R15 ; (r6 = 0x003F)
bis R10,R15
```

12 and 11-Bit Left-Rotations

A 12-bit left-rotation can be achieved through a 1-bit right-rotation operation, as shown in the Figure 3 and implemented with the Assembly code below. The required 1-bit right-rotation can be accomplished by shifting the Least Significant Bit (LSB) to the carry flag and setting the 13th bit of the operand depending on the value of the carry. A 11-bit left-rotation can be achieved by performing two 12-bit left-rotations.

Assembly Routine 6. 12-bit left-rotation

```
      rra    r10
      jnc    done
      bis    r14,r10 ; (r14 = 0x1000)
done:
```

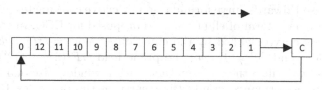

Fig. 3. 12-bit left-rotation

Table 1. Execution times for bitwise left-rotation operations on MSP430F149

Amount of Left-Rotation	# Clock Cycles
1 Bit	3
2 Bits	5
3 Bits	7
4 Bits	9
5 Bits	9
6 Bits	9
7 Bits	9
8 Bits	9
9 Bits	9
10 Bits	9
11 Bits	7
12 Bits	3.5

In the DFT modular multiplication algorithm, all of the above rotation operations, by differing numbers of bits, are used an equal number of times. Our implementation of the rotation operation by different numbers of bits takes 7.4 clock cycles on average. Table 1 shows the number of clock cycles for each bitwise-rotation individually.

4   Implementation Results and Comparisons

We implemented the ECC scalar point multiplication operation on Texas Instrument's low-power 1-series 16-bit microcontroller MSP430F149. We used IAR Embedded Workbench as our development tool and obtained accurate clock cycle counts from it's debugger. Our timings are satisfactory with similar or better results compared to previous implementations on the same family of microcontrollers. Hence, for power efficient implementations of ECC on constrained devices, such as WSN nodes, using frequency domain arithmetic would be desirable. With this work, we present the first ever practical software implementation of ECC in the frequency domain, with promising results for low-power applications of ECC in WSNs.

We present our timing results for ECC point multiplication, for scalar multiplication of random points, in Figure 4 and Table 2. For point addition and doubling, we use Edwards curve formulas in projective coordinates [11]. Edwards curves [16] are a new form of elliptic curves proposed for ECC. In this work, we use an Edwards curve, defined by the equation $x^2 + y^2 = c^2(1 + dx^2y^2)$, with $c = 1$, d random and $dc^4 \neq 1$, over the prime field $GF((2^{13} - 1)^{13})$. For point multiplication, we utilize the NAF method with a window size of 4. In Table 2, we present our performance results (first entry in the table) for 169-bit point multiplication, together with the performance results for related work.

As seen in Table 2, our timing figures for 169-bit ECC scalar multiplication with a random point is only 1.55 s on MSP430F149 running at 8MHz. For the same finite field, and on the faster MC68328EZ running at 20MHz, 169-bit scalar multiplication proposed by Woodbury et al. takes 2.9 s which is almost 2 times slower [32]. Wang et al.'s implementation of 160-bit random point multiplication on MSP430 running at 8MHz (the same as our platform) has a timing performance of 3.51 s [31]. For the much smaller 136-bit field, on the 8-bit 8051 microcontroller running at 12MHz, the implementation of scalar multiplication by Woodbury et al. takes 8.37 s [33]. Finally, a 163-bit scalar multiplication with a random point on a binary field takes 32.5 s in an implementation by Araz et al. on MSP430 running at 8MHz [2].

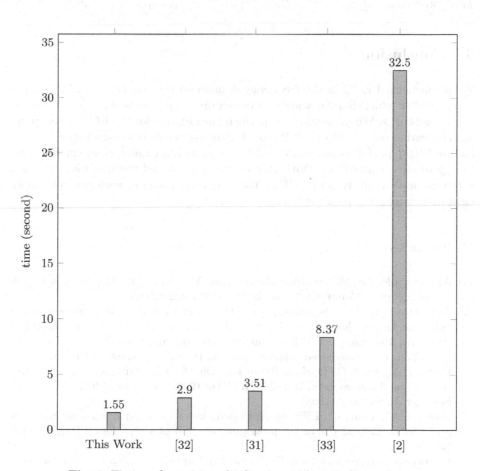

Fig. 4. Timings for point multiplication with a random point

Table 2. Timings for point multiplication with a random point

Platform	Field	Method	Time (s)
MSP430 @8MHz (this work)	$GF((2^{13} - 1)^{13})$	4NAF	1.55
MC68328EZ @20MHz [32]	$GF((2^{13} - 1)^{13})$	Binary	2.9
MSP430 @8MHz [31]	F_{p160}	4NAF	3.51
Intel8051 @12MHz [33]	$GF((2^8 - 17)^{17})$	Binary	8.37
MSP430 @8MHz [2]	$GF(2^{163})$	not specified	32.5

5   Conclusion

We implemented ECC in the frequency domain on the constrained MSP430 microcontroller which is used widely in constrained environments such as wireless sensor networks. We presented our performance figures for the ECC scalar point multiplication operation on MSP430. Our timing result is satisfactory with similar or better performance compared to previous implementations on the same family of microcontrollers. With this work, we proposed the first ever practical software implementation of ECC in the frequency domain, with possible applications in wireless sensor networks.

References

1. Akyildiz, I.F., Su, W., Sankarasubramaniam, Y., Cayirci, E.: Wireless sensor networks: a survey. Computer Networks 38(4), 393–422 (2002)
2. Araz, O., Qi, H.: Load-balanced key establishment methodologies in wireless sensor networks. International Journal of Security and Networks 1(3), 158–166 (2006)
3. Aysu, A., Patterson, C., Schaumont, P.: Low-cost and area-efficient fpga implementations of lattice-based cryptography. In: HOST, pp. 81–86. IEEE (2013)
4. Bailey, D.V., Paar, C.: Optimal Extension Fields for Fast Arithmetic in Public-Key Algorithms. In: Krawczyk, H. (ed.) CRYPTO 1998. LNCS, vol. 1462, pp. 472–485. Springer, Heidelberg (1998)
5. Bailey, D.V., Paar, C.: Efficient Arithmetic in Finite Field Extensions with Application in Elliptic Curve Cryptography. Journal of Cryptology 14(3), 153–176 (2001)
6. Baktir, S.: Frequency Domain Finite Field Arithmetic for Elliptic Curve Cryptography. PhD thesis, Electrical and Computer Engineering Department, Worcester Polytechnic Institute, Worcester, MA, USA (April 2008)
7. Baktir, S., Kumar, S., Paar, C., Sunar, B.: A state-of-the-art elliptic curve cryptographic processor operating in the frequency domain. Mobile Networks and Applications 12(4), 259–270 (2007)

8. Baktir, S., Sunar, B.: Achieving efficient polynomial multiplication in fermat fields using the fast fourier transform. In: Proceedings of the 44th Annual Southeast Regional Conference, ACM-SE 44, pp. 549–554. ACM, New York (2006)

9. Baktır, S., Sunar, B.: Finite field polynomial multiplication in the frequency domain with application to elliptic curve cryptography. In: Levi, A., Savaş, E., Yenigün, H., Balcısoy, S., Saygın, Y. (eds.) ISCIS 2006. LNCS, vol. 4263, pp. 991–1001. Springer, Heidelberg (2006)

10. Baktır, S., Sunar, B.: Optimal extension field inversion in the frequency domain. In: von zur Gathen, J., Imaña, J.L., Koç, Ç.K. (eds.) WAIFI 2008. LNCS, vol. 5130, pp. 47–61. Springer, Heidelberg (2008)

11. Bernstein, D.J., Lange, T.: Faster addition and doubling on elliptic curves. In: Kurosawa, K. (ed.) ASIACRYPT 2007. LNCS, vol. 4833, pp. 29–50. Springer, Heidelberg (2007)

12. Burrus, C.S., Parks, T.W.: DFT/FFT and Convolution Algorithms. John Wiley & Sons (1985)

13. Chen, D.D., Yao, G.X., Koç, Ç.K., Cheung, R.C.C.: Low complexity and hardware-friendly spectral modular multiplication. In: 2012 International Conference on Field-Programmable Technology (FPT), pp. 368–375 (2012)

14. Cheung, R.C.C., Koç, Ç.K., Villasenor, J.D.: A high-performance hardware architecture for spectral hash algorithm. In: ASAP, pp. 215–218 (2009)

15. Diffie, W., Hellman, M.E.: New directions in cryptography. IEEE Transactions on Information Theory 22(6), 644–654 (1976)

16. Edwards, H.M.: A normal form for elliptic curves. Bulletin of the American Mathematical Society, 393–422

17. Göttert, N., Feller, T., Schneider, M., Buchmann, J., Huss, S.: On the design of hardware building blocks for modern lattice-based encryption schemes. In: Prouff, E., Schaumont, P. (eds.) CHES 2012. LNCS, vol. 7428, pp. 512–529. Springer, Heidelberg (2012)

18. Güneysu, T., Lyubashevsky, V., Pöppelmann, T.: Practical lattice-based cryptography: A signature scheme for embedded systems. In: Prouff, E., Schaumont, P. (eds.) CHES 2012. LNCS, vol. 7428, pp. 530–547. Springer, Heidelberg (2012)

19. Kalach, K., David, J.P.: Hardware implementation of large number multiplication by FFT with modular arithmetic. In: Proceedings of the 3rd International IEEE-NEWCAS Conference, pp. 267–270. IEEE (2005)

20. Karatsuba, A., Ofman, Y.: Multiplication of Multidigit Numbers on Automata. Sov. Phys. Dokl. (English translation) 7(7), 595–596 (1963)

21. Koblitz, N.: Elliptic Curve Cryptosystems. Mathematics of Computation 48, 203–209 (1987)

22. Miller, V.S.: Use of Elliptic Curves in Cryptography. In: Williams, H.C. (ed.) CRYPTO 1985. LNCS, vol. 218, pp. 417–426. Springer, Heidelberg (1986)

23. Montgomery, P.L.: Modular multiplication without trial division. Mathematics of Computation 44(170), 519–521 (1985)

24. Pollard, J.M.: The Fast Fourier Transform in a Finite Field. Mathematics of Computation 25, 365–374 (1971)

25. Pöppelmann, T., Güneysu, T.: Towards efficient arithmetic for lattice-based cryptography on reconfigurable hardware. In: Hevia, A., Neven, G. (eds.) LatinCrypt 2012. LNCS, vol. 7533, pp. 139–158. Springer, Heidelberg (2012)

26. Rader, C.M.: Discrete Convolutions via Mersenne Transforms. IEEE Transactions on Computers C-21(12), 1269–1273 (1972)

27. Rivest, R.L., Shamir, A., Adleman, L.: A Method for Obtaining Digital Signatures and Public-Key Cryptosystems. Communications of the ACM 21(2), 120–126 (1978)
28. Saldamli, G., Koç, Ç.K.: Spectral modular exponentiation. In: 18th IEEE Symposium on Computer Arithmetic, ARITH 2007, pp. 123–132 (2007)
29. Tolimieri, R., An, M., Lu, C.: Algorithms for Discrete Fourier Transform and Convolution. Springer (1989)
30. Walters, J.P., Liang, Z., Shi, W., Chaudhary, V.: Wireless sensor network security: A survey. Security in Distributed, Grid, Mobile, and Pervasive Computing 1, 367 (2007)
31. Wang, H., Sheng, B., Li, Q.: Elliptic curve cryptography-based access control in sensor networks. International Journal of Security and Networks 1(3), 127–137 (2006)
32. Woodbury, A.D.: Efficient algorithms for elliptic curve cryptosystems on embedded systems. Master's thesis, Electrical and Computer Engineering Department, Worcester Polytechnic Institute, Worcester, MA, USA (September 2001)
33. Woodbury, A.D., Bailey, D.V., Paar, C.: Elliptic curve cryptography on smart cards without coprocessors. Springer (2000)
34. Yao, G.X., Cheung, R.C.C., Koç, Ç.K., Man, K.F.: Reconfigurable number theoretic transform architectures for cryptographic applications. In: 2010 International Conference on Field-Programmable Technology (FPT), pp. 308–311 (2010)
35. Zhou, Y., Fang, Y., Zhang, Y.: Securing wireless sensor networks: A survey. IEEE Communications Surveys and Tutorials 10(1-4), 6–28 (2008)

Application of Eigensolvers in Quadratic Eigenvalue Problems for Brake Systems Analysis

Sandra M. Aires[1] and Filomena D. d'Almeida[2]

[1] School of Engineering, Polytechnic of Porto, Portugal
sma@isep.ipp.pt
[2] Centro de Matemática and Faculdade de Engenhariada Universidade do Porto, Portugal
falmeida@fe.up.pt

Abstract. We compare, in terms of computing time and precision, three different implementations of eigensolvers for the unsymmetric quadratic eigenvalue problem. This type of problems arises, for instance, in structural Mechanics to study the stability of brake systems that require the computation of some of the smallest eigenvalues and corresponding eigenvectors. The usual procedure is linearization but it transforms quadratic problems into generalized eigenvalue problems of twice the dimension. Examples show that the application of the QZ method directly to the linearized problem is less expensive that the projection onto a space spanned by the eigenvectors of an approximating symmetric problem, this is not an obvious conclusion. We show that programs where this projection version is quicker may not be so accurate. We also show the feasibility of this direct linearization version in conjunction with an iterative method, as Arnoldi, in the partial solution of very large real life problems.

Keywords: eigenvalues, eigenvectors, stability, brake systems, quadratic eigenvalue problems.

1 Introduction

The quadratic eigenvalue problem has many applications in several areas of science and engineering. We were motivated by an application consisting in the study of the stability of brake systems, in particular, in brake squeal analysis. A brake system operates by pressing a set of brake pads against a rotating disc (see Fig.1). The friction between the pads and the disc causes deceleration and may also cause an undesirable squealing effect. The aim is to eliminate this kind of squealing effects ([1], [12]). For that the determination of a certain number of the smallest eigenvalues of a quadratic eigenvalue problem is necessary. As a contribution for this aim we will present a study of efficient eigensolvers for unsymmetric quadratic problems and its implementations. This work extends and completes the preliminary work presented at CMMSE2012 (International Conference on Computational and Mathematical Methods in Science and Engineering) by addressing more extensively the issue of accuracy, by describing the implementation details used to save time and storage, and by treating an example where the mass matrix M and the rigidity matrix K are issued from a real life model of a brake disc, although the damping matrix is artificially created.

B. Murgante et al. (Eds.): ICCSA 2014, Part VI, LNCS 8584, pp. 507–517, 2014.

Fig. 1. Disc brake system of a vehicle

The equations of motion arising in the dynamical analysis of structures discretized by the finite element method are of the form

$$M\ddot{u}(t) + C\dot{u}(t) + Ku(t) = f(t)$$

where M is the mass matrix, C is the viscous damping matrix (energy loss mechanism), K is the stiffness matrix (resistance to displacement), $f(t)$ is a time dependent external force vector and $u(t)$ the displacement vector also time dependent.

The general solution of the homogeneous equation is an important preliminary to study the stability of the system and this solution is given in terms of the solution of a quadratic eigenvalue problem

$$\left(-\lambda^2 M + \lambda C + K\right)x = 0 . \tag{1}$$

Due to friction, the stiffness matrix has specific properties [3] and is defined as

$$K = K_S + \mu K_F$$

where K_S is the structural stiffness matrix (symmetric), K_F the asymmetrical friction induced stiffness matrix and μ the friction coefficient.

The corresponding quadratic eigenvalue problem (in the following designated by QEP) is

$$\left(-\lambda^2 M + \lambda C + \left(K_S + \mu K_F\right)\right)x = 0 . \tag{2}$$

2   Eigenmethods for QEP

Mathematically, the usual procedure for this kind of problems is the linearization ([2],[6],[11]), that is, a transformation of problem (1) into a linear eigenvalue problem with the same eigenvalues and twice the dimension:

$$(A - \lambda B)z = 0 \tag{3}$$

where

$$A = \begin{bmatrix} 0 & I \\ K & C \end{bmatrix} \quad , \quad B = \begin{bmatrix} I & 0 \\ 0 & M \end{bmatrix} \quad \text{and} \quad z = \begin{bmatrix} x \\ \lambda x \end{bmatrix}.$$

In this context it is necessary to study efficient eigensolvers for unsymmetric quadratic problems and its implementations, to take into account the fact that, the damping matrix and part of the rigidity matrix being unsymmetric, the eigenvalues and eigenvectors are in general complex.

We will consider the following options to solve this unsymmetric quadratic eigenvalue problem:

Option 1: linearization of the quadratic equation followed by the QZ method [5] applied to the generalized eigenvalue problem thus obtained in Eq.(3) (in the following designated by GEP).

Option 2: application of a technique similar to the one used in systems modeled by the finite elements method by the software ABAQUS [1]. It consists in solving first the symmetric problem where the damping matrix is ignored as well as the unsymmetric part of K, μK_F, by the Lanczos method [4] or another suitable for symmetric problems, to get a projecting subspace spanned by these eigenvectors. Then the whole problem is projected onto this subspace and the resulting problem is solved by QZ, or, as this is approximately a standard eigenproblem, by QR method [5].

Option 3: solution of the initial quadratic problem directly by the Matlab routine "*polyeig*" ([2], [7]).

3   Implementation Details

3.1   Moderate Size Problems

The options described before were programmed in Matlab, using the Matlab routine "*eig*" to perform QZ method for option 1 and for problems of moderate size.

In option 2, to solve the symmetric eigenvalue problem

$$\left(-\lambda^2 M + K_S \right)\phi = 0 \tag{4}$$

Lanczos method could be used (as in [1]) or a symmetric variant of QZ method (equivalent to QR method applied to the matrix $M^{-1}K_S$ when M is non-singular) as we did in the numerical example in Section 4, by giving the Matlab routine "*eig*" matrices K_S and M, and the option "*chol*". After, the basis ϕ of the invariant subspace of problem (4) is to be normalized thus obtaining a new basis ψ:

$$\psi = \frac{\phi}{\sqrt{\phi^T M \phi}} \ .$$

The projection of the initial QEP onto the subspace spanned by ψ yields a simpler quadratic eigenvalue problem:

$$\left| \begin{array}{l} \psi^T M \psi = I \\ \psi^T C \psi = \hat{C} \\ \psi^T K_s \psi = D \quad \left(= diag\left(\lambda_1^2, \ldots, \lambda_n^2\right)\right) \\ \psi^T K_F \psi = E_F \end{array} \right. \quad \Rightarrow \quad \left(-\lambda^2 I + \lambda \hat{C} + \left(D + \mu E_F\right)\right)\hat{\phi} = 0$$

The linearization of this equation

$$\left(-\lambda^2 I + \lambda \hat{C} + \left(D + \mu E_F\right)\right)\hat{\phi} = 0 \quad \Leftrightarrow \quad \begin{cases} \hat{u} = \lambda \hat{\phi} \\ -\lambda \hat{u} + \hat{C}\hat{u} + \left(D + \mu E_F\right)\hat{\phi} = 0 \end{cases} \quad \Leftrightarrow$$

$$\Leftrightarrow \quad \left(\underbrace{\begin{bmatrix} 0 & I \\ D + \mu E_F & \hat{C} \end{bmatrix}}_{A\_2} - \lambda \underbrace{\begin{bmatrix} I & 0 \\ 0 & I \end{bmatrix}}_{B\_2} \right) \cdot \begin{bmatrix} \hat{\phi} \\ \hat{u} \end{bmatrix} = 0 \quad \Leftrightarrow \quad (A\_2 - \lambda B\_2) \cdot \begin{bmatrix} \hat{\phi} \\ \hat{u} \end{bmatrix} = 0$$

(5)

yields a problem that can be solved by the Matlab routine "*eig*" corresponding to the QR method with matrix A_2, since it is a standard eigenvalue problem (in the following designated by SEP).

However, in practice B_2 may not be the identity, because its computation is affected by the quality of the orthogonality of the computed matrix ϕ and thus of ψ. In that case a GEP must be solved by QZ method, with the Matlab routine "*eig*" with matrices A_2 and B_2.

3.2   Very Large Problems

When the matrices are very large and, thus having a lot of eigenvalues, we can use iterative methods, like Lanczos or Arnoldi (if we have a symmetric or a nonsymmetric problem, respectively) [10] to approximate only some of the smallest in magnitude eigenvalues. In most applications it is not necessary to compute all of them.

So, and because our problem is a nonsymmetric one, we applied Arnoldi method to the linearized problem in option 1 (Eq. (3)) and in the second stage of option 2 (Eq. (5)), instead the QZ or the QR methods using the Matlab routine "*eigs*". For both options 1 and 2, we started by storing the matrices of the linearization in sparse structure. To have consisting measuring times, we use always the same initial vector.

When asking for the smallest eigenvalues in magnitude, the *"eigs"* Matlab routine, that uses the corresponding Arpack routine, may choose a shift and invert mode with shift 0 [8]. In this case, the largest eigenvalues in magnitude v of $A^{-1}B$ are computed and the eigenvalues of the initial problem are $1/v$. This computation is more stable, but requires a factorization of A which may take a long time and a large amount of storage. Alternatively, we can give the routine another inverse mode by noticing that problem (1) is equivalent to

$$\left(M - vC - v^2 K \right) x = 0 \quad , \quad v \text{ being } 1/\lambda \tag{6}$$

and the largest values of V correspond to the smallest values of $\lambda = 1/v$.

Linearizing this inverse problem we obtain

$$\left(E - vF \right) z = 0 \tag{7}$$

with

$$E = \begin{bmatrix} 0 & I \\ M & 0 \end{bmatrix} \quad , \quad F = \begin{bmatrix} I & 0 \\ C & K \end{bmatrix} \text{ and } z = \begin{bmatrix} x \\ vx \end{bmatrix}.$$

We may save time by giving to the *"eigs"* routine directly matrices E and F (instead of A and B) and asking for the largest eigenvalues in magnitude (and corresponding eigenvectors).

4   Numerical Results

In [9] we showed two examples with random matrices but with a special structure since M had 25 blocks (or 50) of dimension 20×20, symmetric positive definite, with known eigenvalues. K and C had the corresponding partition into blocks of random elements. Here, for Example 1, we chose matrices with no particular structure but M is still symmetric positive definite.

4.1   Example 1

To perform some simulations we build four matrices $(M, C, K_F$ and $K_S)$ of dimension 500 with no particular structure, but respecting the main features of the matrices involved in mechanical engineering problems, that is, M is a symmetric positive definite matrix. In this example, C and K_F are random unsymmetric matrices with values between 0 and 1, K_S a random symmetric matrix with values between 0 and 1 and we considered $\mu = 1$. The dimension of the linearized problem becomes 1000, as it was explained above.

We present here comparative results of options 1, 2 and 3 using this test problem. For that purpose, the quadratic equation (2) with option 1, with option 2 with QZ, with option 2 with QR and with option 3 was solved.

As we can see in Table 1, option 1 is more efficient than option 2 with QZ. However, option 2 with QR becomes better in terms of computing time, but is not so safe in accuracy.

Option 3 is the worse in terms of time consumed. It becomes very slow, specially, in problems of large dimension, because "*polyeig*" is a black box routine (it does not required expert information from the user, so it must produce specific parameters from the data).

Then, we applied Arnoldi method instead QZ or QR in options 1 and 2, considering the GEP obtained from linearization and computing only the ten smallest eigenvalues.

We can observe in Table 1 that there is a great reduction in computation time when using the Arnoldi Method instead of QZ in options 1 and 2, but of course, in this case we are only computing ten eigenvalues and eigenvectors.

Table 1. Computation time, in seconds, of option 1, option 2 with QZ and option 3, calculating all the eigenvalues, and options 1 and 2 with Arnoldi, calculating only the ten smallest eigenvalues

Test Problem (n=500)	CPU Times in seconds
Option 1	17.21
Option 2 with QZ	20.08
Option 2 with QR	6.88
Option 3	27.89
Option 1 with Arnoldi	1.12
Option 2 with Arnoldi	3.53

As we do not know the exact solution of this test problem and in order to access the accuracy of the several options, we computed the following estimates of the relative error (REE):

- REE1 is the norm of the eigenvalues of option 1 minus the eigenvalues of option 2 with QZ divided by the norm of the eigenvalues of option 1;
- REE2 is the norm of the eigenvalues of option 1 minus the eigenvalues of option 2 with QR divided by the norm of the eigenvalues of option 1;
- REE3 is the norm of the eigenvalues of option 1 minus the eigenvalues of option 3 divided by the norm of the eigenvalues of option 1;
- REE4 is the norm of the eigenvalues of option 2 with QZ minus the eigenvalues of option 2 with QR divided by the norm of the eigenvalues of option 2 with QZ;
- REE5 is the norm of the eigenvalues of option 2 with QZ minus the eigenvalues of option 3 divided by the norm of the eigenvalues of option 2 with QZ;
- REE6 is the norm of the eigenvalues of option 2 with QR minus the eigenvalues of option 3 divided by the norm of the eigenvalues of option 2 with QR.

We use the inf_norm to have an idea of the worst case and report the corresponding values in Table 2.

Table 2. Relative accuracy of the eigenvalues obtained with option 1, option 2 with QZ, option 2 with QR and option 3

REE1	REE2	REE3	REE4	REE5	REE6
2.7×10^{-15}	9.6×10^{-14}	6.8×10^{-14}	9.9×10^{-14}	6.9×10^{-14}	9.9×10^{-14}

This problem is very sensitive to the precision of the orthogonality of the eigenvector basis. The inf_norm of the difference between the matrix B_2 and the identity is of the order of 6.7×10^{-12} as well as the absolute error between the eigenvalues obtained with B_2 or the identity. However, the relative error estimated in that case is of the order of 9.9×10^{-14} as we can see in Table 2. This explains the need of the investigation that we report here, of using QZ in option 2.

If M has a large condition number, versions of the routine "*eig*" with the option "*qz*" are better than those with the option "*chol*". In option 1, if the condition numbers of matrices K and M are large, then the QEP is almost an ill posed problem and in that case, the computed solution might be inaccurate.

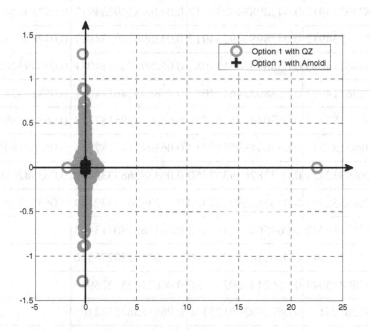

Fig. 2. All eigenvalues computed by option 1 with QZ and the ten smallest eigenvalues computed by option 1 with Arnoldi, with matrices of dimension $n = 500$

In Fig.2 we can see the representation of all 1000 eigenvalues obtained by dense versions (with QR and QZ), and the ten smallest eigenvalues obtained by the sparse version (with Arnoldi). Arnoldi method chooses, in fact, the ten smallest eigenvalues in magnitude (see Table 3).

Table 3 shows the absolute values in ascending order of 25 of the 1000 eigenvalues existing and the smallest 10, 15 and 20 eigenvalues obtained by application of Arnoldi method, to check that Arnoldi is really getting the required eigenvalues.

Table 3. Representation of the absolute values of 25 of the 1000 different eigenvalues existing in the test problem, computed by QZ method (dense version), and the 10, 15 and 20 smallest eigenvalues computed by Arnoldi method (sparse version),all of them in ascending order with four decimal places

Option 1 with QZ (25)	Option 1 with Arnoldi (20)	Option 1 with Arnoldi (15)	Option 1 with Arnoldi (10)
0.023571289145744	0.023571289145747	0.023571289145747	0.023571289145747
0.023571289145744	0.023571289145747	0.023571289145747	0.023571289145747
0.042898349234912	0.042898349234917	0.042898349234917	0.042898349234917
0.042898349234912	0.042898349234917	0.042898349234917	0.042898349234917
0.058893246114083	0.058893246114082	0.058893246114081	0.058893246114082
0.058893246114083	0.058893246114082	0.058893246114081	0.058893246114082
0.063044895187714	0.063044895187713	0.063044895187713	0.063044895187714
0.063044895187714	0.063044895187713	0.063044895187713	0.063044895187714
0.063382689332451	0.063382689332450	0.063382689332449	0.063382689332450
0.063382689332451	0.063382689332450	0.063382689332449	0.063382689332450
0.064568580413447	0.064568580413442	0.064568580413443	
0.066214250275941	0.066214250275940	0.066214250275940	
0.066214250275941	0.066214250275940	0.066214250275940	
0.067379326125173	0.067379326125176	0.069185875332406	
0.069185875332409	0.069185875332409	0.069185875332406	
0.069185875332409	0.069185875332409		

Table 3. *(continued)*

0.072070277845712	0.072070277845711		
0.072070277845712	0.072070277845711		
0.072890551233376	0.072890551233375		
0.072890551233376	0.072890551233375		
0.073269711374474			
0.073269711374474			
0.074564225773052			
0.074564225773052			
0.076542713172203			

4.2   Example 2

Example 2 uses matrices coming from an Abaqus model of a disc analogous to that referred in Section 1 but without pads, so only matrices M and K are obtained by this model. Matrix C is then chosen as $C = 10^{-4} K$. The mesh generated by the finite element method has 1160 of second order elements and 8383 nodes. So, in this example $n=25149$ and thus the dimension of the linearized problem becomes $2n=50298$. Matrix M is a sparse matrix with 984669 nonzero elements (34MB) and matrices K and C are sparse matrices too and have both 2953867 nonzero elements (102MB).

As the dimensions involved are huge we only computed the k smallest eigenvalues, by option 1, inverse formulation in Eq.(7) with Arnoldi method.

The computing times for computing k eigenvalues and eigenvectors are in Table 4 and the ten smallest eigenvalues are represented in Fig.3.

Table 4. Computing times for computing k eigenvalues and eigenvectors

Number of required eigenpairs: k	10	20	30
CPU Times in seconds	15.15	28.94	38.89

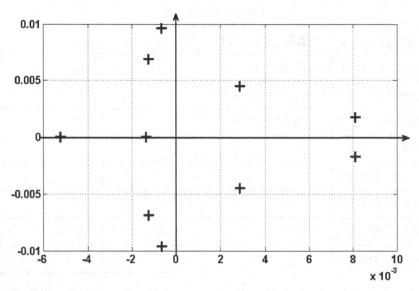

Fig. 3. The $k=10$ smallest eigenvalues computed by option 1 with Arnoldi

5   Conclusions

We can say that Arnoldi method and software packages like Arpack or the Matlab routine *"eigs"* allows us to deal with large dimensional eigenproblems with various computing options very quickly. The quadratic eigenvalue problem must be linearized thus yielding a problem with twice the size of the initial one, and so the possibility of using these packages is very important.

The implementation of Option 2 arises in the hope that the projected problem would not be a generalized problem. But, in practice, we cannot count on it, that is, in practice the projected problem can be a generalized problem. Therefore, QZ method or a conjunction with an iterative method, like Arnoldi, must be applied.

In very large dimension problems the routine *"polyeig"* has no interest in terms of computation time.

The initial problem (1) asking the smallest eigenvalues is equivalent to the inverse problem (6) asking the largest eigenvalues and the latter is better conditioned (Arpack and the routine *"eigs"* do that).

The treatment of these brake systems is difficult due to their large dimension, so the study of the influence of the damping matrix must be done with smaller models asking for a few eigenvalues in order to understand the Physics of the computed values.

Acknowledgments. Financial support provided by the European Regional Development Fund through the programme COMPETE and by the Portuguese Government through the FCT – Fundação para a Ciência e a Tecnologia under the project PEst-C/MAT/UI0144/2011.

The authors thank Jorge Justo and Fernando Ferreira from School of Engineering, Polytechnic of Porto for creating the matrices M and K of the model in Example 2.

References

1. ABAQUS – Brake Squeal Web Seminar, Updated Capabilities in Brake Squeal (February 2006), http://www.abaqus.com/BrakeSqueal/
2. Tisseur, F., Meerbergen, K.: The Quadratic Eigenvalue Problem. SIAM 43(2), 235–286 (2001)
3. Fritz, G., Sinou, J.J., Duffal, J.M., Jézéquel, L.: Investigation of the Relationship Between Damping and Mode-Coupling Patterns in case of Brake Squeal. Journal of Sound and Vibration 307(3-5), 591–609 (2007)
4. Golub, G., Van Loan, C.: Matrix Computations, 3rd edn. The Johns Hopkins University Press, Baltimore (1996)
5. Stewart, G.W.: Introduction to Matrix Computations. Academic Press, New York (1973)
6. Gohberg, I., Lancaster, P., Rodman, L.: Matrix polynomials. Academic Press, New York (1982)
7. MATLAB (General Release Notes for R2012a), The MathWorks (2012)
8. Lehoucq, R.B., Sorensen, D.C., Yang, C.: ARPACK Users' Guide: Solution of Large Scale Eigenvalue Problems with Implicitly Restarted Arnoldi Methods. SIAM, Philadelphia (1998)
9. Aires, S.M., Almeida, F.D.: Comparison of Eigensolvers Efficiency in Quadratic Eigenvalue Problems. In: Vigo-Aguiar, J. (ed.) Proceedings of the 2012 International Conference on Computational and Mathematical Methods in Science and Engineering, Murcia, Spain, vol. IV, pp. 1279–1283 (2012)
10. Saad, Y.: Numerical Methods for Large Eigenvalue Problems, 2nd edn. SIAM (2011)
11. Lancaster, P.: Lambda-Matrices and Vibrating Systems. Pergamon Press, Oxford (1966)
12. Bajer, A., Belsky, V., Zeng, L.: Combining a Nonlinear Static Analysis and Complex Eigenvalue Extraction in Brake Squeal Simulation. SAE Technical Paper 2003-01-3349 (2003), doi:10.4271/2003-01-3349

Modelling Wave Refraction Pattern Using AIRSAR And POLSAR C-Band Data

Maged Marghany

Institute of Geospatial Science and Technology (INSTeG),
Universiti Teknologi Malaysia,
81310 UTM, Skudai, Johor Bahru, Malaysia
maged@utm.my, magedupm@hotmail.com,

Abstract. This study has demonstrated new approach for simulation of wave refraction pattern in airborne radar data. In doing so, the quasi-linear algorithm used to model significant wave height based on new approach of azimuth cut-off algorithm. The study shows that wave refraction pattern can simulate from AIRSAR and POLSAR data with convergence and divergence spectra energy of 0.84 and 0.4 m^2 sec, respectively. In conclusion, modification of conventional azimuth cut-off algorithm can be used to retrieve significant wave height in C_{vv}- band data under circumstance of wave transformation using first order Partial Differential Equation (PDEs).

Keywords: Quasi-Linear model, AIRSAR, POLSAR, Wave refraction, quasi-linear model, first order Partial Differential Equation (PDEs).

1   Introduction

Synthetic Aperture Radar (SAR) has been recognized as a powerful tool for modeling ocean waves and forecasting over an area of 300 km x 300 km. According to Schuler et al., [16] the quantitative descriptors of the ocean surface wave properties such as wave height and wave slope spectra can retrieve from several satellite SAR missions, such as ERS1/2, RADARSAT1/2 and TerraSAR-X(TSX)[4].The retrieving techniques of sea surface quantitative descriptors involve linear and nonlinear techniques. In linear technique, Modulation Transfer Function (MTF) used to relate SAR spectra image with real sea surface spectra. Nevertheless, Schuler et al., [16] agreed with Hasselmann and Hasselmann [3]; Vachon et al.[19]; and Forget and Brochel [1] that the linear techniques produce non accurately quantitative indicators of wave spectra. In contrast, more accurate retrieving of ocean sea surface quantitative spectra descriptors is produced using nonlinear techniques [1],[3],[16],[19],[20]. Currently, several investigations have carried out on the assimilation of SAR wave mode data into wave forecasting models[13],[14],[15]. This is because the SAR image spectrum has turned out to be far removed from the actual wave spectrum and rather complicated post-processing is necessary for extracting quantitative wave information. In this regards, previous studies were carried out by Forget and Brochel [1]; Hasselmann and Hasselmann [3]; and Vachon

B. Murgante et al. (Eds.): ICCSA 2014, Part VI, LNCS 8584, pp. 518–531, 2014.

et al. [19] to develop an inversion algorithm to map SAR wave spectra into ocean wave spectra. Hasselmann and Hesselman [3] introduced a non-linear algorithm which was developed by Vachon et al. [20] to model the significant wave height based on the azimuth cut-off. Vachon et al. [20] defined the azimuth cut-off as the degree to which the SAR image spectrum is constrained in the azimuth direction. Maged [12] modified the azimuth cut-off mode to estimate the significant wave height in the coastal water of Malaysia. The azimuth cut-off is affected by the wind and wave condition in a quasi-linear forward-mapping model [11],[20]. This concept has extended to the complex SAR data and related cross spectra [15]. Further, Schulz-Stellenfleth and Lehner [13] developed an empirical algorithm that named by LISE to derive information of individual wave properties from SAR data. In addition, LISE algorithm is based on the inversion of the quasi-linear model. This model, however, is impacted by the nonlinearities because of wave travelling in azimuth direction. Recently, Schulz-Stellenfleth et al. [14] developed an empirical algorithm CWAVE that based on a quadratic model function to retrieve significant wave height from SAR images. This model does not require a wave model first guess and has the calibrated SAR images as the only source of information[5].

Schuler et al.,[16] developed a new technique based on alpha parameter from the Cloude–Pottier H-A- $\bar{\alpha}$ (Entropy, Anisotropy, and(averaged) Alpha) polarimetric scattering decomposition theorem to measure wave slopes in range direction. Therefore, they used a conventional azimuth cut-off formula to measure the wave spectra slopes in azimuth direction. Recently, Li et al., [4] used TerraSAR-X(TSX) to investigate ocean wave refraction over the coast of Terceira island situated in the North Atlantic with a comparison of X-band marine radar and results of the WAve prediction Model(WAM). Further, they estimated near shore significant wave height from TSX data with the correlation of wave refraction law and the developed XWAVE empirical algorithm. Therefore, they found that The comparison results indicate that the retrieved results using the XWAVE algorithm are reasonable while needing further improvement, including the different sea state. Finally, they concluded that the integration between TerraSAR-X(TSX) and X-band marine radar can use as good tool to observe coastal wave refraction spatial variation.

The question can be raised as to how an integration of quasi-linear algorithm with a first order Partial Differential Equation (PDEs) could use to develop a new approach to observe coastal wave refraction from AIRSAR and POLSAR C_{vv}-band data. The main objective is to modify the conventional azimuth algorithm to model significant wave height in coastal water of Malaysia. The sub-objectives are :(i) to model the physical properties of wave spectra such as wavelength, significant wave height and spectra energy in azimuith direction using C_{vv}-band data; (ii) to simulate wave refection pattern using first order Partial Differential Equation (PDEs); and to model spectra energy variation of wave refraction pattern.

2   Methodology

2.1   Study Area

The study area is located along the coast of Kuala Terengganu in the eastern part of Peninsular Malaysia. The area is approximately 14 km north of Kuala Terengganu coastline, located in the South China Sea between 5° 21' N to 5° 27' N and 103° 10' E to 103° 15' E. Sand materials make up the entire of the shoreline[6],[7],[8], [9],[10],[11],[12]. This area lies in an equatorial region dominated by two seasonal monsoons. The southwest monsoon lasts from May to September while the northeast monsoon lasts from November to March [9]. The monsoon winds affect the direction and magnitude of the waves. Strong waves are prevalent during the northeast monsoon when the prevailing wave direction is from the north (November to March), while during the southwest monsoon (May to September), the wave directions are propagating from the south [10]. According to Maged [11] and Zelina et al. [21] the maximum wave height during the northeast monsoon season is 4 m. The minimum wave height is found during the southwest monsoon which is less than 1 m.

2.2   Input Data

2.2.1 In Situ Wave Collection

The sea wave truth data were collected by wave rider buoy from the Malaysian Meteorological service between latitudes of 5° 18 ' N and 5° 26 ' N and longitude of 103° 32' E 103° 40' E on 6 December 1996 and 19 September 2000 (during that time, the airborne AIRSAR and POLSAR were flown over the study area, respectively). The in situ observation data included wave height and wave direction which were used for wave spectra modulation with AIRSAR and POLSAR data. The wind data were collected at the Meteorological Station at Sultan Mahmed Airport, Kuala Terengganu, at latitudes of 5° 23' N and longitude of 103° 06 'E and obtained by the Malaysia Meteorological Service in Kuala Terengganu (Fig. 1). Wind speed data were used to determine the azimuth cut-off modeled in AIRSAR data. The azimuth cut-off was used to model the significant wave height from AIRSAR SAR data which

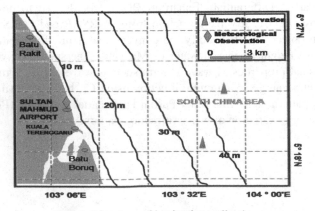

Fig. 1. Locations of in situ data collections

was based on the least squares fit algorithm. A least squares algorithm was applied between azimuth cut-off wavelength and geophysical parameters such as significant wave height [12],[19],[20].

2.3   Wave Spectra Model

According to Pierson and Moskowitz [17] the significant wave height spectrum H_s can be related to the one sided directional wave number spectra density $S(\vec{K},\phi)$ by the following formula

$$H_s = 4[\int\int S(\vec{k},\phi)dk_i dk_j]^{0.5} .\tag{1}$$

where k_i and k_j are the wave numbers in the azimuth and range directions, respectively, and $\vec{K}$ is the wave-number magnitude of k_i and k_j in the azimuth and the range directions, respectively [9],[18]. The ocean wave spectrum $S(\vec{K},\phi)$ was obtained by input of in situ wave parameters (dominant wave number, wave propagation direction ϕ and significant wave height H_s. The actual ocean wave spectrum density $S(\vec{K},\phi)$ can be calculated from the following equation [9],[17]

$$S(\vec{K},\phi) = \frac{\sqrt{S(k_i,k_j)d\vec{K}} \ e^{\left[-2(5.88U^{-2}/\vec{K}\cos\phi)^2\right]}}{\left[\vec{K}^2 - (0.5gU^{-2})\vec{K}\cos\phi + 2.4U^{-2}\right]^2} .\tag{2}$$

where U is the wind speed and ϕ is the wave direction. The actual wave spectra contours have been determined by using equation 2. Furthermore, equation 2 has been used with quasi-linear model to map the AIRSAR and POLSAR wave spectra onto ocean wave spectra.

2.4   AIRSAR and POLSAR Wave Spectra

In this study, a single AIRSAR and POLSAR images frame comprising of 512 x 512 image pixels was extracted from AIRSAR and POLSAR C_{vv}-band (Fig. 2). Each pixel represents a 10 m x 10 m area. The entire image frame of AIRSAR/POLSAR data corresponded to a 2 km x 2 km patch on the ocean surface. AIRSAR and POLSAR images are a two dimensional sampling of the ocean wave field and thus a two-dimensional (2-D) Fourier transfer has to be utilized [9],[18]. When the Fourier transfer was selected, the output domain is the two-dimensional frequency spectrum of the input image [18].

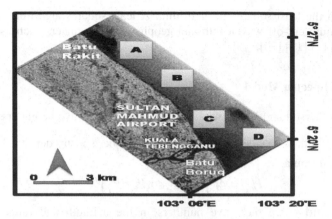

Fig. 2. Location of wave spectra window selections

2.4.1 Quasi-linear Transform and Significant Wave Height

To map observed SAR spectra into the ocean wave spectra, a quasi-linear model was applied. The simplified quasi-linear theory is explained below: according to the Gaussian linear theory, the relation between ocean wave spectra $S(\vec{K},\phi)$ and AIRSAR image spectra $S_Q(\vec{K})$ could be described by tilt and hydrodynamic modulation (real aperture radar (RAR) modulation) [3]. The tilt modulation is linear to the local surface slope in the range direction i.e. in the plane of radar illumination. The tilt modulation in general is a function of wind stress and wind direction for ocean waves and AIRSAR/POLSAR polarization. According to Vachon et al., [19],[20] the tilt modulation is the largest for HH polarization. In addition, hydrodynamic interaction between the scattering waves (ripples) and longer gravity waves produced a concentration of the scatterer on the up wind face of the swell [1],[3]. Following Maged [10] AIRSAR image spectra can map into ocean wave spectra under the assumption of the quasi-linear modulation transfer function $S_Q(\vec{k})$ which is given by

$$S_Q(\vec{K}) = R(K)H(k_i;K_c)\left[\frac{S(\vec{K},\phi)}{2}\left|T_{\text{lin}}(k_{i,j})\right|^2 + \frac{S(-\vec{K},\phi)}{2}\left|T_{\text{lin}}(-k_{i,j})\right|^2\right]. \tag{3}$$

where $H(k_i;K_c)$ is an azimuth cut-off function that depends upon azimuth wave number k_i and range wave number k_j, the cut-off azimuth wave number K_c and R ($\vec{K}$) is the AIRSAR point spread function. The AIRSAR/POLSAR point spread function is a function of the azimuth and the range resolutions [10]. According to Vachon et al., [19], T_{lin} is a linear modulation transfer function which is composed of the RAR (the tilt modulation and hydrodynamic modulation), and the velocity bunching modulation. The RAR modulation transfer function (RAR MTF) is the coherent sum of the transfer function associated with each of these terms, i.e.

$$T_{lin}(\vec{K}) = M_t(\vec{K}) + M_d(\vec{K}) + M_v(\vec{K}). \tag{4}$$

The tilt modulation $M_t(k)$ can be described by Hasselmann and Hasselmann [3] as follow

$$M_t(\vec{K}) = k_j \frac{4\cot\theta}{1 \pm \sin^2\theta} e^{(\frac{i\pi}{2})}. \tag{5}$$

where k_j is the range wave number and θ is the local incident angle of the radar beam. Following Vachon et al., [20], the hydrodynamic modulation transfer function can be given by

$$M_d(\vec{K}) = 4.5\omega K \frac{\omega - i\mu}{\omega + \mu^2} \sin^2\phi. \tag{6}$$

where $\vec{K}$ is long wave number, i is $\sqrt{-1}$, ω is the angular frequency of the long waves, ϕ is the azimuth angle and μ is the relaxation rate of the Bragg waves which is 0.5/s. According to Hasselmann and Hasselmann [3] and Vachon et al. [19,20] velocity bunching can contribute to equation 4 (linear MTF) based on the following equation which was given by Vachon et al.,[20]

$$M_v(\vec{K}) = \frac{R}{V} \omega \left[\frac{k_i}{K} \sin\theta + i\cos\theta \right]. \tag{7}$$

where R/V is the scene range to platform velocity ration which is approximately 32 s in the case of AIRSAR data [10].

In order to estimate the significant wave height from the quasi-linear transform, we adopted the algorithm that was given by Maged [12] to be appropriate for the geophysical conditions of tropical coastal waters:

$$\lambda_c = \beta \left(\int_{H_{s0}}^{H_{sn}} \sqrt{H_s}\,dH_s + \int_{U_0}^{U_n} \sqrt{U}\,dU \right). \tag{8}$$

where λ_c is cut-off azimuth wavelength, H_s and U are the in situ data of significant wave height and wind speed along the coastal waters of Kuala Terengganu, Malaysia. The measured wind speed was estimated at 10 m height above the sea surface. The changes of significant wave height and wind speed along the azimuth direction are replaced by dH_s and dU, respectively. The subscript zero refers to the average in situ wave data collected before flight pass over by two hours while the subscripts n refers to the average of in situ wave data during flight pass over the study area. β is an empirical value which results of R/V multiplied by the intercept of azimuth cut-off (c) when the significant wave height and the wind speed equal zero. A least squares fit was used to find the correlation coefficient between cut-off wavelength and the one calculated directly from the AIRSAR/POLSAR spectra image by equation (8). Then, the following equation was adopted by Maged [9],[12] to estimate the significant wave height (H_{sT}) from the AIRSAR images

$$H_{sT} = \beta^{-2} \int_{\lambda_{c0}}^{\lambda_{cn}} (\lambda_c)^2 d\lambda_c.$$ (9)

where β is the value of $\left(c\, \dfrac{R}{V} \right)^{-1}$ and H_{sT} is the significant wave height simulated

from AIRSAR images. The introduced method (azimuthally cut off) is designed for homogeneous wave fields as waves can be found over the open ocean under deep water condition with homogeneous bathymetry as can be seen in Fig. 1. A linear wave transform model can be used to solve the problem of homogeneous wave fields by simulating the physical wave parameters nearshore.

2.5   Wave Refraction Graphical Method

The wave refraction model over the AIRSAR and POLSAR images is formulated on the basis of wave number and wave energy conversation principle, gentle bathymetry slope, steady wave conditions and only depth refractive (Fig. 3). According to Herbers et al. [2] wave refraction equation takes the following form:

$$\frac{\partial}{\partial x}(H_s^2 c_g \cos\phi) + \frac{\partial}{\partial y}(H_s^2 c_g \sin\phi) = 0.$$ (10)

where the coordinates and the wave angle ϕ are orientated according to the notation of Fig. (3). Equation 10 is a first order Partial Differential Equation (PDEs) in the unknown variables $\phi(x, y)$ and $H_s^2(x, y)$; the group velocity c_g is a known function of the wave period T and the known local depth $h(x,y)$. Following, Herbers et al [2], the notation of Figure 3, the explicit finite difference scheme, centred in x, proposed for the solution of equation 10 takes the form:

(1) Wave angle equation: solved for ϕ_{ij+1}

$$\phi_{ij+1} = \arccos \left[\begin{array}{l} (\dfrac{\Delta y_j}{2}((\dfrac{\sin\phi_{i+1j}}{c_{i+1j}} - \dfrac{\sin\phi_{ij}}{c_{ij}})\Delta x^{-1} + \\[2ex] (\dfrac{\sin\phi_{ij}}{c_{ij}} - \dfrac{\sin\phi_{i-1j}}{c_{i-1j}})\Delta x^{-1}_{i-1}) + \dfrac{\cos\phi_{ij}}{c_{ij}}).c_{ij+1} \end{array} \right].$$ (11)

(2) Significant wave height equation: solved for Hs_{ij+1}

$$H_{s_{ij+1}}^{2} = \frac{1}{c_{gij+1} \sin \phi_{ij+1}} \left[\begin{array}{c} (H^2{}_s c_g \sin \phi)_{ij} - 0.5\Delta y_j (\dfrac{(H^2{}_s c_g \cos \phi)_{i+1j}}{\Delta x_i} - \dfrac{(H^2{}_s c_g \cos \phi)_{ij}}{\Delta x_i} \\[2mm] + \dfrac{(H^2{}_s c_g \cos \phi)_{ij} - H^2{}_s c_g \cos \phi)_{i-1j}}{\Delta x_i - 1}) \end{array} \right]. \quad (14)$$

The boundary conditions completing the model are:

(i)   It is assumed that the parallel depth contours as shown in Fig. 3

(ii)   The ϕ and H_s values are given as initial conditions on the open sea boundary $(j = 1)$.

(iii)   The computation is terminated on the coastal boundaries $(h = 0)$. The wave breaking criterion is applied in shallow waters. The computed significant wave height H_s is compared to $0.78 \ h_{ij}$; if

$$Hs_{ij+1} \rangle 0.78h_{ij}, \text{ then } Hs_{ij+1} = 0.78h_{ij} \quad (15)$$

The spectra energy of significant wave height distribution due to wave refraction is then estimated by using the following formula adapted from Hasselmann and Hasselmann [3]:

$$E(\vec{K}, H_S) - S(k_x, k_y)p(H_S). \quad (16)$$

where $S(k_x, k_y)$ is the distribution for the wave number and $p(H_S)$ is the probability distribution of the significant wave height in the convergence and divergence zone. According to Herbers et al [2],the refraction index (K_r) for a straight coastline with parallel contours can be estimated by using the following equation:

$$K_r = \sqrt{\frac{\cos \theta_d}{\cos \theta_r}} . \quad (17)$$

where θ_d and θ_r are the deep and shallow waves incidence angles

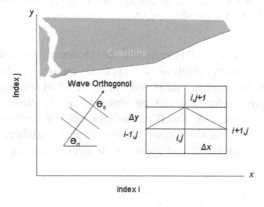

Fig. 3. : Method for wave refraction diagram

3   Results and Discussion

The wave spectral information have been extracted from the average of four sub-images, and each sub-image was 512 by 512 pixels. The average sub-images spectral information were used with the quasi-linear model.   In ease of description, the AIRSAR and POLSAR imagery wavelength spectra are illustrated in polar plots by few circles. These circular areas indicate the wavelength spectra variation. The wavelength values decrease from the outer to inner circle. The distance of the peaks from the center is inversely proportional to its wavelength.   In addition, the angular position of the peaks indicates the wave propagation direction, (Figs. 4 and 5). Indeed, the two-dimensional Fourier transform convert the SAR data   to the domain frequency. The low frequency should be found in the original point and increase towards the outer circles [9]. The domain frequency is inversely proportional to the wavelength. The largest wavelength is found in outer circles and decreased to inner circles (Figs. 4 and 5).   This confirms the study of Maged [9].

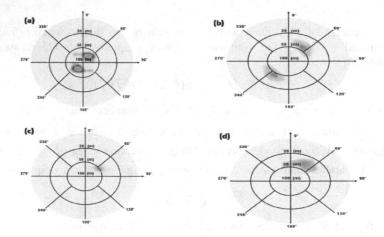

Fig. 4. Average AIRSAR wave spectra (a) Areas A and B, (b) Areas C and D, (c) simulated in situ offshore wave and (d) in situ onshore wave

Close inspection of spectra plots indicated the dominant wavelengths of 20,50 and 100 m, which have been resolved by spectral analysis. The largest average wavelength occurred in areas A, and B with value of 75 m compared to areas C and D. Note that the wave refraction can be observed as the wave directions were a change dramatically in areas A, and B as the approach the shallow water in areas C and D (Figs. 4 and 5). It could further be seen that the offshore wave spectra are traveling with approximately 50° and approach the nearshore zone with incident angle of about 45°.

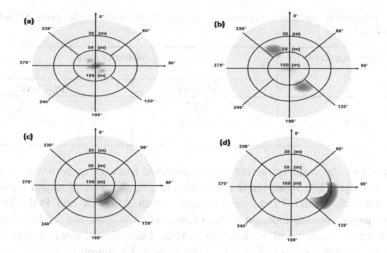

Fig. 5. Average POLSAR wave spectra (a) Areas A and B, (b) Areas C and D, (c) simulated in situ offshore wave and (d) in situ onshore wave

The retrieval wave spectra using the quasi-linear algorithm is in good agreement with in situ measurements (Figs. 4c and 4d). Both AIRSAR and POLSAR data are presented different retrieval wave refraction patterns (Fig. 5). The wavelength was retrieved from POLSAR data is about 80 m and propagated with about 120°. It is interesting to find that the quasi-linear model can retrieve decreasing wavelength of 25 m with direction change to be roughly 140° in the shallow water areas of A and B (Fig. 5b). This means that the retrieval wave refraction pattern from both airborne radar data is agreed with the simulated one using in situ wave measurements. Figure 5d, nevertheless, does not agree with Fig. 5b where the sharp wave refraction pattern is presented in comparison to the quasi-linear model for areas A and B.

Generally, the quasi-linear model predicts only the wave spectra peak which rotates towards the azimuth direction. This means that the quasi-linear model corrects the distortion which had occurred due to the Doppler shift frequency on SAR image. It was noticed that there is a similar pattern of peak locations between observed wave spectra (in situ measurements and AIRSAR and POLSAR spectra) and the retrieval spectra using the quasi-linear model as a result of low R/V i.e. 32s. The retrieval spectra, being characterized by relatively small values for R/V, were less susceptible to the azimuth cut-off, and provide good spectral fidelity. Thus, airborne imagery of ocean waves may be much less susceptible to imaging nonlinearity. This confirms the studies of Forget et al [1]; Hasselmann and Hasselmann [3]; Maged [10];Vachon et al. [19],[20]. Comparison of the retrieval significant wave heights with in situ measurements indicates a positive linear relationship, as shown in Table 1. This indicates by the r^2 value of 0.63, and the probability (p) is less than 0.05. The retrieval significant wave height is ranged between 0.56 and 1.34 m. This agrees with the studies of Maged [9] and Zelina et al. [21].

Table 1. Regression parameters for retrieval significant wave height from AIRSAR and POLSAR data

Statistical Parameters	Airborne data	
	AIRSAR	POLSAR
r^2	0.63	0.62
p	<0.05	<0.05

Fig. 6 shows the wave refraction pattern modeled from the quasi-linear model and in situ wave data. The input quasi-linear wavelength spectra and in situ wavelength spectra were 80 m and 75 m, respectively. Both AIRSAR and POLSAR wave refraction pattern results indicate the refractive index is 2.60 and 2.54 at the Sultan Mahmed Airport station and the location of Batu Rakit station, respectively, showing convergence of wave energy (Fig. 6). At the Batu Rakit station which is close to the river mouth of Kuala Terengganu, the refractive index values are less than 1.00 suggesting divergence of wave energy. In other locations, the refractive index values are close to 0.99, (Fig. 6b) indicating no change in the concentration of wave energy at the coastline. Although the refractive index values for the quasi-linear model differed with those of the in situ wave spectra refraction, the same trend of wave energy dispersion and concentration occur at the coastline. This means that the wave refraction pattern simulated by using the quasi-linear model is similar to the wave refraction simulated from the in situ wave data. The largest refractive index value is observed at the Sultan Mahmed Airport station. This could be attributed to the slight concave shoreline profile which made the incoming north wave energy converge. This result agrees with Maged [10],[11],[12].

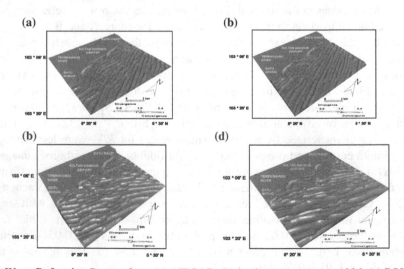

Fig. 6. Wave Refraction Pattern from (a) AIRSAR, (b) in situ measurement 1996, (c) POLSAR and (d) in situ measurement 2000

Fig. 7 shows the wave refraction spectra energy because of convergence and divergence. The convergence spectrum has the sharper peak compared to divergence spectra. The sharp peak of the convergence spectrum is 0.84 m^2 sec (Fig. 7a) while the divergence spectrum peak is less than 0.4 m^2 sec (Fig. 7b). The convergence spectrum peak is located along the azimuth direction. It can be explained that the highest spectra energy propagated close to the azimuth direction is caused by the great influence of the Doppler frequency shift which is produced by convergence. This result agrees with the studies of Maged [12] and Vachon et al., [19],[20].

This study agrees with study of Li et al., [4] that radar satellite or airborne data can use as a good tool to investigate the coastal wave spatial variations. Nevertheless, this result contradicted the study Zelina et al. [21]. Therefore, the largest swell of 250 m occurred in December. Indeed, December is represented the maximum peak of the northeast monsoon season which induces the strongest wind input in the South China Sea 's coastal waters [10],[11],[12]. In December 1996, however, the retrieval wavelength spectra from AIRSAR and POLSAR data is shorter in comparison to the previous study [21].

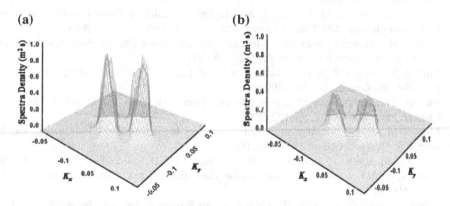

Fig. 7. Wave refraction spectra (a) convergence and (b) divergence

4   Conclusion

This study has demonstrated new approach for simulation of wave refraction pattern in airborne radar data. In doing so, the quasi-linear algorithm used to model significant wave height based on new approach of azimuth cut-off algorithm. The study shows that wave refraction pattern can simulate from AIRSAR and POLSAR data with convergence and divergence spectra energy of 0.84 and 0.4 m^2 sec, respectively. In conclusion, modification of conventional azimuth cut-off algorithm can be used to retrieve significant wave height in C_{vv}- band data under circumstance of wave transformation using first order Partial Differential Equation (PDEs).

References

1. Forget, F., Broche, P., Cuq, F.: Principles of Swell Measurement by SAR with Application to ERS-1 Observations off the Mauritanian Coast. Int. J. Rem. Sen. 16, 2403–2422 (1995)
2. Herbers, T.H., Elgar, C., Guza, R.T.: Directional Spreading of Waves in the Nearshore. J. Geophys. Res. 104, 7683–7693 (1999)
3. Hasselmann, K., Hasselmann, S.: On the Nonlinear Mapping of An Ocean Spectrum and Its Inversion. J. Geophys. Res. 96, 10,713–10,799 (1991)
4. Li, X., Lehner, S., Rosenthal, W.: Investigation of Ocean Surface Wave Refraction Using TerraSAR-X Data. IEEE Tran. Geos. Remote Sens. 48, 830–840 (2010)
5. Li, X.M., Lehnera, S., He, M.X.: Ocean Wave Measurements Based on Satellite Synthetic Aperture Radar (SAR) and Numerical Wave Model (WAM) Data–Extreme Sea State and Cross Sea Analysis. Int. J. Rem. Sen. 29, 6403–6416 (2008)
6. Maged, M.M., Cracknell, A., Hashim, M.: 3-D Visualizations of Coastal Bathymetry by Utilization of Airborne TOPSAR Polarized Data. Int. J. Dig. Ear. 3, 1753–8955 (2010)
7. Maged, M., Cracknell, A., Hashim, M.: 3D Coastal Geomorphology reconstruction Using Differential Synthetic Aperture Radar Interferometry (DInSAR). Int. J. of Com. Sci. and Soft. Tech. 3, 1–4 (2010)
8. Maged, M., Hashim, M., Cracknell, A.: 3D Reconstruction of Coastal Bathymetry from AIRSAR/POLSAR data. Chin. J. Ocean. and Lim. 27(2009), 117–123 (2009)
9. Maged, M.: Velocity Bunching Model for Modelling Wave Spectra along East Coast of Malaysia. J. Ind. Soc. Rem. Sens. 32, 185–198 (2004)
10. Maged, M.: TOPSAR Wave Spectra Model and Coastal Erosion Detection. Int. J. App. Ear. Obs. and Geo. 3, 357–365 (2001)
11. Maged, M.: Operational of Canny Algorithm on SAR Data for Modeling Shoreline Change. Phot. Fer. Geo. 2, 93–102 (2001)
12. Maged, M.: ERS-1 Modulation Transfer Function Impact on Shoreline Change. Int. J. App. Ear. Obs. and Geo. 4, 279–294 (2003)
13. Schulz-Stellenfleth, J., Lehner, S.: Measurement of 2-D Sea Surface Elevation Fields Using Complex Synthetic Aperture Radar Data. IEEE Trans. Geo. and Rem. Sen. 42, 1149–1160 (2004)
14. Schulz-Stellenfleth, J., Koing, T., Lehner, S.: An Empirical Approach for the Retrieval of Integral Ocean Wave Parameters from Synthetic Aperture Radar Data. J. Geo. Res. 112, C03019–C03033 (2007)
15. Schulz-Stellenfleth, J., Lehner, S., Dhoja, D.: A Parametric Scheme for the Retrieval of Two-dimensional Ocean Wave Spectra from Synthetic Aperture Radar Look Cross Spectra. J. Geo. Res. 110, C05004–C05011 (2005)
16. Schuler, D.L., Lee, J.S., Kasilingam, D., Pottier, E.: Measurement of Ocean Surface Slopes and Wave Spectra Using Polarimetric SAR Image Data. Rem. Sen. of Env. 91, 198–211 (2004)
17. Person, W.J., Moskowitz, L.: A Proposed Spectral From Fully Developed Wind Seas Based on the Similarity Theory of S.A. Kitaigorodskii. J. Geo. Res. 69, 5181–5190 (1964)
18. Populus, J., Aristaghes, C., Jonsson, L., Augustin, J.M., Pouliquen, E.: The Use of SPOT Data For Wave Analysis. Rem. Sen. Env. 36, 55–65 (1991)
19. Vachon, P.W., Harold, K.E., Scott, J.: Airborne and Space-borne Synthetic Aperture Radar Observations of Ocean Waves. J. Atm. Oce. 32, 83–112 (1994)

20. Vachon, P.W., Liu, A.K., Jackson, F.C.: Near-shore Wave Evolution Observed by Airborne SAR during SWADE. J. Atm. Oce. 2, 363–381 (1995)
21. Zelina, Z.I., Arshad, A., Lee, S.C., Japar, S., Law, A., Nik Mustapha, R.A., Maged, M.M.: East coast of peninsular Malaysia. In: Sheppard, C. (ed.) Sea at The Millennium: An Environmental Evaluation, Oxford, vol. II, pp. 345–359 (2000)
22. Zebker, H.A.: The TOPSAR Interferometric Radar Topographic Mapping Instrument. IEEE Tran. Geos. Rem. Sen. 30, 933–940 (1992)

Budget Constrained Scheduling Strategies for On-line Workflow Applications

Hamid Arabnejad and Jorge G. Barbosa

LIACC, Departamento de Engenharia Informática
Faculdade de Engenharia, Universidade do Porto
Rua Dr. Roberto Frias, 4200-465 Porto, Portugal
{hamid.arabnejad,jbarbosa}@fe.up.pt

Abstract. To execute scientific applications, described by workflows, whose tasks have different execution requirements, efficient scheduling methods are essential for task matching (machine assignment) and scheduling (ordered for execution) on a variety of machines provided by a heterogeneous computing system. Several algorithms for concurrent workflow scheduling have been proposed, being most of them off-line solutions. Recent research attempted to propose on-line strategies for concurrent workflows but only address fairness in resource sharing among applications while minimizing the execution time. In this paper, we propose a new strategy that extends on-line methods by optimizing execution time constrained to the user budget. Experimental results show a significant improvement of the produced schedules when our strategy is applied.

1 Introduction

With the raising in usage of high-speed internet, grid infrastructures are becoming a novel design structure of distributed computing. The Utility Computing model to deploy those systems has become more popular and widely used to provide a price regulated high performance computing to the users. In the utility model, users are allowed to submit their jobs to different resources, based on the computational cost and jobs deadline. Resources are shared so that the provider can optimize the running costs, but the provider has also to guarantee a high Quality of Service (QoS) by ensuring a fairness access to the shared resources to all users.

Scientific jobs are commonly represented as workflow applications that consist of many tasks, with logical or data dependencies, that can be dispatched to different compute nodes. When scheduling multiple independent workflows that represent user jobs and are thus submitted at different moments in time, a dynamic behaviour is required to redistribute the workload. Most concurrent workflow scheduling algorithms proposed are for static, or off-line, conditions such as [1,2,3]. However, there are some proposed methods which address the problem of scheduling on-line multiple workflows, namely OWM [4], RANK_HYBD [5] and FDWS [6], which target is to minimize the average relative waiting time of the

B. Murgante et al. (Eds.): ICCSA 2014, Part VI, LNCS 8584, pp. 532–545, 2014.

jobs. In this context the common definition of makespan is extended to account for the waiting time and execution time of a given workflow [7].

In the utility model, users consume the services and resources when they need to, and pay only for what they use. Cost and time become the two most important factors that users are concerned about. Thus, the cost/time tradeoff problem for scheduling workflow applications becomes a challenging problem. In this paper we propose a scheduling strategy that extend the former concurrent on-line scheduling algorithms by considering fairness resource sharing constrained to the user defined budget, for each job.

The remainder of this paper is organized as follows: section II describes the scheduling system model; section III reviews related works; section IV presents the proposed algorithm; section V presents the experimental results and discussion; and finally, section VI concludes the paper.

2   Scheduling System Model

The proposed scheduling system model consists of an application model, a utility grid model, budget model and a performance criterion for scheduling.

A typical workflow application can be represented by a Directed Acyclic Graph (DAG), a directed graph with no cycles. In a DAG, an individual task and its dependency is represented by a node and its edges. A dependency ensures that a child node cannot be executed before all its parent tasks finish successfully and transfer the required child input data. The task computation time and communication time is modelled by assigning weight to nodes and edges respectively. A DAG can be modeled by a tuple $G(V, E)$ where V is the set of v nodes and each node $v_i \in V$ represents an application task, and E is the set of communication edges between tasks. Each edge $e(i, j) \in E$ represents the task-dependency constraint such that task v_i should complete its execution before task v_j can be started. In a given DAG, a task with no predecessors is called an *entry task* and a task with no successors is called an *exit task*. We assume that the DAG has exactly one entry task n_{entry} and one exit task n_{exit}. If a DAG has multiple entry or exit tasks, a dummy entry or exit task with zero weight and zero communication edges is added to the graph.

The service-oriented architecture for Utility Grids is shown in Figure 1 [8]. Each user should submit its applications with user QoS requirements into the system. Then, all information about each application is collected into the application Data Base (DB). A utility grid model consists of several Grid Service Providers (GSPs), each of which provides some services to the users. GSPs charge different services by their QoS. Users only execute their jobs in GSPs that satisfy their QoS requirements, and only pay for what they use. To inform and attract users, each GSP should publish their services into Grid market directory (GMD). Ready Tasks Pool component collects tasks which are ready to execute among accepted workflow applications in application DB. A task is ready when all required information are prepared, i.e., its parents are executed. Then, Grid scheduler enquires GMD to query about available services for each task and their

QoS attributes. Also contacts the Grid Service Information to gather detailed information of each service, especially the available time slots for processing tasks. Using these information, the Grid scheduler executes a Task Scheduler algorithm to decide the map allocation based on user QoS parameters for each ready task of a workflow to one of the available services. The Service Executer is used to monitor task assignment on the service and to get notifications of successfully or failure of task execution on GSPs resources.

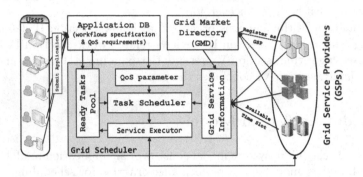

Fig. 1. A General View of Grid Scheduler System

After collecting the job and platform information, the budget model is the third parameter in our model. The available user defined budget for a workflow application must never be exceeded. Generally, the cost model can be divided into two main categories: (1) charge computing resources on hourly or monthly usage (time interval), and any partial hours are rounded up such as Amazon Elastic Compute Cloud (Amazon EC2) [9], (2) charges based on the number of CPU cycles required by a customer's application such as Google AppEngine [10]. In this paper, we consider the second category of pay-as-you-go pricing model.

The general objective is to minimize the completion time, also called *Makespan*. It is denoted by the finish time of the last task of the scheduled workflow application. But, when scheduling various independent workflows that represent user jobs and therefore are submitted at different moments in time, the completion time (or turnaround time) includes both the waiting time and execution time of a given workflow, extending the makespan definition for single workflow scheduling.

Therefore, our performance criteria to assign tasks to processors, from all workflows, considers the minimization of the turnaround time of each workflow, subjected to the user budget limitation imposed to each workflow.

3   Related Work

Generally, the related research in this field can be classified into two main categories: off-line and on-line scheduling. In off-line scheduling, the workflows are available before the execution starts, i.e., at compile time. After a schedule is produced and initiated, no other workflow is considered. This method, although restricted, can be applied in many real-world applications, e.g. when a user has a set of nodes to run a set of workflows. On-line scheduling exhibits dynamic behavior in which users can submit the workflows whenever they need to.

There are some algorithms proposed specifically to schedule concurrent workflows to improve individual QoS, namely, RANK_HYBD [5] and Fairness Dynamic Workflow Scheduling (FDWS) [6]. The first algorithm minimize the average completion time of all workflows. In contrast, FDWS focuses on the QoS experienced by each application (or user) by minimizing the waiting and execution times of each individual workflow. In [7], it is presented a review of on-line and off-line scheduling algorithms, for concurrent workflow scheduling, and a performance comparison for on-line scheduling.

In [5], Yu and Shi proposed a planner-guided strategy, the RANK_HYBD algorithm to address dynamic scheduling of workflow applications that are submitted by different users at different moments in time. The RANK_HYBD algorithm, collects all new arrival workflows submitted by users, and ranks all tasks using the $rank_u$ priority measure [11], which represents the length of the longest path from task n_i to the exit node, including the computational cost of n_i, and it is expressed as follows:

$$rank_u(n_i) - \overline{w_i} + \max_{n_j \in succ(n_i)} \{\overline{c_{i,j}} + rank_u(n_j)\}, \tag{1}$$

where $succ(n_i)$ is the set of immediate successors of task n_i, $\overline{c_{i,j}}$ is the average communication cost of $edge(i,j)$ and $\overline{w_i}$ is the average computation cost of task n_i. For the exit task $rank_u(n_{exit}) = 0$. In each step, the RANK_HYBD algorithm fills the ready tasks pool by all ready tasks from each submitted workflows and selects the next task to schedule based on their rank. If all tasks in the ready tasks pool belong to different workflows, the algorithm selects the task with lowest rank and if they belong to the same workflow, the task with highest rank is selected. Finally, among of all free processor, the processor with lowest finish time for selected task is chosen. The RANK_HYBD algorithm has some weakness points in its strategy. It allows the workflow with the lowest rank (lower makespan) to be scheduled first to reduce the waiting time of the workflow in the system. However, this strategy does not achieve high fairness among the workflows because it always gives preference to shorter workflows to finish first, postponing the longer ones.

To overcome these problems with the RANK_HYBD algorithm, the FDWS algorithm was proposed in [6]. FDWS applied new strategies for selecting the tasks from the ready tasks pool and for assigning the processors to reduce the individual completion time of the workflows, e.g., the turnaround time, including execution time and waiting time. The FDWS algorithm, unlike the RANK_HYBD algorithm, adds only a single ready task with highest priority ($rank_u$) from each workflow to the ready pool. Therefore, instead of scheduling smaller workflows first like RANK_HYBD, it selects and schedules tasks from the longer workflows as well and all workflows have chance to be scheduled in each step. In the Task

Selection phase, FDWS assigns a secondary priority to each task in ready tasks pool. To be inserted into the ready tasks pool, the $rank_u$ was computed for each workflow individually. After filling the ready tasks pool by one ready task from each workflow, $rank_r$ is computed as the secondary priority rank for task n_i belonging to DAG_j, defined by (2). Then, the task with highest $rank_r$ from the ready pool is selected.

$$rank_r(n_{i,j}) = \frac{1}{PRT(DAG_j)} \times \frac{1}{CPL(DAG_j)}. \tag{2}$$

The $rank_r$ metric considers the Percentage of Remaining Tasks value (PRT) of the workflow (DAG) and its Critical Path Length (CPL). The PRT value gives more priority to workflows that are almost completed and only have few tasks to execute. Unlike RANK_HYBD algorithm which uses only the individual $rank_u$ for selecting tasks into ready tasks pool and for execution in each round, in FDWS, the workflow history in the workflow application DB pool is considered to make a scheduling decision.

An alternative strategy is the resource reservation policy, where a static scheduling decision based on resource availability is made when an application workflow is submitted into the system. But this strategy lead us to unfair scheduling because if a workflow application with higher number of tasks and execution time is submitted, it may reserve most of the resources and make higher waiting time for next applications.

But none of these approaches, consider cost as a QoS parameter in their scheduling strategies. In this paper, we propose a strategy for on-line scheduling of concurrent workflows with budget constraints defined by users for each workflow, as described in the next section. We apply our strategy to RANK_HYBD and FDWS in order to consider the budget constraint.

4   The Proposed Strategy

In our model the user specifies the budget constraint by selecting a budget value in the interval limited by the cheapest and highest costs. For the job j the budget is specified as expressed by equation 3:

$$\begin{aligned} BUDGET(j) = Cost_{lowest}(j) + k_{Budget}(j) \times \\ (Cost_{highest}(j) - Cost_{lowest}(j)) \end{aligned} \tag{3}$$

where $Cost_{highest}(j)$ and $Cost_{lowest}(j)$ are the total cost of the assignment produced by the highest and cheapest scheduling, that is obtained by selecting the most expensive and most cheapest processors to execute each task of workflow j. Therefore, for each workflow, it is presented to the user the budget range that is possible to achieve in the selected platform and he/she selects a budget inside that range, that is represented by k_{Budget}. This budget definition was first introduced in [12].

The workflow management system represented in Figure 1 gathers applications information submitted by users, scheduling and runtime monitoring. Next we describe how tasks are selected to the Ready Task Pool and the scheduling strategy.

4.1   Ready Tasks Pool

There is a *ready tasks pool* which is filled by the ready tasks belonging to each submitted and unfinished workflow at each scheduling round. In general, two methods are used to fill the *ready tasks pool*, first, like FDWS algorithm, gather only a single ready task with highest priority ($rank_u$) from each workflow, or insert all ready tasks belonging to each unfinished workflow application into *ready tasks* pool such as RANK_HYBD. But the important key is how to order these ready tasks, i.e., how to assign priority to each ready task based on our QoS parameters to have higher quality solutions and system performance. For selecting a task from *ready tasks pool* to be schedule on resources, we defined a new strategy to assign a secondary priority to each task of the *ready tasks pool*. Because our goal is to execute applications in the lowest turnaround time with its limited budget, the cost factor should be taken into account. To achieve this purpose, for each workflow j, we define the Task Proportion (TP_j), equation 4, which is the ratio of unscheduled number of tasks to the total number of tasks in the workflow (DAG), and Budget Proportion (BP_j), equation 5, which is the ratio of the Remaining Cheapest Budget (RCB_j) to the Remain Budget (RB_j). RCB_j is the remaining *lowest_Cost* for unscheduled tasks and RB_j is the actual remaining budget.

$$TP_j = \frac{unscheduled\ number\ of\ tasks}{Total\ number\ of\ tasks} \tag{4}$$

$$BP_j = \frac{RCB_j}{RB_j} \tag{5}$$

RCB_j is updated in each step after making the processor selection for the selected task belonging to workflow(j) by using equation 6.

$$RCB_j = RCB_j - lowest_{Cost}(T_{sel}) \tag{6}$$

where $lowest_{Cost}(T_{sel})$ is the lowest cost for current task (selected task for scheduling). The initial value for Remaining Cheapest Budget is $RCB_j = lowest_{Cost}$ where $lowest_{Cost}$ is the total cost of the assignment for workflow j produced by the cheapest scheduling. In addition, the initial value for Remain Budget is $RB_j = BUDGET_j$ where $BUDGET$ is user defined budget for application workflow, and it will be updated, by equation 7, after the processor selection phase for the selected task.

$$RB_j = RB_j - Cost(T_{sel}, P_{sel}) \tag{7}$$

where P_{sel} is the processor selected to run the current task (T_{sel}), and $Cost(T_{sel}, P_{sel})$ is the cost of running task T_{sel} on processor P_{sel}.

To assign the secondary priority to each task in *ready tasks pool*, we propose $rank_B$ for each task t_i in the pool, that belongs to workflow j, defined by equation 8. The task with highest $rank_B$ is selected to be schedule.

$$rank_B(t_{i,j}) = \frac{1}{TP_j} \times \frac{1}{BP_j} \tag{8}$$

The $rank_B$ value is the product of two factors: a) the first one is the inverse of the fraction of the workflow j that is remaining in the system; and b) the ratio of the budget value over the remaining cheapest budget. This priority factor gives higher priority to the workflows that have a lower percentage of tasks unscheduled and to workflows that have higher budgets when compared to the cheapest budget for the DAG. The rational for the first factor is to give higher priority to workflows that where submitted earlier, so that a longer workflow with several tasks already executed may have priority over a short and recent workflow. And the rational of the budget factor is that the scheduler will consider first tasks that can spend more budget and therefore they will select more expensive and faster processors, resulting in a lower turnaround time for the workflow.

4.2 Task Scheduler

The Task scheduler is another important key in workflow management system which has responsibility for selecting affordable resource for the current selected task. In this part, we propose a new strategy for processor selection phase based on QoS requirements. The processor to be selected to execute the current task is guided by the following strategy related to cost. To achieve minimum execution time under limited budget, first, the task with highest $rank_B$ is selected from *ready tasks pool*, then the finish time of selected task (t_{sel}) is computed for all processors ($FT(t_{sel}, p_i)$) and these values are one of the factors for processor selection. The variables $FT_{min}(t_{sel})$ and $FT_{max}(t_{sel})$ are defined as lowest and highest finish time for selected task, respectively. The other factor is the cost of executing the task on each processor $Cost(t_{sel}, p_i)$.

The new processor selection strategy is based on the combination of the two factors, time and cost, and therefore two relative quantities are defined, namely, Time quota ($Time_q$) and Cost quota ($Cost_q$) for selected task (t_{sel}) on each processor $p_i \in P$, shown in equations 9 and 10, respectively.

$$Time_q(t_{sel}, p_i) = \frac{FT_{max}(t_{sel}) - FT(t_{sel}, p_i)}{FT_{max}(t_{sel}) - FT_{min}(t_{sel})} \tag{9}$$

$$Cost_q(t_{sel}, p_i) = \frac{Cost_{best}(t_{sel}) - Cost(t_{sel}, p_i)}{Cost_{highest}(t_{sel}) - Cost_{lowest}(t_{sel})} \tag{10}$$

where $Cost_{best}(t_{sel})$ is the cost of selected task on the processor with lowest finish time (processor with $FT_{min}(t_{sel})$). $Time_q$ shows how far is the finish time of selected task on processor p_i, from the worst finish time ($FT_{max}(t_{sel})$). Similarly, $Cost_q$ shows how far the actual cost, on p_i, is from the best cost ($Cost_{best}(t_{sel})$). Both variables are normalized with their highest range.

In addition to these variables that give time and cost relative processor performance, there is a limitation on cost consumption. This constraint is represented by the spare budget, that is defined by the difference between the remaining budget available (RB) and the remaining cheapest assignment (RCB). Once RCB includes the minimum cost of t_{sel}, this quantity is added to the spare budget available, as expressed by equation 11.

$$Cost_{lim}(t_{sel}) = Cost_{lowest}(t_{sel}) + (RB_j - RCB_j) \qquad (11)$$

The Budget Proportion (BP), defined by equation 5, is the ratio between RCB and RB, and gives a measure of how far the cheapest assignment is from the remaining budget available. If BP is near to one, it means that the available budget only allows to select the cheapest assignment. In addition, the $Cost_{lim}$ controls the processor decision to avoid cost consumption higher than user defined budget.

Finally, to select the processor for current task t_{sel} belonging to workflow j, it is computed the Quality value (Q) for each available and free processor $p_i \in P$ as shown in equation 12.

$$\underset{p_i \in P}{Q(t_{sel}, p_i)} \begin{cases} -\infty & \text{if } Cost(t_{sel}, p_i) \\ & > Cost_{best}(t_{sel}) \\ \\ -\infty & \text{if } Cost(t_{sel}, p_i) \\ & > Cost_{lim}(t_{sel}) \\ \\ Time_q(t_{sel}, p_i) + & \text{otherwise} \\ BP_j \times Cost_q(t_{sel}, p_i) \end{cases} \qquad (12)$$

The first two statements guaranty that if the cost of task t_{sel} on processor p_i is higher than the cost on the processor that gives the minimum FT, and if that cost is higher than the available budget for task t_{sel}, then processor p_i cannot be selected. Otherwise, the processor is evaluated considering the time and cost quantities. The processor with higher Quality value is selected.

Algorithm 1 shows the general algorithm used to implement the algorithm versions shown in table 1. The characteristics that differentiate each version are the strategy to select ready tasks from each workflow, the priority assigned to each ready task in the task pool and, the processor selection policy. These policies are parameters of the general algorithm.

5   Experimental Results and Discussion

This section presents performance results obtained with the new strategies applied to two scheduling algorithms. We implemented modified versions of RANK_HYBD [5] and FDWS [6], called Budget RANK_HYBD (B-RANK_HYBD) and Budget FDWS (B-FDWS) to consider budget limitation imposed by users. In the processor selection phase of these two algorithms, cost is not taken into account and there is

---

Algorithm 1. THE GENERAL BUDGET CONSTRAINED SCHEDULING
STRATEGIES FOR ON-LINE WORKFLOW APPLICATIONS

Input:
1. A filling strategy to add ready tasks from each workflow into *Ready Tasks* pool
2. A priority strategy for assigning a rank to each task into *Ready Tasks* pool
3. A processor selection strategy

1 **while** *Application DB* $\neq \phi$ **do**
2   - Fill *Ready Tasks* pool based on the input filling strategy
3   **foreach** $t_i \in$ *Ready Tasks pool* **do**
4   - Assign a rank value for t_i according to the input priority strategy for ready tasks
5   **while** *Ready Tasks* $\neq \phi$ **do**
6   - Select task T_{sel} with highest priority from *Ready Tasks* pool
7   - Select best suitable processor based on the input processor selection strategy
8   - Assign Task T_{sel} to selected Processor
9   - Update the Remain Budget (RB) and the Remain Cheapest Budget (RCB) as defined in Eq.7 and Eq.6
10   - Remove Task T_{sel} from *Ready Tasks* pool

Table 1. Description of the modified algorithms for on-line budget constrained scheduling

Algorithm	Strategies		
Name	Filling Ready Pool	Selecting task to schedule	Processor Selection
B-RANK_HYBD1	for each *workflow* Insert *all ready tasks*	**if all** $t_i \in$ *ready pool* belong to *same workfolow* **then select** t_i with highest $rank_u$, **else** **select** t_i with lowest $rank_u$	select $p_j \in P$ with lowest $FT(t_i, p_j)$ where $cost(t_i, p_j) \leq Cost_{lim}(t_i)$
B-RANK_HYBD2			select $p_j \in P$ with Highest $Q(t_i, p_j)$ value (Eq.12)
B-FDWS1	for each *workflow* Insert *Single ready task* with highest $rank_u$	**select** $t_i \in$ *ready pool* **with** highest $rank_r$ (Eq.2)	select $p_j \in P$ with lowest $FT(t_i, p_j)$ where $cost(t_i, p_j) \leq Cost_{lim}(t_i)$
B-FDWS2			select $p_j \in P$ with Highest $Q(t_i, p_j)$ value (Eq.12)
B-FDWS3		**select** $t_i \in$ *ready pool* with highest $rank_B$ (Eq.8)	select $p_j \in P$ with lowest $FT(t_i, p_j)$ where $cost(t_i, p_j) \leq Cost_{lim}(t_i)$
B-FDWS4			select $p_j \in P$ with Highest $Q(t_i, p_j)$ value (Eq.12)

a possibility to have higher cost than the limited budget defined by the user. So, in modified version, instead of considering all processors to compute the finish time of current task, processors are filtered based on the cost limitation value defined by Equation 11 and we select the processor that allows the lowest finish time among all affordable processors. To evaluate the influence of the new strategy proposed in this study, $rank_B$ for selecting tasks and quality measure value (Q) for selecting processors, we consider several version of the scheduling algorithms as described in Table(1). For all algorithms, we select processors that are free on current time (no reservation policy).

This section is divided into four parts, namely, workflow structure description, the environment scheduling system and hardware parameters, the performance metric, and finally results and discussion are presented.

5.1 Workflow Structure

To evaluate the relative performance of the algorithms, a model of execution time of the tasks on resources is needed. We use the method described in [13] to model the application execution times for our simulation. The model consists of an Expected Time to Compute (ETC) matrix, that contains the estimation execution times of each task on all resources. The parameters of this model can be changed to investigate the performance of algorithms under different heterogeneous computing systems and under different types of tasks. The ETC model is based on three parameters: machine heterogeneity, task heterogeneity and consistency, which allow us to simulate various possible heterogeneous scheduling problems as realistically as possible [14]. The task heterogeneity is defined as the variety among the execution times of the tasks and, Machine heterogeneity, on the other hand, represents the possible variation of the running time of a particular task across all the processors.

Based on the two first parameters, four categories can be proposed for ETC matrix:

1. $Machine_{heterogeneity}$ = High, $Task_{heterogeneity}$ = High
2. $Machine_{heterogeneity}$ = High, $Task_{heterogeneity}$ = Low
3. $Machine_{heterogeneity}$ = Low, $Task_{heterogeneity}$ = High
4. $Machine_{heterogeneity}$ = Low, $Task_{heterogeneity}$ = Low

In our simulation we consider two values 0.2 and 0.8 as low and high heterogeneity values.

The ETC matrix can be further classified into two categories, *consistent* and *inconsistent*. In the consistent model, if for a task t_i a processor p_j has shorter execution time than processor p_k, them the same is true for any other task. In inconsistent model a processor p_j may execute some jobs faster than p_k and be slower for some other jobs. The consistent model can be seen as modelling a heterogeneous system in which the processors differ only in their processing speed and, a inconsistent model may represent a network in which there are different types of machines architectures. In this paper we consider the consistent model.

In order to generate dynamic and concurrent workflows scheduling, we consider 100 workflow applications, in each scenario, that arrive with time intervals that range from 10% to 90% of completed tasks, i.e., a new workflow is inserted when the corresponding percentage of tasks from the last workflow currently in the system is completed. Each workflow application consists of a number of tasks between 10 and 100 tasks. For each workflow, we generate random values for k_{Budget} in the range $(0 \ldots 1)$.

5.2   Simulation Platform

We resort to simulation to evaluate the algorithms from the previous section. It allows us to perform a statistically significant number of experiments for a wide range of application configurations (in a reasonable amount of time). We use the SimGrid toolkit [1] [15] as the basis for our simulator. SimGrid provides the required fundamental abstractions for the discrete-event simulation of parallel applications in distributed environments. It was specifically designed for the evaluation of scheduling algorithms. Relying on a well-established simulation toolkit allows us to leverage sound models of a heterogeneous computing systems, such as the Grid platform considered in this work. The network model provided by SimGrid corresponds to a theoretical *bounded multi-port* model. In this model, a processor can communicate with several other processors simultaneously, but each communication flow is limited by the bandwidth of the traversed route and communications using a common network link have to share bandwidth. In this experiments, we connected all processor over one shared bandwidth.

We consider platforms with 8 and 32 processors as a low and high number of processors, compared to the number of workflows, to analyze the behaviour of the algorithms with respect to the system load. The maximum load configuration is observed for 8 processors and 100 workflows.

5.3   Performance Metric

The metric to evaluate a dynamic scheduler of independent workflows, must represent the individual completion time in order to measure the QoS experienced by the users related to the finish time of each user application. A global measure for the set of workflows would hide relevant delays on shorter workflows.

To evaluate the algorithms we consider the relative improvement on Turnaround Time ($TurnaroundTime_{imp}$) achieved with our strategy, for a given workflow, when compared to the maximum Turnaround Time achieved for that workflow among all strategies. The Turnaround time is the difference between submission and final completion of an application. The ($TurnaroundTime_{imp}$) is obtained by the ratio of the difference of turnaround time for a given workflow G and an algorithm alg_i, and the maximum turnaround time among all algorithms, as shown by equation 13.

$$TurnaroundTime_{imp}(alg_i) = \frac{\max_{alg_k \in alg_{set}}\{TT_G(alg_k)\} - TT_G(alg_i)}{\max_{alg_k \in alg_{set}}\{TT_G(alg_k)\}}, \quad (13)$$

where alg_{set} is the set of algorithms under comparison and $alg_i \in alg_{set}$. This metric for an algorithm alg_i gives the improvement achieved by each algorithm in comparison to the maximum Turnaround Time obtained for a given workflow G. The algorithm that generates more relative improvements is the best algorithm.

---

[1] http://simgrid.gforge.inria.fr

5.4 Results and Discussion

In this section, we compare the modified versions B-FDWS and B-RANK_HYBD in terms of $TurnaroundTime_{imp}$ value. We present results for 1000 instance of the scenario with a set of 100 DAGs that arrive with time intervals that range from 10% to 90% of completed tasks, i.e., a new DAG is inserted when the corresponding percentage of tasks from the last DAG currently in the system is completed. In addition we generate 4 different types based on task heterogeneity and machine heterogeneity. In total we tested 4000 instances. Also we consider two different level of processors (low and high) compared to the number of DAGs, to analyse the behaviour of the algorithms with respect to different system load. We consider 2 sets of processors with 8 and 32 where the maximum load configuration is observed for eight processors. Based on our observations, task heterogeneity does not have a significant effect on results, therefore we do not categorize results based on this parameter.

Results are presented attending to the three strategies referred to in Table 1, namely, Ready Pool Filling strategy, Task Selection strategy and Processor Selection strategy. Figure 2 shows the Turnaround Time percentage improvement achieved for the 6 algorithm's versions.

For low number of CPUs, as we can see in Fig 2(a) and Fig 2(b), the filling policy for adding ready tasks from workflow applications into ready tasks pool, is the strategy that differentiates the two algorithms RANK_HYBD and FDWS. The FDWS filling strategy, which selects a single task from each workflow, leads to higher fairness in scheduling and avoids the postponing of larger workflows as happens with RANK_HYBD, contributing to a better relative turnaround time. The improvements are more significant when we have higher concurrency in the system, i.e. low arrival time, starting on 67% improvement for arrival time interval of 10%, and 30% improvement for arrival time interval of 90%.

As we move to higher arrival time intervals, when comparing the algorithms versions that use the quality measure Q, with the ones that do not use it, we conclude that Q improves the algorithms performance. For instance, B-RANK_HYBD2 has 46% improvment over B-RANK_HYBD1, as well as 27% improvement for B-FDWS2 over B-FDWS1 and 28% for B-FDWS4 over B-FDWS3.

For higher arrival time intervals, which means lower concurrency, using the quality measure Q achieves higher turnaround time percentage improvement on platforms with larger values of machine heterogeneity. We obtained improvements of 58%, 36% and 39% of turnaround time, with arrival time of 90% for B-RANK_HYBD2, B-FDWS2 and B-FDWS4 over B-RANK_HYBD1, B-FDWS1 and B-FDWS3, respectively.

On the other hand, for higher number of CPUs, in addition to filling ready task policy, two other strategies, $rank_B$ and quality measure Q, proposed here, have higher influence in the improvements obtained with both algorithms. Fig. 2(c) and Fig. 2(d) show that, besides the filling ready task policy used by FDWS which improves algorithm performance over RANK_HYBD, quality measure Q always improves the algorithm's performance. The improvements of B-RANK_HYBD2, FDWS2 and FDWS4 over B-RANK_HYBD1, FDWS1 and FDWS3,

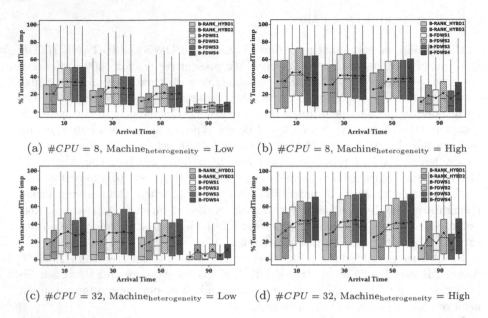

(a) $\#CPU = 8$, Machine$_{heterogeneity}$ = Low   (b) $\#CPU = 8$, Machine$_{heterogeneity}$ = High

(c) $\#CPU = 32$, Machine$_{heterogeneity}$ = Low   (d) $\#CPU = 32$, Machine$_{heterogeneity}$ = High

Fig. 2. Turnaround Time improvement values for 8 and 32 processors and, for low and high machine heterogeneity

respectively, start at 55% for arrival time interval of 10% and increase to 240% for arrival time interval of 90%.

Comparing results of FDWS1 to FDWS3 or FDWS2 to FDWS4, we can conclude that $rank_B$, as the policy for selecting the task from ready tasks pool to be schedule, improves slightly the algorithm performances, in comparison to $rank_r$. The highest improvement observed is 9%.

6   Conclusion

In this paper we present a new strategy for dynamic scheduling of concurrent workflows with two conflicting QoS requirements. To the best of our knowledge, there is no previous research that deal with multiple workflow scheduling that are submitted at different moments in time and that are based on the two conflicting QoS parameters, namely, time optimization and cost constraint at the same time. We propose a new priority rank, called, $rank_B$ for task selection and a quality value Q for the processor selection phase on the workflow management system.

The new strategies allowed to obtain better performances in almost all presented cases. For some other configurations, such as lowest processor number (1/8 of the number of concurrent DAGs) and lower arrival time, the most significant factor is the filling of the ready pool. For the other configurations the most influence factor to improve performance is the usage of the quality factor Q proposed in this paper. Additionally, we apply this strategy to two state of the art algorithms in order to consider the budget constraint, namely, FDWS and RANK_HYBD algorithms. In the future work, we intend to add a deadline constraint as another QoS parameter into our strategies.

Acknowledgments. This work was supported in part by the Fundação para a Ciência e Tecnologia, PhD Grant FCT-DFRH-SFRH/BD/80061/2011.

References

1. Zhao, H., Sakellariou, R.: Scheduling multiple DAGs onto heterogeneous systems. In: Int. Parallel and Distributed Processing Symposium, pp. 1–14. IEEE (2006)
2. N'takpé, T., Suter, F.: Concurrent scheduling of parallel task graphs on multiclusters using constrained resource allocations. In: Int. Parallel and Distributed Processing Symposium, pp. 1–8. IEEE (2009)
3. Bittencourt, L.F., Madeira, E.: Towards the scheduling of multiple workflows on computational grids. Journal of Grid Computing 8, 419–441 (2010)
4. Hsu, C.C., Huang, K.C., Wang, F.J.: Online scheduling of workflow applications in grid environments. Future Generation Computer Systems 27(6), 860–870 (2011)
5. Yu, Z., Shi, W.: A planner-guided scheduling strategy for multiple workflow applications. In: ICPP-W 2008, pp. 1–8. IEEE (2008)
6. Arabnejad, H., Barbosa, J.G.: Fairness resource sharing for dynamic workflow scheduling on heterogeneous systems. In: Int. Symp. on Parallel and Distributed Processing with Applications (ISPA), pp. 633–639. IEEE (2012)
7. Arabnejad, H., Barbosa, J.G., Suter, F.: Fair resource sharing for dynamic scheduling of workflows on heterogeneous systems. In: Jeannot, E., Zilinskas, J. (eds.) High-Performance Computing on Complex Environments, pp. 147–167. John Wiley & Sons (2014)
8. Yu, J., Venugopal, S., Buyya, R.: A market-oriented grid directory service for publication and discovery of grid service providers and their services. The Journal of Supercomputing 36(1), 17–31 (2006)
9. Amazon, http://aws.amazon.com/ec2
10. Google, http://code.google.com/appengine/.
11. Topcuoglu, H., Hariri, S., Wu, M.: Performance-effective and low-complexity task scheduling for heterogeneous computing. IEEE Transactions on Parallel and Distributed Systems 13(3), 260–274 (2002)
12. Sakellariou, R., Zhao, H., Tsiakkouri, E., Dikaiakos, M.: Scheduling workflows with budget constraints. In: Int. Research in Grid Computing, pp. 189–202 (2007)
13. Ali, S., Siegel, H.J., Maheswaran, M., Hensgen, D.: Task execution time modeling for heterogeneous computing systems. In: Heterogeneous Computing Workshop, pp. 185–199. IEEE (2000)
14. Braun, T.D., Siegel, H.J., Beck, N., Bölöni, L.L., Maheswaran, M., Reuther, A.I., Robertson, J.P., Theys, M.D., Yao, B., Hensgen, D., et al.: A comparison of eleven static heuristics for mapping a class of independent tasks onto heterogeneous distributed computing systems. Journal of Parallel and Distributed computing 61(6), 810–837 (2001)
15. Casanova, H., Legrand, A., Quinson, M.: Simgrid: a generic framework for large-scale distributed experiments. In: Int. Conf. on Computer Modeling and Simulation, UKSIM, pp. 126–131. IEEE CS (2008)

Comparing Agility Analysis Techniques: AgilAC Framework versus Ad Hoc Approach

Marcelo Benites Gonçalves[1,*], Débora Maria Barroso Paiva[2],
Valter Vieira Camargo[3], and Maria Istela Cagnin[1]

[1] Department of Computer Systems, University of São Paulo - USP
São Carlos, SP, Brazil, PO Box 668, 13566-590
[2] College of Computing, Federal University of Mato Grosso do Sul (UFMS)
Campo Grande, MS, Brazil, PO Box 549, 79070-900
[3] Computing Department, Federal University of São Carlos (UFSCar)
São Carlos, SP, Brazil, PO Box 668, 13565-905

Abstract. Agile methods have gained ground in academia and industry since their inception. However, identifying whether a development method is in accordance with the agile philosophy has become a challenge because of the diversity and amount of their activities, artifacts and roles. Thus, we defined the AgilAC framework to support agility analysis of development methods. It supports software engineers to better understand the adherence of their methods to agile philosophy. In order to evaluate the applicability of this framework, an experiment was conducted to comparing the AgilAC and the ad hoc analysis, and it is presented in this paper. We adopted two criteria in the comparison: the time spent during the analysis and the correctness of the analysis result. We observed that although the AgilAC requires more time to be applied than the ad hoc analysis, the quality of the analysis results were better. Therefore, it was possible to verify the advantage of the AgilAC over an ad hoc analysis, when considering the correctness of the agility characterization.

Keywords: agility analysis, agile methods, assessing agile methods, software engineering experiment.

1 Introduction

Nowadays, the current scenario of software development is greatly characterized by ever-changing requirements, asking for development processes and methods able to cope with such requirement instability. In this context, agile methods are gaining more follower, especially in the software industry. This new family of methods is lightweight in terms of the number of artifacts that need to be created, and the level of management effort than the traditional development models [1].

* Financial support by CNPq (Brazilian National Counsel of Technological and Scientific Development)

B. Murgante et al. (Eds.): ICCSA 2014, Part VI, LNCS 8584, pp. 546–561, 2014.
© Springer International Publishing Switzerland 2014

There are works reporting the effectiveness of agile methods in certain scenarios as being equal or superior to traditional models [2].

In literature, it is possible to find several development methods that are presented as "agile" and there is also a great heterogeneity among them. It is also possible to find methods that are labelled as "agile" but they cover only part of the agile philosophy [3]. Moreover, for non-expert individuals, many times it is difficult to characterize a method as agile or not. In this sense, it is worthwhile to develop mechanisms to support the evaluation of these methods in relation to what is expected by the agile philosophy.

It is possible to find researches focused on agility analysis [3,4,5]. However, we observed that the criteria for agility analysis used by those works do not provide a complete diagnosis about the agility analysis of the development method (that is, indication of the agility diagnosis complemented by detailed results of the analysis conducted).

So, in order to bridge the identified gaps of the aforementioned approaches, we proposed the AgilAC framework [6] to support agility analysis of development methods. This framework was designed to provide a structured and comprehensive assessment, organized in steps for evaluators that know agile methods but do not have a broad knowledge to perform alone this kind of assessment. The AgilAC is adequate in both when there are an initiative to implement an agile development method and when the goal is to adapt an method already running to agile philosophy. The AgilAC usage is also adequate when the goal is to evaluate which agile practices are present in a development method (for example, regarding a hybrid development method). The main output of the AgilAC is a final report that contains the documentation of all analysis.

The main contribution of this paper is to present an experiment conducted to evaluate the AgilAC framework in relation to ad hoc analysis [1] (referred in the remainder of this paper by ad hoc approach). The experiment results indicated that the analysis using the AgilAC spent more time than the analysis with the ad hoc approach, but our framework provided more accurate and complete results.

The paper is organized as follows: Section 2 presents the AgilAC framework as well as the analysis steps and its flow of use. In Sections 3 and 4, the planning and execution of the experiment conducted according with Wohlin et al. [7] are presented. In Section 5, the related works are presented and in Section 6 we discuss the findings and suggestions for future works.

2   AgilAC Framework

This section presents an overview of the AgilAC framework [6], whose aim is to support the agility analysis of development methods. The main expected contribution of the AgilAC is to allow someone to perform an evaluation without a broad expertise on agile methods, since a basic knowledge concerning agile methods is already enough.

---

[1] Agility analysis performed in a informal and non-systematic way.

Considering that the development methods do not follow a standard structure in their descriptions, it is important to evaluate whether the method documentation is clear enough to conduct the analysis and understand its description and structure. For this, it is necessary to observe all the elements that compose the method documentation (for example, tasks, activities, practices, values, etc). Moreover, the evaluator's knowledge about the method under assessment is a pre-requisite to apply the AgilAC and must be both theoretical and practical. The evaluator must be aware that this knowledge is needed and the detail of the AgilAC's report can be regarded as proportional to this knowledge.

The final result provided by the AgilAC gives directions for improving adherence of development method to agile philosophy. The AgilAC do not have the same formal commitment that can be observed in other evaluation process. For instance, the evaluation process related to maturity models, i. e., SCAMPI for CMMI [8]. Fig. 1 showns the AgilAC steps and the outputs produced for each one. The framework receives as input the documentation of a development method to be evaluated.

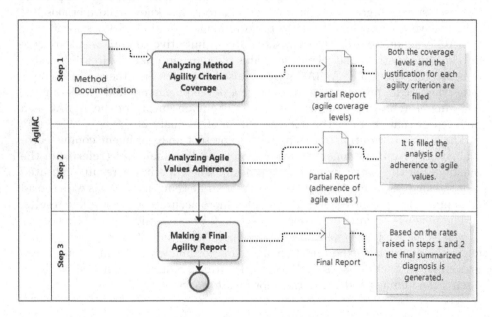

Fig. 1. Steps of the AgilAC

The goal of first step of the AgilAC is verify to what extent the method is consistent with the agility criteria, using coverage levels. The conception of the AgilAC agility criteria was based on well-accepted characteristics for agile methods, according to Abrantes and Travassos [9]. These criteria were grouped according to knowledge areas in project management [10], as shown in Table 1. Due to space limitation, only a few agility criteria of the AgilAC are presented in this table. For example, the criterion "Incremental Development" and "Communicative Teams" are common characteristics to any agile method. For each agility criterion, a question is provided for improving the comprehension and making the identification of the coverage level easier.

Table 1. AgilAC agility criteria (Step 1)

Knowledge Areas [10]	Criteria of Agility
Integration	**Incremental Development:** Does application development involve several short cycles? ...
Scope	**Modularity:** Does the method allow the modularization of its development activities and, when convenient, support the inclusion and exclusion the created modules?
Time	**Emergence:** Are the processes and work structures emerging over the project's lifecycle? ...
Costs	**Leanness:** Does the method contain the minimum set of activities to be effective? Were eliminated all the needless activities?
Quality	**Tests:** Do the test strategies deal with quality problems related to short delivers? Is a testing approach clear in the method? ...
Human Resources	**Reflection/Introspection:** Do the test strategies deal with quality problems related to short delivers? ...
Communication	**Communicative Teams:** Does the method provide a structure that promotes the active participation and communication among all team's members?
Risks	**Convergence:** Are strategies for mitigating risks well-defined in the method? ...
Procurement	**Cooperativity:** Does the method provide a structure that promotes the adequate team's autonomy on purchasing or acquiring products, services, or results from outside the project?
Coverage Results of a Hypothetical Develop. Method	**Percentage**
Level 0	74 % (for example)
Level 1	21 % (for example)
Level 2	5 % (for example)

In context of this work, the term "coverage levels" means to what extent a development method meets each agility criterion defined for the AgilAC. We defined three coverage levels: Level 0 occurs when is not possible to determine elements that meet the requirements established by the criterion; Level 1 is when there are elements in the method that can adequately meet the requirements determined by the criterion; and Level 2 is when there are a heavy amount of elements in the method that can meet the requirements determined by the criterion.

Next, the evaluator should calculate the percentage of each coverage level in relation to all criteria evaluated (coverage level/ total of 19 criteria), as shown at bottom of Table 1. For example, if a total of 14 criteria have obtained the Level 0, the coverage of this level must be 74% (i. e. 14/19); if a total of 4 criteria have obtained the Level 1, the coverage of this level must be 21% (i. e. 4/19) and if a total of 1 criterion has obtained the Level 2, the coverage of this level must be 5% (i. e. 1/19). In this sense, these coverages indicate the predominant level.

In step 2, the idea is to verify the adherence of development method to agile values, performing the "Analysing Agile Values Adherence" step. Basically, the evaluator must check, for each agile value, if the development method is adherent, partially adherent or not adherent to each agile value, give a justification for this and calculate the adherence percentage (AP), as indicated in Table 2. Besides, applying step 2, it is also possible to achieve method aspects that were not observed in step 1. This can occur because they are not directly linked to the agility criteria defined in step 1.

Table 2. Example of evaluation of agile values (Step 2)

Value...	Over...	Adheres to the value? (Yes/No/Partially)	Justification (for example)
Individuals and interactions	Processes and tools	Yes	Daily meeting: regular meetings among team members promote the interactions
Working software	Comprehensive documentation	Yes	It only considers the documentation produced by computer aided automatic generation
Customer collaboration	Contract negotiation	Partially	Specific protocol and documentation to order functions
Responding to change	Following a plan	Yes	Continuous monitoring of changes
Adherence Percentage (AP) of hypothetical development method	**87.5%**		

In step 3, a final report containing a analysis diagnosis (that is, undefined, agile or heavyweight - described in Table 3) of the development method must be obtained based on the data collected from previous steps. This diagnosis is

Table 3. Possibilities of analysis diagnosis provided by the AgilAC

Not agile (undefined)	Agile	Not agile (heavy-weight)
The method can not be considered as agile because it does not achieve the minimum criteria and values considered for agile methods. The improvement for the method should consider the weakness verified in the agility analysis. In this case, the method can be considered as hybrid because has some characteristics of both the agile philosophy and the traditional development	The method can be considered as agile because it is consistent with widely accepted criteria for agile methods and also is adherent to the agile values. However, weaknesses identified in the agility analysis should be considered for improvement.	The method can not be considered as agile because it determines more expensive efforts than the determined agility criteria. Thus, should be considered for optimization the agility critical points observed in the agility analysis.

obtained by comparing the coverage results from the agility criteria (step 1) with the results from the adherence analysis (step 2). The results of steps 1 and 2 can help to identify the causes that make development method analysed not adhering to agile philosophy.

Figure 2 shows the algorithm that must be followed to decide which is the most suitable choice for the agility diagnosis. The percentage limits are based on the scoring mechanisms of ISO/IEC 15504 [11] evaluation method, called SPICE. Moreover, this algorithm prioritizes the adherence of agile values and, if this adherence is satisfied, the agility evaluation follows based on the coverage of agility criteria. Therefore, the step 3 of the AgilAC includes a final report that encompasses all the information gathered in previous steps as well as the agility diagnosis. Considering the example of agility analysis shown in step 1 (Table 1) and step 2 (Table 2), the agility diagnosis for the hypothetical development method is "undefined" after applying the mentioned algorithm.

3 Experiment: Agility Analysis with AgilAC versus Ad Hoc Approach

This section shows the empirical study planned according to the guidelines proposed by Wohlin et al. [7]. The aim of study was to verify whether the agility analysis using the AgilAC has "advantages" versus an ad hoc approach. It is worth noting that the ad hoc analysis is conducted based on the Agile Manifesto (Agile Alliance, 2001) and without following a systematic set of steps.

3.1 Experiment Planning

The experiment was made employing undergraduate students of Computer Science who was taking the "Topics in Information Systems I" course, offered in

```
IF (AP < 75%)
    RETURN "UNDEFINED"
ELSE-IF (Level 1 > Level 0 + Level 2)
    RETURN "AGILE"
ELSE-IF (Level 2 > Level 0)
    RETURN "HEAVY WEIGHT"
ELSE
    RETURN "UNDEFINED"
```

Legend:
Level 0, Level 1 and Level 2 - coverage percentage of the agile criteria levels (see Table 1)
AP - percentage of agile value adherence (see Table 2)

Fig. 2. Algorithm for determining the agility diagnosis

Facom (College of Computing) of UFMS (Federal University of Mato Grosso do Sul).

Formulation of Hypotheses. Table 4 shows the hypothesis as well as the metrics used to evaluate each one. For our proposals, a participant got a correct result when he/she had correctly characterized the agility of the method (as undefined, agile and heavyweight) and had justified his/her decision in a proper way.

Table 4. Hypotheses and metrics

Time spent in the agility analysis	
H_o	There are no differences among subjects that use the AgilAC and the subjects that use the ad hoc approach according to the time to conduct the agility analysis.
H_a	There are differences among subjects that use the AgilAC and the subjects that use the ad hoc approach according to the time to conduct the agility analysis.
Metric	Time spent to make the analysis in each approach.
Correctness of the agility analysis results	
H_1	There are no differences among subjects that use the AgilAC and the subjects that use the ad hoc approach according to the correctness of the results obtained in each approach.
H_{a1}	There are differences among subjects that use the AgilAC and the subjects that use the ad hoc approach according to the correctness of the results obtained in each approach.
Metric	The agility analysis result (composed by the diagnostic of the method on analysis plus the set of reasons for the diagnostic choice).

Variable Selection. Regarding the independent variables, their "factors" are on the input of the experimentation process and are all those that are manipulated and controlled. For this experiment they are the AgilAC, the ad hoc approach and two development methods analysed during the agility analysis. The dependent variables are the output and are those that are being analysed. Their variations must be observed based on changes made in the independent variables. In our case, the dependent variables are the time spent with the analysis conducted and the correctness of the analysis result.

Subject selection: Subjects were selected by random sampling. They were twenty undergraduate students who had a slightly or no experience in agile methods before the provided training.

Training: Prior the experiment, we applied a survey to get the expertise of subjects concerning agile methods, as can be seen in Table 5. The population was characterized as homogeneous and basically formed by subjects with little or no experience with agile methods and also no experience in agility analysis. As the subjects did not have any prior knowledge about agile methods, the following topics were taught in the training phase: (i) agile methods, (ii) AgilAC framework, and (iv) experiment forms - how to fill the analysis forms.

Table 5. Level of experience of the subjects

	No	Little	Average	High
Theoretical knowledge in agile methods	5	8	7	0
Level of practical experience	14	5	1	0
Knowledge of the Agile Manifesto	0	10	4	0
Level of knowledge of the Scrum	5	10	5	0
Level of knowledge of the XP	5	9	6	0
Level of knowledge of the Crystal family	11	4	5	0
Experience with analysis agility	20	0	0	0

Experiment Design. As mentioned, the subjects grouping was conducted with random selection, since the levels of knowledge and experience were homogeneous. So, we formed the following groups: the group 1 with seven subjects, the group 2 with six subjects and the group 3 with seven subjects.

To allow the comparison, we created descriptions of two hypothetical development methods which should be analysed by the subjects. We created a heavyweight method called mA and an agile method called mB. The mA was created by extending the original description of Unified Process [12] with the addition of a set of practices, making this method quite heavy. For example, the addition of a set of artifacts that must be delivered as documentation, practice not followed for agile methods. Thus, the expected result for mA is that the subjects were able to identify that this method is heavyweight. Conversely, the mB was an adaptation of the Scrum [13] without the addition of any activity. Therefore, it was expected that the subjects were be able to identify that it is agile.

During the ad hoc analysis, the subjects were allowed to check the Agile Manifesto [14]. Besides that, the subjects were free to try to follow any strategy for evaluating the description as well as an evaluator that is inexperienced with agile methods can do it, considering a real scenario.

The experiment was designed as shown in Table 6. The groups performed the analysis in different sequences. In the first phase, group 1 analysed the mA method with the AgilAC and in the second phase this group analysed the mB method using the ad hoc approach. At the same way, in the first phase group 2 analysed the mB method using the ad hoc approach and in the second phase this group analysed the mA method with the AgilAC. Also in the first phase, group 3 analysed the mA method using the ad hoc approach and in the second phase this group conducted the analysis of the mB method with the AgilAC.

Table 6. Experiment execution plan

	Group 1	Group 2	Group 3
1st Phase	mA (heavyweight) + AgilAC	mB (agile) + ad hoc	mA (heavyweight) + ad hoc
2nd Phase	mB (agile) + ad hoc	mA (heavyweight) + AgilAC	mB (agile) + AgilAC

Instrumentation. The experiment phases were performed sequentially on the same day. The types of experiment validity threats were based on the classification proposed by Wohlin et al. [7]. The description of each validity threat for the conducted experiment is presented as follow.

Internal validity:

– Learning: since the subjects done two analyses in sequence, one at each experiment phase (according to Table 6), one could argue that the effort in second phase is lower because of knowledge obtained in the first phase. In this context, we consider two distinct possibilities of learning: (1) using the same method in two phases, varying only the approaches used (ad hoc or AgilAC). To eliminate this threat, different methods were provided in each phase. (2) the subject, even with different methods at each stage, could learn or improve their analysis capacity in the second phase with the experience gained in the first phase. In order to mitigate this threat, the groups conducted the experiment using the approaches in a different sequence;
– Productivity under evaluation: As the subjects conducted two analyses in sequence, each one using an approach, may be that the results of second analysis were affected due to lost of productivity. To mitigate this, groups did not perform the analysis following the same sequence of approaches.

Validity by Construction:

- Belief in the hypotheses: to prevent that the experiment's results were influenced by prior knowledge about hypotheses, they were not revealed to subjects;
- Participant expectations: to avoid that subjects carried out the analysis with a favourable expectation in favour of the AgilAC, the hypotheses or objectives of experiment were not revealed for the subjects;
- Development methods provided for agility analysis: if descriptions of existing agile methods were provided for the subjects, they could identify beforehand whether the method was agile or not. To mitigate this, new descriptions of development methods were created. Furthermore, as an agile method and a heavyweight method were provided during the experiment, a method could be analysed more easily than the other. Then, to mitigate this, both development methods were analysed using the two approaches (ad hoc and AgilAC).

External validity:

- Experiment environment versus real environment: in relation to documentations of development methods provided for agility analysis, they are short documentations that were created exclusively for the experiment. Thus, it could be a risk if these methods do not represent real development methods. To mitigate this, the descriptions of development methods were versions adapted of existing methods.

Experiment Execution. The experiment phases (Table 6) were applied sequentially in the same day. Table 7 shows the sequence of steps executed during the agility analysis, using both the AgilAC and the ad hoc approach.

3.2   Analysis of Data Collected

This section presents the data collected and results obtained according to the time spent on the agility analysis, the correctness of agility analysis results, and subjects opinions on the AgilAC and the ad hoc approach.

a) Time spent on the agility analysis

The time was recorded since the beginning of analysis, i.e., after reading the description of the development method (item 2, Table 7), the subjects could start the analysis, and then the time was recorded from this moment until the finalization of writing about the agility diagnosis and its justification (item 3, Table 7). Based on the recorded time, a scatter plot was drawn (Fig. 3), allowing the analysis and elimination of outliers.

Considering subjects who did not conduct the agility analysis as expected, our criteria for eliminating outliers were both: the results with no justification about the agility diagnosis, and the time spent above the average time by others subjects. Thus, subjects 3, 6, 17 and 20, represented in the scatter plot of Fig. 3, were eliminated.

Table 7. Steps executed during agility analysis using the AgilAC and the ad hoc approach

Approach	Steps Sequence
Ad hoc	1.Read the development method description; 2.Analyse the agility of development method in an ad hoc way, with aiding of the reference document for the ad hoc approach (that is, the Agile Manifesto); and 3.Determine the agility diagnosis of development method (such as undefined, agile or heavyweight) and justify it based on the development method description and the reference document for the ad hoc approach.
AgilAC	1.Read the development method description; 2.Analyse the agility of development method using the AgilAC framework, with aiding of the reference document for the AgilAC analysis; and 3.Determine the agility diagnosis of development method (such as undefined, agile or heavyweight) and justify it based on the development method description and the reference document for the AgilAC framework.

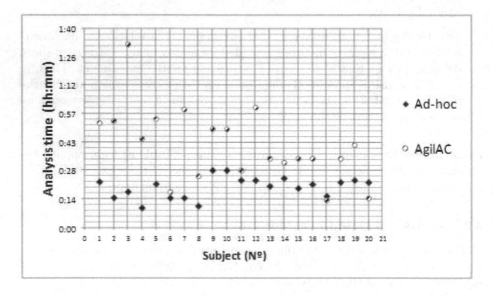

Fig. 3. Dispersion of analysis time per subject

Comparing the time spent in the analysis of each method (Fig. 3), it was observed that the development method did not provide significant influence on the analysis time. However, the approach used had a significant impact on the time spent.

Group 1 and group 2 analysed the mA method with the AgilAC, reaching an average time of 42 minutes. Group 1 spent 47 minutes performing this analysis in the first phase, and group 2 spent 37 minutes, conducting the analysis in the second phase. On the other hand, group 3 analysed the same development method with the ad hoc approach in the first phase, obtaining an average time of 23 minutes. Based on this, it was observed that the agility analysis conducted with the AgilAC framework was 82.6% slower than using the ad hoc approach.

Similar results were found for mB method. When group 1 and group 2 used the ad hoc approach on the second and on the first phases respectively, they had an average time of 19 minutes. Group 1 spent an average time of 16 minutes and group 2 spent 22 minutes. On the other hand, group 3 had an average time of 39 minutes to analyse the same development method (i.e., mB) with the AgilAC. Thus, the analysis with the AgilAC framework had been 105.26% slower than using the ad hoc approach. Therefore, the agility analysis with the AgilAC is, on average, 93.93% slower than performing the analysis with the ad hoc approach. In other words, the evaluator spend nearly twice to conduct the agility analysis with the AgilAC.

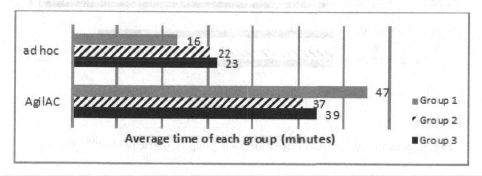

Fig. 4. Average time spent in agility analysis (with outliers removed)

It is possible to observe in Fig. 4 that the analysis conducted with the AgilAC took longer when compared to the ad hoc approach for all groups in all combinations, refuting **H0** null hypothesis and confirming **Ha0** alternative hypothesis. That is, subjects spent more time to perform the analysis using the AgilAC because they were guided by the AgilAC framework; performing a more complete and, consequently, more time consuming analysis. Additionally, an ad hoc analysis considers only "what" the subjects can identify as important to be observed in relation to agility of development methods.

b) Correctness of agility analysis results

Comparing group 1 with group 2 in relation to correctness of agility analysis results, the AgilAC had better results, then it is possible to eliminate the "learning" influence because the AgilAC is used by group 1, in the first phase (Fig. 5), but it is used by group 2 in second phase (Fig. 6), both with the same development method. Thus, group 1 had not chance to learn because it had done anything before.

To mitigate the possibility of one of the methods has favoured the AgilAC, comparing groups 1 and 3, it was observed that the AgilAC achieved better results in relation to the ad hoc approach in both groups, even using the AgilAC

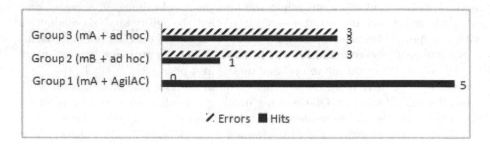

Fig. 5. Number of hits and errors for each group of participants (1st Phase)

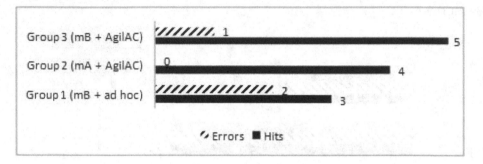

Fig. 6. Number of hits and errors for each group of participants (2nd Phase)

to analyse different development methods. Furthermore, to determine if development methods have not really influenced the correctness of the agility results, we observed that group 2 and group 3 had not significant variations in relation to number of hits at each phase. This means that development methods are equivalent in complexity, since when we compare the planning for these groups only the sequence of methods varies.

In Fig. 7, it is shown the percentage of errors and hits per approach. It is possible to verify that the analyses performed with AgilAC got more correct answers than the analyses performed using the ad hoc approach. As group 1 differs from group 3 just by the execution sequence, we conclude that the null hypothesis **H1** can be refuted and the hypothesis **Ha1** is correct.

c) Subjects opinions on both the AgilAC and the ad hoc approach

As the subjects had no experience with other agility analysis techniques, we applied a questionnaire after experiment execution to subjects point out which of two approaches "was better" in their opinion. This questionnaire has two questions: 1) "Which approach was easier to identify the factors that harm the agile philosophy?" and 2)"Which approach does provide the best support for characterizing the development method agility?".

Besides these questions there was a blank field for additional observations and suggestions. The aim of this field is to help us to identify other needed

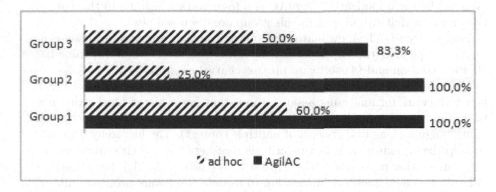

Fig. 7. Hits percentage for group in each analysis approach

improvements in the AgilAC framework. The acceptance rate of 80% for AgilAC in Question 1 showed the subjects thought its usage easier than the ad hoc approach. Regarding the Question 2, 95% of the subjects agreed that AgilAC is better than ad hoc approach. Checking the subjects' observations, we identified that they felt the necessity of a tool for supporting the application of AgilAC. Regarding the difficulties, the subjects reported about the time spent to use the framework, as it took longer than using ad hoc approach – this can also be eased with a tool. Besides, taking into consideration the observations, we were able to improve the descriptions of the agility criteria and also the related questions for guiding the identification of coverage levels.

4 Related Works

There are several mechanisms in the literature for evaluating agility of development methods. It is possible to classify the approaches for analysing agility of development methods in three categories or main groups, according with [15]:

(1) Evaluation Checklists, which are mainly used for agile teams, aiming at estimating the agility degree of a method or team. The evaluation, in this case, is done by evaluating the presence or absence of practices that are considered "agile";

(2) Agile Adoption Frameworks, used for customizing agile methods. We found proposals for modified versions of agile methods, usually taking into consideration issues from organizational contexts. They start from a reference method for supporting the building of a modified version, more suitable to distinct environments. Examples of works are Pikkarainen and Huomo [16], Abrahamsson et al. [17], Soundararajan et al. [18], Russo et al. [19]; and

(3) Agility Measurement Approaches, which are mechanisms whose aim is to provide a diagnosis about a specific method. This diagnosis aims to provide a report or conclusion about the adherence of the method to agile philosophy

adopted by the organization. AgilAC is a framework which fits in this category. Other works that can be put in this group are discussed next.

Framework4-DAT [4] evaluates the agility from four dimensions (1) determining the scope of method application; (2) agility characterization; (3) value-based characterization and (4) software process characterization.

CEFAM (Comprehensive Evaluation Framework for Agile Methodologies) [5] is a framework for analysing agile methods that has a hierarchical structure, in that the method is analysed under different evaluation domains (process, modelling languages, agility, criteria of multiple context). The hierarchy is proposed to help the evaluators in selecting criteria that better suit their evaluation goals.

From another perspective, there are studies (such as [20,21]) that evaluate the adherence of agile methods in relation to models of software processes maturity.

Although there are some mechanisms to help analyse the agility of the development methods, we could not find empirical studies in the literature in order to analyse their applicability and to confirm the benefits of these mechanisms.

5 Conclusion and Future Works

In this paper we give an overview of the AgilAC framework, proposed by us previously [6], and also we describe the experiment to evaluate the applicability of this framework. The experiment evaluated the AgilAC framework from the perspective of how it can support the agility analysis. In other words, the main goal was to evaluate whether the AgilAC is able to improve the capability of evaluators, not experts in agile methods, to elaborate a "correct" and "well justified" analysis.

Based on the data obtained with the experiment, it was possible to verify the advantage of the AgilAC framework over an ad hoc approach, when considering the correctness of results of the agility analysis. Moreover, the experiment's results also provided good feedbacks, which were useful to identify change points for improving the quality of the AgilAC. Another parameter evaluated was the time to perform the analysis, which was bigger when using the AgilAC. This happens because the AgilAC is a framework which needs to be followed; that is the opposite of the ad hoc approach, as it has no systematic steps. However, we claim the improvement in the analysis seems much more important than time spent, since the set of "checkpoints" in a systematic analysis brings the reliability that all important points in the analysis will be covered.

As future works we can mention: (i) to conduct experiments that include experts in the agile philosophy to analyse the AgilAC; (ii) to study the possible refinements in the AgilAC, based on the results of the experiment conducted and further experiments; (iii) to develop computational tool to improve the analysis time, making the process of filing the framework forms quicker; and (iv) to perform case studies with practitioners in real development environments.

References

1. Sommerville, I.: Software Engineering, 9th edn. Addison Wesley (2010)
2. Lappo, P., Andrew, H.C.T.: Assessing agility. In: Eckstein, J., Baumeister, H. (eds.) XP 2004. LNCS, vol. 3092, pp. 331–338. Springer, Heidelberg (2004)
3. Conboy, K., Fitzgerald, B.: Toward a conceptual framework of agile methods: a study of agility in different disciplines. In: 2004 ACM Workshop on Interdisciplinary Soft. Eng. Research, pp. 37–44 (2004)
4. Qumer, A., Henderson-Sellers, B.: A framework to support the evaluation, adoption and improvement of agile methods in practice. Journal of Systems and Soft. 81(11), 1899–1919 (2008)
5. Taromirad, M., Ramsin, R.: CEFAM: Comprehensive evaluation framework for agile methodologies. In: 32nd Annual IEEE Soft. Eng. Workshop, pp. 195–204 (2008)
6. Benites, M., Oliveira, L., Cagnin, M.I.: Agilac: A framework for agile methods evaluation. In: 39th Latin American Computing Conf., pp. 80–95 (2011)
7. Wohlin, C., Runeson, P., Höst, M., Ohlsson, M.C., Regnell, B., Wesslén, A.: Experimentation in software engineering. Springer (June 2012)
8. Software Engineering Institute (SEI): Standard CMMI appraisal method for process improvement (SCAMPI), version 1.1: Method definition doc. (2001)
9. Abrantes, J., Travassos, G.: Common agile practices in software processes. In: 5th Intern. Symp. on Empirical Soft. Eng. and Measurement, pp. 355–358 (2011)
10. PMBok: A Guide To The Project Management Body Of Knowledge (PMBOK Guides). Project Management Institute (2004)
11. ISO/IEC: Software process assessment (SPICE) — part 2 : A model for process management (ISO/IEC 15504) (1995)
12. Jacobson, I., Booch, G., Rumbaugh, J.: The Unified Software Development Process. Addison-Wesley (1999)
13. Schwaber, K., Beedle, M.: Agile Software Development with Scrum, 1st edn. Prentice Hall (2001)
14. Agile Alliance: Manifesto for Agile Software Development. World Wide Web (2001), http://www.agilemanifesto.org (access in January 02, 2013).
15. Soundararajan, S.: A Methodology for Assessing Agile Software Development Approaches. PhD thesis, Virginia Polytechnic Inst. and State University (2011)
16. Pikkarainen, M., Huomo, T.: Agile assessment framework. Technical Report D.4.1 v1.0, Agile VTT (Agile Soft. Develop. of Embedded Systems) Project Publication, Finland (2005)
17. Abrahamsson, P., Warsta, J., Siponen, M.T., Ronkainen, J.: New directions on agile methods: a comparative analysis. In: 25th Intern. Conf. on Soft. Eng., pp. 244–254 (2003)
18. Soundararajan, S., Balci, O., Arthur, J.: Assessing an organization's capability to effectively implement its selected agile method(s): An objectives, principles, strategies approach. In: Agile Conf. (AGILE), pp. 22–31 (2013)
19. Russo, N., Shams, S., Fitzgerald, G.: Exploring adoption and use of agile methods: A comparative case study. In: 19th Americas Conf. on Information Systems, vol. 2, pp. 1565–1572 (2013)
20. Paulk, M.: Extreme programming from CMM perspective. IEEE Software, 1–8 (November 2001)
21. Arimoto, M., Murakami, E., Camargo, V.V., Cagnin, M.I.: Adherence analysis of agile methods according to the MR-MPS reference model. In: 8th Brazilian Simp. on Soft. Quality, pp. 249–263 (2009)

Icon and Geometric Data Visualization
with a Self-Organizing Map Grid

Alessandra Marli M. Morais[1], Marcos Gonçalves Quiles[2],
and Rafael D. C. Santos[1]

[1] National Institute for Space Research
Av dos Astronautas. 1758, CEP 12227-010, São José dos Campos, Brazil
[2] UNIFESP São José dos Campos
Rua Talim, 330, CEP 12231-280, São José dos Campos, Brazil
{alessandra.marli,quiles}@unifesp.br
rafael.santos@inpe.br

Abstract. Data Visualization is an important tool for tasks related to Knowledge Discovery in Databases (KDD). Often the data to be visualized is complex, have multiple dimensions or features and consists of many individual data points, making visualization with traditional icon- and pixel-based and geometric techniques difficult. In this paper we propose a combination of icon-based and geometric-based visualization techniques backed up by a Self-Organizing Map, which allows dimensionality reduction and topology preservation. The technique is applied to some datasets of simple and intermediate complexity, and the results shows that it is possible to reduce clutter and facilitate identification of associations, clusters and outliers.

Keywords: Visualization, Kohonen Self-organizing Maps.

1 Introduction

Recent technological advances in information technology allows the collection and storage of large amounts of data almost effortlessly [1]. Data is collected as much as possible, without much thought at the collecting criteria because it is expected that it may be a potential source of valuable information after analysis [2]. However, finding valuable information from collected datasets (a process known as Knowledge Discovery in Databases or KDD) is often a non-trivial task.

Data visualization techniques has been used in KDD [3] as an attempt to make human beings an integral part of the data analysis process [1], nevertheless, creation of good visual representations which help and maximize data comprehension is not easily achievable. A subjectively good visual representation of a dataset depends largely on the data itself and is still a largely intuitive and ad-hoc process. Many of the well-established visualization techniques found in basic visualization tools (e.g. charts) do not provide adequate support for some large, multidimensional and/or complex datasets [1], being adequate only for summarization of the data.

B. Murgante et al. (Eds.): ICCSA 2014, Part VI, LNCS 8584, pp. 562–575, 2014.

Two of the problems inherent to visualization of large datasets are clutter, which happens due to the need to display many data points at once; and projection/representation, which are techniques required for the visualization of multiple-dimensional data by reducing the number of dimensions for visual representation.

The Self-Organizing Maps (SOM), a neural network algorithm proposed by Teuvo Kohonen [4], is a technique that can be used to attempt to solve these two problems. The SOM is composed of a matrix or lattice of cells, each representing a prototypical data point that can be considered a centroid of a cluster of the original data. The data quantization capabilities of the SOM allows the creation of subsets of the data (reducing clutter). The SOM also creates a view of the data corresponding to a reprojection of the original dimensions into a smaller number of dimensions, usually two, allowing the exploration of the relation between cells, for similarity and dissimilarity identification.

Each cell in a SOM lattice can be used as a basis for well-established visualization techniques or new techniques more adequate to the problem and data. Integration of the SOM features with the particular visualization techniques can improve the identification of patterns in datasets when compared with the traditional visualization techniques alone.

This paper presents techniques for multidimensional and/or large datasets that uses the Kohonen Self-organizing Map as basis to cluster and reorganize data vectors. The paper is organized as follows: Section 2 introduces the main concepts about visualization. Section 3 describes the Self-Organizing Maps algorithm and its use in visualization tasks. Section 4 presents some examples of the technique. Finally, Section 5 presents conclusions and directions for future work.

2   Visualization

Visualization is usually described as the mapping of data to a visual representation. The visual representations, if well constructed, can be useful not only to present the information quickly to users, but also to help and maximize data comprehension [5]. Data visualization can be used for Exploratory Analysis, Confirmatory Analysis and Presentation [6].

A good visual representation enables the processing of several items of information at the same time, thanks to the human visual system and brain which is able to process a huge amount of information simultaneously and which is more efficient to extract information from visual representation than from an amount of numbers, text or a combination of these [5].

According to Tufte [7], a good visual representation is one which shows complex ideas with clarity, precision and efficiency. Graphical excellence is what gives to the viewer the greatest number of ideas in the shortest time with the least ink in the smallest space telling the true and avoiding misinterpretation about data. However, there is no universal process to be done that ensures these requirements.

Defining a mapping that will result in a good visual representation is not a trivial task. The mapping depends largely on the task being considered and it is still a largely intuitive and ad-hoc process. It should feature easy understanding, avoid complex captions and undesirable visualization characteristics like occlusions and line crossings, that might appear as an artifact limiting the usefulness of the visualization [8].

The challenges and issues of a good mapping are significant when the data to be visually represented is large, highly dimensional or complex. The concept of dimensionality and size refers respectively to the amount of characteristics (attributes) which describe a data item of the data set and the total amount of items which comprise this set. There is no precise rule on what characterizes a high-dimensional and large data set [3]. However, despite of the powerful combinations of human brain and visual perception ability and the technological advances, there are some limitations on our capacity to deal with dimensions and the amount of data items that can be displayed.

The human vision is easily able to deal with two and three dimensions, so methods to visualize data with dimensionality greater than three often involves projecting the information into fewer dimensions. This needs to be done with care since some approaches to do it may not maintain the existing topological relationships between data in the original feature space, which is an important feature in the analysis process [9]. For visualization of large datasets, one must also consider a hardware-imposed limitation on the amount of data items that can be displayed on paper or on screen [1].

Even though visualization has an important role in KDD, the major problem that hampers the use of data visualization in KDD is that many of the well-established visualization techniques are not effective when applied to complex data sets [3]. Daniel Keim [10] describes some techniques for visualizing large amounts of multidimensional data. According to the taxonomy described by Keim and Kriegel [10], examples of visualization techniques are: Geometric Techniques (Parallel Coordinates, Scatterplot Matrices and others), Icon Techniques (Chernoff faces, Color Icons and others), Pixel-Oriented Techniques (Recursive Patterns, Pixel Bar Charts and others), Hierarchical (n-Vision, dimensional stacking and treemaps), Graph (Hy+, Margritte and SeeNet), and hybrid approaches.

Oliveira and Levkowitz [3] describe some characteristics considering strengths and weaknesses of those visualization techniques regarding the dimension and size of a given data set, but as mentioned these techniques may not be effective when applied to large datasets or datasets with multiple dimensions. For these cases, techniques that allow dimensionality reduction and grouping to remove clutter may be useful.

3   The Self-Organizing Map

The Self-Organizing Maps (SOM) [4] is a neural network algorithm based on unsupervised learning. It implements an orderly mapping of high-dimensional data into a regular low-dimensional grid, compressing information while preserving the most important topological and metric relationships of the data item on the display [11].

The SOM consist of M units located on a regular low-dimensional grid which represent the map. The grid is usually one- or two-dimensional, particularly when the objective is to use the SOM for data visualization. Each unit j has a prototype vector $m_j = [m_{j1}, ..., m_{jd}]$ in a location r_j, where d represent the dimension of a data item. The map adjusts to the data by adapting the values of its prototype vectors during the training phase. At each training step t a sample data vector $x_i = [x_{i1}, ..., x_{id}]$ is chosen and the distances, usually Euclidean distance, between x_i and all the prototype vectors are calculated to obtain the best-matching unit (BMU), c_i:

$$c_i = argmin_j\{||x_i - m_j(t)||\} \tag{1}$$

Once the closest prototype vector is found its values and the values of its neighborhood prototype vectors are updated, moving them toward x_i. The update rule is:

$$m_j(t+1) = m_j(t) + \alpha(t)h_{c_j}j[x_i - m_j(t)] \tag{2}$$

where t is the training step index, $\alpha(t)$ is the learning rate at moment t and $h_{c_j}j$ is a neighborhood centered winner unit (the best-matching unit). The winner unit is updated with a rate while its neighborhood is also updated with a smaller rate, determined by the distance on the map grid $||r_{c_i} - r_j||$, for example, by a Gaussian:

$$h_{c_j}j(t) = e^{-\frac{||r_{c_j} - r_j||^2}{2\sigma^2(t)}} \tag{3}$$

where r_{c_i} and r_j are positions of units c_i and j on the SOM grid and $\sigma(t)$ is the neighborhood radius.

The algorithm produces two interesting characteristics to complex data set visualization: Quantization and Projection. The quantization result in a tentative to find a set of prototype vectors which reproduce the original data set as well as possible, while the projection try to find low dimensional coordinates that preserve the distances (or the order of distances) between the originally high-dimensional data [12]. The SOM algorithm has proved to be especially good at maintain the topology of the original dataset, meaning that if two data samples are close to each other in the grid, they are likely to be close in the original high-dimensional space data [12].

3.1   Self-Organizing Maps as Visualization Tool – General Ideas

The methods for visualizing data using SOM at literature can be grouped in three categories based on the goal of visualization [12]: methods that get an idea of the overall data shape and detect possible cluster structure, methods that analyze the prototype vectors and methods for analysis of new data samples for classification and novelty detection purposes.

At the first category some projection techniques as Principal Component Analysis (PCA) and Sammon's projection are used to visualize the shape of the SOM in the input space. These approaches plot each prototype vector in a two or three dimension as dots, for example, colored according to its position on the SOM grid and connected with its neighbors by lines, giving an informative picture of the global shape [13] and may exposing the clusters' structures. A more efficient technique to show cluster structure is the distance matrix, being the Unified Distance Matrix (U-Matrix) one of the most used [14]. The U-Matrix is a visual representation of the SOM to reveal cluster structure of the data set. The approach colors a grid according to the distance from each vector prototype and its neighbors: dark colors are chosen to represent large distances while light colors correspond to proximity in the input space and thus represent clusters.

Analysis of prototype vectors is a wide field whose approach is to visualize data attributes using the prototype vectors in order to get some insights about the spreading of values, cluster properties and correlations between attributes. According to Vesanto [12] the component plane plays the key role in the analysis of the prototype vectors. The component plane consist of a SOM grid where each map unit represent one prototype vector attribute colored according to the value of this attribute. The analysis can be done by plotting one component plane or all of them side by side. Other uses are the plotting of vector prototype attributes using some simple visualization technique like scatter plot matrix and by plotting of vector prototype attributes on its grid position.

Methods from the third group can give some insights about which prototype vector corresponds to a given new data vector, how accurate it is and if the new data vector really belongs to the abstract model created by the SOM. In order to detect this the response of the prototype set to the data sample has to be quantified and visualized. Some visual representations based on histograms are described by Vesanto [12].

4   Visualization Using the SOM and Icon and Geometric Techniques

The basic principle of using the SOM grid for visualization is to create a graphical representation associated with each prototype (neuron), then displaying these graphical representations as a matrix. Each component of the matrix can be based on a traditional e.g. icon-, pixel- or geometric-based representations, or specific visualization techniques.

It is expected that the general appearance of the composition of the graphical componentes on the SOM matrix will yield interesting visual clues to the nature of the data and reduce clutter, and at the same time the organization of the components will allow comparison between sets of similar data (in one component) to data that is somehow similar but to a lesser extent (in neighbor components). Due to the SOM properties the visual representation obtained will obey some of the principles posed by Tufte [7], particularly, show the data, induce the viewer to think about the substance rather than about the methodology, avoid distorting what data has to say, present many data in a small space, make large data sets coherent, encourage the eye to compare different pieces of data and reveal the data at several levels of detail.

In order to demonstrate the technique we've developed an API (Application Programming Interface, or a set of classes) using the Java language that allows the creation of a SOM based on input data and the creation of graphical components to represent the data visually.

A very simple example that demonstrate the approach can be seen in Figure 1. In this example, a synthetic dataset with four Gaussian blobs in a two-dimensional dataset is processed by the SOM. In order to visualize the prototype vectors (the neurons) we developed a simple component that displays at the same time the original dataset points (using small dark gray dots); the prototype vector (neuron) as a light, large blue dot, on the position determined by its values. The data points that can be considered as assigned or belonging to that neuron are also displayed as medium blue dots.

Each component displays also the number of data points assigned to that neuron. The visualization components are displayed in a grid, corresponding to the SOM architecture, in this case a 4×4 rectangular lattice.

The plot shown in Figure 1 illustrate the main concepts of using the SOM for visualization: an overview of the grid shows clusters and how the neurons' prototypes approximates the centers of the clusters. A more detailed view shows possible groupings of clusters, e.g. for the large blob in the center of the distribution, which is represented by several neurons which are contiguous due to the topology-preservation capability of the SOM.

It is important to point that this first example is deliberately simple, with two-dimensional data being mapped to two dimensions, but the technique could easily deal with more data and/or more dimensions, provided that the component used for visualization could represent multiple dimensions. That will be demonstrated by other examples in this section.

4.1   Example: Geometric and Icon Techniques Applied to the Display of the Iris Dataset

Although there is no single rule to develop a good visual representation, some well-established visualization techniques may give a first insight about the data and works as a way to construct a better visual representation. For visual representation of data processed by a SOM, geometric and icon techniques can be good starting points.

Geometric techniques consist of mapping the data of the attributes on a geometric space [5], usually by finding an interesting projection of data set dimension [10].

Inselberg's parallel coordinates [15] is an example which maps k-dimensional data to two-dimensional surface by using k equidistant parallel axes where each axes correspond to one dimension and may be scaled according to the minimum

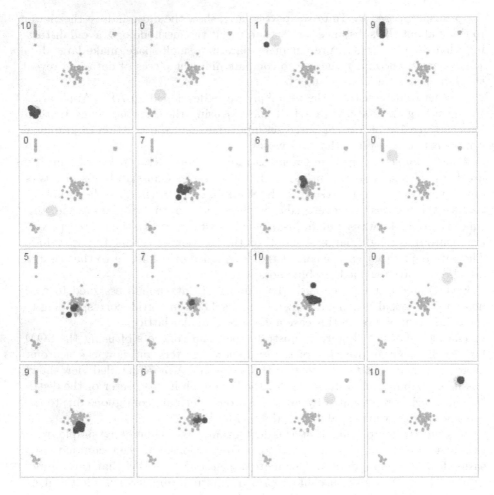

Fig. 1. A visual representation of synthetic data using a 4x4 SOM grid

and maximum value of this dimension; one data item is presented as a polygonal line, intersecting each axes at the point which corresponds to the value of considered dimension. The parallel coordinates technique is effective for detecting outliers and correlation amongst different dimensions [3] but may suffer from overlap and occlusions of lines when used to display large datasets.

To give an example of the capabilities of the proposed API we will explore some visualization techniques applied to the popular Iris dataset [16]. This dataset is formed by 50 samples of each Iris plant species: Iris Setosa, Iris Virginica and Iris Versicolor. Each sample contains four numerical values: sepal length, sepal width, petal length and petal width. This set is known to have two easily separable clusters, being one formed by the samples of Iris Setosa and the other formed by the samples of Iris Virginica and Iris Versicolor. It can not be considered as a complex data set, nevertheless, it is sufficient to demonstrate the concepts.

Figure 2 shows the Iris dataset plotted using a plain Parallel Coordinates plot. Each polygonal line crosses the four vertical axes, each representing a dimension on the dataset.

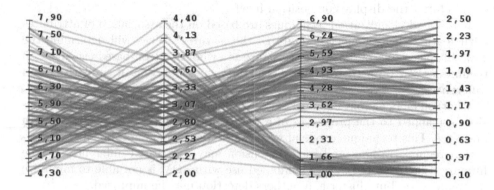

Fig. 2. Iris data set visualized by Parallel Coordinates

Figure 2 represents the 150 data points on the dataset – not a very large amount of data, but even so it is possible to see correlation between some attributes and some grouping, even with minor occlusion problems. Groupings can be made more evident and the occlusion problems can be reduced by preprocessing the data with a SOM and plotting it on a grid, as shown in Figure 3.

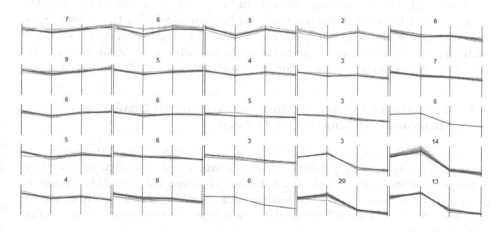

Fig. 3. Iris data set visualized by Parallel Coordinate over a SOM grid

In Figure 3 the prototype vector is displayed by dark polygons while the light ones represent the data items mapped by it. The number on top of each grid unit show the amount of data represented by the neuron at this grid SOM position. It is possible to observe a grouping of data with similar attributes in four elements in the lower right corner (corresponding to the Iris Serosa samples), and that in general that clutter was reduced. It is also important to point out that the component that displays a single parallel coordinate within the SOM was designed to be more concise, i.e. without labels and other information that could clutter the display composition itself.

Icon-based visualization techniques are based on the association of attributes with features in a geometric figure so the features' values will determine the general appearance of the geometric figure as a whole [5]. Icon-based visualization techniques can be used to allow the instinctive comparison of similarities and differences.

Chernoff faces are one example of this technique: in this approach each dimension is mapped to the properties of a face icon, i.e, shape of face, nose, mouth and eyes. This technique can deal with multidimensional data, but its interpretation require training and each dimension may be treated differently by the human visual perception. The combined use with SOM is not able to minimize its weaknesses, but cluster and outliers detection may be improved.

Figure 4 shows the same Iris dataset preprocessed by a SOM and plotted over a grid, with each element on the grid being a Chernoff face representation of the prototype vector (neuron) and data associated to that vector. The prototype vector is displayed int dark lines while the light ones represent the data items mapped by it. The number on top of each grid unit show the number of data vectors represented by the neuron at this SOM grid's position.

In the example shown in Figure 4 the length and width of the sepal are mapped to the shape of mouth and nose and the length and width of petal are mapped to the shape of face and eyes. The cluster (group of neurons) that represents the samples of Iris Setosa can easily be seen in the bottom right corner of the SOM grid. Another interesting feature that can be visualized with the clustering of Chernoff faces is the difference between the prototype vector and data associated with this vector, shown in the neurons in the top left corner, meaning that the data is clustered in those neurons but with some variance on the features' values.

4.2   Example: Geometric Techniques Applied to Display of a Medium-sized Time Series Dataset

As a second example of our approach, we've developed a tool to visualize a collection of time-series representing the accumulated precipitation measured by a sensor network in São Paulo state, Brazil [17]. Each entry on this dataset is composed by a time series with two years of monthly average precipitation as measured by each data collection platform in the sensor network, in a total of 1340 time series.

Fig. 4. Iris data set visualized by Chernoff faces over a SOM grid

Figure 5 shows a parallel coordinates plot of the 1340 time series. In that figure it is possible to identify some seasonal patterns, roughly corresponding to dry and wet seasons. It is also possible to identify several outliers and months with more and less variance on the average precipitation.

Precipitation

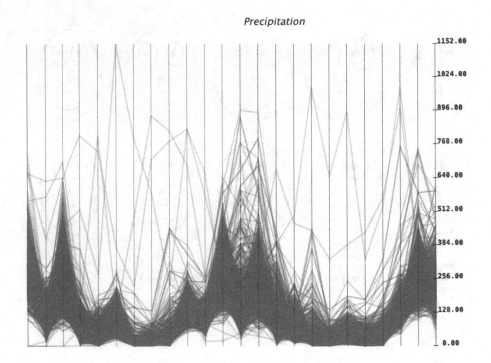

Fig. 5. Visual representation of 1340 time series by a Parallel Coordinates plot

While Figure 5 allows the visualization of some interesting features, it also presents a lot of clutter, making very hard the identification of groups on the dataset. Figure 6 shows the same data, grouped and reprojected into a SOM grid of 7 × 7 neurons.

In Figure 6 we can see the time series clustered into several loosely-defined groups, with the thick blue line representing the prototype vector (neuron) and with the data associated to that vector represented by thinner gray lines. In this example we don't use dimensionality reduction in the visualization, but the neurons are grouped in the grid accordingly to projections of the original data – in other words, similar time series are supposed to be represented by the same neuron or, if not much similar, by neighbor neurons.

The majority of the time series that follow a similar pattern are loosely grouped in the top left corner of the SOM grid shown in Figure 6. Some neurons clearly grouped time series that can be considered outliers, e.g. the neuron on the sixth row, last column and the one on the last row and first column. Reduction of clutter allowed the visualization of the prototypes and variation of the time series assigned to those prototypes, which would be impossible to see in the original Parallel Coordinates plot.

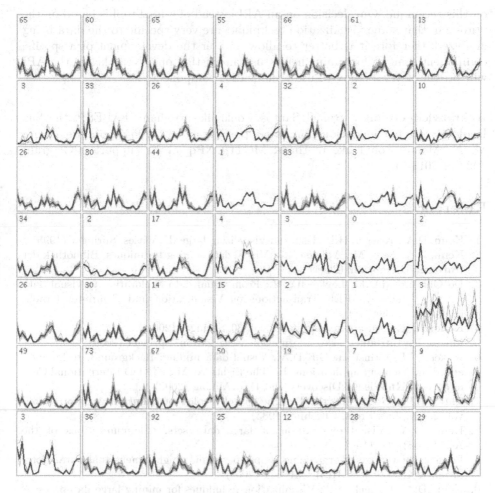

Fig. 6. Monthly rainfall visualized by Parallel Coordinates and SOM grid. The number on top of each grid unit show the amount of data represented by the neuron at this grid SOM position.

5   Conclusions and Future Work

In this paper we presented a technique for data visualization comprised of traditional icon- and geometry-based representations organized in a SOM grid or lattice. This technique allows the visualization of features presented and recognizable in groupings of the dataset, making easier the identification of common patterns, outliers and reducing clutter for data plots with several data points.

Future work will be towards a better, more flexible implementation of the underlying SOM algorithm, particularly allowing the use of hexagonal grids and different distance metrics. We are also working on an API (Application Programming Interface) to facilitate the development of visualization tools based on this technique. Our decision on an API instead of a final tool is based on the argument that some visualization techniques are very specific to the data being analyzed, therefore it is better to allow an user the development of a specific component than to force him/her to use an existing one. Nonetheless the API will provide several ready-to-use visualization components.

Acknowledgements. Rafael Santos would like to thank FAPESP, the São Paulo Research Foundation, for its support (grant 2014/05453-6). Alessandra M. M. Morais would like to thank MCTI/CNPq for its support (PCI grant 302428/2013-5).

References

1. Keim, D.A., Kriegel, H.P.: Issues in visualizing large databases. Springer (1995)
2. Keim, D.A., Sips, M., Ankerst, M.: Visual data-mining techniques. Bibliothek der Universität Konstanz (2004)
3. De Oliveira, M.C.F., Levkowitz, H.: From visual data exploration to visual data mining: A survey. IEEE Transactions on Visualization and Computer Graphics 9(3), 378–394 (2003)
4. Kohonen, T.: Self-organizing maps, vol. 30. Springer (2001)
5. Mazza, R.: Introduction to information visualization. Springer (2009)
6. Ankerst, M., Grinstein, G.K.D.A.: Visual data mining: Background, techniques, and drug discovery applications. In: The Eighth ACM SIGKDD International Conference on Knowledge Discovery and Data Mining (2002)
7. Tufte, E.R., Graves-Morris, P.: The visual display of quantitative information, vol. 2. Graphics Press, Cheshire (1983)
8. Keim, D.A.: Visual exploration of large data sets. Communications of the ACM 44(8), 38–44 (2001)
9. Penn, B.S.: Using self-organizing maps to visualize high-dimensional data. Computers & Geosciences 31(5), 531–544 (2005)
10. Keim, D.A., Kriegel, H.P.: Visualization techniques for mining large databases: A comparison. IEEE Transactions on Knowledge and Data Engineering 8(6), 923–938 (1996)
11. Kohonen, T.: The self-organizing map. Neurocomputing 21(1), 1–6 (1998)
12. Vesanto, J.: Data exploration process based on the self-organizing map. PhD thesis, Helsinki University of Technology
13. Koua, E.: Using self-organizing maps for information visualization and knowledge discovery in complex geospatial datasets. In: Proceedings of 21st International Cartographic Renaissance (ICC), pp. 1694–1702 (2003)
14. Gorricha, J., Lobo, V.: Improvements on the visualization of clusters in georeferenced data using self-organizing maps. Computers & Geosciences 43, 177–186 (2012)

15. Inselberg, A.: Parallel Coordinates – Visual Multidimensional Geometry and Its Applications. Springer (2009)
16. Fisher, R.A.: The use of multiple measurements in taxonomic problems. Annals of Eugenics 7(2), 179–188 (1936)
17. Garcia, J.R.M., Monteiro, A.M.V., Santos, R.D.C.: Visual data mining for identification of patterns and outliers in weather stations' data. In: Yin, H., Costa, J.A.F., Barreto, G. (eds.) IDEAL 2012. LNCS, vol. 7435, pp. 245–252. Springer, Heidelberg (2012)

Analysis of Turing Instability
for Biological Models

Daiana Rodrigues, Luis Paulo Barra, Marcelo Lobosco, and Flávia Bastos

Federal University of Juiz de Fora, Graduate Program in Computational Modelling,
Juiz de Fora, Brazil
{luis.barra,flavia.bastos}@ufjf.edu.br,
daiana.rodrigues@yahoo.com.br, marcelo.lobosco@ice.ufjf.br
http://www.ufjf.br/pgmc/

Abstract. Reaction-diffusion equations are often used to model biological phenomena. This type of system can produce stable spatial patterns from an uniform initial distribution. This phenomenon is known as Turing instability. This paper presents an analysis of the Turing instability for three biological models: a) Schnakenberg model, b) glycolysis model and c) blood coagulation model. The method of lines is used in the numerical solution, and the spatial discretization is done using a finite difference scheme. The resulting system of ordinary differential equations is then solved by an adaptive integration scheme with the use of SciPy, a Python package for scientific computing.

Keywords: Turing instability, Reaction-Diffusion Equations, Schnakenberg model, glycolysis model, blood coagulation model.

1 Introduction

Diffusion is a natural process by which a given system reaches equilibrium [1] and is present in many everyday situations. In diffusive systems composed by more than one substance or specie the diffusion phenomena may be accompanied by chemical reactions or other interactions that locally change concentrations. Systems that include diffusion and such interactions between substances are called reaction-diffusion systems [1,2,3].

Along the pure diffusion of a particular substance, the system tends to reach a homogeneous state as time goes by. Otherwise, reaction-diffusion models can evolve into a stable and heterogeneous spatial pattern from a homogeneous state of spatial equilibrium, due to small perturbations in the initial concentrations. This phenomenon is known as diffusion instability or Turing instability [4,5,6].

This paper presents a stability analysis of three reaction-diffusion models: a) Schnakenberg model, b) glycolysis model and c) coagulation model. The first two models are very known and were used to validate our computational implementation. The coagulation model was recently proposed by Vanegas [7] to describe blood clotting in the dental implant-bone interface. In this work we show that Vanegas' model does not present the Turing instability although the author[7]

B. Murgante et al. (Eds.): ICCSA 2014, Part VI, LNCS 8584, pp. 576–591, 2014.

argues that their coagulation model presents the Turing instability for a given set of parameters.

This paper is structured as follows. First, we give an overview of Turing Instability. Then we present the biological models studied in this work. Our computational implementation is described in section 4, and the results are presented in section 5. Finally, the last section presents our conclusion.

2   Turing Instability

In 1952 Alan Turing wrote the seminal paper [8] proposing that the process of generating structures and shapes in biological systems are based on diffusive chemical processes that occur on them. Three types of instabilities were found: a) oscillatory in time and uniform in space, b) stationary in time and periodic in space, and c) oscillatory in time and space [4]. The second type of instability, which induces the formation of spatial patterns where there was no pattern before, is known as Turing instability [6]. The mathematical analysis of reaction-diffusion equations allows to arrive at conditions that must be verified in order to allow the development of this pattern generation process.

A system describing diffusion and reaction between two species can be generally described by the following equations:

$$\frac{\partial u}{\partial t} = \nabla^2 u + \beta f(u, v) \tag{1}$$

$$\frac{\partial v}{\partial t} = d\nabla^2 v + \beta g(u, v) \tag{2}$$

where u and v are the concentrations of the substances, f and g are the reaction functions and d is the ratio between the diffusion constants of the different species. In the analysis of the Turing instability, usually it is assumed that the fluxes at the boundaries are equal to zero, so the spatial pattern formed suffers no influence from any external inputs.

Turing instability occurs when small spatial perturbations of a homogeneous equilibrium state leads a system described by Equations 1 and 2 to assume a new inhomogeneous steady state.

The analysis is made in two steps. From a homogeneous equilibrium state the conditions for time stability are developed after that small perturbations in the spatial distribution are introduced and the conditions for spatial instability are achieved.

In a homogeneous system, with uniform concentrations in space, the diffusion terms vanish so Equations 1 and 2 can be rewritten as:

$$\frac{\partial u}{\partial t} = \beta f(u, v) \tag{3}$$

$$\frac{\partial v}{\partial t} = \beta g(u, v) \tag{4}$$

A stable state, $(u, v) = (u_0, v_0)$, is found when the time derivative of the concentrations of the substances in the system vanish. From Equations 3 and 4, the following equations must be satisfied:

$$f(u_0, v_0) = g(u_0, v_0) = 0 \tag{5}$$

Considering small perturbations in the concentrations $\mathbf{w} = (w^u(t), w^v(t))^T$ the Equations 3 and 4 can be written for a point, $(u_0 + w^u(t), v_0 + w^v(t))$, in the neighbourhood of the equilibrium state, using the expansion of the Taylor series for functions f and g. Considering only the linear terms, it can be obtained the following equations:

$$\mathbf{w}_t = \beta \mathbf{A} \mathbf{w} \qquad \text{where} \qquad \mathbf{A} = \begin{pmatrix} f_u & f_v \\ g_u & g_v \end{pmatrix} \tag{6}$$

where the subscripts in the definition of the so called stability matrix, $\mathbf{A}$, stand for partial derivatives of f and g computed at (u_0, v_0).

If λ is an eigenvalue of $\beta \mathbf{A}$ with eigenvector $\mathbf{V}$, then:

$$\beta \mathbf{A} \mathbf{V} = \lambda \mathbf{V} \tag{7}$$

Thus $\mathbf{X} = e^{\lambda t} \mathbf{V}$ is the solution of Equation 6.

The original equilibrium state is linearly stable if $Re(\lambda) < 0$, since, in this case, the disturbance $\mathbf{w} \to \mathbf{0}$ when $t \to \infty$. The determination of eigenvalues λ of $\mathbf{A}$ is done, as usual, solving the characteristic equation leading to:

$$\lambda_1, \lambda_2 = \frac{1}{2}\beta[(f_u + g_v) \pm \sqrt{(f_u + g_v)^2 - 4(f_u g_v - f_v g_u)}] \tag{8}$$

From the analysis of Equation 8 it follows that the stability, i.e. $Re(\lambda) < 0$, is guaranteed if:

$$f_u + g_v < 0 \tag{9}$$
$$f_u g_v + f_v g_u > 0 \tag{10}$$

The above conditions guarantee linear stability over time for the original homogeneous state, so without the influence of diffusive terms. In the presence of diffusion, we have the complete reaction-diffusion system so the spatial and temporal perturbations have to satisfy:

$$\mathbf{w}_t = \mathbf{D} \nabla^2 \mathbf{w} + \beta \mathbf{A} \mathbf{w} \tag{11}$$

The solutions of the above system of equations has the form:

$$\mathbf{w}(r, t) = \sum_k c_k e^{\lambda(k)t} \mathbf{W}_k(\mathbf{r}) \tag{12}$$

where the constants c_k are determined by Fourier expansion of the initial conditions in terms of $\mathbf{W}_k(\mathbf{r})$, which are eigenvectors associated with the eigenvalues k of the eigenvalue problem.

If the problem is one-dimensional, defined in $0 < x < a$, $\mathbf{w}$ is a linear combination of terms in the form of $cos(\frac{n\pi x}{a})$ and the eigenvalues, $k = \frac{n\pi}{a}$ (a discrete set of values), are called wave numbers. Substituting the Equation 12 in 11 gives for each k:

$$\lambda \mathbf{W}_k = \beta \mathbf{A} \mathbf{W}_k + D \nabla^2 \mathbf{W}_k \tag{13}$$

Equation 13 can be rewritten as:

$$[\lambda \mathbf{I} - \beta \mathbf{A} - k^2 D] \mathbf{W}_k = 0 \tag{14}$$

Since we are looking for a non-trivial solution for the above system, the determinant of the matrix multiplying $\mathbf{W}_k$ in Equation 14 must be equal to zero. Thus the eigenvalues $\lambda(k)$ must satisfy:

$$\lambda^2 + \lambda[k^2(1 + d) - \beta(f_u + g_v)] + h(k^2) = 0 \tag{15}$$

where

$$h(k^2) = dk^4 - \beta(df_u + f_v)k^2 + \beta^2 \mid \mathbf{A} \mid \tag{16}$$

In order to the steady state be unstable to spatial disturbances, it should be imposed for the solution $\lambda(k)$ of Equation 15 that $Re\lambda(k) > 0$ for some $k \neq 0$. To ensure the existence of such a root, it is necessary that $h(k^2) < 0$ in Equation 15. Furthermore, since the coefficient of k^4 in Equation 16 is positive, negative values of $h(k^2)$ will occur for k^2 between its roots k_1^2 and k_2^2:

$$k_1^2 < k^2 < k_2^2. \tag{17}$$

It can be shown[9] that a necessary condition for obtention of positive roots with d different from 1, is:

$$(df_u + g_v) > 0 \tag{18}$$

On the other hand, in order to obtain two distinct real roots it is necessary a positive discriminant:

$$(df_u + g_v)^2 - 4d \mid \mathbf{A} \mid > 0 \tag{19}$$

In short, the conditions which restrict the space of parameters characterizing the *Turing space* can be summarized with the following set of inequalities, obtained from Equations 9, 10, 18 and 19.

$$f_u + g_v < 0 \tag{20}$$

$$f_u g_v + f_v g_u > 0 \tag{21}$$

$$df_u + g_v > 0 \tag{22}$$

$$(df_u + g_v)^2 - 4d(f_u g_v - f_v g_u) > 0 \tag{23}$$

Besides satisfying the inequalities (20)-(23) above, the two dimensional problems defined in $0 < x < p$, $0 < y < q$, must satisfy the equivalent of Equation 17, i.e., the integer values of the wavenumbers in both directions m and n must satisfy the following inequality:

$$k_1^2 < \pi^2 \left(\frac{n^2}{p^2} + \frac{m^2}{q^2} \right) < k_2^2 \tag{24}$$

3   Biological Models

This section presents three examples of biological phenomena modelled by reaction-diffusion equations: a) Schnakenberg model, b) glycolysis model and c) blood coagulation model.

3.1   Schnakenberg Model

The Schnakenberg model [10] uses partial differential equations to describe an autocatalytic reaction[1] with possible oscillatory behaviour. The Schnakenberg model is represented by the following equations:

$$\frac{\partial u}{\partial t} = \nabla^2 u + \gamma(a - u + u^2 v) \tag{25}$$

$$\frac{\partial v}{\partial t} = d\nabla^2 v + \gamma(b - u^2 v) \tag{26}$$

This model represents the behaviour of two chemical species, usually referred to as activator and inhibitor. If u is the activator chemical, kinetic reaction is such that in Equation 25 the term $u^2 v$ is the production of u in the presence of v. In Equation 26 the same term $(u^2 v)$ represents the consumption of v in the presence of u. The above system is represented in its dimensionless form, where a and b are source terms, d is the diffusion coefficient and γ is a dimensionless constant. All constants a, b, d and γ are positive.

---

[1] Autocatalysis is a type of reaction in which at least one of the reactants is also one of its products [11].

3.2 Glycolysis Model

The process called Glycolysis is a metabolic pathway composed by a set of ten enzyme-catalyzed chemical reactions [12]. This process occurs in the cytosol of the cell.

One glucose molecule has six carbon atoms and when it is broken two molecules are formed, each one with 3 carbons. These molecules are called pyruvate or pyruvic acid. The final products of this process are two molecules of ATP, two molecules of pyruvate and two molecules of NADH+ [2] [14].

The dimensionless form of glycolysis model is given by the following equations:

$$\frac{\partial u}{\partial t} = \nabla^2 u + i - ku - uv^2 \tag{27}$$

$$\frac{\partial v}{\partial t} = d\nabla^2 v + ku + uv^2 - v \tag{28}$$

The biological interpretation is the following: u represents the concentration of glucose, v the concentration of produced pyruvate, $d = \dfrac{D_v}{D_u}$, where D_u and D_v are the diffusion coefficients. The term uv^2 represents the non-linear consumption of u (first equation) and activation of v (second equation). The positive parameter i represents an initial amount of glucose. The parameter k, also positive, is related in the first equation to the natural glucose consumption and in the second one the pyruvate production[achar a citação adequada].

3.3 Blood Coagulation Model

When a blood vessel is injured, the body starts a process to stop the bleeding, called haemostasis. Blood clotting is an important part of haemostasis and is composed by a complex pathway of molecular and cellular events that results in the formation of a fibrin clot. This fibrin clot covers the wall of the injured blood vessel to stop blood loss and to begin the repair of the damaged tissue [15,16].

Three main steps can be used to illustrate in a simplified way the process of clot formation. In the first step, after the rupture of a blood vessel, a complex of substances, called prothrombin activator, is formed. In the second step occurs the conversion of prothrombin into thrombin through the action of prothrombin activator. In the third and final step, thrombin acts as an enzyme to catalyse the conversion of fibrinogen into fibrin filaments, which will retain the erythrocytes, platelets and plasma, forming the fibrin clot [16].

The equations that model the blood coagulation are the following:

$$\frac{\partial u}{\partial t} = \nabla^2 u + \delta - ku - uv^2 \tag{29}$$

---

[2] NAD (nicotinamide adenine dinucleotide) is a coenzyme, in other words, a non-protein organic substance necessary to the operation of some enzymes [13].

$$\frac{\partial v}{\partial t} = d \bigtriangledown^2 v + \gamma + ku + uv^2 - v \tag{30}$$

The above system is written in the dimensionless form, where u is the concentration of thrombin, v is the concentration of fibrinogen, δ is the initial amount of thrombin, k is the initial consumption of thrombin, d is the diffusion coefficient and γ is the initial amount of granules released by platelets.

4   Numerical Implementation

The numerical strategy used to solve the models presented in this work is initially apply the Finite Difference method to discretize the spatial differential operator present in the reaction-diffusion equations. After this discretization, the introduction of the boundary conditions transforms the problem into an initial value problem (IVP), associated with a system of non-linear ordinary differential equations. The technique of transforming a system of reaction-diffusion equations into an IVP is known as method of lines [1].

To solve numerically the models presented in the previous section, we used the *Python* programming language [17]. In the solution of the IVP the *Scipy* library was used [18]. Specifically, in the solution of the IVP the *odeint* package was used [19], as well as the *sparse* package [20], for dealing with the sparsity of the matrix used in the computations.

4.1   Method of Lines

The general form of the equations here studied is the following:

$$\frac{\partial u}{\partial t} = \nabla^2 u(x,t) + f(u,v,x,t) \tag{31}$$

$$\frac{\partial v}{\partial t} = d\nabla^2 v(x,t) + g(u,v,x,t) \tag{32}$$

where it is assumed that that the initial values of u and v are known and the normal derivatives of these quantities vanish at the boundaries.

The first step to solve numerically the above equations is to introduce the approximations for their spatial derivatives. Then the system is said to be in a semi-discrete form and can be written as:

$$\frac{d\mathbf{y}}{dt} = \mathbf{A}\mathbf{y} - \mathbf{f}(\mathbf{y},t) \qquad \text{for} \qquad 0 < t; \qquad \text{with} \qquad \mathbf{y}(0) = \mathbf{y}_0 \tag{33}$$

where here $\mathbf{A}$ is the spatial discretization matrix derived in the next subsection and:

$$\mathbf{f}(\mathbf{y}, t) = \begin{pmatrix} f(u_1, v_1, t) \\ \vdots \\ f(u_N, v_N, t) \\ g(u_1, v_1, t) \\ \vdots \\ g(u_N, v_N, t) \end{pmatrix}, \quad \mathbf{y}(t) = \begin{pmatrix} u_1(t) \\ \vdots \\ u_N(t) \\ v_1(t) \\ \vdots \\ v_N(t) \end{pmatrix} \tag{34}$$

In the form of Equation 33 the problem can be solved efficiently by methods available in the *Scipy* library.

4.2 Space Discretization

In order to approximate the Laplace operator, the spatial discretization was carried out using the second order central finite difference method. Figure 1 illustrates the process.

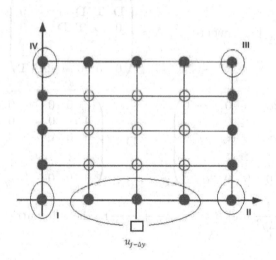

Fig. 1. Illustrative picture of the spatial discretization, modified from [1]. The filled points represent the ones who are located on the boundary.

Using the Laplace operator to represent a generic point j of the domain, we can compute it using the following approximation:

$$\nabla^2 u(x_j) \approx \frac{u_{j+1} - 2u_j + u_{j-1}}{\triangle x^2} + \frac{u_{j+N_{px}} - 2u_j + u_{j-N_{px}}}{\triangle y^2} \tag{35}$$

where N_{px} is the number of points in the x axis and N_{py} is the number of points in the y axis.

The boundary conditions adopted in this work establish that the flow should be zero across the boundary (Neumann boundary conditions). So, for instance, in the lower boundary the points corresponding to $u_{j-N_{px}}$ does not belong to the problem domain and are kwnon as ghost points. Adopting again a second order central difference scheme for the discretized boundary condition it is found that $u_{j+N_{px}} = u_{j-N_{px}}$ and the discretized Laplace operator applied to these points become:

$$\nabla^2 u_j \approx \frac{u_{j+1} - 2u_j + u_{j-1}}{\Delta x^2} + \frac{2u_{j+N_{px}} - 2u_j}{\Delta y^2} \tag{36}$$

The same strategy can be used to find the discretized operator for other parts of the boundary, including corners.

The spatial discretization matrix A in Equation 33 can then be written compactly as:

$$\mathbf{A} = \begin{pmatrix} \bar{\mathbf{A}} & 0 \\ 0 & d\bar{\mathbf{A}} \end{pmatrix} \qquad \text{where} \qquad \bar{\mathbf{A}} = \begin{pmatrix} \mathbf{T} & \mathbf{2D} & 0 & 0 & \cdots & 0 \\ \mathbf{D} & \mathbf{T} & \mathbf{D} & 0 & \cdots & 0 \\ 0 & \mathbf{D} & \mathbf{T} & \mathbf{D} & \cdots & 0 \\ \vdots & \vdots & \vdots & \vdots & \ddots & \vdots \\ 0 & 0 & 0 & \cdots & \mathbf{2D} & \mathbf{T} \end{pmatrix},$$

$$\mathbf{T} = \begin{pmatrix} \beta & 2\alpha_x & 0 & 0 & \cdots & 0 \\ \alpha_x & \beta & \alpha_x & 0 & \cdots & 0 \\ 0 & \alpha_x & \beta & \alpha_x & \cdots & 0 \\ \vdots & \vdots & \vdots & \vdots & \ddots & \vdots \\ 0 & 0 & 0 & 0 & 2\alpha_x & \beta \end{pmatrix}, \qquad \mathbf{D} = \begin{pmatrix} \alpha_y & 0 & 0 & \cdots & 0 \\ 0 & \alpha_y & 0 & \cdots & 0 \\ 0 & 0 & \alpha_y & \cdots & 0 \\ \vdots & \vdots & \vdots & \ddots & \vdots \\ 0 & 0 & 0 & \cdots & \alpha_y \end{pmatrix},$$

$\alpha_x = \dfrac{1}{\Delta x^2}$, $\alpha_y = \dfrac{1}{\Delta y^2}$ and $\beta = 2(\dfrac{1}{\Delta x^2} + \dfrac{1}{\Delta y^2})$. The sub-matrices $\mathbf{T}$ and $\mathbf{D}$ have dimension equal to $N_{px} \times N_{px}$.

4.3   Odeint

The *odeint* package provides functions to solve systems of ordinary differential equations using an adaptive scheme for both the integration step and the convergence order. This routine can solve the problem using either the BDF (Backward differentiation formulas) or the Adams methods[21]. The BDF method is adopted if the system is *stiff* and the implicit Adams method is used if otherwise. The order of convergence of the adopted method is also chosen adaptively along the solution.

5 Numerical Experiments

Using the strategy described in the previous section, a code was implemented using the *Scipy* library. Some simulations were then performed in order to validate the code, using for this purpose the Schnakenberg and the Glycolysis models. The results gave us confidence that our numerical implementation had been correctly implemented. After these tests the Blood Coagulation Model was simulated. Several tests were run with different levels of grid and time discretization refinement. Also the initial disturbance around the equilibrium point was varied. This section presents the most representative results found.

All simulations were performed on a computer equipped with an Intel $i7$ 2.80GHz processor, 8GB of RAM, running the Linux operating system, kernel version 3.9.10, Python interpreter version 2.7, and the *Scipy* library version 0.9.0.

5.1 Results for the Schnakenberg Model

Using the Equations 25 and 26, we assume that the temporal derivatives are equal to zero, $(\frac{\partial u}{\partial t}, \frac{\partial v}{\partial t} = 0)$, which allow us to find the steady state:

$$(u_0, v_0) = \left(a + b, \frac{b}{(a+b)^2} \right) \tag{37}$$

Our simulations were then performed using the parameter described in Table 1. The domain used in this simulations was $0 < x < 1$ and $0 < y < 1$. A mesh of 26×26 points was used in simulations. After time equals to 8, we collected the results. The *odeint* function returned the values for the solution using 60 points equally distributed. For this simulation, we applied a perturbation of 0.9% in the initial condition.

Table 1. Parameters used for simulating the Schnakenberg model

Parameter	Value
a	0.1
b	0.9
d	9.1676
γ	176.72

To obtain the Turing instability, the inequalities defined by Equations 20 to 23 have to be satisfied. Substituting in these inequalities the values given by Table 1, we obtain:

$$f_u + g_v = -35.34 < 0$$
$$f_u g_v + f_v g_u = 31229.96 > 0$$
$$df_u + g_v = 1119.36 > 0$$
$$(df_u + g_v)^2 + 4d(f_u g_v - f_v g_u) = 107748.65 > 0$$

Also, it is necessary non-zero integer values for m and n to represent the wave numbers, respecting the interval defined by Equation 24.

It can be observed that:

$$k_1^2 < \frac{(n^2 + m^2)\pi^2}{L^2} < k_2^2$$
$$4.3717 < n^2 + m^2 < 7.9996$$

In this situation, $m = 2$ and $n = 1$, or vice versa, are values that satisfy the inequality. So the patterns that can be formed are those of type $(2,1)$, as Figure 2 illustrates.

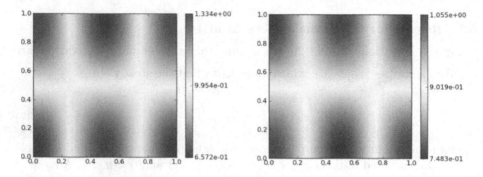

Fig. 2. Concentration of u, on the left, and v, on the right, for the Schnakenberg model on a $2D$ mesh

5.2   Results for the Glycolysis Model

The same strategy used in the previous section is applied to the Glycolysis model. We start calculating the steady state in the absence of any spatial variation:

$$(u_0, v_0) = \left(\frac{\delta}{(\delta^2 + k)}, \delta \right) \tag{38}$$

The parameters used in the two simulations with the Glycolysis model are given in Table 2. Two sets of parameters were used in the simulations. These values are used to solve the inequalities given by Equations 20 to 23; the results obtained are shown in Table 3. Using the first set of values given by Table 3 in Equation 24, we obtain:

$$10.45 < n^2 + m^2 < 60.42$$

Table 2. Parameters used for simulating the Glycolysis model

Parameter	Value for the 1st. test	Values for the 2nd. test
δ	2.8	1.2
k	0.06	0.06
d	0.0125	0.08

Table 3. Values computed for the inequalities given by Equations 20 to 23

Inequality	Values for the 1st. test	Values for the 2nd. test
$f_u + g_v < 0$	$-6.91 < 0$	$-0.58 < 0$
$f_u g_v + f_v g_u > 0$	$7.90 > 0$	$1.50 > 0$
$df_u + g_v > 0$	$0.88 > 0$	$0.80 > 0$
$(df_u + g_v)^2 + 4d(f_u g_v - f_v g_u) > 0$	$0.39 > 0$	$0.16 > 0$

The values that satisfy the inequality are $m = 5$ and $n = 2$, so the patterns that can be formed are those of type $(2, 5)$, as presented in Figure 3. To compute the simulation for this first set of test values, the bi-dimensional domain $0 < x < \pi$ and $0 < y < \pi$ was used. Also, a mesh with 61×61 points and time equal to 500 were used in the simulation that computed 1,000 points equally spaced in the analysed temporal interval. The perturbation level applied in the initial condition used in the simulation was 0.5%.

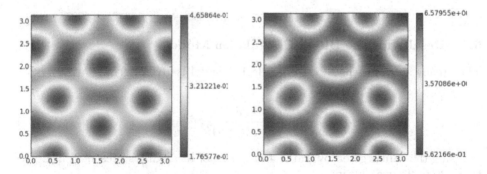

Fig. 3. Concentration of u, on the left, and v, on the right, for the Glycolysis model on a $2D$ mesh for the 1st. set of parameters.

Using the second set of values of Table 3 in Equation 24, we obtain:

$$62.5 < n^2 + m^2 < 187.50$$

Figure 4 shows the patterns formed by the solution of the inequality. To compute the simulation for this second set of test values, the bi-dimensional domain $[0, 5\pi \times 0, 5\pi]$, a mesh with 61×61 points, time equals to 500, and initial perturbation equals to 0.5% were used in the simulation. Again, $1,000$ points equally spaced were computed in the analysed temporal interval.

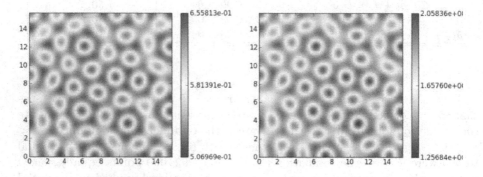

Fig. 4. Concentration of u, on the left, and v, on the right, for the Glycolysis model on a $2D$ mesh for the 2nd. set of parameters.

5.3   Results for the Blood Coagulation Model

The steady state for the blood coagulation model is given by:

$$(u_0, v_0) = \left(\frac{\delta}{k + (\delta + \gamma)^2}, \delta + \gamma \right) \tag{39}$$

Using the values given by Table 4 in the set of inequalities presented in Equations 20 to 23, we obtain:

$$f_u + g_v = -0.6961 < 0$$
$$f_u g_v + f_v g_u = 1.5729 > 0$$
$$df_u + g_v = 0.7509 > 0$$
$$(df_u + g_v)^2 + 4d(f_u g_v - f_v g_u) = 0.0606 > 0$$

Table 4. Parameters used for simulating the Blood Coagulation model

Parameter	Value
δ	1.2
k	0.06
d	0.08
β	1.0
γ	0.03

Analysing the Equation 24, we have:

$$3.15848 < (n^2 + m^2)\pi^2 < 6.2321 \tag{40}$$
$$0.3196 < n^2 + m^2 < 0.6314 \tag{41}$$

As can be observed, we can not found non-zero integer values for m and n that can satisfy Equation 41. The coagulation model described by Equations 29 and 30 was computed in a bi-dimensional domain $[0, 1] \times [0, 1]$. Figure 5 illustrates the value obtained in a spatial mesh composed by 61×61 points, using a final time equals to 1. We have obtained results for 100 points equally spaced in the temporal interval analysed. The perturbation applyed to the initial condition was 0.3%.

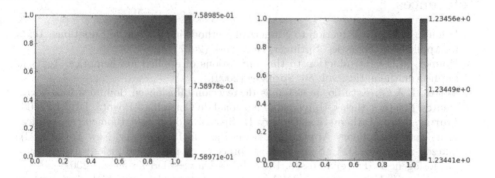

Fig. 5. Concentration of u and v for the coagulation model on a $2D$ mesh

6   Conclusions

This paper presented the study of the Turing instability in biological systems. Three biological models, the Schnakenberg, the Glycolysis and the blood coagulation models were studied. Each biological system can be represented by a simplified mathematical model governed by a system of reaction-diffusion, second order, partial differential equations, which under specific conditions can produce stable spatial patterns from an uniform initial distribution, i.e., the Turing instability.

Several tests were run with different levels of disturbance around the equilibrium point. This paper presented the most representative results found. When instability was detected in the analysis of the conditions to obtain the Turing instability, spatial stable patterns were formed regardless of the value used for the disturbance.

All simulation results obtained when executing the Schnakenberg and the Glycolysis models where very close to those found in the literature, regardless the sets of parameters used in the simulations. Given the non-linearity of the problem, the convergence of these results is a matter that should be taken into account and, although all of these results have not been presented here, distinct spatial refinements were used to obtain them.

For the blood coagulation model, it was not possible to obtain the results presented in the paper where the model was originally proposed[7]. In this paper we presented the linear analysis for this model. Using the original parameters given in the reference [7], a set of non-zero integer values for m and n, that could make possible the Turing instability, was not identified.

Acknowledgments. The authors would like to express their thanks to CAPES, CNPq, FAPEMIG and UFJF for funding this work.

References

1. Holmes, M.H.: Introduction to Numerical Methods in Differential Equations. Texts in Applied Mathematics. Springer, New York (2007)
2. Holmes, M.H.: Introduction to the foundations of applied mathematics. Texts in applied mathematics, vol. 56. Springer (2009)
3. Torres, L.A.G.: Estudio de sistemas de reacción-difusión en dominios fijos y crecientes. Master's thesis, Universidad Nacional de Colombia - Bogotá, UNAL (2008)
4. Murray, J.D.: Mathematical Biology II: Spatial Models and Biomedical Applications. Interdisciplinary Applied Mathematics, vol. 18. Springer, New York (2003)
5. Garzón-Alvarado, D.A.: Simulación de procesos de reacción-difusión: Aplicación a la morfogénesis del tejido óseo. PhD thesis, Universidad Zaragoza (2007)
6. Maini, P.K.: Using mathematical models to help understand biological pattern formation. Comptes Rendus Biologies 327(3), 225–234 (2004)
7. Vanegas-Acosta, J.C., Garzón-Alvarado, P.N.S.L., Mathematical, D.A.: Mathematical model of the coagulation in the bone-dental implant interface. Comp. in Bio. and Med. 40(10), 791–801 (2010)
8. Turing, A.: The chemical basis of morphogenesis. Philosophical Transactions of the Royal Society B 237, 37–72 (1952)
9. Murray, J.D.: Mathematical Biology I. An Introduction. 3 edn. Interdisciplinary Applied Mathematics, vol. 17. Springer, New York (2002)
10. Schnakenberg, J.: Simple chemical reaction systems with limit cycle behaviour. Journal of Theoretical Biology 81(3), 389–400 (1979)

11. TheFreeDictionary.com: Autocatalysis,
    http://medical-dictionary.thefreedictionary.com/Autocatalysis (accessed:
    April 24, 2014)
12. Segel, L.A.: Biological kinetics, vol. 12. Cambridge University Press (1991)
13. TheFreeDictionary.com: Nad,
    http://medical-dictionary.thefreedictionary.com/NAD (accessed: April 24,
    2014)
14. TheFreeDictionary.com: Nadh,
    http://medical-dictionary.thefreedictionary.com/NADH (accessed: April 24,
    2014)
15. Minors, D.S.: Haemostasis, blood platelets and coagulation. Anaesthesia and In-
    tensive Care Medicine 8(5), 214–216 (2007); Paediatrics and blood physiology
16. Hall, J.E.: Pocket companion to Guyton & Hall textbook of medical physiology.
    Elsevier Saunders (2006)
17. Python: Python's homepage, http://www.python.org/ (accessed: April 24, 2014)
18. Scipy: Scipy's homepage,
    http://www.scipy.org/ (accessed: April 24, 2014)
19. Odeint: Odeint's homepage,
    http://docs.scipy.org/doc/scipy/reference/
    generated/scipy.integrate.odeint.html (accessed: April 24, 2014)
20. Sparse: Sparse's homepage,
    http://docs.scipy.org/doc/scipy/reference/sparse.html (accessed: April 24,
    2014)
21. LeVeque, R.J.: Finite difference methods for ordinary and partial differential equa-
    tions - steady-state and time-dependent problems. SIAM (2007)

Weak Equivalents for Nonlinear Filtering Functions

Amparo Fúster-Sabater[1] and Pino Caballero-Gil[2]

[1] Institute of Physical and Information Technologies (CSIC),
Serrano 144, 28006 Madrid, Spain
amparo@iec.csic.es
[2] University of La Laguna, 38271 La Laguna, Tenerife, Spain
pcaballe@ull.es

Abstract. The application of a nonlinear filtering function to a Linear Feedback Shift Register (LFSR) is a general technique for designing pseudorandom sequence generators with cryptographic application. In this paper, we investigate the equivalence between different nonlinear filtering functions applied to distinct LFSRs. It is a well known fact that given a binary sequence generated from a pair (nonlinear filtering function, LFSR), the same sequence can be generated from any other LFSR of the same length by using another filtering function. However, until now no solution has been found for the problem of computing such an equivalent. This paper analyzes the specific case in which the reciprocal LFSR of a given register is used to generate an equivalent of the original nonlinear filtering function. The main advantage of the contribution is that weaker equivalents can be computed for any nonlinear filter, in the sense that such equivalents could be used to cryptanalyze apparently secure generators. Consequently, to evaluate the cryptographic resistance of a sequence generator, the weakest equivalent cipher should be determined and not only a particular instance.

Keywords: Nonlinear filtering function, pseudorandom sequence, LFSR, stream cipher, cryptography.

1 Introduction

A binary additive stream cipher is a synchronous cipher in which the binary output of a keystream generator is added bitwise to the binary plaintext sequence producing the binary ciphertext. The main goal in stream cipher design is to produce random-looking sequences that are unpredictable in an efficient way. From a cryptanalysis point of view, a good stream cipher should be resistant against known-plaintext attacks.

Most known keystream generators are based on Linear Feedback Shift Registers (LFSRs) [10], whose output sequence is the image of a linear function applied to its successive states. Under certain conditions, this structure produces sequences with highly desirable features for cryptographic application. In particular, if its characteristic polynomial is primitive, then the generated sequence, the so-called m-sequence, exhibits certain useful properties such as a large period, good statistical distribution of 0's and 1's or excellent autocorrelation. However, the sequence produced by a LFSR must never be used as keystream sequence in a stream cipher as the inherent linearity of this structure could be easily used to break the cipher.

B. Murgante et al. (Eds.): ICCSA 2014, Part VI, LNCS 8584, pp. 592–602, 2014.

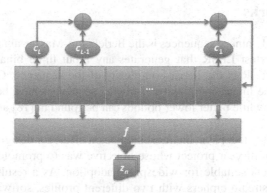

Fig. 1. A Nonlinear Filtering Generator

An interesting LFSR-based keystream generator is the nonlinear filter generator or nonlinear filtering function, which produces the keystream sequence as the image of a nonlinear Boolean function applied to the states of an LFSR. In particular, the nonlinear filter generator here analyzed (see Fig. 1) consists of two parts.

1. A LFSR with length L, characteristic polynomial $P(x) = x^L + c_1 \cdot x^{L-1} + \cdots + c_{L-1} \cdot x + c_L$ with binary coefficients that, from an initial state IS, generates an output sequence $\{a_n\}$.
2. A nonlinear Boolean function $F : GF(2)^L \to GF(2)$, called filter function, whose inputs variables are the successive L-bit states of the LFSR and whose image is the binary keystream sequence $\{z_n\}$.

Although the sequences produced by LFSRs have been well studied, the same do not apply to the sequences obtained from nonlinear filter generators.

This work deals with the relationship between different nonlinear filter generators that produce exactly the same sequence. The main goal is to show that although the study of the generator's properties leads to right conclusions about the properties of the generated sequence, sometimes misleading inferences can be drawn. In particular, this paper shows that two structures with apparently different security levels can produce the same keystream sequence. Indeed, this result can be seen as a proof that the actual security level of a generator is the security level of the weakest element in its corresponding class.

This paper is organized as follows. Section 2 includes a succinct revision of related works. In Section 3 after some necessary preliminaries, the general problem of counting equivalent nonlinear filter generators is addressed as well as the relationship among them is studied. Afterwards, Section 4 provides a brief explanation of the proposal based on the new concept of reciprocal filter generators and introduces a novel method for computing weaker equivalent nonlinear filter generators through a pedagogical example. Finally, Section 5 discusses conclusions and possible future research lines.

2 Related Works

A useful tool to study binary sequences is the Berlekamp-Massey algorithm [14], which determines the shortest LFSR that generates any input finite binary sequence. The length of such a LFSR is the linear complexity of the sequence. General lower and upper bounds on the linear complexity of filtered sequences have been derived in the works [12] and [6], while better lower bounds can be found in [18] and [19] for special cases.

The eSTREAM project [3] is the most significant effort for designing secure stream ciphers. It was a multi-year project whose objective was to promote the design of efficient stream ciphers suitable for widespread adoption. As a result of the project, a portfolio of seven stream ciphers with two different profiles, software and hardware, was published. One of them, the so-called SOSEMANUK, is an LFSR-based generator where the length of the used LFSR is 10 and the content of each stage is an element of $GF(2^{32})$. Such a generator uses design principles similar to the stream cipher SNOW 2.0 that led to the later called SNOW 3G, which forms the heart of the 3GPP confidentiality and integrity algorithms for LTE and LTE-Advanced.

Several references of cryptanalytic attacks on nonlinear filters can be found. The basic correlation attack against the nonlinear filter generator was published in [20], where correlations between the filtered sequence $\{z_n\}$ and the LFSR m-sequence $\{a_n\}$ are used to build an equivalent generator consisting of a nonlinear combination of several LFSRs. The main drawbacks of this attack is the huge amount of time needed for computing the necessary correlations and the requirement that the filter function F must have high correlation to an affine function. After defining the nonlinearity of a Boolean function as the minimum Hamming distance between such a function and an affine function, a practical consequence of the basic correlation attack is that the cryptographic designer has to choose highly nonlinear filter functions for the nonlinear filter generators. Afterwards, the concept of basic correlation attacks was improved by the fast correlation attack described in [15]. Two common disadvantages of the different versions of these attacks are: a) the large number of intercepted keystream bits needed to perform a successful attack and b) the assumption that the filter function is not highly nonlinear.

A general inversion attack was proposed in [9] for any filter function. Consequently, an easy characterization was obtained for filter generators that are resistant against the inversion attack. On the other hand, the works [7] and [5] proposed the so-called decimation attack for any LFSR-based keystream generator. The idea is to consider a decimated sequence of the intercepted keystream sequence so that the decimated sequence can be generated from a decimated LFSR sequence. However, according to [18] if the LFSR's length L is a prime number, then the decimation attack provides no further advantages.

In the last years, several algebraic attacks on stream ciphers have been published. In these attacks, the attacker uses the bits of the intercepted sequence to set a nonlinear system of polynomial equations in terms of the LFSR bits. The main issue regarding these types of attacks is that, as shown in [8], the problem of solving a nonlinear system of multivariate equations is NP-hard even if all the equations are quadratic and the underlying field is $GF(2)$. In order to deal with this, the so-called algebraic method XL [2] was proposed to solve the nonlinear system of quadratic equations for certain nonlinear filter generators. In order to make them resistant against this attack, the filter function should be not only highly nonlinear, but also have a large distance to approximations of low algebraic degree.

The so-called time-memory-data tradeoff attacks [1] can be easily prevented in non-linear filter generators by using LFSRs with large length. There is another interesting attack, the so-called guess and determine attack [16], which exploits the relationship between internal values (such as the recurrence relationship in the LFSR), and the relationship used to construct the keystream sequence from the internal values. As its name indicates, this attack proceeds by guessing some internal values and then using the relationships to determine other internal values. After such an attack, the encryption is considered broken when a complete internal state has been determined from the guessed values. This type of attack can be prevented by choosing adequate polynomials of the LFSRs.

One of the most closely related papers to this one is [13], where the so-called linear transformation attack against the nonlinear filter generator was proposed. The idea behind this attack is to transform the given generator into an equivalent nonlinear filter generator with the same LFSR but a filter function that is better suited for some of the above described attacks.

Another close paper is [17], where the authors define an equivalence class of nonlinear filter generators, showing that a number of important cryptographic properties are not invariant among elements of the same equivalence class. The authors themselves acknowledge that determining the weakest equivalent cipher is a very difficult task because the size of the equivalence class is very large. In this paper we do not deal with the complete nonlinear equivalence class, but only with one of its members, the one we have identified that in many cases leads to a weaker equivalent generator.

In conclusion, each attack against nonlinear filter generators leads to one or more than one conclusion about desirable properties of the LFSR and/or of the filter function. Consequently, one of the main research issues regarding nonlinear filter generators is how to construct a good Boolean function to achieve resistance against all the afore-mentioned attacks. This work deals with this issue because it proves that the properties of the generator not always guarantee security at the output sequences.

3   General Study of Equivalent Nonlinear Filters

In this section, first of all the number of equivalent filters is obtained then the relationship between them is analyzed.

For a filter generator consisting of an LFSR with characteristic polynomial $P_1(x)$ and a filter function $F_1(x)$, it is always possible to generate the same sequence with any other LFSR of the same length and another filter function.

Let α be a root of the characteristic polynomial $P_1(x)$ as well as a primitive element of $GF(2^L)$. In that case, if $gcd(k, 2^L - 1) = 1$, then α^k is also a primitive element of $GF(2^L)$, so there are $\phi(2^L - 1)$ primitive elements of $GF(2^L)$. In particular, the L conjugates of any element (which are the successive square powers), e.g. $\alpha, \alpha^2, \alpha^4, \alpha^8, ..., \alpha^{2^{L-1}}$, are primitive elements of $GF(2^L)$ as well as roots of the same polynomial, which can be computed by the expression $\prod_{i=0}^{L-1}(x - \alpha^{2^i})$ in $GF(2^L)$.

Therefore, there are $\phi(2^L-1)/L$ primitive polynomials of $GF(2^L)$, each one with L roots that are all conjugates of a primitive element. Since each one of these polynomials defines an LFSR of length L, there are $\phi(2^L-1)/L$ different LFSRs of length L, each of them corresponding to a set of conjugates of a primitive element of $GF(2^L)$.

In conclusion, since any binary sequence obtained with a nonlinear filter can be generated by a filter function over each LFSR, then there are $\phi(2^L-1)/L$ different nonlinear filter generators that can be used to generate it.

The relationship between two primitive elements, α and β, roots of two different characteristic polynomials of two different LFSRs of length L is given by the expression $\beta = \alpha^k$ being $gcd(k, 2^L-1) = 1$ and $k \neq 2^i \cdot j \pmod{2^L-1}$ with $i, j > 0$.

This knowledge on the relationship between the characteristic polynomials $P_1(x)$ and $P_2(x)$ of two LFSRs could help to define the relationship between two filter functions $F_1(x)$ and $F_2(x)$, which are part of two equivalent generators that produce the same filtered sequence.

Table 1. Examples of counting of equivalent filters

L	3	4	5	6
$2^L - 1$	7	15	31	63
N. filters	2	2	6	6
k defining filters	1,2,4	1,2,4,8	1,2,4,8,16	1,2,4,8,16,32
	3,5,6	7,11,13,14	3,6,12,24,17	5,10,20,40,17,34
			5,10,20,9,18	11,22,44,25,50,37
			7,14,28,25,19	13,26,52,41,19,38
			11,22,13,26,21	23,46,29,58,53,43
			15,30,29,27,23	31,62,61,59,55,47

As we can see in Table 1, the cases $k = 1$ and $k = 2^{L-1} - 1$ always determine different sets of conjugate roots that define different LFSRs. In fact, the corresponding polynomials for the roots α and $\beta = \alpha^{2^{L-1}-1}$ are always reciprocal.

Any m-sequence $\{a_n\}$ can be written in terms of the roots of the characteristic polynomial of the LFSR through the trace function, so that $a_n = Tr(\alpha^n) = \sum_{i=0}^{L-1} \alpha^{n2^i}$. Consequently, given a sequence $\{a_n\}$ generated by a LFSR with polynomial $P_1(x)$ and root α and another sequence $\{b_n\}$ generated by other LFSR with polynomial $P_2(x)$ and root β such that $\beta = \alpha^k$, we have that $a_n = \sum_{i=0}^{L-1} \alpha^{n2^i}$ and $b_n = \sum_{i=0}^{L-1} \alpha^{kn2^i}$. This is shown with an example in Table 2.

If two filter generators defined by the corresponding polynomials and filter functions $(P_1(x), F_1(x))$ and $(P_2(x), F_2(x))$ generate the same sequence, then we have that:

$$F_1(a_n, a_{n+1}, ..., a_{n+L-1}) = F_1\left(\sum_{i=0}^{L-1} \alpha^{n2^i}, \sum_{i=0}^{L-1} \alpha^{(n+1)2^i}, ..., \sum_{i=0}^{L-1} \alpha^{(n+L-1)2^i}\right) =$$

$$F_2(b_n, b_{n+1}, ..., b_{n+L-1}) = F_2\left(\sum_{i=0}^{L-1} \alpha^{kn2^i}, \sum_{i=0}^{L-1} \alpha^{k(n+1)2^i}, ..., \sum_{i=0}^{L-1} \alpha^{k(n+L-1)2^i}\right).$$

Table 2. Examples of relationships between roots, polynomials and m-sequences

Roots	Polynomial	m-sequence
$\alpha, \alpha^2, \alpha^4, \alpha^8, \alpha^{16}$	$x^5 + x^4 + x^3 + x^2 + 1$	$\{a_n\}$
$\alpha^{15}, \alpha^{30}, \alpha^{29}, \alpha^{27}, \alpha^{23}$	reciprocal= $= x^5 + x^3 + x^2 + x + 1$	$\{b_n\}$ reverse of $\{a_n\}$
$\alpha^3, \alpha^6, \alpha^{12}, \alpha^{24}, \alpha^{17}$	$\prod_{i=0}^{4}(x - \alpha^{3\cdot 2^i}) =$ $= x^5 + x^4 + x^2 + x + 1$	$\{c_n\}$
$(\alpha^3)^{15} = \alpha^{14}, \alpha^{28}, \alpha^{25}, \alpha^{19}, \alpha^7$	reciprocal= $x^5 + x^4 + x^3 + x + 1$	$\{d_n\}$ reverse of $\{c_n\}$
$\alpha^5, \alpha^{10}, \alpha^{20}, \alpha^9, \alpha^8$	$\prod_{i=0}^{4}(x - \alpha^{5\cdot 2^i}) =$ $x^5 + x^3 + 1$	$\{e_n\}$
$(\alpha^5)^{15} = \alpha^{13}, \alpha^{26}, \alpha^{21}, \alpha^{11}, \alpha^{22}$	reciprocal= $x^5 + x^2 + 1$	$\{f_n\}$ reverse of $\{e_n\}$

The Algebraic Normal Form of Boolean functions allows us to write the sequence generated by a filter generator $(P_1(x), F_1(x))$ in terms of a root α of the polynomial $P_1(x)$ and binary coefficients, as follows:

$$F_1(a_n, a_{n+1}, ..., a_{n+L-1}) =$$

$$= c_0 a_n + \cdots + c_{L-1} a_{n+L-1} + c_{0,1} a_n a_{n+1} + \cdots + c_{L-2,L-1} a_{n+L-2} a_{n+L-1} +$$

$$+ \cdots + c_{0,1,...L-1} a_n a_{n+1} \cdots a_{n+L-1} =$$

$$= c_0 \sum_{i=0}^{L-1} \alpha^{n2^i} + \cdots + c_{L-1} \sum_{i=0}^{L-1} \alpha^{(n+L-1)2^i} + c_{0,1} \sum_{i=0}^{L-1} \alpha^{n2^i} \sum_{i=0}^{L-1} \alpha^{(n+1)2^i} + \cdots +$$

$$+ c_{L-2,L-1} \sum_{i=0}^{L-1} \alpha^{(n+L-2)2^i} \sum_{i=0}^{L-1} \alpha^{(n+L-1)2^i} +$$

$$+ \cdots + c_{0,1,...L-1} \sum_{i=0}^{L-1} \alpha^{n2^i} \sum_{i=0}^{L-1} \alpha^{(n+1)2^i} \cdots \sum_{i=0}^{L-1} \alpha^{(n+L-1)2^i}.$$

Thus, if the expression is partitioned into cosets (sets of integers $E \cdot 2^i \ mod(2^L - 1)$ with $0 \le i \le L - 1$), then the function can be expressed as:

$$F_1(a_n, a_{n+1}, ..., a_{n+L-1}) = \sum_{i=0}^{L-1} C_{coset1} \alpha^{n \cdot coset1 \cdot 2^i} + C_{coset2} \alpha^{n \cdot coset2 \cdot 2^i} + \cdots$$

with $C_{cosetj} \in GF(2^L)$.

The weights of the cosets whose coefficients are nonzero in the previous expression provide some information about the function, i.e. its order.

In particular, if the relationship $\beta = \alpha^k$ between two filter generators $(P_1(x), F_1(x))$ and $(P_2(x), F_2(x))$ that generate the same sequence is known, then:

$$F_2(b_n, b_{n+1}, ..., b_{n+L-1}) = \sum_{i=0}^{L-1} D_{coset1}\alpha^{n \cdot coset1 \cdot 2^i} + D_{coset2}\alpha^{n \cdot coset2 \cdot 2^i} + \cdots$$

with $D_{cosetj} \in GF(2^L)$.

Then, the cosets that appear in both expressions must be paired so that for each coset $cosetv$ in the first expression, another coset $cosetw$ exists in the second expression. That is:

$$\sum_{i=0}^{L-1} C_{cosetv}\alpha^{n \cdot cosetv \cdot 2^i} = \sum_{i=0}^{L-1} D_{cosetw}\alpha^{n \cdot cosetw \cdot 2^i}.$$

4   Reciprocal Filters

From the results shown in the previous Section, if two LFSRs with reciprocal polynomials $P_1(x)$ and $P_2(x)$ are considered, two conclusions can be drawn.

1. If the same filter function $F(x)$ is applied to both LFSRs, then different sequences are generated. The Berlekamp-Massey algorithm can be used on the resulting filtered sequences. In fact, from the factorizations of their corresponding characteristic polynomials it can be concluded that they always correspond exactly to the same cosets.
2. In order to generate the same sequence with those LFSRs, two different filter functions $F_1(x)$ and $F_2(x)$ must be used. Since the factorization of the polynomial obtained with the Berlekamp-Massey algorithm corresponds to mirrored complementary cosets in the groups defined by each of the LFSRs, the order of the filter functions are influenced by the weights of those cosets. In particular, it can be concluded that $order(F_i) = max(L\text{-}(\text{weight of each coset linked to the factorization of the polynomial of the sequence}))$.

Thus, if there is a filter generator producing a sequence whose characteristic polynomial factorization only corresponds to cosets of weight $> L/2$, then there is an equivalent filter that is less strong from a cryptographic point of view and has order $< L/2$. Regarding such an equivalent filter, it is a well known fact that the LFSR is the reciprocal of the original one.

If a filter function has order $\sim L/2$, since the order is given by the maximum of the weights of the cosets associated with the factorization, then there is an equivalent filter of order $\geq L/2$ as such a degree is given by the maximum of the weights of the cosets. Consequently, if a reciprocal LFSR is used, then it is known that its weight is at least $L - L/2$. This can be seen as a proof of the known recommendation about using filter functions of order $\sim L/2$.

From all the aforementioned, it can be concluded that for any filter generator, an equivalent filter generator called reciprocal filter can be always obtained to generate the same sequence. In order to determine the reciprocal filter for any known filter generator, the proposed procedure includes the following four basic steps:

1. Determine the relationships between the roots of the characteristic polynomials of the initial LFSR and its reciprocal.

2. Express both m-sequences through the trace function.
3. Compute the coefficients of the cosets in the expression of the filter function.
4. Determine the reciprocal filter function.

This procedure is illustrated through a pedagogical example.

Example:

Given an LFSR of length $L = 5$, characteristic polynomial $P_1(x) = x^5 + x^3 + 1$ and initial state $IS_1 = (1, 0, 0, 0, 0)$, the filter function of order 4

$$F_1(a_0, a_1, a_2, a_3, a_4) = a_0 a_1 a_2 a_4 + a_1 a_2 a_3 a_4 + a_0 a_1 a_3 a_4 + a_0 a_1 a_2 + a_1 a_2 a_3 + a_1 a_2 a_4 +$$

$$+ a_0 a_2 a_4 + a_2 a_3 a_4 + a_0 a_1 a_4 + a_2 a_3 + a_2 a_4 + a_0 a_3 + a_1 a_3 + a_1 a_4 + a_2 + a_4$$

is applied to produce the filtered sequence of period $2^5 - 1$,

$$01011011010110111000010010110.$$

The reciprocal LFSR has characteristic polynomial $P_2(x) = x^5 + x^2 + 1$, whose root β is related to the root α of $P_1(x)$ by means of the expression $\beta = \alpha^{2^{5-1}-1} = \alpha^{15}$. Furthermore, thanks to the modular inverse of $15 \pmod{31}$, the inverse relationship can be obtained $\alpha = \beta^{29}$.

At the same time, the m-sequences $\{a_n\}$ and $\{b_n\}$ obtained from $P_1(x)$ and $P_2(x)$, respectively, can be expressed by means of their trace expressions:

$$a_n = \alpha^n + \alpha^{2n} + \alpha^{4n} + \alpha^{8n} + \alpha^{16n}$$

$$b_n = \beta^n + \beta^{2n} + \beta^{4n} + \beta^{8n} + \beta^{16n}.$$

Consequently, the filter functions F_1 and F_2 can be expressed in terms of the $\phi(2^5 - 1)/5 = 6$ cosets that is: coset 15 (coset of weight 4), coset 11 and coset 7 (cosets of weight 3), coset 5 and coset 3 (cosets of weight 2) and coset 1 (coset of weight 1). In particular, the coefficient C_{15} corresponding to the coset 15 or coset of weight 4 can be obtained by the root presence test [18], while the coefficients C_7 and C_{11} corresponding to the cosets of weight 3 can be computed by grouping terms:

$$C_{15} = \alpha^{21}, C_7 = 1, C_{11} = \alpha^{15}.$$

From these values, it can be concluded that no more cosets of lower weight appear in the expression of the filter function F_1. Thus,

$$F_1 = C_{15}\alpha^{15n} + C_{15}^2 \alpha^{30n} + C_{15}^4 \alpha^{29n} + C_{15}^8 \alpha^{27n} + C_{15}^{16} \alpha^{23n} +$$

$$+ C_7 \alpha^{7n} + C_7^2 \alpha^{14n} + C_7^4 \alpha^{28n} + C_7^8 \alpha^{25n} + C_7^{16} \alpha^{19n} +$$

$$+ C_{11}\alpha^{11n} + C_{11}^2 \alpha^{22n} + C_{11}^4 \alpha^{13n} + C_{11}^8 \alpha^{26n} + C_{11}^{16} \alpha^{21n}.$$

If $\alpha = \beta^{29}$ is substituted in that expression, then the filter function F_2 generating the same sequence can be expressed as:

$$F_2 = C_{15}\beta^{29 \cdot 15n} + C_{15}^2 \beta^{29 \cdot 30n} + C_{15}^4 \beta^{29 \cdot 29n} + C_{15}^8 \beta^{29 \cdot 27n} + C_{15}^{16} \beta^{29 \cdot 23n} +$$

$$+C_7\beta^{29\cdot7n} + C_7^2\beta^{29\cdot14n} + C_7^4\beta^{29\cdot28n} + C_7^8\beta^{29\cdot25n} + C_7^{16}\beta^{29\cdot19n} +$$
$$+C_{11}\beta^{29\cdot11n} + C_{11}^2\beta^{29\cdot22n} + C_{11}^4\beta^{29\cdot13n} + C_{11}^8\beta^{29\cdot26n} + C_{11}^{16}\beta^{29\cdot21n} =$$
$$= C_{15}\beta^n + C_{15}^2\beta^{2n} + C_{15}^4\beta^{4n} + C_{15}^8\beta^{8n} + C_{15}^{16}\beta^{16n} +$$
$$+C_7\beta^{17n} + C_7^2\beta^{3n} + C_7^4\beta^{6n} + C_7^8\beta^{12n} + C_7^{16}\beta^{24n} +$$
$$+C_{11}\beta^{9n} + C_{11}^2\beta^{18n} + C_{11}^4\beta^{5n} + C_{11}^8\beta^{10n} + C_{11}^{16}\beta^{20n}.$$

Consequently, it can be concluded that only the cosets 1, 3 and 5 appear in this expression. Furthermore, their coefficients D_1, D_3 and D_5 are given by:

$$D_1 = C_{15} = \alpha^{21} = \beta^{29\cdot21} = \beta^{20}$$

$$D_3 = C_7^2 = \alpha^0 = \beta^{29\cdot0} = \beta^0 = 1$$

$$D_5 = C_{11}^4 = \alpha^{15\cdot4} = \alpha^{29} = \beta^{29\cdot29} = \beta^4.$$

Since the maximum weight of the cosets in that expression is 2, the nonlinear terms of order 2 in the expression of F_2 are analyzed. As before, for each nonlinear term of order 2 the coefficients D_3 and D_5 can be obtained by the root presence test [18], while the coefficient D_1 corresponding to the coset of weight 1 can be computed by grouping terms, as it is shown in Table 3.

Table 3. Coefficients of the cosets 3, 5 and 1 for all the possible terms of order 1 and 2

	D_3	D_5	D_1
b_0b_1	β^{19}	β^{30}	β^{16}
b_0b_2	β^7	β^{29}	β
b_0b_3	β	β^{19}	β^{17}
b_0b_4	β^{14}	β^{27}	β^2
b_1b_2	β^{22}	β^4	β^{17}
b_1b_3	β^{10}	β^3	β^2
b_1b_4	β^4	β^{24}	β^{18}
b_2b_3	β^{25}	β^9	β^{18}
b_2b_4	β^{13}	β^8	β^3
b_3b_4	β^{28}	β^{14}	β^{19}
b_0	-	-	1
b_1	-	-	β
b_2	-	-	β^2
b_3	-	-	β^3
b_4	-	-	β^4

Now, we can choose in the two first columns of Table 3, corresponding to the cosets of maximum weight, those elements whose sum coincides with the corresponding known coefficient. In particular, the selected coefficients corresponding to the products b_1b_3, b_2b_3, b_0b_2 and b_0b_4 give both values:

$$D_3 = \beta^{10} + \beta^{25} + \beta^7 + \beta^{14} = 1$$

$$D_5 = \beta^3 + \beta^9 + \beta^{29} + \beta^{27} = \beta^4.$$

This result applied on the last column produces that, in order to obtain the final sum $D_1 = \beta^{20}$, the linear elements $b_0 + b_2 + b_3$ have to be included in the filter function F_2:

$$D_1 = \beta^2 + \beta^{18} + \beta + \beta^2 + 1 + \beta^2 + \beta^3 = \beta^{20}.$$

Thus, the final expression of the equivalent filter function is obtained:

$$F_2(b_0, b_1, b_2, b_3, b_4) = b_1 b_3 + b_2 b_3 + b_0 b_2 + b_0 b_4 + b_0 + b_2 + b_3.$$

This function applied on the reciprocal LFSR with characteristic polynomial $P_2(x) = x^5 + x^2 + 1$ and initial state $IS_2 = (1, 0, 0, 1, 0)$, produces the same filtered sequence of the input filter generator

01011011010110111000010010010110.

Recall that F_2 is a function of order 2 with less number of terms of order 2 and 1 than F_1 and in addition without terms of order 3 neither 4. Thus, from a cryptographic point of view, the attacker will find an easier attack against F_2 than against F_1 although both filters generate exactly the same sequence.

Consequently, this example shows that the proposed method can be applied on any known filter generator in order to produce an equivalent filter, which in the case of reciprocal LFSR is of a lower order. This is a proof that some generators apparently secure can have weaker equivalents, and what is more important, that these equivalents can be computed.

5   Conclusions and Future Works

This work has addressed the problem of computing equivalent nonlinear filters that produce the same sequence as a known filter generator. In particular, it analyzes the case in which a reciprocal LFSR is used to define an equivalent nonlinear filter. In fact, under such conditions there are specific relationships between the two filter functions that allow the definition of a specific method for computing the equivalent filter function. The study concludes that the equivalent generator can have a security level that is lower than the one of the original filter. Therefore, the proposed method allows building equivalents that are weaker than the starting filters. In conclusion, this work shows that two structures with apparently different security levels according to their properties, can produce exactly the same keystream sequence, so both generators must be considered as insecure as the weakest one.

Given the difficulty of the subject, there are still many open issues. In particular, one of them is the development of optimum methods for solving the particular knapsack problem that appears in the last phase of the proposed method. Also, a study similar to the one shown in this paper, but on other equivalents that do not correspond to the reciprocal LFSR, could be useful for potential cryptanalitic attacks on nonlinear filter functions.

Acknowledgment. Research supported by the Spanish MINECO and the European FEDER Funds under projects TIN2011-25452 and IPT-2012-0585-370000.

References

1. Biryukov, A., Shamir, A.: Cryptanalytic time/Memory/Data tradeoffs for stream ciphers. In: Okamoto, T. (ed.) ASIACRYPT 2000. LNCS, vol. 1976, pp. 1–13. Springer, Heidelberg (2000)
2. Courtois, N., Klimov, A., Patarin, J., Shamir, A.: Efficient algorithms for solving overdefined systems of multivariate polynomial equations. In: Preneel, B. (ed.) EUROCRYPT 2000. LNCS, vol. 1807, pp. 392–407. Springer, Heidelberg (2000)
3. eSTREAM: the ECRYPT Stream Cipher Project,
   http://www.ecrypt.eu.org/stream/
4. Faugere, J.-C., Ars, G.: An Algebraic Cryptanalysis of Nonlinear Filter Generators using Grobner bases (2003), http://www.inria.fr/rrrt/rr-4739.html
5. Filiol, E.: Decimation attack of stream ciphers. In: Roy, B., Okamoto, E. (eds.) INDOCRYPT 2000. LNCS, vol. 1977, pp. 31–42. Springer, Heidelberg (2000)
6. Fúster-Sabater, A., Caballero-Gil, P.: On the linear complexity of nonlinearly filtered pn-sequences. In: Safavi-Naini, R., Pieprzyk, J.P. (eds.) ASIACRYPT 1994. LNCS, vol. 917, pp. 80–90. Springer, Heidelberg (1995)
7. Games, R.A., Rushanan, J.J.: Blind synchronization of m-sequences with even span. In: Helleseth, T. (ed.) EUROCRYPT 1993. LNCS, vol. 765, pp. 168–180. Springer, Heidelberg (1994)
8. Garey, M.R., Johnson, D.S.: Computers and Interactability. Freeman and Company (1979)
9. Golic, J.D., Clark, A., Dawson, E.: Generalized inversion attack on nonlinear filter generators. IEEE Transactions on Computers 49(10), 1100–1109 (2000)
10. Golomb, S.W.: Shift Register-Sequences. Aegean Park Press, Laguna Hill (1982)
11. Hell, M., Johansson, T., Meier, W.: Grain - A Stream Cipher for Constrained Environments (2005),
    http://www.ecrypt.eu.org/stream/p3ciphers/grain/Grain_p3.pdf
12. Key, E.L.: An analysis of the structure and complexity of nonlinear binary sequence generators. IEEE Transactions on Information Theory 22(6), 732–736 (1976)
13. Lohlein, B.: Design and analysis of cryptographic secure keystream generators for stream cipher encryption. PhD thesis, Faculty of Electrical and Information Engineering, University of Hagen, Germany (2001)
14. Massey, J.L.: Shift-register synthesis and BCH decoding. IEEE Transactions on Information Theory IT-15(1), 122–127 (1969)
15. Meier, W., Staffelbach, O.J.: Fast correlation attacks on stream ciphers. Journal of Cryptology 1(3), 159–176 (1989)
16. Pasalic, E.: On guess and determine cryptanalysis of LFSR-based stream ciphers. IEEE Transactions on Information Theory 55(7), 3398–3406 (2009)
17. Rønjom, S., Cid, C.: Nonlinear equivalence of stream ciphers. In: Hong, S., Iwata, T. (eds.) FSE 2010. LNCS, vol. 6147, pp. 40–54. Springer, Heidelberg (2010)
18. Rueppel, R.A.: Analysis and Design of Stream Ciphers. Springer (1986)
19. Schneider, M.: Methods of generating binary pseudo-random sequences for stream cipher encryption. PhD thesis, Faculty of Electrical Engineering, University of Hagen, Germany (1999)
20. Siegenthaler, T.: Decrypting a class of stream ciphers using ciphertext only. IEEE Transactions on Computers 100(1), 81–85 (1985)

Eliminating Duplicated Paths to Reduce Computational Cost of Rule Generation by Using SDN

Jae-Hwa Sim[1], Sung-Hwan Kim[1], Min-Woo Park[1], and Tai-Myoung Chung[2],*

[1] Department of Electrical and Computer Engineering
Sungkyunkwan University, Suwon, Korea
{jhsim,shkim47,mwpark}@imtl.skku.ac.kr
[2] College of Information and Communication Engineering
Sungkyunkwan University, Suwon, Korea
tmchung@ece.skku.ac.kr

Abstract. Nowadays, SDN(Software Defined Network) has become a powerful technology that has abilities to program network flow paths into flow-table in switches for network control. In SDN architecture, data plane and control plane are decoupled, and then by manipulating flow-table on control plane network control is available. The function of SDN is used to supplement limitation or difficulties of other technologies. For example cloud computing system has difficulties to use existing security devices due to its dynamic and virtualization environment. For the solution, Cloudwatcher framework was designed. In Cloudwatcher, there are four algorithms to take network traffic to certain security node in cloud computing environment by using SDN. However when Cloudwatcher generates paths from source node to destination and security nodes, there is possibility of computational cost according to topology. In this paper, we propose an approach to reduce to the rule generation computation cost by excluding duplicated paths during rule generation time. Our evaluation results show that computation cost can be reduced by considerable amount according to topology environment.

Keywords: Network monitoring, Software Defined Networking, Cloud computing security.

1 Introduction

Recently, collaboration between SDN and other technologies such as network middle-box, network monitoring, and other technologies makes synergy effects in a various of areas in IT industry[1,2,3,4]. At the point, SDN usually complements current issues of other research areas. With such merits, the SDN has become a meaningful research area. As another meaning, SDN is a powerful technology that will replace existing network model. According to the SDNCentral report[5],

---

* Corresponding author.

B. Murgante et al. (Eds.): ICCSA 2014, Part VI, LNCS 8584, pp. 603–613, 2014.

SDN market size will reach 35 billion dollar by 2018. While SDN is rapidly developing, it has already been almost 40 years that traditional network model was first designed, and nowadays the network model has faced some fundamental problems; complexity, scalability and vender dependency. The SDN is deployed to overcome such problems. SDN network model decouples control plane and data plane to manage network. By decoupling both data plane and control plane, administrator can program flow rules on the control plan, and apply the rules to data plane to enforce network traffic to follow the rules.

Cloud computing is one of the prominent technologies in IT industry. Cloud services which are representatively IaaS (Infrastructure as a Service), PaaS (Platform as a service), and SaaS (Software as a service), are widely used in IT environment, and have become more general services for internet user. Anyone who wants to use cloud services just creates an account and uses it, after using services, they just pay for it. However while usage rate of cloud services has increased, security aspects of cloud computing still remain a huge problem. Basically cloud computing environment has structural limits to address security issues. There are two reasons[6]. One is that cloud computing environment is so dynamic that network topology consistently changes. Therefore security devices on cloud computing have difficulties to apply their security rules in dynamic network topology. Specifically, when a cloud network topology changes, security devices is needed to reconfiguration or even move to other place.

Second is virtualization property of cloud computing. In a traditional network, security threats are delivered from outside network, and also security devices are installed to protect outside attack. However in case of cloud network, both inside and outside threats should be considered. For example, in case of multi-tenant cloud network, the tenant and consumer sometimes impose responsibility of security on themselves which could increase the possibility of malware infection of internal hosts/VMs[7]. For example one of the tenants is infected, it could spread out the other tenants/VMs.

Due to those restrictions in cloud computing, network security monitoring is quite complicated, and it is hard to apply security devices. One of the related works for solving the such problems is Cloudwatcher[6] which is a framework to enforce network traffic to reach certain security node. The author proposed four algorithms to do network controlling. In this paper, we propose a more efficient algorithm concerning computation cost based on one of the Cloudwatcher algorithms. By taking account of computation cost of rule generation in huge network topology, traffic jam can be avoid. Organization of this paper is as follows. In section 2 we explain the concept of programmable network which is SDN. Section 3 discusses Cloudwatcher framework, and its algorithms and we suggest an approach to excluding duplicated paths in section 4. Section 5 shows our evaluation. Finally we present future works and conclude this paper in section 6.

2   Programmable Network

Nowadays traditional network environment has some restrictions to evolve. There are some reasons disturbing the network improvement. First is vender-dependency of network devices. Switches of network venders usually include whole data/control plane and other application together, and configuration of switches is dependent on the vender program. Therefore, general users can't program network flow and apply better technologies. Second is network complexity in aspect of management. Traditional network environment is consist of discrete protocols which means each protocol is not managed by one controller, therefore, it causes complexity of the network by a number of protocols or network rules. One of the possible solutions to address the problems is a programmable network.

Network controlling technology using software has sustainedly been developed since Intelligent Network in 1980's. Intelligent Network usually is regarded as the beginning of programmable network, research area of the programmable network has been carried out among many research institutions for about 30 years. Recently, Stanford University fulfill Openflow project which is one of SDN protocol specification [8]. Nowadays, several of companies such as IBM, HP are trying to produce network devices and test-bed related to SDN/Openflow model. With rapid development of Openflow technology, there are also some issues of utilization of Openflow in the fields; different configuration style of venders, compatibility with existing network devices. For this reason, ONF (Open Networking Foundation) for standardizing the Openflow protocol was founded in 2011. ONF manages SDN/Openflow specification and encourages the use of SDN.

Software Defined Network. Compared to a traditional network model, the SDN model decouple existing network model into data plane and control plane. Fig. 1 shows difference between traditional network model and SDN network model. While the control plane doesn't remain in the switch anymore, but is implemented in network operating system such as NOX[9], the data plane is described in flow-table which has a set of rules for traffic forwarding. In a conceptual aspect, SDN should have two characteristic. First, SDN should do software defined forwarding which means that data forwarding in traditional switches/routers should be controlled by open interface and software rules in SDN. Network administrator can write flow-rules in control plane, the rules can be applied into data plane, and then the data plane controls the traffic. Second, SDN environment should be consist of a centralized controller and view of the network. It means that SDN should provide more advanced network management tool in entire network view.

In SDN, control plane uses a particular protocol to communicate with data plan. One of the main protocols is Openflow that we mentioned early. Openflow is used as a typical protocol for the implementation of SDN. Currently, Openflow is deployed between SDN controller and network devices as interface specification. Fig. 2 represents the architecture of SDN. A controller makes set of rules and

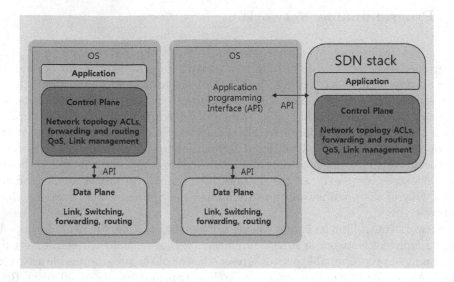

Fig. 1. Comparison between traditional network model and SDN network model[10]

enforces the rules into Openflow-switches/routers through Openflow interface. The rules are commonly applied into flow-table, and when traffic are processed in the flow table pipelining method are deployed. Once traffic is rushing into Openflow routers/switches, packets are compared with flow-table. If there is a matching point, an action that is specified in the table is applied to the traffic.

3   CloudWatcher

As mentioned, cloud computing environment is dynamic and different environment in respects of security due to its virtualization environment, thus existing security devices are hard to be applied into the could computing environment. Therefore, the author proposed a framework that is able to move network traffic to certain node by using network controlling and multi-packet forwarding of SDN. In detail, because traffic in cloud environment is hard to be inspected by security devices, the framework sends simultaneously packets to security devices and destination node on purpose with assistant of SDN.

3.1   Component

Cloudwatcher is consist of three components; Device and Policy, Routing Rule Generator, and Flow Rule Enforcer. First, Device and Policy module is responsible for registration of security devices and security policy information. After the registration, Routing Rule Generator module carries out finding the shortest path from source to security node and destination. The module also makes a graphical topology by sending LLDP(Link Layer Discovery Protocol) to Openflow switches/routers and receiving the response. Finally Flow Rule Enforcer

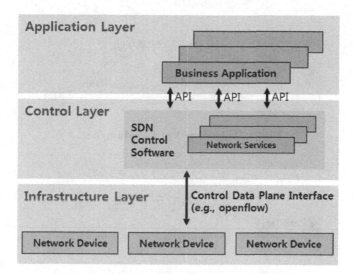

Fig. 2. SDN architecture

delivers rules generated to Openflow switches/routers. Communication between module is done with SLI(Security Language Interface).

3.2 Cloudwatcher's Algorithms

Cloudwatcher suggests four algorithms to find a path from source node to end/security nodes. In this paper, we just represent algorithm 1,2,3. Algorithm 4 has a different topology which is in-line mode while the others operate in passive mode. Packet delivering of all algorithms is based on software defined networking, therefore, network traffic can be routed or forwarded into multiple paths. In the section we demonstrate three algorithms.

Algorithm 1. Algorithm 1 is multipath-naive which means the algorithm finds two types of destination; security nodes and an end node when communication between source node and end node begins. Algorithm 1 takes quite short time to generate a shortest path but it could generate network traffic when there are many security nodes in the topology.

Algorithm 2. Algorithm 2 is shortest-through. In algorithm 2, a network path between source node and end node is established by passing through security node. The process of finding the path is quite complicated because it should consider all intermediate nodes which are included in the path between source and end node.

Algorithm 3. Algorithm 3 is multipath-shortest. In the algorithm, the algorithm finds a shortest path from a source node to a destination node similar

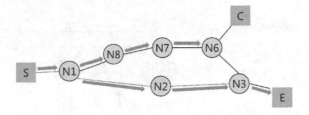

Fig. 3. Algorithm 1 of Cloudwatcher

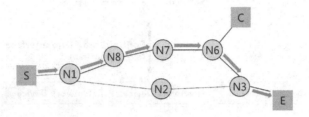

Fig. 4. Algorithm 2 of Cloudwatcher

with algorithm1, and then calculates path cost between each intermediate node of shortest path and security node. However algorithm 3 takes longer time to generate a routing rule than other algorithms when there are many security node and intermediate node while it has better traffic throughput because it should consider all case that it can be.

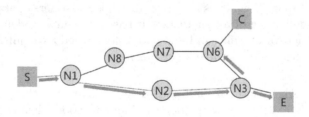

Fig. 5. Algorithm 3 of Cloudwatcher

4   Removing Duplicated Paths

In this section, we suggest a more efficient rule generation algorithm based on algorithm 3 that we mentioned. In algorithm 3, every delivering cost between each intermediate node on a shortest path and security node is calculated, and then the closest node to security nodes is selected. However when there are many security node and intermediate node, rule generation time and computation cost dramatically increases. The point makes network to have possibility of causing network congestion. Thus, we suggest a more efficient rule generation algorithm which considers duplicated paths when looking for a shortest path.

4.1   Duplicated Path

As mentioned earlier, after finding the shortest path between source and end node, every path between each intermediate node on the shortest path and security node is calculated to find the closest node. At that time because all intermediate node are compared and calculated, there could be duplicated paths which makes rule generation time longer and computational cost. Example is as follows. In Fig. 5, when rule generation process is ongoing, a shortest path is calculated, which is N1-N2-N3. After finding shortest path, every path between each intermediate node which are N1, N2, and N3 and security node, which is N6, is calculated to find a shortest path. In case of the N2, it should pass through N1 or N3 to get security node, which is duplicated paths. Actually the paths dont need to be calculated because distance of N2 is bigger than N1 or N3. Such calculation cause computational cost, therefore, it should be excluded during the rule generation time.

4.2   Eliminating Duplicated Path

Below code is part of modified Dijkstra's Algorithm[11] code, we insert pseudo code of checking process to exclude intermediate nodes which pass through other intermediate node on shortest path. When it passes through the intermediate node, it skips the duplicated calculation because it is a duplicated path.

```
include (stdio.h)

int main(void)
{
    int j, k ,p ,start ,min, leng[N+1], v[N+1];
    while(1){

        if (there is an intermediate node on a shortest path){
            skip the computation;
        }

        for (k=1;k<=N;k++) {
            leng[k]=M;v[k]=0;
        }
        leng[start]=0;

        for (j=1;j<=N;j++) {
            min=M;
```

```
                for (k=1;k<=N;k++) {
                    if (v[k]==0 && leng[k]<min) {
                        p=k; min=leng[k];
                    }
                }
                v[p]=1;

                for (k=1;k<=N;k++) {
                    if((leng[p]+a[p][k])<leng[k])
                        leng[k]=leng[p]+a[p][k];
                }
            }
        }
}
```

5   Evaluation

In the section, we evaluate our proposal by showing a example. Below Fig. 6 is a
simple network topology. In this topology, we assume that a shortest path of the
topology is N1-N2-N3-N4-N5. After finding the shortest path, rule generation
process is carried out. Based on the shortest path, each intermediate node on
the shortest path is calculated to find closest node to the security node. Fig. 7
shows all paths that can be generated from the algorithm 3 before eliminating
duplicated paths. The number of paths generated is 16. Fig. 8 represents all
paths generated after excluding duplicated paths by checking the node which
pass through other intermediate nodes. The number of all paths which are generated after eliminating duplicated paths is 3. In Table 1, we show how many
computation paths are excluded by eliminating duplicated paths.

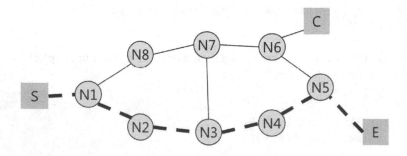

Fig. 6. Exmaple topology

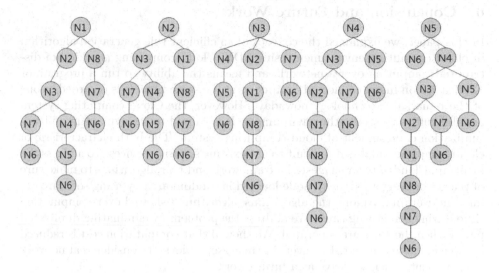

Fig. 7. All paths generated from rule

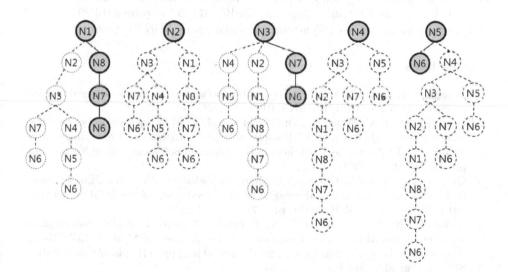

Fig. 8. Eliminated duplicated paths

Table 1. comparison with number of path to be calculated

content	number of path	percentage
original	16	18.7%
eliminated	3	

6   Conclusion and Future Work

In this paper, we explained the concept of an efficient rule generation algorithm in cloud computing environment using SDN. Cloud computing is a kind of distributed computing over a network, and means the ability to run a program or application on many connected computers at the same time. It is regarded as one of the remarkable technology nowadays. However, the Cloud computing system has difficulties to secure the environment. It is caused by its dynamism and virtualization environment of cloud computing system. The both characteristic of cloud computing makes it difficult to use existing security devices. To address the problem, Cloudwatcher suggested a framework and four algorithms to make sure of packet delivery to security node located in random area by using software defined networking. Among the algorithms, algorithm 3 showed over-computation time during generating rules. We address the problem by eliminating duplicated paths when the rules are generated. We showed that computation cost is reduced by considerable amount. However, in this case, it doesn't consider real network environment which we leave as a future work.

Acknowledgements. This research was funded by the MSIP(Ministry of Science, ICT & Future Planning), Korea in the ICT R&D Program 2014[2014044072 003, Development of Cyber Quarantine System using SDN Techniques].

References

1. Ballard, J.R., Rae, I., Akella, A.: Extensible and scalable network monitoring using opensafe. In: Proc. INM/WREN (2010)
2. Fayazbakhsh, S.K., Sekar, V., Yu, M., Mogul, J.C.: FlowTags: enforcing network-wide policies in the presence of dynamic middlebox actions. In: Proceedings of the Second ACM SIGCOMM Workshop on Hot Topics in Software Defined Networking, pp. 19–24. ACM (2013)
3. Qazi, Z.A., Tu, C.C., Chiang, L., Miao, R., Sekar, V., Yu, M.: SIMPLE-fying middlebox policy enforcement using SDN. In: Proceedings of the ACM SIGCOMM 2013 Conference on SIGCOMM, pp. 27–38. ACM (2013)
4. Anwer, B., Benson, T., Feamster, N., Levin, D., Rexford, J.: A slick control plane for network middleboxes. In: Proceedings of the Second ACM SIGCOMM Workshop on Hot Topics in Software Defined Networking, pp. 147–148. ACM (2013)
5. SDNCentral SDN Market Size report,
   http://www.sdncentral.com/sdn-market-size/
6. Shin, S., Gu, G.: CloudWatcher: Network security monitoring using OpenFlow in dynamic cloud networks (or: How to provide security monitoring as a service in clouds?). In: 2012 20th IEEE International Conference on Network Protocols (ICNP), pp. 1–6. IEEE (2012)
7. AsiaCloudForum, Cloud security attacks – are public clouds at risk?,
   http://www.asiacloudforum.com/content/
   cloud-securityattacks-are-public-clouds-risk
8. McKeown, N., Anderson, T., Balakrishnan, H., Parulkar, G., Peterson, L., Rexford, J., Turner, J.: OpenFlow: enabling innovation in campus networks. ACM SIGCOMM Computer Communication Review 38(2), 69–74 (2008)

9. Gude, N., Koponen, T., Pettit, J., Pfaff, B., Casado, M., McKeown, N., Shenker, S.: NOX: towards an operating system for networks. ACM SIGCOMM Computer Communication Review 38(3), 105–110 (2008)
10. Sezer, S., Scott-Hayward, S., Chouhan, P.K., Fraser, B., Lake, D., Finnegan, J., Rao, N.: Are we ready for SDN? Implementation challenges for software-defined networks. IEEE Communications Magazine 51(7) (2013)
11. Johnson, D.B.: A note on Dijkstra's shortest path algorithm. Journal of the ACM (JACM) 20(3), 385–388 (1973)

Towards Application Deployment in Community Network Clouds

Mennan Selimi and Felix Freitag

Department of Computer Architecture
Universitat Politècnica de Catalunya, Barcelona, Spain
{mselimi,felix}@ac.upc.edu

Abstract. Community networks are decentralized communication networks, built and operated by citizens for citizens. Most of these networks originated for enabling Internet access for under-provisioned areas. With the consolidation of cloud technology, there is now also the potential to extend the community networks towards the application level, and deploy within these networks cloud-based applications. In this paper we conduct an experimental deployment of a set of applications on cloud-based infrastructures within community networks. The aim of this study is to understand better the feasibility of running cloud-based applications in the challenging environment of these networks. For our study, we use several nodes inside of two community networks to host our applications and carry out experiments under realistic conditions. The results we obtain with deployments of ownCloud, Tahoe-LAFS and BitTorrent, are a correct functioning of the applications and a promising user experience. Our results could encourage users to contribute and participate in larger scale trials of clouds in community networks, and application developers to envision innovative services based on collective infrastructures.

Keywords: community networks, community cloud, cloud storage.

1 Introduction

Since the first community networks [1] started more than ten years ago, they seem to have become rather successful nowadays. There are several large community networks in Europe having from 500 to 20000 nodes, such as Funk-Feuer[1], AWMN[2], Guifi.net[3], Freifunk[4] and many more worldwide. Most of them are based on Wi-Fi technology (ad-hoc networks, IEEE 802.11a/b/g/n access points in the first hop, long-distance point-to-point Wi-Fi links for the trunk network), but also a growing number of optic fibre links have started to become deployed [2]. Figure 1 shows as example the wireless links and nodes of the Guifi.net community network in the area of Barcelona in Spain. In Figure 2 the map of AWMN nodes in Greece in the area of Athens is shown.

---

[1] http://www.funkfeuer.at
[2] http://www.awmn.gr
[3] http://guifi.net
[4] http://start.freifunk.net

B. Murgante et al. (Eds.): ICCSA 2014, Part VI, LNCS 8584, pp. 614–627, 2014.
© Springer International Publishing Switzerland 2014

Fig. 1. Guifi.net nodes and links in the area around Barcelona in Spain

Fig. 2. Athens Wireless Metropolitan Network (AWMN) nodes in Greece

While the successfully operation of the networking layer of community networks is proved by many of the existing networks, the existence of a large number of applications that run within these networks is far from accomplished. Most users of community networks see the their functionality in providing Internet access to under-provisioned areas. This Internet access is then used consuming valuable up-link capacity among other usages, to access commercial cloud-based applications for storage. Computing and storage resource sharing through cloud computing, now common practice in today's Internet, hardly exists in community networks.

We argue that community networks have now the opportunity to extend the collaborative network building to the next level, that is building collaborative services implemented as community clouds, built, operated and maintained by the community, running on community-owned resources, and being these cloud-based services the ones that are of the community's interest.

In this paper, we contribute with an experimental study of some attractive applications we have deployed within community networks, in order to stimulate with our results further uptake of cloud-based services within these networks.

The slow uptake of application deployment may be related to the fact that these networks are challenging environments, and applications running in there will inherit the challenges of community networks, such as:

- Hardware and software diversity: The network nodes and computers are often inexpensive off the shelf equipment with large heterogeneity in the hardware, software, and capacity.
- Decentralized Management: The network infrastructure and the computers are contributed and managed by the users. They belong to the users and are shared to build the network. There is usually no (or a rather weak) central authority that is responsible for resource provisioning.
- Dynamics: The number of network and computing nodes may rapidly change when members join or leave the network, or when nodes overload or fail.

We select for our study of cloud-based community services three applications with attractive features, whose successful operation we wish to demonstrate should encourage community network users to join and participate, reaching a critical mass which ultimately should become an eco-system. We select as applications to deploy ownCloud [5], a popular open source software which in some features resembles the commercial Dropbox. OwnCloud allows to select among different storage backends. As second application we select to deploy is Tahoe-LAFS [6], a secure and fault-tolerant distributed file systems, potentially suitable for the conditions of community networks, to be used as storage backend of ownCloud. Finally, we deploy BitTorrent for demonstrating a file-sharing option for within the community networks.

The contribution of this paper is an experimental study on the performance and feasibility of applications deployed on cloud-based infrastructures within community networks. Our results aim at opening the door to envision further application deployments and uptake. The examples of the operational applications which we show in this paper should encourage application developers to create additional innovative community network services, and users to contribute resources which enable the hosting and usage of these applications.

2   Applications in Community Networks

Community networks are a kind of eco-system. Participation and contribution of the members is on voluntary basis [3] [4]. In order to raise the interest of users in applications in community networks, these applications must be of value for

---

[5] https://owncloud.org
[6] https://tahoe-lafs.org/trac/tahoe-lafs

the users. A set of carefully selected applications is necessary to motivate these network members start using them. Successful user experience will then drive user participation and collective uptake.

An important requirement for applications to be successful in community networks is that they are open source. The option to be able to customize an application in the source code is important for developers. Secondly, in order to allow an easy application usage by a broad part of society, the applications must have consolidated releases and have a developer community which maintains them. Finally, security, privacy and resilience is important if potentially such applications should be used in the community networks by several kinds of stakeholders.

We have selected ownCloud, Tahoe-LAFS and BitTorrent, whose suitability to fulfil these requirements is explained in the following sections.

2.1 OwnCloud

OwnCloud is an open source cloud Storage as a Service (SaaS) application, which has seen rapid development in the recent years. ownCloud (version 6 at the time of this writing) provides a number of the features synonymous with cloud-based storage solutions including a web-based file access(view/upload/download) and a "desktop sync" client for Windows, OS X and Linux which allows for automated synchronized copies of data on both the client and cloud server. OwnCloud allows users also to mount their own external storage like Dropbox, Google Drive, OpenStack Swift, Tahoe-LAFS, Amazon S3, etc., and use them as ownCloud storage backend. External storage facilities are provided by an app known as "External storage support" which is available on the apps dashboard. OwnCloud with its attractive web-based user interface seems to fit as a case for successful application in community networks.

2.2 Tahoe-LAFS

Tahoe-LAFS is a decentralised storage system with provider-independent security. This features means that the user is the only one who can view or modify disclosed data. The storage service provider never has the ability to read or modify the data thanks to standard cryptographic techniques. The general idea is that the client can store files on the Tahoe-LAFS cluster in an encrypted form using cryptographic techniques. The clients maintain the necessary cryptographic keys needed to access the files. These keys are embedded in read/write/verify "capability strings". Without these keys no entity is able to learn any information about the files in the storage cluster. The data and metadata in the cluster is distributed among servers using erasure coding and cryptography. The erasure coding parameters [5] determine how many servers are used to store each file which is denoted with N, and how many of them are necessary for the files to be available, denoted as K. The default parameters used in Tahoe-LAFS are K=3 and N=10 (3-of-10), which means that each file is shared across 10 different servers, and the correct function of any 3 of those servers is sufficient to retrieve the file. This makes Tahoe-LAFS tolerate multiple storage server failures and attacks.

The Tahoe-LAFS cluster consists a set of storage nodes, client nodes and a single coordinator node called the Introducer. The main responsibility of the Introducer is to act as a kind of publish-subscribe hub. The storage nodes connect to the Introducer and announce their presence and the client nodes connect to the Introducer to get the list of all connected storage nodes. The Introducer does not transfer data between clients and storage nodes, but the transfer is done directly between them. The Introducer is a single-point-of-failure for new clients or new storage peers, since they need it for joining the storage network. We note that for a production environment, the Introducer must be deployed on a stable server of the community network.

When the client uploads a file to the storage cluster, a unique public/private key pair is generated for that file, and the file is encrypted, erasure coded and distributed across storage nodes (with enough storage space). To download a file the client asks all known storage nodes to list the number of shares of that file they hold and in the subsequent round, the client chooses which share to request based on various heuristics like latency, node load etc.

2.3 BitTorrent Application

BitTorrent is an open-source file-sharing application effective for distributing software and media files. As the number of users increases and the file is divided into many small pieces, a client downloads from a swarm of other peers which have the required file. There are several open-source implementations of the components of the BitTorrent system available. While the tracker should run on a stable node, the swarm of BitTorrent peers can grow and shrink, a behaviour that is also typical of many client nodes in a community network. File-sharing within community networks would reduce the traffic on the Internet gateways.

3 Deployment

3.1 The Experimental Environment: Community-Lab

In order to deploy applications in a realistic, community cloud like setting, we use the Community-Lab[7] testbed, a distributed infrastructure provided by the CONFINE[8] project, where researchers can deploy experimental services, perform experiments or access open data traces [6]. We use some of the available nodes of that testbed for deploying the previously introduced applications.

Nodes (i.e. research devices in CONFINE terminology) in the Community-Lab testbed run a custom firmware (based on OpenWRT[9]) provided by CONFINE, which allows running on one node several slivers simultaneously implemented as

---

[7] https://community-lab.net
[8] https://confine-project.eu
[9] https://openwrt.org

Linux containers (LXC). A sliver is defined as the partition of the resources of a node assigned to a specific slice (group of slivers). We can think of slivers as virtual machines running inside of a node. The slivers use the Guifi-Community-Distro (GCODIS) operating system image [7] which contains some of applications, e.g. Tahoe-LAFS, by default.

Our main configuration for the application deployments includes nodes of two geographically distant community networks: Guifi.net in Spain (see Figure 1) and AWMN (Athens Wireless Metropolitan Network) in Greece (see Figure 2). Both community networks (Guifi.net and AWMN) are connected on the IP layer via the FEDERICA (Federated E-infrastructure Dedicated to European Researchers) infrastructure [8], enabling network federation, as illustrated in Figure 3. Therefore, part of the distributed applications in fact is spread over hosts in Guifi.net and other components are in hosts in AMWN. The nodes of our experiments are real nodes in both community networks.

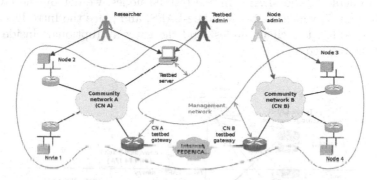

Fig. 3. Network federation in Community-Lab for an experiment over two community networks

For our experiment we create a slice of 20 nodes. From the Guifi.net community network, we use 5 nodes, from AWMN we use 5 nodes, from UPC Campus we use 5 nodes, and 5 nodes are from UPC Lab. Most of the Community-Lab nodes (Guifi.net, AWMN, UPC Campus) consist in terms of hardware of Jetway devices that are equipped with an Intel Atom N2600 CPU, 4GB of RAM and 120GB SSD. A few nodes we add to this slice through UPC Lab are provided as VMs through nodes of a Proxmox[10] cluster, which adds heterogeneity regarding storage space and processing power.

In the first experiment, we deploy an ownCloud instance and Tahoe-LAFS for a distributed cloud storage in community networks. We measure the read and write throughput seen by a Tahoe-LAFS client and the overall functionality of the application.

The second experiments consists of a deployment of BitTorrent in our slice, in order to demonstrate the option of using BitTorrent file-sharing within community networks. For both experiments the nodes from Guifi.net, AWMN and

---

[10] https://www.proxmox.com/

UPC-Campus share 400 MB storage space to our applications, the nodes from UPC-Lab share 5 GB of disk space.

3.2   ownCloud in the Community-Lab Testbed

We deployed an instance of ownCloud in one of the Community-Lab testbed nodes of our slice. We select a stable node since ownCloud is not replicated. Knowing the ownCloud server URL, authenticated users are able to remotely up- and download files over the ownCloud Web interface. To store the files which are uploaded to ownCloud in the community network, we replaced the ownCloud backend with Tahoe-LAFS, as described in more detail in the following.

3.3   Tahoe-LAFS in the Community-Lab Testbed

In order to deploy Tahoe-LAFS in the testbed nodes, we deploy the GCODIS distribution [9] [7] which contains Tahoe-LAFS, and place the Introducer node, the storage nodes, the client nodes, and the gateway (optional) inside of the slivers of the testbed.

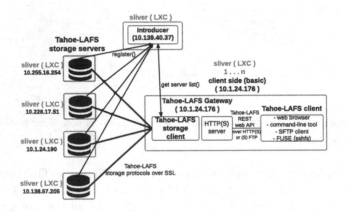

Fig. 4. Tahoe-LAFS deployed in the Community-Lab testbed

Figure 4 shows the resulting Tahoe-LAFS architecture used by our experiments in the Community-Lab testbed. While the Tahoe-LAFS Introducer runs on a different node, the Tahoe-LAFS client is on the same node where ownCloud is also installed. This way the performance of Tahoe-LAFS translates directly to the performance seen at the ownCloud server itself.

Table 1 shows the RTT and number of hops between the client and storage nodes including two community networks (CNs). The client who writes and reads the data is located in a node of UPC Campus. In Table 1 it can be seen that we use 4 types of community network nodes as storage servers of Tahoe-LAFS, having these storage servers to the client different latencies and different numbers of hops. We use the default Tahoe-LAFS erasure coding parameter 3-of-10 which means that for each upload of a file, Tahoe-LAFS uses 10 of the 20 storage servers of our slice, while these storage servers differing significantly in bandwidth and latency.

Table 1. RTT and number of hops between client and storage nodes

Nr. of nodes	CN	RTT	Nr. of hops
5	UPC Lab	0.200-0.400 ms	1
5	UPC Campus	0.300-0.700 ms	1-2
5	Guifi.net	5-13 ms	6-7
5	AWMN	120-150 ms	15-20

3.4 Tahoe-LAFS Experiment

We measure at the Tahoe-LAFS client the upload transfer rate when uploading files of size 20 MB, 50 MB and 100 MB. The upload transfer rate (presented in Mbits/sec) is for storage servers taken from the two community networks and selected by Tahoe-LAFS. Figure 5 shows the obtained measurements. For each file size we measure the throughput achieved in a time window of 7 hours. We measure approximately every 30 minutes. It can be seen that the throughput changes during these 7 hours of the experiment. The lowest throughput achieved is in the midday between 13:00 and 14:00 hour resulting in the upload rate of 1.1 Mbits/sec for 20 MB of file, 1.3 Mbits/sec for 50 MB of file and 1.5 Mbits/sec for 100 MB of file. Since our experiments are carried out in a production community network, the traffic produced in the network by other users influences on the application performance we obtain. The average throughput rates observed during the 7 hours time window are 1.6 Mbits/sec for 20 MB, 1.8 Mbits/sec for 50 MB and 1.9 Mbits/sec for 100 MB of file.

The download transfer rate (presented in Mbits/sec) is shown in Figure 6. As in the upload case, we observe that the lowest throughput is achieved in the midday between 13:00 and 14:00 hour resulting in the download rate of 4.8 Mbits/sec for 20 MB, 4.7 Mbits/sec for 50 MB and 4.9 Mbits/sec for 100 MB of file. The average throughput rates observed during the 7 hours time window are 5.4 Mbits/sec for 20 MB, 5.5 Mbits/sec for 50 MB and 5.9 Mbits/sec for 100 MB of files.

Figure 7 shows the latency of upload and download operation in two community networks comprised of 20 nodes. As it can be seen the download performance is better then upload performance resulting in 2.1 minute for downloading a 100 MB of file, and 6.5 minutes for uploading a 100 MB of file. This is expected in erasure coded systems, since reads (downloads) in fact transfer less data than writes (uploads).

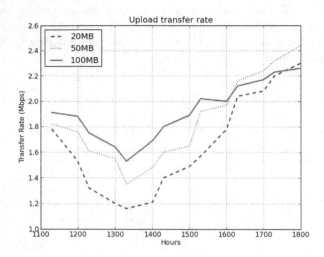

Fig. 5. Upload transfer rate at Tahoe-LAFS client

3.5 BitTorrent Experiment

This experiment aims to show the applicability of BitTorrent in community networks and illustrates in an experiment the file-sharing performance between different geographically-sparse community network nodes while they share a small file using the BitTorrent protocol. The nodes are the same nodes as used in Tahoe-LAFS experiment, in total 20 nodes geographically dispersed in two community networks. We install as a BitTorrent tracker the Opentracker software[11] on a node located in the Guifi network. The transmission BitTorrent client[12] are the peers. A peer which is serving the file (seeder) is located in the AWMN network. The initial seeder provides the complete file of 30 MB and the other peers download this file. The peers that download the file are located in both community networks. The download performance depends on the location of the peers that download the file and the mechanisms of the BitTorrent protocol itself. For the peers located in Guifi.net, we observe that the average download rate achieved is 5.6Mbits/sec resulting in a download latency of 42 seconds for 30 MB of file, and for the peers located in AWMN network download rate achieved is 9.2Mbits/sec resulting in a download latency of 26 seconds for 30 MB of file (Figure 8). All file-sharing operations we experimented with, finished successfully.

3.6 Discussion

Overall we observed that the WAN characteristics have an impact on the performance of the application deployed in a community network cloud. Tahoe-LAFS

---

[11] http://erdgeist.org/arts/software/opentracker/
[12] http://www.transmissionbt.com/

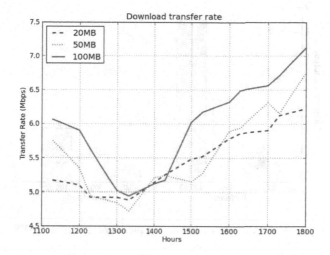

Fig. 6. Download transfer rate at Tahoe-LAFS client

uploads required more bandwidth than downloads[13]. It is explained by the fact that Tahoe-LAFS pushes redundant shares for upload, while for retrieving these are not needed, reducing therefore the download time. The default 3-of-10 encoding uploads produce about 3.3x the traffic of the download of the same file. In practice, these effects may not be significant for the user experience. Users that upload a file do not need a fast upload since they have the file locally. If Tahoe-LAFS is adopted, the traffic produced within community networks during heavy usage of Tahoe-LAFS uploads might reach the limits of the network, but currently the traffic bottleneck of community networks is on the gateways to the Internet, and not in traffic within the network. Erasure coding as done by Tahoe-LAFS adds additional requirement for storage capacity. Using Tahoe-LAFS for large media files might reach the storage capacity limits of clouds in community networks. The usage of Tahoe-LAFS should therefore rather target at smaller personal files, e.g. backup of user data, where security, privacy and resilience are required.

We further observed that the functional performance of Tahoe-LAFS in our experiments was correct. Our experimental study suggests that Tahoe-LAFS is a convenient cloud storage services within community networks where data encryption and fault-tolerance are important. As a result we can state that Tahoe-LAFS seems to be a promising application that should be considered for privacy-preserving, secure and fault-tolerant storage in community network. Though Tahoe-LAFS is not comfortable for users to be handled as stand-alone application, it seems to be particularly suitable in combination with user-friendly front-ends such as ownCloud.

---

[13] https://tahoe-lafs.org/trac/tahoe-lafs/browser/docs/helper.rst

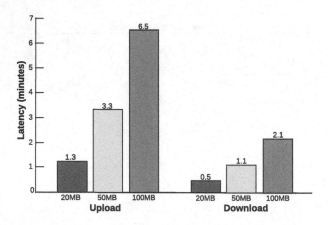

Fig. 7. Upload and download latency at ownCloud/Tahoe-LAFS client

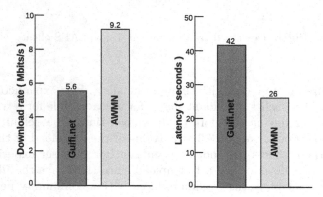

Fig. 8. Average download rate and latency for downloading a 30 MB file

We repeated our experiments during seven hours of a day. We found that the available bandwidth in the community network changes. During some periods of the day, more bandwidth is available, which might be noticed by the end-user in terms of better user experience. Nevertheless, even the lowest bandwidth we observed in our experiments should be well acceptable for the user.

The BitTorrent application deployed in community networks showed fast downloads times when having the file well replicated in a BitTorrent swarm. Our results suggest BitTorrent to be a suitable application for large media file sharing within community networks, where data privacy and security of the file content is not a concern.

4   Related Work

Regarding application deployments related to our scenario, not many systems have been reported, and most of them are not yet available as complete prototypes. The Cloud@Home[14] [10] project proposes to harvest in resources from the community for meeting peaks in resource demands. The system is well described in terms of design and motivation, but a deployed systems seems not yet to be available.

The Clouds@home[15] [11] project focuses on providing guaranteed performance and ensuring quality of service (QoS) even when using volatile volunteered resources connected by Internet. The authors focus on voluntary computing systems, but do not consider the particular context of community networks. The P2PCS[16] [12] project has built an initial prototype implementation of a decentralized Peer-to-Peer cloud system. It uses Java JRMI technology and builds an IaaS system that provides very basic support for creating and managing VMs as a slice (group of VMs). It manages slice information in a decentralized manner using gossip protocols. The system is not completely implemented and integrated.

In terms of providing cloud storage services with Tahoe-LAFS in WAN settings, Chen's paper [13] is the most relevant to our work regarding the study of this application. The authors deployed Tahoe-LAFS, QFS [17] and Swift[18] in a multi-site environment and measured the impact of WAN characteristics on these storage systems. They observed that WAN characteristics have a large impact on the performance of these systems, mainly due to several design choices of these systems. In addition, they found that good system design and extensive optimizations can have a significant effect on performance, as it was seen by the relative difference between Tahoe-LAFS, QFS and Swift. However, the authors deployed their experiments on a multi-site data center with very different characteristics to our scenario. In our work we consider the particular context of community networks with less powerful machines, and heterogeneous and dynamic network conditions.

The work of Tseng et al [14], implements a distributed file system for Apache Hadoop[19]. The original Hadoop distributed file system is replaced with the Tahoe-LAFS cloud storage. The authors investigated the total transmission rate and download time with two different file sizes. Their experiment showed that the file system accomplishes a fault-tolerant cloud storage system even when parts of storage nodes had failed. However in the experiments only three storage nodes and one Introducer node of Tahoe-LAFS were used, and their experiments were run in a local context, which is an unrealistic setting for community network scenario.

From the review of the related work it can be seen that experimental studies of applications deployed in community networks are not well addressed by related work, and the researched systems are usually not deployed in the context of

---

[14] http://cloudathome.unime.it
[15] http://clouds.gforge.inria.fr
[16] https://code.google.com/p/cloudsystem
[17] http://quantcast.github.io/qfs/
[18] http://docs.openstack.org/developer/swift/
[19] http://hadoop.apache.org/

community networks. In our work we put emphasis on studying applications in a real deployment (in production environment) within community network, in order to understand their performance and operation feasibility under real network conditions.

5   Conclusion and Future Work

A set of relevant open source applications were deployed over two community networks and their performance was studied, leveraging a cloud-based infrastructure which provided hosts and realistic conditions to this experiment.

The study aimed to provide a first understanding of the feasibility of application deployments on cloud-based community infrastructures. The applications for the experiment were selected according to their potential relevance for current and potential community network users, covering solutions for different requirements of data sharing and storage. To this end, the popular ownCloud storage application was combined with Tahoe-LAFS, a decentralized secure cloud storage application. The combination of both offers an attractive Web-based GUI through ownCloud and privacy and security through Tahoe-LAFS, since it encrypts data already at the client side and provides by erasure coding fault-tolerance regarding storage node failures. Further to Tahoe-LAFS, BitTorrent was also deployed in the community network. Due to the increased storage and traffic needs of Tahoe-LAFS, it appeared that for large media files BitTorrent is a more appropriate choice, while Tahoe-LAFS finds its application for providing a storage service for private user data that can be stored in a secure way in the community cloud.

Based on the observed successful performance and operation of these applications in community networks, our experimental deployments should now transform into permanently available services open to real users. As the end users will become involved, we will be able to naturally extend our experiments with more storage nodes contributed by real users, more files coming from real data, and real usage. Successful results of that next step could have an important impact on the options available for secure cloud storage, where community clouds might become one of the options user may choose.

Acknowledgement. This work was supported by the European FP 7 FIRE projects CONFINE, FP7-288535, and CLOMMUNITY, FP7-317879. Support is also provided by the Universitat Politècnica de Catalunya BarcelonaTECH and the Spanish Government through the Delfin project TIN2010-20140-C03-01.

References

1. Braem, B., Blondia, C., Barz, C., Rogge, H., Freitag, F., Navarro, L., Bonicioli, J., Papathanasiou, S., Escrich, P., Baig Viñas, R., Kaplan, A.L., Neumann, A., Vilata i Balaguer, I., Tatum, B., Matson, M.: A case for research with and on community networks. SIGCOMM Comput. Commun. Rev. 43(3), 68–73 (2013)

2. Guifi.net: Guifi.net new sections of fiber deployed to the farm (2012), http://en.wikinoticia.com/Technology/internet/122595
3. Bina, M., Giaglis, G.: Unwired Collective Action: Motivations of Wireless Community Participants. In: International Conference on Mobile Business (ICMB 2006), Copenhagen, Denmark, pp. 31–40. IEEE (June 2006)
4. Elianos, F.A., Plakia, G., Frangoudis, P.A., Polyzos, G.C.: Structure and evolution of a large-scale Wireless Community Network. In: 2009 IEEE International Symposium on a World of Wireless, Mobile and Multimedia Networks & Workshops. IEEE (June 2009)
5. Wilcox-O'Hearn, Z., Warner, B.: Tahoe: The least-authority filesystem. In: Proceedings of the 4th ACM International Workshop on Storage Security and Survivability, StorageSS 2008, pp. 21–26. ACM, New York (2008)
6. Neumann, A., Vilata, I., Leon, X., Escrich, P., Navarro, L., Lopez, E.: CommunityLab: Architecture of a Community Networking Testbed for the Future Internet. In: 2012 1st International Workshop on Community Networks and Bottom-up-Broadband (CNBuB 2012), within IEEE WiMob (October 2012)
7. Jimenez, J., Baig, R., Escrich, P., Khan, A., Freitag, F., Navarro, L., Pietrosemoli, E., Zennaro, M., Payberah, A., Vlassov, V.: Supporting cloud deployment in the guifi.net community network. In: Global Information Infrastructure Symposium, pp. 1–3 (2013)
8. FEDERICA: Federated E-infrastructure Dedicated to European Researchers Innovating in Computing network Architectures, http://www.fp7-federica.eu/
9. Jiménez, J., Baig, R., Freitag, F., Navarro, L., Escrich, P.: Deploying PaaS for Accelerating Cloud Uptake in the Guifi.net Community Network. In: International Workshop on the Future of PaaS 2014, within IEEE IC2E, Boston, Massachusetts, USA. IEEE (March 2014)
10. Distefano, S., Puliafito, A.: Cloud@Home: Toward a Volunteer Cloud. IT Professional 14(1), 27–31 (2012)
11. Yi, S., Jeannot, E., Kondo, D., Anderson, D.P.: Towards Real-Time, Volunteer Distributed Computing. In: 11th IEEE/ACM International Symposium on Cluster, Cloud and Grid Computing (CCGrid 2011), Newport Beach, USA, pp. 154–163. IEEE (May 2011)
12. Babaoglu, O., Marzolla, M., Tamburini, M.: Design and implementation of a P2P Cloud system. In: 27th Annual ACM Symposium on Applied Computing (SAC 2012), New York, USA, pp. 412–417 (March 2012)
13. Chen, Y.F., Daniels, S., Hadjieleftheriou, M., Liu, P., Tian, C., Vaishampayan, V.: Distributed storage evaluation on a three-wide inter-data center deployment. In: 2013 IEEE International Conference on Big Data, pp. 17–22 (2013)
14. Tseng, F.H., Chen, C.Y., Chou, L.D., Chao, H.C.: Implement a reliable and secure cloud distributed file system. In: 2012 International Symposium on Intelligent Signal Processing and Communications Systems (ISPACS), pp. 227–232 (2012)

Piecewise Smooth Systems: Equilibrium Points and Application to Gene Regulatory Networks

Marco Berardi[1,2]

[1] Dipartimento di Matematica - Università di Bari,
via Orabona 4, 70125, Bari, Italy
marco.berardi@uniba.it
[2] Istituto di Ricerca sulle Acque - CNR, via De Blasio 5, 70132 Bari, Italy
marco.berardi@ba.irsa.cnr.it

Abstract. After presenting the main issues of piecewise smooth dynamical systems, and defining the different types of equilibrium points, we focus on Piecewise Linear (PWL) systems. A result on the characterization of a particular type of PWL system is given, that relates the *pseudo-equilibrium* point of a PWL system with the solution of a linear algebraic system. Examples are given and an application to Gene Regularoy Networks (GRNs) is provided.

Keywords: Piecewise linear systems, gene regulatory networks.

1 Introduction and Motivation

The study of Piecewise Linear (PWL) systems, as will be defined in Section 4, has a lot of interesting and ongoing applications. Among these, we are mainly interested in applications to Gene Regulatory Networks (GRNs). The activity of these networks can be represented by *rate equations*, expressing the rate of production of a component of the system, as the concentrations of other components. In the context of GRNs, dealing with piecewise linear differential equations means that the regulatory interactions are described by means of step functions, which give to the model a switch like behaviour. We should say that the piecewise linear model, characterized by the jump in the right-hand side of the ODE, can be substituted by a continuous model, where a continuous sigmoid function takes the place of the step function: this is the case of models that obey to Michaelis-Menten kinetics, where the sigmoid function is represented by a first order Hill function; in general, a p-th order Hill function has the following form: $h(x, \theta) = \frac{x^p}{x^p + \theta^p}$, for $x \geq 0$.

The number of papers about different types of GRNs, both from a mathematical point of view, both from a biological one, is quickly increasing: see, for example, [26], [27], [28], [6], [5], [22], [11].

B. Murgante et al. (Eds.): ICCSA 2014, Part VI, LNCS 8584, pp. 628–641, 2014.

2   Introduction to Filippov Theory

A piecewice smooth (PWS) system is generally expressed in the following way:

$$x' = f(x) = \begin{cases} f_1(x) \text{ if } x \in R_1 \\ f_2(x) \text{ if } x \in R_2 \end{cases}, \tag{1}$$

with $x(0) = x_0 \in \mathbb{R}^n$, and $x = x(t)$ and $x' = dx/dt$. In particular, the state space $\mathbb{R}^n$ is split into two subspaces R_1 and R_2 by a surface Σ such that $\mathbb{R}^n = R_1 \cup \Sigma \cup R_2$. The surface is defined by a scalar *event* function $h : \mathbb{R}^n \to \mathbb{R}$, so that R_1, R_2, and Σ, are implicitly characterized as

$$\Sigma = \{x \mid h(x) = 0\}, \quad R_1 = \{x \mid h(x) < 0\}, \quad R_2 = \{x \mid h(x) > 0\}. \tag{2}$$

In equation (1), we will assume that f_i is C^k on $R_i \cup \Sigma$ ($k \geq 1$), for $i = 1, 2$, but we do not assume that f_i extends smoothly also in R_j, for $i \neq j$. We also assume that $h \in C^k$, $k \geq 2$, and that $\nabla h(x) \neq 0$ for all $x \in \Sigma$. Thus, the normal to Σ, $\mathbf{n}(x) = \frac{\nabla(h(x))}{\|\nabla(h(x))\|_2}$, exists and varies smoothly for all $x \in \Sigma$.

In (1), $f(x)$ is not defined if x is on Σ and a standard way to overcome this difficulty is to consider the set valued extension $F(x)$ below:

$$x'(t) \in F(x) = \begin{cases} f_1(x) & \text{when} & x \in R_1 \\ \overline{co} \{f_1(x), f_2(x)\} & \text{when} & x \in \Sigma \\ f_2(x) & \text{when} & x \in R_2 \end{cases} \tag{3}$$

where $\overline{co}(A)$ is the smallest closed convex set containing A. In our case:

$$\overline{co} \{f_1, f_2\} = \{f_\Sigma : \mathbb{R}^n \to \mathbb{R} \mid f_\Sigma = (1 - \alpha)f_1 + \alpha f_2, \ \alpha \in [0, 1]\}. \tag{4}$$

The extension of a discontinuous system (1) into a convex differential inclusion (3) is known as *Filippov convex method*. An absolutely continuous function $x : [0, \tau) \to \mathbb{R}^n$ is a *Filippov solution* of (1) if it holds that $x'(t) \in F(x(t))$ almost eveywhere, being $t \in [0, \tau)$

For the existence of solutions of (3) we refer to [2], [19]. Presently, we only consider the case in which solutions are continuous, though not differentiable in some points. Without loss of generality let us suppose that $x_0 \in R_1$, and consider a trajectory of (1). As long as the trajectory remains in R_1, we have a normal IVP (Initial Value Problem): $x' = f_1(x)$, $x(0) = x_0$. But when the solution reaches a point $x \in \Sigma$, we consider the only two possibilities that guarantee exixtence and uniqueness of solutions: we exit Σ and enter into R_1 or R_2; we remain in Σ with a yet to be defined vector field. Filippov's theory is a very useful tool to decide how the trajectory behaves in these cases, and how to define the vector field in the second case. We summarize it below.

So, let $x \in \Sigma$ and $n(x)$ be the normal to Σ at x. Let $\mathbf{n}^\top(x)f_1(x)$ and $\mathbf{n}^\top(x)f_2(x)$ be the projections of $f_1(x)$ and $f_2(x)$ onto the normal to Σ. We have two main cases.

– *Transversal Intersection.* In case in which, at $x \in \Sigma$, we have

$$[\mathbf{n}^\top(x)f_1(x)] \cdot [\mathbf{n}^\top(x)f_2(x)] > 0 , \tag{5}$$

then we leave Σ. We will enter R_1, when $\mathbf{n}^\top(x)f_1(x) < 0$, and will enter R_2, when $\mathbf{n}^\top(x)f_1(x) > 0$ (see Figure 1). In the former case we will have (1) with $f = f_1$, in the latter case with $f = f_2$. Any solution of (1) reaching Σ at a time t_1, and having a transversal intersection there, exists and is unique.

– *Attracting Sliding Mode.* An attracting sliding mode at Σ occurs if

$$[\mathbf{n}^\top(x)f_1(x)] > 0 \quad \text{and} \quad [\mathbf{n}^\top(x)f_2(x)] < 0, \qquad x \in \Sigma . \tag{6}$$

When we have an attracting sliding mode at $x_0 \in \Sigma$, a solution trajectory which reaches x_0 does not leave Σ, and will therefore have to move along Σ. Filippov's theory provides an extension to the vector field on Σ, consistent with the interpretation in (4), giving rise to *sliding motion.* During the sliding motion the solution will continue along Σ (see Figure 1) with time derivative f_Σ given by:

$$f_\Sigma(x) = (1 - \alpha(x))f_1(x) + \alpha(x)f_2(x) \tag{7}$$

where $\alpha(x)$ is the value for which $f_\Sigma(x)$ lies in the tangent plane T_x of Σ at x, that is the value for which $\mathbf{n}^\top(x)f_\Sigma(x) = 0$. This gives

$$\alpha(x) = \frac{\mathbf{n}^\top(x)f_1(x)}{\mathbf{n}^\top(x)(f_1(x) - f_2(x))}. \tag{8}$$

Observe that a solution having an attracting sliding mode exists and is

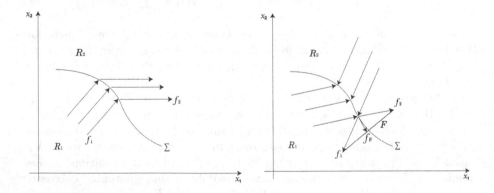

Fig. 1. Example of transversal intersection -left- and sliding mode -right-

unique, in forward time. If the inner products in (6) are of opposite signs we have a *repulsive sliding mode* which does not lead to uniqueness of the solution. Thus, we will not consider repulsive sliding motion in this work.

Remark 1. Numerical methods for solving piecewise-smooth dynamical systems are object of much interest in the last years. For a general overwiev, see, for example, [1], [15].

A lot of different issues arise in the study of piecewise smooth systems. First of all, a couple of different techniques can be used: the *event-driven* method, based on an accurate localization of the event, and the *time-stepping* method, as decribed on [20], based on some estimate of the local truncation error. Moreover, it can happen that f_i, as defined in equation (1), is not defined in R_j, with $i \neq j$. In this case, *one-sided* methods are better (see, for instance, [16], [4], [17]). Also, for example, if the system is quite stiff with regard to function f_1, f_2 as defined in (1), it is advantageous to use semi-implicit methods like Rosenbrock schemes (see [3]). The choice of the sliding vector field when two event functions intersect is still an open problem: different researches have recently focused on this issues (see, for example, [13], [18]); from the numerical point of view, see also [14].

3   Equilibrium Points

An equilibrium point x_1^* of $f_1(x)$ [such that $f_1(x_1^*) = 0$] is said *admissible* if $x_1^* \in R_1$ (i.e. $h(x_1^*) < 0$); while if $x_1^* \in R_2$ (i.e. $h(x_1^*) > 0$) it is called *virtual;* finally, if $x_1^* \in \Sigma$ (i.e. $h(x_1^*) = 0$) it is called a *boundary equilibrium point.* Similar definitions may be given for an equilibrium point x_2^* of $f_2(x)$. Furthermore, x^* is called a *pseudo equilibrium* if it is the zero of the sliding vector field f_Σ in (7). It will be called *admissible* if $0 < \alpha(x^*) < 1$, *virtual* if $\alpha(x^*) \in] - \infty, 0[\cup]1, +\infty[$ (see [12]). A point x^* is said to be a *boundary* equilibrium point if $f_1(x^*) = 0$ or $f_2(x^*) = 0$ and $h(x^*) = 0$. Moreover, if the equilibrium point x_1^* [or x_2^*] belongs to Σ then $\alpha(x_1^*) = 0$ [or $\alpha(x_2^*) = 1$], that is it is also an exit point.

At an admissible pseudo-equilibrium point x^*, we have that the two vector fields f_1 and f_2 become parallel and in the opposite direction, in particular from (7) it follows that :

$$f_1(x^*) = \frac{\alpha(x^*)}{\alpha(x^*) - 1} f_2(x^*) . \tag{9}$$

Example 1. Let us consider the following two-dimensional PWS system:

$$x' = \begin{bmatrix} x_1' \\ x_2' \end{bmatrix} = \begin{cases} f_1(x) & \text{when} \quad h(x) < 0 \\ f_2(x) & \text{when} \quad h(x) > 0 \end{cases}$$

with

$$f_1(x) = \begin{bmatrix} x_2 \\ -x_1 + \frac{1}{1.2 - x_2} \end{bmatrix}, \quad \text{and} \quad f_2(x) = \begin{bmatrix} x_2 \\ -x_1 - \frac{1}{0.8 + x_2} \end{bmatrix},$$

and the discontinuity surface Σ defined by the zero set of $h(x) = x_2 - 0.2$. The normal to Σ is given by $n = (0, 1)^\top$, thus on Σ we have $\mathbf{n}^\top f_1(x) = -x_1 + 1$ and $\mathbf{n}^\top f_2(x) = -x_1 - 1$. which means that there will be an attractive sliding

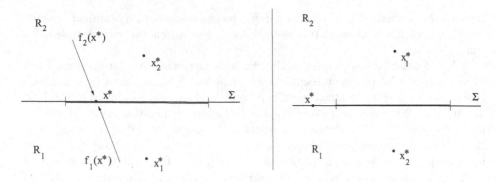

Fig. 2. Admissible and virtual equilibrium and pseudo-equilibrium points. [The sliding region is an interval on Σ.]

mode, on the line $x_2 = 0.2$, when $x_1 \in (-1, 1)$. The sliding vector field is given by $f_\Sigma = \begin{pmatrix} 0.2 \\ 0 \end{pmatrix}$, thus there is no pseudo-equilibrium point. We also notice that the equilibrium point $x_1^* = (1/1.2, 0)$ for f_1 is inside the region R_1 (i.e. $x_2 < 0.2$), [in particular inside the limit cycle of the PWS system], that means that it is an admissible equilibrium point. Moreover, the Jacobian matrix of f_1 at x_1^* is given by $J = \begin{pmatrix} 0 & 1 \\ -1 & \frac{1}{1.44} \end{pmatrix}$, with eigenvalues given by $\lambda_1 = 0.3472 + 0.9378i$ and $\lambda_2 = \bar{\lambda}_1$, with $|\lambda_1| = |\lambda_2| = 1$. Thus the trajectories will leave x_1^* (repulsive equilibrium point) unless we start on it. Furthermore, the equilibrium point of f_2 is $x_2^* = (-\frac{1}{0.8}, 0)$ that is in the region R_1 (virtual equilibrium point). Moreover, the Jacobian matrix of f_2 at x_1^* is given by $J = \begin{pmatrix} 0 & 1 \\ -1 & \frac{1}{0.64} \end{pmatrix}$, with eigenvalues given by $\lambda_1 = 0.7812 + 0.6242i$ and $\lambda_2 = \bar{\lambda}_1$, with $|\lambda_1| = |\lambda_2| = 1$. In the Figure 3 we show the trajectory of the system with initial point $x_0 = (\frac{1}{1.199}, 0)$ [marked with "x"] close to x_1^* and the trajectory starting with $x_0 = (-\frac{1}{0.799}, 0)$ close x_2^*.

Example 2. Let us consider the following two-dimensional PWS system with

$$f_1(x) = \begin{pmatrix} -4x_1 - x_2 + 12 \\ -x_1 - 2x_2 + 10 \end{pmatrix}, \quad f_2(x) = \begin{pmatrix} -5 \\ -1 \end{pmatrix},$$

and the discontinuity surface Σ defined by the line $h(x_1, x_2) = 5x_1 + x_2 = 0$. We notice that the normal vector is $\mathbf{n} = -f_2$. Thus, on Σ, we have

$$\mathbf{n}^\top f_2 = -26; \quad \mathbf{n}^\top f_1(x) = 7(2x_1 + 10), \text{ since on } \Sigma \text{ we have } 5x_1 + x_2 = 0$$

hence, an attractive sliding motion on Σ occurs when $x_1 > -5$, while the trajectory will cross Σ when $x_1 < -5$. From (9), it follows that the pseudo-equilibrium point is such that

$$\begin{pmatrix} 4x_1 + x_2 - 12 \\ x_1 + 2x_2 - 10 \end{pmatrix} = k \begin{pmatrix} 5 \\ 1 \end{pmatrix}$$

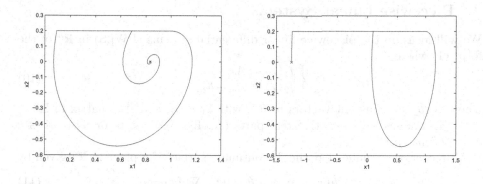

Fig. 3. Trajectories for Example 1 with initial point close x_1^* (left) and x_2^* (right)

with k suitable constant. Thus, by solving the previous system, we have:

$$x_1 = 2 + \frac{9}{7}k , \quad x_2 = 4 - \frac{1}{7}k ,$$

and by imposing this point must be on Σ, it follows that $k = -\frac{49}{22}$, and hence:

$$x^* = \left(-\frac{19}{22}, \frac{95}{22} \right) .$$

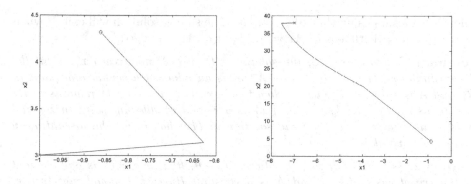

In Figure 2 (left) starting with $x_0 = (-1, 3)$ the trajectory hits Σ then starts to slide on it until to approach the pseudo-equilibrium point. In Figure 5 (right) starting with $x_0 = (-7, 38)$ the trajectory hits Σ then crosses it, until to hit again Σ, then the trajectory starts to slide on Σ until to approach the pseudo-equilibrium point.

4   Piecewise Linear Systems

We will focus on the piecewice linear differential systems (PWLS) in $\mathbb{R}^n$ of the form (1) where

$$\begin{cases} f_1(x) = Ax + b_1 \\ f_2(x) = Ax + b_2 \end{cases}, \tag{10}$$

where b_1, b_2 are constant vectors in $\mathbb{R}^n$, with $b_1 \neq b_2$, and A a matrix in $\mathbb{R}^{n \times n}$ with eigenvalues with negative real parts (i.e. Re $[\lambda(A)] < 0$, or, also, A is a Hurwitz matrix).

Moreover, we assume that the discontinuity surface is a plane in $\mathbb{R}^n$ given by

$$\Sigma : \ h(x) = c^\top x - \theta = 0 ; \quad \nabla h(x) = c ; \tag{11}$$

where c is a constant column vector and θ is a known scalar value.

Assuming initial value $x_0 \in R_1$, the solution in the region R_1 of (1)-(10) is given by:

$$x(t) = \exp(tA)x_0 + [\exp(tA) - I]A^{-1}b_1 , \ t > 0 . \tag{12}$$

Let us suppose $\bar{x}$ is on Σ and $[c^\top A\bar{x} + c^\top b_1] > 0$, if

$$[c^\top A\bar{x} + c^\top b_1][c^\top A\bar{x} + c^\top b_2] > 0 ,$$

then the trajectory will cross Σ (at $\bar{x}$) and enter R_2, while if

$$[c^\top A\bar{x} + c^\top b_1][c^\top A\bar{x} + c^\top b_2] < 0 ,$$

then the trajectory slides on Σ satisfying the following ODE:

$$x' = (1 - \alpha(x))b_1 + \alpha(x)b_2 + Ax , \quad \alpha(x) = \frac{c^\top Ax + c^\top b_1}{c^\top (b_1 - b_2)} , \tag{13}$$

where we are supposing that $c^\top b_1 \neq c^\top b_2$. The sliding solution will exit Σ when $\alpha(x) = 0$ or $\alpha = 1$ [that is $c^\top Ax = -c^\top b_1$ or $c^\top Ax = -c^\top b_2$).

Lemma 1. When $t \to +\infty$, the solution $x(t)$ in (12) may reach x_1^* or hit the discontinuity surface Σ. If $x_1^* = -A^{-1}b_1$ is an admissible equilibrium point in R_1 and A is a diagonal matrix, when $t \to +\infty$, the solution $x(t)$ remains in R_1 and tends asymptotically to x_1^*. If x_1^* is a virtual equilibrium point in R_2 and A is a diagonal matrix, then the solution in (12) has to hit the discontinuity surface at a point $\bar{x}$.

Lemma 2. (see [21].) If in (11) $h(x) = x_s - \theta$, [that is $c = e_s$ is one element of the trivial basis of $\mathbb{R}^n$] and A is a constant Hurwitz diagonal matrix (i.e. $A = \Gamma = \mathrm{diag}(\gamma_1, \ldots, \gamma_n)$), then α in (13) is independent on x:

$$\alpha = \frac{\gamma_s \theta + [b_1]_s}{[b_1 - b_2]_s} . \tag{14}$$

During a sliding motion on Σ, the solution satisfies (13): there is no possibility to leave the discontinuity surface, since α_c is a constant value, and motion will be toward the (admissible) pseudo-equilibrium (since $Re(\lambda(A)) < 0$) which is on the intersection between Σ and the segment $[x_1^*, x_2^*]$.

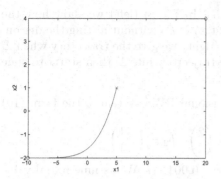 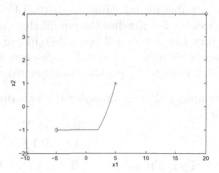

Fig. 4.

Lemma 3. *If $h(x) = c^\top x - \theta$, and A is a constant Hurwitz matrix then we can split $\alpha(x)$ in the following way:*

$$\alpha(x) = \alpha_c + \alpha_v(x) \ , \ \text{with} \ \alpha_c = \frac{c^\top b_1}{c^\top(b_1 - b_2)} \ , \quad \alpha_v(x) = \frac{c^\top A x}{c^\top(b_1 - b_2)} \ . \tag{15}$$

The pseudo-equilibrium solves the system:

$$(1 - \alpha_c)b_1 + \alpha_c b_2 + Ax - \alpha_v(x)(b_1 - b_2) = 0 \ ,$$

and we can prove that

$$\alpha_v(x)(b_1 - b_2) = Wx \ ,$$

where the matrix $W = (w_{ij})$ has the general entry given by:

$$w_{ij} = r_i c^\top A_{*j}, \quad r_i = \frac{[b_1 - b_2]_i}{c^\top[b_1 - b_2]} \ , \tag{16}$$

[A_{*j} denotes the j-th column of A].

Then, the pseudo-equilibrium x^* solves the linear system :

$$[A - W]x = -[(1 - \alpha_c)b_1 + \alpha_c b_2]$$

and it will be admissible if $\alpha(x)$ in (15) belongs to the interval $(0, 1)$ and x^* will be asymptotically stable if $Re[\lambda(A - W)] < 0$.

The pseudo-equilibrium is on the intersection between Σ and the segment $[\tilde{x}_1, \tilde{x}_2]$ where $\tilde{x}_1$ and $\tilde{x}_2$ are the equilibrium points of the perturbed linear systems $x' = [A - W]x + b_1$ and $x' = [A - W]x + b_2$ respectively.

Example 3. Let us consider the two-dimensional PWS system of the form (10) where

$$A = \begin{pmatrix} -0.1 & 0 \\ 0 & -0.5 \end{pmatrix}, \ b_1 = \begin{pmatrix} 2 \\ 2 \end{pmatrix}, \ b_2 = \begin{pmatrix} -1 \\ -1 \end{pmatrix}.$$

We assume initial value $x_0 = (0, 0.1) \in R_2$. In Fig. 4 (left) we show how the trajectory approaches the equilibrium point $x_2^* \in R_2$ without hitting the discontinuity line $\Sigma : x_2 + 3 = 0$; while in Fig. 4 (right) we give the trajectory when Σ is changed in $x_2 + 1 = 0$. In this case, the trajectory hits Σ then starts to slide until to reach the pseudo equilibrium.

Example 4. Let us consider the two-dimensional PWS system of the form (10) where

$$A = \begin{pmatrix} -0.1 & -1 \\ 1 & -0.1 \end{pmatrix}, \quad b_1 = \begin{pmatrix} 2 \\ 2 \end{pmatrix}, \quad b_2 = \begin{pmatrix} -1 \\ -1 \end{pmatrix},$$

with eigenvalues $\lambda_1 = -0.001 + i$ and $\lambda_2 = -0.001 - i$. We assume initial value $x_0 = (0, 0.1) \in R_2$. In Fig. 5 (left) we show how the trajectory approaches the equilibrium point $x_2^* \in R_2$ without hitting the discontinuity line $x_2 + 3 = 0$ while in Fig. 5 (right) we show what happens if we consider a different initial point $x_0 = (0, -10) \in R_1$: in this case the solution hits Σ, crosses Σ then approches the equilibrium point in R_2. We have to observe that the equilibrium point (marked by "diamond") of f_1 is in the region R_2.

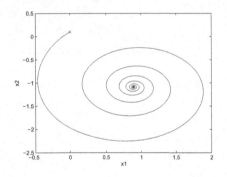 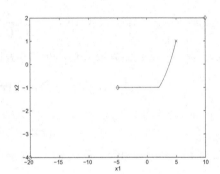

Fig. 5.

In Fig. 6 we consider the same system but with different $\Sigma : x_2 + 2 = 0$ and $x_0 = (0, 0.1) \in R_2$. In this case the trajectory hits the discontinuity line and then slides on Σ until to leave it at $(0.8, -2)$; then the trajectory enters R_2 and hence approaches the equilibrium x_2^*. In Fig. 6 we show the trajectory assuming $\Sigma : x_2 + 0.75 = 0$. In this case the trajectory hits the discontinuity line and starts to slide on it until to reach a pseudo equilibrium on Σ.

Example 5. Let us consider the three-dimensional PWS system of the form (10) (studied in the context of reelay feedback systems [24], [25], [23]) where

$$A = \begin{pmatrix} 0 & 0 & -1 \\ 1 & 0 & -2 \\ 0 & 1 & 2 \end{pmatrix}, \quad b_1 = \begin{pmatrix} 0 \\ -1 \\ 1 \end{pmatrix}, \quad b_2 = -b_1 = \begin{pmatrix} 0 \\ 1 \\ -1 \end{pmatrix},$$

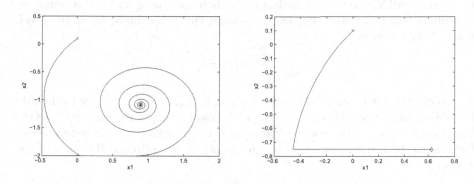

Fig. 6.

where $\Sigma : x_3 = 0$. The eigenvalues of A are given by $\lambda_1 = -1$, $\lambda_2 = -0.5 + 0.8660i$, $\lambda_3 = -0.5 - 0.8660i$. We assume initial value $x_0 = (0.1, 0.1, 0.1) \in R_2$. In Fig. 7 (left) we show how the trajectory, starting with $x_0 = (0.1, 0.1, 0.1) \in R_2$, approaches a limit cycle. We show also the two boundary equilibrium points $x_1^* = (1, -1, 0)$ and $x_2^* = -x_1^*$.

The piecewise linear system in (10) constitutes the basic element in the study of qualitative behaviours in several areas of PWS systems, in particular, in relay feedback systems (see for instance [25]) or in genetic regulatory networks (GRN) (see [8], [7], [10], [5]).

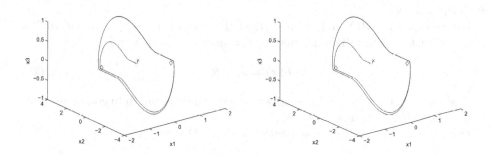

Fig. 7.

5   Piecewise Linear Systems in Gene Regulatory Networks

Equation (10) can model also the dynamics of gene regulatory networks (GRNs): in this case, state variables x_i, for $i = 1, \ldots, n$, typically represent the concentration of proteins, mRNAs or small molecules, thus it has to be $x_i \geq 0$. Assume for example of being in region R_1; let us consider the i-th component of the PWL system (10), let us say

$$x_i' = \sum_{j=1}^{n} a_{ij} x_j + b_i^1. \tag{17}$$

The positive terms in sum (17), say $a_{ij_1} x_{j_1}, a_{ij_2} x_{j_2}, \ldots, a_{ij_k} x_{j_k}$, refer to the rate of synthesis of the proteins $x_{j_1}, \ldots, x_{j_k}$, whereas the negative terms refer to the rate of degradation of the corresponding proteins, and finally the constant terms b_1^j refers to the basal production (or degradation, according to the sign) of the protein x_i.

Here we would like to consider A a Hurwitz matrix in $\mathbb{R}^{n \times n}$ of tridiagonal form:

$$A = \begin{bmatrix} \delta_1 & \beta_2 & 0 & 0 & 0 & 0 & 0 & 0 & 0 \\ \sigma_1 & \delta_2 & \beta_3 & 0 & 0 & 0 & 0 & 0 & 0 \\ 0 & \sigma_2 & \delta_3 & \beta_4 & 0 & 0 & 0 & 0 & 0 \\ 0 & 0 & . & . & . & 0 & 0 & 0 & 0 \\ 0 & 0 & 0 & 0 & . & . & . & 0 & 0 \\ 0 & 0 & 0 & 0 & 0 & . & . & . & 0 \\ 0 & 0 & 0 & 0 & 0 & 0 & \sigma_{n-2} & \delta_{n-1} & \beta_n \\ 0 & 0 & 0 & 0 & 0 & 0 & 0 & \sigma_{n-1} & \delta_n \end{bmatrix} \tag{18}$$

where

$$\delta_i = \sigma_i + \gamma_i + \beta_i, \quad \text{for} \quad i = 1, 2, \ldots, n, \tag{19}$$

with $\sigma_n = \beta_1 = 0$.

For instance in GRN it is well studied the case of $\sigma_i = \beta_i = 0$, for all $i = 1, \ldots, n$, that is A the diagonal matrix $\Gamma = \text{diag}(\gamma_1, \ldots, \gamma_n)$ and

$$h(x) = x_s - \theta \tag{20}$$

The case of A tridiagonal and given by (18) models the diffusion of proteins from a department to another.

During a sliding motion, the solution satisfies (13) where

$$\alpha(x_{s-1}, x_{s+1}) = \frac{\sigma_{s-1} x_{s-1} + \delta_s \theta + \beta_{s+1} x_{s+1} + [b_1]_s}{[b_1 - b_2]_s} = \alpha_c + \alpha_v(x_{s-1}, x_{s+1}) \tag{21}$$

where

$$\alpha_c = \frac{\delta_s \theta + [b_1]_s}{[b_1 - b_2]_s}, \qquad \alpha_v(x_{s-1}, x_{s+1}) = \frac{\sigma_{s-1} x_{s-1} + \beta_{s+1} x_{s+1}}{[b_1 - b_2]_s}.$$

Thus, the trajectory can exit Σ at points where $\alpha(x_{s-1}, x_{s+1}) = 0, 1$.

The sliding vector field will be given by:

$$f_\Sigma(x) = (1 - \alpha_c)b_1 + \alpha_c b_2 + Ax - \alpha_v(x_{s-1}, x_{s+1})(b_1 - b_2) .$$

In this case the matrix W in (16) becomes:

$$W = \begin{bmatrix} 0 \ldots & r_1\sigma_{s-1} & 0 & r_1\beta_{s+1} & \ldots 0 \\ 0 \ldots & r_2\sigma_{s-1} & 0 & r_2\beta_{s+1} & \ldots 0 \\ 0 \ldots & . & . & . & \ldots 0 \\ 0 \ldots & . & . & . & \ldots 0 \\ 0 \ldots & . & . & . & \ldots 0 \\ 0 \ldots & . & . & . & \ldots 0 \\ 0 \ldots & r_{n-1}\sigma_{s-1} & 0 & r_{n-1}\beta_{s+1} & \ldots 0 \\ 0 \ldots & r_n\sigma_{s-1} & 0 & r_n\beta_{s+1} & \ldots 0 \end{bmatrix} , \qquad r_i = \frac{[b_1 - b_2]_i}{[b_1 - b_2]_s} , i = 1, \ldots, n ,$$

then, if $A - W$ is non singular, the pseudo-equilibrium point x^* of the sliding vector must satisfy the linear system:

$$[A - W]x = -[(1 - \alpha_c)b_1 + \alpha_c b_2] ,$$

and if $\alpha(x_{s-1}^*, x_{s+1}^*)$ in (21) is in (0,1) it is admissible and if $\text{Re}(\lambda(A - W)) < 0$ it is attractive.

We notice that the matrix W will have $n - 2$ or $n - 1$ zero eigenvalues.

Example 6. Let us consider the three-dimensional PWS system of the form (10) where

$$A = \begin{pmatrix} -2 & -0.1 & 0 \\ -1 & -1.2 & -1 \\ 0 & -1 & -2 \end{pmatrix} , \quad b_1 = \begin{pmatrix} 1 \\ 0 \\ 2 \end{pmatrix} , \quad b_2 = \begin{pmatrix} 0 \\ 2 \\ -10 \end{pmatrix} ,$$

where $\Sigma : x_3 = 0$. The eigenvalues of A are negative and given by $-0.4775, -2.,$ -2.7225; moreover the matrix A has the form in (18) with the condition (19) satisfied. Instead the matrix W becomes:

$$W = \begin{pmatrix} 0 & -0.0833 & 0 \\ 0 & 0.1667 & 0 \\ 0 & -1. & 0 \end{pmatrix} .$$

The eigenvalues of $-W$ are given by $0, 0, -0.1667$; while the ones of $A - W$ are negative and given by $-2.0253, -1.3414, -2..$ In Fig. 8 (left) we show how the trajectory, starting with $x_0 = (0.1, 0.1, 1) \in R_2$, hits Σ at $\bar{x} = (0.0889, 0.2114, 0)$; then it starts to slide on Σ until to reach the admissible pseudo-equilibrium $x^* = (0.4172, -0.0613, 0)$. In Fig. 8 (right) we show the trajectory of the system with Σ replaced by $\Sigma : x_3 - 3 = 0$. Now the solution remains in R_1 for all $t > 0$ and approaches the admissible equilibrium $x_1^* = (0.6154, -2.3077, 2.1538)$.

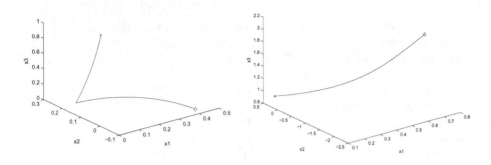

Fig. 8.

6   Conclusions

In this paper, we have considered equilibrium points for Piecewise Linear (PWL) systems: we have given an original characterization for the computation of pseudo-equilibrium points for a particular class of PWL systems. We have also considered a possible application to the analysis of Gene Regulatory Networks. A future work could be to give similar results for a more general class of PWL systems.

References

1. Acary, V., Brogliato, B.: Numerical Methods for Nonsmooth Dynamical Systems. Applications in Mechanics and Electronics. Springer, Berlin (2008)
2. Aubin, J.-P., Cellina, A.: Differential Inclusion. Springer, Berlin (1984)
3. Berardi, M.: Rosenbrock-type methods applied to discontinuous differential systems. Mathematics and Computers in Simulations 95, 229–243 (2014)
4. Berardi, M., Lopez, L.: On the continuous extension of Adams-Bashforth methods and the event location in discontinuous ODEs. Applied Mathematics Letters 25, 995–999 (2012)
5. Casey, R., de Jong, H., Gouzé, J.L.: Piecewise-linear Models of Genetics Regulatory Networks: Equilibria and their Stability. Journal of Mathematical Biology 52, 27–56 (2006)
6. D'Abbicco, M., Calamita, G., Del Buono, N., Berardi, M., Gena, P., Lopez, L.: A Model for the Hepatic Glucose Metabolism based on Hill and Step Functions, preprint (2014)
7. de Jong, H.: Modelling and simulation of Genetic Regulatory Systems: a literature review. Journal of Computational Biology 9, 67–103 (2002)
8. de Jong, H., Gouzé, J.L., Hernandez, C., Page, M., Sari, T., Geiselmann, J.: Qualitative Simulation of Genetic Regulatory Networks Using Piecewise-Linear Models. Bulletin of Mathematical Biology 66, 301–340 (2004)
9. Casey, R., de Jong, H., Gouzé, J.L.: Piecewise-linear Models of Genetics Regulatory Networks: Equilibria and their Stability. Journal of Mathematical Biology 52, 27–56 (2002)

10. de Jong, H., Gouzé, J.L., Hernandez, C., Sari, T., Geiselmann, J.: Dealing with Discontinuities in the Qualitative Simulation of Genetic Regulatory Network. In: Proceedings of the 15th European Conference on Artificial Intelligence, pp. 725–730. IOS Press, Amsterdam (2002)

11. Del Buono, N., Elia, C., Lopez, L.: On the equivalence between the sigmoidal approach and Utkin's approach for piecewise-linear models of gene regulatory networks. To appear on SIAM Journal of Applied Dynamical Systems

12. di Bernardo, M., Budd, C.J., Champneys, A.R., Kowalczyk, P.: Piecewise-smooth Dynamical Systems. Theory and Applications. Springer, Berlin (2008)

13. Dieci, L., Lopez, L.: Sliding motion on discontinuity surfaces of high co-dimension. A construction for selecting a Filippov vector field. Numerische Mathematik 117, 779–811 (2011)

14. Dieci, L., Lopez, L.: Sliding motion in Filippov differential systems: Theoretical results and a computational approach. SIAM Journal of Numerical Analysis 47, 2023–2051 (2009)

15. Dieci, L., Lopez, L.: A survey of numerical methods for IVPs of ODEs with discontinuous right-hand side. Journal of Computational and Applied Mathematics 236, 3967–3991 (2012)

16. Dieci, L., Lopez, L.: Numerical Solution of Discontinuous Differential Systems: Approaching the Discontinuity from One Side. Applied Numerical Mathematics 67, 98–110 (2011)

17. Dieci, L., Lopez, L.: One-Sided Event Location Techniques in the Numerical Solution of Discontinuous Differential Systems. Submitted to BIT Numerical Mathematics

18. Dieci, L., Lopez, L., Elia, C.: A Filippov sliding vector field on an attracting co-dimension 2 discontinuity surface, and a limited loss-of-attractivity analysis. Journal of Differential Equations 254, 1800–1832 (2013)

19. Filippov, A.F.: Differential Equations with Discontinuous Right Hand Sides. Kluwer Academic, Dordrecht (1988)

20. Gear, C.W., Østerby, O.: Solving ordinary differential equations with discontinuities. ACM Transactions on Mathematical Software 10, 23–44 (1984)

21. Gouzé, J.-L., Sari, T.: A class of piecewise linear differential equations arsing in biological models. Dynamical Systems 17, 299–319 (2002)

22. Ironi, L., Panzeri, L., Plahte, E., Simoncini, V.: Dynamics of actively regulated gene networks. Physica D: Nonlinear Phenomena 240, 779–794 (2011)

23. Johansson, K.H., Rantzer, A., Astrom, K.J.: Fast swiyches in relay feedback systems. Automatica 35, 539–552 (1999)

24. Johansson, K.H., Rantzer, A.: Global analysis of third-order relay feedback systems, Technical Report, 1996. IEEE Transations on Automatic Control 247, 1414–1423 (2002)

25. Johansson, K.H., Rantzer, A., Astrom, K.J.: Limit Cycles With Chattering in Relay Feedback Systems. IEEE Transactions on Automatic Control 47, 1414–1423 (2002)

26. Li, J., Kuang, Y., Mason, C.C.: Modeling the glucose-insulin regulatory system and ultradian insulin secretory oscillations with two explicit time delays. Journal of Theoretical Biology 242, 722–735 (2006)

27. Makroglou, A., Li, J., Kuang, Y.: Mathematical models and software tools for the glucose-insulin regulatory system and diabetes: an overview. Applied Numerical Mathematics 56, 559–573 (2006)

28. Wang, H., Li, J., Kuang, J.: Mathematical modeling and qualitative analysis of insulin therapies. Mathematical Biosciences 210, 17–33 (2007)

Adaptive Learning Ant Colony Optimization for Web Spam Detection

Bundit Manaskasemsak, Jirayus Jiarpakdee, and Arnon Rungsawang

Massive Information & Knowledge Engineering Laboratory,
Department of Computer Engineering, Faculty of Engineering,
Kasetsart University, Bangkok 10900, Thailand
{un,arnon}@mikelab.net, jirayusjiar@gmail.com

Abstract. Web spamming is nowadays a serious problem for search engines. It not only degrades the quality of search results by intentionally boosting undesirable web pages to users, but also causes the search engine to waste a significant amount of computational and storage resources in manipulating useless information. In this paper, we present a machine learning approach for spam detection by adopting the ant colony optimization algorithm. We first construct a directed graph corresponding to web hosts and their aggregated hyperlinks. Then, we train a classifier by employing ants to walk along paths in the graph. Each ant will start from an individual non-spam host and afterwards decides to follow a link to the next host with a probability based on both heuristic function and pheromone trail. Relying on the approximate isolation principle of a good set, we reward an ant that can discover a good path, i.e., a sequence of non-spam hosts, by charging energy for its longer walking. In contrast, if the ant instead discovers any spam, it will be penalized by decreasing its walking step. Finally, the classification rules are constructed by choosing common overlapping characteristic features of all non-spam hosts along the discovered paths. Experiments on WEBSPAM-UK2007 dataset show that our approach contributes to more accurately classify spam and non-spam hosts than several rule-based classification baselines.

Keywords: web spam detection, adaptive learning paths, reward distance, penalty distance, ant colony optimization.

1 Introduction

Web search engine has been invented as a tool to serve users' information needs on the World Wide Web. Against billions of web pages, only a first few of search results have high possibility to be clicked and visited by the users. Those web pages earn high impacts on an economic purpose since a high order in ranking provides large free advertising as well as increases web traffic volumes at the same time. In consequence, many web engineers have made hard efforts to boost a ranking order of their web pages. However, most of those attempts are usually backdoors, cloaking, and other cheats that intentionally violate search engine guidelines, all called "web spamming".

B. Murgante et al. (Eds.): ICCSA 2014, Part VI, LNCS 8584, pp. 642–653, 2014.

Web spam has been identified as one of the crucial challenge of search engines [15]. Spam pages containing undeserved content will degrade quality of search results; meanwhile, they cause the search engines to waste amount of computational and storage resources without benefit. Spamming techniques [13] can be considered as three types. First, *content spam* includes techniques that retouch content of target pages, for instance, by inserting a number of keywords that are possibly related to query terms than to their semantic content. Second, *link spam* consists of a creation of link structure to take advantage of link-based algorithms, such as PageRank [21] and HITS [16], in order to boost a ranking score of a target page. Last, *hiding spam*, including cloaking and redirection, attempts to deliver different content to normal web users and search engine web crawlers.

Several studies on web spam detection have been proposed for years in different ways, including content-based techniques (e.g., [11,20]) and link-based ones (e.g., [25,14,26,3,6]), and other learning based on browsing logs (e.g., [18,19]). In this work, we concentrate on link-based spam detection scheme and propose an alternative learning classification. By adopting the ant colony optimization (ACO) algorithm [7,9], the model employs a host graph, constructed from a set of web hosts defined as nodes and their aggregated hyperlinks over web pages defined as edges, and then generates a set of rules from ant trails. In the learning phase using the training dataset from WEBSPAM-UK2007 [5], each web host labeled with "non-spam" is used as a seed for ants. Relying on the approximate isolation principle [14] that good (non-spam) pages—hosts in our case—seldom link to bad (spam) ones, trails on the host graph discovered by ants can be referred to good (non-spam) paths which are subsequently used to generate rules for classifying non-spam hosts from spam ones. Considering each non-spam path, a classification rule is determined by choosing common overlapping characteristic features of all non-spam hosts, and assigned as the "non-spam" class. Experiments conducted on the testing dataset from WEBSPAM-UK2007 reveal that this approach can provide more accurate identification of spam and non-spam hosts than several rule-based classification baselines.

The remainder of this paper is organized as follows. Section 2 reviews some related work and shortly explains the ACO algorithm. Sections 3 and 4 describe our proposed link-based ACO algorithm for spam detection and the extended version, respectively. Section 5 reports performance evaluation. Section 6 finally concludes the paper.

2   Related Work

2.1   Link-Based Spam Detection

Various methods have been proposed to combat web spamming and to detect spam pages. However, we only mention to some of link-based ones that are related to our work here.

One of the important and well-known algorithms is TrustRank [14], which propagates trust from a small selected seed set of good pages to others via personalized PageRank. The algorithm relies on the *approximate isolation* principle, that is "good or non-spam pages seldom point to spam ones", and proceeds as follows. Top-K pages returned from an inverse PageRank computation are first judged by a human expert. Pages annotated as good are included in a seed set. Then, the personalized vector is constructed in which all elements corresponding to good judged pages are assigned to non-zero value. Finally, the trust propagation is computed with a certain iteration. TrustRank was reported to have effective demoting and filtering out web spam.

Opposite to TrustRank, Anti-TrustRank [17] has been proposed in the reverse manner. The algorithm relies on an assumption that pages pointing to spam ones are likely to be spam themselves, and thus propagates distrust from a set of known spam pages on an inverted web graph. The report showed Anti-TrustRank can achieve higher precision in spam detection than TrustRank.

Other common methods have considered web spam detection as a problem of binary classification. Preliminarily, some web pages, referred to a training dataset, are examined and labeled as spam or non-spam by an expert. Then, a classification model is created by adopting any supervised learning algorithm to learn from this training data. Further, the model is used to predict any new web pages as either spam or non-spam. The key issue is what features are defined to represent pages used in both learning and classifying processes. Hence, a number of linked-based features such as values derived from PageRank, TrustRank, and truncated PageRank computation are studied in [3]; the use of content-based features is studied in [20]; and the mixture of both kinds of them is used to combat against web spam in [4].

2.2   Ant Colony Optimization

Naturally, distinct kind of creatures behaves differently in their everyday life. In a colony of social ants, each ant normally has its own duty and performs its own tasks independently from other members of the colony. However, tasks done by different ants are usually related to each other in such a way that the colony, as a whole, is capable of solving complex problems through cooperation [9,22]. For example, many survival-related problems such as selecting the shortest walking path, finding and storing food, which require sophisticated planning, are solved by ant colony without any kinds of supervisor. The extensive study from ethologists reveals that ants communicate with others by means of pheromone trails of which path should be followed. As ants move, a certain amount of pheromone is dropped to make the path with the trail of this substance. Ants tend to converge to the shortest trail (or path), since they can make more trips, and hence deliver more food to their colony. The more ants follow a given trail, the more attractive this trail becomes to be followed by other ants. This process can be described as a positive feedback loop, in which the probability that an ant chooses a path is proportional to the number of ants that has already passed through that path [7,9].

Researchers try to simulate the natural behavior of ants, including mechanisms of cooperation, and devise ant colony optimization (ACO) algorithms based on such an idea to solve the real world complex problems, such as the traveling salesman problem [8], data mining [22]. ACO algorithms solve a problem based on the following concept:

- Each path followed by an ant is associated with a candidate solution for a given problem.
- When an ant follows a path, it drops varying amount of pheromone on that path in proportion with the quality of the corresponding candidate solution for the target problem.
- Path with a larger amount of pheromone will have a greater probability to be chosen to follow by other ants.

In solving an optimization problem with ACO, we have to determine three following functions appropriately to help the algorithm to get faster and better solution. The first one is a problem-dependent heuristic function (η) which measures the quality of items (i.e., attribute-value pairs) that can be added to the current partial solution (i.e., rule). The second one is a principle for pheromone updating (τ) which specifies how to modify the pheromone trail. The last one is a probabilistic transition rule (P), based on the value of heuristic function and on the content of pheromone trail, that is used to iteratively construct the solution.

3   The $\mathcal{LSD}$-$\mathcal{ACO}$ Algorithm

The proposed link-based spam detection using ant colony optimization approach ($\mathcal{LSD}$-$\mathcal{ACO}$) also relies on the *approximate isolation* principle mentioned in [14], i.e., good or non-spam pages seldom point to spam ones. Hence, it can be implied that non-spam hosts usually link to non-spam ones. By adopting ACO on a host graph, our model makes an effort to discover good trails, i.e., path containing non-spam hosts, in order to generate useful classification rules.

3.1   Graph Representation

First of all, an environment employed in ant operations needs to be defined in a way when ants move, they will incrementally construct a solution to the problem. In our case, we thus represent the problem as a host graph $\mathcal{G_H} = (\mathcal{V}, \mathcal{E})$, where $\mathcal{V}$ and $\mathcal{E}$ refer to a set of hosts and their hyperlinks, respectively. A directed edge $e(h_i, h_j) \in \mathcal{E}$ is defined if there is a web page u belonging to host $h_i \in \mathcal{V}$ having a link to a target page v belonging to host $h_j \in \mathcal{V}$. However, multiple edges from h_i pointing to h_j will be collapsed into one edge. Self loops are all omitted.

3.2   Model Training

Given a host graph derived from a training data, a host h_i labeled with *non-spam* will be assigned as a seed for an artificial ant to start walking. In each step, the ant that visits at host h_i will randomly choose a link $e(h_i, h_j)$ and afterwards move to that target h_j. The edge chosen is dependent on the heuristic value and the pheromone information. Let $P_{ij}(t)$ denote a probability assigned to link $e(h_i, h_j)$ at iteration time t. This probability that guides an ant to randomly walk from a current host h_i to the next host h_j is defined as:

$$P_{ij}(t) = \begin{cases} x_j \cdot \frac{\eta_{ij} \tau_{ij}(t)}{\sum_{h_k \in F(h_i)} \eta_{ik} \tau_{ik}(t)} & \text{if } e(h_i, h_j) \in \mathcal{E}, \\ 0 & \text{otherwise,} \end{cases} \tag{1}$$

where η_{ij} and $\tau_{ij}(t)$ are the heuristic value and the pheromone information obtained at iteration time t, respectively. $F(h_i)$ refers to a set of hosts that h_i links to. Last, x_j is an indicator used to avoid a cycle in the trail; in other words, a host h_j will be visited only once. This indicator is defined as:

$$x_j = \begin{cases} 1 & \text{if } h_j \notin \Gamma, \\ 0 & \text{otherwise.} \end{cases} \tag{2}$$

If h_j has been chosen to visit, it will be incrementally included in the trail Γ.

The heuristic value η_{ij} can be defined by two aspects. First, we hypothesize that the number of multiple links pointing out from a non-spam host h_i can heuristically guide ants to fast discover a next non-spam host h_j. The larger number of out-links the non-spam h_i points to h_j, the higher possibility the h_j will be non-spam. Let e_{ij} be the number of hyperlinks aggregated over all web pages belonging to the host h_i pointing to any web pages belonging to the host h_j. Thus, the heuristic value expressed in (1) can be calculated by the proportion of the amount of links pointing out:

$$\eta_{ij} = \frac{e_{ij}}{\sum_{h_k \in F(h_i)} e_{ik}}. \tag{3}$$

Second, in a general situation, an author—non-spammer—of a web page intentionally create a link pointing to a target page which is related to some extent, such as containing similar content or providing further detail. We therefore hypothesize that the similarity of content can heuristically guide the ants to move from a non-spam host to the next non-spam one, as well. In our approach, we use a number of content-based features provided by experts [5], instead of an actual content, to explain a host. Let $\boldsymbol{f}_i$ be a feature vector referred to the host h_i. Then, the heuristic value expressed in (1) can alternatively be calculated by:

$$\eta_{ij} = \frac{Sim(\boldsymbol{f}_i, \boldsymbol{f}_j)}{\sum_{h_k \in F(h_i)} Sim(\boldsymbol{f}_i, \boldsymbol{f}_k)}. \tag{4}$$

Here, $Sim()$ can simply be the cosine similarity function used in the classical IR [2]. Notice that both heuristic functions will assign a constant value to every link; therefore, they can be pre-calculated before starting the ACO learning process.

The other key element of ACO is the pheromone information. Since the ACO algorithm iteratively finds the optimal solution, the pheromone value expressed in (1) need to be changed after each run. The pheromone updating is achieved by

```
 1: function LSD-ACO(G_H)
 2:     for each non-spam host h in G_H do
 3:         initialize heuristics, pheromones, and probabilities of edges
 4:         for each iteration t do
 5:             create a number of ants
 6:             let each ant independently walk from h with a number of hops
 7:             generate rules from ants' paths and choose the (local) best one
 8:             adjust the pheromone levels
 9:             update probabilities of edges
10:             delete all ants
11:         end for
12:         consider all local best rules and choose the (global) best one
13:         collect the global best rule in the answer set
14:     end for
15:     return the sequence of classification rules
16: end function
```

Fig. 1. The LSD-ACO algorithm

the following two fractions: evaporation and reinforcement. The former decreases the pheromone level of each trail by a factor ρ. Typical values for this factor are suggested in the range [0.8,0.99] in $MAX\text{-}MIN$ ant system [23]. The latter increases the pheromone level by a factor σ only to the best ant's path examined among the entire ones after each iteration run. Hence, the pheromone function can be formulated as:

$$\tau_{ij}(t) = \begin{cases} \frac{1}{|\mathcal{E}|} & \text{if } t = 1, \\ \rho \cdot \tau_{ij}(t-1) + \sigma \cdot \tau_{ij}(t-1) & \text{otherwise.} \end{cases} \tag{5}$$

For the first iteration, the pheromone information is initially set to the same value over the entire links in the graph. Further, the reinforcement factor σ, in this work, is defined as the value of confidence metric that measures the quality of classification rules evaluated on the training data. Let CH be a set of hosts covered by the rule, and NH be a set of non-spam hosts. Then, the reinforcement factor is calculated by:

$$\sigma = \frac{|CH \cap NH|}{|CH|}, \tag{6}$$

i.e., the number of hosts covered and correctly classified by the rule divides by the number of all hosts covered by that rule.

After each iteration run, results of the ACO learning can be expressed by ants' paths. These paths are further interpreted and used to generate classification rules. Note that we will address the detail of this process in the next subsection. However, we illustrate here the pseudo-code of our $LSD\text{-}ACO$ in Fig. 1.

As shown in the algorithm, the function requires three pre-defined parameters: the maximum number of iteration runs, ants, and hops, respectively. The first parameter, repeating the learning process with the number of iterations in line 4, is needed for guaranteeing that the model will indeed contribute the best classification rule. The second one, defining the number of ants in line 5, directly affects the number of generated rules. In general, the larger the value of this parameter is set, the higher possibility a good path will be found. The last one, defining the maximum walking distance in line 6, determines how far an ant can move away from a seed. This parameter will affect properties, i.e., either specificity or generality, of a rule.

3.3   Rule Construction

As illustrated the procedure at line 7 of the function $\mathcal{LSD}\text{-}\mathcal{ACO}$ in Fig. 1, all paths discovered by ants are subsequently interpreted and used to generate useful classification rules. The rule is constructed in a simple form; that is, **if** *rule antecedent* **then** *rule consequent*, where the rule antecedent is a conjunction of feature terms and the rule consequent is a class of either spam or non-spam.

In the training dataset, characteristic features and a labeled class of each web host have been already determined by human experts [5]. Given m attribute features $A_1, A_2, \ldots, A_m$ with each feature A_i having n_i possible values $a_{i1}, a_{i2}, \ldots, a_{in_i}$, the construction rule can be expressed by:

$$(A_1 = a_{1j_1}, A_2 = a_{2j_2}, \ldots, A_m = a_{mj_m}) \Rightarrow (Class = \text{either } spam \text{ or } non\text{-}spam),$$

where $j_1, j_2, \ldots, j_m$ are any corresponding indices.

Based on the approximate isolation principle [14], a path discovered by an ant should mainly present a list of non-spam hosts. However, that path may contain some spam ones since, in fact, a non-spam host may possibly have hyperlinks to spam pages by spammers' intentional tricks [13] or the author's mistake. We therefore consider only the non-spam hosts within the path. To generate a rule, a value given for each feature A_i is determined by the common value corresponding to that feature of among all those non-spam hosts. Let $\Gamma_{p(NH)}$ be the p-th ant's path that contains only non-spam hosts, i.e., excluding all spam ones. The rule interpreted from that path is constructed by:

$$\underset{h \in \Gamma_{p(NH)}}{COMMON}(A_1 = a_{1j_1}^h, A_2 = a_{2j_2}^h, \ldots, A_m = a_{mj_m}^h) \Rightarrow (Class = non\text{-}spam), \quad (7)$$

where $a_{ij_i}^h$ presents the same value corresponding to the feature A_i from all non-spam hosts h. Note that if a feature value is given by range, the common value will then be determined by overlapping of those ranges; otherwise, if there is not any values in common, a don't care term "?" will be assigned instead, indicating that feature is not affected in the rule.

Again, the procedure, at line 7 in Fig. 1, will then select the best rule from all generated ones by evaluating them over the training data using the confidence metric illustrated in (6). This best rule is marked as a local candidate. Afterwards, in line 12–13, only one best candidate is selected and accumulated into the model. Eventually, all rules returned at line 15 are sorted by their confidence values, meaning that they are intended to be interpreted in a sequential order in the classification process.

```
1: function walk(nHops)
2:     let an ant starts from the seed host $h_i$
3:     while nHops is not equal to 0 and ant does not reach the end of path do
4:         randomly select a host $h_j$ and move to that target
5:         decrease nHops by 1
6:         if $h_j$ is non-spam then
7:             increase nHops by 1
8:         else if $h_j$ is spam then
9:             decrease nHops by 1
10:        end if
11:    end while
12:    return the ant's path
13: end function
```

Fig. 2. The *walk* function of $\mathcal{LSD}$-$\mathcal{ACO}^+$ algorithm

4   The $\mathcal{LSD}$-$\mathcal{ACO}^+$ Algorithm

One of the important factor of the $\mathcal{LSD}$-$\mathcal{ACO}$ algorithm is the maximum number of hops. This parameter determines how far ants can possibly move away from their seed; the distance of an ant's path can affect the specificity (or generality) of a classification rule.

Instead of using a fixed value of maximum hops employed in the $\mathcal{LSD}$-$\mathcal{ACO}$ algorithm, the key idea of the $\mathcal{LSD}$-$\mathcal{ACO}^+$ algorithm is that this value can be adaptively changed on the way by either *reward distance* or *penalty distance*. The former means increasing value by 1 when an ant can discover a non-spam host; otherwise, the latter will decrease the value by 1 when the ant mistakes by discovering a spam one. We enhance the walk procedure at line 6 illustrated in Fig. 1 with the *walk* function written in Fig. 2. In other words, we let an ant start at seed h_i with an nHops energy level. After moving one step, its energy decreases by one. However, if the ant steps on a good place (i.e., non-spam), it will get reward by a level of energy (see line 7). Otherwise, if the ant steps on a bad one (i.e., spam), it will be penalized with a decay of energy (see line 9).

This adaptive learning mechanism is expected to produce many proper classification rules since if an ant discovers many spams, the function will return a short ant's path. That is, a path containing few non-spam hosts will possibly generate a too specific rule (i.e., many conjunctive terms of the rule antecedent) which is usually useless and should be abandoned.

5   Experiments

5.1   Dataset

We used a publicly available web spam collection [5] based on crawls of the .uk web domain in May 2007. The WEBSPAM-UK2007 dataset includes 105.9 million web pages (in 114,529 hosts) and over 3.7 billion hyperlinks. This collection

Table 1. Statistics of the WEBSPAM-UK2007 collection

	Number of Spam Hosts	Number of Non-spam Hosts
Training dataset	222	3,766
Testing dataset	122	1,933

was annotated by a group of volunteers; a web host was labeled as "spam", "non-spam", or "borderline". However, in our experiments, we restricted the dataset using only hosts labeled with either spam or non-spam. Fortunately, the dataset has already been divided into two subsets for training and testing processes. We report the number of hosts used in the experiments in Table 1. Furthermore, there is a set of pre-computed features over the hosts of this collection, grouped into 96 content-based, 41 link-based, and 138 transformed link-based features, respectively. Since all features have continuous-range values, we therefore used the technique proposed in [10] to first discretize them into multiple intervals.

5.2   Results

We conduct experiments using all non-spam hosts listed in the training dataset as seeds for both $\mathcal{LSD}$-$\mathcal{ACO}$ and $\mathcal{LSD}$-$\mathcal{ACO}^+$ algorithms. In the learning phase, we constantly set the number of maximum iteration runs, ants, and hops, used in both algorithms, to 40, 100, and 12, respectively. Moreover, we separate the experiments by training each algorithm using the heuristic functions expressed in (3) and (4), named *OutDegree* and *ContentSim*, respectively, and only use 96 content-based features for the latter heuristic. Notice that we, however, still need the entire 275 features for identifying the best rule during the training process.

We compare our performance with four rule-based classification algorithms: decision tree ($C4.5$), *RIPPER*, *PART*, and *RandomTree*, respectively, using the same training and testing datasets shown in Table 1. We train those baselines using the WEKA software [24]. For the environment settings, the rule pruning is enabled; all other parameters have been set to default. The experimental results are concluded in Table 2, based on three standard measures: true positive rate (*TPR*), i.e., an ability of identifying spam correctly; false positive rate (*FPR*), i.e., a proportion of incorrect identifying non-spam; and the area under the ROC curve (*AUC*), i.e., the normalized unit over the previous two measures.

As it can be seen in Table 2, both $\mathcal{LSD}$-$\mathcal{ACO}$ and $\mathcal{LSD}$-$\mathcal{ACO}^+$ algorithms yield significant results with much higher *TPR* and *AUC* than ones of all four baselines. However, those baselines can achieve much lower *FPR*. The reason is that since both training and testing datasets, as illustrated in Table 1, contain much larger amount of non-spam hosts than spam ones, the baselines then have recognized most characteristic hosts from the majority class during the training process. Consequently, they possibly predict most unknown hosts as the majority of non-spam, while a few ones as spam. There are thus only few wrong predictions on the testing data. From the results, it can be argued that $\mathcal{LSD}$-$\mathcal{ACO}$ and $\mathcal{LSD}$-$\mathcal{ACO}^+$ have better capability of dealing with the imbalanced data.

Table 2. Performance comparisons on the WEBSPAM-UK2007 collection

Algorithm		Number of Features	TPR	FPR	AUC
$\mathcal{LSD}$-$\mathcal{ACO}$	with *OutDegree*	− / 275	0.844	0.356	0.744
	with *ContentSim*	96 / 275	0.820	0.276	0.772
$\mathcal{LSD}$-$\mathcal{ACO}^+$	with *OutDegree*	− / 275	**0.852**	0.356	0.748
	with *ContentSim*	96 / 275	0.811	0.263	**0.774**
$C4.5$		275	0.082	0.009	0.537
RIPPER		275	0.156	**0.007**	0.575
PART		275	0.164	0.041	0.562
RandomTree		275	0.148	0.036	0.557
$C \cup L \cup LM \cup QL$		291	0.50	0.06	0.76

The bold text highlights the best value.

Furthermore, $\mathcal{LSD}$-$\mathcal{ACO}$ and $\mathcal{LSD}$-$\mathcal{ACO}^+$ using the *ContentSim* heuristic can produce better performance with lower *FPR* and higher *AUC* than ones of the approaches using the *OutDegree* heuristic, indicating that content-based features are still the essential need. Lastly, the comparison of $\mathcal{LSD}$-$\mathcal{ACO}^+$ with $\mathcal{LSD}$-$\mathcal{ACO}$ yields a slight improvement, indicating that the distance of ant's paths can affect the quality of rules.

For our more precise experimental study, we also compare our work with another one well-known spam detection model, named $C \cup L \cup LM \cup QL$ [1], at the last record of Table 2. The model is nearly related to ours, in which the authors have combined the content-based (C), the transformed link-based (L), and the additional new proposed language-model (LM) and qualified-link (QL) features in order to obtain a decision tree ($C4.5$) based classifier. The results illustrated here are excerpted from [1]. It can be seen that both $\mathcal{LSD}$-$\mathcal{ACO}$ and $\mathcal{LSD}$-$\mathcal{ACO}^+$ using the *ContentSim* heuristic can also achieve better performance with a slightly higher *AUC*, while using less features. To our knowledge, the best *AUC* score as reported in "The Web Spam Challenge 2008[1]" is 0.855 [12]; however, those authors have taken many efforts in order to create multiple classifiers using AdaBoost with ensemble random under-sampling (ERUS) strategy.

6   Conclusion

In this paper, we have proposed an alternative approach to detect spam in the Web, based on a classification model adapting the ant colony optimization. In this approach, we follow the approximate isolation principle, introduced in TrustRank, in order to construct non-spam classification rules interpreted from ants' paths. Moreover, we have introduced an adaptive learning technique by giving either a reward or a penalty to ants in the walking procedure for better performance in spam detection.

---

[1] http://webspam.lip6.fr/

We have conducted experiments on the public WEBSPAM-UK2007 dataset and compared performance in detecting spam with several baselines. The results show that our approach gives better performance. In addition, it has been proven that the utilization of content-based features and the application of adaptive learning paths in ACO improve spam classification performance.

There still remains an interesting issue of the distrust propagation concept to be explored. That is, by observations, pages pointing to spam pages are very likely to be spam themselves. For this, in future work, we would like to explore and improve the classification model by employing the reverse path learning, and hope to obtain higher quality set of classification rules.

Acknowledgement. We thank to all research teams who help providing the experimental dataset. We also thank to anonymous reviewers for their comments and suggestions.

References

1. Araujo, L., Martinez-Romo, J.: Web spam detection: New classification features based on qualified link analysis and language models. IEEE Transactions on Information Forensics and Security 5(3), 581–590 (2010)
2. Baeza-Yates, R.A., Ribeiro-Neto, B.A.: Modern Information Retrieval. Addison Wesley, England (1999)
3. Becchetti, L., Castillo, C., Donato, D., Leonardi, S., Baeza-Yates, R.: Link-based characterization and detection of web spam. In: Proceedings of the 2nd International Workshop on Adversarial Information Retrieval on the Web, pp. 1–8 (2006)
4. Becchetti, L., Castillo, C., Donato, D., Leonardi, S., Baeza-Yates, R.: Web spam detection: Link-based and content-based techniques. In: The European Integrated Project Dynamically Evolving, Large Scale Information Systems (DELIS): Proceedings of the Final Workshop, vol. 222, pp. 99–113 (2008)
5. Castillo, C., Donato, D., Becchetti, L., Boldi, P., Leonardi, S., Santini, M., Vigna, S.: A reference collection for web spam. ACM SIGIR Forum 40(2), 11–24 (2006)
6. Castillo, C., Donato, D., Gionis, A., Murdock, V., Silvestri, F.: Know your neighbors: Web spam detection using the web topology. In: Proceedings of the 30th Annual International ACM SIGIR Conference on Research and Development in Information Retrieval, pp. 423–430 (2007)
7. Dorigo, M., Di Caro, G., Gambardella, L.M.: Ant algorithms for discrete optimization. Artificial Life 5(2), 137–172 (1999)
8. Dorigo, M., Gambardella, L.M.: Ant colony system: A cooperative learning approach to the traveling salesman problem. IEEE Transactions on Evolutionary Computation 1(1), 53–66 (1997)
9. Dorigo, M., Maniezzo, V., Colorni, A.: Ant system: Optimization by a colony of cooperating agents. IEEE Transactions on Systems, Man, and Cybernetics 26(1), 29–41 (1996)
10. Fayyad, U.M., Irani, K.B.: Multi-interval discretization of continuous-valued attributes for classification learning. In: Proceedings of the 13th International Joint Conference on Artificial Intelligence, pp. 1022–1027 (1993)

11. Fetterly, D., Manasse, H., Najork, M.: Spam, damn spam, and statistics: Using statistical analysis to locate spam web pages. In: Proceedings of the 7th International Workshop on the Web and Databases, pp. 1–6 (2004)
12. Geng, G.G., Jin, X.B., Wang, C.H.: Casia at web spam challenge 2008 track iii. In: Proceedings of the 4th International Workshop on Adversarial Information Retrieval on the Web (2008)
13. Gyöngyi, Z., Garcia-Molina, H.: Web spam taxonomy. In: Proceedings of the 1st International Workshop on Adversarial Information Retrieval on the Web, pp. 39–47 (2005)
14. Gyöngyi, Z., Garcia-Molina, H., Pedersen, J.: Combating web spam with trustrank. In: Proceedings of the 13th International Conference on Very Large Data Bases, pp. 576–587 (2004)
15. Henzinger, M.R., Motwani, R., Silverstein, C.: Challenges in web search engines. ACM SIGIR Forum 36(2), 11–22 (2002)
16. Kleinberg, J.M.: Authoritative sources in a hyperlinked environment. Journal of the ACM 46(5), 604–632 (1999)
17. Krishnan, V., Raj, R.: Web spam detection with anti-trust rank. In: Proceedings of the 2nd International Workshop on Adversarial Information Retrieval on the Web, pp. 37–40 (2006)
18. Liu, Y., Gao, B., Liu, T.Y., Zhang, Y., Ma, Z., He, S., Li, H.: Browserank: Letting web users vote for page importance. In: Proceedings of the 31st Annual International ACM SIGIR Conference on Research and Development in Information Retrieval, pp. 451–458 (2008)
19. Liu, Y., Zhang, M., Ma, S., Ru, L.: User behavior oriented web spam detection. In: Proceedings of the 17th International Conference on World Wide Web, pp. 1039–1040 (2008)
20. Ntoulas, A., Najork, M., Manasse, M., Fetterly, D.: Detecting spam web pages through content analysis. In: Proceedings of the 15th International Conference on World Wide Web, pp. 83–92 (2006)
21. Page, L., Brin, S., Motwani, R., Winograd, T.: The pagerank citation ranking: Bringing order to the web. Tech. rep., Stanford Digital Libraries (1999)
22. Parpinelli, R.S., Lopes, H.S., Freitas, A.A.: Data mining with an ant colony optimization algorithm. IEEE Transactions on Evolutionary Computation 6(4), 321–332 (2002)
23. Stützle, T., Hoos, H.H.: $\mathcal{MAX}$-$\mathcal{MIN}$ ant system. Future Generation Computer Systems 16(9), 889–914 (2000)
24. Witten, I.H., Frank, E.: Data Mining: Practical Machine Learning Tools and Techniques with Java Implementations, 2nd edn. Morgan Kaufmann, San Francisco (2005)
25. Wu, B., Davison, B.D.: Identifying link farm spam pages. In: Special Interest Tracks and Posters of the 14th International Conference on World Wide Web, pp. 820–829 (2005)
26. Wu, B., Goel, V., Davison, B.D.: Propagating trust and distrust to demote web spam. In: Proceedings of the Workshop on Models of Trust for the Web (2006)

Parallel Implementation of the Factoring Method for Network Reliability Calculation

Denis A. Migov and Alexey S. Rodionov*

Institute of Computational Mathematics and Mathematical Geophysics SB RAS,
prospect Akademika Lavrentjeva 6, 630090 Novosibirsk, Russia
mdinka@rav.sscc.ru, alrod@sscc.ru
http://www.sscc.ru/

Abstract. We consider the problem of a network reliability calculation for a network with unreliable communication links and perfectly reliable nodes. For such networks, we study two different reliability indices: network probabilistic connectivity and its average pairwise connectivity. The problem of precise calculation of both these characteristics is known to be NP-hard. For solving these problems, we propose the parallel methods, which are based on the well-known factoring method. Some optimization techniques for proposed algorithms are proposed for speeding up calculations taking into account a computer architecture.

Keywords: network reliability, probabilistic connectivity, pairwise connectivity, random graph, parallel algorithm.

1 Introduction

The task of calculating or estimating the network reliability is the subject of many researches due to its significance in a lot of applications. As a rule, network reliability is defined as some connectivity measure and the associated problems are NP-hard [1]. Therefore, serious computational obstacles arise while treating practical reliability analysis problems.

Random graphs are commonly used for modeling networks with unreliable elements. It is generally assumed, that the elements failures are statistically independent. The most explored is the case of absolutely reliable nodes and unreliable edges which corresponds to real networks where the reliability of nodes is much higher than reliability of communication links. The transport and wireless networks are good examples.

Various reliability measures, linked to different types of system performance, are used in practice. In particular, the network probabilistic connectivity (NPC) is the probability of all the terminal nodes in a network can keep connected together, given the reliability of each network node and communication link

---

* This research was supported by grants of the Russian Foundation for Basic Research (13-07-00589, 14-07-31069) and by of the Grants Council of the President of Russian Federation "Scientific school 5176.2010.9".

B. Murgante et al. (Eds.): ICCSA 2014, Part VI, LNCS 8584, pp. 654–664, 2014.

in the network. Finding NPC is equivalent to finding if one special node in a network can communicate with all other terminal nodes. This task corresponds to estimating the reliability of monitoring networks in CAM systems or sensor networks, for example. In such networks, one central computer must receive information from all peripheral sensors and computers. The analysis of the network probabilistic connectivity has been subject of considerable research.

Most of the existing calculation methods are based on network decomposition or reduction [2,3,4,5]. The new approach' in this area was introduced in [6]: cumulative update of lower and upper bounds of network reliability for faster feasibility decision. This method allows deciding the feasibility of a given network without performing exhaustive calculation. The approach was further developed with the help of network decomposition [7]. The parallel approach for estimation of network reliability by Monte Carlo technique was studied in [8,9].

Another network reliability measure we consider is the average pairwise connectivity (APC) [10,11]. APC characterize a confidence in possibility of establishment an arbitrary pairwise connection, for example in a peer-to-peer network. The task of calculation of this measure is biunique to the task of obtaining an expected value of a number of disconnected pairs of nodes (EDP) that had been first discussed in [12]. This measure shows how network is reliable from the point of possibility of connection between an arbitrary chosen pair of nodes even if a network is disconnected in a whole.

However, despite the improvements achieved on the efficiency of the computational methods for reliability analysis, they still are ineffective and so their parallel realizations are needed for executing on modern supercomputers. These realizations must take into account particularities of architecture of a specific instrumental computer.

In this paper, we propose the new parallel methods for calculating the network probabilistic connectivity and the expectation of a number of disconnected pairs of nodes in a network. Methods are oriented to using different computer architectures. The proposed methods are based on the well-known sequential factoring method [13] with improvements proposed in [14]. The analysis of the numerical experiments results allowed us to optimize some important parameters of the algorithm, which further increase its speedup.

2   Definitions and Notations

We model the network by an undirected probabilistic graph $G = (V, E)$ whose vertices represent the nodes and whose edges represent the links. We assume that each link succeeds or fails independently with an associated probability. By p_e we denote the probability of an edge $e \in E$ existence. Further on we refer to this probability as edge reliability. We suppose that the nodes are perfectly reliable.

Let us introduce some definitions.

An *elementary event* Q is an individual realization of the graph G defined by existence or absence of each edge. By E_Q we denote the set of all existing edges in Q.

The *probability of an elementary event* Q equals to the product of probabilities of existence of operational edges times the product of probabilities of absence of faulty edges.

$$P(Q) = \prod_{e \in E_Q} p_e \prod_{e \notin E_Q} (1 - p_e), \tag{1}$$

An elementary event Q is *successful* if all nodes in it can be connected by existent edges, i.e. edges from E_Q.

An arbitrary event (an event is a union of elementary events) is called *successful* if it consists only of the successful elementary events.

Probabilistic connectivity of a graph G, $R(G)$, is the probability of a graph G is connected, that is the probability of event which is the union of all successful events and of them only.

There is a similar definition of a more general measure, k-terminal reliability of a graph, which means connectivity only of selected k terminal nodes. Thus, the graph probabilistic connectivity is the all-terminal graph reliability. Below we use two-terminal graph reliability, $R_{st}(G)$, it is the probability of nodes s and t being connected.

Let us define a random variable $Y(Q)$, which is equal to the number of disconnected pairs of nodes in the elementary event Q.

By $N(G)$ we denote the expected value of $Y(Q)$. This value is called *EDP — expectation of a number of disconnected pairs of nodes in the network* and it is the second reliability measure we consider in the present study.

3   Factoring Method for Network Reliability Calculation

Factoring method is the most widely used method for calculating different network reliability measures [2,13]. This technique partitions the probability space into two sets, based on the success or failure of one network's particular element (node or link). The chosen element is called "factored element". So we obtain two graphs, in one of which the factored element is absolutely reliable and in the second one factored element is absolutely unreliable that is, is removed. The probability of the first event is equal to the reliability of factored element; the probability of the second event is equal to the failure probability of factored element. Thereafter obtained graphs are subjected to the same procedure.

The law of total probability gives expression for the network reliability, in the general case for an arbitrary system S with unreliable elements, it takes the following form:

$$R(S) = r_e R(S|e \ works) + (1 - r_e) R(S|e \ fails), \tag{2}$$

where $R(S)$ is the reliability of S and $R(S|e \ works)$ is the reliability of the system S when the element e is in operation, $R(S|e \ fails)$ is the reliability of the system S when the element e is not in operation, r_e is reliability of e.

Figure 1 illustrates the factoring method for probabilistic connectivity of graph with unreliable edges. The corresponding formula takes the following form:

$$R(G) = p_e R(G_e^*) + (1 - p_e)R(G \backslash e), \tag{3}$$

where p_e — reliability of edge e, G_e^* — is a graph obtained by contracting edge e from G, $G \backslash e$ — is a graph obtained by deleting e from G. Recursions continue until either disconnected graph is obtained, or until a graph for which the probabilistic connectivity can be calculated directly is obtained — it can be a graph of a special type or small dimension graph [7].

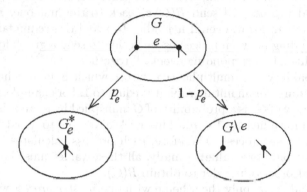

Fig. 1. Factoring procedure

For calculating EDP we can use the factoring method also, the corresponding formula has a similar form:

$$N(G) = p_e N(G_e^*) + (1 - p_e)N(G \backslash e). \tag{4}$$

Recursions go until deriving a graph for which a $N(G)$ is easily obtained. Main differences between these two implementations of the factoring method are different final graphs and different ways for the series-parallel reduction and for other speedup techniques.

On the other hand, we may calculate EDP by the following expression [12]:

$$N(G) = \sum_{u,v \in V,\ u \neq v} (1 - R_{uv}(G)). \tag{5}$$

It leads us to exhaustive search of all pairs of nodes, and for each pair two-terminal network reliability must be calculated.

It seems that the first method is preferable than the second one. However, due to availability of reduction and decomposition methods, calculation of two-terminal network reliability is greatly easier than EDP calculation by factoring method. Therefore, we have an assumption that the second method will be faster if there are enough computational cores.

4   Parallel Algorithm for Network Probabilistic Connectivity Calculation

In this section, we introduce the algorithm with use of MPI for probabilistic connectivity calculation for supercomputers with distributed memory. The main idea of parallel algorithm is the following: during the factoring procedure (3) one graph stays on paternal process while another one is being sent to some idle process.

However, such direct approach has some disadvantages. It leads to constant necessity of sending results from calculating process to a parental process due to the recursion rule (3). For example, if a first process send graph $G \backslash e$ to a second one, the second process will send $R(G \backslash e)$ back to the first one. And the first process must to wait for it, even if it is idle. To avoid this redundant number of operation of sending between processes we decide to send a graph for calculation and a probability of corresponding event all together.

Graph probability is a multiplicative factor which is being changed during factoring procedure, for an inital graph it is equal to 1. For example, in expression (3) the probability of G_e^* is a probability of G multiplied by p_e and the probability of $G \backslash e$ is one multiplied by $1 - p_e$. This trick allows us to avoid back sending during the factoring process. Therefore, each process calculates its own value (most likely not only one value). Finally, all these values must be summarized by some special process in order to obtain $R(G)$.

We have decided to apply the scheme with one *master process* which controls other processes. We call these controlled processes as *guided processes*.

Let us describe the proposed algorithm via the description of the processes interactions.

Master process. This process makes the following:

- monitors a workload of guided processes, i.e. it has an information whether the processes busy or not;
- assigns incoming graphs to idle processes;
- summarizes calculated values coming from guided processes. The final sum is the exact value of the graph probabilistic connectivity that we are searching.

Guided process.

- Initially it is idle and waiting for an incoming graph.
- If a graph is available, then process analyzes it and makes factoring procedure according to (3). Graph G_e^* stays for further calculation in the current process. Another graph ($G \backslash e$) is being attempted to send to some idle process. For this, the request is being sent to the master process. Reply from master process comes as a number k. If $k > 0$ then the current process sends this graph and its probability to the process k. Otherwise, there are no idle processes and both graphs stay for further treating in the current process. Anyway, the next recursion starts on the above-mentioned scheme.
- After the calculation is done, it sends an obtained result to the master process and waits for a next incoming graph with its probability, or for the end message.

Below we represent pseudo code of algorithms for master process and for guided processes. The algorithm for guided processes calls the recursive factoring procedure; pseudo code of this procedure is given separately.

Algorithm for Master Process

1: $R \leftarrow 0$ ▷ "Variable for calculation result"
2: remove attached trees from G; r is obtained factor   ▷ "Preliminary step"
3: **send** 1 to the process No. 1
4: **send** G to the process No. 1
5: process No. 1 becomes busy, other guided processes are idle
6: **while** there is any busy process **do**
7:     **receive** $int\_process\_rank$ from any process
8:     **receive** $float\_value$ from the process No. $int\_process\_rank$
9:     **if** $float\_value = float\_help\_request$ **then**   ▷ "Request for help has been obtained"
10:         **if** there is any idle process, let its No. be $idle\_rank$, **then**
11:             $int\_answer \leftarrow idle\_rank$
12:             **send** int_answer to the process No. $int\_process\_rank$
13:             process No. $idle\_rank$ becomes busy
14:         **else** $int\_answer \leftarrow 0$
15:             **send** int_answer to the process No. $int\_process\_rank$
16:         **end if**
17:     **else** $R \leftarrow R + float\_value$ ▷ "Calculation result has been received"
18:         process No. $int\_process\_rank$ becomes idle
19:     **end if**
20: **end while**
21: **for each** guided process **send** $float\_end\_command$
        return $R * r$

In order to improve performance of the algorithm we have tried to set some important parameters.

First parameter is the lower limit of a dimension of graph that could be assigned to another process. For example, it is meaningless to send a 6-vertex graph to another process because we have the exact formula for 5-vertex graph probabilistic connectivity calculation [7], so the next recursion may be the last and it is better to stay both subgraphs on a parental process. Therefore we introduced parameter N_{Edges} — the minimal number of graph edges required for sending graph to another process.

Another parameter which we introduce — is the necessary amount of recursion before sending graph to another process, N_{Rec}. Perhaps, if we do not try to send each newly formed graph, it will ease the burden on the master process and will help to avoid waste of data communications between processes.

Best values of these parameters highly depend on a graph structure and computer architecture, and can be selected experimentally.

We note that in the case of shared memory there is no need for a master process. Instead, we may use stack for graphs to compute. Let S be the stack of

---

Algorithm for Guided Process

1: **while** $float\_end\_command$ not received **do**
2:     **receive** $float\_value$ from any process
3:     **if** $float\_value \neq float\_end\_command$ **then**   ▷ "Graph's probability's been received"
4:         **receive** $(Graph\_H)$ from the same process
5:         $R \leftarrow$ FACTO_MPI($Graph\_H, float\_value$)
6:         **send** int_my_rank to the master process
7:         **send** R to the master process
8:     **end if**
9: **end while**

Algorithm for Recursive Factoring Procedure with use of MPI

1: **function** FACTO_MPI(G, p)
2:     series-parallel transformation of G; r is obtained factor   ▷ "Preliminary step"
3:     choose edge e for factoring   ▷ "It may be arbitrary edge"
4:     $G \leftarrow G \backslash e$
5:     $q \leftarrow p * r * (1 - p_e)$
6:     $p \leftarrow p * r * p_e$
7:     **if** G is connected **then**
8:         **send** int_my_rank to the master process
9:         **send** $float\_help\_request$ to the master process
10:        **receive** int_answer from the master process
11:        **if** $int\_answer > 0$ **then**
12:            **send** q to the process No. int_answer
13:            **send** G to the process No. int_answer
14:            **return** FACTO_MPI(G_e^*, p)
15:        **else return** FACTO_MPI(G_e^*, p)+FACTO_MPI(G, q)   ▷ "Both graphs stay on paternal process for further factoring"
16:        **end if**
17:    **else return** FACTO_MPI(G_e^*, p)
18:    **end if**
19: **end function**

---

graphs. Initially it contains only one graph. Idle processes appeal to S in order to receive a graph for further factoring. All processes do it alternatively in order to avoid bugs while working with shared memory. When a process receives some graph, this process calculates it directly or makes the factoring procedure. During that procedure one graph stays on a paternal process and another one is being sent to the stack S. When the calculation is completed, process appeals to S again. As in the case of the distributed memory algorithm, it is convenient to send to stack a graph's probability along with a graph. Thus, each process accumulates its own value. When the stack becomes empty, all processes complete their calculations, send accumulated values to the main process that makes global sum, and the whole task is completed with obtaining reliability of the initial graph.

5   Case Studies for the for Probabilistic Connectivity Parallel Algorithm

In this section, we examine scalability of the proposed algorithm. We try to find optimal values for N_{Edges} and N_{Rec} also. It is natural to assume, that with increasing number of computational cores more than some critical amount, parallel realization of the factoring method will not show speedup or even will show slowdown due to enormous amount of data communications. We try to find out this critical amount of cores too.

For numerical experiments we choose the lattice-like graph 7×7 which consists of 49 vertices and 84 edges. In spite of its not large dimension, this graph is very hard for computing its probabilistic connectivity because an inapplicability of various accelerating methods. For example, calculation of its probabilistic connectivity takes more than one day using PC.

The experiments were executed on the computing cluster HKC-30T (Nehalem) of the Siberian Supercomputer Center. This cluster consists of double-blade servers HP BL2220 G6 with Intel Xeon 5540 2.53 GHz CPUs.

During conducting experiments we have obtained that optimal values $N_{Edges} = 30$ and $N_{Rec} = 0$.

Any others values of N_{Rec} make computation slower. For example, for $N_{Rec} = 10$ and optimal N_{Edges} computational time on 8 cores was 16518 second — almost three times greater than for $N_{Rec} = 0$. Thus, our suggestion about inadvisability of sending each newly formed graph to another process was unjustified.

For values $N_{Edges} > 30$ and $N_{Edges} < 30$ computations were slower then in case $N_{Edges} = 30$. Combinations of N_{Edges} and N_{Rec}, in which one or two of them are different from optimums, give worse performance also.

The tables 1, 2 show results of experiments for parallel algorithm with its optimal values of N_{Edges} and N_{Rec}.

The results show that the factoring method works well in parallel implementation. The algorithm shows a superlinear speedup for number of cores, which is less than 128 and a linear speedup for number of cores less than 256. However, it shows deceleration of speedup for number of cores, which is greater than 256.

Table 1. The calculation time in seconds for number of cores from 1 to 64

Number of cores	1	4	8	16	32	64
Calculation time	41455	14122	5972	2791	1346	663

Table 2. The calculation time in seconds for number of cores from 128 to 500

Number of cores	128	192	256	320	400	512
Calculation time	330	219	167	151	156	166

However, it shows deceleration of speedup for number of cores which is greater than 256. Further, it even shows slowdown when number of cores is greater than 320.

6   Parallel Algorithm for Calculating EDP

When large number of computational cores is available, the exhaustive search of all pairs of nodes may be effective for EDP parallel calculation. As in the previous case, a queue to cores pool may be organized, but this time of pairs of nodes for which the 2-terminal probabilistic connectivity must be calculated. In this case we use (5), each R_{uv} is obtained by the factoring method. The serial-parallel reduction [2] and equations for small dimension graphs [15] are used for more efficiency. On the other hand, the adaptation of the parallel Factoring algorithm used for calculating graph's probabilistic connectivity above is possible.

For numerical experiments for EDP calculation by pairs using exhaustive search parallel algorithm, we choose the complete graph that consists of 13 vertices and 78 edges. Like in the section 5, this graph is very hard for computing its reliability, especially for EDP, because an inapplicability of various accelerating methods.

The tables 3, 4 show results of experiments.

Table 3. The calculation time in seconds for number of cores from 1 to 4

Number of cores	1	2	4
Calculation time	1533	770	385

The experiments were done on the computing cluster Jet of the Siberian Sate University of Telecommunications and Information Sciences. It consists of 18 double Intel Quad Xeon E5420 2.5 GHz CPUs

The results show that the proposed EDP calculation method works well in parallel implementation for supercomputers with distributed memory. It shows not a superlinear speedup, as it is with the parallel factoring method for probabilistic connectivity using supercomputers with distributed memory. We guess the main reason of this is the lesser amount of effective accelerating methods applied for EDP calculation. Nevertheless, it shows an acceptable linear speedup.

Table 4. The calculation time in seconds for number of cores from 8 to 64

Number of cores	8	16	32	64
Calculation time	192	96	48	24

7 Conclusion

In this paper we introduce a parallel implementation of the factoring method for exact calculation of a network reliability. Two network reliability measures were considered: network probabilistic connectivity and pairwise connectivity of its nodes. Also we've suggested some techniques which significantly improve the algorithms performance.

Although the parallel algorithm for probabilistic connectivity shows significant speedup in case when amount of cores is less than 256, it also shows a slowdown when amount of cores is greater than 320 due to enormous amount of data communications. Therefore, our primary next goal is to improve scalability of the proposed algorithms. It seems that there are two ways to do this — usage of the several master processes or not using them at all.

Another major problem is to find out whether the pairs exhaustive search parallel algorithm can be faster than the factoring algorithm for EDP parallel calculation with increasing supercomputer performance.

References

1. Colbourn, C.J.: The combinatorics of network reliability. Oxford University Press, New York (1987)
2. Shooman, A.M., Kershenbaum, A.: Exact Graph-Reduction Algorithms for Network Reliability Analysis. In: IEEE Global Telecommunications Conference GLOBECOM 1991, pp. 1412–1420. IEEE Press, New York (1991)
3. Chen, Y., Li, J., Chen, J.: A New Algorithm for Network Probabilistic Connectivity. In: IEEE Military Communications Conference, vol. 2, pp. 920–923. IEEE Press, New York (1999)
4. Migov, D.A., Rodionova, O.K., Rodionov, A.S., Choo, H.: Network Probabilistic Connectivity: Using Node Cuts. In: Zhou, X., et al. (eds.) EUC Workshops 2006. LNCS, vol. 4097, pp. 702–709. Springer, Heidelberg (2006)
5. Migov, D.A.: Computing Diameter Constrained Reliability of a Network with Junction Points. Automation and Remote Control 72(7), 1415–1419 (2011)
6. Won, J.-M., Karray, F.: Cumulative Update of All-Terminal Reliability for Faster Feasibility Decision. IEEE Trans. on Reliability 59(3), 551–562 (2010)
7. Rodionov, A.S., Migov, D.A., Rodionova, O.K.: Improvements in the Efficiency of Cumulative Updating of All-Terminal Network Reliability. IEEE Trans. on Reliability 61(2), 460–465 (2012)
8. Khadiri, M.E., Marie, R., Rubino, G.: Parallel Estimation of 2-Terminal Network Reliability by a Crude Monte Carlo Technique. In: Sixth International Symposium on Computer and Information Sciences, pp. 559–570. Elsevier (1991)

9. Martínez, S.P., Calvino, B.O., Rocco, S.C.M.: All-Terminal Reliability Evaluation through a Monte Carlo Simulation Based on an MPI Implementation. In: European Safety and Reliability Conference: Advances in Safety, Reliability and Risk Management (PSAM 2011/ESREL 2012), Helsinki, pp. 1–6 (2012)

10. Sun, F., Shayman, M.A.: On Pairwise Connectivity of Wireless Multihop Networks. Int. J.Secur. Netw. 2(1/2), 37–49 (2007)

11. Potapov, A., Goemann, B., Wingender, E.: The Pairwise Disconnectivity Index as a New Metric for the Topological Analysis of Regulatory Networks. BMC Bioinformatics 9(1), 1–15 (2008)

12. Rodionov, A.S., Rodionova, O.K., Choo, H.: On the Expected Value of a Number of Disconnected Pairs of Nodes in Unreliable Network. In: Gervasi, O., Gavrilova, M.L. (eds.) ICCSA 2007, Part III. LNCS, vol. 4707, pp. 534–543. Springer, Heidelberg (2007)

13. Page, L.B., Perry, J.E.: A Practical Implementation of the Factoring Theorem for Network Reliability. IEEE Trans. on Reliability 37(3), 259–267 (1998)

14. Rodionova, O.K., Rodionov, A.S., Choo, H.: Network Probabilistic Connectivity: Exact Calculation with Use of Chains. In: Bubak, M., van Albada, G.D., Sloot, P.M.A., Dongarra, J. (eds.) ICCS 2004. LNCS, vol. 3036, pp. 565–568. Springer, Heidelberg (2004)

15. Migov, D.A.: Formulas for Fast Calculating a Reliability of Graphs with Small Dimension. Problems of Informatics 2(6), 10–17 (2010) (in Russian)

A Web System for Solving Real Problems Involving Partial Differential Equations in Generalized Coordinates

Fábio T. Matsunaga, José L. Vilas Boas, Neyva M.L. Romeiro,
Armando M. Toda, and Jacques D. Brancher

Londrina State University (UEL),
P.O.: 10.011, Zip Code: 86.057-970, Londrina-PR, Brazil

Abstract. Numerical simulations based on partial differential equations
are applied to different knowledge areas. There was a motivation for the
use of computational tools to perform real and numerical simulations,
however many of these are restricted to a software installation and limiting processing, mainly in mobile devices. The aim of this study was
to develop a web system which is able to solve real problems involving
partial differential equations in 2D generalized coordinates, by the finite
difference discretization. The system consists of web service for client-
server communication, a database, a module of processing calculation
(back-end) and a graphical user interface (front-end). The system can
be executed on devices regardless their performance, such as personal
computers and mobile devices, since the calculations are made on a remote server. The contribution is the availability of a web architecture
for including other types of differential equations and consequently, for
real problems and situations, such as energy conduction and pollutant
transportation.

Keywords: finite difference discretization, mesh generation, real problems, web architecture.

1 Introduction

Computer simulation processes had a great rise in recent years, and it motivated
researchers to model real world problems in mathematics, physics, chemistry and
engineering. The basis of majority of the simulations are the partial differential
equations (PDEs), which have replaced the descriptions of phenomena instead of
the traditional math formulations [1]. The resolution must be done through the
numerical methods that discretize these equations, by the definition of spatial
and temporal distribution of the problem, such as finite differences, finite volume
and finite element [2,3].

Among the numerical methods, the finite difference discretization (FDD) is
considered the simplest to implement, since it is totally based on algebraic operations. Phenomena and problems can be described symbolically through the
association of global variables and equation parameters, facilitating the algebraic

B. Murgante et al. (Eds.): ICCSA 2014, Part VI, LNCS 8584, pp. 665–680, 2014.
© Springer International Publishing Switzerland 2014

operations and the description of PDEs [1]. For any types of simulation, there are three important steps to be considered: pre-processing (initial parameters description and spatial domain definition), processing (simulation) and post-processing (results evaluation).

Services and applications have been developed to support the demand of simulation processes. However, they use scientific softwares [4], interactive tools such as videos [5], which need some ideals requirements (memory capacity, a setting ideal hardware and software installation). Issues such as the accessibility of mobile services have also been treated [6] and these ideas have emerged due to the need of the simulators have its source codes adapted for each test case and for a specific hardware.

It has become an emerging motivation in the use of web technologies for any application development (weblization process) [7]. The first reason is that the knowledge and learning transfer is practical and allows the reusing of several routines. Another is the use of large data volumes, which requires database and remote server for information storage, which are constantly maintained and easily accessible from anywhere and at any time [8]. The control and version upgrade is automatic and transparent, without any inconvenience to the users. Moreover, the services do not need be reinstalled if any malfunction occurs on the computer.

Some web-based solutions for several real world problems have been developed [7,9]. However, there are few existing web-based solutions that have focused on solving PDEs and simulation of various phenomena described by this model, which most of them focused on developing a web template for a specific case [7,9]. These systems can be explored and extended to multiple situations and solving real problems in general, and as described, these may be modeled mathematically by PDE [1]. In addition, the portability and accessibility of these simulators tools have other important issues to be addressed [10], since many of these devices have restricted execution and the processing hardware is limited.

A web service architecture was proposed in [11], which solved PDEs by FDD method in Cartesian coordinates. However, it should be explored and improved to solve real world problems and situations. Considering the arguments mentioned above, the aim of this work is to develop a system that uses web technologies to solve 2D stationary and time-dependent problems involving PDEs. The numerical method used to solve these equations was the mesh generation and the FDD method in two-dimensional generalized coordinates. For this, we developed a front-end (user interface), a web service communication in runtime between the client and server and a back-end for scalable processing calculations on remote servers.

Thus, the structure of this paper is as follows: Section 2 presents the finite difference discretization method for 2D generalized coordinates. In Section 3, we described the web service architecture to support the demand of finite difference calculation. Section 4 shows some results examples and the discussions. Section 5 is the conclusions and some future works.

2   Finite Difference Method on Generalized Coordinates

Several problems and physical phenomena are described and modeled by PDEs, which are a set of equations containing multiple unknown variables functions along their partial derivatives. Considering this, some studies involving the solution of PDEs by FDD on Cartesians coordinates, for example, a system for heat conduction and wave propagation simulation, have been developed [11]. However, in cases of actual problems, the physical domain (x, y) in most of cases, does not have a regular geometry that coincides with the computational mesh domain. Thus, there is a motivation for the use of methods for mesh generation in irregular geometries, such as the method of FDD in generalized coordinates systems (GCS) [12,13].

2.1   Mesh Generation in 2D Generalized Coordinates

The process of mesh generation by GCS is initially to define the outline of the points of the (x, y) geometry. These points are interpolated by the parametric linear splines method, making possible to know all the points of the geometry boundary. The interpolation method consists in determine a linear equation for each pair of consecutive points (x_i, y_i) and (x_{i+1}, y_{i+1}), where $y = y_i + \frac{y_{i+1} - y_i}{x_{i+1} - x_i}(x - x_i)$ and $x = x_i + \frac{x_{i+1} - x_i}{y_{i+1} - y_i}(y - y_i)$.

After the geometry boundary definition, the transformation of the Cartesian coordinates of the physical domain is performed, as was done at [12] and [13]. This is accomplished through a mapping of the geometry described by the Cartesian coordinates (x, y) for the generalized coordinates or transformed domain (ζ, η). In this case, the dimensions are fixed so that each of the points is mapped to an equivalent in a physical domain.

For the mesh generation, we must initially specify the initial mesh dimensions – N (line numbers in ξ) and M (line numbers in η). In this work, an uniform mesh is assumed, wherein the distance between each node in the mesh is $\Delta\xi = \Delta\eta = 1$. The method used was the elliptical type, described by the governing equations $\xi = \xi(x, y)$ and $\eta = \eta(x, y)$.

The ξ and η lines describes the discretized meshes, which points are the intersections of these lines determined by the governing equations. To solve these, we must firstly define the boundary conditions Γ and the belonging (x, y) points. In this case, the ξ lines start in Γ_1 and end in Γ_3, and the η lines end in Γ_2 and end in Γ_4, as can be seen in Figure 1.

From the Figure 1, all points of the physical domain are mapped to an equivalent point in the transformed domain $\xi(x, y)$ and $\eta(x, y)$. This mapping is performed by metrics denoted by $\alpha = x_\eta^2 + y_\eta^2$, $\gamma = x_\xi^2 + y_\xi^2$ and $\beta = x_\xi x_\eta + y_\xi y_\eta$ and the Jacobian $J = (x_\xi y_\eta - x_\eta y_\xi)^{-1}$, which are responsible for the compensation due to the coordinate system changing. These metrics are applied in the governing equation (1), which calculates a generic point of the mesh grid ϕ described by the (x, y) position.

$$\alpha\phi_{\xi\xi} + \gamma\phi_{\eta\eta} - 2\beta\phi_{\xi\eta} + \frac{1}{J}(P\phi_\xi + Q\phi_\eta) = 0. \tag{1}$$

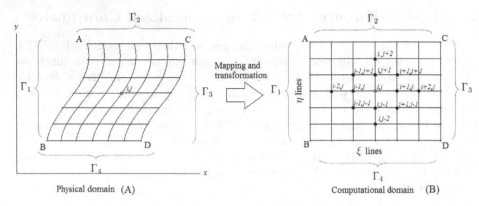

Fig. 1. Computational mesh and the ξ and η lines

From (1), considering x_ξ and y_ξ the partial derivatives of x and y in ξ and x_η and y_η the partial derivatives in η. P and Q source-terms used to attract lines in a ξ and η direction respectively. To calculate each ϕ point, all partial derivatives are approximated via finite difference equations, considering the central differences. The approximations of partial derivatives $x_\xi = \frac{x_{i+1,j}-x_{i-1,j}}{2}$, $x_\eta = \frac{x_{i,j+1}-x_{i,j-1}}{2}$, $y_\xi = \frac{y_{i+1,j}-y_{i-1,j}}{2}$ e $y_\eta = \frac{y_{i,j+1}-y_{i,j-1}}{2}$ and considering a structured mesh ($\Delta\xi = \Delta\eta = 1$), it generates the implicit scheme. This system was solved using the iterative Gauss-Seidel method, whose solution is the value of each grid point x and y (2).

$$\phi_{i,j} = \frac{1}{A_p}(A_e\phi_{i+1,j} + A_w\phi_{i-1,j} + A_n\phi_{i,j+1} + A_s\phi_{i,j-1} + A_{ne}\phi_{i+1,j+1}$$
$$+ A_{se}\phi_{i+1,j-1} + A_{nw}\phi_{i-1,j+1} + A_{sw}\phi_{i-1,j-1}).$$
$$(2)$$

The parameters of the Equation 2 are $A_p = 2\alpha + 2\gamma$, $A_e = \alpha + \frac{P}{2J^2}$, $A_w = \alpha - \frac{P}{2J^2}$, $A_n = \gamma + \frac{Q}{2J^2}$, $A_s = \gamma - \frac{Q}{2J^2}$, $A_{ne} = -\frac{\beta}{2}$, $A_{se} = \frac{\beta}{2}$, $A_{nw} = \frac{\beta}{2}$ and $A_{sw} = -\frac{\beta}{2}$.

2.2   Energy Equation in 2D Generalized Coordinates

For this work, the PDE applied on the system is the 2D transport equation. Depending on the parameterization, the equation can describes various phenomena, such as energy conduction according to the fluid movement [14]. Considering $\sigma = k(\rho C_p)^{-1}$, where k is the thermal conductivity, ρ the specific mass and C_p the specific heat. α, β and γ the transformation metrics with the Jacobian J.

$$\left(\frac{1}{J\sigma}\right)\frac{\partial T}{\partial \tau} - J\alpha\frac{\partial^2 T}{\partial \xi^2} - J\gamma\frac{\partial^2 T}{\partial \eta^2} + \rho u\frac{\partial T}{\partial \xi} + \rho v\frac{\partial T}{\partial \eta}$$
$$= \frac{\partial}{\partial \xi}\left(J\alpha\frac{\partial T}{\partial \xi}\right) + \frac{\partial}{\partial \eta}\left(J\gamma\frac{\partial T}{\partial \eta}\right) - 2\frac{\partial}{\partial \xi}\left(J\beta\frac{\partial T}{\partial \eta}\right)$$
$$(3)$$

On the Equation 3, the ρ value is constant. The u and v describe the fluid movement on x and y direction respectively, generally calculated from Navier-Stokes equation [2]. Besides the parameters described, others are set during pre-processing. Considering Ω the internal mesh domain, to perform the simulation, we must define:

- The maximum simulation time (T_{max}) and the number of time instances (N_t) of $\Delta t = (T_{max})/(N_t - 1)$ length;
- Initial condition $T(\xi, \eta, 0)$ in Ω;
- Dirichlet boundary condition $T(\xi, \eta, t)$ in $\partial\Omega_0 \times [0, T_{max}]$, where $\partial\Omega_0$ is the boundary geometry;
- Neumann boundary condition $\frac{\partial T}{\partial n}$ or $\triangledown T.\overrightarrow{n} = \overline{q}$ in $\partial(\Omega - \Omega_0) \times [0, T_{max}]$, where $\overrightarrow{n}$ is the exterior normal, $\overline{q}$ the energy flow and $\partial(\Omega - \Omega_0)$ the Ω isolated boundary.

The transport equation 2D in generalized coordinates is discretized approximating the PDEs using the FDD method (detailed equations can be found in [14] and [12]). Considering a structured mesh $\Delta\xi = \Delta\eta = 1$, the approximation generates the implicit scheme [12]. This leads to the linear system $Ax = b$ resolution with eight nonzero diagonals, since they consider the eight neighboring points in the region (i, j) – Figure 1. This system was solved using the iterative Gauss-Seidel method, which solution is the value of $T_{i,j}$ in each particular grid point, defined in (4).

$$\frac{1}{A_p}(A_e T_{i+1,j} + A_w T_{i-1,j} + A_n T_{i,j+1} + A_s T_{i,j-1} + A_{ne} T_{i+1,j+1}$$
$$+ A_{se} T_{i+1,j-1} + A_{nw} T_{i-1,j+1} + A_{sw} T_{i-1,j-1} + A_0 T_0).$$

$$(4)$$

From the discretized 2D transport equation (4), we : $A_p = \frac{1}{J_p \sigma \Delta t}$, $A_e = J_p \alpha_p + 0.25(J_e \alpha_e - J_w \alpha_w)$, $A_w = J_p \alpha_p - 0.25(J_e \alpha_e - J_w \alpha_w)$, $A_n = J_p \gamma_p + 0.25(J_n \gamma_n - J_s \gamma_s)$, $A_s = J_p \gamma_p - 0.25(J_n \gamma_n - J_s \gamma_s)$, $A_{ne} = -0.5 J_e \beta_e$, $A_{se} = 0.5 J_e \beta_e$, $A_{nw} = 0.5 J_w \beta_w$, $A_{sw} = -0.5 J_w \beta_w$ e $A_0 = \frac{1}{J_p \sigma \Delta\tau}$.

3   Web System Architecture and Main Modules

The proposed web system consists on four main modules (Figure 2): 1/ a scheme for client-server communication, 2/ graphical user interface (GUI) representing the front-end, 3/ application of calculations and 4/ database module. This system consists in a front-end for client-server communication and interaction and a back-end, responsible for execution of each user requests.

The modules of the web system (Figure 2) supported the construction of the whole system. The front-end consists of the users input web interface as well as the display of results for analysis, exploration and application in a real case. The back-end consists of the database management, the storage and calculations in the field application storage and file of user input. The front-end and back-end communication was implemented by the deployment of a web service that controls/stores user requests and query solutions at the database at runtime.

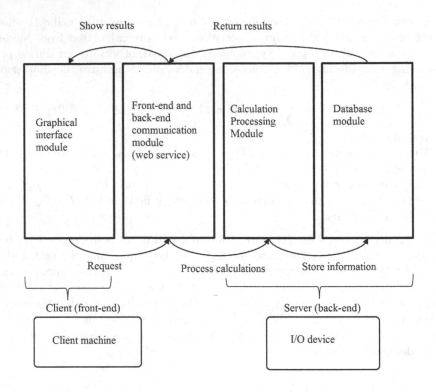

Fig. 2. Main modules of the web system

3.1   Front-End and Back-End Communication

This module is the main way for making the exchange of information input and output calculations, providing a runtime interaction with the user. To communicate with the web service, the SOAP (Simple Object Access Protocol) protocol, which is responsible for the exchange of messages in XML (eXtensible Markup Language) documents format, was used to provide the client-server interaction.

The JAX-WS (Java API for XML Web Services) language was used to implement the web service. The deployment of services was conducted in GlassFish 3.1 server, since this is a servlet container with an application server, which is useful for integrating web services in a service-oriented architecture. The GUI was developed using the Primefaces framework, a useful tool for developing dynamic web pages. The JSF (Java Server Faces) framework was applied to perform session control and connection MVC (Model View Controller) with the interface.

The communication and messaging interface has been described with WSDL (Web Services Description Language), which has several indexes methods to be requested by a web service. The methods range from the search for information in the database, managing user information and requests to the command for executing the application requester.

The developed web service provides a means for communication with the user and the I/O (input-output) device that consists of a remote server that sends and receives data processing. Once this service is available, the communication with the web service through a GUI builds the system web, along with the other modules, establishing a transparency between applications, allowing any user interface to communicate with the web service through XML documents.

3.2   Database Module

The storage method for the user information, calculations and outputs processed by the server is the PostgreSQL database stored in the server. This is accessed by the web service, which sends information to the front-end and the back-end. The entity-relationship diagram of the database is shown in Figure 3.

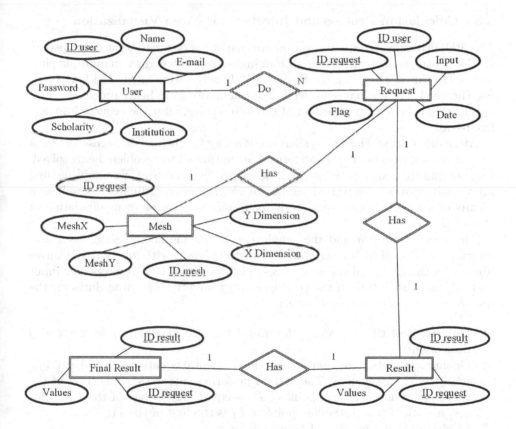

Fig. 3. Entity-relationship diagram of the database composed by five tables

As can be seen in Figure 3, each tuple of the table 'User' is associated with multiple requests ('Request' table) whit the respective results ('Result' table)

and mesh information ('Mesh' table) and a foreign key to the 'Request' table, which contains information for each request calculation. This table also has a relationship with the 'Mesh' table, which contains attributes of mesh ID, corresponding ID request, problem domain and the information of the mesh points stored as 'varchar', both for axis x (MeshX) as for the axis y (MeshY).

The 'Result' table stores results information corresponding to the foreign key in 'Request' table for each time instance, containing the information of the results IDs and a trigger that mirrors the results data to the 'Final Result', which is a table with all problems solved by all the users. This model also allows the sharing of a set of solved problems and a cache storage of previous calculations, where a problem solved need not be calculated again. In addition, the 'Request' and 'Result' tables have features such 'date' and 'time' where any user can access solutions requested in a specific time.

3.3   Calculation Process and Interface for Mesh Visualization

The FDD calculation application on the remote server has been developed in C/C++, one general purpose programming language that runs on multiple platforms. On this application, the PostgreSQL libraries were integrated for recording the results in the database while the calculations are being processed. This application is called by a method of the web service, after the request from the front-end.

After the process, the web system needs a way for the user to access and view the simulation results. This is performed at runtime of the problem being solved. Considering this, an interactive tool, developed in Javascript, was implemented for visualization of discretized meshes in generalized coordinates, as well as a means of user interaction such as query results and access to manipulation of the geometry.

The mesh generator and the simulation results module consists of an user interface developed in Javascript and Primefaces, along with functions of Canvas library for the design of the geometries. The developed GUI receives user input and displays the output of the corresponding geometry. The procedures for the user to use the web system simulator:

1. Dimension of the problem – the user define the maximum size of x and y dimension;
2. Geometry points – the user defines the segmental equation $[(x_0, y_0), (x_1, y_1)]$ with the Γ_n associated. The system performs automatic connection of the segments, where the first point of Γ_n is equals to the last of the Γ_{n-1} (for $1 \leq n \leq 3$). If $n = 4$, the last point of Γ_4 is the first of the Γ_1:
3. Mesh grid size – number of ξ and η lines;
4. Mesh information storage – the calculations processing module stores this information in the 'Mesh' table in the database;
5. PDE input parameters – the calculation process module performs the simulation according to user input. The inputs are the parameters described in section 2.2. The Dirichlet boundary conditions are $(\Gamma, \phi_i, \phi_f, T)_1, (\Gamma, \phi_i, \phi_f, T)_2,$

$(\Gamma, \phi_i, \phi_f, T)_3$ and $\ldots (\Gamma, \phi_i, \phi_f, T)_n$, where Γ is the side of the boundary, ϕ_i and ϕ_f is the interval in $\partial\Omega_0 \times [0, T_{max}]$ and T the value to be assigned in the interval;

6. The calculation processing module performs the simulation for each instance of time. The Poll process, a PrimeFaces artifice, makes constant queries to the database to verify the existence of new results;

7. The GUI module performs the discretized mesh design and distribution of values over that. This procedure consists of the following steps:

   (a) Transforms the string of the x and y grid points extracted from the 'Mesh' table into a set of vectors and sent to the front-end;

   (b) Draws firstly the ξ lines then to draw the η lines using the Path resource (from Canvas), generating the full grid (whole mesh);

   (c) Performs the distribution of values for each node in the grid (intersection of ξ and η lines), by defining the color of the dots according to the color scale calculated on the basis of the higher value of the mesh, generating a legend of chain colors, such as in the Figure 4 (this color scheme is already used on the previous web system version for Cartesian coordinates problems [11]). This procedure is repeated until all instances of time are calculated and updated.

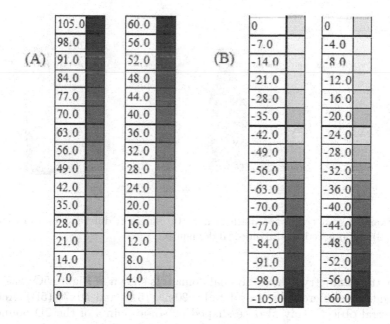

Fig. 4. Example of a legend of chain colors of different ranges of values used to define the PDEs values distribution over the mesh, both positive and (A) to negative (B)

The color chain layout (Figure 4) will be used for filling the geometry to assign a value for each grid points. These are considered for positive and negative values. Regardless of the results obtained from PDE simulated, the same color scheme is used for problems of different distribution of temperature, energy or concentration, which are calculated and distributed over the color tone. The trend is that the warmer colors represent higher and cooler colors represent the lower values.

4   Results and Discussions

For the web system for PDEs simulation testing, some geometry meshes were defined and modeled. Initially, the user specified the coordinates of the physical domain that describes the geometry. In this case, it can make manual measurements for obtaining points or submitting a data file that already contains all the predefined points. The Figure 5 illustrates an example of the generated mesh displayed on the web application developed in Javascript. The geometry boundary points were obtained by millimeter measurements of the silhouette physical plane. In this way, the user do not need to provide all points, since the spline interpolation method applied on the discrete definition calculates the continuous modeling of the whole geometry.

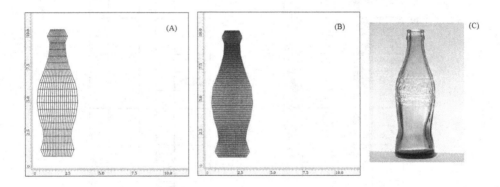

Fig. 5. Example of a generated mesh of a bottle figure in different level of details – (A) and (B) – compared to the physical domain (C)

The bottle geometry, which physical domain is shown in Figure 5C, was modeled considering different level of detail – 20x20 (5A) and 40x70 (5B). Accordingly, several objects may also be shaped by some points of the 2D boundary. Other real geometries may be modeled and simulated, most of them related, for example, to the global disturbances. An example was modeled (Figure 6), where the 2D geometry of Igapo Lake was described in the web mesh generator (size 40x20). Then, it is possible to simulate some phenomena over the mesh (Figure 7), such as the pollutants transportation over the lake surface [15,2].

The following parameters were defined: $T_{max} = 27480$ seconds, $N_t = 100$, $\sigma = 0.055$, mesh size $M = 100$ and $N = 50$, the initial condition $T(\xi, \eta, 0) = 0$ (no initial pollutants over the lake surface), the boundary condition $(1, 2.8, 3.2, 1000)_1$ (representing the specific region of initial launching of a pollutant with 1000mgL^{-1} of concentration) and $q = 0$.

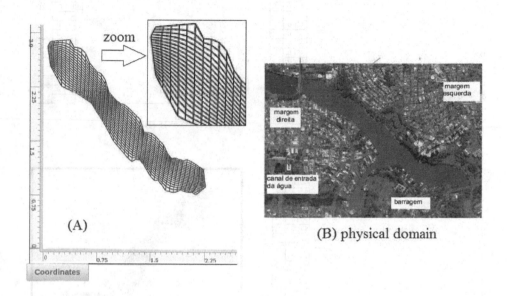

(A)

(B) physical domain

Fig. 6. Igapo Lake 2D mesh generated by the web system (A) compared to the physical domain of the problem (B) obtained from Google Maps. The scale of the problem is measured in Km

In Figure 6, the web system is able to generate the mesh of complex 2D surfaces and to make natural phenomena simulations (Figure 7). However, the flow simulation performed by Romeiro et al. [15] involves the use of flow velocity, obtained from Navier-Stokes equation. In the current work, our web system is limited to 2D transport equation with constant flow velocity, assuming that $u = v = 1$, which creates more uniform flow effect, without vortex effects, compared to the original work (Figure 7D).

The implemented web architecture is able to include different types of PDEs, such as the Navier-Stokes, which enables the calculation of flow velocity. Independent on the equation to be applied and simulated, the architecture of the proposed web-based system is open to modifying each deployed module to adapt the system to solve other kinds of real problems. It is also possible to expand the system to other numerical methods, which can serve the user to select the desired scheme. Thus, the user takes the study of the phenomenon, and determine which types of discretization schemes are best suited (explicit, implicit, Crank-Nicholson), allowing further analysis in the post-processing stage [16].

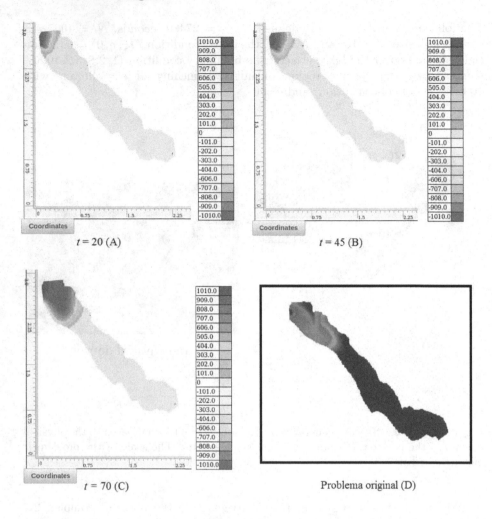

$t = 20$ (A)

$t = 45$ (B)

$t = 70$ (C)

Problema original (D)

Fig. 7. Simulation visualization of pollutant transport in some time instances (A-C) compared to the original solution (D) – obtained from [15]

An example of adaptation and the reusability of modules and routines have already occurred during the actual development of the proposed web-based system, in which the same web architecture was used to solve PDEs in two types of situations: first, to solve problems in Cartesian coordinates [11] and then solve them in generalized coordinates. The modification of some necessary modules allowed the expansion of the first version of the system [11] consisted on create a table of information of grid points in the database module and change the module graphical interface to allow geometries visualization. It was also necessary to integrate the application in Javascript GUI layer with the user (HTML page) and add the methods and algorithms for mesh generation.

A fundamental property of the proposed work is that the web system can run on any device, even mobile (Figures 8), which has limited hardware, where the user needs only a web browser and Internet connection, due to the simple configuration of the web system. Thus, the user need not have knowledge of formulas and resources that are executing the calculations and the local storage of all information, the user enters only with the input parameters and performs interactions with the mesh generator. The performance of the executions and the use of computational resources depends exclusively on the remote server requirements.

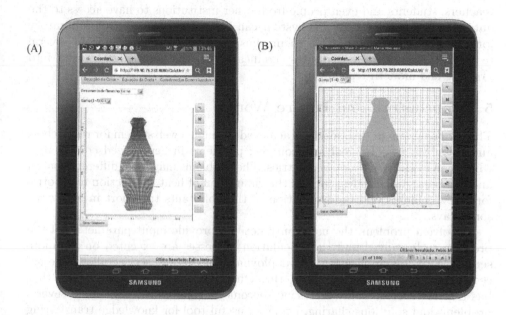

Fig. 8. Mesh generation visualization – pre-processing/sub-processing stage (A) – and resolution of an PDE over the mesh – processing stage (B) – in the web system accessed from a mobile device

Initially, the discrete points of the boundary defined by the user are sent to remote server in text file, through the web service. Then, the remote application read the points, performs the mesh generation and PDE simulation, storing the internal grid points and all results on each time instance on table 'Mesh' and 'Result', respectively. After that, the front-end retrieves the information from database and the Canvas draw the grids of the mesh object. All the onerous procedure are performed on the remote server, while the visible web system, which is apparently local for the client, do only the basic tasks of request sending and results querying. It can be seen that the simulations of different systems of generalized coordinates (Figure 8) are displayed on the screen of the device.

In resolution by FDD method, the pre-processing and post-processing stages are similar among all PDEs, differing only in the processing step. Moreover, the use of a structured discretization allows that all results (grid points and all instances of results) are stored in the same format in the database. The search system can be extended for other purposes, such as the use of the concept of relative error [4] for the system to list a set of similar problems, which results have differences in values that are smaller than the error tolerance, enabling future implementation of the sensitivity analysis of different PDEs [17].

The implemented search system allows that problems solved may be found and listed anywhere at any time, creating a collaborative system for researchers, teachers, students and even people from other institutions to have access to the database of solved PDEs. The system can be applied as a virtual laboratory for physical experiments. This theme has also been a study topic in several researches [18,19] and contributes significantly to the knowledge transfer into practice and modern teaching.

5   Conclusions and Future Works

The proposed aim in this work was achieved through a web system for researchers and scientists to solve real and complex problems in generalized coordinates, which can model real-world geometries. The problems may be of different spatial and temporal magnitudes, such as the simulation of heat conduction in a bottle for a few minutes and the simulation of the pollutants transport in a lake for some days.

To solve a problem, the users only need to provide input parameters of the problem to be solved and the calculation processes are executed on a remote server. With this application, the deployment of the web service allows the access from anywhere and anytime, through a browser or a mobile device that has Internet access. Thus, the service becomes a collaborative system of several problems and solutions sharing, being an useful tool for knowledge transferring through a modern technological learning and education.

The main contribution of this work is the availability of an open web architecture that solves many problems described by PDEs. This system, established by an open and standardized for the simulation of various types of numerical methods architecture model, making it a flexible system for adaptation and adjustment in various real situations, regardless the method to be applied. Furthermore, the database storage solutions can be continuously extended for any user or researcher, making the web proposed system a digital library of real problems.

As a future perspective, we intend to conduct the inclusion of various types of PDEs, such as the Navier-Stokes equations. This equation allows to extend the possibility of solving a wide range of real problems, for example, by calculating the velocity field, needing only the calculations processing module adaptation. Another focus is the great attention on solving three-dimensional problems, which is possible to modify the web interface layer to allow the simulation and visualization of problems, especially in the GUI and processing layer calculations.

Acknowledgements. Thanks to CAPES for the Master's scholarship to the student Fabio Takeshi Matsunaga.

References

1. Tonti, E.: Why starting from differential equations for computational physics? Journal of Computational Physics 257, 1260–1290 (2014)
2. Romeiro, N.M., Castro, R.G., Cirilo, E.R., Natti, P.L.: Local calibration of coliforms parameters of water quality problem at Igapó I Lake, Londrina, Paraná, Brazil. Ecological Modelling 222(11), 1888–1896 (2011)
3. Gupta, P.K., Singh, J., Rai, K., Rai, S.: Solution of the heat transfer problem in tissues during hyperthermia by finite difference-decomposition method. Applied Mathematics and Computation 219(12), 6882–6892 (2013)
4. Reimer, A.S., Cheviakov, A.F.: A Matlab-based finite-difference solver for the Poisson problem with mixed Dirichlet-Neumann boundary conditions. Computer Physics Communications 184(3), 783–798 (2013)
5. Grande, A., Gonzalez, O., Pereda, J.A., Vegas, A.: Educational Computer Simulations for Visualizing and Understanding the Interaction of Electromagnetic Waves with Metamaterials. In: IEEE EDUCON Education Engineering, vol. (1), pp. 543–547 (2010)
6. Dumont, C., Mourlin, F.: A Mobile Computing Architecture for Numerical Simulation. In: International Conference on Mobile Ubiquitous Computing, Systems, Services and Technologies (UBICOMM 2007), pp. 68–74. IEEE (November 2007)
7. Weng, W.C.: Web-based post-processing visualization system for finite element analysis. Advances in Engineering Software 42(6), 398–407 (2011)
8. Isbasoiu, E.C.: Numerical Calculation, Web Services and Principles for Physics Applications. In: 2010 Fourth UKSim European Symposium on Computer Modeling and Simulation, pp. 387–390. IEEE (November 2010)
9. Walker, J.D., Chapra, S.C.: A client-side web application for interactive environmental simulation modeling. Environmental Modelling & Software 55, 49–60 (2014)
10. Wriedt, T.: Mie theory 1908, on the mobile phone 2008. Journal of Quantitative Spectroscopy and Radiative Transfer 109(8), 1543–1548 (2008)
11. Matsunaga, F.T., Vilas Boas, J.L., Romeiro, N.M.L., Brancher, J.D.: CloudPDE: a web solver of differential equations by finite difference discretization. In: IADIS - International Conference on Applied Computing (prelo). (2013)
12. Cirilo, E.R., Bortoli, A.L.D.: Cubic Splines for Trachea and Bronchial Tubes Grid Generation. Semina: Ciências Exatas e Tecnológicas 27, 147–155 (2006)
13. da Silva, W.P., Precker, J.W., Silva, D.D., Silva, C.D., de Lima, A.G.B.: Numerical simulation of diffusive processes in solids of revolution via the finite volume method and generalized coordinates. International Journal of Heat and Mass Transfer 52(21-22), 4976–4985 (2009)
14. Maliska, C.R.: Transferência de Calor e Mecânica dos Fluidos Computacional, 2a edn. Editora Livros Técnicos e Científicos Editora, Rio de Janeiro (2004)
15. Romeiro, N.M.L., Cirilo, E.R., Natti, P.L.: Estudo do escoamento e da dispersão de poluentes na superfície do Lago Igapó I. In: Anais do II Seminário Nacional sobre Regeneração Ambiental de Cidades - Águas Urbanas II, pp. 1–17 (2007)
16. Russell, M., Probert, S.: FDiff3: a finite-difference solver for facilitating understanding of heat conduction and numerical analysis. Applied Energy 79(4), 443–456 (2004)

17. Weibe, A.Y., Huisinga, W.: Error-controlled global sensitivity analysis of ordinary differential equations. Journal of Computational Physics 230(17), 6824–6842 (2011)
18. Singh, V., Dubey, R., Panigrahi, P.K., Muralidhar, K.: An educational website on interferometry. In: 2012 IEEE International Conference on Technology Enhanced Education (ICTEE), pp. 1–10 (January 2012)
19. Perus, I., Klinc, R., Dolenc, M., Dolsek, M.: A web-based methodology for the prediction of approximate IDA curves. Earthquake Engineering & Structural Dynamics 42(April 2012), 43–60 (2013)

Full Body Adjustment Using Iterative Inverse Kinematic and Body Parts Correlation

Ahlem Bentrah[1,2], Abdelhamid Djeffal[1], Mc Babahenini[1], Christophe Gillet[2], Philippe Pudlo[1], and Abdelmalik Taleb-Ahmed[2]

[1] LESIA laboratory, Biskra University, Algeria
[2] LAMIH laboratory, Valenciennes, France

Abstract. In this paper, we present an iterative inverse kinematic method that adjust 3D human full body pose in real time to new constraints. The input data for the adjustments are the starting posture and the desired end effectors positions -constraints-. The principal idea of our method is to divide the full-body into groups and apply inverse kinematic based on conformal algebra to each group in specific order, our method involve correlation of body parts. The paper describes first the used inverse kinematic with one and multiple task simultaneously and how we handle with collision induced by the joints with the objects of the environment. The second part focuses on the adjustment algorithm of the full body using the inverse kinematic described above. Comparison is made between the used inverse kinematic and another inverse kinematic that have the same principle. In this paper we present our preliminary results.

Keywords: Animation, Inverse kinematic, Geometric algebra, Virtual humanoid.

1 Introduction

In various fields such as robotics and animation, there is often a need to control articulated structures in complex way. In the animation case, the control may involve posing the character body or some kinematic chain of the character to respect new constraints. Constraints may be defined as end effectors targets or goal positions. Character body is represented as series of different poses of rigid articulated chain consisting of set of segments connected by joints. The joints correspond to articulations such as elbow, wrist, sterno-clavicular while the segments correspond to the body limbs such as the upper arms, forearms... Each joint have one to three degrees of freedom (DOF) that represent the rotation angle of the joint relative to its joint parent, the character body root is generally in the pelvis. The root position/orientation and the joint rotation represent character pose configuration. To generate new pose with known goal position, inverse kinematics (IK) is one of the most important used techniques. The inverse kinematic methods tend to compute the joints angles in the aim of moving the end effectors as close to the desired position.

B. Murgante et al. (Eds.): ICCSA 2014, Part VI, LNCS 8584, pp. 681–694, 2014.

1.1   Motivation

The focus of this research is aimed to moving away from recent character pose adjustment using the data-driven inverse kinematics which need an important quantity of motions [1,2], towards an online method that can provide realistic character poses. This method can be used in many field such as in robotic to perform task , in the motion editing field to edit the motion to fit some constraints. The main contribution of this paper is to propose pose adjustment by:

1. Dividing the character body into groups and apply IK to each group
2. Using inverse kinematic method based on FABRIK method, our method differ from the original FABRIK method in the following points :
   - Minimizes the number of moved joints to reach targets,
   - Use the priority to avoid conflicts between tasks,
   - Method to avoid joint-environment object collision.
3. involve correlation of body parts.

The present paper is structured as follow: in the first section, we review related work on the inverse kinematic, we discus IK methods, their advantages and drawbacks. In the second section, we present an overview on the proposed method that allows to adjust the full-body to new constraints. In third section, we present the inverse kinematic used in the adjustment method for one and multiple constraints and to avoid collision. In the fourth section, we present our character model and we introduce the steps of proposed adjustment method. In the last section we start with comparison between our proposed inverse kinematic method and another method that have the same principle (FABRIK), next we present preliminary results of our adjustment method.

2   Related Work

The inverse kinematic problem is well studied. To handle simple kinematic chain that contains a few degrees of freedom, the analytical methods are the best choice[3,4]. These methods are fast but they do not give solution to the complex articulated figures case. Numerical methods based on Jacobian matrix [5,6] are proposed to deal with complex articulated figures. These methods tend to be computationally expensive. Moreover, they suffer from singularities due to the matrix using. To overcome the high computational time, heuristic method are proposed such as CCD (Cyclic Coordinate Descent)[7,8] and triangulation [9]. These methods give solution in low computational time. However they do not support multiple tasks control. Kulpa in [10] tried to give solution to this problem by dividing the whole body into groups and apply CCD to each part separately. "Forward and Backward Inverse Kinematic" (FABRIK) is another heuristic method proposed by Aristidou & Lasenby [11] to the hand modeling and traking, it is based on geometric algebra to resolve the inverse kinematic problem. It's a fast method compared to the other methods [11], it can handle with angular limits. However, it presents conflicts when dealing with multiple tasks. FABRIK is for unconstrained environments, hence it gives good results with collisions. Recently, an inverse kinematic methods that work in the distance space are proposed [6]. A general problem of IK algorithms cited above is the difficulty to ensure the naturalness of the generated pose. This is because natural human motion involves correlation of boy parts. Data driven IK systems have been presented [1,2]those methods are based on training data and create model, they can generate natural pose but poses are highly related to the training data and require huge database.

In the character control, it is desired to simultaneously manage multiple tasks [10,11]. Very often, some tasks cannot be satisfied at the same time, whereas they can be so separately. Some solutions of this conflict problem are based on attributing weight to each task to define their relative importance [12,13] and solve the problem as a multi-objective optimization where no task is achieved exactly but the sum of errors is minimized. Priority based method as in [14] tend to sort the tasks by order of priority, in order to satisfy the most important task first, this method can deal with two tasks in the same time. Well know method is PIK [15] where it can handle with many constraints.

The basic IK methods compute virtual human/robot poses in an obstacle free space. However most environments are not obstacle free. Many studies have been reported on the obstacle avoidance problem in the animation and robotic domain. The Path planning methods are mostly used in the animation [16,17]. However, they are moreover computationally expensive. Aside, the path planning method is not useful for the real time motion control.

3   Overview of the Proposed Method

Figure 1 shows an overview of the proposed method in which the the character is divided into six groups that correspond to the legs, trunk, arms and head. The groups are connected by two joints, root and sterno-clavicular. Creating correlation between groups has the advantage to get more realistic poses. In each frame,

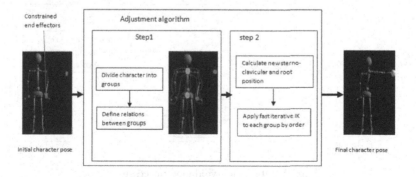

Fig. 1. System overview

our method has as input the character initial pose, the constrained end effectors and their desired positions. In the first step the root and sterno-clavicular positions/orientation are calculated based on the upper and lower body of the constrained end effectors desired positions. in the last step or proposed inverse kinematic is applied in each group based on the root, sterno-clavicular position/orientation and the desired position of the end effectors of each group. Applying the inverse kinematic respect the natural segmental order.

4   Iterative IK

Our inverse kinematic has the same principle as FABRIK [11]. Hence, there is two steps Forward and Backward. New joint position is calculated by keeping constant the length between $joint_i$ and $joint_{i-1}$. Nevertheless, we propose algorithms to handle with environment obstacle and conflict between tasks. The flowing sections represent the proposed idea.

4.1   Obstacle Avoidance

It aim to ensure that the character joints do not collide with static or dynamic objects (obstacles) while it moving to perform task-end effectors desired position. The strategy to avoid collisions must be integrated within the IK algorithm. Our proposition is based on geometric algebra to calculate the joints positions (figure 2). Using this algorithm we keep the rapidity of our IK, which may be useful in the interactive applications and fits with the iterative and local nature of the method used for the inverse kinematic. Our proposed algorithm has two parts: detection and collision response (figure 3).

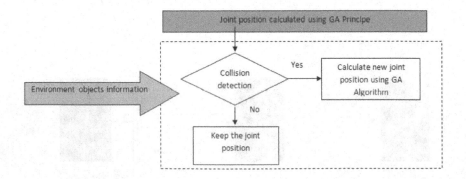

Fig. 2. Obstacle avoidance processes

Collision Detection. In this step we need to detect if the joint collide with the obstacle. Our detection algorithm uses Aligned Bounding Boxes (AABB). It consists to encapsulate the environment objects with simple geometric forms as cubes and spheres, then tests if a specific joint is in intersection with ABB or not. AAB simplifies the resolution of intersection problem of complex forms.

Collision Response. The basic inverse kinematic is position abased local method. The new joint position depend on the previous calculated joint position. Our proposition acts differently: the new joint position does not depend only on the previous calculated joint position, but it depend also on the obstacle position. In the inverse kinematic obstacle avoidance algorithm, we calculate the new $joint_i$ position while keeping the distance D between $joint_i$ and $joint_{i-1}$. We define pos_i as this new position. If the $joint_i$ in $pos_i(x_0, y_0, z_0)$ collide with an object then we calculate new position $pos_i'(x, y, z)$ for this joint, else we keep this position and we pass to the next joint position. pos_i' must respect two distances: D and $dis_{max} + \Delta d$. dis_{max} is the distance between the $joint_i$ and the bounding box edge and Δd represent security value (chosen equal to 0.01). To respect the both distances, we used geometric algebra notions. In this field, to calculate the new joint position where we know the $joint_i$ and $joint_{i-1}$ positions, can be achieved by intersecting two spheres where the first sphere center is $pos_i(x_0, y_0, z_0)$ and its radius is dis $(dis = dis_{max} + \Delta d)$, the second sphere center is $joint_{i-1}$ (x_1, y_1, z_1) and it's radius is D. The two spheres intersection involves to solve the system of equation (1).

$$\begin{aligned} (x - x_0)^2 + (y - y_0)^2 + (z - z_0)^2 - dis^2 \\ (x - x_1)^2 + (y - y_1)^2 + (z - z_1)^2 = D^2 \end{aligned} \qquad (1)$$

The solution of the equation system (1) is a circle. As we need to get one point that will be the new $joint_i$ without collision, we propose to fix the z coordinate, because we have the sum equal to D^2 where $dis^2, (z - z_0) or (z - z_1)$ should not exceeds the radius. We used equation Eq2 to fix z.

$$|z - z_1| < Det |z - z_0| < dis \qquad (2)$$

To find the new position $pos_i'(x, y, z)$ of the $joint_i$ of the actual position $pos_i(x_0, y_0, z_0)$ we must solve the following equation system:

$$\begin{aligned} (x - x_0)^2 + (y - y_0)^2 + (z - z_0)^2 &= dis^2 \\ (x - x_1)^2 + (y - y_1)^2 + (z - z_1)^2 &= D^2 \\ |z - z_1| < D \ et \ |Z - z_0| \qquad &< dis \end{aligned} \qquad (3)$$

The proposed algorithm to find new joint position without collision with an obstacle is summarized in the the following pseudo algorithm:

Obstacle Avoidance Algorithm

```
Input     pos_i(x_0, y_0, z_0 ),  joint_{i-1}(x_1, y_1, z_1)
          Objects list
Output    new joint_i position :  pos'_i(x, y, z)
Check the intersection between joint_i  and the environment object
If there isn't intersection
          Do nothing and keep pos'_i(x, y, z)  as joint_i position
else
          Calculate the distances between joint_i
               and the bounding box corner  (d_1, d_2, ..., d_n)
Calculate  dis_{max} =  max(d_1, ..., d_n)
Calculate z  using equation (3)
Calculate x,y resolving (1) and (2)
```

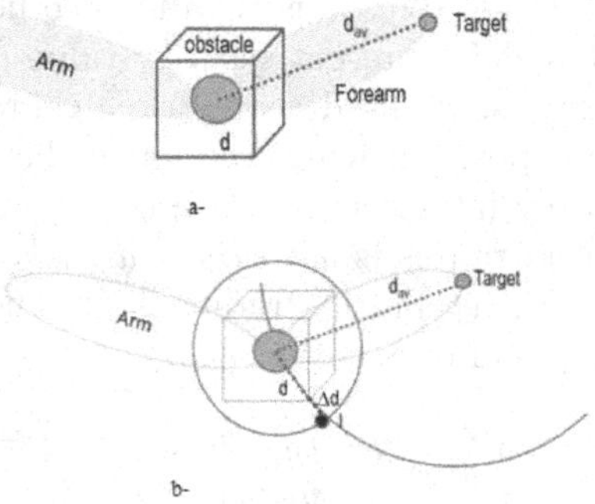

Fig. 3. Obstacle avoidance using geometric algebra

4.2   Multiple Task

It is essential for an IK solver to be able to solve problems with multiple targets. However, without good strategy this may lead to conflict between tasks. In our proposition, the character is subdivided into groups. Those groups are connected between them by sub-base: root and streno-clavicular. Based on the work of Boulic [15] we propose to assign a priority to each task. To get the position of the sub-joints two variants are proposed.

Variant 1. To get the new joints position we apply the forward stage for each group, at the end of this stage, many new sub-base positions are presented as the number of groups. The new sub-base position is calculated using the following equation

$$newpos = \frac{\sum_{i=1}^{i=n} pos_i \times priority_i}{\sum_{i=1}^{i=n} priority_i} \qquad (4)$$

Where Pos_i is the sub-base position for the group i and $Priority_i$ is the task priority of the group i. Using the priority notation, the task that is more important will be achieved before the lowest ones.

Variant 2. In this variant, we sort the tasks by the order of priority. We obtained numerous simple chains, then we apply the previous algorithm for each simple chain, starting by the low priority chain to the high one. Starting by the task with low priority ensure that the task that have the highest priority will be reached much as possible.

We propose two variants, because in some cases we have to verify the height priority task (foot on the ground) and in some other cases it is preferable to reach the highest priority task more then the lower one. We use those variants to calculate the new sub-joints position.

5   Adjustment Using Inverse Kinematic

5.1   Character Model

The structure of our character model consists of rigid kinematic chain (cylinders) connected by angular joints (figure 4).

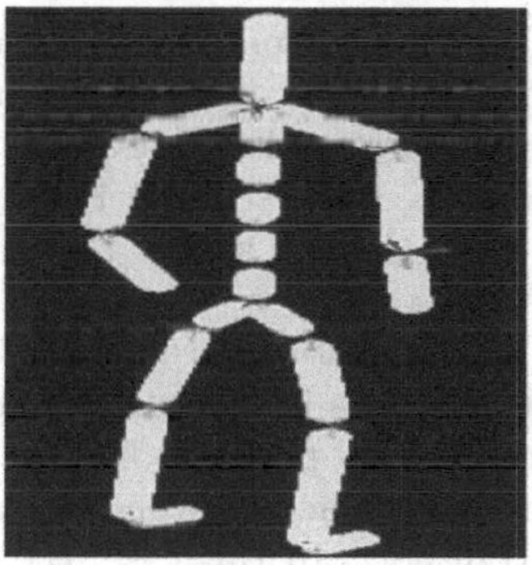

Joint	DOF
Root	6
humeral	2
Stereno-clacvicul	2
elbow	2
wrist	2
hip	3
knee	2
ankle	2

Fig. 4. Character model

As in the work of Shin [18] where the character is subdivided into groups and analytic inverse kinematic is applied to each group, we subdivided the character into six main groups then we apply our inverse kinematic to each group separately. This subdividing makes our character inverse kinematic very fast due to applying few joints; we apply just the joints of the group concerned by the motion. Nevertheless, setting each group separately may yield unrealistic poses. To avoid this problem, the groups are connected by joints "sub-bases".

5.2   Full Body Adjustment

The main idea of our adjustment approach is that the adjustment is solved sequentially using simple iterative IK based on geometric algebra on different parts of the body and in specific order. No motion database is necessary. Based on the desired end-effectors, the root and sterno-clavicular configurations are calculated. The trunk joints orientation are calculated based on the new root and sterno-clavicular configurations . Finally, each of the limbs configurations is determined based on the new clavicle position for the arms, the new root position for the legs and the end-effectors positions with an iterative IK. Algorithm 1 summarizes the proposed method.

Algorithm 1

```
input : starting character pose,  desired end-effectors position
output : new character pose
1-Calculate the new root, sterno-clavicular  position based on
   the desired end-effectors positions
2-Trunk adjustment,  using new sterno-clavicular position
3-Arms adjustment
4-Legs adjustment
end
```

Calculate the New Root, Sterno-Clavicular Position. To reach an object, in real human motion it is preferable to solicit the trunk only when it is impossible to achieve the target by moving just the arm. To achieve the natural pose, we inspired our proposition from the real life. We first verify the contribution of the trunk and the root in the arm or leg motion to satisfy the constraint, and then calculate the new trunk configuration.

To test the contribution of the trunk in the motion, we calculate the distance D between the sterno-clavicular and the target if its lower than the arm length,the sterno-clavicular and the root keep their original positions 5.

In the opposite case, where the target is placed beyond arms length, the trunk chain is integrated in the arm transport. In this case, the new stereno-clavicular position/orientation Pos' in order to get distance between target and this position equal to the arm length.

In some cases we cannot move the trunk in order to be the sterno-clavicular in pos' (the target is very far) one solution is to move the pelvis by error distance to be in pos_{pelvis}. The error distance in the pelvis position is calculated using the distance between the actual position of the pelvis and the actual stereno-clavicular position, and the distance between the actual stereno-clavicular position and Pos'.

In the case where there are two tasks to be achieved, user can choose between the two variants in order to calculate the sub-joint positions. In the case where both arms are constrained the new trunk configuration is calculated considering the sterno-clavicular joint as the sub joint between the two arms. The new sterno clavicular position is calculated by choosing one variant from the variants described in section 4.2. If there is no variant chosen we calculate the new sterno-clavicular position using equation (1). In the default case to calculate the root position where there are constraints on the upper and lower body the root position is calculated using variant two where the lower body constraint have the highest priority.

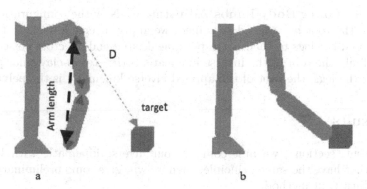

Fig. 5. Reach target in simple chain. Figure a presents the initial arm configuration where we calculate the arm length and the distance target-end effectors, in figure b we move just the arm because the arm lenghth is lower then the distance between the target and the end-effectors

Root Adjustment. Based on the stereno-clavicular and Ankles desired position we determine the new root position. The ankles positions are given high priority. The new root position is calculated using variant two in the sub-section 4.2. we choose this variant because the new root position must verify strictly the ankles position.

Trunk Adjustment. For the trunk adjustment, our inverse kinematic is applied on the trunk chain. The trunk is is constructed from five joints represent the most important spines. The root of the trunk chain is the character body root and the end effectors are the sterno-clavicular. The aim of this step is to find the new joints trunk position/orientation based on the stereno-clavicular configuration.

Algortihm 3

```
Input :target position,  initial  trunk configuration
Output : trunk configuration
1- D1=|sterno-clavicular position-wrist position|
2-D=|sterno-clavicular position-object position|
3-If (D1<D) new stereno-clavicul position is calculated
          Apply IK to find the new trunk configuration
End
```

In the case where both arms are constrained, the new trunk configuration is calculated considering the sterno clavicular joint as the sub joint between the two arms. The new sterno clavicular position is calculated by choosing one variant from the variants described in section 4.2. If there is no variant chosen we calculate the new sterno-clavicular position using equation (1). After getting the new position we apply IK to the trunk chain to get the new trunk orientation.

Upper and Lower Body Limbs Adjustment. Now the configuration of the trunk and the root has been determined, we apply inverse kinematic to adjust the arms and the legs to their corresponding desired end effectors positions. For the arm limb, the root of the inverse kinematic is the sterno-clavicular position. While for the legs, the root of the applied inverse kinematic is the pelvis.

6   Results

In the result section , we first compare our inverse kinematic with FABRIK method that have the same principle, then we will give some preliminary results of our adjustment method.

6.1   Comparison of Our IK to FABRIK

Priority Inverse Kinematic. The figure Figure 6 represent three chains connected by sub-base with two end effectors and two desired positions. In the character case, those chains represent arms and trunk. This figure aim to represent how deal our inverse in the presence of two target in the same time compared to FABRIK. Figure 6.a represents the result when applying FABRIK to the chains, as shown there is conflict between tasks. When we apply our inverse kinematic to the chains in 6.b the height priority target is achieved without conflict.

The Variant1 result depends on the priority value of each task, thus the choice of the priority value is very important, a wrong choice may achieve the high priority task but in high computational time. While in the second variant the result depends only on the high priority task. As conclusion, variant2 is used when the high priority task must be completely achieved.

Obstacle Avoidance. In this section and to prove the efficacy of the obstacle avoidance algorithm , figure 7-a represents chain with five joints that represent an arm , the arm end-effectors have to reach the target (red object) while there are obstacles in the environment (blue and green objects). When we apply FABRIK to reach the target the joints arm collide with the other objects in the environment Figure 7-b. But when applying our inverse kinematic with the obstacle avoidance test, figure 7-c represents the result: the end-effectors reach the target without collision with the obstacles.

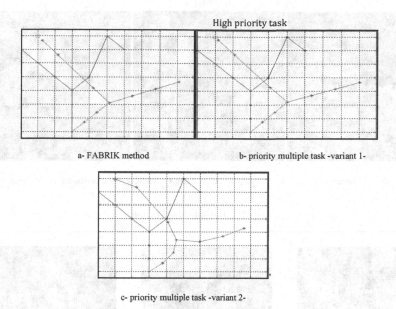

High priority task

a- FABRIK method   b- priority multiple task -variant 1-

c- priority multiple task -variant 2-

Fig. 6. Multiple tasks case

6.2   Preliminary Adjustment Method Results

Reaching Objects. In the first two result subsections, we tested the reliability of our inverse kinematic in independent chains. Finally, in this section we test the adjustment of the full body method. Figure 8.a represents the character initial pose, the task is to reach the red object. Applying the algorithms 1 and 2 we reconstruct the full body pose. In the first case 8.b, the target is close to the character so the trunk keep its initial configuration and just the arm changes its configuration. This is natural pose because in real life when the target is close we use just the arm to reach it. We made another test where the target (8.c) is far from the character, in this case the trunk moves to help the arm to reach the task .

Multiple Constraints. To discuss the performance of our method, in figure 9, we compare the adjustment method results to a natural captured pose. In figures 9 (b,d) the right character represents natural captured pose while the left character represents the result pose of our adjustment method. The initial pose of our characters are presented in figure 9 (a,c), the desired position of each end effector is represented by sphere. Taking the example of figure 9 (c) representing initial pose of our character (left one), we constrain four end effectors (upper and lower body) where the red spheres represent the desired position of the character foot and green and blue ones represent the desired position of each hand. In the natural case, the result must be like the left character. Applying the adjustment method (algorithms 1 and 2), we reconstruct the character pose to

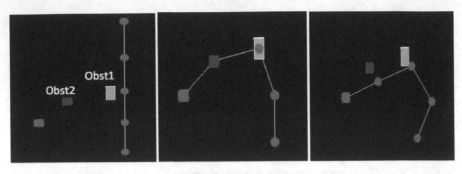

a- initial configuration   b- final configuration using FABRIK c- final configuration using our method

Fig. 7. obstacle avoidance

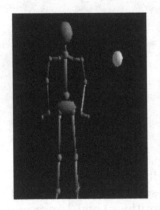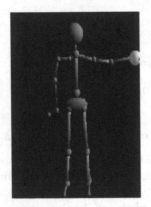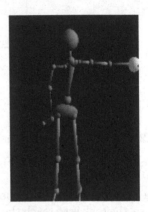

a- initial configuration   b- reach target just by moving arm   c- including the trunk to reach target

Fig. 8. Reaching objects

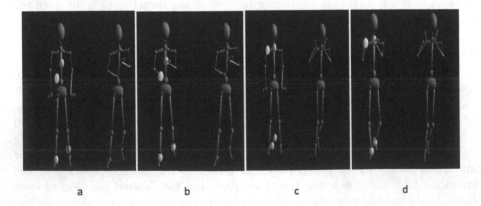

a   b   c   d

Fig. 9. Multiple constraints

fit the constraints, the result is shown in figure 9(d) of left character. Comparing the left character pose that represents our result to the right character pose that represents the natural captured pose , we find that our result is very close to the natural pose.

7 Conclusion

In this paper, we propose a new fast method for pose adjustment to new constraints. This method can be used in the motion editing to edit pose to fit new constraints, in the motion retargeting to adapt motion to new character in fast way. The principal idea of the method is to divide the character body into groups connected between them with sub-joint, to keep the correlation between body parts and apply an inverse kinematic to each group in a specific order. The used inverse kinematic method is based on FABRIK. To improve the reliability of our method, based on inverse kinematic, we improved first the reliability of the used inverse kinematic. Our inverse kinematic controls multiple end-effectors without conflict by assigning priority to each task. The proposed inverse kinematic provides result without collision in the case of the presence of obstacles in the environment. We compared the results of our inverse kinematic to the result of FABRIK. After improving the IK reliability, we tested the full body character to new constraints using proposed algorithms. Dividing the character body into groups has the advantage to make easy the computation of a solution. Selecting the root and stereno-clavicular as connection between the body part and applying the inverse kinematic in specific order allows to respect the segmental order and gives more natural pose. As a future work, we plan to maintain the pose balancing to get realistic poses by combining the inverse kinematic and the kinetic in the same algorithm and test the adjustment algorithm in specific problems cases as in motion editing.

8 Thanks

This work is under the Hubert Curien Partnership (PHC) of the Tassili scientific cooperation and is funded by the French Ministry of Foreign Affairs (MFA) and the Algerian Ministry of Higher Education and Scientific Research. We thank them all for their help.

References

1. Grochow, K., Martin, S.L., Hertzmann, A., Popović, Z.: Style-based inverse kinematics. ACM Transactions on Graphics (TOG) 23, 522–531 (2004)
2. Shin, H.J., Lee, J.: Motion synthesis and editing in low-dimensional spaces. Computer Animation and Virtual Worlds 17(3-4), 219–227 (2006)
3. Craig, J.J.: Introduction to robotics, vol. 7. Addison-Wesley, Reading (1989)
4. Tolani, D., Goswami, A., Badler, N.I.: Real-time inverse kinematics techniques for anthropomorphic limbs. Graphical Models 62(5), 353–388 (2000)

5. Buss, S.R.: Introduction to inverse kinematics with jacobian transpose, pseudoinverse and damped least squares methods. IEEE Journal of Robotics and Automation 17 (2004)
6. Kenwright, B.: Responsive biped character stepping: When push comes to shove. In: 2012 International Conference on Cyberworlds (CW), pp. 151–156. IEEE (2012)
7. Welman, C.: Inverse kinematics and geometric constraints for articulated figure manipulation. PhD thesis, Simon Fraser University (1993)
8. Wang, L.C., Chen, C.C.: A combined optimization method for solving the inverse kinematics problems of mechanical manipulators. IEEE Transactions on Robotics and Automation 7(4), 489–499 (1991)
9. Muller-Cajar, R., Mukundan, R.: Triangualation-a new algorithm for inverse kinematics (2007)
10. Kulpa, R., Multon, F., et al.: Fast inverse kinematics and kinetics solver for humanlike figures. In: Humanoids, pp. 38–43 (2005)
11. Aristidou, A., Lasenby, J.: Fabrik: a fast, iterative solver for the inverse kinematics problem. Graphical Models 73(5), 243–260 (2011)
12. Badler, N.I., Manoochehri, K.H., Walters, G.: Articulated figure positioning by multiple constraints. IEEE Computer Graphics and Applications 7(6), 28–38 (1987)
13. Zhao, J., Badler, N.I.: Inverse kinematics positioning using nonlinear programming for highly articulated figures. ACM Transactions on Graphics (TOG) 13(4), 313–336 (1994)
14. Siciliano, B., Slotine, J.J.: A general framework for managing multiple tasks in highly redundant robotic systems. In: Fifth International Conference on Advanced Robotics, Robots in Unstructured Environments, ICAR 1991, pp. 1211–1216. IEEE (1991)
15. Baerlocher, P., Boulic, R.: An inverse kinematics architecture enforcing an arbitrary number of strict priority levels. The Visual Computer 20(6), 402–417 (2004)
16. Kallman, M., Mataric, M.: Motion planning using dynamic roadmaps. In: Proceedings of the 2004 IEEE International Conference on Robotics and Automation, ICRA 2004, vol. 5, pp. 4399–4404. IEEE (2004)
17. Kallmann, M.: Scalable solutions for interactive virtual humans that can manipulate objects. In: AIIDE, pp. 69–75 (2005)
18. Shin, H.J., Lee, J., Shin, S.Y., Gleicher, M.: Computer puppetry: An importance-based approach. ACM Transactions on Graphics (TOG) 20(2), 67–94 (2001)

Privacy-Preserving Kriging Interpolation on Distributed Data

Bulent Tugrul[1] and Huseyin Polat[2]

[1] Computer Engineering Department, Ankara University, 06100 Ankara, Turkey
btugrul@eng.ankara.edu.tr
[2] Computer Engineering Department, Anadolu University, 26470 Eskisehir, Turkey
polath@anadolu.edu.tr

Abstract. Kriging is one of the most preferred geostatistical methods in many engineering fields. Basically, it creates a model using statistical properties of all measured points in the region, where a prediction value is sought. The accuracy of the kriging model depends on the total number of measured points. Acquiring sufficient number of measurement requires so much time and budget. In some scenarios, private or governmental institutions may collect geostatistical data for the same or neighbor region. Collaboration of such organizations may build better models, if they join their data sets. However, due to financial and privacy reasons, they might hesitate to collaborate.

In this study, we propose a solution to build kriging model using distributed data while preserving privacy of each data owners and the client that requests prediction. The proposed scheme creates a kriging model on joint data of all parties who wants to collaborate. We analyze our solution with respect to privacy, performance, and accuracy. Our solution has extra costs; however, they are not that critical. We conduct experiments on real data sets to show that our scheme gives better result than the model created on insufficient measured data.

Keywords: Privacy, kriging, distributed data, prediction, geostatistics.

1 Introduction

Geostatistical modeling is widely used for interpolation. In addition to interpolation, such modeling can also be used for spatial data analysis, variograms modeling, exploratory variography, and so on. Kriging is a widely used geostatistical modeling in many applications [1]. Kriging, as described below, has two main steps [2]. First step finds a variogram model, which represents the statistical properties of the region in which prediction requested. For this purpose, distances between all points in the data set and semi-variance values are calculated. The second step finds the weights for each point used for prediction calculation. The concept of kriging was first studied by Krige [3]. In traditional kriging method, there are two parties. One party, which holds all coordinates and measurement data, is called server (S). The other party, which needs a prediction for a specific location, is called client (C).

B. Murgante et al. (Eds.): ICCSA 2014, Part VI, LNCS 8584, pp. 695–708, 2014.

In traditional kriging process, there is no privacy. Data collected for geostatistics purpose are assumed as private data. The server spends so much time and budget to obtain such valuable information. The economic future of server depends on keeping such data secret. In addition to this, the client does not want to reveal the location for which it will make investment in the future. Third parties may be interested in such valuable information to gain advantage.

As mentioned above, kriging uses all coordinates and measurements to figure out the statistical properties of the region. Therefore, if the server does not have sufficient data, the accuracy of the kriging model may be questionable. In some situations, it is not possible to obtain sufficient data due to time and budget constraints. However, multiple servers may collect geostatistics data for the same or neighbor region. If a model can be created on joint data of such parties, the accuracy and reliability of the model will be better. However, the data held by each server and the prediction coordinate of the client are accepted as private data of each party. Therefore, they may hesitate to join such collaboration. Privacy-preserving solutions encourage them to integrate their data to produce better results.

We propose a method in order to perform kriging interpolation on distributed data between multiple parties while preserving their confidentiality and the clients' privacy. We focus on ordinary kriging. However, our proposed solution can be easily extended to simple kriging. Our scheme makes it possible to offer accurate predictions on distributed data-based kriging efficiently without jeopardizing data holders' privacy. Our analysis show that the proposed method is able to preserve privacy and estimate predictions with decent accuracy. Although it causes some additional costs, they are not critical for the overall success.

The paper is organized as follows. Next section is devoted for related studies in the literature. Third section explains how to produce a kriging prediction in brief. Section 4 explains details of our proposed solution. Performance, privacy, and accuracy analysis of the proposed scheme appears in Section 5. The results of experiments on real data are presented in the same section. We finally discuss conclusions and future works.

2   Related Work

Applying various data mining methods while preserving privacy has become a hot topic for researchers. Clifton et. al [4] propose a toolkit to achieve privacy for various data mining tasks. Xu and Yi [5] classify privacy-preserving distributed data mining protocols. They put each protocol into four different categories like data partitioning model, data mining algorithms, secure communication model, and privacy preservation techniques. Lu et al. [6] propose a secure framework for collaborative data mining tasks for cloud computing model with distributed data. A new approach for privacy-preserving distributed data mining tasks without using secure computation or perturbation methods is studied by Gurevich and Gudes [7]. Emekci et al. [8] build a privacy-preserving decision tree over distributed databases.

Data collected for data mining methods may be stored by various databases or companies. As in geostatistics case, data is a valuable asset of companies. They do not want to share their knowledge with other companies. However, insufficient data may lead to inaccurate data mining models. Therefore, collaboration of companies to apply data mining methods on their joint data has very high importance. In the literature, there are various data partitioning schemes including horizontal, vertical, and arbitrary partitioning. Kaleli and Polat [9] propose a privacy-preserving method to provide recommendations on horizontally distributed data. An efficient solution for privacy-preserving distributed clustering protocol for horizontally partitioned data is proposed by Kaya et al. [10]. An efficient solution for k-NN queries for vertically partitioned data is proposed by Amirbekyan and Estivill-Castro [11]. Yang and Wright [12] study how to compute a privacy-preserving Bayesian network on databases vertically partitioned among two parties. Yakut and Polat [13] propose privacy-preserving recommender system for arbitrary distributed data. A solution of how to construct a privacy-preserving back-propagation algorithm over arbitrarily partitioned databases is presented by Bansal et al. [14]. Secure two party computation methods allow computing output of a function f with private inputs of two parties while preserving privacy. Yao [15] defines secure two party computations. Goldreich et al. [16] propose algorithms for multi-party computations. Zhan et al. [17] develop a secure association rule mining protocol based on homomorphic encryption (HE) on joint data of multiple parties without revealing their secret data each other.

One of the most widely applied protocols in privacy-preserving data mining (PPDM) is oblivious transfer (OT) protocol. In this protocol, there are two parties, sender and receiver. Sender sends private information to the receiver and receiver computes the outputs of all data sent by sender, but sender gets only one of the outputs. An adaptive OT protocol is proposed by Green and Hohenberger [18]. They claim that their solution satisfies all necessary requirements of OT protocol. Camenisch et al. [19] propose a secure solution to access databases or file system in multi-user environment using OT. Corniaux and Ghodosi [20] claim that OT protocols are based on a semi honest model; therefore, there can be a malicious server that can disrupt the protocol. HE is also used in PPDM solutions. HE is based on public key encryption. In addition to this, it has another property that allows calculating addition and multiplication on cipher-text. Pailier HE system system is one of the well-known and applied HE schemes [21]. Guajardo et al. [22] propose a protocol, where they claim that their protocol computes mean, median, and variance efficiently for multi-party computation using Pailier HE system. Agrawal and Srikant [23] use randomization method to achieve privacy. Basically, randomization adds noise to the original data to perturb its true value. Thus, it protects privacy of data owner. A random number value, chosen from a uniform or Gaussian distribution with zero mean and standard deviation, can be added true value stored in the data set. Although there are various methods that are utilized to preserve confidentiality, randomization is widely used in PPDM for achieving data privacy.

Tugrul and Polat [24] study how to estimate kriging-based predictions on central data with privacy. In that study, the authors assume that data held by a single party. They propose a scheme to offer predictions using kriging without violating privacy. In another study proposed by Tugrul and Polat [25], the authors propose a privacy-preserving scheme in order to estimate inverse distance weighted-based interpolations on central data while preserving confidentiality. Again, they assume that data held by a single party. They utilize inverse distance weighted interpolation method to produce predictions with privacy. Both studies consider central data. However, our work here considers distributed data between more than two companies. Tugrul and Polat [26] propose different schemes in order to provide kriging-based predictions on partitioned data between two parties only while preserving privacy. They assume that data are partitioned between two servers only while our work here considers that data are distributed between multiple parties or more than two companies.

3   Background: Kriging Interpolation

In this section, we briefly explain kriging. Producing a prediction using kriging method requires creating a variogram model on all coordinates and measurements held by the server S and calculating the prediction value for the location, where the client C needs a prediction value. The steps of kriging are as follows:

1. S calculates all distances for any two locations i and j using the Euclidean distance formula. The distance (d_{ij}) between i and j is found as follows, where x and y values show coordinates:

$$d_{ij} = \sqrt{(x_i - x_j)^2 + (y_i - y_j)^2}. \tag{1}$$

2. Similarly, S calculates semi-variances (s_{ij} values) for any two locations i and j as follows, where P_i and P_j values show the related measurements:

$$s_{ij} = 0.5 \times [P_i - P_j]^2. \tag{2}$$

3. Then, S computes average semi-variances and distance for a specific bin.
4. S plots the average semi-variances versus the average distances and finds a formula to estimate the semi-variance at any given distance. Semi-variances can be denoted as follows:

$$Semi - variance = f(distance), \tag{3}$$

where function f shows the relationship between semi-variances and distances.

5. Γ matrix, which is a $(G + 1) \times (G + 1)$, is created by S. Both last row and column of Γ matrix are filled with 1s, except the diagonal entry, which is set to 0.
6. Next, S computes the Γ^{-1} matrix.
7. C sends the prediction coordinate q (x_q, y_q) to S.

8. S computes the distances between q and each measured location using Eq. (1). It then creates the matrix g, which is a $(G+1) \times 1$ matrix.
9. S then solves the equation given below and finds weights (λ matrix) as follows:

$$\lambda = \Gamma^{-1} * g. \tag{4}$$

10. Finally, S produces the prediction value for the location (referred to as P_q) by multiplying the weight for each measured location and the related measurement value. If we consider λ and P as vectors of length $G+1$, then P_q can be estimated by finding the scalar product of λ and P as follows:

$$P_q = \lambda \cdot P = \sum_{i=1}^{G+1} \lambda_i P_i. \tag{5}$$

4   Private Kriging on Distributed Data

In a traditional kriging interpolation, there is one server that holds coordinate and measurement information. The client asks a prediction for a specific location and the server does all the required calculations and returns an estimated value as a prediction. In such scheme, there is no privacy. The client does not will to disclose the coordinate, where it is interested in. In addition, in some region, there might be more than two servers that collect data for kriging purposes. As mentioned before, accuracy of geostatistical methods depends on the number of sample points. Therefore, the servers may want to collaborate to provide better services. This scheme is called distributed data-based method (PKDD), where there are m servers $S_1, S_2, \ldots, S_m$ hold private data. The servers do not want to share their private data in order to survive in the future. The proposed solution gives servers an opportunity to come together and create more accurate geostatistical models.

In the following section, the steps of the protocol are described. The computations can be grouped as off-line and on-line computations, where on-line performance is not that critical.

Off-line Phase: Calculations like distance, semi-variance, binning, and creating kriging model are performed in this phase.

A. Distance and Semi-variance Calculation: In order to create a kriging model, all distance and semi-variance values for locations, where the servers have measurements have to be calculated. In the proposed protocol, there is a total number of $G(\sum_{i=1}^{m} G_{Si})$ sample points in region **A**. The servers $S_1, S_2, \ldots, S_m$ hold $G_{S_1}, G_{S_2}, \ldots, G_{S_m}$ sample points, respectively. There are two cases like points i and j are held by the same server and points i and j are held by two different servers.

Case I: Any two points i and j are held by same server: The servers do not need to collaborate with others to calculate distance and semi-variance values for points, which are stored in its database. Therefore, each server calculates distance and semi-variance values using Eq. (1) and Eq. (2).

1. Each server calculates distances between points i and j, where $i = 1, 2, \ldots,$ $G_{S_k} - 1$, $j = i + 1, i + 2, \ldots, G_{S_k}$ and $k = 1, 2, \ldots, m$.
2. Each server calculates semi-variance values for points i and j, where $i = 1, 2, \ldots, G_{S_k} - 1$, $j = i + 1, i + 2, \ldots, G_{S_k}$ and $k = 1, 2, \ldots, m$.

Case II: Each of any two locations i and j is held by different server: The servers need to collaborate with the other servers to find distance and semi-variance values for points, which are held by different servers. The distance equation, Eq. (1), can be expanded as follows:

$$d_{ij} = \sqrt{x_i^2 + y_i^2 + x_j^2 + y_j^2 + (-2)x_i x_j + (-2)y_i y_j}. \tag{6}$$

The server, which possesses point (x_i, y_i), calculates the values x_i^2 and y_i^2 by itself. The corresponding server, which holds point (x_j, y_j), calculates x_j^2 and y_j^2. However, they need to collaborate to calculate values $(-2)x_i x_j$ and $(-2)y_i y_j$. Similarly, they can calculate semi-variance value. The Eq. (2) can be expanded as follows:

$$s_{ij} = \frac{P_i^2}{2} + \frac{P_j^2}{2} + (-1)P_i P_j. \tag{7}$$

The server, which possesses point (x_i, y_i), calculates value $P_i^2/2$ by itself. The corresponding server, which holds point (x_j, y_j), calculates $P_j^2/2$. However, they need to collaborate to calculate the values $(-1)P_i P_j$.

The steps of how to calculate distance and semi-variance values are explained in terms of S_k. The corresponding server will be called as S_m ($m = 1, 2, \ldots, k - 1, k + 1, \ldots, m$).

1. S_k calculates the values $\xi_{K_{Sk}}(-2x_i), \xi_{K_{Sk}}(-2y_i)$, and $\xi_{K_{Sk}}(-P_i)$ using an HE method and its public key K_{Sk}. S_k sends all encrypted values to S_m.
2. S_m masks its coordinates using randomization technique to prevent S_k from learning S_k's private data. To do so, S_m creates two random arrays (r and v), which has uniform distribution with zero mean and σ. In each array, there are G_{S_m} numbers of random numbers. Then, S_m adds random numbers to its coordinates to find $x_j + r_j$ and $y_j + v_j$. After calculating perturbed coordinates, S_m finds $dp'_{ij} = \xi_{K_{Sk}} \left[(x_j^2 + y_j^2) - 2x_i x_j - 2y_i y_j - 2x_i r_j - 2y_i v_j \right]$ for $j = i + 1, i + 2, \ldots, G_{Sm}$.
3. Similarly, S_m produces a random array (u) using uniform distribution with zero mean and σ. To hide its measurement, S_m adds these random numbers. S_m finds $\xi_{K_{Sk}}((sp_{ij})) = \xi_{K_{Sk}}[(P_j^2)/2 + (-P_i)P_j + z_{ij}] = \xi_{K_{Sk}}(sp_{ij} + z_{ij})$ and sends to S_k.

4. S_k decrypt $\xi_{K_{Sk}}(dp_{ij} + u_{ij})$ and $\xi_{K_{Sk}}(sp_{ij} + z_{ij})$ values using its private key and gets $dp_{ij} + u_{ij}$ and $sp_{ij} + z_{ij}$.
5. S_k has all necessary values to compute distances and semi-variance value for points, which are held by S_m.
6. S_k repeats these process described above with other servers.
7. The servers change their roles and all servers get distance and semi-variance values for all points in the region $\mathbf{A}$ to create a kriging model.

B. Binning and Model Creation: After obtaining all required distance and semi-variance values to create a kriging model, the servers execute the following steps:

1. They agree on the binning methodology.
2. Next, they calculate average distance and semi-variance values for each bin.
3. Each server then plots average distance and semi-variance using one of the functions used in kriging models. They come up with a formula to describe the relationship between distance and semi-variance. Due to random data, which are used to hide coordinate and measurement values, each server might end up with a slightly different formula. $Semi - variance_{S_1} = f_{S_1}(distance), Semi-variance_{S_2} = f_{S_2}(distance), \ldots, Semi-variance_{S_m} = f_{S_m}(distance)$, respectively.
4. Each server calculates Γ matrices using the semi-variance formula and distance values for all points $(G = \sum_{i=1}^{m} G_{Si})$. Γ is a $(G+1) \times (G+1)$ symmetric matrix. The last row and column are filled with 1s except the diagonal element. The value of diagonal element is 0.
5. They finally find Γ^{-1} matrices to use in estimation phase, as explained below.

Online Phase: In this phase, the client C sends the coordinate for which it needs a prediction to the master server (MS). It is assumed that S_1 acts as MS, but any server may act as MS. The coordinate value is considered as private data of the client. The servers, which have measurement values in the same region, calculate kriging weights and produce the final estimation collaboratively.

A. Estimating Weights: The servers and the client C calculates weights, as described below:

1. The C utilizes OT to hides its true location q. Therefore, it produces $n - 1$ bogus coordinate values and stores true location in $n - 1$ bogus coordinate. Then, it permutes n locations and sends them to the MS.
2. The MS forwards all n coordinate values to other servers.
3. For each location $z = 1, 2, \ldots, n$, the servers perform the followings:
   (a) Each server finds distances between z and each coordinate in their database using Eq. (1).
   (b) The servers $S_1, S_2, \ldots, S_m$ encrypt each distance using an HE scheme with their public key K_{S_i} and send them to other servers.

(c) The *MS* finds the related semi-variance values for distances received from servers $S_2, S_3, \ldots, S_m$ utilizing $Semi-variance_{S_1} = f_{S_1}(distance)$ in the encrypted form using an HE scheme.

(d) The *MS* then creates the matrix g, which is a $(G+1) \times 1$ matrix including the semi variances in encrypted form estimated between z and each measured location.

(e) Since the *MS* has the matrix $\Gamma_{S_1}^{-1}$ and the matrix g (including encrypted values), it can estimate the kriging weights (λ matrix) by employing an HE scheme. Notice that λ is a $(G+1) \times 1$ matrix including encrypted weights, which are encrypted with public keys of the servers $K_{S_2}, K_{S_3}, \ldots, K_{S_m}$.

B. Prediction Estimation: For each location $z = 1, 2, \ldots, n$, the servers perform the followings:

1. The *MS* sends corresponding weights, which are encrypted using public keys to servers $S_2, S_3, \ldots, S_m$. The servers decrypt encrypted values using their private key and gets $\lambda_{S_{mj}}$.

2. They perform scalar dot products to calculate partial aggregate of the prediction for location z (referred to as $PP_{S_{mz}}$). $PP_{S_{mz}}$ can be calculated as follows:

$$PP_{S_{mz}} = \lambda_{S_m} \cdot P_{S_m} = \sum_{j=1}^{G_{S_m}} \lambda_{S_{mj}} \times P_{S_{mj}}. \tag{8}$$

3. Then, they compute $\xi_{KC}(P_{S_{mz}})$ using client's public key KC and send to the *MS*.

4. The *MS* similarly calculates partial aggregate of the prediction for location z (referred to as $PP_{S_{1z}}$) in encrypted form by performing a scalar dot product using an HE scheme with client's public key KC. $PP_{S_{1z}}$ can be calculated as follows:

$$\xi_{KC}(PP_{S_{1z}}) = \xi_{KC}(\lambda_{S_1} \cdot P_{S_1}) = \xi_{KC}\left[\sum_{j=1}^{G_{S1}} \lambda_{S_{1j}} \times P_{S_{1j}}\right]. \tag{9}$$

5. The *MS* computes $\xi_{KC}(PP_z) = \xi_{KC}(PP_{S_{1z}}) + \xi_{KC}(PP_{S_{2z}}) + \ldots + \xi_{KC}(PP_{S_{mz}})$ when it gets all prediction values from all servers.

6. The *C* gets the final prediction value for its real location q using OT. It decrypts using its private key and obtains prediction P_q.

5   Performance and Privacy Analysis

In this section, PKDD scheme is analyzed with respect to performance and privacy. PKDD scheme brings extra storage, communication, and computation cost to assure privacy between collaborating parties. These costs can be divided into two parts: off-line and online. In real life scenarios, off-line costs are less crucial than online costs. Moreover, these online constraints in kriging are not too rigid as compared to other online protocols such as recommender systems. In the last subsection, PKDD is analyzed in terms of privacy.

5.1   Storage Cost Analysis

PKDD scheme uses randomization, OT, and encryption to provide privacy. Randomization method increases storage cost. Parties use random numbers to disguise coordinate and measurement values. Each party creates three arrays storing random numbers for x_i, y_i, and P_i. Each array has a length of G_{Si}, where $i = 1, 2, \ldots, m$. In total, randomization requires $3 \times \sum_{i=1}^{m} G_{Si} = 3 \times G$ extra storage area. In other words, storage costs for random numbers are in the order of $O(G)$. In addition to randomization, parties need to store kriging model parameters. For Γ matrix, they need $m \times (G + 1) \times (G + 1)$ and for λ vector, they need $m \times (G + 1) \times 1$ storage area. In total, extra storage requirement is $3 \times G + m \times (G + 1) \times (G + 1) + m \times (G + 1) \times 1$, which is in order of $O(m \times G^2)$, where m is much smaller than G.

5.2   Communication Cost Analysis

In real life applications, kriging requires two communications between client and server. The client sends the coordinate, where it needs a prediction and the server does all the required calculations and sends back a prediction value to the client. If privacy is necessary between collaborating parties, the servers are supposed to establish extra communications. During off-line phase, they have to establish $2 \times (m - 1) \times \sum_{i=1}^{m} G_{Si} = 2 \times (m - 1) \times G$, which is in order of $O(mG)$ communications between each other. In online phase, the client performs OT, which can be conducted in poly-logarithmic time (in n). Therefore, the servers have to make $m \times n + 2n \times (m - 1) \times G + n \times G$, which is in order of $O(mnG)$ communications. In total, number of communications in PKDD are in the order of $O(mnG)$. G is always much bigger than both m and n. Therefore, it requires $O(G)$ extra communications.

5.3   Computation Cost Analysis

The proposed scheme has two phases: off-line and online. Accordingly, it is scrutinized with respect to two phases. The costs of some operations such as random number generation, subtractions, and additions are negligible. In the off-line phase, order of $O(G_{S_1}, G_{S_2}, \ldots, G_{S_m})$ encryption should be conducted. In the similar manner, the decryption is in the order of $O(G_{S_1}, G_{S_2}, \ldots, G_{S_m})$. Moreover, additional computations occur due to model creation for each server. If privacy is not a concern, a model creation is adequate to provide a prediction; however, the proposed method requires m different model creations for m parties. Thus, computation costs increase by n times. The online phase demands order of $O(nG^2)$ encryptions. Likewise, number of decryptions is in the order of $O(nG^2)$. Additionally, number of multiplications increases by $O(n)$ times because predictions are estimated for n bogus locations rather than just one. In the final step of the protocol, an OT is conducted. However, computation cost of OT can be omitted.

5.4   Privacy Analysis

The proposed scheme should verify that the private data of both client and the servers should be kept secret. Locations and corresponding measurements held by the servers and the coordinate for which the client asks prediction and result of interpolation method are assumed as confidential data. It is also assumed that the servers and the client are semi-honest. They follow the protocol steps; however, they try to learn as much information as possible about each other's private data.

The client C composes $n - 1$ bogus coordinates and hides its true location among them and uses OT. The probability of guessing its true coordinate is 1 out of n. If n is chosen a bigger value, it increases privacy, but causes more computation cost. A reliable n value can be chosen to provide adequate privacy. The final result is also private data of the client. Therefore, the servers should not have an idea of such value. Since encryption enforces private key must be known by the client only, the servers cannot decrypt the cipher-text, which is encrypted by client's public key. Hence, the servers cannot learn the final prediction value. The servers should hide their coordinates and measurements from other servers and the client. The client gets an aggregate value, which does not give information about the servers' private data. The servers use random perturbation and encryption methods to hide confidential data. Random perturbation methods change the coordinate value for distance calculation. Random values generated from a uniform distribution over a reliable range are added to coordinate and measurements. The servers cannot find the exact coordinate and measurements of other servers. Finally, aggregate results also prevent the servers from deriving useful information about each other's data.

5.5   Experiments-Accuracy Analysis

We conducted two sets of experiments using two real geostatistics data sets to evaluate our proposed scheme with respect to prediction accuracy. (i) The first set of experiment show how collaboration between multi parties affects accuracy. For varying party sizes and G values, we presented Root mean squared error (RMSE) and mean absolute error (MAE) values. (ii) As mentioned in our solution, we utilized random perturbation, HE, and OT methods. HE and OT methods do not effect accuracy; however, randomized perturbation method may worsen accuracy. The second set of experiments present how distribution range of the random numbers affects prediction results.

Data Sets and Evaluation Metrics. For evaluation purposes, we used two real data sets obtained from the U. S. National Geochemical Survey Database (http://tin.er.usgs.gov/geochem/select.php) in our experiments. The data sets contain the sodium (**Na**) content of soil in Illinois and Colorado State. For Illinois data set, there are 1,331 measurements for different coordinates along the state. The measured minimum and maximum **Na** contents are 0.018 and 1.273, respectively. The mean and median values of the data set are 0.602 and 0.597, respectively. The second data set includes 1,150 measurements along the state. The minimum and maximum values of the set are 0.063 and 3.230, respectively. The mean and median of the set are 0.9986 and 0.9115, respectively. Histogram analysis verifies that both data sets have a normal distribution.

There are various evaluation metrics used to compare geostatistics methods. RMSE and MAE are the two of such metrics used to compare the accuracy of the models. Smaller RMSE and MAE values imply better accuracy. RMSE and MAE can be formulated as follows:

$$RMSE = \left[\frac{1}{n} \sum_{i=1}^{n} \left(\hat{Z}(i) - z(i) \right)^2 \right]^{1/2}$$

$$MAE = \left[\frac{1}{n} \sum_{i=1}^{n} \left| \hat{Z}(i) - z(i) \right| \right]$$

in which n represents number of measurements, $\hat{Z}(i)$ is the prediction value for point i, and $z(i)$ is the measured value for that point.

Methodology. We used leave-one-out method to calculate RMSE and MAE values. This method separates a point from the data set and creates a model with remaining points. The difference between prediction and measured value is computed as error of the model for that point. After calculating error for each point in the data set, RMSE and MAE values are computed. The average of several experiments are displayed in the tables.

Experiment I: Effects of Collaboration. It is hypothesized that collaboration between parties provide more accurate and dependable predictions. In order to prove this hypothesis, different sets of experiments are conducted using real data. It is also assumed that there are one, three, four, or five parties. In one party case, all measurements are held by one party. In three-party case, there are three parties that hold one third of all measurements. The distribution is held randomly. The same methodology is followed for the four and five-party cases. For all cases, varying G values are used. The results of RMSE and MAE values are presented for both Illinois and Colorado data set in Table 1 and Table 2. The results shown in the tables verify the hypothesis. Collaboration provides more accurate predictions. In each column, accuracy is getting worse for varying G values. Thus, if geostatistics methods are based on more measurements, they produce better results with respect to accuracy. The RMSE value improves 12% and the MAE value increases 13% for the Illinois data set when G is 15.

Experiment II: Overall Performance of the Proposed Scheme. We also analyze overall effects of randomization method on both coordinates and measurement values. PKDD scheme utilizes randomization method to disguise coordinate values. However, randomization may affect accuracy due to adding random numbers. On the contrary, HE and OT do not change accuracy. The overall performance of the proposed scheme is presented in Table 3 and Table 4.

As expected, randomization makes accuracy worse. Adding a random number over the range $[-0.05, 0.05]$ does not change the RMSE value for the Colorado data set when G is 15. If the range is chosen over the range $[-0.25, 0.25]$, accuracy decreases 9% for the same G value. However, the RMSE value is less than five-party and the MAE value is equal to five-party case. Therefore, if the range is chosen less than 0.25, PKDD method will produce better result than five-party case. In conclusion, adding random numbers, if they are chosen from an acceptable range, does not significantly affect accuracy.

Table 1. Effects of collaboration on RMSE with varying G values

	Illinois Data Set				Colorado Data Set			
G	5	15	30	50	5	15	30	50
Integrated	0.1207	0.1185	0.1183	0.1208	0.3165	0.3092	0.3098	0.3113
3-Party	0.1353	0.1321	0.1323	0.1326	0.3382	0.3306	0.3307	0.3317
4-Party	0.1380	0.1333	0.1329	0.1336	0.3237	0.3192	0.3194	0.3199
5-Party	0.1404	0.1345	0.1333	0.1340	0.3499	0.3362	0.3394	0.3456

Table 2. Effects of collaboration on MAE with varying G values

	Illinois Data Set				Colorado Data Set			
G	5	15	30	50	5	15	30	50
Integrated	0.0905	0.0885	0.0882	0.0905	0.2260	0.2201	0.2205	0.2207
3-Party	0.1005	0.0982	0.0986	0.0991	0.2419	0.2359	0.2362	0.2369
4-Party	0.1036	0.1006	0.1007	0.1012	0.2505	0.2426	0.2449	0.2464
5-Party	0.1063	0.1014	0.1012	0.1017	0.2528	0.2462	0.2488	0.2544

Table 3. Overall performance on MAE with varying δ values

	Illinois Data Set			Colorado Data Set		
	0.05	0.15	0.25	0.05	0.15	0.25
Integrated	0.1200	0.1424	0.1633	0.3092	0.3228	0.3376
3-Party	0.1313	0.1527	0.1828	0.3369	0.3502	0.3615
4-Party	0.1328	0.1529	0.1859	0.3380	0.3527	0.3639
5-Party	0.1334	0.1568	0.1899	0.3394	0.3531	0.3647

Table 4. Overall performance on MAE with varying δ values

	Illinois Data Set			Colorado Data Set		
	0.05	0.15	0.25	0.05	0.15	0.25
Integrated	0.0901	0.1102	0.1266	0.2205	0.2339	0.2487
3-Party	0.0986	0.1174	0.1432	0.2414	0.2543	0.2656
4-Party	0.1018	0.1194	0.1470	0.2470	0.2567	0.2685
5-Party	0.1022	0.1224	0.1503	0.2487	0.2583	0.2699

6   Conclusions and Future Work

To sum up, insufficient measurements lead to inaccurate model creation. If the kriging model generated using less data, as understood from the results of the experiments, accuracy of the model is not reasonable. Therefore, companies should be encouraged to collaborate to join their data. However, their financial futures depend on such valuable data. The proposed solution fills this gap. It enables companies to create a better kriging model based on measurements of three or more parties without revealing their private data. The privacy of the client is also taken into consideration. The coordinate and final prediction values are accepted as private data of the client. The servers cannot learn the coordinate for which the client needs prediction.

There are several geostatistics methods described in the literature. Future direction is to explore each geostatistics method. Therefore, there remains work to propose solutions for each one of the geostatistics methods. As in this study, our new solutions will also be analyzed with respect to performance, privacy, and accuracy. Each study will be verified with experiments on real data sets. We are also planning to scrutinize whether our proposed scheme can be used to deal with spatial nonstationarity or not.

References

1. Johnston, K., Ver Hoef, J.M., Krivoruchko, K., Lucas, N.: Using ArcGIS geostatistical analyst. Environmental Systems Research, Redlands, CA, USA (2001)
2. Isaaks, E.H., Sristava, M.R.: An Introduction to Applied Geostatistics. Oxford University Press, USA (1990)
3. Krige, D.G.: A statistical approach to some basic mine valuation problems on the Witwatersrand. Journal of the Chemical, Metallurgical and Mining Society of South Africa 52(6), 119–139 (1951)
4. Clifton, C., Kantarcioglu, M., Vaidya, J., Lin, X., Zhu, M.Y.: Tools for privacy preserving distributed data mining. SIGKDD Explorations Newsletter 4(2), 28–34 (2002)
5. Xu, Z., Yi, X.: Classification of privacy-preserving distributed data mining protocols. In: The 6th International Conference on Digital Information Management, Australia, pp. 337–342 (2011)
6. Lu, Q., Xiong, Y., Gong, X., Huang, W.: Secure collaborative outsourced data mining with multi-owner in cloud computing. In: IEEE 11th International Conference on Trust, Security and Privacy in Computing and Communications, Liverpool, UK, pp. 100–108 (2012)
7. Gurevich, A., Gudes, E.: Privacy preserving data mining algorithms without the use of secure computation or perturbation. In: The 10th International Database Engineering and Applications Symposium, Delhi, India, pp. 121–128 (2006)
8. Emekci, F., Sahin, O.D., Agrawal, D., El Abbadi, A.: Privacy preserving decision tree learning over multiple parties. Data & Knowledge Engineering 63(2), 348–361 (2007)
9. Kaleli, C., Polat, H.: Privacy-preserving SOM-based recommendations on horizontally distributed data. Knowledge-Based Systems 33, 124–135 (2012)

10. Kaya, S.V., Pedersen, T.B., Savaş, E., Saygıýn, Y.: Efficient privacy preserving distributed clustering based on secret sharing. In: Washio, T., Zhou, Z.-H., Huang, J.Z., Hu, X., Li, J., Xie, C., He, J., Zou, D., Li, K.-C., Freire, M.M. (eds.) PAKDD 2007. LNCS (LNAI), vol. 4819, pp. 280–291. Springer, Heidelberg (2007)

11. Amirbekyan, A., Estivill-Castro, V.: Privacy-preserving k-NN for small and large data sets. In: The 7th IEEE International Conference on Data Mining-Workshops, Omaha, NE, USA, pp. 699–704 (2007)

12. Yang, Z., Wright, R.N.: Privacy-preserving computation of Bayesian networks on vertically partitioned data. IEEE Transactions on Knowledge and Data Engineering 18(9), 1253–1264 (2006)

13. Yakut, I., Polat, H.: Arbitrarily distributed data-based recommendations with privacy. Data and Knowledge Engineering 72, 239–256 (2012)

14. Bansal, A., Chen, T., Zhong, S.: Privacy preserving Back-propagation neural network learning over arbitrarily partitioned data. Neural Computing and Applications 20(1), 143–150 (2011)

15. Yao, A.C.: Protocols for secure computations. In: The 23rd Annual Symposium on Foundations of Computer Science, Chicago, IL, USA, pp. 160–164 (1982)

16. Goldreich, O., Micali, S., Wigderson, A.: How to play any mental game. In: The 19th Annual ACM Symposium on Theory of Computing, New York, NY, USA, pp. 218–229 (1987)

17. Zhan, J., Matwin, S., Chang, L.: Privacy-preserving collaborative association rule mining. Journal of Network and Computer Applications 30(3), 1216–1227 (2007)

18. Green, M., Hohenberger, S.: Practical adaptive oblivious transfer from simple assumptions. In: Ishai, Y. (ed.) TCC 2011. LNCS, vol. 6597, pp. 347–363. Springer, Heidelberg (2011)

19. Camenisch, J., Dubovitskaya, M., Neven, G., Zaverucha, G.M.: Oblivious transfer with hidden access control policies. In: Catalano, D., Fazio, N., Gennaro, R., Nicolosi, A. (eds.) PKC 2011. LNCS, vol. 6571, pp. 192–209. Springer, Heidelberg (2011)

20. Corniaux, C.L.F., Ghodosi, H.: A verifiable distributed oblivious transfer protocol. In: Parampalli, U., Hawkes, P. (eds.) ACISP 2011. LNCS, vol. 6812, pp. 444–450. Springer, Heidelberg (2011)

21. Paillier, P.: Public-key cryptosystems based on composite degree residuosity classes. In: Stern, J. (ed.) EUROCRYPT 1999. LNCS, vol. 1592, pp. 223–238. Springer, Heidelberg (1999)

22. Guajardo, J., Mennink, B., Schoenmakers, B.: Modulo reduction for paillier encryptions and application to secure statistical analysis. In: Sion, R. (ed.) FC 2010. LNCS, vol. 6052, pp. 375–382. Springer, Heidelberg (2010)

23. Agrawal, R., Srikant, R.: Privacy-preserving data mining. In: The 2000 ACM SIGMOD International Conference on Management of Data, Dallas, TX, USA, pp. 439–450 (2000)

24. Tugrul, B., Polat, H.: Estimating kriging-based predictions with privacy. International Journal of Innovative Computing, Information and Control 9(8), 3197–3209 (2013a)

25. Tugrul, B., Polat, H.: Privacy-preserving inverse distance weighted interpolation. Arabian Journal for Science and Engineering (2013b), doi:10.1007/s13369-013-0887-4

26. Tugrul, B., Polat, H.: Privacy-preserving kriging interpolation on partitioned data. Knowledge-Based Systems (2014), doi:10.1016/j.knosys.2014.02.017

A Brain Computer Interface for Enhancing the Communication of People with Severe Impairment

Osvaldo Gervasi[1], Riccardo Magni[2], and Stefano Macellari[1]

[1] University of Perugia, Dept. of Mathematics and Computer Science, Perugia, Italy
[2] Pragma Engineering SrL, Perugia, Italy

Abstract. We present a novel approach for providing people with severe impairment the possibility of collecting feedbacks and communication prompts using a Brain Computer Interface (BCI). In particular we implemented an experimental apparatus, based on MindWave; the latter is a low cost and well known BCI device released by NeuroSky®, based on a dry single electrode and designed for enhancing Human-Computer Interaction (HCI), especially for videogames[8]. Even if BCI is an emerging research area and appears to be still relatively immature, the expected future impact on HCI is really considerable.

The experimental apparatus allows users to interact with the computer only through their brain biological signals, without the need for using muscles. The BCI system is based on the Steady–State Visual Evoked Potential (SSVEP). The SSVEP is generated in presence of a repetitive visual stimuli; in our experiment two different visual stimuli are available: when the patient looks at one of the stimuli the SVVEP signal is generated and she/he can express a decision or respond to a question.

Even if the experiment has been tested only on a very limited patient set, the results are extremely promising.

1   Introduction

The idea of exploiting BCI based methods to enhance communication of people with severe impairment is not new [14,12,10]. Several methods have been tested and implemented by means of multi-electrode systems (wet and dry) in experimental setup. The results are encouraging about the chance to put in place extensive utilization of such a mean to collect feedbacks and communication prompts. However the level of investigation has been limited to the level of proof of the concepts, working in a well-defined experimental setting. As a consequence, such systems have been implemented only on test cases, with a limited number of patients.

Our work originated from a collaboration with the Counseling Center for Technological Aids in Umbria (ONLUS Assoc. COAT) with the aim of demonstrating a feasible way to introduce BCI devices to enhance the communication ability of Amyotrophic Lateral Sclerosis (ALS) patient in realistic scenarios. We are

B. Murgante et al. (Eds.): ICCSA 2014, Part VI, LNCS 8584, pp. 709–721, 2014.
© Springer International Publishing Switzerland 2014

focusing our attention on the sustainability of the adopted solution, considering the overall cost of the system and all concrete implications.

Amyotrophic Lateral Sclerosis, also referred to as Lou Gehrig's disease, is a debilitating disease caused by progressive weakness, muscle atrophy, difficulty speaking (dysarthria), swallowing (dysphagia) and difficulty breathing (dyspnea). The origin of such a pathology is not clear and there are several theories about the progressive damage of moto-neurons of the brain: genetic causes (as observed in family related situations), toxic causes coming from body functioning (oxidation of melatonin) or from environmental factors (such as some classes of pesticides). ALS is classified as a rare disease: the incidence of the disease can be evaluated in 3 cases per 100.000 and the prevalence in 10 per 100.000. The progression of the disease produces a total paralysis of the patient which can survive by means of external aids like percutaneous endoscopic gastrostomy (PEG) for nutrition and pulmonary mechanical ventilator for breathing. At this stage of impairment, patient has no chance to communicate verbally, but only by means of assistive technologies adapted to her/his conditions. Although a reported percentage (5-15%) of cognitive decline caused by the pathology, in most cases cognitive capacities remain preserved.

In this work we describe the preliminary results obtained under such assumptions, in order to implement a low cost solution, to be made available to all interested patients.

The paper is structured as follows: in section 2 the most relevant works on BCI in neuroscience and biomedicine are discussed; in section 3 the experimental apparatus and the adopted methodology are presented; in section 4 the obtained results are presented and then discussed in section 5; in section 6 some conclusions are drawn and the future work is sketched.

2   Related Work

The neuronal activity of the central nervous system has been studied since the mid-70s by researchers in neuroscience and biomedicine. At later stage the possibility of developing BCI as a means of interaction with computer applications was investigated. Immediately, the possibility of application in the field of assisted technologies for people with severe physical paralyzes appeared clear. In fact, the use of BCI does not require muscle movements and/or the application of physical forces: in some cases it is sufficient the movement of eyes in a state of consciousness to express a choice.

There are three main ways to read the neuronal activity signals: invasive, when the electrodes are inserted into the human brain cortex; partially invasive, when electrodes are placed under the scalp without penetrating the skull; non-invasive, when external sensors are used. Non-invasive system are obviously preferred both for the easy of use and the acceptableness by patients.

Non-invasive BCI signals are intercepted by sensors of type EEG (Electroencephalogram) or fMRI (functional magnetic resonance) or fNIRS (functional near-infrared spectroscopy). In EEG applications, sensors consist of electrodes reading the electrical potentials (approximately $30\mu V$) that emerge from the skull, corresponding to the activity of brain areas.

To increase the conductivity between the electrode and the scalp, wet electrodes are often applied by means of saline solutions, gels, etc. that increase the sensitivity and require periodic recharging.

In fMRI applications the detection of the degree of oxygenation of the different brain areas allows to reveal the activity of the various brain areas.

In fNIRS applications the optical properties (in the Near InfraRed region) of the various cortex regions reveal the related brain activity. In fact, the optical properties vary as a function of the oxygen present in the brain region.

EEG is the most widespread method for the implementation of BCI, since a good response time is obtained[13]. A series of periodic brain activities may be electrically measured and can be distinguished by the specific frequency bands related to the brain states:

- *Delta Waves* (0.5-4Hz) related to sleep (non-REM) and unconsciousness
- *Theta Waves* (3-8 Hz) associated with a state of relaxation or dream
- *Alpha Waves* (6-13 Hz) related to a state of consciousness without special attention
- *Beta Waves* (12-30 Hz), especially at the level of the frontal parietal region and related to a state of active attention
- *Gamma Waves* (30-50Hz) related to the process of multi-sensory stimuli, Mu Rhythm, associated with motor activities.

The presence and intensity of the rhythms vary according to the different brain areas; this is the reason why many BCI systems are using multi-electrode configurations (up to 256) whose signals are analyzed and correlated. In Figure 1 the electrode positions are shown[1].

The complexity depends on the number of electrodes and on the number of samples (one hour of EEG recording using 128 electrodes, collecting 500 samples per second and using 16 bits, produces approximatively 0.5 GB of data to be analyzed). Even with a small number of electrodes, BCI systems are able to detect three fundamental mental states through the analysis of the prevalence of the rhythms (Waves): concentration, attention (response to a stimulus) and imagination-dream. The response to a stimulus can occur either due to a single stimulus (a light is turned on), either due to an oscillatory stimulus (with a characteristic frequency). A stimulus of this type can be both auditory (Steady State Auditory Evoked Potential, SSAEP) and visual (Steady State Visually-Evoked Potential, SSVEP). Thanks to its ease of implementation and its responsiveness in the application, many experiments have been conducted based on SSVEP. Applications with a greater degree of complexity are using multi-electrode sensors. Such systems perform a double mapping of the response to visual stimuli: in terms of frequency (oscillation on the basis of the excitation frequency) and spatial property (based on the excited area of the brain)[4]. The area of greatest excitation appears to be the occipital lobe of the head: different experiences have shown that even a reduced number of electrodes located in the occipital area produces a significant response in relation to the excitation frequency[11,3].

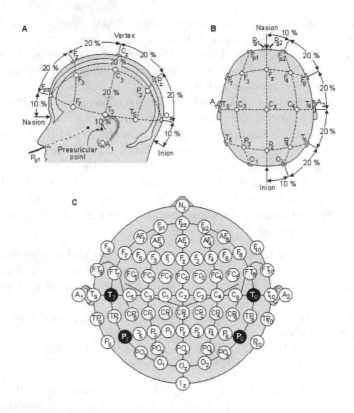

Fig. 1. The electrode positions for BCIs implementation

The range of excitation frequencies goes from 6Hz to 20Hz approximately; the source of stimulation is a source of illumination at different frequencies (in the case of multiple stimuli) in non-harmonic relation: such a condition guarantees the recognition of the excitation frequency in the signal without EEG ambiguity.

Several attempts have been made to produce visual stimulation systems based on CRT and LCD screens: unfortunately the refresh time and the relative low frequency (60Hz) of the screen, compared to the frequency of the signal to be monitored, have produced frequency unstable stimuli and a relatively uncertain signal[6,2]. In most cases, the stimuli were implemented through LED sources connected to generators that would ensure sufficient frequency stability[14].

Data analysis depends on the complexity of the setup configuration (i.e., number of electrodes and their position on the scalp): to this purpose various recognition systems with an increasing degree of complexity have been developed[7]. Such systems require more computational resources or an off-line operation, such as systems based on the analysis of spectral density-mediated[6], more suitable for on-line systems. Depending on the complexity of the algorithm, the number, type and position of the electrodes, the percentage of confidence of the responses obtained may vary from 65% to 88%, in the case of off-line processing[5].

3   Materials and Methods

3.1   Device

In order to plan the deployment of a low cost device for impaired patients, we considered commercial off-the-shelf systems, essentially developed for game applications. The Neurosky® Mindset[9] shown in Figure 2 is a very simple device with amplification and data sampling on-board and designed with a single dry electrode to be placed on the frontal region. The device has been designed by Neurosky® as a bio-sensor measuring the Electroencephalogram (EEG) activity, with the dry electrode positioned in the forehead (see Figure 3). In this way NeuroSky® is able to measure the level of attention, the meditation states and eye blinking activity. The device is mainly oriented to games activity, however it may used in several applications as an emerging interaction device.

Wearing procedures are facilitated and easy-to-learn thanks to the similarities of the Mindset device with headphones. Even if ALS patients are forced to stay on bed because of the severe disability, the Mindset device is easy to wear and remove. Our intent is to provide ALS patients who can not express verbally because of the mechanical ventilation and/or movement impairment with a new communication mean using their eyes movements. The movement of eyes is the last ability that ALS patients loose. There are several systems designed to intercept such movements; unfortunately, they fail in tracking the eye movement when she/he weeps or in presence of glasses or eyelid ptosis. Since the visual evoked potentials are preserved even in such severe impairment conditions, we decided to investigate brain signals in presence of visual and audio stimuli.

After an initial attempt to interpret via spectrum analysis the signals coming from the device in the standard position (the electrode is placed in the user's forehead, as shown in Figure 3), we decided to use the device in a non-conventional

Fig. 2. The MindWave system

Fig. 3. The MindWave system worn based on Neurosky® indications (image from NeuroSky® website[8])

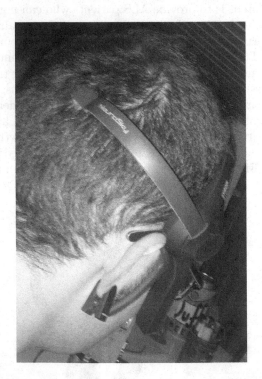

Fig. 4. The experimental setup of the MindWave system

way, in order to catch SSVEP signals. Since we have to capture the brain signals from the occipital-ocular cortex, the electrode arm was moved on the opposite side, as shown in Figure 4.

The evoked potential is captured by the single dry electrode, amplified internally and, finally, converted by Mindset digitally. The raw signal is then sent via a Bluetooth connection to a computer which interprets it. Using the drivers provided by the manufacturer the resulting sampling rate was approximatively 250Hz, which guarantees a good waveform reconstruction till 30 Hz.

The control panel implemented on the PC side (Figure 5) shows the output of the program that interprets the biological sensors output, which may be read as "Yes" (green button), "No" (red button) and "I don't know" (orange button). From such a control panel a series of detailed information may be provided to the medical staff to facilitate the interpretation of the program output.

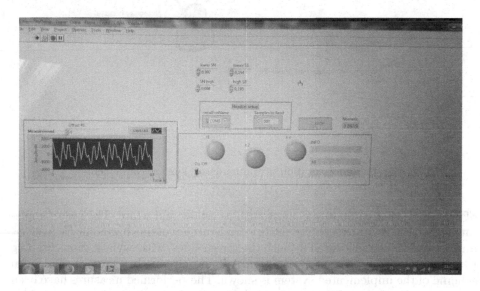

Fig. 5. The control panel implemented on the PC side

3.2 Stimuli

The visual stimuli have been generated from two separated light sources using two "Arduino Uno"[1] circuits (positioned at a given distance) and two (non harmonic) frequencies.

The visual stimuli has been determined in order to optimize the overall system performance in the given configuration, taking into account the distance. The approach to visual stimuli has been driven by tests and considerations about the device performance in the given configuration: the aim is to consider the overall performance of the system (intended as stimuli sensitivity+electrode location and efficiency+transducer+bandwidth). Tests made with several subjects in different conditions have selected a range of pulsed frequencies in the band range

---

[1] Arduino Uno is an Open Source Electronics prototyping platform based on flexible and easy-to-use Hardware and Software.

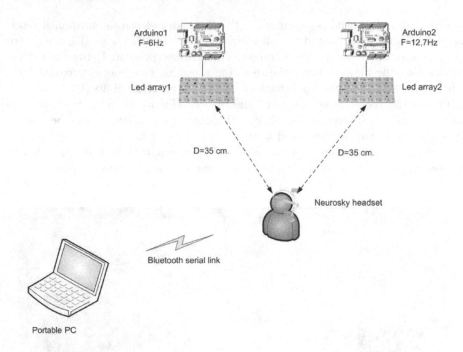

Fig. 6. Representation of the hardware system components

of 6-13 Hz, using composite colours such as yellow and white. They were implemented on the PC screen (LCD, 60Hz refresh rate) avoiding harmonic frequencies and with led clusters located at both the screen sides, with asynchronous flashing rate generation, implemented using an Arduino circuit. In Figure 6 the logical scheme of the implemented system is shown. The performed measures have been done at a distance of 35cm to the light sources. It has been demonstrated[14] that led stimulation has been preferred to other sources, because of the stability and reproducibility of the overall signal.

3.3   Software

The Arduino boards are programmed in Processing language in order to produce settable frequencies.

The acquisition software is based on Neurosky® drivers (serial bluetooth profile). Signal processing has been realized with LabView tools: for sake of simplicity the first experience exploited signal processing functions supplied in LabView libraries. The dominant frequency function invoked by the decision engine guarantees an high computational speed and the rejection of the spurious stimuli. Furthermore, when the subject is looking away from the imposed visual stimuli, the system is able to identify a "I don't know" decision.

In Figure 7 the block diagram related to the main software components is shown.

4   Results

The system has been tested on healthy people by now, since we are interested in exploiting the feasibility of such an approach and trying to measure the reliability of the system in different operating conditions. The preliminary tests highlight that the SVVEP signal is more stable when the visual stimulus occurs through led based lights, against stimulating through the computer screen. The effect is revealed by the stability of the implemented decision algorithm.

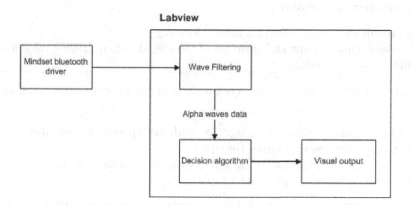

Fig. 7. Logical scheme of the Labview software system

4.1   The Test Phase

The test phase was conducted in a controlled environment both at the level of visual and noise stimuli; all sources of electromagnetic interference (e.g., mobile devices) were removed in an area of about 15 square meters. During each test the following protocol was adopted:

- *Self confidence phase*: the subject is asked to wear the device and to observe light stimuli intermittently. During this period the electrode is repositioned to get a better response, in terms of the received signal. The signal must be noise-free and the shape of the wavelength well defined and repeatable during the application of the stimulus. The duration of this phase was approximatively 20 minutes.
- *Test phase*: The tests were carried out with the assistance of an operator who asked for the observation of one of the two stimuli. The tests consisted of sessions of 10 exposures. Between two exposures the subject observed 3 minutes of rest (wearing the headset).

The tests were performed in different conditions:

- *Silence and natural light*: the ideal state for the synchronization of excitations;
- *Silence and artificial light* (neon light): to verify the level of disorder induced by pulsed light on system reliability;
- *Intelligible spoken* (same language as the subject) *and natural light*: in order to verify the possibility of using such a system at the level of interpersonal communication (orthogonality of the stimuli).

For each test we measured:

a) the required time to obtain a stable decision
b) the delay time for the stabilization of the signals when passing from a stimulus to the succeeding.

The characteristics of the subjects involved in the experiment are the following:

Subject 1: male, Caucasian, age 28, with no specific visual problems, no diagnosis, wearing eyeglasses (myopic);
Subject 2: male, Caucasian, age 49, with no specific visual problems, no diagnosis, wearing eyeglasses (myopic).

The experimental results obtained measuring the responses of the two subjects in different operating conditions are shown in Table 1. The sample considered is so limited that the results should be considered only qualitative.

Table 1. Experimental results of tests

	correct decision	std. dev.	delay time (sec)	std. dev. (sec)
Natural light	79%	2,36%	8	2,62
Neon	72%	2,67%	14	5,12
Natural light + acoustic stimulus	79%	2,83%	9	3,15

5   Discussion

A preliminary analysis of the data shows a good stability of the output. The observed uncertainty might be considered acceptable, if the system is properly interfaced to a communication system.

The experience suggests that the main sources of uncertainty are:

- *The characteristics of the subject wearing the headset*: we have to consider the physical characteristics of the subject and the relative uncertainty due to the way the electrode is positioned on the scalp;
- *The ambient light*: it impacts on the measure because of the different sensitivity of the eye to the pulsed stimuli;

The presence or absence of intelligible acoustic stimuli (human speech) does not seem to add a large variability: as expected it is orthogonal to the SSVEP stimuli.

The times of convergence to a stable output vary based on the following aspects:

- The features related to the positioning of the electrode and its physical magnitude may affect the sensitivity of the system in the signals reading;
- The physical conditions of the person (in particular the fatigue) tend to modify the system responsiveness.

6 Conclusions

Our work was devoted to investigate the following issues:

1. Is it possible to use a commercial, single electrode, BCI device to read SVVEP signals simply changing the position of the single dry electrode?
2. How much reliable such a system is?
3. Is it possible to characterize the global behavior of the implemented system?

Even if performed on a very limited set, the tests clearly show a positive response to the first question. The reliability of the system is influenced by several factors. In particular, the stress and fatigue conditions of the person who is making the test is the most important limiting factor. Furthermore the ambient light and the environmental conditions are other crucial aspects which may downgrade the efficiency of the system. The response to the third question requires more intensive tests, already planned as immediate future work.

In conclusion we may assert that the overall performances of the BCI system are encouraging, though not sufficient for its immediate integration, as a sensor, in a system for symbolic communication for people with severe disabilities.

We identified the following actions to increase the reliability and the efficiency of the system:

- at the level of *stimulation*: introducing more sophisticated methods of generating waveforms characteristics, compared to the on/off currently used, and reducing the size of the array of LEDs and their mechanical integration
- at the level of the *decision algorithm*: introducing the possibility to isolate the evoked potential even in presence of high level of noise, reducing the time for the stabilization of the output;
- at the level of *system customization*: custom calibration of the excitation frequencies and management of the power of the output light.

As per a wide experimentation of the system, we will apply for the authorization of the Ethics Committee of the USL (regional Public Health Authority) to proceed with the application (on a voluntary basis) of the system in subjects with a range of neurological diseases (diagnosis of ALS, MS, Stroke).

Acknowledgments. We acknowledge the scientific support of the Counseling Center for Technological Aids in Umbria (ONLUS Assoc. COAT).

References

1. Website source of the image, http://www.bci2000.org (accessed on February 28, 2014)
2. Cecotti, H., Volosyak, I., Graser, A.: Reliable visual stimuli on LCD screens for SSVEP based BCI. In: Proc. of the 18th European Signal Processing Conference (EUSIPCO 2010). p. 5 pages. Aalborg, Danemark, This research was fully supported within the 6th European Community Framework Program by a Marie Curie European ToK grant BrainRobot, MTKD-CT-2004-014211 and within the 7th European Community Framework Program by a Marie Curie European Re-Integration Grant RehaBCI, PERG02-GA-2007-224753 and by an EU ICT grant BRAIN, ICT-2007-224156 (August 2010),
   http://hal.archives-ouvertes.fr/hal-00536125
3. Guneysu, A., Akin, H.: An ssvep based bci to control a humanoid robot by using portable eeg device. In: 2013 35th Annual International Conference of the IEEE Engineering in Medicine and Biology Society (EMBC), pp. 6905–6908 (July 2013)
4. Kuś, R., Duszyk, A., Milanowski, P., Łabęcki, M., Bierzyńska, M., Radzikowska, Z., Michalska, M., Zygierewicz, J., Suffczynski, P., Durka, P.J.: On the quantification of ssvep frequency responses in human eeg in realistic bci conditions. PLoS ONE 8, 1–9 (2013)
5. Lin, Y.P., Wang, Y., Jung, T.P.: A mobile ssvep-based brain-computer interface for freely moving humans: The robustness of canonical correlation analysis to motion artifacts. In: 35th Annual International Conference of the IEEE Engineering in Medicine and Biology Society, EMBC 2013 (2013)
6. Liu, Y., Jiang, X., Cao, T., Wan, F., Mak, P.U., Mak, P.I., Vai, M.I.: Implementation of ssvep based bci with emotiv epoc. In: 2012 IEEE International Conference on Virtual Environments Human-Computer Interfaces and Measurement Systems (VECIMS), pp. 34–37 (July 2012)
7. Luo, A., Sullivan, T.J.: A user-friendly ssvep-based brain computer interface using a time-domain classifier. Journal of Neural Engineering 7(2), 026010 (2010),
   http://stacks.iop.org/1741-2552/7/i=2/a=026010
8. The neurosky biosensors website,
   http://neurosky.com/products-markets/eeg-biosensors/ (accessed on February 28, 2014)
9. The neurosky mindset website, http://developer.neurosky.com/docs/
   doku.php?id=mindset_instruction_manual (accessed on February 28, 2013)
10. Resalat, S.N., Setarehdan, S.K.: An improved ssvep based bci system using frequency domain feature classification. American Journal of Biomedical Engineering 3, 1–8 (2013)
11. Seyed, N.R., Seyed, K.S.: An improved ssvep based bci system using frequency domain feature classification. American Journal of Biomedical Engineering 3, 1–8 (2013)
12. Stamatto Ferreira, A.L., Cunha de Miranda, L., Gomes Sakamoto, S.: A survey of interactive systems based on brain - computer interfaces. SBC Journal on 3D Interactive Systems 4, 3–12 (2013)

13. Wang, Y., Gao, X., Hong, B., Gao, S.: Practical designs of brain computer interfaces based on the modulation of eeg rhythms. In: Graimann, B., Pfurtscheller, G., Allison, B. (eds.) Brain-Computer Interfaces, The Frontiers Collection, pp. 137–154. Springer, Heidelberg (2010), http://dx.doi.org/10.1007/978-3-642-02091-9_8
14. Zhu, D., Bieger, J., Molina, G.G., Aarts, R.M.: A survey of stimulation methods used in ssvep-based bcis. Intell. Neuroscience 2010, 1:1–1:12 (2010), http://dx.doi.org/10.1155/2010/702357

Micro Simulation to Evaluate the Impact of Introducing Pre-signals in Traffic Intersections

António Vieira, Luís S. Dias, Guilherme B. Pereira, and José A. Oliveira

University of Minho, Campus Gualtar, 4710-057, Braga, Portugal
`antonio6vieira@gmail.com, {lsd,gui,zan}@dps.uminho.pt`

Abstract. The resolution of problems related to the saturation of traffic intersections usually consists in the construction of infrastructure such as bridges or tunnels. These represent the most costly type of solutions. As such, it becomes necessary to ponder other types of solution, of lower cost. Thus, this paper intends to provide a way to significantly improve the performance of a traffic intersection, by using pre-signals, on its approaches. For this purpose, a traffic micro simulation model was developed, using Simio. The simulation experiments show that the implementation of pre-signals result in an increase of the intersection flow's upper ceiling in over 10%, a decrease in the average waiting time per vehicle in 1 minute and a decrease of the queues average size. In addition, it was also found that there is always gain in the space occupied by the queues, taking into consideration the space investment needed to implement pre-signals.

Keywords: Traffic Intersections, Pre-signals, Micro simulation, Simio.

1 Introduction

Since the motor vehicle has become the main means of transport of the human being, we have been witnessing a growing number of vehicles circulating on traffic lanes. This will eventually cause problems related to traffic congestion. To overcome these problems, generally 2 types of solutions are used: "signal control optimization approach and geometric channelization approach" [26]. The signal optimization is so important that many researchers believe that it can help solving problems related to traffic congestion and, thereby, avoid having to make changes to the infrastructures [13], [26]. Geometric channelization usually involves the expansion of the intersection or the construction of bridges/tunnels, representing the most onerous type of solution [23]. Thus, signal optimization is the optimum type of solution.

For signal optimization, we often witness attempts to solve congestion problems by increasing the duration of cycles. In fact, "in order to maximally satisfy the demand of the passage of motor vehicles", traffic control departments usually increase number of lanes and extend the cycle length with the excuse that long cycle length leads to short loss time in each cycle, so that the capacity of motor vehicles would be improved" [13]. These increases can go up to 240 or 300 seconds. However, it will only lead to

B. Murgante et al. (Eds.): ICCSA 2014, Part VI, LNCS 8584, pp. 722–745, 2014.

an increase of the waiting time of the pedestrians and of the vehicles on the remaining lanes. Thus, very long signal cycles should be avoided [13].

In the perspective of using a low cost technique and as a way to improve the performance of a traffic intersection, this paper intends to analyse the impact of introducting pre-signals on its approaches. For this purpose, it is necessary to provide some dozens of meters on the approaches of the intersection that will work as a kind of "launch pad". Thus, with the introduction of this technique, it is expected that the wasted time between the instant at which the last vehicle of an approach passes through the intersection and the instant at which the first vehicle of the next approach passes through the intersection is minimized; and the impact of the acceleration from the rest of the vehicles is also minimized.

To analyse this possibility and due to the fast development of computers, the risk of implementing solutions, without properly analysing its impacts, no longer exists. Simulation enables the visualization of the results from a modification made to a particular system, without the need to modify the reality of the system itself. From the several simulation tools on the market, the choice fell on a very recent that uses an object-oriented paradigm: Simio.

This paper is organised as follows. Chapter 2's main objective is to make the usual literature review on the topics adjacent to the main theme of this paper. Chapter 3 is dedicated to the different forms of data collection used on this project. In chapter 4 the various phases of modelling carried out for the elaboration of the simulation model are covered. Chapter 5 is related to the simulation experiments conducted in order to evaluate the system with and without pre-signals. Lastly, the final chapter summarizes the study, conclusions and includes directions for future research. Furthermore, exemplification videos of the model in execution were recorded and placed online at the following address: http://pessoais.dps.uminho.pt/lsd/pre_semaforos/. This document is also accessible, in pdf format, in the same address.

2   Literature Review

2.1   Pre-signals

Despite only having been first documented in 1991 in the UK [15], pre-signals were already in use in several European cities [27]. Its implementation is becoming significative in some cities of the United Kingdom and, in fact, until 1993, only in London, 14 pre-signals were implemented and a further 20 to 25 pre-signals were planned for the coming years [27].

The implementation of pre-signals can have many goals. One of these is "to give buses priority access into a bus advance area of the main junction stop line so as to avoid the traffic queue and reduce bus delay at the signal controlled junction" [27]. Conversely, in [6] Hanzhou and Wanjing used pre-signals to avoid losses of capacity on the lanes that can't discharge completely during its green phases, due to the existence of turning lanes. More recently, Xuan et al. in [28] were pioneers on the utilization of pre-signals to increase the capacity of a traffic intersection. In their study the approaches receives "2 green sub-phases: one for protected left turns only, and the other exclusively for through movements and right turns" [28]. In [30], Zhou and Zhuang seized this idea and proposed "an integrated model for lane assignment and signal timing optimization at tandem intersections. The model aimed to minimize the average delay that vehicles experienced in the pre-signal and main signal.

2.2   Simio

The majority of the studies that use simulation to model problems related to the traffic congestion use packages of micro simulation tools like VISSIM or AIMSUN. However, for this study, a discrete event simulation software was used. Since the number of tool options can be very high, simulation tool comparison becomes a very important task. However, most of scientific works related to this subject "analyse only a small set of tools and usually evaluating several parameters separately avoiding to make a final judgement due to the subjective nature of such task" [4].

In [9], Hlupic and Paul compared a set of simulation tools, distinguishing between users of software for educational purpose and users in industry. In his turn, Hlupic developed "a survey of academic and industrial users on the use of simulation software, which was carried out in order to discover how the users are satisfied with the simulation software they use and how this software could be further improved" [8]. In [4] and [19], Dias and Pereira et al. compared a set of tools based on popularity on the internet, scientific publications, WSC (Winter Simulation Conference), social networks and other sources. "Popularity should never be used alone otherwise new tools, better than existing ones would never get market place, and this is a generic risk, not a simulation particularity" [4]. However, a positive correlation may exist between popularity and quality, since the best tools have a greater chance of being more popular. According to the authors, the most popular tool is Arena. Nonetheless, the good classification of the "newcomer" Simio is noteworthy.

Simio, developed in 2007 [25], is based on intelligent objects [16], [18] and [22]. These "are built by modellers and then may be used in multiple modelling projects. Objects can be stored in libraries and easily shared" [17]. Unlike other object-oriented systems, in Simio there is no need to write any programing code, since the process of creating a new object is completely graphic [16], [18], [22]. The activity of building an object in Simio is identical to the activity of building a model. In fact there is no difference between an object and a model [16], [18]. A vehicle, a costumer or any other agent of a system are examples of possible objects and, combining several of these, one can represent the components of the system in analysis. Thus, a Simio model looks like the real system [16], [18]. This fact can be very usefull, particularly while presenting the results to someone non-familiar to the concepts of simulation.

In Simio the model logic and animation are built in a single step [16], [18]. This feature is very important, because it makes the modelation process very intuitive [18]. Moreover, the animation can also be useful to to reflect the changing state of the object [16]. In addition to the usual 2D animation, Simio also supports 3D animation as a natural part of the modelling process [22]. To switch between 2D and 3D views the user only needs to press the 2 and 3 keys of the keyboard [22]. Moreover, Simio provides a direct link to Google Warehouse, a library of graphic symbols for animating 3D objects [18], [22].

Simio offers 2 basic modes for executing models: the interactive and the experimental modes. In the first it is possible to watch the animated model execute, which is useful for building and validating the model. In the second, it is possible to define one or more properties of the model that can be changed, in order to see the impact on the system performance [22].

Although Simio incorporates a number of innovative features in pursuit of this goal, "only time will tell if this tool has bridged the many practical issues that must be addressed to trigger a widespread paradigm shift in the way practitioners build models" [16].

Currently there are not many studies that use Simio for modelling systems. In fact, regarding traffic congestion problems, no studies were found that used this tool. Even so, it is possible to find some studies that used this tool for other types of problems. In [1], Akhtar et al. studied the role of consanguineous marriages in causing congenital defects. In [12], Li and Wang developed a micro simulation model to evaluate the performance and service level of a ticket office. In [25], Vik et al. used Simio to model a logistic system design of a cement plant. In [3], Brown and Sturrock also used this tool to improve a set of production processes. Lastly, in [10], Kai et al. used Simio to explore simulation of casualty treatment in wartime.

3   Data Collection

For the purpose of building a model consistent with the real system, data related to real traffic situations was gathered and some literature was collected and analysed.

1)   Literature Considered

• **Cycle times of the traffic lights:** In [13], Maolin et al. stated that when the signal cycle length is around 100 seconds, the waiting time of vehicles is minimal.
• **Safety distances kept while driving:** Drivers that travel at a speed next to 50 km/h maintain a safety distance of about 16 meters [21]. Nevertheless, in [20], Pipes considered a lower value of about 1 meter for each unity of velocity (m/s).
• **Space occupied by a vehicle in a queue:** The analysed studies indicated that a stopped vehicle in a queue can occupy a distance of about 7.62 meters [2], [14], [31] or 7.89 meters [2], [7].
• **Start-up acceleration:** In [31], Zhu gathered and analysed several studies regarding this matter. The author developed a polynomial acceleration model characterized by expression (1). Since in Simio it is not yet possible to implement the acceleration of entities, it was necessary to use the correspondent velocity expression provided by the same authors (2).

$$a = 2,46 - 0,24t + 0,006t^2 \tag{1}$$

$$v = 2,66 + 2,46t - 0,12t^2 + 0,002t^3 \tag{2}$$

- **Reaction time of drivers on the first position of a queue:** Some authors considered that the first vehicle of a queue normally wastes 2 seconds to initiate the start-up acceleration process after the traffic light changes to green [2], [14]; others, considered the wasted time lies between 1.5 and 2 seconds [2], [5].
- **Reaction time of drivers on the remaining positions of a queue:** In [2], Bonneson gathered and analysed several studies regarding this matter and the values he collected were 1 second per vehicle, 1.22 seconds or 1.3 seconds.
- **Discharge rate of a queue:** In [2], Bonneson concluded that generally, this phenomenon takes place at a rate of approximately one vehicle every 2 seconds. Tough, some authors considered lower values like 1.97 [11] or 1.92 seconds [29].
- **Instant speed when crossing the stop line of an intersection:** In [2], Bonneson stated that the velocity of each vehicle increases until the fourth or fifth vehicle. From that number, the velocity of the vehicles tends to stabilize.

2)   Data Gathered on the Ground

Real traffic data was gathered from 2 distinct places. The first was Avenida 31 de Janeiro in Braga. In this place, data related to the the reaction time of 30 drivers to the change of the signal of the traffic light to green was collected and analysed. On average the drivers wasted 2.08. The second place was Avenida Marechal Humberto Delgado in Vila Nova de Famalicão. This street has a road on which the vehicles, after having passed the traffic light, have several dozen of meters, in a straight line, to accelerate. This enables the drivers to adopt a similar behaviour to the one that's supposed to be modelled. Data was collected from 30 drivers and it was found that, on average, drivers take 7 seconds to travel 40 meters, starting from a rest state.

4   Model Development

The whole project includes 3 models: Intersection, TrafficLight and Automobile. The first is the main one that models the traffic lights and the major part of the vehicle's behaviour. The design of the intersection is also made in this model. Automobile models part of the behaviour of the vehicles. Lastly, TrafficLight's only goal was to animate the change of the signals of the traffic lights.

4.1   Modelling the Traffic Lights Behaviour

The system consists of an X-shaped intersection (4 approaches). Each approach has 2 traffic lanes. Also, on each approach there are 2 traffic lights: one is the main and the other is the pre-signal. The signal cycles are processed through the regular repetition of green, yellow and red lights. As for the order, the traffic lights process on a counter clockwise direction.

Each traffic light has a state that on each instant can only have the value 0 (red), 1 (green) or 2 (yellow). These values are changed through processes. Since these processes work the same way for any approach, only the processes that change the signals of the main traffic light and the pre-signal of one approach will be illustrated, through Figure 1 and Figure 2.

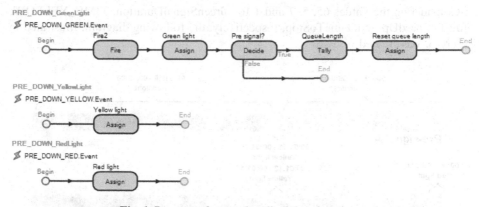

Fig. 1. Processes for pre-signal's lights changing

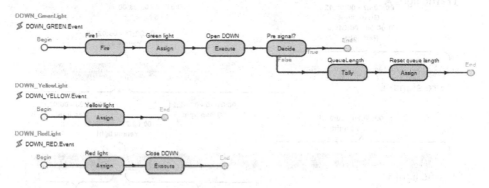

Fig. 2. Processes for main traffic light's changing

The first process of Figure 2 changes the state that indicates the signal of the pre-signal to 1, the second changes the same state to 2 and the last to 0. These changes are made by the Assign steps "Green light", "Yellow light" and "Red light". The remaining steps will be discussed later. Similarly, on Figure 1, the first process changes the state that indicates the signal of the main light to 1, the second to 2 and the last to 0, through equivalent steps.

These processes are event-triggered. Thus, they are associated to events, i.e., when a certain event is fired, the process associated to that event initiates its execution. These events are fired by timers at certain moments specified by their time offset and time interval properties. The data structures that define these properties are:

- **GreenSignalDuration:** Property that defines the duration of the green signal of the main traffic lights;
- **TIME_YELLOW:** State that defines the duration of the yellow signal;
- **TimeToSpeedUp:** Function that returns the time the main lights take to change their sign to green, after the pre-signal changed to the same colour;
- **TimeToStop:** Function that returns the time the main lights take to change their sign to yellow or red, after the pre-signal changed to the same colour;

Considering the values 65, 5, 7 and 1 for GreenSignalDuration, TIME_YELLOW, TimeToSpeedUp and TimeToStop, respectively, the following diagram, illustrated in Figure 3, was created.

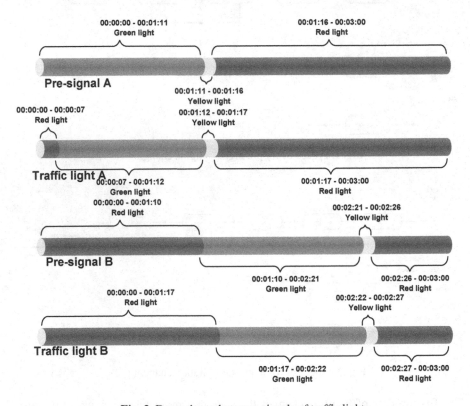

Fig. 3. Dependency between signals of traffic lights

Pre-signals act in two phases. First, their job is to minimize the impact of the reaction time of the drivers and the start-up acceleration of the drivers, by turning to green prior to the main traffic lights (TimeToSpeedUp). Very high values for this interval will result on the vehicles starting to reduce their speeds, or even having to stop because when they're approaching the main light the signal is still red. Conversely, very low values will result on a very high distance between the first vehicle and the intersection, when the main light turns to green [27]. Thus, the impact of this type of signalization would be lost. Approaching the end of the green phase, the role of the pre-signals is to allow as many vehicles as possible to pass through the intersection, while trying to avoid that no vehicle stays retained in the area between the pre-signal and the main light (TimeToStop). Very high values for this interval, would result on vehicles passing through the intersection even after the main traffic light's signal has changed to red. Conversely, very low values would increase the possibility of the vehicles becoming retained in the area between the pre-signal and the main light. Obviously, these intervals are dependent of the distance between traffic lights of the same approach. Therefore, changing the distance must result in a change of the intervals. Some examples can be seen in Table 1. PRE_SIGNAL_LaneLength represents the distance between traffic-lights of the same approach.

Table 1. Influence of the distance between traffic-lights on their signal's synchronization

PRE_SIGNAL_LaneLength	TimeToSpeedup	TimeToStop
10	4	0
20	5	0
40	8	0
50	8	1
60	9	1
70	10	2

4.2   Modelling Vehicles Behaviour

To model the behaviour of the vehicles it was necessary to create a large number of processes, functions and states both in the model Intersection and Automobile, in order to model all the situations through which the vehicles pass. Though, in this paper, only some of the processes will be illustrated.

The project was built in order to run in one of two possible ways: with or without pre-signals. For this purpose, MODE property was created. Thus if the value 1 is saved on this property, the model will run with pre-signals, otherwise it will run without pre-signals. Therefore, when Simio initiates the model execution, the process illustrated in Figure 4 is executed.

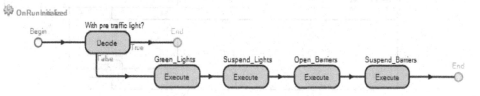

Fig. 4. Process OnRunInitialized

This process evaluates the value on the MODE property. Since the model is prepared to execute with pre-signals, if the value is 1, then the process terminates. If the value is different than 1, it is necessary to assign the green colour to all the pre-signals (step Execute "Green_Lights"), suspend all the processes that change the colour of the pre-signals (step Execute "Suspend_Lights"), open all the safety barriers (step Execute Open_Barriers) and suspend all the processes that close them (Suspend_Barriers). Either the project is supposed to run with or without pre-signals, the entities that enter the model have their correspondent tokens execute the process illustrated on Figure 5.

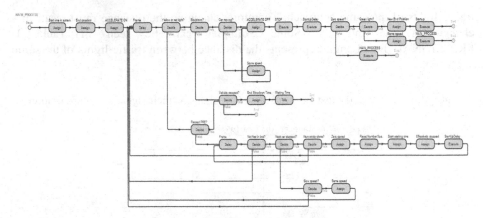

Fig. 5. Process MAIN_PROCESS

This process is responsible for modelling all the situations that can happen to the vehicles since the moment they enter the system to the moment they leave it. However, to model the safe distance kept by them it was necessary to create an additional process at the Automobile model, illustrated in Figure 6.

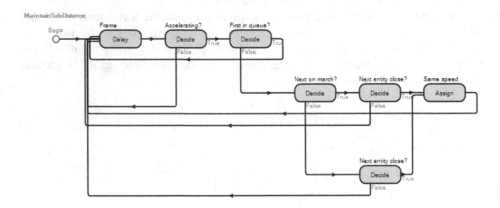

Fig. 6. Process MaintainSafeDistance

To provide greater realism to the model, 3D models of road segments, traffic lights, vehicles and safety cancels were downloaded from Google Warehouse. To visualize the change of the signals model TrafficLight was developed and to visualize the safety cancels opening and closing, the correspondent processes were developed. Lastly, some sample videos of the model in execution were recorded and placed online at the following address: http://pessoais.dps.uminho.pt/lsd/pre_semaforos/. The remaining functions, events, states and processes can be consulted on [24].

4.3   Validation of the Model

In order for the built model to correspond to the system in analysis the validation is a very important step. To that end, only the measures that can influence the KPI were validated. The considered measures were: space occupied by each vehicle in a queue; safety distance kept by vehicles; start-up acceleration; reaction time of vehicles in the first position of each queue to the change of the traffic light's signal to green; reaction time of vehicles of the remaining positions of the queue to the traffic light's signal change to green; queue discharge rate; instant speed of each vehicle when crossing the stop line of an intersection. Data related to these measures was collected and compared to the data analysed in chapter 3. For the majority of the measures the vehicles respond equally whether the model is executing with or without pre-signals. Yet, for some of them this isn't true. The measures in question are:

• **Queue discharge rate:** Simio was used to create Graph 1 and Graph 2. These indicate the Interarrival time of vehicles to the intersection.

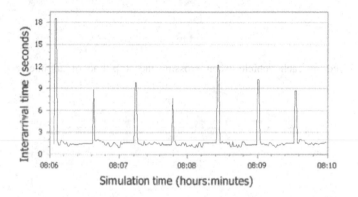

Graph 1. Interarrival time (without pre-signals)

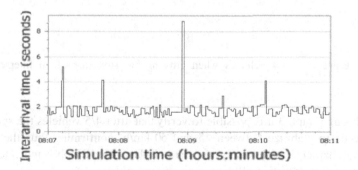

Graph 2. Interarrival time (with pre-signals)

As can be seen, the interarrival time of vehicles observed in Graph 1 is slightly lower than 2 seconds, confirming the data reffered in chapter 3. The data from Graph 2 indicates that the higher interarrival times of vehicles to the intersection in this case are lower than the values displayed in Graph 1.

• **Speed of each vehicle when crossing the stop line of an intersection:** Graph 3 and Graph 4 were created to display the instant speed of each vehicle when crossing the stop line of an intersection.

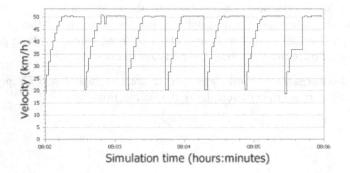

Graph 3. Speed of each vehicle when crossing the stop line of an intersection without pre-signals

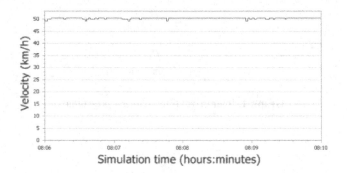

Graph 4. Instant speed of vehicles when crossing the stop line of an intersection with pre-signals

By analysing Graph 3 it is possible to verify that after 4/5 vehicles the speed of the vehicles tends to stabilize between 45 and 50 km/h, confirming thus the data data presented in chapter 3. Conversely, the data displayed by Graph 4 indicates that the vehicles pass the intersection allways at a constant speed.

5   Simulation Experiments

One of the major benefits of using Simio is the possibility of conducting simulation experiments on the models. A simulation experiment is a set of scenarios, one which each one executes the model with different values on the properties of the model in cause. Thus, to use simulation experiments it is necessary to define the properties of the model that we can change to see the impact on the system performance (response of the KPI). In this context, average waiting time, average queue size and flow were defined as KPI. The properties of the model are: MODE, GreenSignalDuration, ExponencialMean and PRE_SIGNAL_LaneLength. ExponencialMean defines the mean of the interarrival time of entities to the system. All the results of the simulation experiments can be consulted on [24].

In order to ensure that the results do not contain irrelevant data, as a result of the time needed for the system to achieve a "full-operating status", it is very important to define an accurate warm-up period. In this context, a warm-up period of 360 seconds was defined because, on the several tests conducted, it was found that from this time on, the KPI values achieved a more stable status. Figure 7 illustrates the described situation.

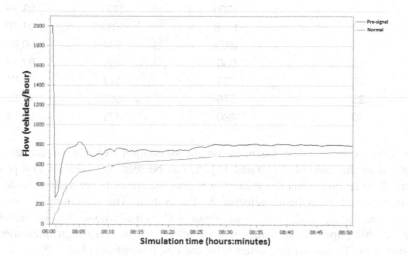

Fig. 7. Evolution of the flow KPI without warm-up period

5.1   Traffic Intensities

Traffic intensities are defined by interarrival times of entities to the system. Thus, the ExponencialMean property defines the intensity of traffic and, in order to determine the values to consider for this property, different values were used and the maximum possible flow was determined. Thereafter, a simulation experience, composed by a scenario for each of the considered values of ExponencialMean, was executed. All the scenarios were executed with 30 seconds to GreenSignalDuration, 50 for

PRE_SIGNAL_LaneLenght, 1 for MODE, 1 hour for simulation time and 5 replications were used. The observed flows were compared with the maximum possible flows and Table 2 was created.

Table 2. Flow values for different traffic intensities

Average interarrival time	Maximum flow rates	Flow rates in SIMIO	Utilization %
1	14400	2056	14,28
2	7200	2058	28,58
3	4800	2059	42,90
4	3600	2054	57,04
5	2880	2027	70,38
6	2400	1965	81,86
7	2057	1740	84,57
8	1800	1555	86,39
9	1600	1411	88,21
10	1440	1294	89,83
11	1309	1174	89,65
12	1200	1125	93,79
13	1108	1011	91,30
14	1029	934	90,80
15	960	895	93,19
20	720	677	94,05
25	576	562	97,52
30	480	471	98,21

The values chosen to be executed on the simulation experiments are the ones highlighted on the table (4, 7, 10 and 15). As can be seen, the values 1 to 4 produce the same flow results. Thus the value 4 was selected, since lower values would only result in many vehicles being rejected by the system. The remaining values were selected considering a difference of about 400 vehicles/hour. In the remaining sections of this paper, these values will be referenced as: very high intensities (4), high intensities (7), normal intensities (10), and low intensities (15).

5.2   Validation of the Simulation Experiences

• **Determination of the simulation time:** For this determination, only one simulation experience with 4 scenarios was executed. All the scenarios executed one replication and with the values of 30 to GreenSignalDuration, 50 to PRE_SIGNAL_ LaneLength and 1 to MODE. The only difference between scenarios is the values for ExponencialMean (4, 7, 10 and 15). This experience was executed for the values 5, 10, 15, 20, 25, 30, 45, 60, 120, 180 and 240 minutes of simulation time. All the values obtained for the KPI were compared, showing that the system tends to stabilize from 45 minutes of simulation time.

- **Determination of the number of replications:** For this purpose, 4 simulation experiences were executed: one for each intensity. Each experience executes 10 scenarios (each with a different number of replications from 1 to 10). All the scenarios executed 45 minutes of simulation time and with the values of 30 to GreenSignalDuration, 50 to PRE_SIGNAL_LaneLength and 1 to MODE. The obtained results show that the number of replications needed for the system to stabilize is 6.

5.3   Experiences with Green Light's Duration

The main purpose of these experiences was to discover the optimum range of green light signal duration. Thus 12 simulation experiences were conducted, all with the value of 50 meters for PRE_SIGNAL_LaneLength. Each experience consisted of 8 scenarios: 4 with the value 1 and 4 with the value 0 for MODE. In its turn, each 4 scenarios (with and without pre-signal) executed a different traffic intensity. Each simulation experience executes a different value for GreenSignalDuration (from 10 to 120 seconds). Based on the results Graph 5 until Graph 16 were prepared.

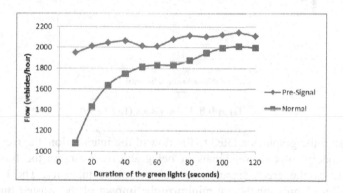

Graph 5. Flow values (very high)

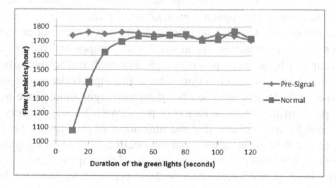

Graph 6. Flow values (high)

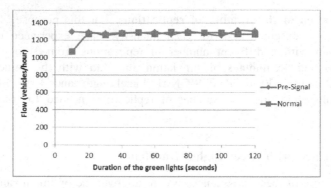

Graph 7. Flow values (medium)

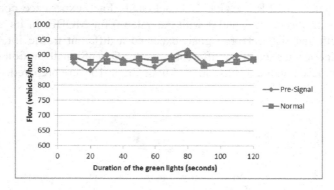

Graph 8. Flow values (low)

By analysing the graphics related to the flow of the intersection, the fact that stands out is the capacity of the pre-signals of being able to maintain the flow constant, independently of the green signal time and the traffic intensity. The fact that the implementation of pre-signals can minimize the impact of the wasted time between the last passage of a vehicle of an approach and the passage of the first vehicle of the next approach as well as the start-up acceleration can justify this situation. Conversely, in the case of the regular signalization, the same only applies for medium and low traffic intensities (Graph 7 and Graph 8). This is somewhat expected because the approaches to the intersection are always nearly empty and the probability of a queue being emptied by a green phase is high. For high traffic intensities (Graph 5 and Graph 6) the flow of an intersection without pre-signals decreases sharply when values of 50 seconds or less are used for the green light time. Once again, this is related to the perks of the implementation of pre-signals. Lastly, for very high traffic intensities (Graph 5), it is notable that the utilization of pre-signals can increase the upper ceiling of the flow of vehicles in an intersection with regular signalizing. Also in the same graph it is possible to verify that the flow improves when higher green light times are used in an intersection without pre-signals. This is also expected because start-up accelerations and wasted times of the drivers are minimized with the implementation of pre-signals. Nevertheless, the graphs related to the average waiting time (Graph 9 to Graph 12) show that very high green light durations should not be used, since waiting time reaches very high values.

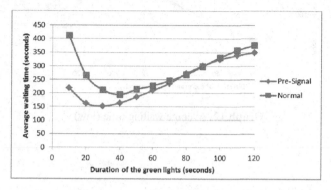

Graph 9. Average waiting time (very high)

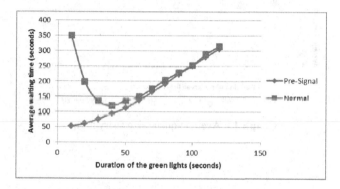

Graph 10. Average waiting time (high)

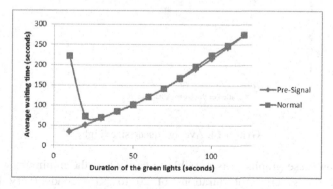

Graph 11. Average waiting time (medium)

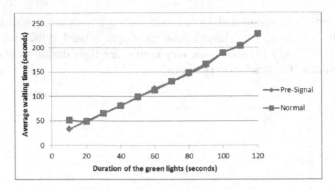

Graph 12. Average waiting time (low)

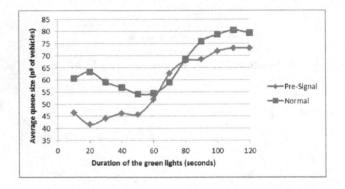

Graph 13. Average queue size (very high)

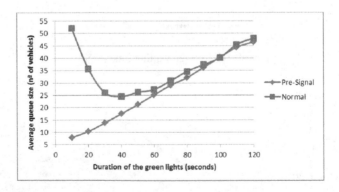

Graph 14. Average queue size (high)

Focusing on these graphs, it is possible to verify that the minimum waiting times are achieved for green light durations of 20 to 30 seconds (very high traffic intensities), totalizing a cycle time between 100 to 140 seconds. This confirms what had been stated by Maolin et al., in [13]. Nevertheless, this only applies in intersection without pre-signals. For intersections with pre-signals the average waiting time continues to decrease when lower green light times are being used. In fact, for high traffic intensities (Graph 10), differences of about 5 minutes per vehicle are registered when green light time of 10 seconds are being used.

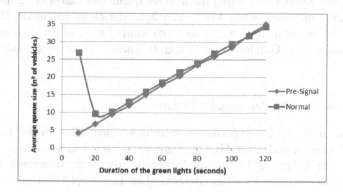

Graph 15. Average queue size (medium)

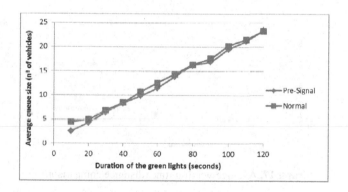

Graph 16. Average queue size (low)

By analysing the graphs related to the average queue size, gains or losses are hardly noted for the implementation of pre-signals. In fact, in Graph 14, Graph 15 and Graph 16 differences are only noted for low green light times.

A correct analysis to the present problem should not be made by comparing the KPI of the 2 signalling types with the same green light duration because, as it was evidenced by the analysed data, both types perform better for different green light durations. In that sense, it is important to determine what values of green light time should be used on each signalization type. For vintage intersections, the green light time that provides the best results is not immediately determined, since increasing the duration will also increase the flow, the average queue size and the average waiting time, i.e. there is no optimum point or direction on the various graphs where it is possible to get the maximum flow and minimum average queue size and waiting times. Thus, a balance has to exist. Therefore 40, 50, 60 or 70 seconds seems to be the durations that can provide the best performance for a vintage intersection. Conversely, an intersection with pre-signals can use low green signal times and maintain the flow at a steady level without decreasing the average queue size and waiting time. Hence, its performance increases when low green light durations are used. Though, for durations of 10 seconds, some queue size and waiting time values increase for very high traffic intensities (Graph 13 and Graph 9). Thence, 20 or 30 seconds should be used.

5.4   Experiences with Pre-signal's Distance

In this phase, 6 simulation experiences were conducted, all with the value 20 for GreenSignalDuration and 1 for MODE. Each experience had 4 scenarios, each one with a different value for ExponencialMean. Lastly, each simulation experience executed a different value for PRE_SIGNAL_LaneLength (10 to 70 meters). Based on the results Graph 17 to Graph 19 were created.

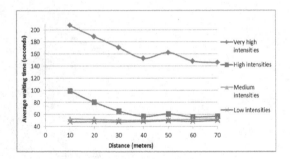

Graph 17. Average waiting time (distance comparison)

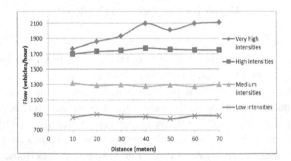

Graph 18. Average queue size (distance comparison)

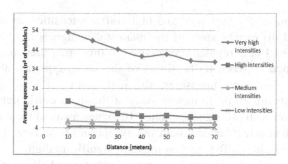

Graph 19. Flow values (distance comparison)

As the graphs illustrate, the performance of the intersection doesn't suffer significant changes for medium and low traffic intensities. For high and very high intensities, the graphs show that the intersection performs better for distances equal or greater than 40 meters. Though, the lower the distance the better, since it: increases the safety levels (vehicles don't have enough space to over speed), increases the effectiveness of the control of the time vehicles take to travel the distance and the space required to implement pre-signals is minimized.

5.5 Comparison of the Signalization Types

In the perspective of comparing the performance of a vintage intersection (with green light duration of 60 seconds) with one with pre-signals (green light duration of 20 seconds) and to ascertain if, for distances lower than 40 meters, the performance of the intersection still justifies the implementation of pre-signals Table 3 was elaborated.

Table 3. Comparison of the signalization types

Traffic intensities		Very high		High		Medium		Low	
Signalization (Green light duration in seconds)		Pre-Signals (20)	Regular (60)	Pre-Signals (20)	Regular (60)	Pre-Signals (20)	Regular (60)	Pre-Signals (20)	Regular (60)
	Distance								
Flow rates (vehicles/hour)	10	1763		1697		1318		867	
	20	1859	1833	1731	1728	1285	1274	905	882
	30	1932		1744		1295		876	
	40	2097		1776		1272		875	
Average waiting time (seconds)	10	207		99		52		47	
	20	189	227	80	150	52	121	48	113
	30	171		66		50		48	
	40	153		57		50		48	
Average queue size (nº of vehicles)	10	53		18		7		5	
	20	49	55	14	27	7	18	5	13
	30	44		11		7		4	
	40	40		10		7		4	

As the table shows, for very high and high traffic intensities, an intersection with pre-signals placed 40 meters away of the main traffic signals has better KPI, when compared to the KPI of a regular intersection. In this intensities the decrease of the KPI as the distance between traffic lights of the same approach also decreases is notable. For the remaining intensities, the KPI don't suffer significant changes whether pre-signals are 10 or 40 meters away of the main traffic lights. In this case, even an intersection with pre-signals situated 10 maters away of the main traffic lights performs better than a regular one, with the exception of the flow. For this KPI the different displayed values differ only on the third significant digit.

As it can be noticed, considering a 40 meters distance and all the traffic intensities, on average, an intersection with pre-signals has queue size values with 13 less vehicles. Regarding average waiting times, on average one vehicle saves around 1 minute and 15 seconds. As for the last KPI, an increase of the upper ceiling of about 15% can be verified.

Despite the advantages of the utilization of pre-signals, to proceed to its implementation it is necessary to "invest" some dozens of meters. In order to help the managers of an intersection making a decision on whether to implement pre-signals or not, Table 4 was prepared. In it, the average earnings recorded regarding the space occupied by a queue of vehicles are displayed.

Table 4. Space gained in the approaches to the intersection

Traffic intensities	Very high	High	Medium	Low	
Space requirements (meters)	Profited space (meters)				Average (meters)
10	5	59	74	51	47
20	26	79	64	41	52
30	54	92	54	39	60
40	74	90	44	29	59

The values in this table were obtained by subtracting the space "invested" to the gain recorded in Table 3, after converting the queue size to space occupied by vehicles on a queue (using the data from chapter 3). As can be seen, regardless the traffic intensity, there is always gains on space after providing 10 to 40 meters to the implementation of pre-signals on a traffic intersection.

6   Conclusions

The construction of certain kinds of infrastructures (e.g. bridges and tunnels) represents the most obvious and costly type of solution, for the resolution of traffic saturation problems [23]. In order to propose a less costly solution, this paper aimed to investigate whether it is possible to significantly **improve the performance of a traffic intersection by implementing pre-signals** in its approaches. In this context, a **micro simulation** traffic model was developed, using Simio, where it is possible to modify: the type of intersection (with or without pre-signals); the distance between the pre-signals and the respective main traffic lights of each approach; the green signal time and the traffic intensity. The **input data** for modelling the system (e.g.

reaction times, safety distances) were gathered through field observations and literature reviewed. Thus, the developed model allows the proper modelling of the system in analysis. Some sample videos of the model in execution were recorded and placed online at the following address: http://pessoais.dps.uminho.pt /lsd/pre_ semaforos/. Average waiting time, average queue size and the flow rate were defined as **KPI** (Key Performance Indicators). Lastly, the green light duration, the traffic intensity, the distance between traffic lights of the same approach and the mode of execution (with or without pre-signals) were defined as the properties of the model.

Simio's **simulation experiments** were used in order to analyse the impact that changes to the properties of the model produce in the KPI. The several experiments conducted show that the average waiting time and the average queue size reaches minimum values for lower green signal times (Graph 9 until Graph 16). Though, for lower values of green signal time the flow of vehicles diminishes (Graph 5 until Graph 8) being impossible to achieve a high flow rate, without harming the average queue size and the average waiting time. Conversely, the main objective of the implementation of pre-signals is to: **reduce the time** elapsed **between the passage** of the last vehicle of an approach and the passage of the first vehicle from the next approach (Graph 1 and Graph 2) and lessen the impact of the start-up acceleration by the vehicles on the flow (Graph 3 and Graph 4). Hence, the implementation of pre-signals enables the utilization of low green signal times in order to obtain better results for the average waiting time and the average queue size, without affecting the flow of vehicles through the intersection. The simulation experiments indicate that the appropriate green signal time for the traffic lights of a traffic intersection with pre-signals is around 20 seconds.

In a second phase, the impact of the **distance between** the **pre-signals** and the respective main traffic lights was analysed. It was found that the best distance lies around 40 meters for very high or high traffic intensities (Graph 17, Graph 18, Graph 19 and Table 3) - See considerations on traffic intensities in chapter 5. For the remaining intensities, it's still possible to verify gains with the utilization of pre-signals, however, it is noted that the distance does not influence the performance of the intersection.

Finalized the simulation experiments, it was verified that the implementation of pre-signals results in an **increase** of the **upper ceiling** of the **flow** of the intersection in about **15%**, a **decrease** in the average **waiting time** in approximately **1 minute** and 15 seconds and a **decrease** of the average size of the **queues** in around **60 meters**. In addition, it was also found that there is always gain in the space occupied by the queues, taking into consideration the "investment" needed to implement pre-signals (Table 4).

6.1   Future Work

After the conclusion of the project it was possible to verify that, for low traffic intensities, there are some fluctuations in the values of the KPI. This can be explained by the arrival of a different number of vehicles on distinct simulation times, whether the traffic intersection is being modelled with or without pre-signals. Since few vehicles arrive at the system, these differences become significant. In an attempt to try to solve this problem, the entire model could be duplicated and whenever a new entity is created, it also had to be duplicated and sent to both models. Nevertheless, this solution would make the model very extense and its modelling ineffective, because changes would have to be made on both models. Another solution could be to increase the number of replications and simulation time. This was not possible due to the limitations of the machine where the model was developed. It would also be important to analyse the viability of the implemented acceleration model, since there are differences between an acceleration in a straight line and an acceleration in a straight line followed by a curve. Lastly, it would be interesting to evaluate if this new technique could contribute for the reduction of noise and environmental pollution.

Acknowledgements. This work has been supported by FCT – Fundação para a Ciência e Tecnologia in the scope of the project: PEst-OE/EEI/UI0319/2014.

References

1. Akhtar, N., Niazi, M., Mustafa, F., Hussain, A.: A discrete event system specification (DEVS)-based model of consanguinity. Journal of Theoretical Biology 285, 103–112 (2011)
2. Bonneson, J.A.: Modeling Queued Driver Behavior at Signalized Junctions. Transportation Research Record 1365, TRB, Washington, D.C., pp. 99–107 (1993)
3. Brown, J.E., Sturrock, D.: Identifying Cost Reduction and Performance Improvement Opportunities Through Simulation. In: Rossetti, M.D., Hill, R.R., Johansson, B., Dunkin, A., Ingalls, R.G. (eds.) Proceedings of the 2009 Winter Simulation Conference (2009)
4. Dias, L., Pereira, G., Rodrigues, G.: A Shortlist of the Most Popular Discrete Simulation Tools. Simulation News Europe 17, 33–36 (2007)
5. Georgia, E.T., Heroy, F.M.: Starting Response of Traffic at Signalized Intersections Traffic Engineering, pp. 39-43 (1966)
6. Hanzhou, X., Wanjing, M.: Simulation-based study on a pre-signal control system at isolated intersection with separate left turn phase. In: 2012 9th IEEE International Conference on Networking, Sensing and Control (ICNSC), April 11-14, pp. 103–106 (2012)
7. Herman, R., Lam, T., Rothery, R.W.: The Starting Characteristics of Automobile Platoons. In: Proc., 5th International Symposium on the Theory of Traffic Flow and Transportation, pp. 1–17. American Elsevier Publishing Co., New York (1971)
8. Hlupic, V.: Simulation software: an Operational Research Society survey of academic and industrial users. In: Simulation Conference, 2000. Proceedings, vol. 2, pp. 1676–1683 (Winter 2000)
9. Hlupic, V., Paul, R.: Guidelines for selection of manufacturing simulation software. IIE Transactions 31, 21–29 (1999)
10. Kai, Z., Ruichang, W., Jie, N., Xiaofeng, Z., Haijian, D.: Using Simio for wartime casualty treatment simulation. In: 2011 International Symposium on IT in Medicine and Education (ITME), December 9-11, pp. 322–325 (2011)
11. Lee, J., Chen, R.L.: Entering Headway at Signalized Intersections in a Small Metropolitan Are. In: Transportation Research Record 1091,TRB, National Research Council, Washington, D.C. pp.17–126 (1986)

12. Li, J., Wang, L.: Microscopic simulation on ticket office of large scale railway passenger station. In: 7th Advanced Forum on Transportation of China (AFTC 2011), pp. 41–47 (October 22, 2011)
13. Maolin, P., Sheng, D., Jian, S., Keping, L.: Microscopic Simulation Research on Signal Cycle Length of Mixed Traffic Considering Violation. In: 2010 International Conference on Intelligent Computation Technology and Automation (ICICTA), May 11-12, pp. 674–678 (2010)
14. Messer, C.J., Fambro, D.B., Texas Transportation, I.: National Research Council. Transportation Research Board, M, Effects of Signal Phasing and Length of Left Turn Bay on Capacity, Texas Transportation Institute, Texas A & M University (1997)
15. Oakes, J., Thellmann, A.M., Kelly, I.T.: Innovative bus priority measures. In: Proceedings of Seminar J, Traffic Management and Road Safety, 22nd PTRC European Transport Summer Annual Meeting, University of Warwick, U.K, vol. 381, pp. 301–312 (1994)
16. Pegden, C.D.: Simio: A new simulation system based on intelligent objects. In: 2007 Winter Simulation Conference, December 9-12, pp. 2293–2300 (2007)
17. Pegden, C.D.: Intelligent objects: the future of simulation (2013)
18. Pegden, C.D., Sturrock, D.T.: Introduction to Simio. In: Proceedings - Winter Simulation Conference, Phoenix, AZ, pp. 29–38 (2011)
19. Pereira, G., Dias, L., Vik, P., Oliveira, J.A.: Discrete simulation tools ranking: a commercial software packages comparison based on popularity (2011)
20. Pipes, L.A.: An Operational Analysis of Traffic Dynamics. Journal of Applied Physics 24, 274–281 (1953)
21. Qiang, L., Lunhui, X., Zhihui, C., Yangou, H.: Simulation analysis and study on car-following safety distance model based on braking process of leading vehicle. In: 2011 9th World Congress on Intelligent Control and Automation (WCICA), June 21-25, pp. 740–743 (2011)
22. Sturrock, D.T., Pegden, C.D.: Recent innovations in Simio. In: Proceedings - Winter Simulation Conference, Baltimore, MD, pp. 21–31 (2010)
23. Treiber, M., Helbing, D.: Microsimulations of Freeway Traffic Including Control Measures. at - Automatisierungstechnik 49, 478 (2001)
24. Vieira, A.: Micro simulation to evaluate the impact of introducing pre-signals in traffic intersections. MSc Dissertation. University of Minho (2013)
25. Vik, P., Dias, L., Pereira, G., Jos, O., Abreu, R.: Using simio for the specification of an integrated automated weighing solution in a cement plant. In: Proceedings of the Winter Simulation Conference, Baltimore, Maryland (2010)
26. Wei, H., Perugu, H.C.: Oversaturation inherence and traffic diversion effect at urban intersections through simulation. Jiaotong Yunshu Xitong Gongcheng Yu Xinxi/ Journal of Transportation Systems Engineering and Information Technology 9, 72–82 (2009)
27. Wu, J., Hounsell, N.: Bus Priority Using pre-signals. Transportation Research Part A: Policy and Practice 32, 563–583 (1998)
28. Xuan, Y., Daganzo, C.F., Cassidy, M.J.: Increasing the capacity of signalized intersections with separate left turn phases. Transportation Research Part B: Methodological 45, 769–781 (2011)
29. Zegeer, J.D.: Field Validation of Intersection Capacity Factor. In Transportation Research Record 1091, TRB, National Research Council, Washington, D.C., pp.67–77 (1986)
30. Zhou, Y., Zhuang, H.: The optimization of lane assignment and signal timing at the tandem intersection with pre-signal. Journal of Advanced Transportation (2013)
31. Zhu, H.: Normal Acceleration Characteristics of the Leading Vehicle in a Queue at Signalized Intersections on Arterial Streets, Oregon State University (2008)

An Approach for Automatic Expressive Ontology Construction from Natural Language

Ryan Ribeiro de Azevedo[1,2], Fred Freitas[2], Rodrigo Rocha[1,2],
José Antônio Alves Menezes[1], and Luis F. Alves Pereira[1,2]

[1] Computer Science, UAG/Federal Rural University of Pernambuco, Garanhuns, Brazil
{ryan,rodrigo,jam,lfap}@uag.ufrpe.br
[2] Center of Informatics, Federal University of Pernambuco, Recife, Brazil
{rra2,fred,rgcr,lfap}@cin.ufpe.br

Abstract. In this paper, we present an approach based on Ontology Learning and Natural Language Processing for automatic construction of expressive Ontologies, specifically in OWL DL with ALC expressivity, from a natural language text. The viability of our approach is demonstrated through the generation of descriptions of complex axioms from concepts defined by users and glossaries found at Wikipedia. We evaluated our approach in an experiment with entry sentences enriched with hierarchy axioms, disjunction, conjunction, negation, as well as existential and universal quantification to impose restriction of properties. The obtained results prove that our model is an effective solution for knowledge representation and automatic construction of expressive Ontologies. Thereby, it assists professionals involved in processes for obtain, construct and model knowledge domain.

Keywords: Description Logic (DL), Ontology, Ontology Learning, PLN.

1   Introduction

One of the subfields of Ontology that has been standing out along the last decade is Ontology Engineering. Its purpose is to create, represent and model knowledge domains, most of which are not trivial, such as Bioinformatics and e-business, among others. However, as pointed out by [14], the task of Ontology Engineering still consumes a big amount of resources even when principles, processes and methodologies are applied. Besides financially expensive, the ontology design is also an arduous and onerous task [6]. Thus, new technologies, methods and tools for overtaking these technical and economic challenges are necessary. This way, the need for highly specialized personnel and the manual efforts required can be minimized.

For this purpose, a research line that is gaining importance through the past two decades is the extraction of domain models from text written in natural language, using Natural Language Processing (NLP) techniques. The process of modeling a knowledge domain from text and the automated design of ontologies, through an analysis of a set of texts using NLP techniques, for example, is known as Ontology Learning and was first proposed by [11]. Even so, as affirmed by [17], despite the

B. Murgante et al. (Eds.): ICCSA 2014, Part VI, LNCS 8584, pp. 746–759, 2014.

increasing interest and efforts taken towards the improvement of Ontology Learning methods based in NLP techniques [15][16][1][2][3], the notable potential of the techniques and representations available to the learning process of expressive ontologies and complex axioms has not yet been completely exploited. In fact, there are still gaps and unanswered questions that need viable and effective solutions. Among them, these stand out [13][16] [15]:

- The necessity of combining the knowledge of specialists of domain with the competencies and experience of ontology engineers in a single effort: there are scarce specialized resources and demand for professionals. This obstacle reduces the use of semantic ontologies by users and specialists in general.
- There is a considerable amount of tools and *Frameworks* of *Ontology Learning* that have been developed aiming at the automatic or semi-automatic construction of ontologies based on structured, semi-structured or unstructured data. Nonetheless, although useful, the majority of these tools used in Ontology Learning are only capable of creating informal or unexpressive ontologies.
- Evaluating the consistency of ontologies automatically: it is necessary that the automatically created ontologies be assessed by the time of their development, minimizing the amount of errors committed by the ones involved in the development phase and verify whether or not the ontology is contradictory and free of inconsistencies.

All the questions and issues abovementioned justify the approach hereby proposed. It is based in a translator, which consists in the utilization of a hybrid method that combines syntactic and semantic text analysis both in superficial and in-depth approaches of NLP. Demonstrating that a translator for creating ontologies that formalizes and codifies knowledge in OWL DL $\mathcal{ALC}$ [7] from sentences provided by users is a viable and effective solution to the process of automatic construction of expressive ontologies and complete axioms.

The rest of this paper is organized as follows: Section 2 presents the assumptions and a detailed description of our approach. We report our experimental setup, results and discussions in Section 3. Section 4 presents related work. Finally, Section 5 concludes this paper.

2   The Approach and Example

One of the goals of this work consists in demonstrating that a translator, through the processing of sentences in natural language provided by users, is capable of creating – automatically and according to the discourse interpreted $\mathcal{ALC}$ [7] ontologies with minimal expressivity. An overview of the translator's architecture and function flow diagram are depicted in Figure 1 and described as follows.

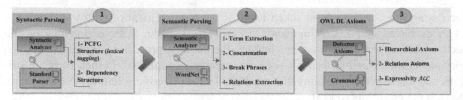

Fig. 1. Translator's architecture and function flow diagram

The architecture of our approach is composed by 3 modules: the **Syntactic Parsing Module (1)**, **Semantic Parsing Module (2)** and the **OWL DL Axioms Module (3)**. The activities executed in the respective modules and their functions are presented in the following sections.

2.1   Syntactic Parsing Module

The syntactic analysis of the sentences inserted by users takes place in the **Syntactic Parsing Module (1)**, this module uses a Probabilistic Context-Free Grammars (PCFGs). For this purpose, we used the Stanford Parser 2.0.5[1]. A PCFG consists of:
1. A context-free grammar $G = (N, \Sigma, S, R, q)$ as follows:
    a. N is the set of all non-terminals seen in the trees $t_1 \ldots t_m$.
    b. Σ is the set of all words seen in the trees $t_1 \ldots t_m$.
    c. The start symbol S is taken to be S.
    d. The set of rules R is taken to be the set of all rules $\alpha \rightarrow \beta$ seen in the trees $t_1 \ldots t_m$.
    e. The maximum-likelihood parameter estimates are
    $$qML(\alpha \rightarrow \beta) = \frac{Count(\alpha \rightarrow \beta)}{Count(\alpha)}$$
    where $Count(\alpha \rightarrow \beta)$ is the number of times that the rule $\alpha \rightarrow \beta$ is seen in the trees t1 ... tm, and $Count(\alpha)$ is the number of times the non-terminal α is seen in the trees $t_1 \ldots t_m$.
2. A parameter $q(\alpha \rightarrow \beta)$. For each rule $\alpha \rightarrow \beta \in R$. The parameter $q(\alpha \rightarrow \beta)$ can be interpreted as the conditional probabilty of choosing rule $\alpha \rightarrow \beta$ in a left-most derivation, given that the non-terminal being expanded is α. For any $X \in N$, we have the constraint:

$$\sum_{\alpha \rightarrow \beta \in R: \alpha = X} q(\alpha \rightarrow \beta) = 1$$

In addition we have $q(\alpha \rightarrow \beta) \geq 0$ for any $\alpha \rightarrow \beta \in R$. Given a parse-tree $t \in \mathcal{T}_G$ containing rules $\alpha 1 \rightarrow \beta 1, \alpha 2 \rightarrow \beta 2, \ldots, \alpha n \rightarrow \beta n$, the probability of t under the PCFG is:

$$p(t) = \prod_{i=1}^{n} q(\alpha i \rightarrow \beta i)$$

---

[1] http://nlp.stanford.edu/software/lex-parser.shtml

Two activities are executed by this module, the lexical tagging and the dependence analysis (Using PCFG). The results obtained by this module are shown in Fig. 2 and 3. We used the sentence (**S1**): "A self-propelled vehicle is a motor vehicle or road vehicle that does not operate on rails" to illustrate the results obtained by the translator's modules.

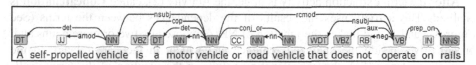

Fig. 2. Lexical tagging and dependence structure

Each word of the sentence (S1) above (Fig. 2) is grammatically classified according to their lexical categories and the dependence between them is attributed.

> (NP (DT A) (JJ self-propelled) (NN vehicle)) | (**VP** (VBZ is) (NP (DT a) (NN motor) (NN vehicle) | (CC or) (NN road) (NN vehicle)) | (**SBAR** (**WHNP** (WDT that)) | (**VP** (VBZ does) (RB not) (**VP** (**VB** operate) (PP (IN on) (NP (NNS rails))

Fig. 3. Classification in syntagmatic or sentential categories

Syntagmatic categories are in red and, in black, the lexicon to which each category pertains (See Fig. 3).

2.2 Semantic Parsing Module

The results of the activities carried out by the systems of the **Syntactic Parsing Module (1)** are used by the systems of the **Semantic Parsing Module (2)**, which carries out the activities shown in Figure 4 and are detailed as follows.

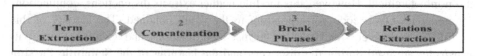

Fig. 4. Activities carried out in the Semantic Parsing Module

This module initiates its activities by assessing the entry sentence and the referred result of the syntactic analysis obtained in the previous module and then starts the extraction of terms (**Term Extraction**) that are fit to be concepts of the ontology (Activity (1)). In this phase, terms classified as prepositions (IN), conjunctions (CC), numbers (CD), articles (IN, CC, RB, DT+PDT+WDT) and verbs (EX+MD+VB+VBD+VBG+VBN+VBP+VBZ) are discarded, and the terms classified as nouns (NN+NNS+NNP+NNPS) and adjectives (JJ+JJR+JJS) are indicated as possible concepts of the ontology, therefore, the terms extracted, who are fit to be concepts were: motor/NN, vehicle/NN, road/NN, vehicle/NN, self-propelled/JJ, vehicle/NN and rails/NNS as presented in Figure 5.

Fig. 5. Result of the extraction of terms

After the term extraction activity is done, the Activity (2), called **Concatenation** is enabled. This activity uses the results of the dependences between the terms (See Figures 2 and 3) and makes the junction of NPs composed by two or more nouns and/or adjectives inside the analyzed sentence and which, in fact, are related. In the example sentence ($S1$), the concatenation results in the junction of the terms (self-propelled/JJ ↔ vehicle/NN), (motor/NN ↔ vehicle/NN) and (road/NN ↔ vehicle/NN) into an only term, because they are dependent of one another, resulting in just 3 terms: *self-propelled-vehicle*, *motor-vehicle* e *road-vehicle*, and no longer 6 terms, as in the initial phase of the **Term Extraction** activity.

In Activity (3), **Break Phrases,** every time terms or punctuation marks like comma (,), period (.), *and*, *or*, *that*, *who* or *which* (what we call sentence breakers) are found, the sentences are divided into subsentences and analyzed separately, the result for (S1) was:

A self-propelled-vehicle is a motor-vehicle | or road-vehicle | that does not operate on rails

The last activity to take place in the **Semantic Parsing Module** is Activity (4), **Relations Extraction**. The relations between the terms are verified and validated through verbs found in the sentences and patterns observed in the translator's inner grammar. The verbs are separated and the terms dependent on verbs are extracted, resulting in:

Self_propelled_vehicle **is a** motor_vehicle | self-propelled_vehicle **is a** road_vehicle | self_propelled_vehicle **operate** on rails

This module detects the terms and the relations between them, both hierarchical and nonhierarchical. However, this module neither extracts disjunctions, conjunctions nor generates OWL code corresponding to the result obtained. The activity of this module is exclusively for detecting terms, their relations and validity. Some patterns/rules are used during the discovery and the learning of the **Semantic Parsing Module (2).** The patterns/rules used for learning hierarchical axioms are showed in Table 1.

The generated result for the above patters are the same, *i.e.*, {SN_1, SN_2, SN_3, SN_n} ⊑ SN, notice that the **and** conector for the mentioned patterns play the role of conector of concepts, establishing among them a dependency relation. The patterns above also are recognized as axioms OR, AND, and NOT and may be represented in the following way (See Table 2):

Table 1. Patterns/Rules for transforming the hierarchical axioms of terms

Construction Patterns of hierarchical axioms of terms

(1) SN_1 (*is a* | *is an* | *is*) SN

(2) SN_1 (*are a* | *are an* | *are*) SN

(3) SN_1 *and* SN_2 *and* SN_3 *and* SNn (*are a* | *are an* | *are* | *is a* | *is an* | *is*) SN

(4) SN_1, SN_2 , SN_3, SN_n (*are a* | *are an* | *are* | *is a* | *is an* | *is*) SN

(5) SN_1, SN_2 *and* SN_3 (*are a* | *are an* | *are* | *is a* | *is an* | *is*) SN

(6) SN_1, SN_2 *and* SN_3, SN_4 (*are a* | *are an* | *are* | *is a* | *is an* | *is*) SN

(7) SN_1 (*is a* | *is an* | *is*) SN and/that/who/which (**is a** | **is an** | **is**) SN and (**is a** | **is an** | **is**) SN...

(8) SN_1 SNn (*are a* | *are an* | *are* | *is a* | *is an* | *is*) SN and SNn (*are a* | *are an* | *are* | *is a* | *is an* | *is*) SN_1 and SN_n...

(9) SN_1 *and* SN_2 *and* SN_3 *and* SNn (*are a* | *are an* | *are* | *is a* | *is an* | *is*) SN and (*are a* | *are an* | *are* | *is a* | *is an* | *is*) SN_1 and SN_n...

(10) SN_1, SN_2 *and* SN_3, SN_n (*are a* | *are an* | *are* | *is a* | *is an* | *is*) SN

(11) SN_1 (*are a* | *are an* | *are* | *is a* | *is an* | *is*) SN_2 *or/and* SN_3

Table 2. Patterns/Rules for transforming the hierarchical axioms of relations using Inserction (and), Conjuction (or) and Nagation (not)

Construction Patterns using verbs, intersection (and), conjuction (or) and negation (not)

(1) SN_1 (*are a* | *are an* | *are* | *is a* | *is an* | *is*) SN *or/and* SN (***That/Who/Which***) (***has* not**) SN

(2) SN_1 (*are a* | *are an* | *are* | *is a* | *is an* | *is*) SN *or/and* SN (***That/Who/Which***) (***has***) SN

(3) SN_1 (*are a* | *are an* | *are* | *is a* | *is an* | *is*) SN *or/and* SN (***That/Who/Which***) (***Verb***) SN

(4) SN_1 (Verbo) SN (*or/and*) SN

(5) SN_1 (Verb) SN (*or/and*) (Verb) SN

(6) SN_1 *has* SN_n *and* SN_n

(7) SN_1 (*has not*) SN_n

(8) SN_1 (*does not* | *doesn't* | *is not* | *isn't*) SN_n

(9) SN_1 (*does not* | *doesn't* | *is not* | *isn't*) (Verb) SN_n

(10) SN_1 *has* SN_n *and has not* SN_n

(11) SN_1 (*are a* | *are an* | *are* | *is a* | *is an* | *is*) SN (***That/Who/Which***) (Verbo) (*or*) (Verbo) SN

(12) SN_1 (*are a* | *are an* | *are* | *is a* | *is an* | *is*) SN *or/and* SN (***That/Who/Which***) (Verbo) (*or/and*) (Verbo) SN

(13) SN_1 (Verbo) SN (*or/and*) (Verbo) SN (*or/and*) (Verbo) SN...

One should notice that new construction patterns of hierarchical axioms and relations may be inserted to the internal grammar of the described approach by human intervention. All the patterns presented in Table 2 composes disjunction ($\sqcup$), conjunction ($\sqcap$) and negation ($\neg$) rules beyond the axioms with $\forall r.C$: universal restriction and $\exists r.C$: existencial restriction. In the next section, the operation of the **OWL DL Axioms Module** is described and the obtained results using its processing is showed.

2.3   OWL DL Axioms Module

The function of the **OWL DL Axioms Module** is to symbolically find/learn axioms that prevent ambiguous interpretations and limit the possible interpretations of the discourse, enabling systems to verify and disregard inconsistent data. The process of discovering the axioms is the hardest part of the process of creating ontologies. Here, the axioms discovered correspond to DL $\mathcal{ALC}$ expressivity. The module recognizes coordinating conjunctions (*OR* and *AND*), labeled CC, indicating the union (disjunction) and intersection (conjunction) respectively for concepts and/or properties, recognizes linking verbs followed by negations, like *does not* (or *doesn't*), *has not* (or *hasn't*), and *is not* (or *isn't*) for negation axioms ($\neg$), besides generating universal quantifiers ($\forall$) and existential quantifiers ($\exists$). In this module, we used Protégé-OWL API 3.5[2] and OWL API[3].

It also recognizes *is* and *are* as taxonomic relations ($\sqsubseteq$ - hierarchical). The transformations occur in four steps and make use of the results obtained by the previous modules:

Step (1): construction of taxonomic/hierarchical relations. The basic pattern used here is **<NPs> <VP> <NPs> where <VP> in this case is a (*is a/an, is* or *are*).** Other patterns are possible (See Table 1). For all the transformations, the patterns are automatically chosen by the translator. For our example, the results of Step (1) were:

self_propelled_vehicle is a motor_vehicle → *self_propelled_vehicle $\sqsubseteq$ motor-vehicle*
self_propelled_vehicle is a road_vehicle → *self_propelled_vehicle $\sqsubseteq$ road-vehicle*

By subsumption reasoning the implicit hierarchical relations (motor_vehicle $\sqsubseteq$ vehicle) and (road-vehicle $\sqsubseteq$ vehicle) are discovered and created automatically.

Step (2): construction of nonhierarchical relations. The pattern used here is **<NPs> <VP> <NPs> where <VP>** in this case is a verb other than **(is a/an, is or are).** Other patterns are possible (See Table 2).

*self_propelled_vehicle **operate** on rails* →
self_propelled_vehicle $\equiv \exists operate.rails$

---

[2] http://protege.stanford.edu/plugins/owl/api/
[3] http://owlapi.sourceforge.net/

Step (3): verification of conjunctions and disjunctions. The conjunctions OR and AND are verified and analyzed. They can be associated with concepts and/or properties. The pattern **<NPs>** *is a/an* or *are* **<NPs> <CC> <NPs>**, where **<CC>** is the conjunction *Or* or *And* that links two or more **<NPs>** is one of the patterns associated with union and intersections of concepts, and is chosen by the translator resulting in:

> A *self_propelled_vehicle is a motor_vehicle* **or** *road_vehicle* →
> *self_propelled_vehicle* ≡ *(motor-vehicle* ⊔ *road-vehicle)*

Step (4): detection of negations. The fourth analysis detects the negations, its dependences and classifies the sentence to apply the patterns. Two negations are possible: negations and disjunctions of concepts and negations of properties. Two patterns or a junction of these patterns are taken into consideration in the process of extraction of negation axioms for hierarchies: **<NPs> is not <NPs>** and the pattern **<NP>does not<VP><NP>** for negation of properties (See Table 2). For (S1), the following result was obtained:

> *self_propelled_vehicle that does not operate on rails* →
> *self_propelled_vehicle* ⊓ ¬∃*operate.rails*

The final result, after the integration of the partial results obtained by the three modules, for (S1) in OWL 2 code, was:

(S1): "*A self-propelled vehicle is a motor vehicle or road vehicle that does not operate on rails*"

> *self_propelled_vehicle* ≡ *(motor-vehicle* ⊔ *road-vehicle)* ⊓ ¬∀*operate.rails*
> *(motor_vehicle* ⊓ *road_vehicle)* ⊑ *vehicle*

Our approach generated 4 axioms, 2 of which being hierarchical axioms, 1 being the union between concepts and 1 other of negation of properties (Universally Restricting Property Values). The approach proposed by us is effective in patterns like this and makes correct or approximately correct interpretations of what the user desires. This statement can be verified by observing the early results in the following section.

3   Results and Discussions

3.1   Data Set

In order to validate the translator, sentences from various knowledge domains were used. The data set utilized in the experiments contains a total of 120 sentences and in all of them there were negation axioms, conjunction or disjunction axioms and/or two and/or three types of axioms in the same sentence, as well as axioms with definition of terms hierarchy. We opted for sentences found in *Wikipedia* glossaries because they offered in principle a controlled language without syntactic and semantic errors, besides providing a great opportunity for automatic learning. Some examples of sentences having negation, union and conjunction axioms used in the experiments and the respective results generated by the translator, along with a discussion on these results are shown as follows.

3.2   Generating OWL DL $\mathcal{ALC}$

Processed Sentence (1): juvenile is an young fish or animal that has not reached sexual maturity.

Result: $\rightarrow$ *juvenile* $\sqsubseteq$ *(young_fish* $\sqcup$ *young_animal)* $\sqcap$ $\neg$ *$\forall$hasReached.Sexualmaturity* | $\rightarrow$ *young_fish* $\sqsubseteq$ *fish* | $\rightarrow$ *young_animal* $\sqsubseteq$ *animal*

Discussion: The result of the analysis of the sentence is different from the results of the processing performed by the LExO system [16] [15]: Juvenile $\equiv$ (young $\sqcap$ (Fish $\sqcup$ Animal) $\sqcap$ $\neg\exists$reached.(Sexual $\sqcap$ Maturity). The compared system (LExO) classifies *young, fish* and *animal* as distinct terms, however, by the interpretation in natural language of the sentence in analysis, the word *young* is an adjective of the *fish* concept, thus, our approach classifies and represents '*young fish*' as a composite noun, that is, composing a single concept (*young_fish*). The same occurs for *sexual maturity*, these two terms are classified as a single concept in the same way as the classification of the previous concept (*young_fish*). We can also observe the creation of two axioms, one of union of concepts and one of negation of property. By subsumption reasoning (in the reasoning activities, we used the inference machine Pellet 2.2.2[4] [5]) the hierarchical axioms: young_fish $\sqsubseteq$ fish and young_animal $\sqsubseteq$ animal were discovered and automatic created by our approach. Figure 6 shows part of the resultant OWL DL axiom constructed automatically for sentence (**1**).

```
▼<owl:Class rdf:about="#?juvenile">
  ▼<rdfs:subClassOf>
    ▼<owl:Restriction>
      ▼<owl:onProperty>
        <owl:ObjectProperty rdf:ID="has_reached"/>
        </owl:onProperty>
      ▼<owl:allValuesFrom>
        ▼<owl:Class>
          ▼<owl:complementOf>
            <owl:Class rdf:about="#?sexual_maturity"/>
          </owl:complementOf>
        </owl:Class>
      </owl:allValuesFrom>
      </owl:Restriction>
    </rdfs:subClassOf>
  ▼<rdfs:subClassOf>
    ▼<owl:Class>
      ▼<owl:unionOf rdf:parseType="Collection">
        <owl:Class rdf:about="#?young_fish"/>
        <owl:Class rdf:about="#?young_animal"/>
        </owl:unionOf>
      </owl:Class>
    </rdfs:subClassOf>
  </owl:Class>
```

Fig. 6. Fragment of the OWL DL files for sentence (1)

---

[4] http://clarkparsia.com/pellet/

Processed Sentence (2): vector is an organism which carries or transmits a pathogen.

Result: → *vector* ⊑ (*organism* ⊓ (∃*carries.Pathogen* ⊔ ∃*transmits.Pathogen*))

Discussion: The result obtained in (2) was also compared with results generated by LExO [16]: Vector ≡ (Organism ⊓ (*carries* ⊔ ∃*transmit.Pathogen*)). The verb *to carry* was not correctly classified as an existential restriction when analyzed by LExO, whereas in our approach, the sentence was coherently classified, the existential quantifier was created and the disjunction of the relations created was performed, where carriesPathogen and transmitsPathogen are disjoint (⊔), which evidences the accurate interpretation of the sentence in natural language. Figure 7 shows part of the resultant OWL DL axiom constructed automatically for sentence (2).

```
▼<rdfs:subClassOf>
  ▼<owl:Class>
    ▼<owl:unionOf rdf:parseType="Collection">
      ▼<owl:Restriction>
        ▼<owl:onProperty>
          <owl:ObjectProperty rdf:ID="carries"/>
          </owl:onProperty>
        ▼<owl:someValuesFrom>
          <owl:Class rdf:about="#?pathogen"/>
          </owl:someValuesFrom>
        </owl:Restriction>
      ▼<owl:Restriction>
        ▼<owl:onProperty>
          <owl:ObjectProperty rdf:ID="transmits"/>
          </owl:onProperty>
          <owl:someValuesFrom rdf:resource="#?pathogen"/>
        </owl:Restriction>
      </owl:unionOf>
    </owl:Class>
  </rdfs:subClassOf>
```

Fig. 7. Fragment of the OWL DL files for sentence (2)

Processed Sentence (3): whale is an aquatic animal and mammal.

Result: → whale ⊑ (aquatic_animal ⊓ aquatic_mammal) | → *aquatic_animal* ⊑ *animal* | → aquatic_mammal ⊑ mamal | → whale ⊑ *animal* | → whale ⊑ mammal

Discussion: In this sentence, the concept *whale* forms a hierarchy with the other two concepts (*aquatic animal* and *aquatic mammal*). Besides, it generates an intersection of both terms, meaning that individuals pertaining to the concept *whale* pertain to the set of individuals of both concepts at the same time. By subsumption reasoning the hierarchical axioms: aquatic_animal ⊑ animal and aquatic_mammal ⊑ mammal were discovered and created automatically by our approach. Futhermore, our approach discovered, by deduction, the following axioms: whale ⊑ animal and whale ⊑ mammal, in this case, distinctly of sentence in (2), and as expected, the OWL 2 code associated to this deduction is not generated. The user is only informed that the translator was able to deduce according to the sentence processing and it is showed that it performs a reasoning although there are limitations. Figure 8 shows part of the resultant OWL DL axiom constructed automatically for sentence (3).

```
▼<owl:Class rdf:about="#?whale">
  ▼<rdfs:subClassOf>
    ▼<owl:Class>
      ▼<owl:intersectionOf rdf:parseType="Collection">
          <owl:Class rdf:about="#?aquatic_animal"/>
          <owl:Class rdf:about="#?aquatic_mammal"/>
        </owl:intersectionOf>
      </owl:Class>
    </rdfs:subClassOf>
  </owl:Class>
```

Fig. 8. Fragments of the OWL DL files for sentence (3)

Processed Sentence (4): fish is an aquatic animal and isn't a mammal.

Result: → fish $\sqsubseteq$ (aquatic_animal $\sqcap \neg$ mammal) | → *aquaic_animal* $\sqsubseteq$ *animal*

Discussion: The expected result was generated, the *fish* concept is subclass of *aquatic animal* concept and disjoint of *mammal* concept. Furthermore, the hierarchical relation *aquatic_animal* $\sqsubseteq$ *animal* was discovered by subsunction reasoning. This complement alone is useless, however when combined with an intersection, it can be really useful in reasoning tasks. Figure 9 shows part of the resultant OWL DL axiom constructed automatically for sentence **(4)**.

```
▼<owl:Class rdf:about="#?fish">
    <rdfs:subClassOf rdf:resource="#?aquatic_animal"/>
    <owl:disjointWith rdf:resource="#?mammal"/>
  </owl:Class>
```

Fig. 9. Fragments of the OWL DL files for sentence (4)

Processed Sentence (5): *human is a rational animal and oganic being.*

Result: → *human* $\sqsubseteq$ (*rational _animal* $\sqcap$ *organic being*) | *rational _animal* $\sqsubseteq$ *animal*
| *organic_being* $\sqsubseteq$ *being.*

Discussion: The concept domain (*human*) is subclass of an intersection between the concept elements (*rational_animal* $\sqcap$ *organic_being*). It points that each individual of the *human* concept is also a member of the concept equivalent to the intersection generated. Furthermore, the *being* and *animal* concepts, and also the axioms *rational _animal* $\sqsubseteq$ *animal* and *organic_being* $\sqsubseteq$ *being* were created by subsunction reasoning. Figure 9 shows part of the resultant OWL DL axiom constructed automatically for sentence **(5).** Figure 10 shows part of the resultant OWL DL axiom constructed automatically for sentence **(5).**

```
▼<owl:Class rdf:about="#?human">
  ▼<rdfs:subClassOf>
    ▼<owl:Class>
      ▼<owl:intersectionOf rdf:parseType="Collection">
          <owl:Class rdf:about="#?rational_animal"/>
          <owl:Class rdf:about="#?oganic_being"/>
        </owl:intersectionOf>
      </owl:Class>
    </rdfs:subClassOf>
  </owl:Class>
```

Fig. 10. Fragments of the OWL DL files for sentence (5)

3.3   Validation

The objective of planning and performing the experiments was to verify the model constructed and test it in order to answer the following research questions: is the translator capable of constructing minimal-expressivity $\mathcal{ALC}$ ontologies and represent knowledge? And, is the translator capable of identifying rules and axioms inherent to $\mathcal{ALC}$ expressivity based on the texts produced from dialogues with users? In order to answer these questions, the following hypothesis was formulated to assess the translator's performance during the experiment:

- $H_{a,1}$: The approach constructs the expressive ontology correctly in more than half of the cases;
- $H_{0,1}$: The approach does not construct the expressive ontology in more than half of the cases.

Analyzing all the 120 sentences, we obtained the following results: in 75% of the sentences analyzed (90 sentences), the translator detected and created coherently the axioms, whereas in 30 sentences (25%, also including the ones the translator could not possibly solve in any way). The translator did not detect the axioms coherently and committed errors; however, in 24 out of those 30 sentences, the translator created axioms and made possible the recreation of the ontologies through the proceeding of inserting new definitions. The right unilateral binomial test was applied and the following results obtained.

Limiar → 50% | Level of significance α (5%) | p-value = 1.886e-08

3.4   Sustention

As p-value is lower than the significance of the null hypothesis ($H_{0,1}$: The translator does not construct the ontology in more than half of the cases), it is not accepted, that is, the translator statistically constructs the ontology correctly in more than half of the cases. Using the same binomial test we confirm that this success ratio is statistically superior to 67% (0.67 < success ratio < 1; p-value=0.03636). Therefore, we conclude that the success ratio presented by our approach, 75%, is statistically higher than 67% and, in fact, significant.

4   Related Works

Our system conceiving was based in previous works proposed in the literature, despite some of them have distinct focus and detailing levels, all of them deal with the problem of generating expressive ontologies automatically and assistance to ontology engineering activities. Among the researches that contributed to maturing and development of our proposed approach, we should highlight the following works.

The LExO system LExO [16][15] published satisfactory results in state-of-the-art, it is also a comparative point for other systems which proposes automatic generators of expressive ontologies from natural language (NL) texts. The LExO[5] does not uses the Stanford Parser at syntactic analysis phase, but the MINIPAR [12]. Some published works pointed a slighted advantage to Stanford Parser in relation to others. This way, we would conclude that systems which uses better parsers obtain more accurate results. Despite LExO transforms and represents accordingly the processed text in OWL DL code, it does not performs subsumption reasoning and does not discovers implicit relations in the analyzed propositions.

Another excellent system related to our approach is the ACE[6] (Attempto Controlled English) [8][9][10]. It transforms the given definitions in OWL DL code. However, it uses a specific syntax for its input propositions, then users not familiar with this syntax in NL will face problems in using it. We highlight that ACE represents accordingly in OWL DL the propositions given by the user, however it do not present subsumption reasoning and it fails checking inconsistencies in the classes defined and transformed in OWL DL. The ACE works with ABOX and TBOX which may cause problems in DL representation, once it models individuals as concepts and concepts as individuals. The system does not performs deduction and subsumption reasoning.

Finally, we present the TextOntoEx [4] system. The authors point that the system discovers relations between terms, however it does not generate OWL DL code, and seems to create only RDF codes. Despite it has been validated, fragments of code automatically generated are not shown and PLN techniques used are not presented. Furthermore, the system does not performs subsumption reasoning, does not discovers other relations, does not deduces from a given set of propositions and does not check for inconsistencies.

Our approach uses the Stanford Parser, performs subsumption reasoning and checks inconsistencies in the created classes during the development, this way it is an advantageous system in relation to others previously cited here. It processes texts in NL without properly syntaxes, however it requires a formal language in the conception of propositions given to the system. We only work with TBOX to eliminate confusions between individuals and concepts in definition representations. Nevertheless, when a concept is classified as an individual, the user is alerted to fix it.

5   Conclusions and Future Works

In this paper, we describe an approach to automatic development of expressive ontologies from definitions provided by users. The results obtained through the experiments evidence the need of automatic creation of expressive axioms, sufficient to creating ontologies with $\mathcal{ALC}$ expressivity, besides the success in the identification of rules and axioms pertaining to $\mathcal{ALC}$ expressivity. We also conclude that the translator can aid both experienced ontology engineers and developers and inexperienced users just starting to create ontologies. As future works, we include the integration of our approach with other existing approaches in the literature, the creation of a module for automatic inclusion of unprecedented patterns in the translator and one module for automatic insertion of individuals for terms of ontologies created by the translator.

---

[5] https://code.google.com/p/lexo/
[6] http://attempto.ifi.uzh.ch/site/description/

References

1. Buitelaar, P., Cimiano, P.: Ontology learning and population: Bridging the gap between text and knowledge. In: Frontiers in Artificial Intelligence and Applications Series, vol. 167, IOS Press (2008)
2. Buitelaar, P., Cimiano, P., Magnini, B. (eds.): Ontology learning from text: Methods, applications and evaluation, pp. 3–12. IOS Press (2005)
3. Cimiano, P., Völker, J.: Text2Onto. NLDB, 227–238 (2005)
4. Dahab, M.Y., Hassan, H.A., Rafea, A.: TextOntoEx: Automatic Ontology Construction from Natural English Text. Expert Systems with Applications 34(2), 1474–1480 (2008)
5. Sirin, E., Parsia, B., Grau, B.C., Kalyanpur, A., Katz, Y.: Pellet: A practical OWL-DL Reasoner. J. Web Sem. 5(2), 51–53 (2007)
6. Gómez-Pérez, A., Fernández-López, M., Corcho, O.: Ontological Engineering with examples from the areas of Knowledge Management, e-Commerce and the Semantic Web. In: Advanced Information and Knowledge Processing, Springer, London (2004)
7. Horrocks, I., et al.: OWL: a Description-Logic-Based Ontology Language for the Semantic Web. In: The Description Logic Handbook: Theory, Implementation and Applications, 2nd edn., pp. 458–486. Cambridge University Press (2007)
8. Kuhn, T.: A Survey and Classification of Controlled Natural Languages. In: Computational Linguistics - Association For Computational Linguistics (2014)
9. Kuhn, T.: A Principled Approach to Grammars for Controlled Natural Languages and Predictive Editors. Journal of Logic, Language and Information 22(1) (2013)
10. Kuhn, T.: Controlled English for knowledge representation. Ph.D. thesis, Faculty of Economics, Business Administration and Information Technology of the University of Zurich (2010)
11. Madche, A., Staab, S.: Ontology learning for the semantic web. IEEE Intelligent Systems 16(2), 72–79 (2001)
12. Lin, D.: Dependency-based evaluation of MINIPAR. In: Proceedings of the Workshop on the Evalution of Parsing Systems at the 1st International Conference on Language Resources and Evaluation, LREC (1998)
13. Pease, A.: Ontology: A Practical Guide. Published by Articulate Software Press, USA (2011)
14. Simperl, E., VTempich, C.: Exploring the Economical Aspects of Ontology Engineering synthetic. In: Handbook on Ontologies, International Handbooks on Information Systems, pp. 337–358. Springer, Berlin (2009)
15. Völker, J.: Learning Expressive Ontologies. No. 002. In: Studies on the Semantic Web. AKA Verlag / IOS Press (2009) ISBN: 978-3-89838-621-0
16. Völker, J., Hitzler, P., Cimiano, P.: Acquisition of OWL DL Axioms from Lexical Resources. In: Franconi, E., Kifer, M., May, W. (eds.) ESWC 2007. LNCS, vol. 4519, pp. 670–685. Springer, Heidelberg (2007)
17. Zouaq, A.: An Overview of Shallow and Deep Natural Language Processing for Ontology Learning. In: Ontology Learning and Knowledge Discovery Using the Web: Challenges and Recent Advances. ch. 2, pp. 16–37. IGI Global, EUA (2011) ISBN 978-1-60960-625-1 (hardcover)

A Mobile Visual Technique to Support Civil Protection in Risk Analysis

Luca Paolino[1,*], Monica Sebillo[1], Genoveffa Tortora[1],
Giuliana Vitiello[1], and Marco Romano[2]

[1] Dipartimento di Studio e Ricerca Aziendale,
Management & Information Technology, University of Salerno
Via Giovanni Paolo II, 84084 Fisciano(SA), Italy
{lpaolino,msebillo,gvitiello}@unisa.it
[2] Departamento de Informtica
universidad carlos III de Madrid Madrid, Spain
mromano@inf.uc3m.es

Abstract. The importance of the territory defense is becoming a more and more important activity of national civil defense agencies. In our research we describe a participatory design process involving real stakeholders from the emergency management field, which has led to the development of a mobile application supporting on-site operators. The application was based on the requirements we derived by interviewing operators on a real scenario. The utility of the application is successively described on three typical scenarios.

Keywords: mobility, digital society, risk analysis.

1   Introduction

Vulnerable sites enlarge upon one or more dangerous or vulnerable geographic features inside a specific area. Experts consider a geographic feature as "dangerous" if, due to human or natural causes, it can induce the partial or total failure or loss of functionality of other features in its neighborhood [12]. Examples of dangerous sites are portions of land subject to landslides, chemical plants, dumps, etc. Differently, the presence of dangerous features, which can be subject to partial or total failure or loss of functionality, makes a geographic feature "vulnerable". Some examples of vulnerable features are civil infrastructures and human settlements. Anyway, the same geographic feature can be considered both as dangerous and vulnerable.

Examples of dangerous sites are portions of land subject to landslides, chemical plants, dumps, etc. Differently, the presence of dangerous features, which can be subject to partial or total failure or loss of functionality, makes a geographic feature "vulnerable". Some examples of vulnerable features are civil infrastructures and human settlements. Anyway, the same geographic feature

---

[*] Corresponding author.

B. Murgante et al. (Eds.): ICCSA 2014, Part VI, LNCS 8584, pp. 760–769, 2014.
© Springer International Publishing Switzerland 2014

can be considered both as dangerous and vulnerable. In this context, it is very important that nations are able to locate and evaluate vulnerability and dangerousness of their own territories also by taking into account human or natural factors. In order to reach this aim, every nation has established special departments and agencies whose goal is the territory control with all aspects of the issue, through several instruments and techniques functional to this aim. In particular, the diversity of emergency situations which originated from natural disasters occurred on the Earth in recent years, have raised several ICT research issues covering different areas, including data visualization, (geo)visual analytics, advanced (mobile) interfaces, communication technology and collaborative environments. The research efforts put to address those issues have come to highlight that reliable information and good communications are crucial requirements to achieve effectiveness in emergency response preparedness and management.

In this paper we describe a participatory design process involving real stakeholders from the emergency management field, which has led to the development of a mobile application supporting on-site operators for dynamic, advanced analysis of heterogeneous data. The application exploits the analytics power of a visual technique named Framy [1,8,9] meant to facilitate the analysis of an area surrounding a user in terms of qualitative and quantitative parameters. In order to elicit the application requirements, we performed a contextual inquiry on the territory of Salerno, Italy, which involved the Civil Defense Department of forest rangers. We interviewed subjects on the issue of vulnerable areas, on the kind of activities they commonly perform when dealing with an emergency and on the extent to which mobile devices (including cell phones and personal digital assistants, PDAs) are adopted for continuous information exchange between on-site operators and the emergency operating center. The goal of the study was to analyze current practices and gain insight on both the activities requiring the use of computers, as they really occur during the risk analysis phase, and the activities that could benefit from the adoption of (mobile) computer applications. The derived requirements guided the subsequent design of a mobile application addressing the need to dynamically analyse spatial data on small sized screen.

The paper is organized as follows. In Section 2, we describe the rangers' current practices addressed to patrol, identify and analyze vulnerable areas. Then, we describe the requirements that an information system should present to support this kind of users during their operations. Section 3 illustrates the mobile application designed on the basis of the requirements and three typical scenarios of usage which better illustrate the advantages coming from the adoption of Framy in the field of civil defense. Some final remarks conclude the paper.

2   Practices and Requirements of Emergency Management by Civil Defense Departments

Operators of the Civil Defense Department of Salerno are in charge to evaluate the risk level of vulnerable areas of the province. Some evaluations can be directly performed by the command center through advanced software that analyzes data stored in a database. Some others have to be performed in situ because the presence of an operator is essential. Moreover, the majority of data stored in the central database can be collected only in situ, therefore the involved operators have to explore the vulnerable areas to gather information and evaluate the risks. These task forces are continuously in contact with both the command center and the other rangers by means of radios, pocket radios and mobile phones. They use professional GPS modules to accurately annotate positions to orient themselves. The kinds of areas that they can explore are different, they can be in charge of patrolling urban areas, forests or vulnerable technological parks such as dams. During the exploration they need data coming from other groups of operators to make decisions about the risk level of an area. As an example, the water level of a river that flows near a urban area is too high, a violent storm is coming from a near region towards the city; these two pieces of information, if combined, can produce a high risk level for the urban area.

Another reason to combine data is to allow operators to make decisions about the route to be followed during the exploration. As an example, a team can alert about obstacles in their area that can make the route hard, such as fires and landslides. Decision makers, on the basis of such an information, are able to choose the most convenient route.

The proposed scenario of existing working practices gave us the opportunity to reason about the key requirements emerging from the context. At this stage, we were primarily interested in deriving a list of requirements that could be taken into account throughout the design process. Summing up all the considerations and discussions we had upon the fieldwork completion, and reasoning on the derived scenario, we were therefore able to elicit an initial set of requirements divided into five categories according to the classification described by Preece [11].The initial set of requirements and the corresponding rationale for the design of an application supporting rangers patrolling rural areas or estimating risk levels are specified as follows.

Functional requirements:

1. the application allows the exchange of heterogeneous data between operators working both on vulnerable sites in mobility and at the command center;
2. the application analyzes data in run-time to establish the risk level.

- Rationale (R): by sharing valuable information about natural phenomena and morphological data, users would be able to make better decisions during the risk analysis and the exploration of vulnerable site.
- Environmental and contextual Requirements(ECR): the application is used on the go.
- R: users are mobile, they need light instruments to nimbly move from a place to another.
- ECR: the application may be used in online mode.
- R: some environments could not be reached by communication networks.
- ECR: Users can receive a training provided by experts.

- R: Users are members of the staff of a professional organization that can provide for an adequate training .

Data requirements (DR):

1. the application accesses data related to the distribution of geographical elements, such as the extension of mountain areas, rivers and forest;
2. it accesses the elements that present both geographic and temporal components, such as storms, fires, natural disasters, that is, elements that are only momentarily present in the area.

- R: Operators are interested to get information about geographic elements all around them and natural phenomena that have temporally a geographical position.
- DR: Data must be accurate and frequently updated.
- R: Users make crucial decisions on the basis of provided data.

User profiles:

1. the range of the instruction level varies between Ordinary Level and Master Degree;
2. users already use some technological instruments during the exploration of a vulnerable area, such as GPS module, phone, and radio.

- R: All the information is carried out by the initial survey directly conducted in situ.
- Usability requirements (UR): the application should be easy to learn and should require a little training effort.
- R: The application is frequently used by operators working directly on vulnerable sites.
- UR: The user interface should be effective: it should provide users with a simple management of users mistakes.
- R: The application supports a critical task on the basis of data updated directly by users. Therefore, it is paramount to reduce the number of possible unintentional user mistakes.

3   Application Design and Usage Scenarios

Starting from the previous set of requirements we decided to exploit the visual analytics power of the technique named Framy [9] to provide users with a usable mobile application addressed to analyze the risk levels of the area surrounding the operators. Framy has proven a useful technique to address different problems related to the management of spatial data on small sized screens [6], including augmented reality applications [10] and multimodal interfaces [2] [4]. to support decision makers working directly on the disaster sites during the critical phases of the crisis management. This inspired us in reusing the technique in a different context where the mobility and the advanced analysis of heterogeneous data play the main roles.

Framy is explicitly targeted to enhance geographic information visualization on small-sized displays by providing for information clues about features of either the on-screen or the off-screen space. The rationale behind it is to display semi-transparent colored frames along the border of the device screen. Each frame is then partitioned and colored with a saturation index featuring the Hue-Saturation-Value model. The intensity of such an index is proportional to the result of an aggregate function which either aggregates a property of objects located in the corresponding map sector, either inside or outside the screen, such as sum and count, or calculates the distance between the map focus and a point of interest (POI) in that portion. Thus, for instance, a frame may indicate the amount of facilities located within some specific sectors which satisfied a given requirement. If several aggregates are displayed on the map, nested frames may be visualised along the borders, each one corresponding to a different aggregate, with a different color.

Formally, the colour intensity of a frame portion C_i can be expressed as
$$Intensity(C_i) = f(g(S_i)) \; \forall i \in \{1, \ldots, n\}$$
where f is a monotonic function; g is a function which aggregates a set of spatial data; and given U_i (invisible) which represents the map portion of a sector which lies out-side the screen and V_i (visible) which represents the map portion which is inside screen, S_i is one of either U_i or V_i or $U_i \cup V_i$ (Figure 1(a)). Figure 1(b)

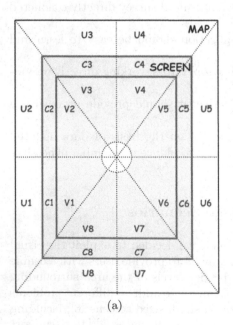

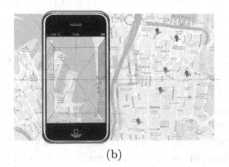

(a)   (b)

Fig. 1. (a) An example of (on/off-)screen subdivision accomplished by Framy, (b) Application running Framy

illustrates an application running the Framy technique. The frame is partitioned into eight coloured portions. The intensity of each portion is proportional to the number of POIs located around the map focus.

Details about the aggregate values associated to each sector, such as the number of POIs, the POI names and their distances from specified locations, may be required by tapping on the corresponding sector.

In order to allow users to easily distinguish colours and correctly interpret their meanings, the Framy visualisation technique has been supplied with colour rules derived from Itten's [7] theory which provides for a way to select highly contrasted colours. In particular, the rules we have adopted concern the colour combination of a frame and the map background, the colour combination of a frame and the corresponding features and the colour combination of nested frames. As for the territory analysis and the evaluation of its vulnerability and dangerousness, the application can be directly used during the exploration activity by exploiting the integrated localization sensors, data gathered on the go, and data being in other locations.

As Figure 2(a) shows, the applied technique is used, first of all, to analyze data surrounding the operators, such as the presence of forests, rivers, urban settlements and natural phenomena, as storms and snowfalls, and then to visual aggregate this data by using the visual metaphor of the frames. Combined elements generate a risk level: the higher the level, the more intense the color of the corresponding frame portion. Therefore, just by looking the frame, the operators immediately understand the directions where the vulnerable areas are and their estimated risk levels.

The criteria to combine elements can be directly set by the operators. For example, they can decide that an area is dangerous if at least three elements are present, such as a storm, a river and an urban area. Finally for each kind of combination the operator can set a different risk level.

The evaluation of the risk is performed directly by the mobile application. The choice of not involving a remote server is due to possibility that the operators may be in areas out of the communication networks and can work with off-line data. The application allows users to send updated information to the server of the command center. Then, it updates the current situation to the rangers in the interested area.

In the example depicted in Figure 2(a), icons represent the elements that can generate a risk. Circle shapes indicate the territorial extension of those elements. In northwestern areas snowfalls can cause serious damages to the routes through the mountains and forests, while in northeastern areas the presence of a river near to a urban area during the storm causes a higher risk level even.

The usage of the frame can be fully personalized. As an example, the user can set different nested frames; each frame can have a different hue that is related to different risk evaluations (Figure 3(a)). Another possibility is to exploit the analytical power given by the application of two aggregate functions to the same frame. The first function modifies the saturation while the second the brightness. It is formally expressed as follows:

- $saturation(C_i) = f_1(g_1(S_i, L_j)) \; \forall i \in \{1, \ldots, n\}$ and $j \in \{1 \ldots m\}$
- $brightness(C_i) = f_2(g_2(S_i, L_j)) \; \forall i \in \{1, \ldots, n\}$ and $j \in \{1 \ldots m\}$

where f_1, f_2 are monotonic functions; g_1, g_2 are functions which aggregate a set of spatial data of specific layer.

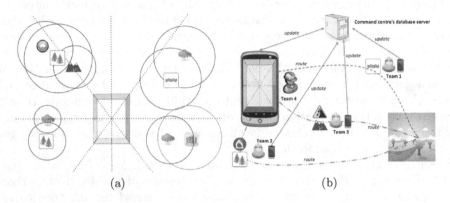

(a)   (b)

Fig. 2. (a) The frames display the analysis of the natural phenomena combined with the human structures, (b) Identification of the easiest path to follow

3.1   Scenario 1: Identification of a Safe Path in a Rural Area

In Figure 2(b), a typical scenario is depicted where the proposed application may be exploited. Four teams are working in patrolling a quite huge vulnerable area. The area is a natural park that is often at fire risk during the summer period. Moreover, the whole mountain range of the park is at risk of landslide.

Team 1 is in charge of patrolling an area beside the river. This area has been often flooded during the last winter; then it is necessary to check the status of the ground. Team 2 is exploring a forest area and notices a fire beginning caused by a pyromaniac. Team 3 is assigned to a mountain area at risk of landslide. Team 4 is the last team to start the mission and is in charge of monitoring the most distant area. In order to get that area they can follow three different routes across the mountains, namely the river and the forest. The first three teams send the information gathered during the monitoring activities to the command center server that informs all the teams about the current situation. The team 4 calculates the risk level for the different areas surrounding it through the proposed application. By analyzing the frame intensities, the decision maker takes in account the high difficulty level related to the routes crossing the forest affected by the fire and the mountain that presents some landslides. Meantime, as the team 1 reported that the status of the area surrounding the river is good, the frame portion related to that area presents a low intensity level. Therefore, the decision maker decides to follow the route across the river monitored by the team 1.

3.2 Scenario 2: Arsons Prevention

In this subsection a typical scenario is described where the application can be used to support the evaluation task of arson risks. A decision maker of a civil defense task force is planning to patrol a huge area in order to prevent arsons. He sets the application with two personalized frames. As Figure 3(a) shows, the first frame (the green one) represents the distribution of the areas vulnerable to the arsons. The risk is calculated on the basis of the vulnerable elements present in those areas. More in detail , the areas in the South, Southwest and Southeast are considered more vulnerable than the others, because of the high presence of forest. In this case, the intensities of the corresponding frame portions are very high. The urban areas, represented by the motorway and the city, are considered less vulnerable and the intensities of the corresponding frame portions are significantly lower. The areas on the West are considered not particularly vulnerable because the river makes those areas not easily accessible. The second frame (the red one) represents the distribution of the arsons during the last two years. In particular, the portion indicates that the areas with the highest number arsons in the past are those near the motorway and the mountain. In the Southern area the number of arsons is quite small, while they are totally absent near the river and the city. Therefore, the red frame, on one hand, confirms that the difficult accessibility of the Eastern areas protects them from the arsons, and it also highlights that the area around the motorway is the most exposed to the risk of arsons, even if the green frame portion corresponding to it has not a valuable color intensity. This is probably due to the easy access by pyromaniacs to that area.

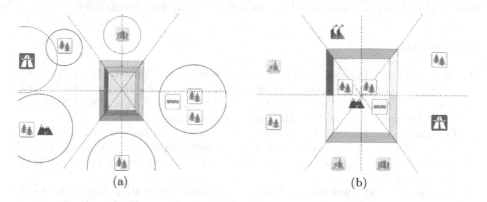

(a)   (b)

Fig. 3. (a) The usage of the technique to analyze the risks of fires, (b) The usage of Framy to evaluate the industrial pollution

3.3 Scenario 3: Pollution Control

The last scenario shows how a single frame can be exploited to analyze different kinds of information. The civil defense is evaluating the risks of contamination coming from the industrial and urban areas situated all around a natural park.

As it is shown in Figure 3(b), the single frame displays the level of produced pollution through the saturation, while the brightness indicates the distance between the origin of the pollution and the park. In particular, although the motorway does not produce a high level of pollution, its distance from the park is really short, thus increasing the level of contamination risk. The saturation of the corresponding portion is low, but its brightness is really high. Another example is given by the industrial park in the North. The pollution level is very high and the distance from the natural park is short. In this case both the brightness and the intensity level are quite high.

4   Conclusions

In the paper we have briefly described the design of a mobile application which allows to visualize information and aggregation of results on small screen devices. The scenario-based method we adopted was especially useful for participatory design . The participation of stakeholders and potential users in the development process started from the requirement analysis and continued throughout the design and prototyping phases, relying on their expertise in the domain and their feedback on all crucial choices. Interaction scenarios were also used to provide for evidence of the appropriateness of the proposed system for on-site emergency response, preparedness and management. At present we are planning a usability evaluation study for the developed application. It will involve emergency operators and allow us to identify where and how some improvements could be applied. In particular, the goal is to understand the extent to which the effectiveness of the Framy visualization technique can be enhanced.Finally, we are planning to extend the proposed metaphor for supporting visual data integration [5] [3].

References

1. De Chiara, D., Paolino, L., Romano, M., Sebillo, M., Tortora, G., Vitiello, G.: LINK2U: Connecting social network users through mobile interfaces. In: Qiu, G., Lam, K.M., Kiya, H., Xue, X.-Y., Kuo, C.-C.J., Lew, M.S. (eds.) PCM 2010, Part II. LNCS, vol. 6298, pp. 583–594. Springer, Heidelberg (2010)
2. De Chiara, G., Paolino, L., Romano, M., Sebillo, M., Tortora, G., Vitiello, G., Ginige, A.: The framy user interface for visually-impaired users. In: 2011 Sixth International Conference on Digital Information Management (ICDIM), pp. 36–41 (September 2011)
3. Deufemia, V., Moscariello, M., Polese, G.: Visually integrating databases at conceptual level. In: to appear in Proceedings of International Working Conference on Advanced Visual Interfaces. ACM, New York (2014)
4. Deufemia, V., Paolino, L., de Lumley, H.: Petroglyph recognition using self-organizing maps and fuzzy visual language parsing. In: 2012 IEEE 24th International Conference on Tools with Artificial Intelligence (ICTAI), vol. 1, pp. 852–859 (November 2012)
5. Deufemia, V., Giordano, M., Polese, G., Tortora, G.: A visual language-based system for extraction-transformation-loading development. In: to appear in Software: Practice and Experience (2013)

6. Ginige, A., Romano, M., Sebillo, M., Vitiello, G., Di Giovanni, P.: Spatial data and mobile applications: General solutions for interface design. In: Proceedings of the International Working Conference on Advanced Visual Interfaces. AVI 2012, pp. 189–196. ACM, New York (2012)
7. Itten, J.: The Art of Color, 1st edn. John Wiley & Sons, Inc., New York (1961)
8. Paolino, L., Romano, M., Sebillo, M., Tortora, G., Vitiello, G.: Audio-visual information clues about geographic data on mobile interfaces. In: Muneesawang, P., Wu, F., Kumazawa, I., Roeksabutr, A., Liao, M., Tang, X. (eds.) PCM 2009. LNCS, vol. 5879, pp. 1156–1161. Springer, Heidelberg (2009)
9. Paolino, L., Sebillo, M., Tortora, G., Vitiello, G.: Framy-visualising geographic data on mobile interfaces. Journal of Location Based Services 2(3), 236–252 (2008)
10. Paolino, L., Sebillo, M., Tortora, G., Vitiello, G., Laurini, R.: Phenomena: A visual environment for querying heterogenous spatial data. Journal of Visual Languages and Computing 20(6), 420–436 (2009)
11. Preece, J., Rogers, Y., Sharp, H.: Interaction Design, 1st edn. Apogeo, New York (2004)
12. Della Rocca, B., Fattoruso, G., Locurzio, S., Pasanisi, F., Pica, R., Peloso, A., Pollino, M., Tebano, C., Trocciola, A., De Chiara, D., Tortora, G.: SISI project: Developing GIS-based tools for vulnerability assessment. In: Sebillo, M., Vitiello, G., Schaefer, G. (eds.) VISUAL 2008. LNCS, vol. 5188, pp. 327–330. Springer, Heidelberg (2008)

Soft Computing Approach in Modeling Energy Consumption

Haruna Chiroma[1], Sameem Abdulkareem[1], Eka Novita Sari[2], Zailani Abdullah[3], Sanah Abdullahi Muaz[4], Oguz Kaynar[5], Habib Shah[6], and Tutut Herawan[7,8]

[1] Department of Artificial Intelligence
[4] Department of Software Engineering
[7] Department of Information system
University of Malaya
50603 Pantai Valley, Kuala Lumpur, Malaysia
[2] AMCS Research Center, Yogyakarta, Indonesia
[3] School of Informatics & Applied Mathematics
Universiti Malaysia Terengganu
Gong Badak, Kuala Terengganu, Malaysia
[6] Faculty of Computer Science and Information Technology
Universiti Tun Hussein Onn Malaysia
86400 Parit Raja, Batu Pahat, Malaysia
[5] Department of Management Information System
Cumhuriyet University, 58140, Turkey
[8] AMCS Research Center, Yogyakarta, Indonesia
hchiroma@acm.org, {sameem,tutut}@um.edu.my, eka@amcs.co,
zailania@umt.edu.my, samaaz.csc@buk.edu.ng,
habibshah.uthm@gmail.com, okaynar@cumhuriyet.edu.tr

Abstract. In this chapter, we build an intelligent model based on soft computing technologies to improve the prediction accuracy of Energy Consumption in Greece. The model is developed based on Genetic Algorithm and Co-Active Neuro Fuzzy Inference System (GACANFIS) for the prediction of Energy Consumption. For verification of the performance accuracy, the results of the propose GACANFIS model were compared with the performance of Backpropagation Neural network (BP-NN), Fuzzy Neural Network (FNN), and Co-Active Neuro Fuzzy Inference System (CANFIS). Performance analysis shows that the propose GACANFIS improve the prediction accuracy of Energy Consumption as well as CPU time. Comparison of the results with previous literature further proved the effectiveness of the proposed approach. The prediction of Energy Consumption is required for expanding capacity, strategy in Energy supply, investment in capital, analysis of revenue, and management of market research.

Keywords: Genetic algorithm, Co-Active Neuro Fuzzy Inference System, Energy Consumption, Prediction.

B. Murgante et al. (Eds.): ICCSA 2014, Part VI, LNCS 8584, pp. 770–782, 2014.
© Springer International Publishing Switzerland 2014

1   Introduction

Human's require energy for their daily activities to properly function. Almost all activities in the society require energy [1]. The consumption of energy has significantly improved in the last decade across the globe as a result of population increase as well as economic development. In both social and economic development, energy is viewed as a critical factor which impact on the wealth of individuals [2]. The world energy demand increases by 2.5% annually and it will continue to increase in the future as expected. In Greece, energy consumption includes the energy delivered to the industrial, transportation, household, among others. In the last sixteen years (16), from 1992 to 2007 energy uses in Greece increase from 14, 079,000 to 22,552,000 Tones of Oil Equivalent (TOE). In terms of percentage, the energy increases by 60% with about 4.1% yearly increment [1]. The prediction of energy consumption are required for expanding capacity, strategy in energy supply, investment in capital, analysis of revenue, and management of market research. Though, high level of uncertainty characterized the prediction of long term energy consumption, in some instances the prediction covers thirty (30) years into the future. This attracted the interest of scientist to propose novel methods for a relatively reliable and accurate prediction of long-term Greek Energy Consumptions (GEC) [2].

GEC was forecast using a model developed based on Backpropagation neural networks (BP-NN) [2]. Hernandez et al. [3] proposes a model based on NNs for the forecasting of electricity load for microgrids. An NNs model was build to predict the monthly energy consumption of Iran [4]. However, NNs is well known to be slow and can easily be trapped in local minima. In another study, the demand for fossil fuels in Turkey was estimated using Genetic Algorithm (GA) model [5]. The GA can effectively improve the performance of NNs [6] which motivated Azadeh et al. [7] to hybridized NNs through GA searches to build a model for the prediction of electrical energy consumption in Turkey. Padmakumari et al. [8] applied fuzzy NNs (FNN) since it combined the power of NNs and fuzzy logic to create more effective hybrid model to build a synergistic model for the prediction of long-term energy consumption. Experimental evidence in [9] indicated that Co-active neuro fuzzy inference system (CANFIS) performs better than the FNN. Yet, CANFIS has been trained using the backpropagation algorithm (BP) which is susceptible to the limitations earlier mentioned. The GA can be applied for training without being trapped in the local minima and is faster than the BP in convergence to the optimal solution [10]. Fuzzy regression has a unique advantage of modeling using a small amount of datasets [11].

To improve the performance of CANFIS and considering the relatively few experimental data at our disposal, we propose to genetically optimize the parameters of CANFIS to build a GACANFIS model for the prediction of GEC.

The rest of this chapter is organized as follows. Section 2 presents a detailed description of the soft computing methodologies. Section 3 presents the results and a discussion of the study before concluding remarks and further works in Section 4.

2   Proposed Method

2.1   Genetic Algorithm

The GA operates by the initialization of the random generation of chromosomes in a population believe to have the solution to the problem. Fitness for each of the chromosomes in the population is evaluated to select those with the best fitness value for mating. The iteration is terminated when the criteria set for termination is reached. For example, if the error is reduced to a predefine threshold, number of generations exceed predefine limits, predefine time, exceed predefine number of generations without improvement, etc. The chromosomes in the population are selected based on their fitness, many selection methods exist, e.g. Roulette wheel, ranking, Boltzmann etc. Genetic operators such as crossover, mutation rate, selection generation gap, elitism etc. is applied to reproduce the next generation. The reproduction continues in an iterative manner until the predefine stoppage criteria is reached and the optimal chromosome typically in the current population is returned as the best solution to the problem [12].

2.2   Co-Active Neuro Fuzzy Inference System

Jang *et al.* [13] CANFIS is a more general category of neuro fuzzy inference systems. It can be used as a universal approximation model for the mapping nonlinear function of any type. The capability and strengths of CANFIS lie on the advantages of hybridizing modular neural network and fuzzy inference systems. Mostly, the strength of CANFIS comes from the weights between the consequent layer and fuzzy association layer. The major constituent of CANFIS is fuzzy neurons through which membership function (MFs) is applied to the CANFIS inputs that are fed externally to the network. There are two commonly MFs that are used in the literature, namely, generalized bell and Gaussian [14]. The output of CANFIS is expanded by normalized axon within the range of 0 to 1 and the modular network supply function rules to inputs. The number of experts contain in the modular network correspond to the number of outputs and the number of neurons in each expert correspond to the number of MFs. This hybrid intelligent system has a combination of axon which applies MFs outputs to the modular neural network outputs. In the final stage of CANFIS operations the ensemble outputs pass through the last output layer. The error in the output is propagated to MFs as well as the modular neural network, every node in the first layer is the membership grade of the fuzzy set and the degree to which the input vectors belongs to one of the fuzzy set. The product of all outputs from the first layer is computed and transmits to the second layer. The upper, and lower components of the third layer, apply MFs to every input and represent modular neural network that carry out summation computation for each output [15].

2.3   Design of the Proposed GACANFIS for Modeling the Greek Energy Consumption

2.3.1   Greek Energy Consumption Datasets

The data required for the modeling of GEC in this research are collected from different sources. The installed power capacity (IPC) and the residence yearly/electricity consumption (RYEC) data were collected from [16]. The yearly ambient temperature (YAT) data were extracted from [17] and [1] supply the Greek gross domestic product (GGDP) and the GEC data. Though, there are other factors that determine the energy consumption, such as the amount of CO_2 pollution, number of air conditioners, electricity price, installation of renewable energy technologies among others, but the record of the historical data are not available as pointed out in [2].

Table 1. Descriptive statistics of the GEC

	Observation	Minimum	Maximum	Mean	Std Deviation
GEC	22	2	72	31.95	20.37

Therefore, we decided to use only the variables with available data since it was argued in [18] that data availability is one of the criteria's for the selection of independent variables for modeling purposes.

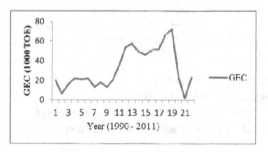

Fig. 1. Fluctuation of GEC from 1990 to 2011

The dependent variable is the GEC because its fluctuation as depicted in Fig. 1 depends on the independent variables, the descriptive statistics on the GEC data is presented in Table 1, in which the standard deviation computed based on Eq. (1) shows the dispersion of distribution in the data.

$$\sigma = \sqrt{\frac{1}{n}\sum_{i=0}^{n} x_i^2 - \left(\frac{1}{n}\sum_{i=0}^{n} x_i\right)^2} \,, \tag{1}$$

where σ, n and x_i represent the standard deviation, observations in the sample dataset, and real number of the random variable respectively.

Fig. 1 clearly shows the increase in the demand of energy consumption in Greece and a sudden fall, which could likely be attributed to the financial crises, but the details is not within the focus of this research. The raw data were not normalized to prevent the destruction of the original pattern in the historical data, a practice suggested by [19].

2.3.2   Design of the Proposed GACANFIS Model

The GACANFIS approach proposes to model the GEC. The structure of the CANFIS used in this study comprised of four input vectors representing RYEC, YAT, GGDP, and IPC. The output layer contained the output neuron which produces the GEC. The objective of the GA is to turn the CANFIS in order to determine the optimal parameter settings of fuzzy weights, mean and standard deviation of the MFs since they significantly affect the prediction performance of the CANFIS model. The datasets described in sub-section 2.3.1 are partitioned into 80% (1990 - 2008) for training and 20% (2007 - 2011) for testing the efficacy of the propose GACANFIS model. The CANFIS fitness function is the accuracy between the predicted values of GEC and the actual values. In this research, we chose Mean Square Error (MSE) computed using Eq. (2) as the fitness function because [20] argued that the MSE is more preferable than other measurement metrics such as normalized mean square error, mean absolute error, relative mean absolute error among others in measuring the accuracy of several algorithms on the same datasets.

$$MSE = \frac{\sum_{j=0}^{k} \sum_{i=0}^{n} (a_{ij} - x_{ij})^2}{nk}, \tag{2}$$

where

k = Neurons in the output layer
n = Observations in the dataset
x_{ij} = CANFIS output for observation i at neuron j
a_{ij} = Actual GEC for observation i at neuron j

The initial population of the chromosomes is randomly selected, typically fifty (50) chromosomes in the population based on the GA standard parameter values proposes by De Jong in 1975, reported in [21]. Each of the chromosomes in the population is evaluated using MSE.

The GA stop execution if the criteria set for termination is reached, otherwise the new generation of the chromosomes is created using crossover and mutation operators. The new population is created from chromosomes selected based on their fitness values. The chromosomes with the best fitness values are selected for mating whereas those with poor fitness values were rejected. Roulette wheel was the selection technique after trials with Boltzmann and rank selection. The crossover and mutation rates use in this simulation are 0.6 and 0.001 respectively, adopted from [21]. The value of generation gap, generation gap strategy, and chromosome length were selected after experimentations. The children resulted from the mating replace the older population of the chromosomes and iteratively continue searching the problem space. The iteration typically results to improve performance of the GACANFIS fuzzy weights, mean and standard deviation of the MFs. The generations of the population in our simulation was set to terminate when the GA runs for thirty (30) generations without improvement in the best fitness value found in the last population. Subsequently, the optimal chromosome is returned as the candidate solution, in our case the best GACANFIS model with the best fuzzy weights, mean and standard deviation of the MFs is returned as the final model for the prediction of GEC. The conceptual framework for the propose methodology is presented in Fig. 2. For comparison, BP-NN, FNN, Autoregressive Integrated Moving Average (ARIMA), and CANFIS were applied to predict the GEC. The models of the comparison methods were realized through trial and error technique to obtain the best model with the optimal performance.

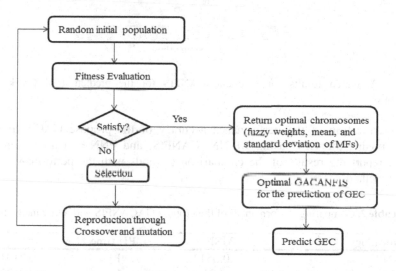

Fig. 2. The optimization of CANFIS through GA operation to build GACANFIS

3   Results and Discussion

3.1   The Performance Accuracy of the Propose GACANFIS Model and Its Evaluation

The approach proposed in the study was implemented in GeneHunter, PB-NN implemented in MATLAB 2013a Neural Network ToolBox and CANFIS was implemented in Neurosolution on a computer system (HP L1750 model, 4 GB RAM, 232.4 GB HDD, 32-bit OS, Intel (R) Core (TM) 2 Duo CPU @ 3.00 GHz).

   The optimal parameters of the GA that build the GACANFIS are: the GA operators - crossover rate was set to 0.6, mutation rate 0.001, population size fifty (50), roulette wheel was the selection technique. The generation gap was 0.76 with elitist strategy, chromosome length was 32 bits. The GA was set to terminate evolution when the best fitness runs for thirty (30) generations without improvement. The structure of the CANFIS was Gaussian. Fig. 3 is the simulation of the GA during building the GACANFIS, the straight line at the end of the learning curve shows convergence to the optimal solution and the GACANFIS with the optimal fuzzy weights, mean and standard deviation of the MFs was returned as the best model for the prediction of GEC.

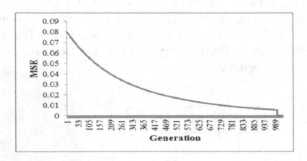

Fig. 3. GA Search results for modeling CANFIS for the prediction of Greek energy consumption

For verification of the performance accuracy of the propose GACANFIS, this chapter predict the GEC using BP-NN, CANFIS, and FNN as earlier mentioned. Table 2 report the results of the comparison methods with the performance of our propose method.

Table 2. Comparing performances of the propose GACANFIS with other methods

Methodology	MSE	CPU time	r
BP-NN	0.05132	13	0.7191
CANFIS	0.0156	16	0.8137
ARIMA	0.593	21	0.5321
FNN	0.02311	15	0.8691
Propose GACANFIS	**0.0000627**	**9**	**0.9213**

Table 2 shows the performances of the soft computing methods applied for the prediction of GEC. The results suggest that the propose GACANFIS significantly improve the performances of the methods previously utilized in the literature for the modeling of energy consumption. The performance criteria's display in Table 2 clearly indicates the robustness and effectiveness of the propose method. Also, the propose GACANFIS was found to be faster than the other comparison methods in converging to the optimal solution. The probable reason for this performance could be as a result of the GA capability to avoid being trapped in a local minimum, and searches for a very large space to obtain the optimal fuzzy weights, mean and standard deviation of the MFs which have not been possible without the GA. The performance of the our proposal is more evident when considering MSE and CPU time as performance criteria's whereas in terms of correlation coefficient (r) between the GACANFIS output (y) and actual GEC (a) computed using Eq. (3), though the GACANFIS has the best values, but is close to the r values of the other methods.

$$r = \frac{\dfrac{\sum_i (y_i - \overline{y})(a_i - \overline{a})}{n}}{\sqrt{\dfrac{\sum (a_i - a)^2}{n}} \sqrt{\dfrac{\sum_i (y_2 - \overline{y})^2}{n}}} \tag{3}$$

This is not surprising as r indicates directional movement of the predicted and observed GEC. On the other hand, MSE shows error between predicted and observed values of GEC. This means that MSE can be changed without r being affected and vice versa. Among the other methods, including BP-NN, FNN, and CANFIS, their respective MSE values are close to each other. The worst performance was recorded by ARIMA model the probable reason for this performance can likely be caused by its linear nature which makes it unsuitable for solving complex and nonlinear problems including modeling of energy consumption. Though, the results generated by ARIMA are not surprising because experimental evidence in [22] among other literature with similar findings prove the superiority of soft computing techniques over statistical methods.

Fig. 4 display the performances of the propose GACANFIS and the comparative methods on the GEC test dataset to show the generalization ability of the methods in the prediction of GEC. The propose GACANFIS performs better than the BP-NN, CANFIS, and FNN as the predicted values of GEC from 2007 to 2011 are more closer to the actual GEC values than the GEC values predict by BP-NN, CANFIS, and FNN. Based on the promising values of GEC predict by the GACANFIS and the significant value of r as shown in Table 2. Therefore, the propose GACANFIS model can comfortably be applied to predict long-term GEC. The results of the study can favorably be compared with the results in the literature.

Table 3. The percentage error between actual GEC and GEC predict by GACANFIS

Year	% Error
2011	0.27
2010	0.14
2009	0.22
2008	1.10
2007	0.12

$$\%\mathrm{Error} = \frac{\mathrm{Actual\,GEC - GACANFIS\,predicted\,GEC}}{\mathrm{Actual\,GEC}} \times 100 . \qquad (4)$$

We found only the work of [2] which predicts the GEC uses soft computing techniques in the context of Greece. Thus, we select it for fair comparison since we also applied soft computing techniques to predict GEC in the context Greece. The percentage error of the propose GACANFIS presented in Table 3 computed using Eq. (4) indicates performance improvement over the 2% percentage error reported in [2]. The difference in the results could be attributed to the effectiveness of our approach which probable reason was earlier explained.

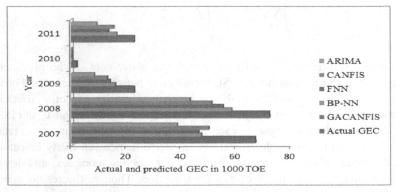

Fig. 4. Comparing GEC prediction performances of the soft computing technologies

Fig. 4 display the performances of the propose GACANFIS and the comparative methods on the GEC test dataset to show the generalization ability of the methods in the prediction of GEC. The propose GACANFIS performs better than the BP-NN, CANFIS, and FNN as the predicted values of GEC from 2007 to 2011 are more closer to the actual GEC values than the GEC values predict by BP-NN, CANFIS, and FNN. Based on the promising values of GEC predict by the GACANFIS and the significant value of R^2 as shown in Table 2. Therefore, the propose GACANFIS model can comfortably be applied to predict long-term GEC. The results of the study can favorably be compared with the results in the literature.

3.2   Statistical Test

Non parametric statistical test was deployed for assessing the performances of the propose GACANFIS, FNN, BP-NN, ARIMA, and CANFIS on the GEC test datasets. The analysis of variance (ANOVA) was employed due to multiplicity of the independent datasets predict by each of the five (5) models which t-test cannot explore the significant difference among those models as t-test is limited to two independent samples. The ANOVA is a statistical test among several means under the hypothesis that they are equal. The ANOVA test result is determined based on *p-values* typically 0.05 at 95% confidence interval. The result based on *p-value* is that, if the critical value is less than the *p-value, then* the null hypotheses are rejected otherwise the research hypothesis is accepted.

Table 4. Post hoc multiple comparison test results

(I) Group	(J) Group	Mean Difference (I-J)	P	0.05 Confidence Interval Lower Bound	Upper Bound
1.00	2.00	.07708	1.000	-49.8262	49.9804
	3.00	9.73140	.990	-40.1719	59.6347
	4.00	11.55070	.978	-38.3526	61.4540
	5.00	11.42729	.979	-38.4760	61.3306
	6.00	17.50530	.883	-32.3980	67.4086
2.00	1.00	-.07708	1.000	-49.9804	49.8262
	3.00	9.65432	.990	-40.2490	59.5576
	4.00	11.47362	.979	-38.4297	61.3769
	5.00	11.35022	.980	-38.5531	61.2535
	6.00	17.42822	.885	-32.4751	67.3315
3.00	1.00	-9.73140	.990	-59.6347	40.1719
	2.00	-9.65432	.990	-59.5576	40.2490
	4.00	1.81929	1.000	-48.0840	51.7226
	5.00	1.69589	1.000	-48.2074	51.5992
	6.00	7.77390	.996	-42.1294	57.6772
4.00	1.00	-11.55070	.978	-61.4540	38.3526
	2.00	-11.47362	.979	-61.3769	38.4297
	3.00	-1.81929	1.000	-51.7226	48.0840
	5.00	-.12340	1.000	-50.0267	49.7799
	6.00	5.95460	.999	-43.9487	55.8579
5.00	1.00	-11.42729	.979	-61.3306	38.4760
	2.00	-11.35022	.980	-61.2535	38.5531
	3.00	-1.69589	1.000	-51.5992	48.2074
	4.00	.12340	1.000	-49.7799	50.0267
	6.00	6.07801	.999	-43.8253	55.9813
6.00	1.00	-17.50530	.883	-67.4086	32.3980
	2.00	-17.42822	.885	-67.3315	32.4751
	3.00	-7.77390	.996	-57.6772	42.1294
	4.00	-5.95460	.999	-55.8579	43.9487
	5.00	-6.07801	.999	-55.9813	43.8253

Actual GEC (1), GACANFIS (2), BP-NN (3), FNN (4), CANFIS (5), and ARIMA (6).

The ANOVA [F (Sig. = 0.86, df = 5, 29, P > 0.05) = 0.374], and Tukey results tabulated in Table 4 shows that there is no significant difference among the GEC values predicted by propose GACANFIS, FNN, BP-NN, ARIMA, CANFIS and the original GEC values. However, the mean difference of the propose GACANFIS predicted GEC and the Original GEC is very low and very close to the *p-value* (column 3 of Table 4) whereas others are very high. Implies that the propose GACANFIS predicted GEC values are promising than other GEC values produce by the comparison models. The means difference between the GEC values predicted by ARIMA model and the original once is very high and more than all other methods which signify poor performance.

3.3   Long – Term Prediction of Energy Consumption

The performance accuracy of the propose GACANFIS is promising and robust. The sigficant r values of original GEC and predicted once shows that the model has a capability of representing a real life system. Therefore, the model can be relied upon and applied to predict future values of GEC considering its performance and improvement made over approaches in the literature. As pointed out in [23], a model with a significant r constitute a true representation of the real system. Thus, we applied the propose GACANFIS to predict the long term GEC presented in Figure 5.
Fig. 5 is a long term prediction of the GEC from 2014 up to 2023, the trend depicted by the figure indicated that the GEC will continue to fluctuate in the future but as a whole, the GEC demand will experience growth. This implies there will be economic development in Greece since energy consumption is positively correlated with economic development as reported in [1]. The policy makers in Greece can fine our propose GACANFIS model useful in the decision making process, especially on issues related to energy consumption and economic development.

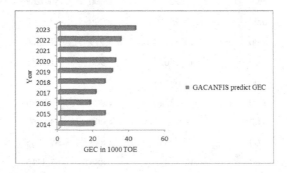

Fig. 5. Long term GEC predicts by GACANFIS

4   Conclusion and Future Work

The research presented in this chapter is to improve the prediction accuracy of Greek energy consumption. The soft computing technologies applied for modeling the Greek energy consumption comprised of GA [24], fuzzy logic [25-27], inference system, and modular NNs. Specifically, in hybrid form is referred to as GACANFIS. Datasets were collected from various sources and use to build the propose GACANFIS model for the prediction of Greek energy consumption. For evaluation purposes, FNN, BP-NN, ARIMA, and CANFIS were also applied to predict Greek energy consumption for comparing the results with that of the propose GACANFIS. Simulation of the comparative analysis indicates that the propose GACANFIS outperform the FNN, BP-NN, ARIMA, and CANFIS in terms of MSE, R^2, and CPU time. We take additional steps to compare the results with existing results in the literature and it was found to perform better than the results reported in the literature. The prediction of energy consumption is required for expanding capacity, strategy in energy supply, investment in capital, analysis of revenue, and management of market research. These can be applied in the context of Greece based on our propose model. Further research will be conducted to predict energy consumption of six (6) European countries and six (6) African countries based on data mining techniques selection criteria's proposed by Chiroma et al. [28]. Uncertainties will be considered in the future work by modifying the framework of Chiroma et al. [29] to suit the prediction of GEC while considering uncertainties.

Acknowledgments. This work is supported by University of Malaya High Impact Research Grant no vote UM.C/625/HIR/MOHE/SC/13/2 from Ministry of Education Malaysia.

References

1. European Statistics, http://ec.europa.eu/eurostat
2. Ekonomou, L.: Greek long-term energy consumption prediction using artificial neural networks. Energy 35(2), 512–517 (2010)
3. Hernandez, L.S., Baladrón, A.J.M., Carro, B., Sanchez-Esguevillas, A.J., Lloret, J.: Short-term load forecasting for microgrids based on artificial neural networks. Energies 6(3), 1385–1408 (2013)
4. Azadeh, A., Ghaderi, S.F., Sohrabkhani, S.: A simulated-based neural network algorithm for forecasting electrical energy consumption in Iran. Energ Policy 36(7), 2637–2644 (2008)
5. Canyurt, O.E., Ozturk, H.K.: Application of genetic algorithm (GA) technique on demand estimation of fossil fuels in Turkey. Energ Policy 36(7), 2562–2569 (2008)
6. Samanta, B.: Gear fault detection using artificial neural networks and support vector machines with genetic algorithms. Mech. Syst. Signal Pr. 18(3), 625–644 (2004)
7. Azadeh, A., Ghaderi, S.F., Tarverdian, S., Saberi, M.: Integration of artificial neural networks and genetic algorithm to predict electrical energy consumption. Appl. Math. Comput. 186(2), 1731–1741 (2007)
8. Padmakumari, K., Mohandas, K.P., Thiruvengadam, S.: Long term distribution demand forecasting using neuro fuzzy computations. Int. J. Elec. Power 21(5), 315–322 (1999)
9. Chiroma, H., Abdulkareem, S., Abubakar, A., Zeki, A., Gital, A.Y.U., Usman, M.J.: Co—Active neuro-fuzzy inference systems model for predicting crude oil price based on OECD inventories. In: 2013 International Conference on Proceedings of the IEEE Research and Innovation in Information Systems (ICRIIS), pp. 232–235 (2013)

10. Furtuna, R., Curteanu, S., Leon, F.: An elitist non-dominated sorting genetic algorithm enhanced with a neural network applied to the multi-objective optimization of a polysiloxane synthesis process. Eng. Appl. Artif. Intell. 24, 772–785 (2011)
11. Chan, K.Y., Kwong, C.K., Tsim, Y.C.: Modelling and optimization of fluid dispensing for electronic packaging using neural fuzzy networks and genetic algorithms. Eng. Appl. Artificial Intell. 23(1), 18–26 (2010)
12. Season, D.: Non – linear PLS using genetic programming. PhD thesis submitted to school of chemical engineering and advance materials, University of Newcastle (2005)
13. Jang, J.S.R., Sun, C.T., Mizutani, E.: Neuro-fuzzy and soft computing. Prentice-Hall, New Jersey (1997)
14. Principe, C.J., Euliano, N.R., Lefebvre, W.C.: Neural and adaptive systems: fundamentals through simulations. Wiley, London (2000)
15. Ishak, S., Trifiro, F.: Neurol networks, transportation research circular E-C113. Artif. Intell. Transp. 17–32 (2007)
16. Public Power Corporation SA, http://www.dei.gr
17. Greek national meteorological service, http://www.hnms.gr
18. Khazem, H.A.: Using artificial neural network to forecast the futures prices of crude oil. PhD dissertation, Nova south Eastern University, Florida (2007)
19. Jammazi, R., Aloui, C.: Crude oil forecasting: experimental evidence from wavelet decomposition and neural network modeling. Energ. Econ. 34, 828–841 (2012)
20. Peter, G.Z., Patuwo, B.E., Hu, M.Y.: A simulation study of artificial neural networks for nonlinear time-series forecasting. Comp. Oper. Res. 28, 381–396 (2001)
21. Grefenstette, J.J.: Optimization of control parameters for genetic algorithms. IEEE T. Syst. Man Cy. 16(1), 122–128 (1986)
22. Chiroma, H., Abdulkareem, S., Abubakar, A.I., Sari, E.N., Herawan, T.: A Novel Approach to Gasoline Price Forecasting Based on Karhunen-Loève Transform and Network for Vector Quantization with Voronoid Polyhedral. In: Linawati, Mahendra, M.S., Neuhold, E.J., Tjoa, A.M., You, I. (eds.) ICT-EurAsia 2014. LNCS, vol. 8407, pp. 257–266. Springer, Heidelberg (2014)
23. Azar, A.T.: Fast neural network learning algorithms for medical applications. Neural Comp. Appl. 23(3-4), 1019–1034 (2013)
24. Tan, L., Taniar, D.: Adaptive estimated maximum-entropy distribution model. Information Sciences 177(15), 3110–3128 (2007)
25. Kalia, H., Dehuri, S., Ghosh, A.: A Survey on Fuzzy Association Rule Mining. International Journal of Data Warehousing and Mining 9(1), 1–27 (2013)
26. Wan, R., Gao, Y., Li, C.: Weighted Fuzzy-Possibilistic C-Means Over Large Data Sets. International Journal of Data Warehousing and Mining 8(4), 82–107 (2012)
27. Kwok, T., Smith, K.A., Lozano, S., Taniar, D.: Parallel Fuzzy c-Means Clustering for Large Data Sets. In: Monien, B., Feldmann, R.L. (eds.) Euro-Par 2002. LNCS, vol. 2400, pp. 365–374. Springer, Heidelberg (2002)
28. Chiroma, H., Abdul-Kareem, S., Abubakar, A.: A Framework for Selecting the Optimal Technique Suitable for Application in a Data Mining Task. In: Park, J.J(J.H.), Stojmenovic, I., Choi, M., Xhafa, F. (eds.) Future Information Technology. LNEE, vol. 276, pp. 163–169. Springer, Heidelberg (2014)
29. Chiroma, H., Abdulkareem, S., Gital, A.Y.: An Intelligent Model Framework for Handling Effects of Uncertainty Events for Crude Oil Price Projection: Conceptual Paper. In: Proceedings of the International Multi Conference of Engineers and Computer Scientists, vol. I, pp. 58–62 (2014)

Mining Indirect Least Association Rule from Students' Examination Datasets

Zailani Abdullah[1], Tutut Herawan[2,3], Noraziah Ahmad[4],
Rozaida Ghazali[5], and Mustafa Mat Deris[5]

[1] School of Informatics & Applied Mathematics
Universiti Malaysia Terengganu
Gong Badak, Kuala Terengganu, Malaysia
[2] Faculty of Computer Science & Information Technology
University of Malaya
50603 Pantai Valley, Kuala Lumpur, Malaysia
[3] AMCS Research Center, Yogyakarta, Indonesia
[4] Faculty of Computer System and Software Engineering
Universiti Malaysia Pahang
Lebuh Raya Tun Razak, Gambang, Kuantan, Pahang, Malaysia
[5] Faculty of Science Computer and Information Technology
Universiti Tun Hussein Onn Malaysia
86400 Parit Raja, Batu Pahat, Johor, Malaysia
zailania@umt.edu.my, tutut@um.edu.my,
noraziah@ump.edu.my, {rozaida,mmustafa}@uthm.edu.my

Abstract. Association rule mining (ARM) is one of the most important and well researched area in data mining. Indirect association rule, a part of ARM, provides a different perspective in identifying the most useful infrequent patterns. Specifically, it refers to the property of high dependencies between two items that are rarely appeared together but indirectly occurred through another items. Besides generating nontrivial information, it also can implicitly reveal a new fact of relationship which cannot be directly determined using the typical interestingness measures. Therefore, in this paper we applied our novel algorithm called Mining Lease Association Rule (MILAR) and our measure called Critical Relative Support (CRS) to mine the indirect least association rule from the students' examination datasets. The experimental results show that the numbers of extracted indirect association rules are reduced when the threshold value of CRS is increased. This number is also lesser than the least association rule. In addition of decreasing the number of uninteresting rules, the obtained information also can be used by educators as a basis to improve their teaching and learning strategies in the future.

Keywords: Data mining, Indirect, Least association rule, Algorithm.

1 Introduction

Data mining, an interdisciplinary subfield of computer science, is the process of extracting some new nontrivial information from large data repository. It is an

B. Murgante et al. (Eds.): ICCSA 2014, Part VI, LNCS 8584, pp. 783–797, 2014.
© Springer International Publishing Switzerland 2014

extension of traditional data analysis and statistical approaches that incorporates with analytical techniques such as numerical analysis, machine learning, etc. In other definitions, data mining is about making analysis convenient, scaling analysis algorithms to large databases and providing data owners with easy to use tools in helping the user to navigate, visualize, summarize and model the data [1]. Regardless of the various interpretation of data mining, it ultimate goals is to extract the high-level knowledge from the low-level data. One of the important models and extensively studies in data mining is known as association rule mining (ARM).

Since Agrawal *et al.* [2] introduced ARM in 1993; it has been extensively studied in various domain applications by many researchers [3-11]. It is also known as market basket analysis or affinity analysis. The universal aim of ARM is at discovering the most potential interesting relationship among a set of items that frequently occurred together in transactional database [12]. Association rule is an implication expression of A → B (such as Milk and Diaper → Coke), where A and B are itemsets. An item is said to be frequent if it appears more than a minimum support threshold. Thus under this concept, infrequent or least items are automatically classified as uninteresting and pruned out during rules generation. However in certain domain and critical applications, least items may also provide a useful insight about the data such as competitive product analysis [13], text mining [14], web recommendation [15], biomedical analysis [16], radar system [28], etc. The classical form of association rule which is derived from the frequent items is typically known as direct association rules. As contradiction from the classical one, indirect association rule [17] refers to a pair of items that are rarely occurred together but their existences are highly depending on the presence of mediator itemsets. It was first proposed by Tan *et al.* [13] as a new way of interpreting the value of infrequent patterns and effectively pruning out the uninteresting infrequent patterns. Mining indirect association rule is quite challenge because it involves with the least items. In many series of ARM algorithms, lowering the value of minimum support can assist in capturing the least items from the database. The problem is, it will indirectly force to produce abundant numbers of undesired rules. As a result, the process of determining which least items that are very useful becomes more tedious and complicated. In addition, the lowering the minimum support will also proportionally intensify the memory consumption and its complexity. Therefore, due to the difficulties in the algorithms [18] development, measures formulation and it may require excessive of computational cost, there are very limited efforts have been put forward.

From the literature, the problem of indirect association mining has become more and more important because of its various domain applications [17,19-23]. Generally, the studies on indirect association mining can be divided into two categories, either focusing on proposing more efficient mining algorithms [14,17,21] or extending the definition of indirect association for different domain applications [5,17,19]. Therefore, from the mentioned problems, there are three main contributions of this paper. First, LP-Tree data structure is employed to minimize memory consumption during generating the least items. Second, we applied Mining Indirect Least Association Rule (MILAR) [20] algorithm and Critical Relative Support (CRS) measure [24]. On top of that, the pervious MILAR algorithm is defined using simple flowchart and it has been improved by extending into pseudocode. CRS has been widely employed in measuring the least association rule [24, 29-37]. A range of CRS is always in 0 and 1. The more CRS value reaches to 1, the more interestingness of that particular rule. Besides that, the itempair support and mediator support are also used as complimentary of CRS to capture the indirect least association rules. Third, Student Examination Result dataset and Student Examination Anxiety dataset have been used in the experiment. Resulting from the experiments is very important to evaluate the performance of LP-Tree data structure, MILAR algorithm and CRS measure.

The rest of the paper is organized as follows. Section 2 describes the related work. Section 3 explains the proposed method. This is followed by performance analysis through two experiment tests in section 4. Finally, conclusion and future direction are reported in section 5.

2   Related Work

Indirect association is closely related to negative association, they are both dealing with itemsets that do not have sufficiently high support. The negative associations' rule was first pointed out by Brin et al. [25]. The focused on mining negative associations is better that on finding the itemsets that have a very low probability of occurring together. Indirect associations provide an effective way to detect interesting negative associations by discovering only frequent itempairs that are highly expected to be frequent.

Until this recent, the important of indirect association between items has been discussed in many literatures. Tan et al. [13] proposed INDIRECT algorithm to extract indirect association between itempairs using the famous Apriori technique. There are two main steps involved. First, extract all frequent items using standard frequent pattern mining algorithm. Second, find the valid indirect associations from the candidate indirect association from candidate itemsets.

Wan et al. [14] introduced HI-Mine algorithm to mine a complete set of indirect associations. HI-Mine generates indirect itempair set (IIS) and mediator support set (MSS), by recursively building the HI-struct from database. The performance of this algorithm is significantly better than the previously developed algorithm either for synthetic or real datasets. IS measure [26] is used as a dependence measure.

Lin et al. [27] proposed GIAMS as an algorithm to mine indirect associations over data streams rather than static database environment. GIAMS contains two concurrent processes called PA-Monitoring and IA-Generation. The first process is to set off when the users specify the required window parameters. The second process is activated once the users issues queries about current indirect associations. In term of dependence measure, IS measure [26] is again adopted in the algorithm.

Chen et al. [17] proposed an indirect association algorithm that was similar to HI-mine, namely MG-Growth. The aim of MG-Growth is to discover indirect association patterns and its extended version is to extract temporal indirect association patterns. The differences between both algorithms are, MG-Growth used the directed graph and bitmap to construct the indirect itempair set. The corresponding mediator graphs are then generated for deriving a complete set of indirect associations. In this algorithm, temporal support and temporal dependence are used in this algorithm.

Kazienko [15] presented IDARM* algorithm to extracts complete indirect associations rules. In this algorithm, both direct and indirect rules are joined together to form a useful of indirect rules. Two types of indirect associations are proposed named partial indirect association and complete ones. The main idea of IDARM* is to capture the transitive page from user-session as part of web recommendation system. A simple measure called Confidence [2] is employed as dependence measure.

Lin *et al.* [38] presented EMIA-LM algorithm for mining indirect association rules over web data stream. EMIA-LM uses a mediator-exploiting search strategy in the process of generating the rule. It also adopts a compact data structure, alleviates unnecessary data transformation processes and minimizes the usage of memory. The preliminary experiments also showed that EMIA-LM is better than HI-mine* for static data in term of computational speed and memory consumption.

Liu *et al.* [39] suggested FIARM (Filtering-Based Indirect Association Rule Mining) algorithm to analyze gene microarray data.It is a Apriori-based algorithm. The algorithm can determine indirect gene associations to assist the biologists in finding a new insightful knowledge. FIARM-Measure is also introduced to help in discovering indirect association rules from the rules that have a negative correlation. In the analysis, Gene Ontology is employed to verify the accuracy of the relationships.

Koh and Dobbie [40] proposed a novel indirect rule mining algorithm with weighting mechanism called sociability to mine social networks for collaboration recommendation. In their experiment, they are using the collaborative network from the digital community DBLP Computer Science Bibliography dataset and a frequent mining dataset. The result shows that the number of indirect rule generated from their proposed algorithm and weighting mechanism are inversely proportional to the minimum confidence threshold.

Herawan *et al.* [41] introduced an Indirect Pattern Mining Algorithm (IPMA) in an attempt to mine the indirect patterns from data repository. IPMA embeds with a measure called Critical Relative Support (CRS) measure rather than the common interesting measures. The finding shows that IPMA is successful in generating the indirect patterns with the various threshold values.

3   The Proposed Method

3.1   Association Rule

Throughout this section the set $I = \{i_1, i_2, \cdots, i_{|A|}\}$, for $|A| > 0$ refers to the set of literals called set of items and the set $D = \{t_1, t_2, \cdots, t_{|U|}\}$, for $|U| > 0$ refers to the data set of transactions, where each transaction $t \in D$ is a list of distinct items $t = \{i_1, i_2, \cdots, i_{|M|}\}$, $1 \leq |M| \leq |A|$ and each transaction can be identified by a distinct identifier TID.

Definition 1. *A set* $X \subseteq I$ *is called an itemset. An itemset with k-items is called a k-itemset.*

Definition 2. *The support of an itemset* $X \subseteq I$, *denoted* $\text{supp}(X)$ *is defined as a number of transactions contain X.*

Definition 3. *Let* $X, Y \subseteq I$ *be itemset. An association rule between sets X and Y is an implication of the form* $X \Rightarrow Y$, *where* $X \cap Y = \phi$. *The sets X and Y are called antecedent and consequent, respectively.*

Definition 4. *The support for an association rule* $X \Rightarrow Y$, *denoted* $\text{supp}(X \Rightarrow Y)$, *is defined as a number of transactions in D contain* $X \cup Y$.

Definition 5. *The confidence for an association rule* $X \Rightarrow Y$, *denoted* $\text{conf}(X \Rightarrow Y)$ *is defined as a ratio of the numbers of transactions in D contain* $X \cup Y$ *to the number of transactions in D contain X. Thus*

$$\text{conf}(X \Rightarrow Y) = \frac{\text{supp}(X \Rightarrow Y)}{\text{supp}(X)}$$

Definition 6. (Least Items). *An itemset X is called least item if* $\text{supp}(X) < \alpha$, *where* α *is the minimum support (minsupp)*

The set of least item will be denoted as Least Items and

$$\text{Least Items} = \{X \subset I \mid \text{supp}(X) < \alpha\}$$

Definition 7. (Frequent Items). *An itemset X is called frequent item if* $\text{supp}(X) \geq \alpha$, *where* α *is the minimum support.*

The set of frequent item will be denoted as Frequent Items and

$$\text{Frequent Items} = \{X \subset I \mid \text{supp}(X) \geq \alpha\}$$

3.2   Indirect Association Rule

Definition 8. *An itempair* $\{X \, Y\}$ *is indirectly associated via a mediator M, if the following conditions are fulfilled:*

1) $\text{supp}(\{X, Y\}) < t_s$ (itempair support condition)
2) There exists a non-empty set M such that:
   a) $\text{supp}(\{X\} \cup M) \geq t_m$ and $\text{supp}(\{Y\} \cup M) \geq t_m$ (mediator support condition)
   b) $\text{dep}(\{X\}, M) \geq t_d$ and $\text{dep}(\{Y\}, M) \geq t_d$, where $\text{dep}(A, M)$ is a measure of dependence between itemset A and M (mediator dependence measure)

The user-defined thresholds above are known as itempair support threshold (t_s), mediator support threshold (t_m) and mediator dependence threshold (t_d), respectively. The itempair support threshold is equivalent to *minsupp* (α). Normally, the mediator support condition is set to equal or more than the itempair support condition $(t_m \geq t_s)$

The first condition is to ensure that (X,Y) is rarely occurred together and also known as least or infrequent items. In the second condition, the first-sub-condition is to capture the mediator M and for the second-sub-condition is to make sure that X and Y are highly dependence to form a set of mediator.

Definition 9. (Critical Relative Support). *A Critical Relative Support (CRS) is a formulation of maximizing relative frequency between itemset and their Jaccard similarity coefficient.*

The value of Critical Relative Support denoted as CRS and

$$\mathrm{CRS}(A,B) = \max\left(\left(\frac{\mathrm{supp}(A)}{\mathrm{supp}(B)}\right),\left(\frac{\mathrm{supp}(B)}{\mathrm{supp}(A)}\right)\right) \times \left(\frac{\mathrm{supp}(A \Rightarrow B)}{\mathrm{supp}(A) + \mathrm{supp}(B) - \mathrm{supp}(A \Rightarrow B)}\right)$$

CRS value is between 0 and 1, and is determined by multiplying the highest value either supports of antecedent divide by consequence or in another way around with their Jaccard similarity coefficient. It is a measurement to show the level of CRS between combination of the both Least Items and Frequent Items either as antecedent or consequence, respectively. Here, Critical Relative Support (CRS) is employed as a dependence measure for 2(a) in order to mine the desired Indirect Association Rule.

3.3   Algorithm Development

Determine Minimum Support. Let I is a non-empty set such that $I = \{i_1, i_2, \cdots, i_n\}$, and D is a database of transactions where each T is a set of items such that $T \subset I$. An itemset is a set of item. A k-itemset is an itemset that contains k items. From Definition 6, an itemset is said to be least (infrequent) if it has a support count less than α.

Construct LP-Tree. A Least Pattern Tree (LP-Tree) is a compressed representation of the least itemset. It is constructed by scanning the dataset of single transaction at a time and then mapping onto a new or existing path in the LP-Tree. Items that satisfy the α (Definition 6 and 7) are only captured and used in constructing the LP-Tree.

Mining LP-Tree. Once the LP-Tree is fully constructed, the mining process will begin using bottom-up strategy. Hybrid 'Divide and conquer' method is employed to decompose the tasks of mining desired pattern. LP-Tree utilizes the strength of hash-based method during constructing itemset in support descending order.

Construct Indirect Patterns. The pattern is classified as indirect least association rule if it fulfilled with the two conditions. The first condition is elaborated in Definition 8 where there are another three sub-conditions. One of them is mediator dependence measure. CRS from Definition 9 is employed as mediator dependence measure between itemset in discovering the indirect patterns. Fig. 1 shows Mining Indirect Least Association Rule algorithm (MILAR).

```
1.  Input   : Dataset, vMinSupp, vMedSupp, vCRS
2.  Output : ILAR, LAR, AR
3.  WHILE ( Dataset ≠ eof ) DO
4.      FOR ( Itemset ∈ Dataset ) DO
5.          ItemSupp ← Compute ( ∀ Itemset )
6.          IF ( ItemSupp > vMinSupp) THEN
7.              FItem ← Item
8.          ELSE
9.              LItem ← Item
10.         END IF
11.         FLItems ← ( FItem ∪ LItem )
12.     END FOR
13. END WHILE
13. FLItems_DO ← SortDescOrder ( FLItems )
14. FOR ( LineOfTrans ∈ FLItems_DO ) DO
15.     ItemsPath ← ( FLItems_DO ∩ LineOfTrans )
16.     LP-Tree ← Update ( ItemsPath )
17. END FOR
18. FOR ( FPT_ItemsPath ∈ LP-Tree ) DO
19.     CondItems ← DetermineCondItems ( FPT_ItemsPath )
20.     CondFPT ← DetermineCondFPT ( CondItems )
21. END FOR
22. PatternSet ← ConstructPattern ( CondFPT )
23. FOR ( PS ∈ PatternSet ) DO
24.     X, Y, M ← SplitSingleItem ( PS )
24.     IF ( Support(X,Y) ≥ vMinSupp) ) THEN
25.         IF ( Support(X ∪ M) ≥ vMedSupp & Support(Y ∪ M) ≥ vMedSupp ) THEN
26.             IF ( Support(X,M) ≥ vCRS & Support(Y ∪ M) ≥ vCRS ) THEN
27.                 ILAR ← ConstructIndirectLeastAR ( X,Y,M )
28.             ELSE
29.                 LAR ← ConstructLeastAR ( X,Y,M )
30.             END IF
31.         END IF
32.         AR ← ConstructAllAR ( X,Y,M )
33.     END IF
34. END FOR
```

Fig. 1. MILAR Algorithm

4   Experiment Test

In this section, the analysis is made by comparing the total number of different classification of association rules being extracted based on the predefined thresholds. In the analysis, three items are involved in forming a complete association rule; two items as an antecedent and one item as a consequence. The mediator is appeared as a part of antecedent. We conducted our experiment using two datasets. The experiment has been performed on Intel® Core™ 2 Quad CPU at 2.33GHz speed with 4GB main memory, running on Microsoft Windows Vista. All algorithms have been developed using C# as a programming language.

The first dataset is students' examination anxiety dataset. The dataset was taken from a survey on exploring examination anxiety among engineering students at Universiti Malaysia Pahang (UMP) [3]. The respondents were 770 students, consisting of 394 males and 376 females. They are undergraduate students from five engineering based faculties, i.e., 216 students from Faculty of Chemical and Natural Resources Engineering (FCNRE), 105 students from Faculty of Electrical and Electronic Engineering (FEEE), 226 students from Faculty of Mechanical Engineering (FME), 178 students from Faculty of Civil Engineering and Earth Resources (FCEER), and 45 students from Faculty of Manufacturing Engineering and Technology Management (FMETM). To this, we have a dataset comprises the number of transactions (student) is 770 and the number of items (attributes) is 5. Table 1 displays the mapped of original survey dimensions, Likert scale with a new attribute id.

Table 1. The mapping between survey dimensions of students' examination anxiety, Likert scale and a new attribute Id

Survey Dimensions	Likert Scale	Attribute Id
Lack of preparation	1 – 5	1
Feel depressed after test	1 – 5	2
Lost concentration during exam	1 – 5	3
Prepared for exam	1 – 5	4
Do not understand the test question	1 – 5	5
Important exam	1 – 5	6
Take a surprise test	1 – 5	7

Item is constructed based on the combination of survey dimension and its likert scale. For simplicity, let consider a survey dimension "Lack of preparation" with likert scale "1". Here, an item "11" will be constructed by means of a combination of an attribute id (first characters) and its survey dimension (second character). Different Interval Supports were employed for this experiment.

Fig. 2 shows the performance analysis against Students' Examination Anxiety dataset. Minimum Support (*minsupp or* α) and Mediator Support Threshold (t_m) are set to 20% and 10%, respectively. Varieties of minimum CRS (min-CRS) were employed in the experiment. During the performance analysis, the total number of association rules that filtered by *minsupp* threshold was 1641. The maximum number of least association rules and indirect least association rules are 396 and 904, respectively. The percentage of total indirect least association rule and least association rule as compared to typical association rule are 12% and 29%, respectively. The total number of indirect least association rule is the lowest with the different of 17% against least association rule. The general trend is, the total of indirect least association rule is kept reducing when the values of min-CRS is kept increasing. However, both minimum number of least association rules and indirect least association are 0 when min-CRS value is set to 0.5. Table 2 depicts example of candidate mediator from association rules that have been successfully extracted. The main focus here is more on the candidate mediator of 13 which represent the lack of preparation with the likert scale of 3 (fair).

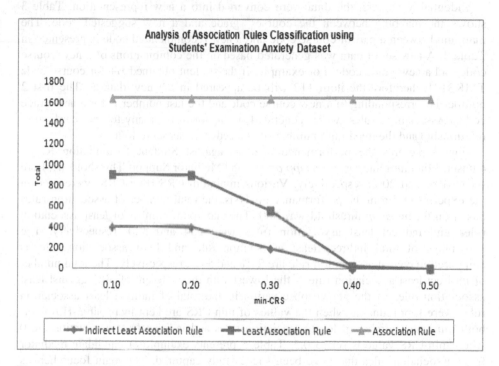

Fig. 2. Analysis of the Generated Association Rules against Students' Examination Anxiety Dataset with variation of min-CRS

Table 2. Example of Candidate Mediator from Association Rule for Students' Examination Anxiety Dataset

Association Rule	Support	Candidate Mediator	CRS
33 -> 13	10.00		0.32
44 -> 13	10.39		0.35
33 13 -> 44	5.19	13	0.47
44 13 -> 33	5.19	13	0.44

The second dataset is Students' Examination Result dataset. The dataset was obtained from the student examination result in computer science programme for intake July 2007/2008. This programme has 30 subjects in computer science and the period of study is 3 years. There were 80 students involved and their identities were removed due to the confidentiality agreement. The data was obtained from IT Centre, Universiti Malaysia Terengganu in Microsoft excel format. The original data was given in the horizontal format which is only suitable for reporting purposes. It consists of 12 attributes: student status, metric number, name (firstname, middle name and last name), session (July or December), result (pass or fail), CGPA, GPA, course code, course name, credit hour, level (elective or compulsory) and grade. Due to the confidentiality matters, the data were converted into a new representation. Table 3 shows the mapping between the courses' grade and a new suggested code. The mapping between a part of course code and a new recommended code is presented in Table 4. A new set of data was generated based on the combinations of a new course code and a new grade code. For example, if the student obtained B+ for course code TMK3101, therefore the item 112 will be appeared in the new dataset. The first 2 number is corresponding to a new course code and the last number is for a new grade code. Association rules were generated in a form of many-to-one cardinality relationship and the maximum number of antecedents was set to four.

Fig. 3 presents the performance analysis against Student Examination Result dataset. Minimum Support (*minsupp or* α) and Mediator Support Threshold (t_m) are set to 40% and 20%, respectively. Various minimum CRS (min-CRS) were used in the experiment. From the performance analysis, the total number of association rules based on the *minsupp* threshold was 660. The maximum number of least association rules and indirect least association rules were 254 and 520, respectively. The percentage of total indirect least association rule and least association rule as compared to typical association rule are 24% and 54%, respectively. The total number of indirect least association rule is the lowest with the different of 30% against least association rule. At the glance observation is, the total of indirect least association rules were kept reducing when the values of min-CRS are kept increasing. However, both minimum number of least association rules and indirect least association are 0 when min-CRS value is set to 0.5. Table 5 presents example of candidate mediator from association rules that have been successfully captured. The main focus here is more on the candidate mediator of 112 which represent the course code with the category of 2 (good).

Table 3. Mapping of students' examination grade and a new code

Category	Grade	Total Marks	Code
Excellent	A, A-	Above 75	1
Good	B+, B, B-	60 - 70	2
Fair	C+, C, C-	45 - 60	3
Weak	D, F	Less than 50	4

Table 4. Mapping of a part of students' examination course and a new code

Course Code	Course Name	Code
TMK3101	Computer Systems and Internet	11
TMK3102	Computer Programming	12
⋮	⋮	⋮
TMK4999B	Final Year Project	48

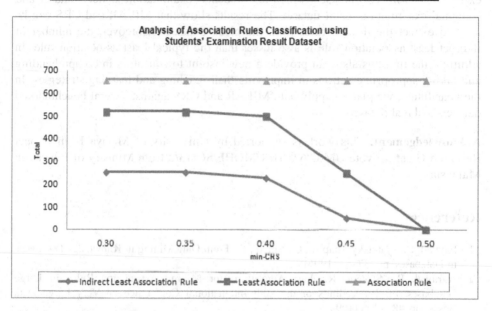

Fig. 3. Analysis of the Generated Association Rules against Students' Examination Result Dataset with variation of min-CRS

Table 5. Example of Candidate Mediator from Association Rule for Examination Result Dataset

Association Rule	Support	Candidate Mediator	CRS
153 -> 112	21.11		0.37
172 -> 112	25.75		0.49
153 112 -> 172	13.17	112	0.54
172 112 -> 153	13.17	112	0.40

5 Conclusion

In data mining, association rule mining (ARM) is considered as one of the most prominent research areas and it has been applied in various domain applications

[42-46]. Indirect association rule, a subset of ARM, has been broadly studied to deal with the rarity cases which cannot be captured from the common rules. The main advantage is, it capable in analyzing the dependency among the least items from association rules. However, finding this information is very difficult as compared to the common rules. In fact, a scalable algorithm and appropriate measure are became necessity. Therefore, in this paper we applied our novel algorithm Mining Indirect Least Association Rule (MILAR) and employed our scalable measure called Critical Relative Support (CRS) to capture the indirect least association rules. We conduct our experiments with two datasets which are students' examination anxiety dataset and students' examination result dataset. The results show that MILAR and CRS can be used to extract the desired rules from the given datasets. Moreover, the number of indirect least association rule is also lesser than the typical least association rule. In addition, the findings also can provide a new insight to educators in comprehending and taking appropriate actions in improving their teaching and learning strategies. In the near future, we plan to apply both MILAR and CRS against several benchmarked datasets and real datasets.

Acknowledgement. This work is supported by University of Malaya High Impact Research Grant no vote UM.C/625/HIR/MOHE/SC/13/2 from Ministry of Education Malaysia.

References

1. Fayyad, U., Piatetsky-Shapiro, G., Smyth, P.: From Data Mining to Knowledge Discovery in Databases, pp. 37–54 (1996)
2. Agrawal, R., Srikant, R.: Fast Algorithms for Mining Association Rules in Large Databases. In: Proceedings of the 20th International Conference on Very Large Data Bases, pp. 487–499 (1994)
3. Mannila, H., Toivonen, H., Verkamo, A.I.: Discovery of Frequent Episodes in Event Sequences. Data Mining and Knowledge Discovery 1, 259–289 (1997)
4. Park, J.S., Chen, M.S., Yu, P.S.: An Effective Hash-based Algorithm for Mining Association Rules. In: Proceedings of the ACM-SIGMOD Int. Conf. Management of Data (SIGMOD 1995), pp. 175–186. ACM Press (1995)
5. Savasere, A., Omiecinski, E., Navathe, S.: An efficient algorithm for mining association rules in large databases. In: Proceedings of the 21st International. Confenference on Very Large Data Bases (VLDB 1995), pp. 432–443. ACM Press (1995)
6. Fayyad, U., Patesesky-Shapiro, G., Smyth, P., Uthurusamy, R.: Advances in Knowledge Discovery and Data Mining. MIT Press, MA (1996)
7. Bayardo, R.J.: Efficiently Mining Long Patterns from Databases. In: Proceedings of the ACM-SIGMOD International Conference on Management of Data (SIGMOD 1998), pp. 85–93. ACM Press (1998)
8. Zaki, M.J., Hsiao, C.J.: CHARM: An efficient algorithm for closed itemset mining. In: Proceedings of the 2002 SIAM Int. Conf. Data Mining, pp. 457–473. SIAM (2002)
9. Agarwal, R., Aggarwal, C., Prasad, V.V.V.: A tree projection algorithm for generation of frequent itemsets. Journal of Parallel and Distributed Computing 61, 350–371 (2001)

10. Liu, B., Hsu, W., Ma, Y.: Mining Association Rules with Multiple Minimum Support. In: Proceedings of the 5th ACM SIGKDD International Conference on Knowledge Discovery and Data Mining, pp. 337–341. ACM Press (1999)
11. Abdullah, Z., Herawan, T., Deris, M.M.: Scalable Model for Mining Critical Least Association Rules. In: Zhu, R., Zhang, Y., Liu, B., Liu, C. (eds.) ICICA 2010. LNCS, vol. 6377, pp. 509–516. Springer, Heidelberg (2010)
12. Leung, C.W., Chan, S.C., Chung, F.: An Empirical Study of a Cross-level Association Rule Mining Approach to Cold-start Recommendations. Knowledge-Based Systems 21(7), 515–529 (2008)
13. Tan, P.N., Kumar, V., Srivastava, J.: Indirect Association: Mining Higher Order Dependences in Data. In: Proceedings of the 4th European Conference on Principles and Practice of Knowledge Discovery in Databases, pp. 632–637. Springer, Heidelberg (2000)
14. Wan, Q., An, A.: An Efficient Approach to Mining Indirect Associations. Journal Intelligent Information Systems 27(2), 135–158 (2006)
15. Kazienko, P.: Mining Indirect Association Rules for Web Recommendation. International Journal of Applied Mathematics and Computer Science 19(1), 165–186 (2009)
16. Tsuruoka, Y., Miwa, M., Hamamoto, K., Tsujii, J., Ananiadou, S.: Discovering and Visualizing Indirect Associations between Biomedical Concepts. Bioinformatics 27(13), 111–119 (2011)
17. Chen, L., Bhowmick, S.S., Li, J.: Mining Temporal Indirect Associations. In: Ng, W.-K., Kitsuregawa, M., Li, J., Chang, K. (eds.) PAKDD 2006. LNCS (LNAI), vol. 3918, pp. 425–434. Springer, Heidelberg (2006)
18. Zhou, L., Yau, S.: Association Rule and Quantitative Association Rule Mining among Infrequent Items. In: Proceedings of the 8th International Workshop on Multimedia Data Mining, ACM SIGMOD, pp. 1–9 (2007)
19. Cornelis, C., Yan, P., Zhang, X., Chen, G.: Mining Positive and Negative Association from Large Databases. In: Proceedings of the 2006 IEEE International Conference on Cybernatics and Intelligent systems, pp. 1–6. IEEE (2006)
20. Abdullah, Z., Herawan, T., Deris, M.M.: Mining Indirect Least Association Rule. In: Herawan, et al. (eds.) DaEng 2013. LNEE, vol. 285, pp. 159–166. Springer, Heidelberg (2014)
21. Kazienko, P., Kuzminska, K.: The Influence of Indirect Association Rules on Recommendation Ranking Lists. In: Proceeding of the 5th International Conference on Intelligent Systems Design and Applications, pp. 482–487 (2005)
22. Tseng, V.S., Liu, Y.C., Shin, J.W.: Mining Gene Expression Data with Indirect Association Rules. In: Proceeding of the 2007 National Computer Symposium (2007)
23. Wu, X., Zhang, C., Zhang, S.: Efficient Mining of Positive and Negative Association Rules. ACM Transaction on Information Systems 22(3), 381–405 (2004)
24. Abdullah, Z., Herawan, T., Noraziah, A., Deris, M.M.: Mining Significant Association Rules from Educational Data using Critical Relative Support Approach. Procedia Social and Behavioral Sciences 28, 97–191 (2011)
25. Brin, S., Motwani, R., Ullman, J., Tsur, S.: Dynamic itemset counting and implication rules for market basket data. In: Proceedings of the International ACM SIGMOD Conference, pp. 255–264. ACM Press (1997)
26. Tan, P., Kumar, V., Srivastava, J.: Selecting the Right Interestingness Measure for Association Patterns. In: Proceedings of the 8th International Conference on Knowledge Discovery and Data Mining, pp. 32–41 (2002)

27. Lin, W.-Y., Wei, Y.-E., Chen, C.-H.: A Generic Approach for Mining Indirect Association Rules in Data Streams. In: Mehrotra, K.G., Mohan, C.K., Oh, J.C., Varshney, P.K., Ali, M. (eds.) IEA/AIE 2011, Part I. LNCS, vol. 6703, pp. 95–104. Springer, Heidelberg (2011)
28. Wu, X.C., Cheng, Y., Yan, L.L., Xue, F.X.: A Radar Intelligence Simulation Method Based on Data Mining Techniques. Applied Mechanics and Material 263-266, 277–282 (2013)
29. Herawan, T., Vitasari, P., Abdullah, Z.: Mining Interesting Association Rules of Students Suffering Study Anxieties Using SLP-Growth Algorithm. International Journal of Knowledge and Systems Sciences 3(2), 24–41 (2012)
30. Abdullah, Z., Herawan, T., Noraziah, A., Deris, M.M.: Detecting Critical Least Association Rules in Medical Databasess. International Journal of Modern Physics: Conference Series, World Scientific 9, v464–v479 (2012)
31. Herawan, T., Abdullah, Z.: CNAR-M: A Model for Mining Critical Negative Association Rules. In: Li, Z., Li, X., Liu, Y., Cai, Z. (eds.) ISICA 2012. Communications in Computer and Information Science, vol. 316, pp. 170–179. Springer, Heidelberg (2012)
32. Herawan, T., Noraziah, A., Abdullah, Z., Deris, M.M., Abawajy, J.H.: IPMA: Indirect Patterns Mining Algorithm. In: Nguyen, N.T., Trawiński, B., Katarzyniak, R., Jo, G.-S. (eds.) Adv. Methods for Comput. Collective Intelligence. SCI, vol. 457, pp. 187–196. Springer, Heidelberg (2013)
33. Herawan, T., Vitasari, P., Abdullah, Z.: Mining interesting association rules of student suffering mathematics anxiety. In: Zain, J.M., Wan Mohd, W.M.b., El-Qawasmeh, E., et al. (eds.) ICSECS 2011, Part II. CCIS, vol. 180, pp. 495–508. Springer, Heidelberg (2011)
34. Abdullah, Z., Herawan, T., Deris, M.M.: Efficient and Scalable Model for Mining Critical Least Association Rules. Journal of The Chinese Institute of Engineer, Taylor and Francis 35(4), 547–554 (2012); special issue from AST/UCMA/ISA/ACN (2010)
35. Abdullah, Z., Herawan, T., Noraziah, A., Deris, M.M.: Extracting Highly Positive Association Rules from Students' Enrollment Data. Procedia Social and Behavioral Sciences 28, 107–111 (2011)
36. Abdullah, Z., Herawan, T., Noraziah, A., Deris, M.M.: Mining Significant Association Rules from Educational Data using Critical Relative Support Approach. Procedia Social and Behavioral Sciences 28, 97–101 (2011)
37. Abdullah, Z., Herawan, T., Mat Deris, M.: An Alternative Measure for Mining Weighted Least Association Rule and Its Framework. In: Zain, J.M., Wan Mohd, W.M.b., El-Qawasmeh, E., et al. (eds.) ICSECS 2011, Part II. CCIS, vol. 180, pp. 480–494. Springer, Heidelberg (2011)
38. Lin, W.-Y., Chen, Y.-C.: A Mediator Exploiting Approach for Mining Indirect Associations from Web Data Streams. In: IEEE Xplore in the 2nd Intl. Conference on Innovations in Bio-inspired Computing and Applications (IBICA), pp. 183–186 (2011)
39. Liu, Y.-C., Shin, J.W., Tseng, V.S.: Discovering Indirect Gene Associations by Filtering-Based Indirect Association Rule Mining. International Journal of Innovative Computing, Information and Control 7(10), 6041–6053 (2011)
40. Koh, Y.S., Dobbie, G.: Indirect Weighted Association Rules Mining for Academic Network Collaboration Recommendations. In: Proceedings of the 10th Australasian Data Mining Conference, pp. 167–174 (2012)
41. Herawan, T., Noraziah, A., Abdullah, Z., Deris, M.M., Abawajy, J.H.: IPMA: Indirect Patterns Mining Algorithm. In: Nguyen, N.T., et al. (eds.) Advanced Methods for Computational Collective Intelligence, vol. 285, pp. 159–166. Springer (2013)
42. Kalia, H., Dehuri, S., Ghosh, A.: A Survey on Fuzzy Association Rule Mining. International Journal of Data Warehousing and Mining 9(1), 1–27 (2013)

43. Priya, R.V., Vadivel, A.: User Behaviour Pattern Mining from Weblog. International Journal of Data Warehousing and Mining 8(2), 1–22 (2012)
44. Daly, O., Taniar, D.: Exception Rules Mining Based on Negative Association Rules. In: Laganá, A., Gavrilova, M.L., Kumar, V., Mun, Y., Tan, C.J.K., Gervasi, O. (eds.) ICCSA 2004. LNCS, vol. 3046, pp. 543–552. Springer, Heidelberg (2004)
45. Taniar, D., Rahayu, W., Lee, V.C.S., Daly, O.: Exception rules in association rule mining. Applied Mathematics and Computation 205(2), 735–750 (2008)
46. Ashrafi, M.Z., Taniar, D., Smith, K.A.: Redundant association rules reduction techniques. International Journal of Business Intelligence and Data Mining 2(1), 29–63 (2007)

Towards a New Generation ACO-Based Planner

Marco Baioletti, Andrea Chiancone, Valentina Poggioni,
and Valentino Santucci

Department of Mathematics and Computer Science,
University of Perugia, Perugia, Italy
{baioletti,andrea.chiancone,poggioni,valentino.santucci}@dmi.unipg.it

Abstract. In this paper a new generation ACO-Based Planner, called ACOPlan 2013, is described. This planner is an enhanced version of ACOPlan, a previous ACO-Based Planner [3], which differs from the former in the search algorithm and in the implementation, now done on top of Downwards. The experimental results, even if are not impressive, are encouraging and confirm that ACO is a suitable method to find near optimal plan for propositional planning problems.

1 Introduction

The basic principle of the first generation of ACO-Based planner ACOPlan, described in [3,1,2], was to use the well known *Ant Colony Optimization* metaheuristic *(ACO)* [6] to solve planning problems with the aim of optimizing the quality of the solution plans. The approach was based on the strong similarity between the process used by artificial ants to build solutions and the way used by state–based planners to find solution plans. Therefore, we had defined an ACO algorithm which handles a colony of planning ants with the purpose of solving planning problems by optimizing solution plans with respect to the overall plan cost.

ACO is a metaheuristic inspired by the behavior of natural ants colony which has been successfully applied to many *Combinatorial Optimization* problems. Being ACO a stochastic incomplete algorithm, there is no guarantee that optimal solutions are ever found, but in many CO problems ACO is able to find very good or near optimal solutions, sometimes being competitive with state-of-arts algorithms.

In this paper a new generation of planners based on ACO search algorithm is introduced. The most important change lies in the search process performed by the ants. In particular the ants start from the most promising states reached in the previous generations and perform a variable number of steps. A smaller number of steps L is used to enhance exploitation, while a larger number of steps B gives more importance to exploration. The parameters L and B are tuned by means of an auto-adaptive process. Other new features are described in Section 4.

The paper is structured as follows. In the two first sections a brief introduction to the metaheuristic ACO, and the previous generation of ACOPlan algorithm are described, while, in the next section the characteristics and peculiarity of new

B. Murgante et al. (Eds.): ICCSA 2014, Part VI, LNCS 8584, pp. 798–807, 2014.

generation of ACO based algorithms are presented. The experimental results are described in Section 5, while some conclusions are drawn in Section 6.

2   Ant Colony Optimization

ACO is a well–known metaheuristic to tackle Combinatorial Optimization problems introduced since early 90s by Dorigo et Al. [6]. It is inspired by the foraging behavior of natural ant colonies. When walking, natural ants leave on the ground a chemical substance called *pheromone* that other ants can smell. This stigmergic mechanism implements an "indirect communication way" among ants, in particular when looking for the shortest path to reach food.

ACO is usually applied to optimization problems whose solutions are composed by discrete components. A Combinatorial Optimization problem is described in terms of a *solution space S*, a set of (possibly empty) *constraints Ω* and an *objective function* $f : S \rightarrow \mathbb{R}^+$ to be minimized (maximized).

The colony of artificial ants builds solutions in an incrementally way: each ant probabilistically chooses a component to add to a partial solution built so far, according to the problem constraints. The random choice is biased by the *artificial* pheromone value τ related to each component c and by a heuristic function η. Both terms evaluate the desirability of each component. The probability that an ant will choose the component c is

$$p(c) = \frac{[\tau(c)]^\alpha [\eta(c)]^\beta}{\sum_x [\tau(x)]^\alpha [\eta(x)]^\beta} \tag{1}$$

where the sum on x ranges on all the components which can be chosen, and α and β are tuning parameters which differentiate the pheromone and heuristic contributions.

The pheromone values represent a kind of memory shared by the whole ant colony and are subject to *update* and *evaporation*. In the most applications only the best solutions are considered in the pheromone update phase: the global best solution found so far (*best–so–far*) and/or the best solution found in the current iteration (*iteration–best*). Moreover, most ACO algorithms use the following update rule [4]:

$$\tau(c) \leftarrow (1 - \rho) \cdot \tau(c) + \rho \cdot \sum_{s \in \Psi_{upd} : c \in s} F(s) \tag{2}$$

where Ψ_{upd} is the set of solutions involved in the update, F is the so called *quality function*, which is a decreasing function of the objective function f (increasing if f is to be maximized), and $\rho \in]0,1[$ is the pheromone evaporation rate. ρ is a typical ACO parameter which was introduced to avoid a premature convergence of the algorithm towards sub–optimal solutions.

The simulation of the ant colony is iterated until a satisfactory solution is found, a termination criterion is satisfied or a given number of iterations is reached.

3   First Generation ACO Based Planners

According to the main features of ACO the ants–planners of the colony are stochastic and heuristic–based.

Each ant–planner executes a forward search, starting from the initial state $\mathcal{I}$ and trying to reach a state in which the goal $\mathcal{G}$ is satisfied. The solution is built step by step by adding components. At each step, the search process performs a randomized weighted selection of a solution component c which takes into account both the pheromone value $\tau(c)$ associated to the component and the heuristic value $\eta(a)$ computed for each action a executable in the current state and related to the chosen solution component. Once an action a has been selected, the current state is updated by means of the effects of a.

The construction phase stops when at least one of the following termination condition is verified

1. a solution plan is found, i.e. a state where all the goals are true is reached;
2. a dead end is met, i.e. no action is executable in the current state;
3. an upper bound L_{max} for the number of execution steps is reached.

Variants of the system was been presented and experimented, for example in [3,1,2]. These variants differ in several point, for example in terms of pheromone models, evaluation functions and implementation, but they share the general algorithm presented in Fig. 1. The experimental evaluation presented in [3] shows that the approach was competitive and comparable with the state of the art. There was also an unsuccessful version that due to some unexpected bugs run at the IPC 2011 with terrible results.

Let $(\mathcal{I}, \mathcal{G}, \mathcal{A})$ be the planning problem, the optimization process is iterated for a given number N of iterations, in which a colony of n_a ants build plans with a maximum number of steps L_{max}. At each step, each ant chooses an action among the executable ones by the *ChooseAction* function that encodes the transition probability function previously described. When all ants have completed the search phase, the best plan π_{iter} of the current iteration is selected and the global best plan π_{best} is possibly updated. Finally, the pheromone values of the solution components are updated by means of the function *UpdatePheromone* that implements the updating rules 2. Relevant parameters are c_0 which denotes the initial value for the pheromone, ρ which represents the evaporation rate and σ that is a parameter of the pheromone update rule (see 3).

3.1   Pheromone Models

The effectiveness of an ACO algorithm firstly depends on the choice of the pheromone model and its data structures. A good model should be simple to compute but enough informative to characterize the context in which an ant–planner can choose a specific action. Moreover it should allow them to distinguish the context of most successful choices from the worst ones. On the other hand the characterization of the component should not be too much detailed in order to allow the pheromone to deposit in a significant quantity.

Algorithm 1. The algorithm ACOPlan

1: $\pi_{best} \leftarrow \emptyset$
2: $InitPheromone(c_0)$
3: **for** $g \leftarrow 1$ to N **do**
4: **for** $m \leftarrow 1$ to n_a **do**
5:   $\pi_m \leftarrow \emptyset$
6:   $s \leftarrow \mathcal{I}$
7:   $A_1 \leftarrow$ executable actions in $\mathcal{I}$
8:   **for** $i \leftarrow 1$ to L_{max} **while** $A_i \neq \emptyset$ **and** $\mathcal{G} \not\subseteq s$ **do**
9:    $a \leftarrow ChooseAction(A_i)$
10:    extend π_m with a
11:    $s \leftarrow Res(s, a)$
12:    $A_{i+1} \leftarrow$ executable actions on s
13:   **end for**
14: **end for**
15: find π_{iter}
16: update π_{best}
17: $UpdatePheromone(\pi_{best}, \pi_{iter}, \rho, \sigma)$
18: **end for**

In [3] several pheromone models have been proposed and empirically compared by systematic experiments.

State-State (SS): A component is defined by the current state s. This is one of the most expensive pheromone model from the space complexity point of view, because the number of possible states is exponential with respect to the problem size.

State-Action (SA): The pheromone value τ depends on the current state s and on the action a to be executed. This model is even more expensive than SS, because for each state s there can exist several actions executable (and chosen) in s. On the other hand, the pheromone values can be interpreted in terms of a preference policy: $\tau(a, s)$ represents how much it is desirable, or it has been useful, to execute a in state s.

Action-Action (AA): In this model a notion of local history is introduced: the pheromone depends both on the action a under evaluation and on the last executed action a', i.e. the pheromone is a function $\tau(a, a')$. Considering only the previous action is the simplest way in which the action choice can be directly influenced by history of previous decisions. AA allows a manageable representation and defines a sort of local first order Markov property.

Fuzzy Level-Action (FLA): The basic idea underlying this model is to associate the action under evaluation with the plan time step, i.e. the planning graph level, where it is executed. Since the limited number of levels, such approach has a more tractable space complexity with respect to the SA model. On the other hand a pure Level-Action model would present the drawback that the pheromone of an action at a time step t cannot be used in other close time steps, while it is often likely that an action desirable at certain time step, say 2, will also be desirable at close time steps, say 1 and 3. To solve this problem the FLA model which fuzzifies the Level-Action representation just described has been introduced: when pheromone is distributed over an action a executed at time step t is also spread in decreasing quantity over the same action in the close time steps, conversely the pheromone level computed for an action a to be executed at time t is computed as the weighted average of the pheromone values $\tau(a,t)$, where $t = t - W, ..., t + W$ computed with the Level-Action model, where weights, i.e. the spread distribution, and the time window W are parameters of the model. The two models showing the best performances were AA and FLA.

3.2   The Heuristic

The heuristic function is a key feature of ACOPlan because it directly affects the transition probability function (1) used to synthesized the solution plan. The heuristic value for a component c is defined by

$$\eta(c) = \frac{1}{h(s_c)}$$

where h is an heuristic function which evaluates the state s_c resulting from the execution of the action a_c associated to the component c in the current state.

The ACOPlan variants used both the heuristic function FF and its variant FFAC for actions costs. This function was presented in [3]; to note that FFAC does not have static costs as in the heuristic proposed by Keyder and Geffner in [10] but dynamic costs depending on the level at hand; it is similar to the heuristic function used in SAPA [5] with the sum propagation for action cost aggregation.

3.3   Plan Comparison and Pheromone Updating

A critical point of the optimization process is the ability of comparing plans found by the colony of planner ants. This is particularly important in pheromone updating phase where the best plan of the iteration must be selected, as well as in the general ACOPlan which returns the best plan found in all the iterations.

Any comparison criteria should obviously prefer a *solution plan* to a *non solution plan*. On the other hand comparison of two *solution plans* π, π' can be easily based on actions costs. A comparison criteria cannot be easily defined when both plans π and π' are not solution plans. In this case a plan π is evaluated by a combination of the heuristic value on the best state s ever reached by π with the cost of reaching s.

The *pheromone update phase* evaporates all the pheromone values and increases the pheromone value of the components belonging to π_{iter} and π_{best} according to the formula

$$\tau(c) \leftarrow (1 - \rho) \cdot \tau(c) + \rho \cdot \Delta(c) \tag{3}$$

where

$$\Delta(c) = \begin{cases} \sigma & \text{if } c \text{ belongs to } \pi_{iter} \\ 1 - \sigma & \text{if } c \text{ belongs to } \pi_{best} \\ 1 & \text{if } c \text{ belongs to both} \\ 0 & \text{otherwise} \end{cases}$$

and σ is usually set to $\frac{2}{3}$ to give more influence to the exploration.

4   Second Generation ACO Based Planners

The new version of ACOPlan, called ACOPlan 2013, differs from the previous implementation in several points.

A first important difference is that ACOPlan 2013 is now implemented in top of Fast Downwards [8], as other planners, like Lama [13]. In this way we can exploit of efficient routines for parsing the planning problem, translating into a finite domain representation, finding landmarks and computing heuristic functions. Nevertheless, our search engine is independent on the actual representation of states and only uses some Downward internal procedures, like the successor generator procedure.

The most important algorithmical difference lies in the choice of starting point and the length of the exploration performed by each ant. In the previous version of ACOPlan, each ant starts from the initial state and begins to build a plan for at most a certain number N of steps.

Since each ant always starts from the initial state, all the states close to the initial state are explored very often, thus wasting a lot of computation time. The repeated exploration of some states is a major problem in this kind of algorithms. LAMA and other similar planners avoid at all to explore a state s more than once, except when s is reachable from the initial state by a path shorter than before. In ACOPlan, a state could be explored several times, in different iteration or by different ants, and forcing the algorithm to avoid already visited states would make the search process incomplete. The use of a cache data structure, where information about already visited states are stored, only mitigates the computation efforts of re-exploring states.

In ACOPlan2013 the choice of the starting point is made in two steps.

First of all, an already found plan p, among the best-so-far and the iteration-best, is selected at random with equal probabilities. Then a state s is randomly chosen in the following way. With equal probability, s is the last state reached by p, the best state in the whole p, or the best state reached in the last 10 steps in p.

Then the ant starts its exploration from s and its initial plan is the prefix of the plan p, truncated at the step corresponding to s.

In this way, an important drawback of ACOPlan is overcome, i.e. that each generation tries to build plans starting from scratch, thus discarding many information acquired by the previous generations (except for the pheromone contribution). Indeed, in ACOPlan 2013, the ants start from the best states reached in the previous generations.

The bound N on the plan length was another major problem of ACOPlan. It is obvious that N should be large enough to allow the ant to find a plan, for instance $N \geq PL$, where PL is the length of a solution plan, which should be estimated somehow from the problem to be solved (for instance as a multiple of a relaxed plan starting from the initial state) or fixed to a large enough value.

In ACOPlan 2013, the exploration is performed using two different bounds, denoted by L (Little) and B (Big), on the number of steps from s. Usually $L < B$. The bounds L and B are used in alternation, and the sequence of generation where L is used is called "Little" phase, while in those where L is used is called "Big" phase.

In the "Little" phase, the ants explore for a small number of steps, and in this way they can exploit and continue, as much as possible, the exploration performed in the previous generation. The aim of a "little" number of steps is to make small, but progressive enhancements.

In the "Big" phase, the ants explore for a greater number of steps, thus the possibility of finding completely different solutions and the exploration of a larger part of the search space are possible.

Hence, the alternation between "Little" and "Big" phases then corresponds to having phases in which a different weight is given to the exploitation and to the exploration aptitudes.

The bounds L and B start with the value $L = 10$ and $B = 100$ and are changed in a dynamic way, making them converging to a common intermediate value.

In this way, the algorithm dynamically adapts the search process to give more importance to the exploration or to exploitation, according to the number of improvements of the heuristic function and the number of solutions found.

Another important new feature is the particular use of two heuristic functions, h_{FF} (i.e. the FF heuristics) and h_{LM} (i.e. the Landmarks count heuristics), both used also by LAMA [13].

In Lama each state is evaluated with both heuristic functions. In the ACOPlan approach, the simultaneous use of both of them, although possible, is a little bit problematic. While the formula for the transition probabilities can be easily extended to take into account of two heuristic functions, the major difficulty is that states (and plans) are evaluated by using heuristic values, hence the presence of two heuristic functions can cause situations of incomparable states.

Moreover, since the number of state evaluations required by ACOPlan is large, a double heuristic evaluation can be computationally heavy.

Therefore, we decide to mostly use one heuristic function and the decision is automatically taken during the search process seeing the progress obtained by using all the two heuristics for some generations each. If both heuristic functions produces comparable results, they are kept for the next sequence of generations. On the other hand, if a heuristic function h appears to work better than the other h', a "last chance" is given to h'. After this period, if h is still better than h', h' is removed and only h is used in the remaining generations.

Table 1. Results for plan quality

(a) **Barman** domain

Problem	Lama score	FDss-1 score	ACO score	best cost
p01	0.92	0.99	0.70	279
p02	0.98	0.87	0.66	259
p03	0.85	0.99	0.75	274
p04	0.86	0.99	0.70	281
p05	0.86	0.92	0.69	297
p06	0.90	0.83	0.72	322
p07	0.98	0.99	0.63	305
p08	0.86	0.83	0.73	290
p09	0.91	0	0.58	348
p10	0.93	0.94	0.70	323
p11	0.86	0	0.70	354
p12	0.82	0.75	0.54	334
p13	0.90	1.00	0.71	396
p14	0.85	0.72	0.60	372
p15	0.96	0.99	0.64	387
p16	0.87	1	0.65	386
p17	0.87	1	0.62	383
p18	0.73	0.62	0.65	380
p19	0.90	1	0.65	387
p20	0.89	0.91	0.70	356
total	17.70	16.34	13.3	

(b) **Elevator** domain

Problem	Lama score	FDss-1 score	ACO score	best cost
p01	0.52	0.75	0.73	191
p02	0.46	0.71	0.75	417
p03	0.45	0.62	0.76	464
p04	0.92	0.75	0.59	256
p05	0.48	0.61	0.80	253
p06	0.47	0.68	0.67	513
p07	0.48	0.66	0.63	409
p08	0.38	0.68	0.58	505
p09	0.45	0.73	0.74	671
p10	0.40	0.61	0.68	602
p11	0.58	0.56	0.53	635
p12	0.48	0.53	0.57	691
p13	0.53	0.72	0.88	992
p14	0.55	0.61	0.62	804
p15	0.57	0.42	0.69	923
p16	0.50	0.62	0.53	891
p17	0.53	0.64	0.52	1066
p18	0.47	0.59	0.57	1148
p19	0.41	0.41	0.56	1417
p20	0.64	0.61	0.85	1386
total	10.28	12,52	13.25	

(c) **Parcprinter** domain

Problem	Lama score	FDss-1 score	ACO score	best cost
p01	1.00	1.00	1.00	1383121
p02	1.00	1.00	1.00	1852217
p03	0.93	1.00	0.96	2490322
p04	1.00	0.78	0.91	2754187
p05	1.00	1.00	1.00	1216462
p06	1.00	1.00	1.00	1270874
p07	0.91	1.00	0.98	2121255
p08	1.00	1.00	1.00	1681282
p09	1.00	1.00	0.97	2387265
p10	1.00	1.00	0	2021893
p11	0.86	1.00	1.00	1891203
p12	1.00	0.88	0.97	2828340
p13	0.83	0.83	0.99	3335367
p14	1.00	0.74	0.87	3119803
p15	1.00	0.74	0.85	3160821
p16	0.83	0.70	0.81	3526437
p17	0.99	1.00	0	1556448
p18	0.99	1.00	0	2376643
p19	0.98	1.00	0.98	3072626
p20	1.00	1.00	0	2308715
total	19.32	18.68	15.29	

5   Experiments

This section presents and discusses the results of experiments held with this new generation of ACOPlan optimizing the overall plan execution costs.

The benchmarks domain problems from the last planning competition IPC2011 [] has been used in the experiments. The results presented here refers to the domains *Barman*, *Elevators*, and *Parcprinter*.

In these experiments the pheromone model *FLA* has been used.

Since the behavior of ACOPlan depends on many parameters, a preliminary phase of parameters tuning has been held by systematic tests in order to establish the ACOPlan general setting: 10 planner–ants, $\alpha = 3$, $\beta = 3$, $\rho = 0.10$. Moreover, since ACOPlan is a not deterministic planner, the results collected for each single problem instance are the median values obtained over 15 runs. The experiments were run on the EGI grid.

Comparisons have been made with LAMA 2011 [13] and FDss-1[7] respectively the winner and the runner-up planners at the Sequential Satisficing track of the last planning competition IPC2011.

For each tested domains a table showing the score obtained by each planner with respect the best cost solution (the known best cost) is built and presented in Table 1. The score is computed as in the planner competition, that is the ratio between the solution cost and the cost of the best solution known so far, $score = cost/best\_cost$.

The tests show ACOPlan 2013 obtain comparable results with respect to the other planners, even if it is apparent that ACOPlan 2013 is not competitive with them. Anyway its results on Elevators are better than those of LAMA and FDss-1.

6   Conclusions and Future Works

In this paper we have described ACOPlan 2013, a new version of the planner ACOPlan, which is based on a new search algorithm and a new implementation. This new planner is then compared with some state-of-arts planners, obtaining encouraging results. Although ACOPlan 2013 is not competitive, it appears to be comparable with them and the improvements obtained with respect to the results obtained by ACOPlan are impressive both from the quality of the solutions found and the number of problems solved whith the time limit. Hence, we believe that ACO still remains a good method to solve planning problems.

Following this line, it is apparent that for this approach there is room for improvement and it can be made competitive with the state-of-art by using, for instance, learning or auto-adaptivity methods to tune the parameters of the algorithm.

As a future work we are also planning to study new pheromone models, by exploiting, for instance, the finite domain representation computed by Fast Downwards [8] and to apply the model to other planners, real world applications and problems, like for example [11,14,15,12,9].

Acknowledgements. The research leading to the results presented in this paper has been possible thanks to the grid resources and services provided by the European Grid Infrastructure (EGI), the Italian Grid Infrastructure (IGI) and the National Grid Initiatives that support the Virtual Organization (VO) COMPCHEM.

References

1. Baioletti, M., Milani, A., Poggioni, V., Rossi, F.: An ACO approach to planning. In: Cotta, C., Cowling, P. (eds.) EvoCOP 2009. LNCS, vol. 5482, pp. 73–84. Springer, Heidelberg (2009)
2. Baioletti, M., Milani, A., Poggioni, V., Rossi, F.: Ant search strategies for planning optimization. In: Proc of the International Conference on Planning and Scheduling, ICAPS 2009 (2009)
3. Baioletti, M., Milani, A., Poggioni, V., Rossi, F.: Experimental evaluation of pheromone models in acoplan. Ann. Math. Artif. Intell. 62(3-4), 187–217 (2011)
4. Blum, C.: Ant colony optimization: Introduction and recent trends. Physics of Life Reviews 2(4), 353–373 (2005)
5. Do, M.B., Kambhampati, S.: Sapa: A multi-objective metric temporal planner. Journal of Artificial Intelligence Research (JAIR) 20, 155–194 (2003)
6. Dorigo, M., Stuetzle, T.: Ant Colony Optimization. MIT Press, Cambridge (2004)
7. Fawcett, C., Helmert, M., Hoos, H., Karpas, E., Roeger, G., Seipp, J.: Fd-autotune: Domain-specific configuration using fast downward. In: ICAPS 2011, PAL Workshop, Runner-up in Learning Track, IPC (2011)
8. Helmert, M.: The fast downward planning system. J. Artif. Intell. Res (JAIR) 26, 191–246 (2006)
9. Intelligent Social Media Indexing and Sharing Using an Adaptive Indexing Search Engine. Community of scientist optimization: An autonomy oriented approach to distributed optimization. ACM TIST 3(3) (2012)
10. Keyder, E., Geffner, H.: Heuristics for planning with action costs revisited. In: Proc. of ECAI 2008, pp. 588–592 (2008)
11. Cialdea, M., Limongelli, C., Poggioni, V., Orlandini, A.: Linear temporal logic as an executable semantics for planning languages. Journal of logic, language and information 16, 63–89 (2007)
12. Milani, A., Santucci, V.: Community of scientist optimization: An autonomy oriented approach to distributed optimization. AI Commun. 25(2) (2012)
13. Richter, S., Westphal, M.: The lama planner: Guiding cost-based anytime planning with landmarks. J. Artif. Intell. Res (JAIR) 39, 127–177 (2010)
14. Tasso, S., Pallottelli, S., Ciavi, G., Bastianini, R., Laganà, A.: An efficient taxonomy assistant for a federation of science distributed repositories: A chemistry use case. In: Murgante, B., Misra, S., Carlini, M., Torre, C.M., Nguyen, H.-Q., Taniar, D., Apduhan, B.O., Gervasi, O. (eds.) ICCSA 2013, Part I. LNCS, vol. 7971, pp. 96–109. Springer, Heidelberg (2013)
15. Milani, J.L.A., Cheng, V.C., Leung, C.H.C.: Probabilistic aspect mining model for drug reviews. IEEE Transactions on Knowledge and Data Engineering 99 (2014)

Author Index

Printed in the United States
By Bookmasters